PROCEEDINGS

World Conference on
Biotechnology
for the Fats and Oils Industry

Edited by
Thomas H. Applewhite
(retired)

Kraft, Inc.
Research & Development
801 Waukegan Rd.
Glenview, Illinois USA

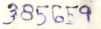

American Oil Chemists' Society ©1988

Library of Congress Catalog Card Number: 88-7540

ISBN 0-935315-21-7

Printed in the United States of America

Contents

Contents

Contents

Foreword

T. H. Applewhite, Chairman

These proceedings cover another "first" for the American Oil Chemists' Society. In cooperation with the Deutsche Gesellschaft für Feltwissenschaft, we embarked on a venture to examine the past, present and future aspects of biotechnology as related to the fats and oils industry. In some respects, this is an exciting new area of scientific endeavor; and in others, it is as old as agriculture itself. As the reader will find, our aim is to educate. And, to that end, the steering committee selected recognized leaders from around the world to cochair the sessions that were designed to cut across the entire field of biotechnology and the fats and oils industry. These chairmen then selected speakers that are recognized experts in the respective fields to deliver the plenary lectures that make up the bulk of these proceedings. To accommodate the many scientists who labor in this field but could not be invited to speak, we also provided poster sessions each day so that volunteer papers could be presented. Under the able direction of Richard F. Wilson, who chaired the poster sessions, 41 papers were offered during the meeting. Manuscripts were solicited and those submitted and judged acceptable as to content and format by Dr. Wilson are included in these proceedings.

Another important aspect of such a meeting are the discussion sessions. To assure that the key ideas and exchanges from these were retained, E.G. Perkins agreed to be the recorder and has prepared reports that are included with each session.

We believe the material included here will provide an excellent guide to biotechnology for the fats and oils industry, and we must acknowledge the contributions of all who participated in the conception, development and execution of this world conference.

Plant Lipid Biotechnology Through The Looking Glass

P.K. Stumpf

Department of Biochemistry and Biophysics, University of California, Davis, CA 95616

The biosynthesis of economically important fatty acids has been resolved in recent years. The individual enzymes have been isolated, purified and characterized; the specific compartmentations of these enzymes in the cell have been determined; the flow of precursors required to supply the substrates for fatty acid synthesis has been defined. With these data at hand, a new thrust has been brought forth, namely the input of the concepts of molecular biology to the solution of a number of problems, which until now have not been resolved. If and when these problems are elucidated, application of these new solutions will have a profound effect on the agroeconomics involved in the annual production and consumption of over 50 million metric tons (m MT) of vegetable oil throughout the world. This discussion focuses on some of the problems facing plant scientists in their attempts to understand and control plant lipid biosynthesis. Some possible solutions will be suggested, and the impact of these solutions on the economy of nations will be examined.

"'The time has come' the Walrus said,
'To talk of many things:
Of shoes-and ships-and sealing wax-
Of cabbages-and kings-
And why the sea is boiling hot-
And whether pigs have wings.'"
Alice's Adventures in Wonderland, Lewis Carroll

INTRODUCTION

Rather then talk of many things, I would like to limit myself to some important implications concering biotechnology and the future direction in which plant lipid research will move during the next decade. I think it is quite safe to say that the next decade will bring remarkable new discoveries in the plant sciences. These discoveries will have a direct bearing not only in giving the research investigator a much better understanding of what transpires in the intact plant but also will make possible a dramatic change in how industry faces its problems and these problems will be resolved effectively and efficiently.

Biotechnology is a new name for fundamental processes that occur continuously in nature. It has been practiced for thousands of years in complete ignorance of any concepts of molecular biology by honey bees, primitive farmers and primitive breeders. The new twist is the explosive growth of knowledge in both the biochemistry and molecular biology of plants, spilling out in ever-increasing volumes. As a result, the application of new knowledge will have a tremendous impact on the efficiency of crop production for food and industrial uses, will force the improvement of the quality of crops (1) and even now is stimulating the dcevelopment of environmentally friendly crop chemicals such as glyphosate.

However, up to now, the vegetable oil industry has benefited largely from the employment of conventional agricultural technologies coupled with an increasingly sophisticated science of plant breeding. As a result in less than 20 years vegetable oil production has been doubled from about 25 million tons (MT) in 1969 to over 50 MT in 1987.
The purpose of this conference, I would assume, is to explore how genetic engineering and associated technologies will interface with the current technologies employed in the production of oil crops and how these will be used to modify and improve the quality of vegetable oils, yet still meet the old as well as the new demands of industry.

THE BASIC CONCEPTS

The ultimate source of carbon in a plant is atmospheric CO_2. We now have a reasonably full understanding of the biochemical events that are involved in the conversion of carbon dioxide to vegetable oil. As depicted in Figure 1, a remarkable number of reactions are required to channel the carbon of CO_2 to the final product, a triglyceride. At least 50 enzymes function in this sequence. First of all, atmospheric carbon dioxide is fixed exclusively in the chloroplast compartment of the leaf canopies of all plants. The key enzyme, ribulose 1,5 bisphosphate carboxylase/oxygenase, converts carbon dioxide to organic carbon, which then enters the Calvin Cycle to be converted eventually to a triose phosphate, a simple sugar phosphate. This intermediate is translocated out of the chloroplast into the cytosolic compartment where enzymes convert this compound to sucrose.

Nature was very wise in selecting sucrose as the carbon carrier; it is the ideal substance. It is very soluble, relatively metabolically inert, neutral and thus readily transported from the leaf canopy to a variety of sinks via the plant vascular systems. In one type of sink, namely the developing pea or corn seed, sucrose is converted to starch, the major storage metabolite. In developing rapeseed, sunflower seed, soya seed, etc., sucrose is converted to acetyl-CoA, a very active metabolite used almost exclusively for fatty acid synthesis. When the fully developed oil seed germinates, the stored fatty acids (as triglycerides) are broken down to acetyl-CoA by β-oxidation and then channeled via a set of enzymically catalyzed reactions to sucrose, which is then translocated to appropriate growing tissues of the new plant. There it is converted into proteins, nucleic acids, etc. Quite differently, in developing fruit mesocarp tissue such as the oil palm fruit and the avocado, fatty acids are synthesized and stored as triglycerides. However, these stored lipids have no physiological function. The ripened fruit drops to the ground where it is degraded by bacteria and fungi to carbon

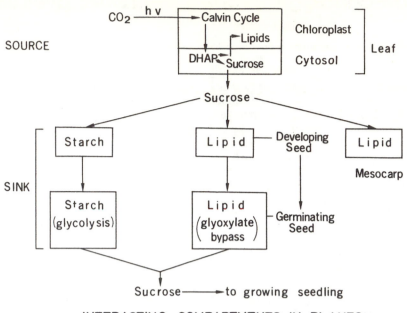

INTERACTING COMPARTMENTS IN PLANTS

FIG. 1. The interaction of various parts of the plant in the biosynthesis and utilization of lipids.

dioxide if it is not sent to the mills for extraction, etc. In summary, the leaf cell traps carbon dioxide and converts it to sucrose. In addition, the leaf cell has the capacity to synthesize those fatty acids and complex lipids that are essential for the construction and maintenance of its membranes (2). The fatty acids found in leaf membrane lipids always consist of palmitic acid and a large amount of linoleic and linolenic acids. It is important to note that these fatty acids, as complex polar lipids, are found in the membrane lipids of all green plants; that is to say, through millions of years of evolution, nature has settled on these acids as essential in the make-up of complex polar lipids that must play key roles in the structure of membranes involved in photosynthesis.

It is in the triglycerides or neutral lipids of the seed and fruit where the fatty acid composition varies according to the genetics of the particular plant. For example, although the leaf lipids of the castor bean plant have the predicted complement of palmitic, oleic, linoleic and linolenic acids, the triglyceride composition of the fully developed seeds contain as much as 90% ricinoleic acid. This acid is associated exclusively with the seed triglycerides but is completely excluded from the lipids of the leaf and the seed membranes. Thus, in developing new varieties the genetic engineer must manipulate a set of genes so that all membrane lipid compositions are conserved. In contrast, the storage lipids can be altered to any desired composition providing that in the germination of the seeds the pathways essential for the conversion of lipid to sucrose remain intact.

For the actual synthesis of fatty acids, the plant cell must convert sucrose to a very active metabolite called acetyl-CoA, its carboxylation product, malonyl-CoA, and then seven or eight enzymes utilize these sub-

strates for the construction of the typical fatty acid (3). In sharp contrast to the bacterial and animal cells, all the plant enzymes are localized exclusively in organelles. In the leaf cell, the organelle is the chloroplast, a highly efficient multifunctional plastid that also houses all the enzymes required for photosynthesis. In the developing seed or fruit mesocarp cell, the organelle is called the proplastid, and this contains all the enzymes for fatty acid synthesis and the ancilliary enzymes required to convert sucrose to acetyl-CoA. Another interesting facet of the plant fatty acid synthase system (PFAS) is that all the enzymes in this system are individual proteins that make-up what is called the nonassociated system, a very similar system also is found in all bacteria. In sharp contrast all animal and yeast FAS systems consist of multifunctional polypeptides with each of the enzyme activities represented by a domain or a stretch of amino acid residues that constitute the active enzymic site scattered up and down the polypeptide chain(s). This difference in molecular architecture is important because it allows the biochemist to separate each of the seven plant enzymes from each other, thus permitting reconstruction experiments in which one enzyme concentration can be varied while all the other six enzyme concentrations are held constant. This type of experiment obviously cannot be done with the animal FAS complex (3). Furthermore, because the nonassociated FAS systems of bacteria and plants are so similar, the genetic engineer might even consider transferring key bacterial FAS enzymes into the plant genome to improve the kinetics of lipid synthesis.

Figure 2 summarizes the present knowledge concerning the synthesis of fatty acids in plants. Suffice it to say that by a series of reactions, acetyl-CoA and malonyl-CoA are condensed in an ordered series of reactions to form palmitic acid, which is elongated to

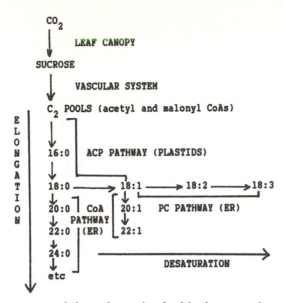

FIG. 2. A summary of the pathways involved in the conversion of acetyl-CoA to the final product, the localization of these pathways in the seed and the substrate specificities.

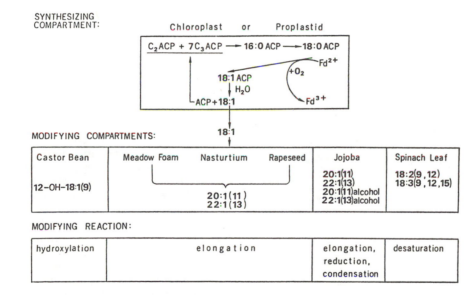

CENTRAL ROLE OF OLEIC ACID

FIG. 3. Central role of oleic acid. A plant cell has a synthesizing compartment and modifying compartments (indicated here as ER, i.e. endoplasmic reticulum, etc.). Modifying reactions are indicated in the figure. Fd, ferredoxin; C_2ACP, C_3ACP, malonyl-ACP.

stearic acid and then desaturated to the mono-, di- and tri- enoic acids (3). Figure 3 emphasizes these reactions as they relate to the various compartments.

TARGET REACTIONS

With these comments as a backdrop, what can be said about the rational application of molecular biology to the improvement of world oil production? As I have indicated, at least 50 enzymes participate in the conversion of carbon dioxide to fats and oils. The source of energy for these reactions is from photosynthetic reactions, which make possible the fixation of carbon dioxide, evolution of oxygen and the generation of adenosine triphosphate (ATP), the immediate driving force in all biosynthetic reactions. Which of these 50 enzymes can be altered or modified? Obviously, over several billion years plants that make possible the production of the present vast tonnage of vegetable oil have evolved. What is there left for the plant lipid scientist to tackle? As we learn more about the detailed biochemistry of higher plants, possible opportunities in which we can accelerate the very slow process of evolution arise. Thus, genetic engineering can function at several levels: (a) the introduction of a single gene that is expressed at all times throughout the entire plant, (b) the introduction of a single gene that is expressed only in a specific developing seed cell at a given time, (c) the introduction of controlling genes that regulate expression at temporally different stages of development, (d) the insertion of genes that control multiple traits such as drought, temperature or water stress (4) and (e) in the future, a procedure that would allow the deletion of a normally functioning gene and its replacement with a gene specifically designed to alter a specific reaction could be developed.

Before the genetic engineer can carry out these changes, the biochemist must be able to identify target reactions that regulate the synthesis of plant fatty acids. The very difficult problem the biochemist faces is determining which of the 50 or more enzymes fit that bill. I have selected a few reactions that should be examined carefully by the plant biochemists and the plant molecular biologist as possible target reactions.

Carbon dioxide fixation. The key enzyme in all green plants for carbon dioxide fixation is ribulose 1,5 bisphosphate carboxylase/oxygenase (also called rubisco) (5). The most abundant soluble protein in the world, it is located solely in the chloroplast compartment, and in all higher plants it consists of eight small polypeptide subunits (S) and eight large polypeptide subunits (L) held together as the active complex L8S8. The "L" subunit contains the catalytic site and is encoded by chloroplast DNA while the "S" subunit is nuclear-encoded but its function is defined poorly. The enzyme being bifunctional, catalyzes the reactions:

(1) ribulose 1,5 bisphosphate + CO_2 $\longrightarrow$
$\qquad$ 2 phosphoglyceric acids
$\qquad$ a carboxylation reaction

(2) ribulose 1,5 bisphosphate + O_2 $\longrightarrow$
$\qquad$ phosphoglyceric acid + phosphoglycolic acid
$\qquad$ an oxygenation reaction

Reaction 2, the oxygenation reaction, deprives the Calvin Cycle of one phosphoglyceric acid because phosphoglycolic acid is formed, becoming a substrate for photorespiration and ultimate conversion to carbon dioxide. One of the features of rubisco is its low rate of fixation. As Andrews and Lorimer (5) stated recently, "Given that for more then 3.5×10^9 years, nature has been conducting a selection experiment with rubisco . . . . is rubisco in higher plants already perfect in the sense that no further increases in these parameters are possible?" This would suggest that if rubisco has not improved over these many years, a barrier too wide to be bridged by natural evolution must exist in the enzyme structure. Perhaps molecular biology, coupled with a detailed knowledge of the structure of rubisco,

can be employed to design an enzyme that has suppressed the oxygenase step and increased its affinity for CO_2 and, hence, the rate of carboxylation. If genetic engineering can redesign rubisco, the final product of CO_2 fixation, namely sucrose, would be increased and indirectly lead to great yields of triglycerides. It should be pointed out here that an equally important candidate for the control of sucrose synthesis involves the regulation of the synthesis of another molecule, fructose 2,6 bisphosphate, which recently has been implicated as playing a key role in controlling sucrose synthesis in the leaf cell. Whether these and other reactions can be considered targets of control awaits much more research.

Acetyl-CoA carboxylase. The biotinyl enzyme synthesizes the key substrate necessary for fatty acid synthesis:

acetyl-CoA + CO_2 + ATP $\longrightarrow$
$\qquad$ malonyl-CoA + ADP + Pi

The enzyme has been investigated in plant tissue by a number of workers (3). Although the structure appears to be rather complex, its regulation also is poorly understood. Nevertheless, low levels of malonyl-CoA in plant extracts lead to the synthesis of a number of fatty acids ranging from C8 to C18 fatty acids, high levels narrow the range of fatty acid chain lengths. It seems that a much broader understanding of the carboxylase gene(s) presumably encoded for by nuclear DNA would allow a sharper definition of the subunit structure of this important enzyme and its regulation at either the transcriptional or translational level. Perhaps gene transfer of a more active bacterial carboxylase gene into the plant DNA also could be explored.

Fatty acid synthesis enzymes. Before we can discuss what, if any, enzymes could be modified or replaced to improve the fatty acid composition of a commercial vegetable oil, a basic observation must be made. In general, there are at least two types of genes: one that can be labeled as housekeeping genes and is expressed in all cells; their expression is required for the orderly functions common to all cells. The second type is organ- or tissue-specific and is expressed in a temporal or developmental mode. These genes are switched on and off as a function of the development of the cell or tissue, and their gene products or enzymes are involved in synthesizing compounds uniqe to a specific tissue such as pigments, alkaloids or unique fatty acids such as ricinoleic or erucic acids. Thus, although the mechanism of fatty acid synthesis may be identical in the seed or leaf cell, the end product control may be unique in the seed cell. The conclusion one must accept is that in the attempts by the molecular biologist to modify the type and amount of fatty acids in the plant, care must be taken to include in gene transfers suitable regulatory regions that will make possible a normal seed developmental sequence and also the specific assignment of fatty acids to the different sites in the cell.

There are a number of target reactions in fatty acid synthesis that could be considered as potential regulatory reactions:

(1) Developmental control systems are possible regulatory actions. At a critical time after fertilization in the developing seed, a coordinated and intense increase occurs in the activity of at least 40–50

enzymes required for the conversion of glucose phosphate to fatty acids. FAS activity reaches a maximum value and then falls rapidly as the seed approaches dormancy. Evidence strongly suggests that the transcription of specific genes encoded in nuclear DNA takes place when a switch is turned on, and as a result a rapid synthesis of mRNA occurs. Ohlrogge has evidence that this occurs in the formation of acyl carrier protein, which is critical for fatty acid synthesis (6). In addition, all the ancilliary enzymes responsible for the conversion of sucrose to acetyl-CoA and for the formation of the plastid structure also must be synthesized. The fundamental dilemma here is that very little is known about the switching mechanism involved in turning on and off the genes that are responsible for the specific gene products. A complete understanding of this switching mechanism would permit the molecular biologist to manipulate these "on-off" controls with possible amplification of important regulatory enzymes. An even more interesting approach would be the isolation or synthesis of an inducer that could be sprayed onto the whole plant; the inducer then would turn on the capacity of the plant to synthesize triglycerides, for example, in all tissue. The plant would die, of course, but its entirety could be harvested for oil. With no inducer, the plant would go through its normal life cycle to produce the necessary seeds.

(2) Of the seven enzymes involved directly in fatty acid synthesis, at least two or three may have important regulatory effects on the overall formation of fatty acids (3,7,8). These include:

(a) acetyl CoA:ACP transacylase:
   acetyl CoA + ACP $\longleftrightarrow$ acetyl ACP + CoA

(b) β-ketoacyl ACP Synthase I:
acyl ACP + malonyl ACP $\longrightarrow$
β-ketoacyl ACP + ACP + CO$_2$

(c) β-ketostearoyl ACP Synthase II:
palmitoyl ACP + malonyl ACP $\longrightarrow$
β-ketostearoyl ACP + ACP + CO$_2$

All these enzyme have low specific activities in a wide variety of seed extracts and thus are candidates for regulatory functions. In reconstitution experiments with highly purified PFAS enzymes, in which one enzyme concentration was varied while the others were kept constant, it was observed that (a) increasing the concentration of the transacylase markedly decreased the chain length of the newly synthesized fatty acids, (b) Synthase I is involved in the synthesis of fatty acids only up to palmitic and (c) Synthase II was responsible for the conversion of palmitic to stearic and was terminated there.

Therefore, one could suggest that a manipulation of the transacylase by recombinant DNA technology may alter greatly the end-product of fatty acid synthesis, favoring the formation of the shorter-chain fatty acids, and that deletion of Synthase II would result in a plant that will synthesize only C16 fatty acids. Deletion of Synthase I, I would imagine, would prove lethal to the plant.

In summary, a number of possible target reactions may be vulnerable to genetic engineering, providing the precautions I listed earlier are taken into account. Needless to say, to the imaginative plant lipid scientist, a number of other enzymes can be mentioned but those listed above could be of primary importance.

MODIFYING REACTIONS

A number of plants accumulate fatty acids that are quite different from the normal C16 and C18 fatty acids. The production of these have considerable potential in new industrial oleochemical uses. These include the Cuphea, the jojoba and the Brassica, to list just a few. All these plants have a mechanism that terminates chain elongation at a given chain length. With the exception of the Cuphea, we now are able to explain the mechanisms that the plants employ to achieve the synthesis of the fatty acids characteristic of the species. In these cases, we return to the basic concepts outlined earlier, namely the synthesizing compartment in which acetyl-CoA is used to construct the C16 and C18 fatty acids on the ACP track and then the transfer of the end product, be it stearic or oleic acid, to the cytosol for further modifications. Thus, in the developing jojoba seed oleic acid is synthesized in the proplastid compartment, translocated to the cytosol where it then is elongated on the CoA track to C20 and C22 moneoic acids, reduced to the corresponding alcohols and then these are condensed to form the valuable oxygen esters. In Brassica, oleoyl-CoA and stearoyl-CoA are ineffective as primers (9). In leek epidermal tissue, palmitoyl-CoA is the primer for the formation of the very long chain saturated acids and oleoyl-CoA is ineffective as a primer (10). Thus, the specificity of the final product is in large part dictated by the nature of the primer substrate.

leek: *palmitoyl-CoA* + malonyl-CoA
β-ketostearoyl-CoA + CoA   etc.

Brassica: *oleoyl-CoA* + malonyl-CoA
β-ketoeicosenoatoyl-CoA + CoA   etc.

THE FUTURE TREND—THE LOOKING GLASS

We have discussed a number of target reactions that are being investigated in University and, presumably industrial laboratories. The new techniques of molecular biology have given an enormous impetus to the exploration of the modification of these target reactions. Robbelen, Theimer, Stobart, Harwood, Knowles, Scowcroft, Horsch, Jones and Ohlrogge and others will describe in much greater detail what their researches are beginning to show. In addition, it now is possible to engineer genetically a number of plants that are resistant to virus infections, herbicides and pesticides.

What disadvantages result from these new approaches? For example, if a plant such as rapeseed or soya could be altered genetically to produce any desired fatty acid composition that industry would require—a number of problems would arise. One would be the further drift from genetic diversity to genetic uniformity (11). The next step would be genetic vulnerability, that is vulnerability to pests, disease and weather.

History very dramatically has documented the problem of genetic uniformity. For example, as recently as 1985, a new strain of a crop pathogen, citrus canker, threatened to wipe out the susceptible citrus varieties grown in Florida.

In addition, a versatile oil crop could affect greatly the economy of an entire nation. The oil palm is the principal agronomic crop in Malaysia, Indonesia and some African countries. If a genetically designed rapeseed or soya seed could produce the same type of triglycerides as economically what now is produced by the oil palm, then the oil palm industry would collapse, and the palm oil producing countries would suffer. Conversely, if the oil palm industry would apply the same techniques to the oil palm that were used to alter rapeseed or soya, then the oil palm would become the prime source of vegetable oils. When one considers the average yield of various oil-producing crops, the oil palm is prodigiously productive with an average yield of 4000 kg oil/ha whereas soya is much lower with only about 400 kg oil/ha. Of course, the soya bean is grown primarily for its protein; the oil is a side product!

Thus, as biotechnology enters the 21st century, the explosive techniques of molecular biology coupled with the equally explosive needs of an ever-growing population will revolutionize the current modes of agriculture and will have a profound effect on the economies of the world. Achieveing these goals will require (a) adequate funding in research from both federal and industrial sources, (b) properly trained biochemists and molecular biologists to develop the basic knowledge needed to solve technical problems, (c) an exchange of ideas not only at the academic level but also at the industrial level. The question arises as to the impact secrecy would have on the development of fruitful research.

(d) A solution to the problem of genetic uniformity of plant material. (e) Political considerations to control trade barriers that even now are appearing above the horizon. For example, restrictions placed on oil imports because of nutritional dogma, and (f) competition between temperate zone high technology countries vs third world tropical countries with their evolving agriculture and technology.

REFERENCES

1. *New Directions for Biosciences Research in Agriculture*, National Academy Press, Washington D.C., 1985; *Genetic Engineering of Plants*, National Academy Press, Washington D.C., 1984.
2. Stumpf, P.K. in *The Biochemistry of Plants*, Vol. 4, edited by P.K. Stumpf and E.E. Conn, Academic Press, New York, 1980, p. 177.
3. Stumpf, P.K. in *The Biochemistry of Plants*, Vol. 9, edited by P.K. Stumpf and E.E. Conn, Academic Press, New York, 1987, p. 121.
4. Goodman, R.M., H. Hauptli, A. Crossway and V.C. Knauf, *Science 236*:48 (1987).
5. Andrews, T.J., and G.H. Lorimer in *Biochemistry of Plants*, Vol. 10, edited by P.K. Stumpf and E.E. Conn, Academic Press, New York, 1987, p. 209.
6. Ohlrogge, J.B., and T.M. Kuo, *Plant Physiol. 74*:622 (1984).
7. Shimakata, T., and P.K. Stumpf, *Proc. Natl. Acad. Sci, USA 79*:5808 (1982).
8. Shimakata, T., and P.K. Stumpf, *J. Biol. Chem. 258*:3592 (1983).
9. Agrawal, V.P., and P.K. Stumpf, *Lipids 20*:361 (1985).
10. Agrawal, V.P., and P.K. Stumpf, *Arch. Biochem. Biophys. 240*:154 (1985).
11. Plucknett, D.L., N.J.H. Smith, J.T. Williams and N.M. Anishetty, *Gene Banks and the Worlds Food*, Princeton University Press, Princeton, NJ, 1987.

Lipid Biotechnology: A Wonderland for the Microbial Physiologist

Colin Ratledge
Department of Biochemistry, University of Hull, Hull HU6 7RX, UK

Many definitions of biotechnology have been advanced but all agree that biotechnology encompasses the use of microorganisms or their component parts for the manufacture of products or service systems. Plant and animal culture systems are included in most definitions, thus biotechnology can be seen as playing a vital part in some of the new advances that are occurring in agricultural and medical sciences. The application of biotechnology to the oils and fats industry, as just one sector of the very large agrochemical industry, is such that already in this decade we have seen the American Oil Chemists' Society (AOCS) sponsor several biotechnology sessions at annual conferences, a monograph on the subject (1), a major conference in 1985 that was devoted to emerging technologies (2), partially including the contributions of biotechnology, and now a major conference devoted exclusively to biotechnology.

This awakening of interest has stemmed from the rapid advances that occurred in the major biological disciplines and, in particular, in the advances that have occurred in molecular genetics.

The preceding contribution from Stumpf has set out the background and growth points for the contribution of the plant geneticist and biochemist to the oils and fats industry. Of course, plants are the cornerstone of the industry; the vast majority of oils and fats are produced by agricultural means. Animal fats constitute an important but nevertheless minor sector of the market, and microbial oils and fats contribute nearly nothing. Biotechnological applications to animal fat research and development are still very much in their infancy, though occasional thoughts have been raised about the possibility of changing the degree of unsaturation of the fatty acids in animals. However, such changes—even if nutritionally desirable (3)—would not be brought about by genetic engineering techniques but by more conventional means of animal breeding or even by control of the animals' dietary intake of fatty acids. Genetic engineering to modify the fat of an animal might be regarded as a project for the next century, though, this is only 13 years away.

Why then, if microorganisms currently contribute so little to the oils and fats industry, should we now be so enthusiastic about them? First, we can view microorganisms as potential sources of oils and fats, though for obvious economic reasons these oils should be of the higher value-added types rather than resembling the bulk low-priced commodities such as soybean oil, sunflower oil, etc. Second, microorganisms provide excellent means of carrying out numerous biotransformation reactions using either the whole microbial cell or one or more of its component enzymes. Third, microorganisms can provide useful models for studying the more intricate aspects of lipid biochemistry, metabolic control and function, thus giving invaluable leads to plant and animal lipid biochemists.

Each of the above three aspects will be covered in this short review; aspects of microbial biotransformations largely will be left to the succeeding presentation of Yamane for full coverage. However, it should be admitted that the study of microbial lipids has always been the "Cinderella" of the subject with regard to lipids and to microbiology. Even when microbiologists have chosen to address themselves to potentially interesting problems in this field, such as the biosynthesis of fatty acids, they have tended to use a very restricted range of microorganism: usually *Escherichia coli* for bacterial work and *Saccharomyces cerevisiae* for yeast work. Although much of this work has had considerable elegance, it has had only an indirect bearing on what happens in the oil-bearing seeds of plants or even in the adipose tissue of animals. Thus, the more applied aspects of lipid microbiology, what now would be called lipid biotechnology, often has been an "orphan" subject. Advances have failed to keep pace with developments elsewhere. We still do not know, for example, what controls the degree of unsaturation of a fatty acid; yet, almost every undergraduate biochemistry student is familiar with the intimate details of DNA organization with its introns and exons, TATA boxes and leader sequences. Thus, for a microbial physiologist such as myself to find a relatively unexplored area of biochemistry in the last quarter of the 20th century suggests that this is a wonderland just waiting for exploration. Therefore, I hope this review gives a reflection of how some of these perambulations have been proceeding in recent years.

MICROORGANISMS

Microorganisms range from bacteria without a defined nucleus, procaryotes, to yeasts and molds that have such a nucleus, eucaryotes. They also include the photosynthetic microorganisms, which may be both pro- and eucaryotic, thus covering the cyanobacteria, previously known as the blue-green algae, and the larger eucaryotic algae that range from the microscopic individual cells to the macroscopic large, brown algae recognized as seaweeds throughout the world.

Microorganisms can be seen to have the following advantages: a prodigious growth rate, 1 g of bacteria can generate a second gram of biomass in 20 minutes, a yeast cell takes a little longer—two to four hr; an omnivorous appetite in that almost any carbon-containing substrate can be used to sustain growth; and they can be manipulated by the geneticist or the physiologist to produce a wide range of products or to have the yield of a minor component amplified considerably. A chemist achieving a 0.1% yield for a reaction in the laboratory probably would give up and look in another direction for the right process. That is not so with the microbiologist. As witnessed by most of the antibiotics, there are many examples in which the initial yield of a product was extremely low; by the application of genetical mutation and physiological control of the growth process, the yield has been multiplied many-fold.

The technology for sustaining large-scale cultivation of microorganisms has advanced over many years. Fermenters capable of producing up to 10^5 ton of biomass per year now are not uncommon. Microorganisms, according to the process, may be grown batch-wise or in a continuous mode. However, the real problem comes in deciding on what to grow the cells; for photosynthetic algae, the choice is obvious: CO_2. For other microorganisms, an organic source of carbon is necessary; for large scale fermentation processes, the price and availability of the starting substrate may be a key factor in determination of the process' economic viability (4).

PROSPECTS FOR MICROBIAL OILS

Types capable of production. Probably the most expensive triacylglycerol oils being produced commercially today are those containing the essential γ-linolenic acid (6,9,12-octadecatrienoic acid). These oils are found in the seeds of *Oenothera* (Evening Primrose) and *Ribes* species (especially blackcurrant and gooseberries) (see Table 1). There is a small volume market (5200–500 ton/year) which is growing steadily for these oils. Although the exact nutritional value of exogenously supplied γ-linolenic acid is challenged, sometimes the role of this acid in vivo, which is to act as a precursor of the prostaglandins (PGE_1) series, is undisputed. As these oils command prices of $50/kg and beyond, there is little wonder that several groups have seen this as an obvious target for biotechnological innovation and exploration.

γ-Linolenic acid has been known as a fungal fatty acid since 1948 (5) and was established by Shaw (6) to be confined to the family of molds known as the *Mucorales*. Species of *Mucor* and other related organisms also have been known for many years to be capable of accumulating considerable amounts of lipid (7,8). Thus, by virtue of extensive screening for strains of

fungi accumulating high amounts of γ-linolenic acid (9–12). Because of the prior publications in this area, these patent applications may not all be recognized or granted. Today, there are at least two commercial processes for the production of this oil. One in Japan is based on *Mortierella* and will be described later by Suzuki (13); the other uses *Mucor javanicus* and has been developed by J. & E. Sturge, Ltd., in the UK in conjunction with work in our own laboratories. This latter process is now at the 220,000 liter level of production with opportunities for expansion of up to four or five times by using similarly sized fermenters.

The fatty acid profile of this oil is distinct from that of the plant oils (see Table 1) but, interestingly, bears a strong resemblance to that of the milk fat from humans. The lower content of linoleic acid (18:2) in the *Mucor* oil than in plant oils makes it an attractive starting point for producing purified γ-linolenic acid; such as an oil is easier to fractionate than one containing many polyunsaturated fatty acids.

The prospects for producing other dietarily important polyunsaturated fatty acids (PUFA) including arachadonic acid (20:4), eicosopentaenoic acid (20:5) and docosahexaenoic acid (22:6) have been examined in microorganisms. Although these PUFA have been reported sometimes in high concentrations, they have been studied only in algae (18–20). Although algae require much less financial investment for their cultivation than do yeasts and molds, the yields of algae per liter of medium (or even expressed per m²) are very much lower than yeasts or molds. Maximum yields of algae (*Chlorella*, *Spirulina* and *Scenedesmus*) under optimal outdoor conditions are unlikely to exceed 25 g/m²/day (21). Thus, to produce 100 tons of biomass it would require four hectares of algal lagoons illuminated for 100 days with continuous sunshine and a temperature between 28–30 C. On the other hand, with yeasts and molds capable of reaching 80–100 g of cells/l in a 60 hr fermentation, a modest 200 m³ fermenter would produce well over 1500 tons of biomass in one year, irrespective of sun, wind or rain and even allowing for a 30 hr turn-around between fermenter runs.

The oil produced in algae also is not as attractive as those of other microorganisms; it usually is comprised of numerous types in which the triacylglycerol fraction

TABLE 1

Spectra of Fatty Acids in Various Oils and Fats Containing γ-linolenic Acid

Source of oil	Oil content (% by wt)	Principal fatty acyl constituents (% W/W) (14–17)							
		16:0	16:1	18:0	18:1	18:2	γ-18:3	α-18:3	18:4
Oenothera biennis[a]	16	7	tr	2	9	74	8	0.2	—
Ribes nigrum[b]	30	6	tr	1	10	48	17	13	3
Ribes rubrum[c]	25	4	—	1	14	41	4	30	3
Ribes uvacrispa[d]	18	7	—	1	15	40	11	19	4
Borago officinalis[e]	30	10	—	4	18	42	23	0.4	—
Human breast milk		23	—	9	35	11	2*	0.2	—
Mucor javanicus[f]	18	23	tr	7	39	11	18	0.2	—

[a]Evening Primrose; [b]Blackcurrant; [c]Redcurrant; [d]Gooseberry; [e]Borage; [f]Production fungus (J. & E. Sturge, UK).

*Includes other polyunsaturated fatty acids.

TABLE 2

Effect of Sterulic Acid on Formation of Stearic in Selected Oleaginous Yeasts (31).

Yeast		Sterulic acid* added (ml l⁻¹)			
		0	0.02	0.10	1.0
Candida 107 (= NCYC 911)					
Lipid in cells		37	18	25	33
Rel. %	16:0	26	28	28	24
	18:0	5	24	33	44
	18:1	35	12	10	10
	18:2	21	30	23	14
Trichosporon cutaneum					
Lipid in cells		29	31	30	31
Rel. %	16:0	29	30	32	35
	18:0	7	19	24	23
	18:1	50	33	27	25
	18:2	13	16	14	13
Rhodosporidium toruloides					
Lipid in cells		30	30	34	32
Rel. %	16:0	16	15	15	13
	18:0	4	17	32	41
	18:1	42	19	23	18
	18:2	29	17	15	15
	18:3	5	9	7	6

*Added as sterculia oil, which contains 50% sterulic acid ($\triangle^9$ cyclopropene 17:1) and 5% malvalic acid ($\triangle^9$ cyclopropene 18:1).

may be only a minor component. Thus, extraction and processing of an algal oil becomes more complex than that of a microbial oil. There also is the problem of the co-extracted chlorophyll with which to deal, otherwise the algal oil will have an unwanted green color.

Nevertheless, in spite of these problems algae potentially are useful sources of arachadonic acid (20:4) (18) and eicosapentaenoic acid (20:5) (19), as other symposium presentations clearly indicate (22,23). The advantage of using algae to produce PUFA rather than to rely on fish oils is that the algae could be induced to produce high concentrations of a single fatty acid (18,19) rather than an array of fatty acid types. There also may be the possibility of producing these fatty acids as an adjunct to the algal production of carotenoids, especially β-carotene, which now is a commercial reality in several locations throughout the world (24).

Other high-value microbial lipids that are under current consideration include production of a cocoa butter fat or even a fat with superior specifications to cocoa butter. Cocoa butter currently sells for between $4,500–$5,000 per ton (25,26), although prices over the past six years often have exceeded this. Annual production of cocoa butter is about 800,000 tons (27), and a number of reformulated vegetable oils currently are

produced as cocoa butter equivalents (CBE). The potential market for these CBE could be up to 10% of the total cocoa butter market.

Cocoa butter is characterized by a high content of stearic acid (30–35%) and has as its principal triacylglycerol 1-palmitoyl-2-oleoyl-3-stearoylglycerol. Unfortunately, most oleaginous microorganisms have stearic acid contents of 10% or less, and so efforts have been made to increase this level either by feeding stearic acid or its esters to yeasts (28–30) or by adding an inhibitor of stearate $\triangle^9$-desaturase (31). This latter approach, using the naturally occurring cyclopropene fatty acid sterculic acid, has led to up to 40% increases of stearic acid in selected oleaginous yeasts (see Table 2). Comparisons of this yeast fat with cocoa butter have indicated a striking similarity between the two lipids (25).

Prospects for being able to delete by genetic mutation and selection the $\triangle^9$-desaturase enzyme in oleaginous yeasts also should not be forgotten as a possible third route to achieving biotechnological production of CBE. However, as will be appreciated from Table 2, inhibition (or potential deletion) of the $\triangle^9$-desaturase does not seem to greatly diminish the formation of linoleic acid (18:2), which would be necessary to achieve if the new cocoa butter substitute was to be entirely satisfactory.

Other lipids that currently are under consideration for biotechnological production include a number of glycolipids as potential surfactants, various carotenoids and, if we can include as a lipid the unusual bacterial polyester, poly-β-hydroxybutyrate. Such opportunities have been reviewed recently by several researchers (32–34). It seems unlikely, however, that wax ester substitutes for jojoba oil or spermaceti could be produced economically by microbial means.

COST

The increase in activity towards producing a microbial oil has led a number of the interested parties to suggest possible economies for potential Single Cell Oil (SCO) process.

Clearly, with a very high priced oil such as those containing γ-linolenic acid the costs look attractive. Prescott (35) has calculated that a single 220 m³ fermenter can produce as much oil in four days as 30 acres of a good crop of Evening Primrose or Borage does in a year. Furthermore, the annual output of two such fermenters probably could satisfy the entire current world demand for this oil. Approximate cost for this process compared with agricultural costs are given in Table 3. It should be pointed out that these estimates are for average yield models for both plant and fungal oils; they do not take into account losses of oil or costs at the extraction stage. Even so, the lower costs of the biotechnological route are quite evident.

Costs for larger scale productions of oils and fats have been made by Davies (36,37), Moreton (25), Floetenmeyer et al. (38) and by Moreton and Norris (39) for fermentation processes in general. Interested readers are referred to these reports for further details as a complete discussion of the various points that must be considered are too many for a short review like this.

The important factors that have a bearing on the eventual costs include:

TABLE 3

Comparisons of Variable Costs of GLA from Different Sources (35).

Source	High margin	Low margin	Percentage of GLA in oil	Minimum cost of GLA
Evening Primkrose	£10,000	£5,755	9%	£64,000/ton
Borage	£9,333	£5,608	20%	£28,040/ton
M. javanicus	£2,000*		16%	£12,500/ton

*Cost per ton of oil in cells; i.e. does not include processing or extraction costs (but these are likely to be approximately equal in all three cases).

- *Cost of substrate.* Even the most efficacious oleaginous organism cannot convert substrate to lipid with a yield much better than 22%. Thus 4.5 to five ton of substrate are needed to produce one ton of oil. Year-round availability of substrate also is important.
- *Scale of operation.* The larger the process, the cheaper the unit costs become.
- *Fermenter design and operation.* Different economic costs depend on whether a batch or continuous process is chosen (or is desirable from a product viewpoint). Stirred tank reactors may not be as effective as airlift fermenters.
- *Down-stream processing and oil extraction.* This is the area that still requires the most work. There are obvious differences between a yeast-based process and a mycelial fungal process; the latter is easier to harvest and extract but poses more problems during the fermentation phase, and continuous operation almost is impossible. Extraction of cells with hexane is a standard practice with plant oilseeds, and a similar technology can be applied to yeast cells (40). Improvements in this process to avoid complete and expensive drying of the cells have been attempted (37), and the use of cross-flow filtration rather than centrifugation has been advocated (41).
- *Capital costs.* Returns on capital invested must take into account discount cash flow to provide a proper evaluation of the financial viability of the project (39).
- *Revenue.* There is no doubt that a good quality oil of almost invariable composition and properties can command a premium price. Single Cell Oils (SCO) can be produced year-round with high constancy of composition. Supplies of the oil are not influenced by the vagaries of the weather; agreed supplies at agreed costs are attractive propositions to potential customers of the SCO. The cell residue after oil extraction also is recognized as potential value as an animal fodder equivalent to soybean meal.

Davies (37), in a very detailed series of cost estimates, has calculated that it would cost approximately $3 million to build a process plant capable of producing from whey 1,000 tons of oil plus 1,800 tons of protein residue that is salable as animal fodder. To make the process attractive to potential financial companies, requiring a 20% rate of return after taxation,

the oils would have to sell for more than $1,000/ton and the protein for $275/ton. However, costs do not include any credit for abatement of a potential pollution problem by using the whey.

Moreton (25) has indicated that the lowest reasonable cost of a yeast SCO is more likely to be $2,500/ton than $1,000/ton. Floetenmeyer et al. (38), however, have suggested that given a zero-cost substrate this value could be as low as $680/ton of lipid, but they did not take into account the extraction costs of the oil. Moreton's calculations do include both the cost of substrate (at $240/ton) and extraction costs of about $1000/ton of lipid.

Thus, by a variety of calculations we probably can conclude that production of SCO will be limited only by economic considerations to the more expensive "up-market" oils and fats. However, given that a number of processes exist throughout the world for the production of Single Cell Protein (SCP), which utilize waste (zero- or negative-cost) substrates, it is fairly easy to show that production of SCO would be a better financial proposition than continuing with SCP (41). A fuller review than this of the microorganisms potential for oil production recently has been published (42). This covers bacteria, algae, yeasts and fungi, and it summarizes the literature from 1980 to 1985.

MICROBIAL TRANSFORMATIONS

The ability of microorganisms to carry out numerous transformation reactions is well-known and forms one of the cornerstones of microbial technology. Transformations with lipids may be carried out by using either whole cells or their component enzymes (Table 4). A brief outline of the current activities and future prospects of this topic follows.

The simplest of the transformation reactions attempted with lipids has been feeding fats or fatty acids to selected microorganisms in the hope that the microorganisms would upgrade the oil quality by desaturating, or even saturating, the component fatty acids. In most cases (43–45), the composition of the fatty acyl groups recovered from the microorganisms has proved to be similar to those on which it was obliged to grow. The reason for this conservation of lipid structure is that the yeast (it usually has been yeasts that have been tried) obviously is able to satisfy its own requirements for fatty acids, either for the structural phospho-

TABLE 4

Agents for Biotransformation Reactions

Order of complexity (and cost)		
	1. *Whole cells*	– growing
		– nonproliferating
		– permeabilized by organic solvents (two-phase liquid systems)
	2. *Mutant cells*	– blocked to cause product accumulation
	3. *Crude enzyme preparations* (extracellular or intracellular)	– free
		– immobilized: single phase or liquid two-phase systems
	4. *Partially purified enzymes*	– free
		– immobilized
	5. *Enzymes with co-factors*	– coupled enzymes for co-factor regeneration

lipids or storage triacylglycerols, with a wide range of acyl groups. Thus, there is no environmental pressure exerted upon the microorganism to change the fatty acid substrate with which it is presented. The growth of yeasts on various fats and oils has been reviewed recently (46). However, much of the current interest in this area appears to be towards complete utilization of the unwanted fat to produce yeast salable as animal fodder material.

If changes in the fatty acids of a substrate are required, it usually will be necessary to mutate the parent microbial cells. Mutation can be used to block degradation of the fatty acid via the β-oxidation cycle, thus preventing undue losses of the substrate, or it can block the further oxidation of the product before it is linked into β-oxidation. Consequently, if fatty acids are presented now to such cells for transformation, partial oxidation products of fatty acids should be found (47,48). Such products would include hydroxy fatty acids and dicarboxylic acids.

Most of the metabolisms for fatty acids have not been worked out using fatty acids; they have stemmed from extensive studies on alkane metabolism in a variety of microorganisms. These studies elegantly have demonstrated the power of using mutants to produce greatly elevated levels of intermediate products. Conversions of alkanes to α,ω-dioic acids in yield up to 70% and with amounts up to 60 g/l have been reported with mutants of *Candida cloacae* and *C. tropicalis* (49,50). 3-Hydroxyalkanedioic acids also have been recovered from such transformation experiments (50).

Beyond the use of whole cells lies the possibility of using cells held in solvents; this obviously would kill the cells but allows the intrinsic enzyme activity to be retained and even enhanced. Such a technique is practiced with some sterol transformations in which the solvent, often chloroform, allows penetration of the substrate into the cells and migration of the product out of the cells (51,52). The advantage of this technique is that it does not require the physical isolation of

unstable enzymes, such as the hydroxylases, to be involved in the initial attack on alkanes or fatty acids. Furthermore, the enzymes can retain activity for a considerable time and can transform many times their own weight from substrate into product (52). Cells used in such systems may be the original parent (wild-type) strain or mutant produced to prevent further reactions of the product (53). Recent examples of biocatalysis occurring in two-phase systems include epoxidation of 1,7-octadiene by whole cells of *Pseudomonas putida* (54) and stereospecific hydrolysis of *dl*-menthyl acetate by cells of *Bacillus subtilis* (55).

The next degree of complexity for biotransformation is the use of isolated enzyme systems, and these may be recovered from the extracellular growth medium or from within the cells. The less purification that has to be carried out, the cheaper the eventual product. Consequently, considerable effort currently is being spent trying to isolate new microorganisms with enhanced enzyme activity or, if this fails, even resorting to genetic engineering to enhance the enzyme complement of the cells. For lipids, isolated-enzyme technology usually means lipases that now are featured so prominently in this and previous symposia organized by the AOCS (56–58). Lipases, besides being either nonspecific or specific and being able to carry out transesterification reactions, may also act in a synthetic mode (57,58) but only if the water content of the reaction mixtures is kept extremely low. Once again, we see that enzymes are not to be thought of as just water-soluble agents; they can, and do, perform certain reactions better in a nonaqueous environment than in water.

Lipases have much to offer to the oils and fats industry (56,59), and one of their main potentials is catalyzing the formation of cocoa butter-type triacylglycerols from palm oil and stearic acid by interesterification (59). Unfortunately, commercialization of this has not been realized yet. The difficulty appears that although purified vegetable oils, such as palm oil, work extremely well with lipases in the laboratory, the enzyme does not retain its longevity when operating with commercial grades of oil at pilot-scale levels (60). Thus, activity is lost prematurely; this adversely affects the overall economics of the process.

Lipases used in these interesterification reactions are immobilized to prevent their loss from the reactor. Unfortunately, the number of commercially successful immobilized enzymes still is extremely small (about four or five), in spite of considerable early promise that this would be a major activity of biotechnology by the 1980s. The lack of success is due principally to many enzymes requiring specific biological (and hence expensive) co-factors or requiring an input of energy into the reaction to make it proceed. These constraints mean that it is the hydrolytic or isomerizing enzymes that are potentially the most useful for commercial processes but there are only a limited number of such reactions that are interesting commercially. In spite of these drawbacks, considerable effort currently is underway to circumvent some of these problems; it is likely that the next decade will see coupled enzymes being used for co-factor recycling or even an energy input into reactions. However, these reactions probably will operate on a relatively small scale and will deal with high value-added products.

Because lipases catalyze the hydrolytic type of reaction and require neither co-factor nor energy input, they would seem ideal candidates for exploitation. However, applications of them are very narrow; it has to be remembered that chemists have been very successful in devising all of the reactions that are used in the oleochemical industry. Thus the biotechnologist has to devise some novel enzymological reactions that the chemist would find very hard to emulate. Biotransformation reactions undoubtedly will find a place in the oils and fats industry, but the biotechnologist must not expect the chemist to sit idle, waiting for his reactions to be superseded.

MICROORGANISMS AS MODEL SYSTEMS FOR STUDYING LIPID BIOCHEMISTRY

"I should like to have it explained," said the Mock Turtle. Alice's Adventures in Wonderland, Lewis Carroll.

Plants and animals are complex; microorganisms, in comparison are simple. Most biochemical pathways that are common to all living cells usually have been first recognized in microorganisms. Lipid biochemistry follows this generalization, though there are some exceptions.

Pathways for the biosynthesis of fatty acids and from fatty acids to triacylglycerols, phospholipids and glycolipids have been recognized in most microorganisms, and the complexity of the component enzymes has been analyzed in some detail (61,62). Even biosynthesis of some of the more complex lipids that often are thought of as animal lipids, such as the sphingolipids, which includes the cerebrosides and sphingomyelins, also have been clarified in yeasts (see ref. 63 for review).

By the mid 1970s, it seemed the whole of lipid biochemistry in microorganisms probably had been elucidated and that microbial systems having played their part could be relegated to a minor position while more

interesting projects involving plant or animal tissues could proceed. Since then, however, we have had to revise completely our ideas of how the fatty acid synthetase complex is organized in yeast and in animals. We now are faced with the enormous task of trying to understand how such a complex process as fatty acid assembly is carried out by only two proteins when eight separate reactions, some repeated seven times with different substrates, are required for the conversion of acetyl-CoA to palmitoyl-CoA. The approach to this problem now requires a combination of genetics and biochemistry, as eloquently described by Schweizer (64). Ultimately, this work should unravel not only the process in Saccharomyces cerevisiae (baker's yeast) but also in animal systems that share properties with the eucaryotic microbial cell. For assistance with plant lipid biochemistry, we again can look to microorganisms for guidance; although here it is the procaryotic bacterial cell that shows the greatest similarities.

It also was considered in the mid 1970s that yeasts and fungi would be the same in regards to lipid biosynthesis and that by understanding the process in one organism (S. cerevisiae), the process would be understood in all organisms. However, this ignored the real differences that obviously occurred between different microorganisms, namely that a few species had the ability to accumulate considerable amounts of lipid in their cells whereas other cells, even when placed under exactly the same conditions, did not. These fat-accumulating microorganisms, referred to as the oleaginous species (65), clearly had some unusual features of their biochemistry to account for this accumulation. A series of investigations (66) led us to the conclusion that it was not that these cells had a hyperactive system for synthesizing fatty acids but that they possessed a system for producing acetyl-CoA, the building units for fatty acid biosynthesis, which was not present in the nonoleaginous cells.

For once, the tables were turned on the microbial physiologist. It was the animal biochemists who had pioneered this area, though even here the animal biochemists considered their system for acetyl-CoA unique, thus completely absent from microbial cells (66).

The key to efficient acetyl-CoA production in oleaginous yeasts (and molds) was found to be ATP:citrate lyase (67), an enzyme catalyzing the cleavage of citrate in the presence of CoA and ATP to acetyl-CoA and oxaloacetate. However, the mere presence of this enzyme is not sufficient to explain oleaginicity, and one then must begin to work backwards to answer a series of questions:

Questions: How is the citrate provided for the ATP:citrate lyase, which is in the cytoplasm?
Answer: By transport out of the mitochondrion in exchange for malate (68).

Question: Why does the citrate accumulate and not get metabolized via the citric acid cycle?
Answer: Because its metabolism, at the level of isocitrate dehydrogenase in the mitochondrion, is blocked (67,69).

Question: How does the isocitrate dehydrogenase become blocked?
Answer: Because it requires AMP for activity (67,69,70)

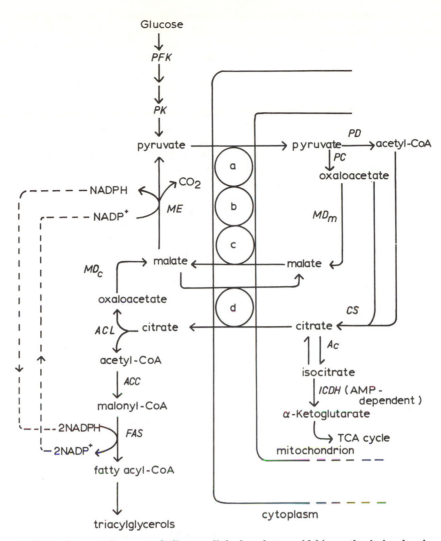

FIG. 1. Intermediary metabolism as linked to fatty acid biosynthesis in oleaginous microorganisms. Mitochondrial transport process: a, b, c, interlinked pyruvate-malate translocase systems; d, citrate -malate translocase. Enzymes: AAC, acetyl-CoA carboxylase; AC, aconitase; CL, ATP:citrate lyase; CS, citrate synthase; FAS, fatty acid synthetase complex; ID, isocitrate dehydrogenase; MD$_c$, malate dehydrogenase (cytosolic); MD$_m$, malate dehydrogenase (mitochondrial); ME, malic PC, pyruvate carboxylase; PD, pyruvate dehydrogenase; PFK, phosphofructo PK, pyruvate kinase.

this co-factor quickly disappears when the cells run out of nitrogen and are ready to begin the accumulation of lipid (71).

Question: How does the AMP deaminase become activated?
Answer: Not known.

Question: What makes the AMP disappear?
Answer: AMP deaminase, which becomes activated when the nitrogen supply is exhausted (72).

Question: How does the malate get out of the mitochondrion so it can exchange with the citrate, which is inside the mitochondrion?
Answer: Probably in exchange for pyruvate (72).

Question: How is the supply of pyruvate then ensured?
Answer: By the key enzymes of glycolysis phospho-

fructokinase (PFK) and pyruvate kinase (PK) being fully active (73,74).

Question: Where does the malate come from in the first place?
Answer: From oxaloacetate, which comes from the pyruvate entering the mitochondrion.

Question: When the citrate is cleaved, the acetyl-CoA goes for fatty acid biosynthesis but what happens to the other product, oxaloacetate?
Answer: This is converted to malate.

Question: Is this the malate, which then exchanges for citrate?
Answer: Only in part as more malate arrives outside the mitochondrion than is taken in exchange for citrate.

Question: What happens to this extra malate?
Answer: It is converted to pyruvate, which then can

begin the process over again by malic enzyme, an enzyme which simultaneously produces NADPH (72).

Question: What happens to the NADPH?
Answer: This is used for fatty acid biosynthesis.

Question: Is this a complicated scheme?
Answer: Look at Figure 1.

Question: How is the whole process coordinated to achieve the right balance of acetyl-CoA and NADPH production?
Answer: Not known.

The main aspects for a biochemical explanation for oleaginicity thus now are known (Fig. 1). It would seem more than likely that further study will answer some of the outstanding problems and, because the process is similar to that seen with lipid accumulation in animal cells, it may be possible to suggest a universal hypothesis. The accumulation of lipids in plant seeds probably is by the same mechanism. Certainly, ATP:citrate lyase is present in the cytosol of germinating endosperm (75), developing soybean cotyledons (76), ripening mango fruit (77) and pea leaves (78), though it has yet to be demonstrated in plant tissues synthesizing lipid reserves. The process used for NADPH production for fatty acid biosynthesis is less certain and though malic enzyme

does occur (79), it may be linked only to NADH formation and not NADPH.

Microorganisms not only provide excellent models for studying pathways of biosynthesis and degradation but also for the regulation of these pathways. Techniques such as continuous culture are denied to animal and plant biochemists but such systems enable stringent control to be held over the environmental conditions and, most importantly, also to produce cells being held at a constant growth rate in a steady state system. Therefore, it becomes impossible by using a chemostat (i.e. continuous culture fermenter) to calculate from observed enzyme activities if these are sufficient to account for the rates of cell growth and even for the overall rates of synthesis of individual components. In this way, it has been possible to show that ATP:citrate lyase in yeasts is probably the rate-limiting step in the whole of lipid biosynthesis (71). Thus, our view that this enzyme is probably the key regulatory enzyme because it is so sensitive to feedback inhibition by long chain fatty acyl-CoA esters is confirmed (80). However, the CoA esters, also inhibit equally strong the efflux of a citrate from the mitochondria (81), and it is possible that the enzyme could be coordinated and controlled, though this yet has to be established.

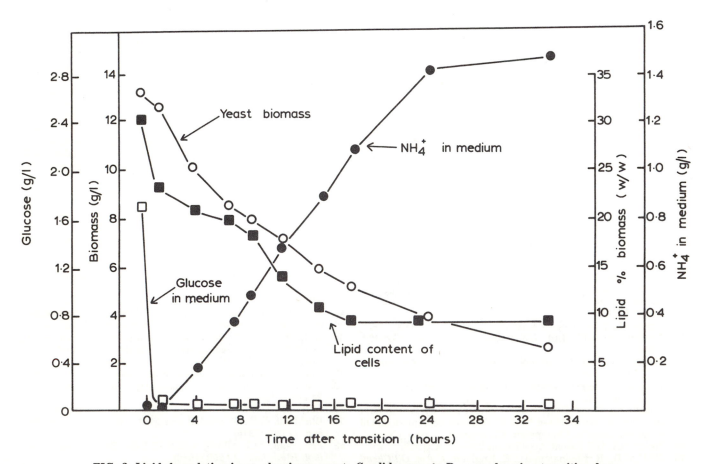

FIG. 2. Lipid degradation in an oleaginous yeast, *Candida curvata* D, on undergoing transition from steady-state lipogenic growth in a carbon-excess, nitrogen-limiting chemostat to a carbon-starvation environment while still in continuous culture at a dilution rate of 0.05 h⁻¹. Lipid mobilization immediately begins the transition is made (achieved by switching media entering the chemostat). (Taken from refs. 82 and 84.)

Just as we have been able to use chemostat cultures to follow the events leading up to lipid accumulation, we have been able to use the same techniques to follow lipid degradation (82,83). Again, the utilization of accumulated lipid reserves is an important process in animals, plants and microorganisms, and yet little is known about the events that trigger lipid mobilization and its subsequent degradation.

Initiation of lipid utilization can be a rapid event, as shown when steady-state, lipid-rich yeasts are switched from a high-fat state in carbon-excess growth to a carbon-starvation environment, a change that can be achieved in less than five minutes. The enzymes and organelles (peroxisomes) that are associated with the degradation of stored lipid then rapidly make their appearance. Utilization of lipid can be recognized readily following the switch to carbon starvation, as shown in Fig. 2 (82,83). It also has been relatively easy to show that the carbon within the stored triacylglycerol does indeed become incorporated into other cell components during starvation, thus proving that the lipid genuinely is used to generate new cells.

To achieve similar control over lipid degradation in other, more complex systems would not be possible. Though, by examining the microorganism as a model system, it should be possible to know what events might be investigated. Therefore, microorganisms can tell us much very quickly, and it is often prudent to study the simpler systems before trying to unravel the more complex.

"A likely story indeed!" said the Pigeon. Alice's Adventures in Wonderland, Lewis Carroll.

REFERENCES

1. *Biotechnology for the Oils and Fats Industry*, edited by C. Ratledge, P.S.S. Dawson and J.B.M. Rattray. American Oil Chemists' Society, Champaign, IL (1984).
2. *World Conference on Emerging Technologies in the Fats Industry Proceedings*, edited by A.R. Baldwin, American Oil Chemists' Society, Champaign, IL (1986).
3. Gurr, M.I., *J. Dairy Technol.*, in press.
4. Ratledge, C., *Ann. Rep. Ferment. Proc.*, 1:49 (1977).
5. Bernhard, K., and L. Albrecht, *Helv. Chim. Acta*, 31:977 (1948).
6. Shaw, R., *Adv. Lipid Res.*, 4:107 (1966) and *Comp. Biochem. Physiol.* 18:325 (1966).
7. Blinc, M., and M. Bojec, *Arch. Mikrobiol.* 12:41 (1941).
8. Bernhauer, K., A. Niethammer and J. Rauch, *Biochem. Z.*, 319:94 (1948).
9. Efamol Ltd, European Patent Application 0153134A (1985).
10. Nisshin Oil Mills Ltd, UK Patent Application 2152042A (1985).
11. Suzuki, O., and T. Yokochi, European Patent Application 0155420A (1984).
12. Suzuki, O., and T. Yokochi, European Patent Application 0125764A (1983).
13. Suzuki, O., *World Conference on Biotechnology for the Fats and Oils Industry*, edited by T.H. Applewhite, American Oil Chemists Society, Champaign, IL, in press.
14. Nestlé SA, UK Patent Application 2118567A (1983).
15. Wolf, R.B., R. Kleiman and R.E. England, *J. Am. Oil Chem. Soc.* 60:1858 (1983).
16. Meers, J.L., and W. Prescott, *Chemspec. Europe 87 BACS Symp.*, 1987, pp. 1)3.
17. Gibson, R.A., and G.M. Kneebone, *Amer. J. Clin. Nutr.* 34:200 (1981).
18. Ahern, T.J., *J. Am. Oil Chem. Soc.* 61:1754 (1984).
19. Seto, A., H.L. Wang and C.W. Hesseltine, *J. Am. Oil Chem. Soc.* 61:892 (1984).
20. Mangold, H.K., *Chemy. Ind.*, pp. 260)267 (1986).
21. Prokop, A. and Fekri, M., *Biotech. Bioeng.* 26:1282 (1984).
22. Kyle, D.J., in *World Conference on Biotechnology for the Fats and Oils Industry*, edited by T.H. Applewhite, American Oil Chemists' Society, Champaign, IL, in press.
23. Yamada, H., in *World Conference on Biotechnology for the Fats and Oils Industry*, edited by T.H. Applewhite, American Oil Chemists' Society, Champaign, IL, in press.
24. Borowitzka, M.A., *Microbiol. Sci.* 3:372 (1986).
25. Moreton, R.S., in *World Conference on Biotechnology for the Fats and Oils Industry*, edited by T.H. Applewhite, American Oil Chemists' Society, Champaign, IL, in press.
26. Sinden, K.W., *Enz. Microb. Technol.* 9:124 (1987).
27. Moreton, R.S., *Single Cell Oil* edited by R.S. Moreton, Longmans, London, 1987.
28. Fuji Oil Co. Ltd., UK Patent 1555000 (1979).
29. CPC International Inc., UK Patent Application 2091286A (1982).
30. Noguchi, Y., M. Kame and H. Iwamoto, *Yukagawa* 31:431 (1982).
31. Moreton, R.C., *Appl. Microbiol. Biotechnol.* 22:41 (1985).
32. Falbe, J., and R.D. Schmid, *Fette Seifen Anstrichmittel* 88:203 (1986).
33. Edited by N. Kosaric, W.L. Cairns and N.C.C. Gray, *Biosurfactants and Biotechnology*, Marcel Dekker, New York (1987).
34. Ratledge, C., in *Biotechnology—A Comprehensive Treatise*, Vol. 4, edited by H. Pape and H.-J. Rehm, 1986, pp. 185–213.
35. Prescott, W., *Edible Oil Processing Symposium*, Institute of Chemical Engineers, Hull, UK (1987).
36. Davies, R.J., *Food Technol. N.Z.* (June) pp. 33–37 (1984).
37. Davies, R.J., in *Single Cell Oil*, edited by R.S. Moreton, Longmans, London, in press.
38. Floetenmeyer, M.D., B.A. Glatz and E.G. Hammond, *J. Dairy Sci.* 68:633 (1985).
39. Moreton, R.S. and J.R. Norris, in *Developments in Food Microbiology*, edited by R.K. Robinson, Elsevier, in press.
40. Simon Rosedowns Ltd., British Patent 1466853 (1977).
41. Bell, D.J., and R.J. Davies, *Biotech. Bioeng.* 29:1176 (1987).
42. Ratledge, C., in *Proceedings: World Conference in Emerging Technologies in the Fats and Oils Industry*, edited by A.R. Baldwin, American Oil Chemists' Society, Champaign, IL, 1986, p. 318.
43. Bati, N., E.G. Hammond and B.A. Glatz, *J. Am. Oil Chem. Soc.* 61:1743 (1984).
44. Montet, D., R. Ratomahenina, P. Galzy, M. Pina and J. Graille, *Biotechnol. Lett.*, 7:733 (1985).
45. Yamauchi, T., T. Kimura, K. Umezawa and Y. Ohtaki, *Nippon Shokuhin Kogyo Gakkaishi* 33:256 (1986).
46. Ratledge, C., and K.H. Tan, in *Yeast: Biotechnology-Biocatalysis*, edited by H. Verachtert and De Mot, Marcel Dekker, in press.
47. Moissdorfer, F., in *World Conference on Biotechnology for the Fats and Oils Industry*, edited by T.H. Applewhite, American Oil Chemists' Society, Champaign, IL, in press.
48. Soda, K., in *World Conference on Biotechnology for the Fats and Oils Industry*, edited by T.H. Applewhite, American Oil Chemists' Society, Champaign, IL, in press.
49. Uchio, R., and I. Shiio, *Agric. Biol. Chem.*, 36:1389 (1972).
50. Hill, F.F., I. Venn and K.L. Lukas, *Appl. Microbiol. Biotechnol.* 24:168 (1986).
51. Lilly, M.D., *J. Chem. Technol. Biotechnol.*, 32:162 (1982).
52. Buckland, B.C., P. Dunhill and M.D. Lilly, *Biotechnol. Bioeng.* 17:815 (1975).
53. Fish, N.M., D.J. Allenby and M.D. Lilly, *Eur. J. Appl. Microbiol. Biotechnol.* 14:259 (1982).
54. Harbron, S., B.W. Smith and M.D. Lilly, *Enz. Microb. Technol.* 8:85 (1986).
55. Brookes, I.K., M.D. Lilly and J.W. Drozd, *Enz. Microb. Technol.* 8:53 (1986).
56. Macrae, A.R., in *Proceedings: World Conference in Emerging*

Technologies in the Fats and Oils Industry, edited by A.R. Baldwin, American Oil Chemists' Society, Champaign, IL, 1986, p. 7.

57. Lazar, G., A. Weiss, R.D. Schmid, *Ibid.*, p. 346.
58. Baratti, J., G. Buono, H. Deleuze, G. Langrand, M. Secchi and C. Triantaphylides, *Ibid.*, p. 355.
59. Macrae, A.R., and R.C. Hammond, *Biotechnol. Gen. Eng. Rev. 3*:193 (1986).
60. Wisdom, R.A., P. Dunhill and M.D. Lilly, *Biotech. Bioeng. 29*:1081 (1987).
61. *Fatty Acid Metabolism and its Regulation*, edited by S. Numa, Elsevier, Amsterdam (1984).
62. Wakil, S.J., J.K. Stoops and V.C. Joshi, *Annu. Rev. Biochem. 52*:537 (1983).
63. Ratledge, C., and C.T. Evans, in *The Yeasts*, 2nd ed. Vol. 4, edited by A.H. Rose and J.S. Harrison, Academic Press, in press.
64. Schweizer, E., in *World Conference on Biotechnology for the Fats and Oils Industry*, edited by T.H. Applewhite, American Oil Chemists' Society, Champaign, IL, in press.
65. Thorpe, R.F., and C. Ratledge, *J. Gen. Microbiol. 72*:151 (1972).
66. Srere, P.A., *Curr. Top. Cell. Regln. 5*:229 (1972).
67. Botham, P.A., and C. Ratledge, *J. Gen. Microbiol., 114*:361 (1979).
68. Evans, C.T., A.H. Scragg and C. Ratledge, *Eur. J. Biochem. 130*:195 (1983).

69. Evans, C.T., A.H. Scragg and C. Ratledge, *Eur. J. Biochem. 32*:609 (1983).
70. Evans, C.T., and C. Ratledge, *Can. J. Microbiol. 31*:845 (1985).
71. Boulton, C.A., and C. Ratledge, *J. Gen. Microbial. 129*:2871 (1983).
72. Evans, C.T., and C. Ratledge, *Can. J. Microbiol. 31*:1000 (1985).
73. Evans, C.T., and C. Ratledge, *Gen. Microbiol. 130*:3251 (1984).
74. Evans, C.T., and C. Ratledge, *Can. J. Microbiol. 31*:479 (1985).
75. Fritsch, H., and H. Beevers, *Plant Physiol. 63*:687 (1979).
76. Nelson, D.R., and R.W. Rinne, *Plant Physiol. 55*:69 (1975).
77. Mattoo, A.K., and V.V. Modi, *Biochem. Biophys. Res. Commun. 39*:885 (1970).
78. Kaethner, T.M., and R. ap Rees, *Planta, 163*290 (1985).
79. ap Rees, T., J.H. Bryce, P.M. Wilson and J.H. Green, *Arch. Biochem. Biophys. 227*:511 (1983).
80. Boulton, C.A., and C. Ratledge, *J. Gen. Microbiol. 129*:2863 (1983).
81. Evans, C.T., A.H. Scragg and C. Ratledge, *Eur. J. Biochem. 132*:617 (1983).
82. Holdsworth, J.E., and C. Ratledge, *J. Gen. Microbiol.*, in press.
83. Holdsworth, J.E., and C. Ratledge, *Proc. 11th Intern. Spec. Symp. on Yeasts*, 1986, p. 66.
84. Holdsworth, J.E., *Aspect of Lipid Metabolism in Oleaginous Yeast*, Ph.D. thesis, University of Hull, UK (1987).

Enzyme Technology for the Lipids Industry: An Engineering Overview

Tsuneo Yamane

Laboratory of Bioreaction Engineering, Department of Food Science and Technology, School of Agriculture, Nagoya University, Nagoya 464, Japan

Enzymes useful in the lipids industry, i.e. lipases, phospholipases and lipoxygenases, are surveyed as to source, pH optimum, specificity and so on. Some useful biochemical reactions catalyzed by these enzymes are discussed: hydrolysis of fats and oils by lipases, transesterifications (acidolysis, alcoholysis, interesterification and aminolysis) of fats and oils by lipases, hydrolysis of lecithin by phospholipase A_2 and transphosphatidylation of phospholipids by phospholipase D. Research and development activities in these fields in the academic and industrial sectors of Japan are discussed.

With reference to the lipolytic enzymes' applications, forms or states with which enzymes and microorganisms are used in microaqueous solvent systems, i.e. in low water-activity media or in nearly anhydrous solvents, are summarized. Some configurations of reactors for the microaqueous biosystems are shown, and some engineering problems involved in the systems are identified. The importance of optimal moisture content control is emphasized.

In this review, lipids cover not only edible fats and oils but also natural phospholipids and glycolipids. Among interesting and promising aspects of old and new biotechnologies relevant to the lipids industry, this overview will be confined to enzyme (isolated and whole cells as an container) technology and engineering.

Many enzymes are involved in the catabolic and anabolic metabolisms of lipids but there are not many enzymes and the types of reactions that are or will be considered to apply to the lipids' industry. They are mostly lipolytic enzymes such as lipases and phospholipases. This should not be surprising in consideration of the present state of the art of enzyme technology and engineering. How many enzymes have been adopted in industry? Only some hydrolases, isomerases and lyases are sharing success stories among thousands of enzymes that are known to exist in life forms. The lipolytic enzymes (lipases and phospholipases), an oxidase (lipoxygenase) and some engineering aspects in their industrial applications are discussed here. Recent trends in biochemical and biotechnological studies on these enzymes and R & D in both academic and industrial sectors of Japan also will be introduced briefly.

LIPASES AND ESTERASES

Variety of lipases. Generally, the properties of enzymes, even though they catalyze the same type of reaction, differ with their origins. Nature is surprisingly profound and of great variety in this respect—far beyond man's imagination. This is valid completely with lipase, which is a single enzyme numbered as triacylglycerol acyl-

hydrolase E.C. 3.1.1.3. The enzyme has a very wide range of properties, depending on its source, with respect to positional specificity, fatty acid specificity, pH optimum and thermostability. This suggests that one can find a suitable lipase (or lipase-producing organism) from nature that fits a given application. It often is stated that the lipases can be placed into two groups according to their positional specificity, 1,3-specific and nonspecific. However, the author agrees with H. Machida's comment that the positional specificity of lipases is not divided clearly into two but it continuously changes from very distinctly specific to very weakly specific or completely nonspecific (1). The situation is made more complicated due to nonenzymatic acyl migration from β- to α-position in glycerides (2).

According to the endoplasmicreticulum (EC of IUB, there is another enzyme, monoacylglycerol lipase (E.C. 3.1.1.23, glycerol-monoester acylhydrolyase), which is different from genuine lipase. This often is found in animal tissues and also is produced by a microorganism. It is difficult to differentiate this enzyme from what is called "esterase."

Fatty acid specificity of lipases is ambiguious. The substrate specificity of a lipase usually is reported as the relative hydrolysis rate of a single triglyceride vs the number of carbons in the fatty acid. The spectrum diverges significantly, which indicates that lipases do not have strict substrate specificity. The substrate specificity is made more vague by the fact that the rate of hydrolysis is affected not only by the substrate but also by physical factors (solid or liquid, particle or emulsion size, degree of turbulence, solubility in water in case of low carbon-numbered triglyceride).

As with other enzymes, each lipases has its own optimal pH, ranging from acid to neutral to alkaline. Development of alkaliphilic lipase aims at two applications: its addition in laundry detergents to enhance cleaning and a substitute for pancreatic lipase in digestive medicine.

Thermostability of lipases also varies considerably with their origins. Animal and plant lipases usually are less thermostable than microbial extracellular lipases. Relatively thermostable lipases produced by *Pseudomonas* species (3,4) and by *Humicola* sp. (5) have been reported in Japan.

Microbial lipases produced industrially in Japan and their properties are summarized in Table 1. Other lipases would be available on request but their scale of production may be small.

Biochemical reactions catalyzed by lipases. Lipases and esterases catalyze three types of reaction (Fig. 1). The catalytic action of lipases is reversible. It catalyzes ester synthesis in a microaqueous system (see Section 5.1). However, in view of biotransformation in oleochemical industry yielding value-added products,

TABLE 1

Microbial Lipases Produced Industrially in Japan

Strain	Manufacturer	Thermostability (remaining activity)	Optimal pH	Positional specificity	Molecular weight
Candida cylindracea *(Candida rugosa)*	Meito Sangyo	40% (50 C, 10 min)	7.0	α, β, α'	55,000
Aspergillus niger	Amano Pharmaceutical	50% (60 C, 15 min)	5.6	α, α'	38,000
Rhizopus niveus	Amano Pharmaceutical	60% (50 C, 30 min)	7.0	α, α'	—
Pseudomonas fluorescens	Amano Pharmaceutical	60% (60 C, 30 min)	7.0	α, α'	31,000
Rhizopus japonicus	Osaka Saikin Kenkyusho	50% (55 C, 30 min)	5.0	α, α'	30,000

In addition to these, Sapporo Breweries, Ltd. is producing *Pseudomonas fragi* lipase and Toyo Jozo Co., Ltd. is producing *Chromobacterium viscosum* lipase.

the transesterification action seems more worthwhile than hydrolysis and ester synthesis. The difference in free energy involved in triglyceride hydrolysis is quite small and the net free energy of transesterification is zero. Consequently, transesterification reactions take place easily. Transesterification is categorized into four subdivisions according to the chemical species with which the ester reacts (Fig. 1). Some researchers designate these four types of reaction by "interesterification" but the author prefers transesterification to interesterification as the technical term covering all four types of reaction because in biochemistry transfer of a group from one chemical species to another is called "trans", such as transglycosylation, transpeptidylation, transphophatidylation. Therefore, the author confines the term interesterification only to the type 3 reaction (ester exchange).

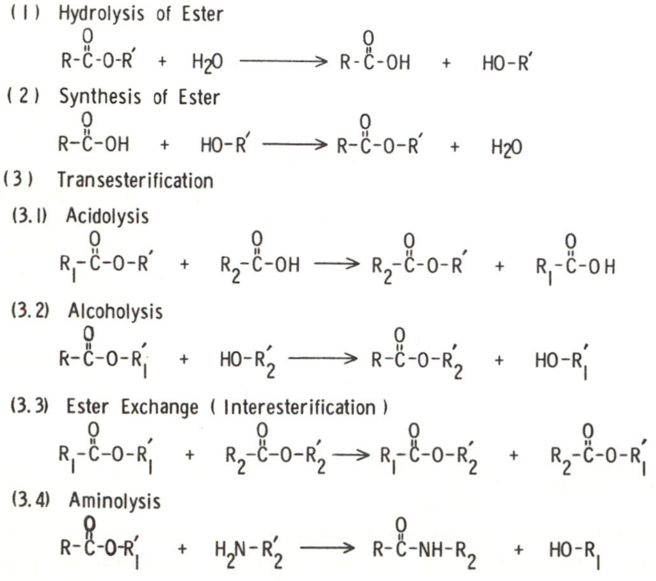

R & D activities on lipase applications in Japan. Research on microbiology, biochemistry and biotechnology of lipases and on their industrial applications are quite active in Japan. Division of Biochemistry, Osaka Municipal Technical Research Institute has a long history of research on microbial lipases. They currently are interested in gene cloning of *Geotrichum candidum* lipase and its crystallographic structure. Extensive microbiological and biochemical studies on *Saccharomycopsis lipolytica (Candida paralipolytica)* have been carried out by Y. Otha et al. (6). Development of polyethylene glycol (PEG)-modified lipases (organic solvent-soluble lipases) and their application to a number of biochemical reactions in organic solvent media have been reported by Y. Inada and his associates (7). The use of hydrophobic gels for the immobilization of lipase was reported by S. Fukui, A. Tanaka and coworkers (8). Lipase-catalyzed synthesis of a microcyclic lactone (cyclopentadecanolide) was reported by Y. Yamada et al. (9). This will be useful in the production of synthetic musks. Hydrolysis of fats and oils dissolved in isooctane with aqueous lipase solution was studied by J. Takahashi and coworkers (10). Esterification of high acid-value rice bran oil has been studied with a thermostable lipase immobilized in cationic resin beads by Y. Kosugi et al. (11). Enzymic sugar ester syntheses were attempted by H. Seino (12) and by S. Nagai (13), both in collaboration with Dai-ichi Kogyo Seiyaku Co. Ltd. The author and coworkers have studied lipase-catalyzed glycerolysis of fats and oils (14).

The chemical process of fat-splitting currently being performed in industry is very efficient in energy recycling, and it is unlikely that enzymatic fat-splitting will compete economically with the conventional chemical process in the near future. The interest of people in the enzymatic hydrolysis is shifting from bulky fatty acid products to high value-added products or to unstable fatty acids that may be decomposed through

(1) Hydrolysis of Ester

$$R-\overset{O}{\overset{\|}{C}}-O-R' + H_2O \longrightarrow R-\overset{O}{\overset{\|}{C}}-OH + HO-R'$$

(2) Synthesis of Ester

$$R-\overset{O}{\overset{\|}{C}}-OH + HO-R' \longrightarrow R-\overset{O}{\overset{\|}{C}}-O-R' + H_2O$$

(3) Transesterification

(3.1) Acidolysis

$$R_1-\overset{O}{\overset{\|}{C}}-O-R' + R_2-\overset{O}{\overset{\|}{C}}-OH \longrightarrow R_2-\overset{O}{\overset{\|}{C}}-O-R' + R_1-\overset{O}{\overset{\|}{C}}-OH$$

(3.2) Alcoholysis

$$R-\overset{O}{\overset{\|}{C}}-O-R'_1 + HO-R'_2 \longrightarrow R-\overset{O}{\overset{\|}{C}}-O-R'_2 + HO-R'_1$$

(3.3) Ester Exchange (Interesterification)

$$R_1-\overset{O}{\overset{\|}{C}}-O-R'_1 + R_2-\overset{O}{\overset{\|}{C}}-O-R'_2 \longrightarrow R_1-\overset{O}{\overset{\|}{C}}-O-R'_2 + R_2-\overset{O}{\overset{\|}{C}}-O-R'_1$$

(3.4) Aminolysis

$$R-\overset{O}{\overset{\|}{C}}-O-R' + H_2N-R'_2 \longrightarrow R-\overset{O}{\overset{\|}{C}}-NH-R_2 + HO-R'$$

FIG. 1. Types of reaction catalyzed by lipase.

the superheated steam splitting procedure. Industrial preparations of optical active alcohols related to synthetic pyrethroids insecticides with an *Arthrobacter* lipase has been realized by H. Hirohara and S. Mitsuda working with Sumitomo Chemical Co. Ltd. (15). Enrichment or isolation of some long-chain polyunsaturated fatty acids from refined fish oil through lipase action are getting much attention in Japanese fats and oils companies because of their potential for reducing the incidence of thrombosis and related diseases.

Production of cocoa butter substitutes through acidolysis of inexpensive fats by lipase is under active R & D in several Japanese companies (Fuji Oil, Ashai Denka Kogyo, Kanegafuchi Chemical Industry). Isolated lipase immobilized in an appropriate carrier or dry microbial cells having lipase are used in packed bed bioreactors. Moisture content control is the key factor of the acidolysis reaction.

Characterization and application of *Pseudomonas fragi* lipase are being investigated by T. Nishio working with Sapporo Breweries Ltd. (16). A unique monoacylglycerol lipase produced by *Penicillium* sp. has been commercialized recently from Amano Pharmaceutical Co., Ltd., with the aim of monoglyceride synthesis from free fatty acid and glycerol (17). Enzymatic production of monoglyceride still is too costly to compete with a chemical process (a glycerolysis reaction).

Phospholipases. Five kinds of phospholipases exist in nature according to their sites of attack of ester bonds of phospholipids. Only a phospholipase A_2 ("Lecitase" from Novo Industrie A/S) manufactured from porcine pancreatic glands and a phospholipase D (PLD) excreted from *Streptomyces chromofuscus* (Toyo Jozo Co., Ltd.) are available commercially on large scale. The class of phospholipids, substrates for phospholipases, is synonymous with lecithin, and is isolated in various grades from soybean and egg yoke. They are essential constituents of biomembrane and have surface active properties. Lecithins have been utilized as one of five emulsifiers permitted in the food industry. Patents claim that when soybean lecithin is partially hydrolyzed by phospholipase A_2, the resulting phospholipids (containing lyzolecithin) increase in hydrophilicity and have an enhanced power of o/w type emulsification (18). The modified lecithin is on sale now though the production scale is unclear. Phospholipids also find application in medical and pharmaceutical fields. Liposomes, made of various kinds of phospholipids, are expected to be good drug delivery systems.

PLD not only hydrolyzes phospholipids but also catalyzes a transphosphatidylation reaction in which exchange of base takes place as indicated in Figure 2. Attempts are being made by Yakurt Central Institute to convert soybean lecithin to phosphatidylglycerol with the transphosphatidylation action of PLD. Phosphatidylglycerol is claimed to be an effective emulsifier even in the presence of calcium ion. The author and others are studying this reaction with the gaol of producing each component of phospholipid in pure state (19). Kinetic estimation of transphosphatidylation activities of several PLD from different sources reveals that the activity varies considerably with the enzyme sources (unpublished data).

Lipoxygenase. The major or sole source of lipoxy-

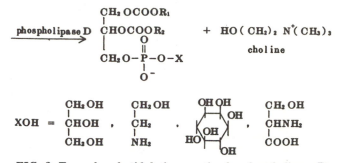

FIG. 2. Transphosphatidylation reaction by phospholipase D.

genase now utilized is soybean. However, lipoxygenases from *Fusalium oxysporum* were discovered and investigated enzymatically by K. Arima et al. (20). Chemical modification of lipoxygenase has been attempted by H. Hirata et al. of the National Chemical Laboratory for Industry, MITI, as a national project of the Research Association for Biotechnology. If the products, fatty acid hydroperoxides or their reduced derivatives (hydroxyconjugated fatty acid), find useful industrial application, lipoxygenase-catalyzed oxidation of unsaturated fatty acids will be realized commercially.

GENETIC AND PROTEIN ENGINEERING

The production costs of industrial lypolytic enzymes are high as compared with amylases and proteases. The cost seems to hamper wider industrial application of the enzymes; it depends heavily on the degree of purity. There is a hope that the cost can be reduced by gene technology such as gene amplification in addition to a traditional random mutation. The first and essential step of genetic manipulation is cloning of genes involved in the enzyme's biosynthesis. Recently, genes of bacterial lipases (21,22), a mammalian phospholipase A_2 (23) and a bacterial sphingomyelinase (24) have been cloned. A number of genes of lypolytic enzymes will be cloned rapidly in the coming years. Many useful reactions catalyzed by the enzymes in oleochemistry are conducted in microaqueous organic media (see the next section). In this respect, protein engineering approaches will help the elucidation of mechanism of solvent-denaturation and will create novel enzyme proteins that are more resistant to the organic solvents.

BIOCATALYSIS IN MICROAQUEOUS ORGANIC SOLVENT SYSTEM

What is a microaqueous biosystem? As mentioned above, many industrially intriguing reactions catalyzed by the lipolytic enzymes are carried out in organic media. However, one should realize, that the organic media are not completely anhydrous. Recent research at many laboratories in the world have shown that a complete depletion of water from the system results in no biochemical reaction. Water is essential for the enzyme to display its full catalytic activity because it is a protein. With reference to biocatalysts' application in organic media, some technical phrases have been used: "in low water-activity media," "in nearly anhydrous solvent," "in an organic solvent containing a little amount of water," etc., to show that the reaction systems are not absolutely anhydrous. To emphasize the importance of water for biocatalysis in organic solvents and to cover all possible forms or states of biocatalysts utilization in organic media, would like to propose a shorter but clearer term "microaqueous," which means that the system is not aqueous, nonaqueous or anhydrous. "Microaqueous biosystem" is defined roughly as a biochemical or biological system in which moisture content is low as compared with a conventional one. That is, microaqueous biochemical systems are made up of water-soluble organic substances or water-insoluble organic solvents.

Optimal control of the moisture content is the keystone of the microaqueous biocatalysis system as it affects the reaction rate, product yield, product selectivity and operational stability. A rough profile of effects of moisture content on the last two variables is drawn in Figure 3, which suggests that at lower moisture content the yield of the product may be high, but the reaction rate may be lower. While at the greater moisture content the rate becomes higher but the yield drops. In between, there is an optimal moisture content from an engineering viewpoint. The effects of moisture content on the biochemical reactions in organic media are exemplified by glyceride synthesis (25) and glycerolysis reaction (14) both carried out by the author's group using acidolysis (26) and by reaction in reversed micelles (27). These are only several among a number of reactions that follow the scheme shown in Figure 3.

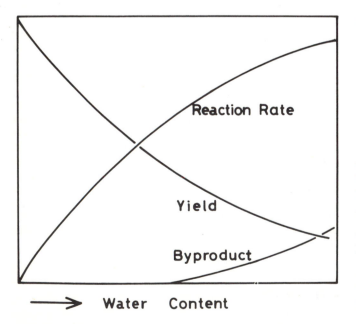

FIG. 3. Effect of moisture content on a reaction conducted in a microaqueous organic medium.

Enzyme forms used in microaqueous solvent systems. In the past several years, a broad range of forms or states have been developed with which enzymes may be used for diversed reactions in microaqueous solvent systems. These can be classified tentatively into six forms:

(1) If a substrate is water-soluble, its highly concentrated aqueous solution becomes microaqueous. Enzymes are dissolved in the concentrated substrate solutions. Carbohydrates like glucose and sucrose are examples. Alternately, enzymes are dissolved in a water-soluble organic solvent that has some amount of water. Methanol, ethanol, ethylene glycol, glycerol, dimethylformamide, dimethylsulfoxide and acetone are some such solvents. In our glyceride synthesis and glycerolysis reaction, lipases are dissolved in microaqueous glycerol (14,25).

(2) Solid enzyme is suspended in a water-immiscible or water-insoluble organic solvent (9,28,29). If the solid enzyme is not suspended easily, it is adsorbed on particles of support like diatomaceous earth prior to its suspension.

(3) An enzyme molecule is confined in a reversed micelle solubilized in a water-insoluble organic solvent (30,31). The formation of reversed micelles using a surfactant, AOT, in aliphatic hydrocarbons such as heptane, octane and isooctane is well-known. Sharp dependence of the activity of enzyme in reversed micelles on the water content as expressed by $w_0 = [H_2O]/[AOT]$ generally is recognized.

(4) An enzyme is solubilized in an aromatic solvent (benzene, toluene, etc.) by binding it with PEG via a triazine ring. It was demonstrated that some amount of water is needed for the PEG-bound enzyme to exhibit its activity (32).

(5) An enzyme is entrapped within a gel whose degree of hydrophobicity can be controlled. The gel-entrapped enzyme is suspended in water-immiscible organic solvent (8).

(6) Porous particles are impregnated with an aqueous solution of the enzyme. The particles are suspended or packed in a water-immiscible organic solvent. The enzyme can be used in free and immobilized states. This is a sort of w/o emulsion but the proportion of water phase to the organic phase is lowered greatly in such a system, and the system is more convenient technologically than the free w/o emulsion (33).

The forms (1), (2), (4) and (5) are microaqueous at molecular level, while the form (6) is microaqueous at the phase level. The form (3) lies between molecular and phase levels. It may be called microaqueous at pseudophase level. Thus people have obtained a variety of forms of enzyme workable in organic media. It is difficult to predict which form is most suitable for industrial purposes but economics eventually will decide this.

Microorganism forms used in microaqueous solvent systems. Microbial cells are used in one of the following forms in microaqueous solvent.

(1) Wet or semi-dried cells are suspended in water-immiscible organic solvent (34). Wet microbial cells contain 70–80 wt % of water. If they are dispersed in the solvent, the whole system is microaqueous

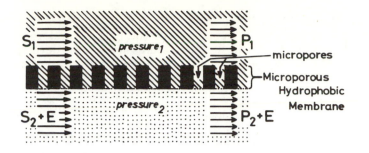

pressure$_1$ < pressure$_2$

FIG. 4. The neighborhood of the membrane in the microporous hydrohobic membrane bioreactor system performed continuously. The substrates, S$_1$ and S$_2$, and the products, P$_1$ and P$_2$, for four types of reaction are shown below. Enzyme is dissolved in S$_2$ phase or adsorbed on the S$_2$-side surface of the hydrophobic membrane.

	Type of reaction	S$_1$	S$_2$	P$_1$	P$_2$	Ref.
1.	Hydrolysis	Fat	Water	Fatty acid	Glycerol	37
2.	Glyceride synthesis	Fatty acid	Glycerol	Glyceride	Water	25
3.	Glycerolysis	Fat	Glycerol	Partial glyceride	—	14
4.	Trans-phosphatidylation	Phosphatidyl-choline in ether	Glycerol in water	Phosphatidyl-glycerol	Choline in water	38

because water is confined only within the cells.

(2) Semi-dried mycellium cells packed in a column (35).

(3) Wet or semi-dried cells immobilized by entrapping in gels having an appropriate hydrophobicity-hydrophilicity balance (36) or by holding them in a sponge-like support.

As usual, the use of whole cells having the enzyme under consideration seems more economical than the use of the enzymes isolated from their cultures.

REACTOR SYSTEMS FOR MICROAQUEOUS BIOCATALYST

The reactor configurations for microaqueous biocatalyst may not differ much from those for conventional aqueous media. Batch reactors (stirred tank reactors) and continuous reactors such as packed bed (PBR), stirred tank (CSTR) and fluidized bed (FBR) are conceivable. Membrane reactors with either batch or continuous operation are also a choice. We have developed a novel bioreactor system having a microporous hydrophobic membrane. This is based on the principle of contacting two immiscible liquid phases at the membrane surface so the reactor is suitable for two-liquid phase biocatalysis system as shown by

Substrate 1 (phase 1) + Substrate 2 (phase 2)

  biocatalyst

      Product 1 (phase 1) + Product 2 (phase 2)

In this bioreactor, reaction and phase separation can be achieved simultaneously. The representations of the membrane in the bioreactor for the four cases are depicted schematically in Figure 4.

One of the advantages of the microaqueous bioreactor system is a lack of microbial contaminations. This should be emphasized in view of the bioreactor commercialization system.

The substrate solution fed to a microaqueous bioreactor system must contain a definite amount of water. If it is anyhdrous, biocatalysts in the reactor would lose moisture gradually, which would result in loss of activity. However, moisture content will be undesirable in terms of yield and selectivity of product. When water is formed through the reaction as in a condensation reaction, it must be removed by an appropriate method such as purging the liquid with bubbles of dry inert gas, reducing the pressure or adsorption with molecular sieves.

For monitoring and automatic control of moisture content, a moisture sensor is required. Several choices are available for this purpose: refractometer, electric conductivity meter, water activity meter, dielectric constant meter and dew point meter (thin aluminum film sensor). The last sensor can detect 0–400 ppm of free water dissolved in hydrocarbon on line. We have found that 0–5% moisture content of glycerol could be detected by an electric conductivity electrode (39). One of these sensors will enable control of the moisture content of an organic medium at an optimal level.

DISCUSSION

Research and development concerning enzyme applications to the lipids industry has not been as active as those to the areas of carbohydrate, protein and amino acids. This is mainly due to the following facts:

(1) The number of high value-added products in the oleochemical industry is limited.

(2) The enzymes involved in the biotransformations for the lipids industry are costly.

(3) There are many difficulties in solving engineering problems because of heterogeneous and/or micro-aqueous natures of oleochemical bioreactions. Basic quantitative data, especially data on operational stability of enzymes, are scarce.

The future prospects in enzyme technology for the lipids industry will be widened by further endeavor in these three subjects. Fortunately, we are witnessing various and enthusiastic activities of R & D all over the world, and we are optimistic enough to see several new enzyme-catalyzed processes in the lipids industry in the near future.

REFERENCES

1. Machida, H., *Yukagaku* (in Japanese) *33*:691 (1984).
2. Mattson, F.H., and R.A. Volpenhein, *J. Lipid Res.*, *3*:281 (1962).
3. Sugiura, M., T. Oikawa, K. Hirano and T. Inukai, *Biochem. Biophys. Acta.* *488*:353 (1977).
4. Kosugi, K., H. Suzuki, A. Kanbayashi, T. Funada, M. Akaike and K. Honda, Japanese Patent 57-38239 (1982).
5. Liu, W.-H., T. Beppu and K. Arima, *Agric. Biol. Chem.* *37*:2493 (1973).
6. Gomi, K., Y. Ota and Y. Minoda, *Agric. Biol. Chem.* *50*:2531 (1986).
7. Yoshimoto, T., K. Takahashi, H. Nishimura, A. Ajima, Y. Tamaura and Y. Inada, *Biotechnol. Letters* *6*:337 (1984).
8. Yokozeki, K., S. Yamanaka, K. Takinami, Y. Hirose, A. Tanaka, K. Sonomoto and S. Fukui, *J. Appl. Microbiol. Biotechnol.* *14*:1 (1982).
9. Makita, A., T. Nihira and Y. Yamada, *Tetrahedron Letters* *28*:805 (1987).
10. Mukataka, S., T. Kobayashi and J. Takahashi, *J. Ferment. Technol.* *63*:461 (1985).
11. Kosugi, K., H. Igusa and N. Tomizuka, *Abstract of Annual Meeting of Soc. Agric. Biol. Chem.*, Japan, 1985, p. 657.
12. Seino, H., T. Uchibori, T. Nishitani and S. Inamasu, *J. Am. Oil Chem. Soc.* *61*:1761 (1984).
13. Nakata, H., K. Jinno, Y. Chikazawa, I. Morita, N. Nishio, M. Hayashi and S. Nagai, *Abstract of Annual Meeting of Soc. Ferment. Technol.*, Japan, 1986, p. 78.
14. Yamane, T., M.M. Hoq, S. Itoh and S. Shimizu, *J. Jpn. Oil Chem. Soc.* *35*:625 (1986).
15. Hirohara, H., S. Mitsuda, E. Ando and R. Komki, *Biocatalysts in Organic Synthesis*, edited by J. Tramper, H.C. van der Plas and P. Linko, Elsevier, Amsterdam, 1985, p. 119.
16. Nishio, T., T. Chikano and M. Kamimura, *Agric. Biol. Chem.* *51*:181 (1987).
17. Yamaguchi, S., T. Mase, S. Asada, European Patent 0191217 (1986).
18. Kyowa Hakko Kogyo Co., Ltd., Japanese Patent 58-51853 (1983).
19. Juneja, L.R., N. Hibi, T. Yamane and S. Shimizu, *Appl. Microbiol. Biotechnol.*, in press.
20. Matsuda, Y., T. Beppu and K. Arima, *Agric. Biol. Chem.* *43*:1179 (1979).
21. Odera, M., K. Takeuchi and A. Toh-e, *J. Ferment. Technol.* *64*:363 (1986).
22. Kugimiya, W., Y. Otani, Y. Hashimoto and Y. Takagi, *BBRC* *141*:185 (1986).
23. Seilhamer, J.J., T.L. Randall, M. Yamanaka and L.K. Johnson, *DNA* *5*:519 (1986).
24. Tsukagoshi, N., A. Yamada, M. Tomita, H. Ikezawa and S. Udaka, *Abstract of Annual Meeting of Soc. Agric. Biol. Chem.* Japan, 1987, p. 31.
25. Hoq, M.M., T. Yamane, S. Shimizu, T. Funada and S. Ishida, *J. Am. Oil Chem. Soc.* *61*:776 (1984).
26. Macrea, A.R., *Biocatalysis in Organic Synthesis*, edited by J. Tramper, H.C. van der Plas and P. Linko, Elsevier Amsterdam, 1985, p. 195.
27. Barbaric, S., and P.L. Luisi, *J. Am. Chem. Soc.* *103*:4239 (1981).
28. Klibanov, A.M., *Biocatalysis in Organic Media*, edited by C. Laane, J. Tramper and M.D. Lilly, Elsevier, Amsterdam, 1987, p. 115.
29. Ooshima, H., H. Mori and Y. Harano, *Biotechnol. Letters* *7*:789 (1985).
30. Edited by Luisi, P.L. and B.E. Straub, *Reverse Micelle*, Plenum Press, New York (1984).
31. Han, D., and J.S. Rhee *Biotechnol. Bioeng.* *28*:1250 (1986).
32. Takahashi, K., H. Nishimura, T. Yoshimoto, M. Okada, A. Ajima, A. Matsushima, Y. Tamaura, Y. Saito and Y. Inada, *Biotechnol. Letters* *6*:765 (1984).
33. Martinek, K., and I.V. Berezin, *J. Solid-Phase Biochem.* *2*:343 (1977).
34. Buckland, B.C., P. Dunnill and M.D. Lilly, *Biotechnol. Bioeng.* *17*:815 (1975).
35. Bell, G., J.R. Todd, J.A. Blain, J.D.E. Patterson and C.E.L. Shaw, *Biotechnol. Bioeng.* *23*:1703 (1981).
36. Fukui, S., A. Tanaka and T. Iida, *Biocatalysis in Organic Media*, edited by C. Laane, J. Tramper and M.D. Lilly, Elsevier, Amsterdam, 1986, p. 21.
37. Hoq, M.M., T. Yamane, S. Shimizu, T. Funada and S. Ishada, *JAOCS* *62*:1016 (1985).
38. Lee, S.Y., N. Hibi, T. Yamane and S. Shimizu, *J. Ferment. Technol.* *63*:37 (1985).
39. Yamane, T., J.S. Rhee, Y. Ohta and S. Shimizu, *J. Jpn. Oil Chem. Soc.* *36*, in press.

Biosynthesis of Triglycerides in Plant Storage Tissue

Gareth Griffiths[a], **Sten Stymne**[a] and **Keith Stobart**[b]

[a]Department of Plant Physiology, Swedish University of Agricultural Sciences, S-750 07 Uppsala, Sweden, and [b]Department of Botany, The University, Bristol BS8 1 UG, U.K.

This review deals with our work on the synthesis of C18-polyunsaturated fatty acids and triacylglycerol assembly in linoleate–rich oil seeds. Comparative studies with the developing cotyledons of turnip rape (Brassica campestris) are described that show: (1) that the enzymes involved in channeling oleate to phosphatidylcholine for desaturation are a magnitude lower in activity than similar enzymes in linoleate-rich oil species, (2) Erucoyl–rich oils are synthesized by some modification of the basic triacylglycerol pathway that is envisaged for most C18–rich species. Consideration is given to those reactions that may regulate the acyl quality of the oil.

In recent years, the interest in fatty acid synthesis and its relationship to the assembly of triacylglycerols in oleaceous seeds has increased appreciably. This is due largely to the realization that eventually it may become possible to manipulate acyl quality in a precise fashion and so produce a more desirable commodity. Research in this area has received some impetus with the advent of recombinant-DNA technology and the expectation of successful interspecies gene transfer. Unfortunately, the basic understanding of triacylglycerol biosynthesis and the mechanisms that govern acyl quality, which are all necessary to underpin future ventures into the genetic engineering of plant oils, are lacking somewhat, and a great deal of fundamental research still is required. This is particularly so in those species that can accumulate oils rich in the more uncommon fatty acids such as erucate and γ-linolenate. Our work has been concerned largely with seeds rich in linoleate in order to elucidate the formation of triacylglycerol and the polyunsaturated fatty acid biosynthesis and to provide a model system for comparison with other species. This article deals with some of our findings and where appropriate, emphasis is given to those reactions that may regulate the acyl quality of the final oil.

OIL ACCUMULATION

Triacylglycerols are stored in the seed tissues in discreet oil bodies of ca. $1 \mu m$ in diameter (1). The deposition of the oil in the developing cotyledons of the seed follows defined kinetics. The cotyledons of safflower (2,3), for instance, commence oil deposition after a lag of 12–14 days after pollination and actively accumulate triacylglycerol for 8–10 days more. In safflower, almost 70% of the oil is laid down in a relatively narrow window for 16–20 days after pollination. Although oil deposition in other species may occur over a longer time period, it is essential for biochemical studies to establish its kinetics to take experimental material in a precise and defined state (4). Oil deposition during the development of the seed warrants further investigations to determine the influences that regulate oil quality. Of particular interest is the role of plant hormones in seed development and the factors that govern the supply of photosynthate from mature leaves to the developing seed cotyledons. Of further importance is establishing the details of cell division and expansion in the maturing cotyledons and the geneity of the tissue during development. This is not important only for the interpretation of biochemical studies but is essential for an appraisal of the status of the cells for future genetic engineering work.

Microsomal membrane preparations obtained from young cotyledons at the early phase of oil deposition generally are most active at synthesizing triacylglycerol in vitro from exogenously supplied acyl-CoA and sn-glycerol 3- phosphate. Preparations from the developing cotyledons of the species studies (safflower, sunflower, linseed, soy and rape) so far contain all the necessary enzymes for the acylation of glycerol phosphate and the operation of the Kennedy pathway and can catalyze the assembly of triacylglycerols at respectable rates. In fact, in vitro preparations from many of the above species are so active that under the correct conditions visible oil droplets will accumulate in the incubation mixture; by electron microscopy, these can be observed to originate from the microsomal membrane vesicles (5). The triacylglycerol arises as oil droplets that are much smaller in size than the oil bodies observed in vivo in the maturing cells. The oil droplets appear stabilized by the bovine serum albumin in the reaction mixture and are not surrounded by the half-unit membrane that has been described for the oil body (6). Despite the physical differences of the oil body in vivo and the oil droplet in vitro, the microsomal preparations provide an excellent model system to follow the ontogeny of the droplet in the membrane vesicle and the mechanism of its release into the surrounding medium. The oil body appears to be surrounded by an array of distinct proteins (1). Huang and coworkers (7), based on rather elegant studies, suggest that the boundary proteins of the lipid bodies are synthesized on the polyribosomes of the rough endoplasmic reticulum (ER) and become integrated into the half-unit membrane of the lipid body as it forms and/or is liberated in the bilayers of the ER. It may become possible to study this further in isolated rough ER membranes from oil seeds by providing m-RNA transcripts for some of the oil body proteins together with the necessary cofactors and perhaps follow the simultaneous formation of the triacylglycerol oil droplet with a half-unit membrane.

TRIACYLGLYCEROL ASSEMBLY

Therefore, microsomal membrane preparations from the developing cotyledons of most oil seed species contain all the enzymes for the assembly of the triacylglycerol and at specific activities similar to the situation in vivo (8). This is particularly so in preparations from the maturing cotyledons of safflower, and this species has been favored in our attempts to establish the fundamental aspects of triacylglycerol biosynthesis and its relationship to the formation of the C18-polyunsaturated fatty acids. We will describe these aspects and later will relate our observations to more recent studies with the developing cotyledons of rape. The fatty acids in the triacylglycerol molecule generally are arranged in a nonrandom fashion (2,9) with the more saturated fatty acids predominating at position 1, whereas the mono- and polyunsaturated C18 acids are the only fatty acids at position 2. On the other hand, the fatty acids at position 3 have a more general mix although in some plants a particular fatty acid can predominate (Table 1). It also is of significance that an asymmetry in fatty acid distribution exists in the phosphatidylcholine molecule with saturated and unsaturated species at position sn-1 and only the C18-unsaturates at position 2. Therefore, a biochemical understanding of triacylglycerol biosynthesis has to account fully for these observations. Microsomal membrane preparations from most oil seeds have proved particularly active in the acylation of sn-glycerol 3-phosphate to yield phosphatidic acid. Generally, phosphatidic acid accumulates in the membranes, particularly in the presence of EDTA. Hence, it can be purified and analyzed readily for its acyl group distribution. Acyl specificity and selectivity (10) experiments show that the saturated substrates, palmitate and stearate, are esterified to position 1 whereas the unsaturated C18-acids become associated with position 2 during the concerted two–step acylation of sn-glycerol 3-phosphate:

sn-glycerol 3-P + acyl- CoA → sn-1-acyl-lysophosphatidate
sn-1-acyl-lysophosphatidate + acyl-CoA → phosphatidic acid

Thus, the properties of the enzymes involved in catalyzing the acylation of sn- glycerol 3-phosphate control the asymmetric distribution of saturated and unsaturated acyl species at positions sn-1 and sn-2 of the triacylglycerol molecule.

TABLE 1

Stereospecific Distribution of Fatty Acids in Triacylglycerol from the Developing Cotyledons of _Borago officinalis_

sn-carbon	Fatty acid (mol %)					
	16:0	18:0	18:1	18:2	γ18:3	20:1
sn-1	23	6	16	51	0	4
sn-2	0	0	17	39	44	0
sn-3	5	0	7	31	53	4

The enzyme phosphatidase catalyzes the cleavage of phosphatidic acid to yield diacylglycerol. The oil seed enzyme is inhibited by EDTA; this can be overcome with Mg^{2+} (10). In microsomal preparations, the activity of the phosphatidase is somewhat rate limiting with phosphatidic acid accumulating to some extent when the membranes are incubated with acyl-CoA and sn-glycerol 3-phosphate. Interestingly, in this context the enzyme appears to play a regulatory role in controlling the flow of intermediates towards triacylglycerol in mammalian tissues (11).

The ultimate step in the synthesis of triacylglycerol is the acylation of diacylglycerol at position sn-3. This is catalyzed by a diacylglycerol acyltransferase, the only enzyme considered unique to triacylglycerol biosynthesis. This reaction is responsible for one third of the oil quality of the triacylglycerol; its importance previously has been somewhat neglected. In earlier studies with

safflower, it was considered that the enzyme showed little selectivity (12). However, more recently with the use of an improved assay system the enzyme from a number of sources had some specificity for acyl species (13). That the diacylglycerol acyltransferases may exhibit some acyl specificity is indicated, for example, in the fatty acid distribution in the triacylglycerols of borage _(Borago officinalis)_ seeds. Here (Table 1), the γ-linolenate shows some enrichment at the sn-3 position.

ROLE OF PHOSPHATIDYLCHOLINE IN TRIACYLGLYCEROL BIOSYNTHESIS

In many cases, it is the C18-unsaturated and polyunsaturated fatty acid components that regulate the quality and commercial application of the seed oil. Therefore, much work has been devoted to understanding their formation in relation to triacylglycerol assembly. In vivo radiolabeling studies with developing oil seeds suggested that oleate entered phosphatidylcholine and there was desaturated to linoleate (14,15). This now has been confirmed convincingly in microsomal preparations from developing seeds in which [^{14}C] oleoyl-phosphatidylcholine can be generated, and the formation of [^{14}C] linoleoyl-phosphatidylcholine monitored upon the addition of NADH (8). In the last few years, particular emphasis has been placed on understanding the mechanisms by which oleate enters phosphatidylcholine for desaturation and how the polyunsaturated products are made available for triacylglycerol synthesis. [^{14}C] Oleate experiments in vivo (16) and in vitro (17) show that the substrate very efficiently enters position sn-2 of sn-phosphatidylcholine. This suggested that perhaps a 1-acyl lyso-phosphatidylcholine species was the oleoyl acceptor and that the reaction was catalyzed by the well-documented enzyme, acyl-CoA: lyso-phosphatidylcholine acyltransferase. The regeneration of the acyl acceptor, lyso-phosphatidylcholine, originally was considered to arise perhaps by the action of a phospholipase A$_2$. Microsomal preparations from a number of species (8), however, catalyze the transfer of oleate to position 2 of sn-phosphatidylcholine and the concomitant return of the fatty acids from this position back to the acyl-CoA pool. In microsomes, no evidence could be obtained for the involvement of a phospholipase, and we consider it more plausible that the acyl exchange between acyl-CoA and position 2 of sn-phosphatidylcholine occurs by the concerted forward and reverse reactions of the acyltransferase (18). Hence, the acyl-CoA: lyso-phosphatidylcholine acyltransferase in developing oil seeds controls the entrance of oleate into phosphatidylcholine for desaturation and the return of the C18-polyunsaturated products to the acyl-CoA pool for utilization in triacylglycerol assembly. In vivo studies also indicate that the acyl exchange perhaps is the most important mechanism in oil seeds that is involved in the channeling of substrate into the phosphatidylcholine molecule (16). The bias required to move oleate to phosphatidylcholine may be aided by the selectivity of the glycerol–acylating enzymes for the linoleate in the acyl-CoA pool that is being produced via the acyl exchange (10).

In vitro and in vivo experiments also indicate that glycerol backbone with associated acyl components readily is incorporated into phosphatidylcholine in developing oil seeds (19). This may be catalyzed by a cholinephosphotransferase (20) although the enzyme appears to be somewhat novel and may be involved in equilibrating diacylglycerol and phosphatidylcholine during triacylglycerol formation (21). This would result in the return of glycerol backbone to phosphatidylcholine, perhaps containing some of the oleate used in the acylation of glycerol phosphate. Such a mechanism will allow the entry of C18-substrate into phosphatidylcholine for desaturation with subsequent enrichment of the molecule with C18-polyunsaturated fatty acids. The "topping-up" of the glycerol backbone with C18-polyunsaturates also is aided by the presence of a Δ12-desaturase enzyme that also operates with substrate at position 1 of the sn-phosphatidylcholine (17,19). Whether this is the same enzyme that is responsible for the desaturation of oleate at the sn-2 position of phosphatidylcholine remains to be established. However, it is noteworthy that in recent studies on γ-linolenic acid synthesis in the developing seeds of Borago officinalis, the Δ6-desaturase almost was totally specific for the linoleate at position sn-2 of phosphatidylcholine (22,23). Generally, this may indicate that the desaturase enzymes that utilize the fatty acids in phosphatidylcholine may be specific for the substrate at positions sn-1 and 2.

To conclude this section, our proposals for basic triacylglycerol formation and the relationships to C18-polyunsaturated synthesis are illustrated in Scheme 1 and are summarized here: (1) Acyl-CoAs (largely oleate with some palmitate and stearate) are made available from the plastid. The saturated fatty acids are utilized largely in the acylation of position sn-1 of sn-glycerol 3-phosphate. The acylation of position sn-2 has a strong preference for linoleate and excludes the saturated acids completely. (2) Oleate is transferred to position 2 of sn-phosphatidylcholine (reaction A) for desaturation, and the polyunsaturated products are returned to the acyl-CoA pool for utilization in the acylation of sn-glycerol 3-phosphate. (3) Phosphatidate is hydrolyzed to diacylglycerol (Reaction C). (4) Diacylglycerol can interconvert with phosphatidylcholine (Reaction D) during triacylglycerol formation, bringing about further enrichment of the glycerol backbone with C18-polyunsaturated fatty acids. (5) Diacylglycerol is acylated at sn-3 to yield triacylglycerol (Reaction E).

The acyl exchange coupled with the selectivity properties of the sn-glycerol 3-phosphate–acylating enzymes and the activity of the cholinephosphotransferase govern the asymmetric distribution of fatty acids at positions sn-1 and 2 of phosphatidylcholine and triacylglycerol.

SOME OBSERVATIONS WITH RAPE SPECIES

Many problems still relate to the biosynthesis of triacylglycerols in species of rape and in particular with the assembly of oils that contain the longer chain fatty acids such as erucate (C22:1). Because of the importance of rape oils in the agricultural economy of many countries, it is essential to understand more fully the situation in such species. Recently, we have obtained for the first time microsomal preparations from turnip rape (Brassica campestris) that equate in activity with the other oil-synthesizing systems studied. Such membranes acylate sn-glycerol 3-phosphate and form triacylglycerols in a fashion similar to that observed in other oil seed species (23). However, the microsomal membranes appear somewhat deficient in the activities of those enzymes responsible for the channeling of oleate into phosphatidylcholine. Even with rates of triacylglycerol formation of 17 nmol/100 nmol microsomal phosphatidylcholine/60 min (compared with 45 nmol/100 nmol phosphatidylcholine in safflower), the incorporation of [14C] glycerol 3-phosphate into phosphatidylcholine is low. Also, the specific activity of the acyl-CoA; lyso-phosphatidylcholine acyltransferase is usually ca. 20 nmol oleate incorporated/min per mg protein, a magnitude less than observed in the highly polyunsaturated species such as safflower (300 nmol/min per mg protein). An estimate of the activity of the acyltransferase in the reverse direction was only ca. 1 nmol fatty acid transferred/mg protein per min; again, this is lower than found for safflower (20 nmol fatty acid/mg protein per min). Thus, the enzymes in rape that are involved in trafficking acyl substrate through phosphatidylcholine for desaturation (acyl-CaA; lyso-phosphatidylcholine acyltransferase and cholinephosphotransferase) are relatively less efficient than found in the highly polyunsaturated plant species. Therefore, the acyl quality of the oil in rape may be controlled not only by the activity of the Δ12 and Δ15 desaturase systems but also by those enzymes that channel acyl substrate to the sites of desaturation in the phosphatidylcholine of the ER. Although the modes of entry of oleate into phosphatidylcholine in rape are somewhat limiting compared to other oil seed species, it still is possible to assess their impor-

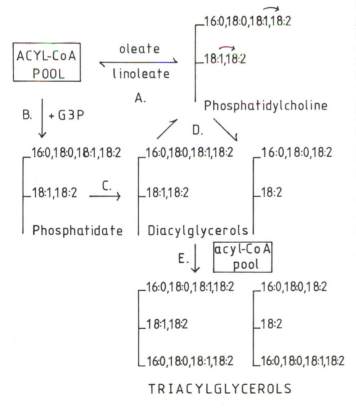

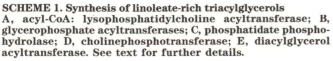

TRIACYLGLYCEROLS

SCHEME 1. Synthesis of linoleate-rich triacylglycerols
A, acyl-CoA: lysophosphatidylcholine acyltransferase; B, glycerophosphate acyltransferases; C, phosphatidate phosphohydrolase; D, cholinephosphotransferase; E, diacylglycerol acyltransferase. See text for further details.

tance relative to each other. Tissue-slices of rape cotyledons (*Brassica campestris*) were incubated with [^{14}C] oleate and the appearance of radioactivity monitored in positions *sn*-1 and 2 of phosphatidylcholine. The results show that over the early incubation times (Table 2) the radioactivity in position *sn*-1 is relatively high and accounts for over 30% of the total activity. Comparison with safflower (Table 2), a developing seed with a highly active acyl-CoA; lyso-phosphatidylcholine acyltransferase, shows that relatively smaller amounts of [^{14}C] oleate appear in position 1 and that position 2 can account for over 90% of the activity. The results indicate that in rape, the diacylglycerol-phosphatidylcholine interconversion may play a proportionately greater role than found in other oil seeds in providing substrate to phosphatidylcholine for desaturation.

TABLE 2

Positional Distribution of Radioactivity in Phosphatidylcholine in Cotyledons Incubated with [^{14}C] Oleate

Incubation time	Radioactivity (% distribution)	
	sn-1	*sn*-2
Safflower		
10 min	7	93
20 min	6	94
40 min	10	90
Rape		
10 min	27	73
20 min	33	67
40 min	37	63

Cotyledon slices were incubated in [^{14}C] oleate. After extraction, the phosphatidylcholine was purified and the distribution of radioactivity in positions *sn*-1 and 2 determined after treatment with phospholipase A2.

However, triacylglycerol assembly with saturated and C18-polyunsaturated fatty-acids in the ER of rape appears not to be the whole story. Erucoyl–rich oils may be synthesized in a different fashion and quite possibly in a separate compartment. The microsomal glycerol acylating enzymes in rape actually are very poor acceptors of erucoyl-CoA compared to the more conventional fatty acids (Table 3). Experiments with radioactive acyl-CoAs and nonlabeled *sn*-glycerol 3-phosphate gave similar results. However, a small but significant amount of [^{14}C]-

TABLE 4

Fatty Acids in Complex Lipids in the Developing Cotyledons of *Brassica campestris* (var. Bele)

Complex lipid	Fatty acid (mol%)							
	16:0	18:0	18:1	18:2	18:3	20:1	20:2	22:1
Phosphatidylcholine	12	2	33	34	14	6	tr	tr
Diacylglycerol	9	5	39	20	10	8	<1	9
Triacylglycerol	4	1	30	18	10	13	<1	23

erucate was found associated with the triacylglycerol. That there are mechanisms for excluding erucate from the phosphatidylcholine of the ER is illustrated by an

TABLE 5

Stereospecific Distribution of Fatty Acids in Triacylglycerol from Developing Cotyledons of *Brassica campestris* (var. Bele)

sn-carbon	Fatty acid (mol%)						
	16:0	18:0	18:1	18:2	18:3	20:1	22:1
sn-1	6	3	34	9	2	17	29
sn-2	0	0	54	36	10	tr	tr
sn-3	3	0	0	11	19	18	49

analysis of the fatty acids present in the complex lipids of the developing cotyledons (Table 4). Although phosphatidylcholine contains some C20:1, little or no erucate is found. The diacylglycerol and particularly the triacylglycerol contain appreciable amounts of erucate.

We also have determined the stereospecific distribution of fatty acids in the triacylglycerol of *B. campestris* var. Bele. The results (Table 5) show that the saturated fatty acids and C20:1 and C22:1 are notable by their absence at position *sn*-2 of the triacylglycerol molecule and that this position is particularly rich in C18-unsaturated and polyunsaturated fatty acids. This may indicate that the acyl constituents at position *sn*-2 of triacylglycerol are governed by similar mechanisms to those in other oil seed species such as safflower. Further experiments with rape have assessed the utilization of [^{14}C] oleate in vivo and the synthesis of its desaturation and elongation products. The results (Table 6) show that the oleate enters phosphatidylcholine and there is desaturated to linoleate.

TABLE 3

Kennedy Pathway in Microsomal Preparations of *Brassica campestris* (var. Bele)

Substrate		^{14}C-glycerol incorporated (nmol)				
	Phosphatidic acid	Phosphatidylcholine	Diacylglycerol	Triacylglycerol	Rest	Total
Linoleate	19	4	4	16	5	48
Erucate	0	0	0	0	0	0.2
Linoleate + Erucate	7	2	2	7	2	20

Microsomes (equiv. 174 nmol phosphatidylcholine) were incubated with either [^{14}C] glycerol 3-phosphate (400 nmol) or acyl-CoAs. The reaction mixture contained BSA (10 mg), Mg Cl$_2$(10 μmol) in P-buffer, pH 7.2. After a 95-min incubation, the radioactivity in the complex lipids was determined.

TABLE 6

Utilization of [^{14}C]-oleate in the Developing Cotyledons of
Brassica campestris (var. Bele)

Lipid and incubation time	Radioactivity (dpm x 10^{-3} per 30 cotyledon pairs)[a]			
	18:1	18:2	20:1	22:1
Triacylglycerol				
40 min	205(68)	tr	54(18)	42(14)
200 min	363(63)	tr	92(16)	121(21)
Phosphatidylcholine				
40 min	247(95)	13(5)	0	0
200 min	229(82)	42(5)	8(3)	0

[a]Figures in parenthesis are the relative distribution of radioactivity in the complex lipid.
Developing cotyledons were incubated with [^{14}C]-oleate, and after the desired time interval the radioactivity
in the component fatty acids of the phosphatidylcholine and triacylglycerol were determined.

OIL QUALITY (FIG. 1)

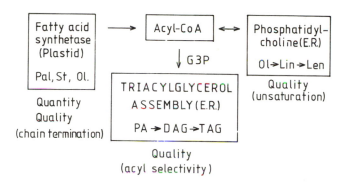

FIG. 1. Compartmentation of the reactions that affect acyl quality in the developing seed.

Some C20:1 at longer incubation times also is associated with the phosphatidylcholine but no erucate is present. On the other hand, the triacylglycerol contains [^{14}C] oleate and substantial amounts of radioactive C20:1 and erucate, even at the shorter incubation times. Therefore, in rape there appears to be two perhaps competing systems that operate for the exogenously supplied substrate oleate (24,25). One is concerned with the desaturation of oleate to C18-polyunsaturated fatty acids, and the other its further elongation to eurcate. The synthesis of erucate from [^{14}C] oleate appears to predominate over desaturation in experiments in vivo. It is essential to understand how these, possibly two assembly sites, are related and how C18-polyunsaturated fatty acid synthesis is integrated with that of erucic acid. The acyl-remodeling of triacylglycerols and diacylglycerols (26,27) is a possibility, and good active in vitro preparations are required to substantiate these proposals. Other possibilities, of course, exist and require investigation. For instance, the primary acyl acceptor and/or acyl donor species may differ from that required for the basic triacylglycerol assembly that utilizes the saturated and C18-unsaturated fatty acids. The site(s) of erucate synthesis also requires further elucidation. It has been suggested that the elongation of oleate to erucate can occur in a particulate fraction that contained chloroplasts (28). However, other studies (29) suggest that some of the reactions in erucate formation may reside in the oil body itself.

A most important asset will be the ability to regulate the C18-unsaturated and polyunsaturated fatty acid content of seed oil. In this context, rape varieties lacking in linolenic acid or even rich in oleate and palmitate (similar to cocoa butter) would be most valuable. With this in mind, it is worth considering some of the factors that appear to regulate the desaturation of oleate and linoleate. The oleoyl desaturase(s) in varieties of safflower and sunflower that are rich in linoleate utilize the oleate substrate at positions *sn*-1 and 2 in *sn*-phosphatidylcholine. Kinetic studies on the desaturation of oleate in phosphatidylcholine by microsomal preparations of safflower indicate that the complete conversion of oleate to linoleate never is achieved. The rate of desaturation of oleate may depend on its concentration in the phosphatidylcholine and rapidly ceases when the substrate approaches amounts that are similar to the levels found in the endogenous lipid. This may be due to the affinity of the desaturase(s) for its substrate and/or a product control of activity. In either case, this could be a factor in determining the ratio of oleate and linoleate in the triacylglycerol oil. It also is possible that the desaturase enzymes are specific positionally for the acyl substrate at the *sn*-1 and 2 positions in *sn*-phosphatidylcholine. The Δ6-desaturase in borage only is active with the linoleate at *sn*-2 and is ineffective with substrate at the *sn*-1 position (22,23). On the other hand, the Δ12-desaturase is active on the oleate at both positions. It would seem that different desaturase proteins may be involved in recognizing the acyl topography in the phosphatidylcholine molecule. If this were so, then such observations are pertinent particularly to programs attempting to manipulate the C18-polyunsaturated quality of vegetable oils. Recently, varieties of linseed that are produced by mutagenesis have been found to contain intermediate and almost zero levels of α-linolenate (30). This success generally bodes well for the continued application of plant breeding and genetics in the quest for important new varieties. The results also are interesting; they show that the capacity to synthesize α-linolenic acid for incorporation into triacylglycerol can be removed almost entirely without affecting oil quantity, seed viability and, very importantly, the structural membrane lipids in the other organs of the plant (31). How far one can go with this approach remains to be fully established. However,

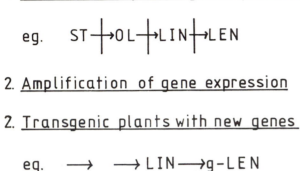

1. PLANT BREEDING
versus
2. GENETIC ENGINEERING

1. Reduction in specific gene expression

eg.   ST ┤→OL ┤→LIN ┤→LEN

2. Amplification of gene expression

2. Transgenic plants with new genes

eg.   ⟶   ⟶ LIN ⟶ g-LEN
                    <u>Δ 6</u>

FIG. 2. Possible manipulation of acyl quality through plant breeding and genetic engineering.

it is noteworthy that varieties of safflower (32) and sunflower (33) that are rich in oleate and soya bean (34) varieties exist and are available with an increased stearate content. Even varieties of rape whose triacylglycerols are enriched particularly with oleate now have been produced (35).

Another area worth considering (Fig. 1) in the manipulation of fatty acid quality is at the level of glycerol phosphate acylation. It eventually may be possible to regulate the properties of these enzymes and control their acyl selectivity. In this context, it is notable that chocolate derives its texture from the specific distribution of the fatty acids in the triacylglycerols of cocoa butter. It is possible that these properties could be developed in other oil crops such as rape to produce a more useful oil. The design of new varieties that will form triacylglycerols with short chain fatty acids (C12 and C14) and uncommon fatty acids (γ-linolenate) will be of considerable agricultural importance. The possibility of using rape as a vehicle for such oils is receiving attention. It now is possible with the in vitro membrane systems from rape to study the assembly of triacylglycerols with foreign fatty acids that normally are absent in the seed and so establish some of the strategies required for acyl manipulation through genetic engineering. It is most worthwhile to continue investment in the search for new oil crop varieties through sustained programs of mutagenesis and plant breeding rather than depend at the moment on speculative proposals in genetic engineering. Certainly, the reduction in specific gene expression for example in the synthesis of the C18-series (Fig. 2) by present–day techniques demonstrably is possible. On the other hand, the amplification of gene expression and the production of transgenic plants making new fatty acids (Fig. 2) will require recombinant-DNA technology. To this end, a great deal of fundamental research still is required to understand the regulatory mechanisms involved in the biosynthesis of triacylglycerols with their component fatty acids. A major need is to establish the key reactions involved and the purification of their enzymes (often insoluble and membrane bound). In the longer term, it is to be hoped that the genes for these proteins will be cloned and as the techniques become more widely applicable for successful gene transfer and expression, it then may become feasible to tailor oil-producing systems that satisfy specific industrial and nutritional requirements.

ACKNOWLEDGMENTS

The authors are grateful to the following for financial support: Karlshamns Oil Company, Sweden; National Swedish Board for Technical Development; Royal Society, U.K.; Science and Engineering Research Council, U.K.; Swedish Council for Forestry and Agricultural Research; Swedish Natural Science Research Council. We thank Claire Griffiths for typing the manuscript.

REFERENCES

1. Slack, C.R., W.S. Bertaud, B.D. Shaw, R. Holland, J. Browse and H. Wright, *Biochem. J. 190*:551 (1980).
2. Ichiara, K., and M. Noda, *Phytochem. 19*:49 (1980).
3. Slack, C.R., P.G. Roughan, J.A. Browse and S.E. Gardiner, *Biochim. Biophys. Acta 833*:438 (1985).
4. Griffiths, G., A.K. Stobart and S. Stymne, in *The Metabolism, Structure and Function of Plant Lipids*, edited by P.K. Stumpf, J.B. Mudd and W.D. Nes, Plenum Press Publ., New York, NY, 1987, p. 361.
5. Stobart, A.K., S. Stymne and S. Hoglund, *Planta 169*:33 (1986).
6. Wanner, G., H. Formanek and R.R. Theimer, *Planta 151*:109 (1981).
7. Huang, A.H.C., R. Qu, S. Wang, V.B. Vance, Y. Cao and Y. Lin, in *The Metabolism, Structure and Function of Plant Lipids*, edited by P.K. Stumpf, J.B. Mudd and W.D. Nes, Plenum Press Publ., New York, NY, 1987, p. 239.
8. Stymne, S. and A.K. Stobart, in *The Biochemistry of Plants*, Vol. 9, edited by P.K. Stumpf, Acad. Press Publ., New York, NY, 1987, p. 175.
9. Gurr, M.I., in *The Biochemistry of Plants*, Vol. 4, edited by P.K. Stumpf and E.E. Conn, Acad. Press Publ., New York, NY, 1980, p. 205.
10. Griffiths, G., A.K. Stobart and S. Stymne, *Biochem. J. 230*:379 (1985).
11. Brindley, D.N., *Prog. Lipid Res. 23*:115 (1984).
12. Ichihara, K., and M. Noda, *Phytochem. 21*:1895 (1982).
13. Cao, Y.Z., and A.H.C. Huang, *Plant Physiol. 82*:813 (1986).
14. Dybing, C.D., and B.M. Craig, *Lipids 5*:422 (1970).
15. Slack, C.R., P.G. Roughan and N. Balasingham, *Biochem J. 170*:421 (1978).
16. Griffiths, G., S. Stymne and A.K. Stobart, *Planta*, in press.
17. Slack, C.R., P.G. Roughan and J. Browse, *Biochem. J. 179*:649 (1979).
18. Stymne, S., and A.K. Stobart, *Biochem. J. 223*:305 (1984).
19. Stobart, A.K., and S. Stymne, *Planta 163*:119 (1985).
20. Slack, C.R., P.G. Roughan, J.A. Browse and S.E. Gardiner, *Biochim. Biophys. Acta 833*:438 (1985).
21. Stobart, A.K., and S. Stymne, *Biochem. J. 232*:217 (1985).
22. Stymne, S., and A.K. Stobart, *Biochem. J. 240*:385 (1986).
23. Stymne, S., G. Griffiths and A.K. Stobart, in *The Metabolism, Structure and Function of Plant Lipids*, edited by P.K. Stumpf, J.B. Mudd and W.D. Nes, Plenum Press Publ., New York, NY, 1987, p. 405.
24. Pollard, M.R., and P.K. Stumpf, *Plant Physiol. 66*:641 (1980).
25. Pollard, M.R., and P.K. Stumpf, *Ibid. 66*:649.
26. Mukherjee, K.D., *Plant Physiol. 73*:929 (1983).
27. Mukherjee, K.D., and I. Kiewitt, *Phytochem. 23*:349 (1984).
28. Agrawal, V., and P.K. Stumpf, *Lipids 20*:361 (1985).
29. Wright, H.C., and M.Sc., Thesis, Massey University, New Zealand, 1980.
30. Green, A.G., *Canad. J. Plant Sci. 66*:499 (1986).

31. Tonnet, M.L., and A.G. Green, *Arch. Biochem. Biophys. 252*:646 (1987).
32. Knowles, P.F., *J. Am. Oil. Chem. Soc. 49*:27 (1972).
33. Purdy, R.H., *J. Am. Oil Chem. Soc. 63*:1062 (1986).
34. Graef, G.L., W.R. Fehr and E.G. Hammond, *Crop Sci. 25*:1076 (1985).
35. Jonsson, R., in *Svalof 100 year Jubilee Symposium*, 1986, p. 101.

Fatty Acid Synthesis in Plant Cells[a]

J.L. Harwood

Department of Biochemistry, University College, Cardiff, CF1 1XL, U.K.

Plants catalyze the formation of fatty acids by the use of a large number of different enzymes. Two systems, acetyl-CoA carboxylase and fatty acid synthetase, are responsible for de novo formation. The plant acetyl-CoA carboxylase appears to be a high molecular weight, multifunctional protein and the fatty acid synthetase is a dissociable Type II complex. A specific condensing enzyme allows the elongation of palmitoyl-acyl carrier protein (ACP) for stearate synthesis. De novo synthesis of long chain fatty acids and desaturation of stearoyl-ACP to oleoyl-ACP take place in plastids and the elongation of long chain to very long chain ($>$C18) fatty acids uses various elongation systems located on the endoplasmic reticulum. Linoleate and linolenate formation are complex, using acyl lipid substrates and entailing the cooperation of different subcellular compartments within the plant cell.

Acetyl-CoA is regarded generally as the starting point for fatty acid biosynthesis. Ultimately, this acetyl-CoA comes from the photosynthetic fixation of CO_2 although, in developing seeds, sucrose usually will be the carbon precursor actually used. By the use of different pathways of intermediary metabolism (1), pyruvate is formed, and the pyruvate dehydrogenase/decarboxylase complex can then form acetyl-CoA. The site of the pyruvate dehydrogenase/decarboxylase complex and its role in the supply of acetyl-CoA for fatty acid synthesis have been discussed (2,3).

THE ACETYL-CoA CARBOXYLASE COMPLEX

Acetyl-CoA carboxylase catalyzes the ATP-dependent formation of malonyl-CoA from acetyl-CoA and bicarbonate. The reaction can be regarded as the first committed step for de novo fatty acid formation. Moreover, the product malonyl-CoA also is used for elongation reactions. The first acetyl-CoA carboxylase to be studied in detail was that from the gram negative bacterium, *E. coli*, where two proteins catalyzing the partial reactions (biotin carboxylase and the transcarboxylase) and the biotin carboxyl carrier protein could be isolated independently. By contrast, mammalian acetyl-CoA carboxylase was found to be a high molecular weight multifunctional protein. The situation in plants has been rather unclear until recently. Early work has been described well (4). Since 1980, purification methods for the enzyme have improved considerably with the use of affinity chromatography utilizing avidin, which binds tightly to biotin (and, therefore, biotin-containing proteins). Our knowledge of proteinase inhibitors also has advanced recently, and it now is possible to purify acetyl-CoA carboxylase rapidly and with minimal loss of activity from almost any plant tissue. A number of preparations from different plants have been made which show molecular weights of 210-240 kDa on SDS-polyacrylamide gel electrophoresis. In some cases, the native enzyme may have a molecular weight of 420-500 kDa (3). Although it is too early to make an unequivocal statement applying to all species and tissues, it seems most probable that the plant acetyl-CoA carboxylase is a multifunctional protein that has at least three functional domains. Whether a fourth regulatory domain also is present must await sequence analysis and further biochemical studies.

Some properties of plant acetyl-CoA carboxylases are shown in Table 1. A slightly alkaline pH optimum has been found and, moreover, the activity-pH curve is usually quite sharp so that small changes in intracellular pH could have a significant effect on acetyl-CoA carboxylase activity. The Km's for the three different substrates are all rather high with the values for bicarbonate indicating poor binding (Table 1).

TABLE 1

Comparison of the Properties of Acetyl-CoA Carboxylases from Different Plants

Plant source	Castor bean	Maize	Parsley	Rapeseed
pH optimum	8.0	8.4	8.0	8.5
Km Acetyl-CoA	50μM	100μM	150μM	74μM
Km ATP	100μM	*var.	70μM	38μM
Km HCO$_3$	2mM	2mM	1mM	3mM

*Dependent on Mg^{++} concentrations. See (3) for a detailed discussion of the properties of acetyl-CoA carboxylases.

It is well known that in many other tissues, acetyl-CoA carboxylase activity may be rate-limiting for fatty acid and fat formation. In addition, the control of this activity by processes such as phosphorylation/dephosphorylation, activation by tricarboxylic acids (such as citrate) or inhibition by acyl-CoA's has been described well in mammals. In some situations, the activity of plant acetyl-CoA carboxylase also increases dramatically, a good example being the illumination of leaves, and the mechanism of this activation has been investigated. No evidence has been found for stimulation by tricarboxylic acids or for regulation by phosphorylation/dephosphorylation (3). Instead, the plant enzyme may be controlled by changes in a number of factors. For example, in chloroplasts the changes that are known to occur in vivo upon illumination have been correlated with a potential increase of about 24-fold in acetyl-CoA carboxylase activity (5). These changes include increases in pH and in Mg^{++} and ATP levels while the concentration of ADP declines. Similar methods of control also have been postulated for other plant acetyl-CoA carboxylases although the different responses of this enzyme as isolated from various sources (3) makes the description of generalized control mechanisms difficult at this time.

---

[a]The editor is indebted to Dr. J.L. Harwood for substituting for J.B. Mudd on short notice when Dr. Mudd was unable to speak due to ill health.

FATTY ACID SYNTHETASE IS A DISSOCIABLE COMPLEX

Since the early observations by Stumpf and his coworkers that acyl carrier protein (ACP) could be isolated as a heat-stable, low molecular weight protein from several plant tissues (6), it has been clear that the plant fatty acid synthetase belongs to the Type II dissociable complexes (3). However, it is only in the last five years that enzymes catalyzing the various partial reactions of fatty acid synthetase (acetyl-CoA:ACP transacylase, malonyl-CoA:ACP transacylase, β-ketoacyl-ACP synthetase, β-ketoacyl-ACP reductase, β-hydroxyacyl-ACP dehydrase, enoyl-ACP reductase) have been isolated and examined individually. For details of these purifications and details of the properties of isolated enzymes the reader is referred to (3). Space does not allow much detail here, instead reference will be to a few of the interesting results that have accrued from such experiments.

Localization of all detectable ACP in the chloroplast stroma of spinach leaves (8), showed that this organelle was the site of most, if not all, de novo fatty acid synthesis in leaves as previously suggested by Weiare and Kekwick (9). All the individual proteins of fatty acid synthetase, together with the enzymes responsible for palmitate elongation and stearate desaturation are soluble and, therefore, localized in the stroma. Nevertheless, although they are soluble they may be associated loosely with the thylakoid membranes in vivo. Evidence for this comes from the loose association of fatty acid synthetase with isolated chloroplast thylakoids (10) and the localization of ACP on thylakoid membranes as determined by electron microscopy (11). Thus, the products of de novo synthesis can be transferred easily to membrane lipids that would release ACP for further participation in fatty acid formation.

ACP has been purified from a number of plants and, in particular, has been studied in detail from rapeseed and barley and spinach leaves (8). Two isoforms have been found in spinach leaves of which the most abundant, ACP-1, has been sequenced completely. Interestingly, only ACP-2 is found in spinach seeds, a finding that has been confirmed for other plants. There is considerable sequence homology between the various ACP isoforms of spinach and barley as well as with ACP's from E. coli or animal systems, particular in the region of the prosthetic group 4' phosphopantotheine. The reason for ACP isoforms has been investigated. It does not appear that one form is needed for fatty acid synthetase and another for other acyl-ACP dependent reactions or that ACP-1 is the only form in plastids (8). However, two acyl-ACP-utilizing enzymes have been shown to have different specificities towards acyl-ACP substrates made with different ACP isoforms. Thus, oleoyl-ACP thioesterase is most active with ACP-1 and glycerol 3-phosphate:acyl-ACP acyltransferase is most active with ACP-2 (8,12,13). These results mean that by altering the relative levels of plant ACP isoforms, one may be able to alter the pattern of fatty acid and lipid products in plant tissues. However, the proportions of the ACP isoforms in vivo do not support the idea of such a role in nature (3,8).

Acetyl-CoA:ACP transacylase is the partial reaction that usually has the lowest activity in vitro and evidence has been produced that increasing the enzyme's amount or by-passing the formation of acetyl-ACP can augment fatty acid formation (14). Furthermore, alteration of the enzyme's activity may alter the pattern fatty acid products (3,15). These observations both obviously are of relevance to biotechnology. Malonyl-CoA:ACP transacylase has been purified from several plants and the cyanobacterium *Anabaena variabilis*. In soybean leaves, isoforms have been detected; and in leek, the malonyl-CoA:ACP transacylases from epidermal and parenchymal cells had molecular weights of 38kDa and 45kDa, respectively (3).

TABLE 2

Some Features of the Individual Proteins of Plant Fatty Acid Synthetase

Protein	Purified from?	Characteristics
ACP	Barley and spinach leaves, rapeseed and other tissues	Low molecular weight. Isoforms found in leaves that have different activities with some acyl-ACP reactions. Nuclear-coded. Synthesized as larger precursor and 4' phosphopantetheine added before transport into chloroplast.
Acetyl-CoA:ACP transacylase	Barley, spinach and *Brassica campestris* leaves	Probably the slowest reaction of fatty acid synthetase. Isoforms found in *B. campestris*. Sensitive to arsenite. Ratio of acetyl-ACP: malonyl-ACP may alter the chain length of fatty acyl products.
Malonyl-CoA:ACP transacylase	Barley, leek, soybean, spinach leaves and avocado fruits	Isoforms found in soybean leaves and in different cell types from leek leaves.
β-Ketoacyl-ACP synthetase	Spinach and barley leaves, parsley cultures and rapeseed	Synthetase I from spinach is sensitive to cerulenin and uses C2-C14 substrates while synthetase II is sensitive to arsenite and uses C16 substrate.
β-Ketoacyl-ACP reductase	Spinach leaves, avocado fruit, rape and safflower seeds	Isoforms separated from avocado using NADH or NADPH as reductant. In most tissues, NADPH activity is more important.
β-Hydroxyacyl-ACP dehydrase	Spinach leaves and safflower seeds	Tetrameric in spinach. Stereospecific for the D-β-hydroxyacyl-ACP substrate.
Enoyl-ACP reductase	Safflower, castor bean and rape seeds, spinach and other leaves, avocado fruit	Isoforms usually found. Type 1 uses NADH while Type 2 prefers NADPH. Exists as a tetramer.

The reason for isoforms of this enzyme is not known.

Evidence that palmitate elongation was carried out by an enzyme system distinct from that forming palmitate originally was suggested by experiments with arsenite (16) and later confirmed by experiments with cerulenin (3,7). When two condensing enzymes were isolated from spinach, their chain-length specificity and inhibition characteristics confirmed that palmitate was elongated by a specific enzyme (3,7). The β-ketoacyl-ACP synthetase I will utilize substrates in the C2-C14 range and is sensitive to cerulenin. In contrast, β-ketoacyl-ACP synthetase II is active with palmitoyl-ACP and is inhibited by arsenite (Table 2). Alteration in the relative activity of β-ketoacyl-ACP synthetase II may be another target for gene manipulation, controlling, as it does, the ratio of C16/C18 products.

The remaining enzymes of fatty acid synthetase have been purified from spinach leaves and the two reductases from several other tissues as well. Two forms of the β-ketoacyl-ACP reductase, one using NADPH and the other NADH, have been resolved in fractions from avocado but the NADPH-form seems to be quantitatively more important in several other plant tissues (3). Two forms of the enoyl-ACP reductase have been detected (Table 2). Type I is NADH-specific, while Type II prefers NADPH over NADH as the source of reductant. Both types are found in castor bean, rape and safflower seeds, but only Type I could be detected in avocado fruits or leaf tissue (3). Again, the reason for isoforms of the two reductases is unknown at present.

Regulation of the chain length of the products of fatty acid synthesis is of importance in the edible oil industry. However, the process is poorly understood at present, and more experiments are clearly needed with tissues such as *Cuphea* spp. or coconut where medium-chain products predominate. Several possible regulatory factors have been identified by in vitro experiments (3,7,15), but their physiological relevance has yet to be proved.

ELONGATION OF FATTY ACIDS

Stearate can be elongated to the very long chain fatty acids (>C18) that are so important as precursors or the surface layers of plants: cutin, suberin and wax. The enzymes responsible for such elongations use malonyl-CoA and NADPH and are located on the endoplasmic reticulum (3,17). Similar systems presumably are used for the elongation of monoenoic acids, such as in erucic-containing rape seed, although saturated, very long chain fatty acid synthesis has been much better studied because of its widespread relevance to all plants. The elongation systems appear to be chain-length specific and, in the case of potato, different proteins are required for formation of C20, C22 and C24 acids. Evidence for this came from experiments in which the elongation systems were induced successively by ageing potato slices and by blocking such induction with cycloheximide, a protein-synthesis inhibitor ([18], Table 3). The presence of chain-length specific elongases in plant tissues is supported by genetic observations and by the isolation of a C18-specific elongase from leek (3). For most systems, acyl-CoA substrates seem to be used (3,17).

TABLE 3

Effect of Sequential Treatment of Aging Potato Slices with Cycloheximide

Treatment	Aging time	Distribution of radioactivity (% total)				
	(hours)	C_{16}	C_{18}	C_{20}	C_{22}	C_{24}
Control	0	24	69	1	n.d.	n.d.
	2	27	57	7	6	n.d.
	4	25	56	9	8	n.d.
	6	25	51	10	11	3
	8	23	48	10	11	5
+Cycloheximide	8	24	70	1	n.d.	n.d.
Control period + cycloheximide	1+1	19	75	3	n.d.	n.d.
	2+2	24	71	2	1	n.d.
	3+3	24	68	4	2	n.d.
	4+4	28	59	5	5	tr.

n.d., non detected; tr., trace. For further details see (18).

FORMATION OF UNSATURATED FATTY ACIDS

In higher plants the majority of the total unsaturated fatty acids are represented by just three molecules: oleic, linoleic and α-linolenic acids. The last two predominate in leaf tissues, while oleate and linoleate are major components of most economically-important edible oils (19). As all of these acids have 18 carbons, it is no surprise that they are made by the successive desaturation of stearate. However, whereas the water-soluble strearoyl-ACP is the substrate for the first desaturation, further removal of hydrogens usually is thought to involve complex lipid substrates.

Stearoyl-ACP was identified as the substrate for oleate formation in spinach chloroplasts and *Chlorella vulgaris*. The Δ9-desaturase has been best studied using fractions from developing safflower seeds. The enzyme is soluble and was purified about 200-fold by a combination of ion-exchange and affinity chromatography (7). Although the purified desaturase would use palmitoyl-ACP as substrate, the kinetics were such that this substrate had only about 1% of the activity of stearoyl-ACP. This explains well why palmitate is much more prevalent than palmitoleate and also why oleate is the major monoenoic acid in plants (3,7). Although all the enzymes of the fatty acid synthetase complex and the Δ9-desaturase are soluble proteins of the plastid stroma, their reactions are tightly coupled. Thus, if the labeling pattern of oleate made during incubations of chloroplasts with [1-^{14}C]acetate is examined, the acid is found to be made de novo (3).

Once oleoyl-ACP has been formed a major branch-point of fatty acid synthesis is reached. If the thioester is used by glycerol 3-phosphate acyltransferase then oleate becomes part of the diacylglycerol structure used for chloroplast lipid synthesis (e.g., monogalactosyldiacylglycerol, digalactosyldiacylglycerol, sulphoquinovosyldiacylglycerol, phosphatidylglycerol). Here, oleate will accumulate at the *sn*-1 position while palmitate occupies the *sn*-2 position. Thus, chloroplast lipids characteristically may contain a saturated acid at the *sn*-2 position and an unsaturated acid at the *sn*-1 position. This is the opposite of a typical mammalian acyl lipid. Where the above fatty acid arrangement is found, then the chloroplast lipid is referred to as having a prokaryotic distribution. In contrast, chloroplast lipids (and seed oil triacylglycerols) that have a high concentration of C18 unsaturated fatty acids at the *sn*-2 position are called eukaryotic to denote the participation of the endoplasmic reticulum in their formation (below).

A more complete discussion of the prokaryotic and eukaryotic pathways is given in (20), but a short description follows: hydrolysis of oleoyl-ACP by a thioesterase yields an unesterified acid that can be converted to oleoyl-CoA in the plastid envelope. Provision of oleoyl-CoA then allows other extra-plastidic lipids to become esterified. However, the *sn*-2 position of phosphatidylcholine is particularly utilized. This phosphoglyceride is the major extra-chloroplast lipid and it generally is thought that the endoplasmic reticulum is the major site for esterification. Oleate at the *sn*-2 position of phosphatidylcholine is a particularly good substrate for Δ12-desaturation (21) and the linoleate produced thus may be further desaturated to α-linolenate (or γ-linolenate in *Borago officinalis*) while still attached to phosphatidylcholine in oil-accumulating seeds (3). However, seeds do not usually produce a large amount of α-linolenate, which is the major fatty acid of chloroplast membranes (19). In leaves, there is a general consensus that linoleate must be transferred back to the monogalactosyldiacylglycerol of chloroplasts in order to be further desaturated to α-linolenate (3). The mechanisms of lipid transfer between organelles in plant cells is unknown at present but may involve some of the lipid-transfer proteins that have been isolated and purified from different plant tissues (22). The overall pathway of unsaturated fatty acid synthesis in plants is summarized in Figure 1.

Some aspects of Δ12- and Δ15-desaturation have been delineated in plants but, in general, our knowledge of these reactions is poor (3,7). The membrane-bound enzymes have proved to be difficult to solubilize or often are poorly active in many subcellular fractions. Even such questions as the source of reductant for linoleate desaturation is unclear (3). However, it is becoming increasingly clear that definite statements that embrace all different plant tissues are impossible to make with regard to such aspects as the exact pathway of desaturation used. Thus, other acyl lipids, and even oleoyl-CoA, have been suggested as substrates for oleate desaturation in various tissues. Moreover, the relative proportion of phosphatidylcholine or monogalactosyldiacylglycerol used for linoleate desaturation is not defined completely. Nonetheless, the pathways shown in Figure 1 represent a good summary of the presently accepted picture which applies to plants in general if not absolutely (3). Further exciting developments are to be expected in this area in the next few years.

ACKNOWLEDGMENTS

The author's research on fatty acid synthesis has been supported by the S.E.R.C., A.F.R.C. and N.E.R.C. as well as Unilever and I.C.I.p.l.c. I am indebted to Prof. Paul K. Stumpf who first excited (and has ever since encouraged) my interest in plant lipids.

REFERENCES

1. Stumpf, P.K., in *Proceedings of the World Conference on Biotechnology for the Fats and Oil Industry*, edited by T.H. Applewhite, The American Oil Chemists' Society, Champaign, IL, 1988, pp. 1–6.
2. Liedvogel, B., in *The Metabolism, Structure and Function of Plant Cells*, edited by P.K. Stumpf, J.B. Mudd and W.D. Nes, Plenum Press, New York, NY, 1987, p. 509.
3. Harwood, J.L., *Annual Revs. Plant Physiol*, in press.
4. Stumpf, P.K. in *The Biochemistry of Plants*, edited by P.K. Stumpf and E.E. Conn, Academic Press, New York, NY, Vol. 4, 1980, p. 177.
5. Nikolau, B.J., and J.C. Hawke, *Arch. Biochem. Biophys. 228*:86 (1984).
6. Simoni, R.D., R.S. Criddle and P.K. Stumpf, *J. Biol. Chem. 242*:573 (1967).
7. Stumpf, P.K., in *Fatty Acid Metabolism and its Regulation*, edited by S. Numa, Elsevier, Amsterdam, 1984, p. 155.
8. Ohlrogge, J.B., in *The Biochemistry of Plants*, edited by P.K. Stumpf and E.E. Conn, Academic Press, New York, NY, Vol. 9, 1987, p. 137.
9. Weaire, P.J., and R.G.O. Kekwick, *Biochem. J. 146*:425 (1975).
10. Walker, K.A., and J.L. Harwood, *Biochem. J. 226*:551 (1985).
11. Slabas, A.R., J. Harding, P. Roberts, A. Hellyer, C. Sidebottom, C.G. Smith, R. Safford, J. de Silva, C. Lucas, J. Windust, C.M. James and S.G. Hughes, in *The Metabolism, Structure and Function of Plant Cells*, edited by P.K. Stumpf, J.B. Mudd and W.D. Nes, Plenum Press, New York, NY, 1987, p. 697.
12. Ohrogge, J.B., in *Proceedings of the World Conference on Biotechnology for the Fats and Oils Industry*, edited by T.H. Applewhite, The American Oil Chemists' Society, Champaign, IL, 1988, pp. 102–109.
13. Guerra, D.J., in *Proceedings of the World Conference on Biotechnology for the Fats and Oils Industry*, edited by T.H. Applewhite, The American Oil Chemists' Society, Champaign, IL, 1988, pp. 39–42.
14. Shimakata, T., and P.K. Stumpf, *J. Biol. Chem.* 258:3592 (1983).
15. Pollard, M.R., and S.S. Singh, in *The Metabolism, Structure and Function of Plant Cells*, edited by P.K. Stumpf, J.B. Mudd and W.D. Nes, Plenum Press, New York, NY, 1987, p. 455.

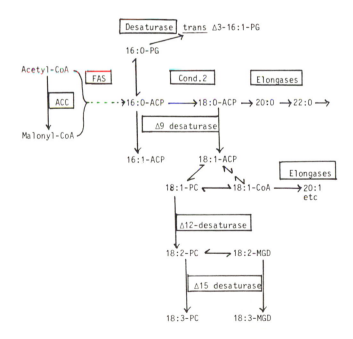

FIG. 1. Major pathways of fatty acid synthesis in plants. ACC, Acetyl-CoA Carboxylase; FAS, fatty acid synthetase; Cond. 2, β-ketoacyl-ACP II. The elongases for saturated fatty acid synthesis work mainly with acyl-CoA substrates. Δ15-Desaturase activity uses mainly phosphatidylcholine (PC) as substrate in developing seeds and mainly monogalactosyldiacylglycerol (MGD) in leaf tissues.

16. Harwood, J.L., and P.K. Stumpf, *Arch. Biochem. Biophys.* 142:281 (1971).

17. Cassagne, C., R. Lessire, J-J Bessoule and P. Moreau, in *The Metabolism, Structure and Function of Plant Cells*, edited by P.K. Stumpf, J.B. Mudd and W.D. Nes, Plenum Press, New York, NY, 1987, p. 481.

18. Walker, K.A., and J.L. Harwood, *Biochem. J.* 237:41 (1986).

19. Harwood, J.L., in *The Biochemistry of Plants*, Vol. 4, edited by P.K. Stumpf and E.E. Conn, Academic Press, New York, NY, 1980, p. 1.

20. Roughan, P.G., in *The Metabolism, Structure and Function of Plant Lipids*, edited by P.K. Stumpf, J.B. Mudd and W.D. Nes, Plenum Press, New York, NY, 1987, p. 247.

21. Stobart, A.K., in *Proceedings of the World Conference on Biotechnology for the Fats and Oils Industry*, edited by T.H. Applewhite, The American Oil Chemists' Society, Champaign, IL, 1988, pp. 22–29.

22. Mazliak, P., and J.C. Kader in *The Biochemistry of Plants*, Vol. 4, edited by P.K. Stumpf and E.E. Conn, Academic Press, New York, NY, 1980, p. 283.

Recent Advances in Oil Crops Breeding

P.F. Knowles
Department of Agronomy and Range Science, University of California, Davis, CA 95616

Plant breeding began some 5,000 to 10,000 years ago when primitive societies began the process of crop domestication. Though the process was slow, it yielded most of our oil crops essentially in their modern form. Plant breeding now provides cultivars that are better adapted to different environments, higher in oil content, changed in oil quality through changes in fatty acid composition, more resistant to diseases, insects and other pests, and more resistant to pesticides. Plant breeding procedures and technology that will facilitate achieving those objectives are germplasm collection, screening and preservation programs, development of hybrid cultivars, use of mutagens to change genes, particularly those affecting adaptation, oil quality and resistance to pesticides and herbicides, and among several genetic engineering techniques, the use of doubled haploids developed through microspore and anther culture, thus abbreviating the duration of breeding programs. To be most effective in providing benefits to the fats and oil industry, plant breeding and modern biotechnology must be merged closely. Increasingly, breeding augmented by genetic engineering is being done in the private sector. This leads to increased financial investment and with it more rapid and abundant cultivar development, increased protection being sought for both newly developed cultivars and for plant breeding procedures, decreased germplasm and information exchanges, and increased involvement of the public sector in research basic to plant breeding programs.

By way of introduction, I would emphasize that plant breeding is a form of biotechnology. In fact, when I began my career in plant breeding, I considered myself an applied geneticist, or a genetic engineer involved in the construction of superior genotypes of crop plants.

The major advances in the breeding of all crops were made 5,000 to 10,000 years ago when crop plants including most oil crops were domesticated from wild species to the forms we know today. The process was slow but effective.

My remarks are general in nature, and I shall confine my discussion to crops grown as annuals and shall draw to some extent on my experience with safflower. Three general areas will be covered: major breeding objectives, breeding strategies and technology and public and private sectors in breeding.

MAJOR OBJECTIVES IN BREEDING OIL CROPS

Major objectives that will be considered are: better adaptation to different environments, higher yields, higher oil content, changed and improved oil quality, resistance to pests and resistance to pesticides.

Better Adaptation to Different Environments

Examples from a few crops will illustrate where in recent years the challenge of the environment has been met and where it has not.

For many years in India, the American soybean cultivars were successful in research trials but not in commercial production. The reason was that the seeds had fragile seed coats and readily were damaged during local harvesting procedures; poor germination led to thin stands and low yields. Now (G.V. Ramanamurthy, personal communication), from crosses of American cultivars to indigenous types with smaller seeds and tough seed coats, cultivars are available with resistance to seed damage. Soybean now occupies a large area in India. Similar results were obtained in Mississippi with lines derived from a cross of the American cultivar Forrest with a wild soybean species (*Glycine soja* Sieb. & Zucc.) (1). After 42 days of adverse weather at harvest time, the average viability of seed of Forrest was 10% and line D81-9760 was 84%.

On the other hand, soybean has not been too successful in the arid environment of California, primarily due to its susceptibility to damage from spider mites (*Tetranychus* spp.). It has not become a crop of major importance in arid regions of the Middle East.

Safflower has shown most promise in arid environments if adequate moisture is available. This has limited its distribution to California, Arizona and parts of the Western Great Plains in the United States and to areas with a similar environment in other countries. Production in most areas is prohibited by an array of foliar diseases that are favored by high humidities. With resistance to those diseases, safflower, because of its high yield potential, could be grown over a much wider area.

Higher Yields

Major improvements have been made in the yields of oil crops. The situation in the United States (Table 1) is an example. Over the periods presented, with the first 10 years of data as a base, increases in yields were peanut, 279%; soybean, 54%; sunflower, 21%; and flaxseed, 40%. Obviously, the increases were due in part to improved cultural practices. Nevertheless, improved cultivars played a major role. Only in the case of sunflower has the availability of hybrid cultivars been a major factor in yield increases.

Higher Oil Contents

Two changes in oilseeds can lead to higher oil content. The first is a reduction in the amount of hull or seed coat, both fibrous, that cover the embryo and, if present, the endosperm. This was the basis for much of the increase in oil content of safflower (Table 2). It is likely that partial hull discovered by Urie (2) will become a feature of future safflower cultivars that will raise oil contents to 50%. Obviously, by reducing the hull and/or seed coat, protein contents also will increase.

TABLE 1

Yields of Major Oilseed Crops of United States (10–Year Averages, 1941–1980, 5–Year Average, 1981–1985, kg/ha)[a]

Years	Flaxseed	Peanut	Soybean	Sunflower
1941–1950	577	780	1283	
1951–1960	505	1110	1426	
1961–1970	662	1758	1661	998[b]
1971–1980	676	2579	1851	1227
1981–1985	811	2960	1980	1204
Increase, %[c]	40	279	54	21

[a]USDA Agricultural Statistics.
[b]1962–1970.
[c]Increase of 1981–1985 yields over those in 1941–1950 (1962–1970 for sunflower) in %.

TABLE 2

Average Percentage of Hull, Oil, Protein and Fiber in Safflower Seed of Three Hull Types (3)

Hull Type[a]	Hull	Oil	Meal	
			Protein	Fiber
Normal	40	39	20	41
Striped	25	46	34	25
Thin	20	47	40	17

[a] Genotypes: Normal, *ThThStpStp*; Striped, *ThThstpstp*; Thin, *ththStpStp*.

The second change is raising the oil content of the embryo or both the embryo and endosperm without changing the hull or seed coat, in which case protein content will be reduced. Pushing oil contents above 55% probably will lead to problems during seed handling because of fragility of the protective hull or seed coat or fragility of the seed contents.

Changes in Oil Quality

Safflower provides an interesting array of changes in fatty acid composition (Table 3), all of them occurring naturally. Only the standard type with high levels of linoleic acid and the high oleic acid type available for about 20 years (5) are grown commercially. The original source of the high oleic acid type appears to be Bangladesh, where collections I made in 1964 near Dacca were of that type.

Other examples of changes in C-18 fatty acid composition are shown in Table 4. All of them resulted from the use of mutagens. Of increasing commercial interest is the high oleic acid sunflower type because of the value of the oil for both edible and industrial purposes. Of equal interest is the very low linolenic acid type of flaxseed developed by Green (7) in Australia.

I stress that many of these new fatty acid types are essentially new crops, requiring isolation from other types during production, transportation, storage, processing and marketing. They are the first steps in the development of an array of cultivars in a single species each with

high levels of a specific fatty acid. For the different types there will be a need for plant markers such as leaf shape, flower color, pollen color and seed appearance.

TABLE 3

Fatty Acid Compositions of the Seed Oil of Safflower Types (4)

Type[a]	Fatty acid content, %			
	Palmitic	Stearic	Oleic	Linoleic
Very high linoleic	3-5	1-2	5-7	87-89
High linoleic	6-8	2-3	16-20	71-75
Intermediate oleic	5-6	1-2	41-53	39-52
High oleic	5-6	1-2	75-80	14-18
High stearic	5-6	4-11	13-15	69-72

[a]Very high linoleic from Portugal; high linoleic from the cultivar Gila; intermediate oleic from Iran; high oleic from the cultivar UC-1; and high stearic from Israel.

The stability of fatty acid composition is of concern to the oils and fats industry. It is well known that linoleic and oleic acid contents of sunflower oil are sensitive to temperature; the oleic/linoleic ratio increases with high temperatures and decreases with low temperatures. In flax, Green (8) found that levels of oleic and linoleic acids in the very low linolenic acid genotype showed a similar response. The same was true of a safflower genotype with about equal amounts of oleic and linoleic acids (9) but not true of genotypes with high levels of either linoleic or oleic acids. Soybean appeared to be responsive to selection for stability of oleic/linoleic acid ratios (10).

Resistance to Pests

For the most part, breeders of oil crops have kept pace with the major needs for resistance to pests. The exceptions, as mentioned above, are those in which pests have been features of environments that have prohibited successful production of some oil crops. Less success has been achieved for resistance to insects than for resistance to disease, in part because agriculture has depended heavily on chemicals for control of insects. With increasing concerns about pesticides in the environment expressed by the development of integrated pest management programs, oil crops breeders will increase efforts to develop insect-resistant cultivars. For example, Kilen and Lambert (11) report that for soybean they were planning to combine the partial resistance of three genotypes to the velvetbean caterpillar (*Anticarsia gemmatalis* Hubner) into one gene pool that should provide a higher level of resistance. Modern biotechnology will play a major role in transferring genes for resistance to many pests from one species to another.

Resistance to Pesticides

A major area of interest by biotechnology and chemical companies is in the development of genotypes and cultivars of oil crops with resistance to herbicides. In a review, Sun (12) points out that instead of producing herbicides to fit a crop plant, companies will produce cultivars to fit the herbicide. Possible benefits for the companies involved are an expansion and increased control of the market for the chemical and the resistant cultivars, for the farmer less tillage and hand hoeing, reduced soil erosion and the use of one herbicide for most or all crops in a rotation, and for the environmentalist the use of efficient new herbicides with little mammalian toxicity and that are quickly biodegradable. Resistance to triazine was found to be controlled by genes in the chloroplasts of a weedy version of turnip rape. Through backcrossing the cytoplasm with the chloroplasts was transferred to rapeseed, resulting in the development of the cultivar Triton (13). While this was done using conventional breeding techniques, modern biotechnology in the future will play a major role in the transfer of genes for herbicide resistance to different species.

BREEDING STRATEGIES AND TECHNOLOGY

There is nothing unique about breed strategies used for oil crops to achieve breeding objectives. These briefly will be considered: germplasm collection, screening and preservation, use of hybrid cultivars, use of mutagens and anther and microspore culture.

Germplasm Collection, Screening and Preservation

Increasingly, breeding programs in both the public and private sector have become international in scope leading to a decrease in total variability. For many crops, much of the total germplasm traces back to a rather small number of cultivars. More often than not, the value of germplasm collections are realized long after they were made. For example, resistance in safflower to Fusarium wilt (14) and Verticillium wilt (15) were identified several years after collections were made. It is imperative that germplasm be collected and preserved not only of all oil crops and their related wild relatives but also of other species, particularly endangered species, with seed oils of possible future value.

Hybrid Cultivars

Cytoplasmic male sterility has provided the basis for hybrid cultivars of sunflower to the extent that they provide much of the production in the United States, Europe and elsewhere. With sunflower, hybrid cultivars have two major advantages over nonhybrid cultivars: they are higher in yield and they are more uniform. The first steps in the introduction of hybrid rapeseed have been made, and undoubtedly most cultivars will be hybrids in the near future. The same will be true of turnip rapeseed and other *Brassica* species. A.B. Hill of Cargill, Inc. (Minneapolis, MN)(personal communication) reports on the development of hybrid cultivars of safflower. Hybrid cultivars of castor long have been grown commercially, though all-female flowered plants, instead of cytoplasmic male sterile plants, have provided the basis for hybrid seed production. An added advantage of hybrid cultivars is that they provide the developer with biological control of seed production; the hybrid cultivar will not breed true, necessitating that the grower go back each year to the developer for seed.

Mutagens

Sixty years ago, Muller (16) reported that the mutation rate in *Drosophila* could be increased by treatment with x rays. Since then, several ionizing agents, ultraviolet radiation and chemicals have been used. The plant breeder has been attracted to mutagens primarily in two situations: one in which a character expression such as an increased level of a particular fatty acid (Table 4) is unavailable in a species and a second in which a cultivar is superior in all characteristics except one such as a low level of resistance to one disease. Three features of mutagens discourage their widespread use. First, mutagens cause more adverse than beneficial effects, and adverse

TABLE 4

Fatty Acid Composition of the Seed Oil of Different Crops Induced by Mutagens

Crop and identity or type	Fatty acid composition, %					Reference
	Palmitic	Stearic	Oleic	Linoleic	Linolenic	
Soybean						(6)
FA8077[a]	9.1	4.3	44.2	37.5	5.0	
A6[b]	7.7	30.4	21.1	35.4	5.5	
Sunflower						Sungene
Standard	5.1	3.5	40.7	50.7		
High oleic	3.2	2.0	87.8	6.9		
Flaxseed						(7)
Glenelg[a]	7.0	3.7	35.1	14.1	40.1	
M1589[b]	7.2	5.0	44.1	24.6	19.1	
M1722[b]	8.4	4.8	35.5	27.9	23.4	
F$_2$ plants[d]	8.7	5.5	33.5	50.7	1.6	

[a]Standard type.
[b]Mutant type.
[c]Personal communication, Sungene Technologies Corporation.
[d]Average of seven F$_2$ low linolenic plants form a cross of M1589 x M1722.

effects may be combined with desired effects, requiring breeding procedures to separate them. Second, the changed genes usually are recessive and are not manifested until the second generation from treated seed. And third, very large populations must be examined.

Anther or Microspore Culture

Anther or microspore culture is a technique to exploit the haploid stage of plant development in accelerating breeding programs that involve hybridization of different cultivars. In conventional hybridization programs, the plant grown from the crossed seed is highly heterozygous. Usually four or more generations from that plant are required to provide an adequate number of true breeding plants. However, the heterozygous plant grown from the crossed seed will produce microspores, which develop into pollen grains that are extremely variable in their genetic constitution. These haploid microspores are treated to encourage their development into embryos and then into plants. The haploid plant normally will be sterile. However, if doubling of the chromosome number of the haploid occurs naturally or following appropriate treatment, the resulting plant will breed true. This means that the plant breeder can begin evaluation of the progenies of doubled haploids about two years after a cross instead of five or six years later. While accelerating a breeding program, a great deal of skilled labor is involved in haploid breeding, such that the breeder may elect to use the technique only where time is critical. *Brassica* species have proved to be very responsive not only to microspore culture techniques (17) but to other genetic engineering technology.

Actually, haploids have occurred naturally in breeding populations and have been doubled by chemical treatments to give true breeding plants. The progeny of some have become commercially grown cultivars.

THE PUBLIC AND PRIVATE SECTORS IN OIL CROPS BREEDING

Plant breeding, in terms of the generation of cultivars, increasingly is moving from the public to the private sector. In the private sector this has led or will lead to: (1) Increased investments in breeding, leading to increased sizes of conventional breeding programs and increasing integration into them of modern biotechnology. (2) Increased protection being sought for newly developed cultivars through biological means or through systems of registration and plant breeding procedures through patents. (3) Decreased exchange of germplasm, particularly that with novel or valuable characteristics. (4) Decreased exchange of information, particularly that relating to new technology.

In the public sector, with declining involvement in cultivar development the emphasis has been or will be on long–term projects that are basic to plant breeding.

Among these are: (1) Increased attention to germplasm collection, evaluation and preservation. (2) Increased development of germplasm with characteristics of value in breeding programs, characteristics such as resistance to pests, changed fatty acid composition and cytoplasmic male sterility necessary for the production of hybrid cultivars. (3) Increased involvement in research necessary to more efficient plant breeding and genetic engineering. (4) Increased attention to the development of new oil crops of interest for edible and industrial purposes. (5) Continued involvement in training of plant breeders. (6) Continued involvement in extension of information on all phases of oil crops production and utilization.

It is obvious from the foregoing that plant breeding will benefit in many ways from advances in biotechnology, particularly in its contribution to the manipulation of genes and expansions in the boundaries of variability within a species.

In cultivar development, modern biotechnology will not succeed without a first class breeding program using tested conventional breeding technology. Modern biotechnology is becoming increasingly expensive; it could reduce seriously resources allotted to breeding programs.

Plant breeding is much more than generation of variability. In fact, the hard part of plant breeding is the identification of superior genotypes, the putting together of genes that will result in a cultivar well adjusted to the environment, high in yield, resistant to pests and producing products superior in quality.

REFERENCES

1. Hartwig, E.E., and H.C.Potts, *Crop Sci. 27*:506 (1987).
2. Urie, A.L., *Ibid. 26*:493 (1986).
3. Rubis, D.D., First Research Conference on Utilization of Safflower. Agric. Res. Service, USDA, Report 74-43, 1967, pp. 23-28.
4. Futehally, S., *Inheritance of Very High Levels of Linolenic Acid in the Seed Oil of Safflower (Carthamus tinctorius L.)* Master of Science Thesis, University of California, Davis, CA (1982).
5. Knowles, P.F., A.B. Hill and J.E. Ruckman, *California Agric. 19 (12)*:15 (1965).
6. Graef, G.L., W.R. Fehr and E.G. Hammond, *Crop Sci. 25*:1076 (1985).
7. Green, A.G., *Can. J. Plant Sci. 66*:499 (1986).
8. Green, A.G., *Crop Sci. 26*:961 (1986).
9. Knowles, P.F., *J. Am. Oil Chem. Soc. 46*:27 (1972).
10. Carver, B.F., J.W. Burton, T.E. Carter Jr. and R.F. Wilson, *Crop Sci. 26*:1176 (1986).
11. Kilen, T.C., and L. Lambert, *Ibid. 26*:869 (1986).
12. Sun, M., *Science 231*:1360 (1986).
13. Beversdorf, W.D., and D.H. Hume, *Can. J. Plant Sci. 64*:1007 (1984).
14. Knowles, P.F., J.M. Klisiewicz and A.B. Hill, *Crop Sci. 8*:636 (1968).
15. Urie, A.L., and P.F. Knowles, *Ibid. 12*:545 (1972).
16. Muller, H.J., *Science 66*:84 (1927).
17. Keller, W.A., and G.R. Stringham, Production and Utilization of Microsphore-Derived Haploid Plants, in *Frontiers of Tissue Culture 1978*, edited by T.A. Thorpe, International Association for Plant Tissue Culture, Univ. Calgary, Calgary, Alberta, Canada.

Plant Fatty Acid Synthesis: Sites of Metabolic Regulation and Potential for Genetic Engineering

Daniel J. Guerra and **Larry Holbrook**
Biotechnica Canada, Inc., 170, 6815-8th Street N.E., Calgary, Alberta, Canada, T2E 7H7

Fatty acid synthesis in plants is under developmental and tissue-specific control. The molecular genetics of plant lipid metabolism essentially are uncharacterized. However, the biochemical framework has been described adequately. In photosynthetic tissues, fatty acid synthesis (FAS) de novo is localized in the chloroplast. The central cofactor/cosubstrate for C_8 to C_{18} fatty acid synthesis is acyl carrier protein (ACP). Evidence from photosynthetic tissue suggests that acyl chain length and in situ distribution may be regulated by ACP isoform structure. In developing oilseed, the plastid is most likely the site of ACP-mediated FAS. To potentially engineer triacylglycerol synthesis, we first must understand the relationships between ACP-mediated de novo FAS and acyl chain length. Several research laboratories are involved with aspects of this complex developmental system. ACP gene manipulation in oilseed crops has become a major focus. A general overview and some specifics of research target agendas will be described.

Plant fatty acid synthesis (FAS) is intracellularly compartmentalized and developmentally regulated (1). In oilseed crops such as soybean and rape, a seed–specific triacylglycerol biosynthetic pathway is coupled to a plastid–bound FAS de novo via a cytoplasmic pool of neutral and phospholipids (2). The pathway leading to C_{16} and C_{18} fatty acids in seed plastids probably is homologous with chloroplast FAS. The initial entry of carbon arises from different metabolic routes. These differences are analogous to physiological source-sink relationships. Acetyl-CoA ultimately is from carbon fixation in green tissues through the production of dihydroxyacetone phosphate and glyceraldehyde-3 phosphate. These phosphorylated sugars may traverse the chloroplast envelope and enter glycolysis in the cytosol or remain plastid-bound where an active glycolytic pathway has been reported (3). Through the activity of pyruvate dehydrogenase (chloroplast, mitochondrial), acetyl-CoA is formed. Malonyl-CoA is synthesized de novo by direct CO_2 incorporation via acetyl-CoA carboxylase (1).

In developing castorbean endosperm (4), sucrose transport and cytosolic hydrolysis (via invertase) yield hexose phosphates. A portion of the oxidative pentose phosphate pathway (OPP) is believed to function in the proplastid (5). OPP intermediates are shuttled through glycolysis and the pyruvate dehydrogenase complex to yield acetyl-CoA (5).

Acetyl-CoA and malonyl-CoA formed from the preceding pathways enter de novo FAS via their respective acyl carrier proteins (ACP) transacylases. The condensation of acetyl-ACP and malonyl-ACP by the β-ketoacyl-ACP synthetase I yields acetoacetyl-ACP, CO_2 and free ACPSH. Acetoacetyl-ACP is reduced by NADPH H^+ to form 3-hydroxybutyryl-ACP, which is dehydrated to crotonyl-ACP. A subsequent reduction with either NADH or NADPH H^+ yields butyryl-ACP, which becomes the 4-carbon substrate for subsequent condensation with a new molecule of malonyl-ACP (1,6). The cycle proceeds to a family of acyl chain lengths (C_8-C_{16}), depending upon plant species and tissue differentiation. There are certain steps in FAS in which regulation of acyl chain length and distribution may occur. The elucidation of these enzymatic steps provides important targets for genetic engineering. Techniques of gene isolation, vector construction and plant transformation may lead to molecular modifications of triacylglycerol composition. We will describe the current understanding of regulatory steps in FAS and some approaches to oilseed modification.

EXPERIMENTAL PROCEDURES

The research reported here is from our own and other laboratories. Specific protocols for experiments may be found in the original research papers cited within the text.

ACETYL-CoA AND MALONYL-CoA:ACP TRANSACYLASES

The initial commitment of carbon to de novo FAS is ACP-mediated. Acetyl-CoA:ACP transacylase (ACT) is the first reaction in the pathway. ACT activity in vivo has not been characterized adequately because of low specific activity in tissue extracts (7,8). Although little is known about the kinetics of this reaction, it has been noted (8) that enzyme level regulation of de novo FAS must be coordinated through ACT activity. In a detailed account of ACT reactivity on in vitro plant FAS, Shimakata and Stumpf (8) showed that the addition of isolated ACT caused synthesis of shorter chain fatty acids. These authors suggested that plants may regulate de novo FAS through the activity of ACT. The enzyme appeared to be rate-limiting for FAS in vitro when the ratio of substrates (ACP/Acetyl-CoA) was held constant at 1:6. Furthermore, these authors demonstrated that all other enzymes in the pathway served only to increase the synthesis of long chain fatty acids (e.g., 16:0, 18:0). The difference in control mechanisms suggested that β-ketoacyl-ACP synthetase was rate-limiting when hexanonyl-ACP was used as reactant instead of acetyl-CoA (8).

Malonyl-CoA:ACP transacylase (MCT) has been purified to near homogeneity and considerable information has accumulated on enzyme kinetics and tissue-specific expression (9–13). The Km and Vmax of semi–purified MCT have been reported but precise information has been lacking. Km values for both substrates (malonyl-CoA, ACP) have been within reported concentrations of these moieties in vivo (micro-molar) (11,13).

Two distinct forms of MCT were found in leaf tissue and one form in developing soybean seed (13). Although these isozymes were similar in reaction rate and substrate specificity, differences were noted in degree of hydrophobicity and inhibition analyses (13). The major questions regarding MCT isoform expression have not been resolved. Indeed, only one form of MCT has been reported for spinach and barley leaf tissue (11,12). Due to the central role of malonyl-ACP in acyl chain-lengthening coordination of MCT, expression must be linked to the in vivo rate of FAS.

Our laboratory has recently characterized MCT from *B. napus* and *E. coli* (14). Table 1 shows the relationship of the ACP/malonyl-CoA molar ratio on *B. napus* MCT activity. When the substrates are present in micro-molar concentration and the ratio of ACP-I to malonyl CoA is increased to $\doteq 2$, the reaction rate is linear. Further increases in the molar ratio resulted in stationary reactivity followed by a steady decline (data not shown). The molecular basis for these complex kinetics is not known. It is possible that ACP substrate may inhibit the reaction through allosteric interactions or that the absolute amount of each substrate plays a key role in the reaction. These substrate relationships not only reveal a higher order of enzyme regulation but also may point to a direct regulation of de novo FAS by MCT. *E. coli* MCT reactivity was similar to *B. napus* MCT (data not shown).

TABLE 1

Effect of Substrate Ratio on the Reactivity of *B. napus* malonyl-CoA:ACP Transacylase

	MCT activity
[a]ACP/malonyl-CoA	(pmoles ^{14}C malonyl-ACP/min/μl)
0.5	10.0
1.6	23.0
3.2	28.0
4.8	32.0

[a]ACP concentration was increased from 4.0 μm to 36.0 μm with [2-^{14}C] malonyl-CoA fixed at 8.0 μm. ACP-I was purified from transformed *E. coli*.

The distinct regulatory nature of both ACT and MCT in FAS de novo awaits a more careful examination of those enzymes with varying substrate concentration.

PLASTID–BOUND ELONGATION/DESATURATION

The reactions leading to the synthesis of stearoyl-ACP from palmitoyl-ACP and malonyl-ACP is distinct from the C_4 to C_{16} FAS pathway (1). β-Ketoacyl-ACP synthetase II was shown to have very low activity in five examined plant tissues (8). This elongation system is situated at an important branch-point between C_{16} and C_{18} pathways. The substrate 16:0-ACP must show different reaction rates with the various enzymes competing for it. The observation that a specific synthase is utilized for C_{16} to C_{18} chain elongation targets this reaction for regulatory control in vivo.

Stearoyl-ACP formed from the β-Ketoacyl ACP synthetase II does not accumulate in plant tissues but is desaturated to oleoyl-ACP by a specific desaturase (15,16). This enzyme has been purified to some extent, and a characterization has been made (17). In leaf and seed tissue, the enzyme required a modified electron transport system including O_2, NAD(P)H and ferrodoxin. McKeon and Stumpf (17) purified the desaturase 200-fold from safflower using an ACP affinity column. Initial characterization suggested that the enzyme was a dimer, and that less than 5% activity was observed when stearoyl-CoA or palmitoyl-ACP was used as substrate.

BRANCH-POINT AND END PRODUCT METABOLISM

As described above the major end product of plant FAS de novo is oleoyl-ACP ($\Delta 9$) with palmitoyl-ACP being the next most significant. Various branch-point reactions may occur to these acyl-ACP substrates. A plastid–bound glycerol 3-phosphate acyl-ACP transferase has been isolated from spinach leaves, which shows a marked preference for oleic acid as the acyl substrate. Frentzen et al. (18) have shown that the enzyme is soluble and that the 1-position of *sn*-glycerol 3-phosphate is acylated preferentially by its reaction with oleoyl-ACP. Guerra et al. (19) further have shown that ACP-II isoform from spinach leaf tissue is the preferred cosubstrate over ACP-I. This specificity for position, acyl functional group and ACP structure have caused us to suggest that this reaction is highly regulated from core components of plant FAS (20).

The monoacylglycerol 3-phosphate acyl-ACP transferase also has been characterized. Andrews et al. (21) have shown the enzyme to be located at the chloroplast envelope and that the preferred substrates are 1-oleoyl lysophosphatidic acid and palmitoyl-ACP. Thus, the sequential formation of phosphatidic acid for subsequent galactolipid biosynthesis in higher plants has been characterized and distinct specificity for reactants described.

Another important branch-point reaction in higher plant plastids is the hydrolysis of oleoyl-ACP. Ohlrogge et al. (22) showed the specificity of the soluble thioesterase, and McKeon and Stumpf (17) reported a 770–fold purification from safflower using an ACP affinity column. Guerra et al. (19) reported a 10-fold lower Km value for ACP-I as cosubstrate when compared to ACP-II. The complementary data for the glycerol 3-phosphate acyl-ACP transferase and oleoyl-ACP thioesterase allowed us to suggest a switching mechanism for 18:1 ACP utilization in higher plant plastids (20). The overall implication was that ACP isoform expression may regulate partially the partitioning of acyl chains among subcellular compartments.

RECOMBINANT ACYL CARRIER PROTEIN

Beremand et al. (23) recently reported the complete chemical synthesis of spinach ACP-I gene. From the amino acid sequence of authentic spinach ACP-I and with the use of plant codon usage tables, the gene was assembled, annealed and finally ligated into M13 vectors. Subcloning procedures resulted in an ACP-I gene construct within a bacterial expression vector under the *trc* promoter. Expression of the recombinant protein in competent *E. coli* allowed us to purify the recombinant ACP-I gene product to homogeneity (14).

Sequence analysis verified the structure of the plant ACP, and a series of characterizations indicated that ACP-I structure imparted unique reaction kinetics to both plant and bacterial enzyme systems. This new source of ACP-I should prove invaluable to researchers interested in evaluating potential targets for enzyme regulation in plant FAS de novo.

As described above, the ratio of ACP to cosubstrate was shown to affect the rate of MCT activity. Somewhat different results were obtained when total FAS was analyzed in a cell-free system (14). Table 2 shows the effect of altering the ratio of ACP-I to ^{14}C malonyl-CoA in such a system.

TABLE 2

Effect of ACP/^{14}C malonyl-CoA Ratio on *B. napus* FAS In Vitro

^{a}ACP/malonyl-CoA	FAS activity (pmoles/min/μl)
0.5	0.87
0.6	0.96
1.0	0.27
2.0	0.15
3.0	0.04
5.0	0.06

^{a}Concentration of ACP-I kept constant at 3.0 μm and concentration of [2-^{14}C] malonyl-CoA was decreased from 7.8 μm to 0.78 μm.

Increasing the ratio of ACP to malonyl-CoA above 0.6 caused a steady decline in de novo FAS from a *B. napus* chloroplast preparation (14). These data show the complexity of ACP cosubstrate/cofactor involvement in overall FAS from acetyl-CoA and malonyl-CoA. Apparently, the total amount of free ACP flux through the system can alter greatly the rates of acyl chain accumulation. The implication is that one or more of the initial reactions (e.g. acetyl-CoA:ACP transacylase, malonyl-CoA:ACP transacylase, β-Ketoacyl-ACP synthase) are sensitive to the free pool of ACP and/or malonyl-ACP. Any endeavor to manipulate FAS in vivo through genetic engineering should address the regulatory control over this anabolic pathway. The stoichiometric ratio of each of the proteins of FAS reactions may be more critical than the absolute amount of any one regulatory element. An identical relationship of de novo FAS vs ACP/malonyl-CoA ratio was obtained using a crude *E. coli* FAS preparation (14). These data suggest a similar control mechanism for both plant and bacterial FAS.

The recombinant ACP-I has great potential as a molecular probe. Isolation of polyadenylated mRNA can allow for northern hybridization analyses of ACP message levels in developing oilseeds or actively growing leaf tissue. Using the techniques of genetic engineering, cDNA copies of hybrid selected mRNA will provide a series of clones that may possess the authentic ACP-I structural gene.

Genomic libraries prepared from restriction digests of developing oilseeds or seedlings may be probed with the synthetic ACP-I gene for analysis of full-length genomic clones. Genomic clones provide an added advantage in that 5′ (or 3′) flanking regions may be found that possess promoter- or enhancer-like elements for gene transcription.

Finally, recombinant ACP-I may be used in the context of in vitro and in vivo reaction kinetics as described (14). A clearer understanding of ACP structure/function relationships should provide avenues for protein engineering of novel isoforms of ACP. The use of site-directed mutagenesis on ACP-I synthetic gene structure (24) also will allow for analyses of rates of transcription, translation and transgenic expression.

OILSEED MODIFICATION VIA GENETIC ENGINEERING

As outlined above, much information has accumulated as to the enzymology of FAS de novo in plants. With the successful synthesis, cloning and expression of a spinach

ACP-I gene and the isolation of a full-length cDNA clone, a new dimension of study is developing. The transfer of foreign genes from one plant to another should provide a new resource base for novel lipid expression. In collaboration with conventional plant breeding techniques, transgenic expression may become stable and the diversity of oil quality greatly expanded. Plant triacylglycerols rich in specific fatty acids (e.g. hydroxy, epoxy, n-3) may be a new source for biotechnological development.

Specific strategies include the use of the Ti plasmid from *Agrobacterium tumefaciens* (25). Using either nopaline or octopine borders for gene constructs in cointegrate vector systems provides transformation of many agronomic species. To date, Ti plasmids are most successful as plant vectors in dicotyledenous plants. Some of these species are major oilseed crops (e.g. soybean, rapeseed and cotton) with an existing market niche. The successful transformation of crops with unique FAS genes will provide new oilseed products for both the producers and distributors.

Other techniques of genetic engineering include electroporation (electronic-pulsed current of naked DNA into plant protoplasts or isolated organelles) and microinjection (direct injection of naked DNA into plant nuclei and protoplasts). All of these technologies are in the early developmental stages: it would be premature to evaluate their direct involvement in oilseed modification. The potential for these techniques and others is quite substantial, and research funding in this area is encouraged.

From a more biological perspective, the use of microspores (male gametes) for oilseed modification deserves attention. Using chemical mutagens such as ethylmethane sulfonate (EMS) microspores may be mutagenized chemically to grow on selected media. Herbicide tolerance, agronomic performance and oilseed quality all may be affected by these mutations. Propagation of plantlets from mutagenized microspores provide a haploid genome possessing modified chemical or physical characteristics. These plants then may be used in conventional breeding schemes to develop new lines of oilseed crops.

The phenomenon of somaclonal variation (somatic genome random mutation in cultured plant cells) also has been suggested as a source of novel germplasm (26). Plantlets derived from bulk culture can be screened for chemical or physical characteristics much in the same way as mutagenized microspores. Desirable mutation can be developed into a conventional breeding program for development and marketing.

The end product of successful genetic engineering is the expression of foreign gene function in vivo. A goal for plant biotechnology is successful production of oil-modified transgenic varieties. *Brassica* spp. transformation has been achieved but the gene systems involved have not addressed oilseed modification (27,28). The current method employed for gene transfer is by the *Agrobacterium tumefaciens* pTi. Although many *Brassica* spp. produce tumors with both octopine and nopaline strains of *A. tumefaciens*, *B. napus* is more responsive to the nopaline phenotype (29). This host plant selectivity probably is a function of the VIR region on the pTi plasmid.

A commonly used selectable marker for *B. napus* transformation with pTi is kanamycin resistance derived from transposons possessing the neomycin phosphotransferase (NPT) gene (27). Methotrexate resistance via dihydro folate reductase mutation also has been employed for selection of *B. napus* transformants (28).

One of the advantages of pTi transformation is the ability to infect tissue explants. Given the proper culturing conditions, this transformed tissue may undergo organogenesis (shoot and root formation). Plants regenerated from this system will show a minimum of ploidy amplification because of the rapid differentiation rate. An epidermal thin cell layer from *B. napus* internode has proven quite successful (30).

The combined techniques of gene isolation, tissue transformation and plant regeneration may be applied to oilseed modification. Recombinant ACP-I and other FAS proteins may be introduced to tissue explants via the appropriate structural gene(s) and pTi construct. The resultant biochemical effect may be ascertained by lipid fractionation and FAME analyses. In this way, experiments conducted in vitro (e.g. MCT, de novo FAS) may be compared directly with transformed fatty acid metabolism in vivo.

There are several sites in the FAS de novo pathway where initial and branch-point reactions may be regulated. The initial acylation of acetyl-CoA and malonyl-CoA with acyl carrier protein provide for the synthesis of C_6 to C_{18} fatty acids. The condensation reaction of palmitoyl-ACP with malonyl-ACP is another key reaction for acyl chain distribution. Acyl transferases and thioesterases provide endpoint regulation of the products of FAS de novo. In all instances, acyl carrier protein served as cosubstrate/cofactor. The use of acyl carrier protein genes for in vivo manipulation of FAS should provide an avenue for exploring the molecular biology of the system.

REFERENCES

1. Stumpf, P.K., in *Fatty Acid Metabolism and Its Regulation*, edited by S. Numa, Elsevier Press, Amsterdam, 1984, pp. 155–179.
2. Stymne, S., and K. Stobart, in *The Biochemistry of Plants*, edited by P.K. Stumpf, Vol. 10, Academic Press, New York, NY (1986).
3. Dennis, D.T., and J.A. Miernyk, *1982 Annual Review of Plant Physiol 33*:27 (1982).
4. Yamada, M., and Y. Nakamura, *Plant Cell Physiol 16*:151 (1975).
5. Simcox, P.D., E.E. Reid, D.T. Canvin and D.T. Dennis, *Plant Physiol 59*:1128 (1977).
6. Harwood, J.L., *Prog. Lipid Res. 18*:55 (1979).
7. Shimakata, T., and P.K. Stumpf, *Arch. Biochem. Biophys. 217*:144 (1982).
8. Shimakata, T., and P.K. Stumpf, *J. Biol. Chem. 258*:3592 (1983).
9. Joshi, V.C., and S.J. Wakil, *Arch. Biochem. Biophys. 143*:493 (1971).
10. Lessire, R., and P.K. Stumpf, *Plant Physiol. 73*:614 (1983).
11. Stapleton, S.R., and J.G. Jaworski, *Biochem. Biophys. Acta. 794*:240 (1984).
12. Hoj, P.B., and I. Svendsen, *Carlsberg Res. Commun. 48*:285 (1983).
13. Guerra, D.J., and J.B. Ohlrogge, *Arch. Biochem. Biophys. 246*:274 (1986).
14. Guerra, D.J., K. Dziewanowska, J.B. Ohlrogge and P.D. Beremand, in press.
15. Nagai, J., and K. Bloch, *J. Biol. Chem. 243*:4626 (1968).
16. Jaworski, J.G., and P.K. Stumpf, *Arch. Biochem. Biophys. 162*:158 (1974).
17. McKeon, T.A., and P.K. Stumpf, *J. Biol. Chem. 257*:12141 (1982).
18. Frentzen, M., E. Heinz, T.A. McKeon and P.K. Stumpf, *Eur. J. Biochem. 129*:629 (1983).
19. Guerra, D.J., J.B. Ohlrogge and M. Frentzen, *Plant Physiol. 82*:448 (1986).
20. Guerra, D.J., J.B. Ohlrogge and M. Frentzen, *Proc. Intl. Lipid Symposium*, Academic Press, New York, NY, in press.
21. Andrews, J., J.B. Ohlrogge and K. Keegstra, *Plant Physiol. 78*:459 (1985).
22. Ohlrogge, J.B., W.E. Shine and P.K. Stumpf, *Arch. Biochem. Biophys. 189*:382 (1978).
23. Beremand, P.D., D.J. Hannapel, D.J. Guerra, D.N. Kuhn and J.B. Ohlrogge, *Arch. Biochem. Biophys. 256*, in press.
24. Beremand, P.D., D.J. Guerra, D.J. Hannapel, D.N. Kuhn and J.B. Ohlrogge, *J. Cellular Biochemistry*, in press.
25. Caplan, A., L. Herrera-Estrella, D. Inze, E. Van Haute, M. Van Montagu, J. Schell and P. Zambryski, *Science 222*:815 (1983).
26. Orton, T.J., in *Gene Manipulation in Plant Improvement*, edited by J.P. Gustafson, Plenum Press, New York, NY (1984).
27. Charest, P.J., L. Holbrook, J. Gabard, V.N. Iyer and B.L. Miki, *Theor. App. Gen.*, in press.
28. Pua, E.C., A. Mehra-Palta, F. Nagy and N.H. Chua, *Bio/Technology*, in press.
29. Holbrook, L., and B.L. Miki, *Plant Cell Reports 4*:329 (1985).
30. Klimaszewska, K., and W. Keller, *Plant Cell Tissue and Organ Culture 4*:183 (1985).

Genetic Transformation for the Improvement of Canola

R.B. Horsch, J.E. Fry, A. Barnason, S. Metz, S.G. Rogers and **R.T. Fraley**

Monsanto Company, 700 Chesterfield Village Parkway, St. Louis, MO 63198

Transformed *Brassica napus* plants have been produced by means of a simple stem segment transformation and regeneration method that couples the gene transfer mechanism of *A. tumefaciens* with kanamycin selection and de novo shoot regeneration. Surface-sterilized stem segments were inoculated with an *A. tumefaciens* strain containing our vector and cultured for two days. The explants then were transferred to regeneration medium containing kanamycin. Shoot regeneration occurred within three to six wk, and transformants were confirmed by a leaf callus assay on medium containing kanamycin. Progeny of the transgenic plants inherited the new genes in the expected Mendelian patterns of 3:1 or 15:1. This method should be adaptable to many species that are within the host range of *A. tumefaciens* and where regeneration of plants from culture is possible.

The ability to alter genes and then reintroduce them into plants opens the door to unprecedented structure-function analysis in genetics, biochemistry and developmental biology. The first generation studies have examined gene structure and expression and quickly are being followed by analysis of altered expression of proteins. The next generation of experiments will likely examine the effects of altered proteins on biochemical pathways and will lead to insights into physiological systems. We now can monitor expressions of genes at different times and places during plant development and soon will be able to change the time and place of expression of such genes by changing promoters. Ultimately, directed changes in genes will lead to a better understanding of how molecules give rise to the growth and development of plants. This understanding and technical capability will revolutionize how we can alter/improve plants for human benefit. One of the first traits that may be possible to change will be oil quality.

AGROBACTERIUM

A. tumefaciens causes crown gall disease by transferring a defined segment of DNA (T-DNA) from its tumor-inducing (Ti) plasmid into the nuclear genome of cells in an infected wound on many dicotyledonous plants (for recent reviews, see [1-3]). The T-DNA contains genes that encode enzymes that catalyze the synthesis of the phytohormones IAA (4-6) and IPA (7), an auxin and a cytokinin. This overproduction of phytohormones is responsible for the proliferation of wound callus into a gall that can harbor a population of the bacteria. Other T-DNA genes cause synthesis of unique metabolites called opines that *Agrobacterium* can catabolize but that cannot be utilized by many other microorganisms or plants. One strain of *A. tumefaciens* (A208) that we have used produces nopaline synthase, an enzyme that produces nopaline from arginine and α-ketoglutarate.

The movement of the T-DNA is mediated by genes in another region of the Ti plasmid. These virulence genes (8) are not themselves transferred with the T-DNA, but act in *trans* to cause the transfer (9,10). The T-DNA is defined by a 25-base sequence, the border sequence, which is present as a direct repeat outside the ends of the T-DNA (11). Recent evidence suggests that the border sequence is the site of specific nicking of the DNA by one of the vir gene products (12). It has been reported that a bacterial origin of transfer can substitute for the border sequence when compatible mobilization functions also are present (13). None of the genes or DNA sequences within the T-DNA are required for the transfer process (14).

These properties of the *Agrobacterium* DNA transfer system are invaluable for developing a powerful vector system for plant transformation. It is possible to remove all of the original T-DNA that is responsible for the disease and replace it with any DNA as long as the border sequence is maintained (15). The DNA is inherently stable once in the plant genome because neither the border nor the virulence genes are transferred. Finally, because the borders define a discrete T-DNA segment the frequency of cotransfer of the entire segment is very high. This means that the favored genes usually are transferred along with the selectable marker used to identify the transformed cells or plants.

VECTORS

To harness this mechanism, we designed a shuttle vector (16) that can replicate in *E. coli* where recombinant manipulations are easily handled, and then can be transferred into *A. tumefaciens* in preparation for transfer into plants. There are two basic modes of maintenance of the shuttle vectors in *Agrobacterium*: either by integration into the Ti plasmid by recombination at a region of DNA homology (15,17,18) or by autonomous replication in *trans* to the Ti plasmid (19-21). The former type of vector is referred to as a *cis* or integrating vector, while the latter is called a *trans* or binary vector. In most cases, they accomplish the same goals of shuttling genes from *E. coli* to *A. tumefaciens* in a T-DNA package that then can be transferred to plants.

The essential components of our vectors are shown in Figure 1. The binary vector pMON505 (22) is shown as an example. They include plasmid functions for replication (pBR322, RK2 Replicon) and/or integration (LIH, not present in binary vector) in the bacteria as well as a bacterial selectable marker for spectinomycin and streptomycin resistance (Spc/Str^R) for genetic manipulations with the bacteria. They also include a chimeric selectable marker for plant antibiotic resistance such as neomycin phosphotransferase (NOS-NPTII-NOS), which makes transformed plant cells resistant to kanamycin, and a border sequence (NRB) to identify the end of the new T-DNA. There is also a scoreable marker, the gene for nopaline synthase (NOS), that provides a powerful genetic tool for monitoring the presence of the T-DNA in plants and their progeny. The final feature is the most important, a multicloning site to add the favored gene.

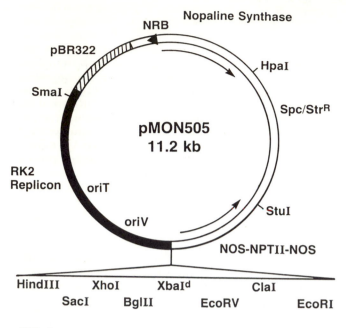

| HindIII | XhoI | XbaI^d | ClaI |
| SacI | BglII | EcoRV | EcoRI |

FIG. 1.

SELECTABLE MARKERS

Because only a small proportion of cells are transformed by *A. tumefaciens*, it is important to have a selectable marker to recognize the transformed cells and suppress the growth of wild-type cells. The first marker to be used was the neomycin phosphotransferase gene from *E. coli*. This gene has been used successfully in many systems, including mammalian cells, fungi and *Dictyostellium*. In solanacious plants such as petunia and tobacco it provides an excellent, unambiguous selectable marker.

The features of a good selectable marker are high-level resistance of the transformed cells to concentrations of the drug that completely inhibit wild-type growth and efficient selectability of rare transformants from a large excess of inhibited wild-type cells that surround them. Kanamycin-inhibited petunia cells do not appear to interfere with the growth of resistant cells in any way, and actually may facilitate their growth by acting as a nurse tissue. Single resistant clones can be obtained from petunia leaf discs transformed with attenuated mutants of *A. tumefaciens* (23), demonstrating the efficacy of selection for this marker.

A second selectable marker that we have used is the mouse gene for dihydrofolate reductase (DHFR) that encodes an altered protein that does not bind to methotrexate, a potent inhibitor of this essential enzyme. Petunia cells that contain and express this gene are able to grow on concentrations of methotrexate that completely inhibit the control cells (24). Selectability also appears to be good for methotrexate resistance, although the efficiency is dependent on the strength of expression of the DHFR gene. A simple trick to improve the selectability of resistant cells is wait two or three days after transformation to add selective pressure, thus allowing time for the cells to accumulate resistant DHFR enzyme.

A third marker we have used is a bacterial gene for hygromycin phosphotransferase that confers resistance to hygromycin (25). This marker works reasonably well

in solanacious species as well as *Arabidopsis thaliana*, in which it is the marker of choice. The selectability of hygromycin resistance is poor, probably due to the extreme toxicity of the drug. For species such as *A. thaliana* in which kanamycin resistance functions poorly, hygromycin can be a good alternative. However, where kanamycin resistance works, it is clearly a better choice.

LEAF DISC TRANSFORMATION

The simple and powerful interface between *A. tumefaciens* and plant cells is illustrated by the leaf disc transformation system (26) where gene transfer, selection and regeneration are coupled together in an efficient process. Tobacco is an excellent host for *A. tumefaciens*, and also responds exceedingly well in culture. Although the technique is most easily practiced with tobacco, it has been applied to a number of other species (Table 1).

TABLE 1

Transgenic Plants Produced with *A. tumefaciens*-mediated Transformation

Species
Nicotiana plumbaginifolia (23)
petunia (26)
tobacco (29)
tomato (30)
potato (31)
lettuce (Michelmore, personal communication)
poplar (32)
Arabidopsis thaliana (25)
Medicago varia (33)
flax (34)
Brassica napus (35)
sunflower (Everett et al., in press)
cotton (36)

Surface sterilized stem segments of *B. napus* are infected with the appropriate strain of *A. tumefaciens* carrying the vector of choice and cocultured on regeneration medium for two or three days. During this time, the virulence genes in the bacteria are induced, the bacteria bind to the plant cells around the wounded edge of the explant, and the gene transfer process occurs. One empirical observation we have made is that for species other than tobacco, a nurse culture of tobacco cells increases the transformation frequency when used during the coculture period. This may be due to more efficient induction of the virulence genes or some other factor not yet understood.

After the transformation has occurred, the explants are transferred to regeneration/selection medium. This contains 500 µg/ml carbenicillin to kill the bacteria and the appropriate antibiotic to inhibit untransformed plant cells, usually kanamycin. During the next three wk, the transformed cells grow into shoots via organogenesis. Between three and six wk, the shoots develop enough to remove them from the explant and to induce rooting in preparation for their transfer to soil.

There are a number of technical issues that require careful attention, especially when attempting to apply this technique on species other than petunia or tobacco. First, some of the shoots that grow in the presence of kanamycin do not express and/or contain the foreign DNA. The reasons for these apparent escapes are not clear but may include loss of expression or loss of DNA during plant development, or incomplete selection due to cross-protection of wild-type cells by transformed cells nearby. Whatever the reasons, the problem can be dealt with by a second selection for ability to root in the presence of kanamycin (26), and subsequent callus induction assays or opine assays for continued expression of the new genes.

Position effects, the effects of the surrounding DNA or chromatin structure, result in differences in expression of the T-DNA genes in different transformants. The extreme of this is complete loss of expression during differentiation of the plantlet. This occurs in about a quarter to a third of the shoots that grew initially in the presence of 300 μg/ml kanamycin. One possible reason for failure to select against these escapes is that the kanamycin is not a good herbicide, and shoots can continue growth in its presence once they are large enough. It usually is necessary to screen several independent transgenic plants to identify the best expression of the favored gene that sometimes but not always is correlated with the expression of the selectable marker.

INHERITANCE

Progeny of transgenic plants produced using this process always have inherited the T-DNAs found in the parent plant. Progeny usually inherit the T-DNA according to Mendelian rules for dominant genes. In cases in which there is only one T-DNA, selfed progeny show a 3:1 ratio for nopaline or kanamycin resistance, while backcrossed progeny show a 1:1 ratio. This is also true for multiple T-DNAs that are present in a tandem array in which they are all genetically linked. In a few cases, there are multiple T-DNAs, some that are unlinked. In these cases, selfed ratios are usually 15:1 (two unlinked loci) or higher (three or more unlinked loci). In nearly all cases, the selected and unselected markers on a single T-DNA are coinherited.

The T-DNA thus becomes a permanent, stable part of the genome of transgenic plants and behaves just as the other endogenous genes. The availability of plants with T-DNA inserts mapped in chromosomal locations brings a valuable new tool to plant genetics. These transformants have an easily scoreable marker that can be used to map new mutants and a selectable chromosome tag that might facilitate breeding or selection of interesting genes nearby.

ENGINEERING HERBICIDE TOLERANCE

The herbicide glyphosate is a potent inhibitor of an enzyme in the shikimate pathway in higher plants. The herbicide competitively blocks EPSPS (5-enolpyruvylshikimate-3-phosphate synthase) and kills plants as well as tissue cultures. We have engineered plants to be tolerant to otherwise lethal doses of glyphosate by a strategy of overexpression of EPSPS in transgenic plants. The method for achieving this goal started in tissue culture to obtain the EPSPS protein, proceeded through molecular biology to obtain and modify the gene, and back into tissue culture to put the gene back into plants.

A glyphosate tolerant suspension culture of petunia hybrida was selected by sequential step-up from 0.1 mM glyphosate to 20 mM glyphosate over a period of several months. The resistant cell line (MP4G) produces about 20 times more EPSPS than the sensitive parent culture (MP4). The EPSPS protein was purified from the MP4G culture, and this pure protein was partially sequenced to obtain an amino acid sequence from its NH$_2$-terminus. A series of oligonucleotide probes was synthesized (in three families) that would cover all possible DNA sequences for the amino acid sequence. These families then were used to probe a northern blot of poly(A)+mRNA from the two cell lines. One of the families of probes hybridized to a specified mRNA from the MP4G culture about 20 times more strongly than to the same size mRNA from the MP4 culture. This strongly suggested that this probe was specific for EPSPS and that the MP4G culture was overproducing the mRNA for this enzyme. The gene then was isolated from a cDNA library prepared from the MP4G culture. The cause of the overproduction of the EPSPS mRNA was an amplification of the gene during the sequential adaptation of the culture to glyphosate (27).

Thus, the suspension culture had provided an enriched source for purification of the protein and a means to identify which probe (from a family of redundant probes) was the proper one to use to screen a cDNA library for the EPSPS gene. The culture also was used to prepare a cDNA library that was enriched for the EPSPS sequence.

A derivative of the pMON505 vector, pMON530, was used for *Agrobacterium*-mediated transfer of a chimeric EPSPS gene back into plant cells. It contains the NOS-NPTII-NOS gene for kanamycin resistance and an expression cassette consisting of the cauliflower mosaic virus (CaMV) 35S promoter and a nopaline synthase (NOS) polyadenylation signal. The complete open reading frame of the petunia EPSPS was cloned into pMON530 to create a chimeric EPSPS gene capable of high level expression in plant cells (pMON546). The vector then was introduced into *A. tumefaciens* for transfer to plant cells.

Leaf discs transformed with vectors pMON505 and pMON546 both produced large amounts of kanamycin resistant callus. Small pieces of this callus were transferred to medium containing 0.1, 0.25, or 0.5 mM glyphosate. The callus transformed with pMON546 continued to grow at all concentrations of glyphosate while the callus transformed with pMON505 was inhibited. Glyphosate tolerance also could be demonstrated by direct selection for growth from infected leaf discs on the same concentrations of glyphosate. Four independent transgenic plants containing pMON546 were selected on kanamycin and grown in soil before spraying with a lethal dose of glyphosate. These plants were sprayed with formulated glyphosate (containing a surfactant) at a rate equal to 0.8 pounds per acre, twice the dose needed to kill all control plants. All four transformed plants survived and continued to grow with sight chlorosis at the growing points, from which they later recovered. These experiments demonstrate that EPSPS is the prime target of glyphosate in plant cells and whole plants because over expression of that one gene confers tolerance to the herbicide. These results have been extended to tobacco and tomato using the petunia gene construction in pMON546. In the near future, we expect to extend these results to other crops such as *B. napus*.

Herbicide resistance (especially resistance to a broad spectrum compound like glyphosate) will likely have a major impact on the methods and costs of weed control in the future. Canola is an excellent commercial target for glyphosate resistance because current methods of weed control are inadequate over much of its growing area.

DISCUSSION

Molecular biology and gene transfer technology have opened the door to unprecedented experimental power for understanding and manipulating plants. This power will build on itself as it is used to discover the mechanisms of gene expression, protein structure and function and the biochemistry and physiology of plants. This knowledge will rapidly increase the ability to intelligently predict beneficial changes to be introduced into crops, as well as the means to do it. The results of faithful expression of soybean seed storage genes in heterologous species or with chimeric gene constructions (28), imply that genes for changing oil production also can be expressed correctly in the desired tissues and at the desired times of development in transgenic plants.

REFERENCES

1. Bevan, M., M.D. Chilton, *Ann. Rev. Genet. 16*:357 (1982).
2. Depicker, A., M. Van Montagu and J. Schell, in *Genetic Engineering of Plants: an Agricultural Perspective*, edited by T. Kosuge, C. Meredith and A. Hollaender, Plenum Press, New York, NY, 1983, pp. 143-176.
3. Nester, E., M. Gordone, R. Amasino and M. Yanofsky, *Ann. Rev. Plant Physiol. 35*:387 (1984).
4. Schroder, G., S. Waffenschmidt, E. Weiler, J. Schroder, *Eur. J. Biochem. 138*:387 (1984).
5. Thomashow, L., S. Reeves, M. Thomashow, *Proc. Natl. Acad. Sci. USA 81*:5071 (1984).
6. Akiyoshi, D., H. Klee, R. Amasino, E. Nester and M. Gordon, *Proc. Natl. Acad. Sci. USA 81*:5994 (1984).
7. Barry, G., S. Rogers, R. Fraley and L. Brand, *Proc. Natl. Acad. Sci. USA 81*:4776 (1984).
8. Klee, H., F. White, V. Iyer, M. Gordon and E. Nester, *J. Bacteriol. 153*:878 (1983).
9. deFramond, A., K. Barton, M.D. Chilton, *Bio/Technology 1*:262 (1983).
10. Hoekema, A., P. Hirsch, P. Hooykaas and R. Schilperoort, *Nature 303*:179 (1983).
11. Barker, R., K. Idler, D. Thompson and J. Kemp, *Plant Mol. Biol. 2*:335 (1983).
12. Stachel, S.E., B. Timmerman and P. Zambryski, *Nature 322*:706 (1986).
13. Buchanan-Wollaston, V., J.E. Passiatore and F. Cannon, *Nature 328*:172 (1987).
14. Garfinkel, D., R. Simpson, L. Ream, F. White, M. Gordon and E. Nester, *Cell 27*:143 (1981).
15. Zambryski, P., H. Joos, C. Genetello, J. Leemans, M. Van Montagu and J. Schell, *Eur. Mol. Biol. Org. J. 2*:2143 (1983).
16. Fraley, R.T., S.G. Rogers, R.B. Horsch, P.R. Sanders, J.S. Flick, S.P. Adams, M.L. Bittner, L.A. Brand, C.L. Fink, J.S. Fry, G.R. Galluppi, S.B. Goldberg, N.L. Hoffman and S.C. Woo, *Proc. Natl. Acad. Sci. USA 80*:4803 (1983).
17. Comai, L., C. Schilling-Cordaro, A. Mergia and C. Houck, *Plasmid 10*:21 (1983).
18. Fraley, R.T., S.G. Rogers, R.B. Horsch, D.A. Eichholtz, J.S. Flick, C.L. Fink, N.L. Hoffman and P.R. Sanders, *Bio/Technology 3*:629 (1985).
19. Bevan, M., *Nucleic Acids Res. 12*:8711 (1984).
20. An, G., B. Watson, S. Stachel, M. Gordon and E. Nester, *Eur. Mol. Biol. Org. J. 4*:277 (1984).
21. Klee, H.J., M.F. Yanofsky and E.W. Nester, *Bio/Technology 3*:637 (1985).
22. Horsch, R.B., and H.J. Klee, *Proc. Natl. Acad. Sci. 83*:4428 (1986).
23. Horsch, R.B., H.J. Klee, S. Statchel, S.C. Winans, E. Nester, S.G. Rogers and R.T. Fraley, *Proc. Natl. Acad. Sci. 83*:2571 (1986).
24. Eichholtz, D.A., S.G. Rogers, R.B. Horsch, H.J. Klee, M. Hayford, N.L. Hoffman, S.B. Braford, C. Fink, J. Flick, K.M. O'Connell and R.T. Fraley, *Somatic Cell Mol. Genet. 13*:67 (1987).
25. Lloyd, A.M., A.R. Barnason, S.G. Rogers, M.C. Byrne, R.T. Fraley and R.B. Horsch, *Science 234*:464 (1986).
26. Horsch, R.B., J. Fry, N.L. Hoffman, M. Wallroth, D. Eichholtz, S.G. Rogers and R.T. Fraley, *Science 227*:1229 (1985).
27. Shah, D.M., R.B. Horsch, H.J. Klee, G.M. Kishore, J.A. Winter, N.E. Tumer, C.M. Hironaka, P.R. Sanders, C.S. Gasser, S. Aykent, N.R. Siegel, S.G. Rogers and R.T. Fraley, *Science 233*:478 (1986).
28. Beachy, R.N., P. Abel, M. Oliver, B. De, R.T. Fraley, S.G. Rogers and R.B. Horsch, in *Biotechnology in Plant Science: Relevance to Agriculture in the Eighties*, edited by M. Zaitlin and A. Hollander, Academic Press, New York, NY, in press.
29. DeBlock, M., L. Herrera-Estrella, M. Van Montagu, J. Schell and P. Zambryski, *Eur. Mol. Biol. Org. J. 3*:1681 (1984).
30. McCormick, S., J. Niedermeyer, J. Fry, A. Barnason, R. Horsch and R. Fraley, *Plant Cell Reports 5*:81 (1986).
31. Shahin, E., and R. Simpson, *Hort. Sci. 21*:1199 (1986).
32. Fillatti, J.J., J. Sellmer, B. McCown, B. Haissig and L. Comai, *J. Mol. Gen. Genet. 13*:67 (1987).
33. Deak, M., G.B. Kiss, C. Koncz and D. Dudits, *Plant Cell Reports 5*:97 (1986).
34. Basiran, N., P. Armitage, R.J. Scott and J. Draper, *Plant Cell Reports 6*:396 (1987).
35. Fry, J., A. Barnason and R.B. Horsch, *Plant Cell Reports 6*:321 (1987).
36. Umbeck, P., G. Johnson, K. Barton and W. Swain, *Bio/Technology 5*:263 (1987).

Commercial Development of Oil Palm Clones

L.H. Jones
Unilever Research Laboratory, Colworth House, Sharnbrook, Bedford, MK44 1LQ, UK

Commercial development of clonal oil palms produced by tissue culture requires optimization and the up-scale of laboratory methods for cost–effective plant multiplication. In addition, effective selection of elite plants for clonal multiplication requires detailed palm recording in the most up-to-date progenies from palm–breeding programs. Finally, clones must be fully field tested in a range of environments before being selected for commercial production. Abnormal flowering leading to bunch failure in some clones selected for large–scale multiplication has delayed commercial production while the cause is investigated until safe procedures can be adopted.

Methods for clonal propagation of oil palm using tissue culture were developed in the 1970s, and the first experimental plants from tissue culture were field-planted in Malaysia 10 years ago. Activity since then has been concerned primarily with developing a commercially viable process from the initial laboratory technique. This paper discusses some of the factors that have to be considered in this process.

ESSENTIAL ELEMENTS IN DEVELOPMENT OF NEW CLONES

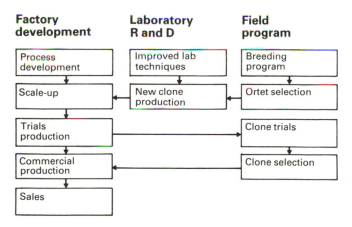

FIG. 1

The essential requirement is for a cost effective process producing reliable uniform plants of guaranteed high performance. Figure 1 shows the main elements that have to be put together. The development of a laboratory technique is only one step in a series of operations requiring development. The route from top to bottom of the diagram outlines the steps of optimization of the methodology to provide a cost-effective production system. However, the value of the plants and their marketability depends upon their potential as oil producers, which is determined by the genetic make-up of the stock being propagated. This requires careful ortet selection from a well–founded breeding program and intensive clone testing and reselection. Each of the headings in Figure 1 contains a multitude of activities that I only can review briefly here.

LABORATORY TECHNIQUE

The majority of plant propagation systems in commercial use rely on micropropagation from proliferating axillary buds and avoid the production of callus because of the consequent problems of plant regeneration and possible genetic instability.

In the case of the palms, all reports of successful propagation from culture have involved the preliminary development of callus. In the oil palm, we outlined the sequence in 1974 (1) and in 1976 Rabechault and Martin (2) described a similar system in more detail starting from leaf tissue. A fuller description of the French work has been published by Ahee et al. (3). This follows our own observations very closely. Tisserat's detailed work on date palm (4,5) also reflects oil palm experience in many respects, and similar findings were reported by Reynolds and Murashige (6) in 1979 for several palm species. The essential features of the process are shown in Figure 2 as described by Lioret (7).

OPTIMIZATION AND UPSCALE

Optimization of the laboratory process includes the purely technical requirements of the culture process itself, such as fine tuning of hormonal requirements for different genotypes, determination of optimum transfer times and environmental requirements of light and temperature. It also includes minimizing all the elements of cost required to produce a plant ready for sale. Development of the laboratory process to a commercial scale requires careful planning of the logistics to establish a cost-effective operation.

Process optimization requires consideration of the following costs: capital costs are the costs of provision of adequate facilities of aseptic transfer and culture incubation to provide for the cultures and the workforce at minimum cost. Process costs cover the ergonomic study to minimize operator time per plant produced, e.g. medium preparation, culture inspection, operator transfer rate, quality control, plantlet handling. Materials and energy costs are costs of the medium components, culture vessels, energy inputs, transport, nursery supplies. Insurance costs include safeguards against contamination, loss of regeneration potential, genetic drift, impure chemicals, power failures, human error, strikes.

The operation is divided into two stages: preproduction and production. The preproduction stage covers the primary isolation of callus from explants, early stages of callus multiplication, initiation of embryogenesis and early embryoid multiplication. Once sufficient embryogenic cultures have been produced for further multiplication to become routine, the new embryogenic line is passed on to production.

In production, choices are made between further multiplication and shoot production in appropriate culture regimes. The main production cycle comprises embryoid

PROPAGATION OF OIL PALM BY TISSUE CULTURE

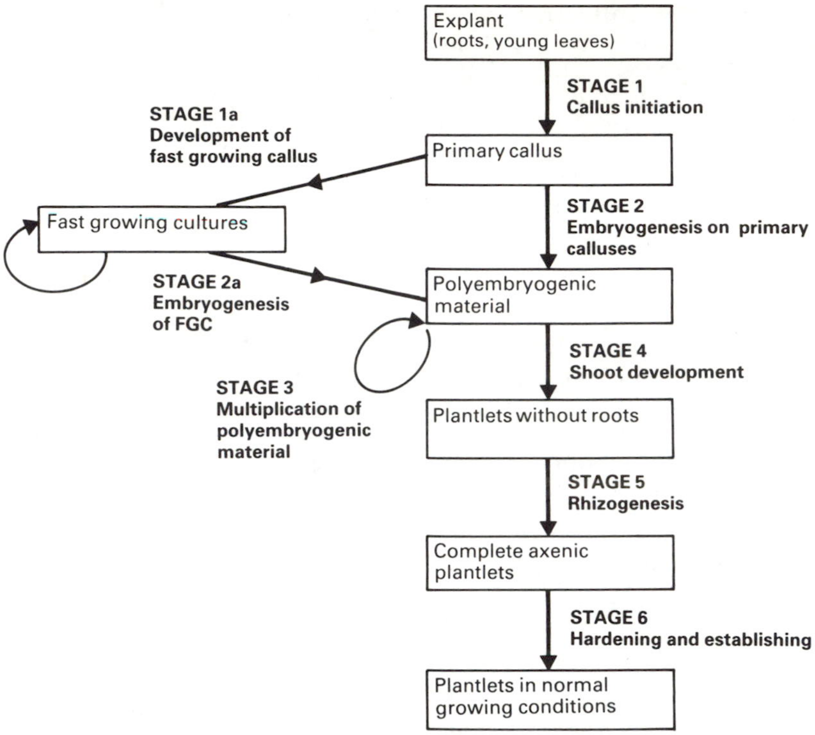

```
                                          ┌─────────────────────┐
                                          │ Explant             │
                                          │ (roots, young leaves)│
                                          └─────────────────────┘
                                                    │ STAGE 1
                                                    │ Callus initiation
  STAGE 1a                                          ▼
  Development of                          ┌─────────────────────┐
  fast growing callus                     │ Primary callus      │
                                          └─────────────────────┘
         ┌─────────────────────┐                   │ STAGE 2
         │ Fast growing cultures│                   │ Embryogenesis on primary
         └─────────────────────┘                   │ calluses
            STAGE 2a                                ▼
            Embryogenesis            ┌─────────────────────┐
            of FGC                   │ Polyembryogenic     │
                                     │ material            │
                   STAGE 3           └─────────────────────┘
                   Multiplication of          │ STAGE 4
                   polyembryogenic            │ Shoot development
                   material                   ▼
                                     ┌─────────────────────┐
                                     │ Plantlets without roots│
                                     └─────────────────────┘
                                              │ STAGE 5
                                              │ Rhizogenesis
                                              ▼
                                     ┌─────────────────────┐
                                     │ Complete axenic     │
                                     │ plantlets           │
                                     └─────────────────────┘
                                              │ STAGE 6
                                              │ Hardening and establishing
                                              ▼
                                     ┌─────────────────────┐
                                     │ Plantlets in normal │
                                     │ growing conditions  │
                                     └─────────────────────┘
```

FIG. 2.

multiplication, singling of individual shoots and subsequent rooting, and hardening prior to dispatch.

In both preproduction and production, there are at any time a large number of clones in various stages of developments for production of plants for commercial trials. Once major commercial production begins on selected clones, there will be relatively few clones in large-scale production but always with a background of new clones under development and new isolates of old clones for stock replenishment.

A key operation is the provision of adequate supplies of the appropriate media for each operation. Computer stock control is a vital part of the smooth running of the operation. In designing a factory operation, careful thought must be given to the entire system from the outset. A preliminary choice must be made between reusable glass tubes or jars that can be autoclaved after dispensing the media or nonautoclavable, disposable plastic containers requiring aseptic dispensing.

Choice of container then dictates design of racks, trolleys, trays, shelving and the basic operating system of the factory. Once a major capital commitment has been made to one system, it becomes difficult to change without far-reaching implications for the whole network of factory operations.

Hygiene in the factory requires constant vigilance. In addition to the hazards of microbial contamination during the culture transfer operation, there are additional hazards from pest infestation in the culture room. Mites are a fairly frequent pest, and seasonal infestations with thrips can occur (8) particularly during the summer grain harvest. The presence of mites or insects usually can be detected by the development of microbial colonies in the cultures in characteristic lines of footprints. Thrips have been known to form axenic breeding colonies on cultured shoot systems, causing considerable damage to the leaf tissues without otherwise contaminating the media.

Contamination of cultures during transfer can be kept to a very low level by good operators but remains a probable event. Careful inspection and rejection of any tubes showing visible levels of contamination provide a primary safeguard against spreading contamination by subculture of contaminated tubes. However, contaminating organisms can be present at subvisible levels, and culture lines that have been subcultured over long periods frequently build up unacceptable levels of microbial contaminants. We have not found antibiotic treatments to be effective in disinfecting infected cell lines except on a small scale for valuable material. Since long–term cultures also deteriorate in their quality of shoot production, good practice requires regular reisolation of fresh cell lines either from the original ortet or from proven clonal ramets.

NURSERY AND PLANT ESTABLISHMENT

After rooting and hardening, the ramets are packaged into nylon sachets with roots wrapped in moist tissue and dispatched to the tropics in polystyrene-lined boxes. In these conditions, they show survival for periods up to a week before planting out in polybags. Initially, the plants are kept misted under polythene tunnels. After an initial acclimatization period, the polythene is removed but plants are kept under light shade with overhead spray irrigation for several months until they are large enough to be repotted into large polybags and transferred to the nursery. Here they are treated in the same way as conventional seedlings. They are field-planted into their permanent positions after about nine months in the nursery, perhaps 15 months after removal from the culture tube.

ORTET SELECTION

Clearly, it is essential to identify the best available palms for multiplication. Their performance must be determined genetically and not due simply to their growing in a favored environment. The selection of vigorous individual palms from unrecorded commercial plantings is unlikely to be much better than a random "lucky dip" into the gene pool. Only when comparative records are available over several years in well–planned trials is it possible to select ortets with a good chance of giving outstanding performance. In practice, we select from the best progenies of proven parents in well-recorded progeny trials. Criteria for selection are not based solely on fresh fruit bunch yield, which is known to be very dependent on environmental conditions, but on other more highly heritable characters such as bunch index (ratio of bunch weight to total weight gain) and oil to bunch ratio (9).

Resistance (tolerance) to *Fusarium* wilt is an important selection criterion for clones to be grown in areas affected by *Fusarium* wilt, for example Zaire, or Cameroon. Although ortets in non-wilt endemic areas cannot be rated for *Fusarium* tolerance, all clones are tested at the ramet stage for reaction to *Fusarium*. Other factors to be taken into account include agronomic factors such as drought tolerance, fertilizer economy and ease of harvesting. Oil quality in terms of carotene content and fatty acid composition show a considerable range of variation and could be included in criteria for selection (10).

There is interest in oils with higher than average linoleic acid content for its nutritional quality and in more liquid oils (11). However, selection for oil composition has not been applied seriously yet because of the commercial implications. A clone producing a special oil would have to be planted on a scale large enough to produce a commercially useful tonnage and be concentrated close to dedicated processing mills in which separate storage and transport facilities would have to be provided. We still are doubtful of the wisdom of planting large areas with single clones, and producers are not prepared to develop what would, in effect, be a new oil crop unless they can see a clear price premium over standard palm oil over at least a 20–year period.

CLONE TESTING

Increasing numbers of clone trials have been planted as clonal material has become available from the laboratory program. The majority of early trials were planted in Malaysia, although some material was planted in Lobe in Cameroon. The early clones were derived from un-

selected seedlings of Cameroon origin that were used in culture experiments at the Colworth House. Later clones were produced in Malaysia from selected ortets but only the most recent material, which only has just begun fruiting, is derived from rigorously selected ortets. Corley et al. (12) have analyzed the performance of clones in Malaysia and Cameroon planted up to 1981.

The first clone trials were designed to establish the phenotypic uniformity of clonal material (13). It was clear that clones were significantly less variable for both vegetative and bunch characteristics than comparable seedling populations.

In more recent trials, clones are being considered for performance relative to seedling controls for future commercial planting. The design and execution of clone trials also is being considered.

Results from clonal plantings made up to 1981 show relative performances of various clones in different years. This may be due to differential age–dependent performance or various responses to weather in different seasons. Proper evaluation of clone performance will require at least five years yield–recording although if several trials are considered together a shorter period might suffice. There also are clone x-location interactions demonstrating the importance of selecting clones for performance in specific regions (Tables 1 and 2).

TABLE 1

Yield of Oil in the First Two Years of Production from Nine Trials with Clones 31A, 54A, 90A and 115E(12)

Trial	Clone Months recorded	31A	54A	90A	115E
			Oil yield (t/ha yr)		
Inland soils					
PCT 11	31-54	3.26	3.94	2.47	3.11
PCT 16	31-56	1.77	2.82	1.78	2.52
HCT 6A	28-56	2.09	3.14	1.64	3.26
HCT 6B	28-56	3.03	3.24	1.94	2.59
HCT 8	28-56	2.48	2.77	1.75	3.03
Mean (inland trials)		2.53	3.18	1.92	2.90
Coastal soils					
HCT 5A	30-55	2.69	3.34	4.61	3.05
HCT 5B	30-55	2.84	4.98	5.55	5.29
HCT 7	30-55	2.11	3.49	4.06	3.93
HCT 9	36-56	5.62	7.15	7.80	6.85
Mean (coastal trials)		3.32	4.74	5.50	4.78

TABLE 2

FFB Yield (Kg/palm yr) in PCT 3 and Lobe Trial 79/1 over First Four Years of Production (12)

Clone	Yield years 1 & 2		Yield years 3 & 4	
	PCT 3	Lobe 79/1	PCT 3	Lobe 79/1
926	63	35	123	74
924	61	34	124	74
975	72	9	129	62
997	74	57*	145	132*
949	62	10*	133	84*
Mean	66	29	131	85

*Clones 997 and 949 planted adjacent to trial 79/1.

Selection of comparison standards in clone trials is not straightforward. The objective is to select clones of superior performance to seedling material currently available. Because all seedlings and every seedling progeny are different, there is no standard against which comparison can be made. One approach is to use good quality current commercial seed for comparison but it must be remembered that by the time clone trials have been completed and the clone selected for commercial production, a new generation of seed progenies will be available for commercial planting. A more stringent test of clonal material, which may go some way to discounting the steady improvement in seed quality, is to compare clones with the most recent selection within a breeding or seed production program, which may yield 10% more than the average.

In some trials, we have attempted to use a standard clone for comparison. This will become more useful as we learn more about the potential of different clones but the evidence of significant clones x-location interactions suggests that we will need a portfolio of standard clones. The need to include such a variety of controls in clone trials increases the area required for each trial.

Since the normal planting density for oil palms is about 140 palms/hectare and about 50 palms are required for each clone (say five replicates of 10 palm plots), it is apparent that even modest trials can occupy large areas and incur considerable management costs over several years if full recording is to be done.

In plantings made since 1983, particularly in clones subject to large-scale production, a high proportion of palms have produced abnormal flowers resulting in bunch abortion (14). This phenomenon temporarily has halted full-scale commercial development while the cause of the problem is identified. The problem is being pursued actively in Unilever, IRHO, PORIM and presumably in all clonal oil palm laboratories. Our current view is that the problem is due to an epigenetic change in endogenous growth regulator balance resulting from a particular change in the culture protocol. We are confident that we can avoid the particular treatment but do not understand yet why the cultures responded this way. Our major problem now is to devise an early-detection method to use as a quality control method at the culture stage and to rebuild confidence in use of clonal palms.

REFERENCES

1. Jones, L.H., *Oil Palm News 22*:2 (1974).
2. Rabechault, H., and J.P. Martin, *C.R. Acad. Sci. Paris. 283*:1735 Serie D. (1976).
3. Ahee, J., P. Arthuis, G. Cas, Y. Duval, G. Guenin, J. Hanower, D. Lievoux, C. Lioret, B. Malurie, C. Pannetier, D. Raillot, C. Varechon and L. Zuckerman, *Oleagineaux 36*:113 (1981).
4. Tisserat, B., *J. Exptl. Bot. 30*:1275 (1979).
5. Tisserat, B., *Date Palm Tissue Culture*, Advances in Agricultural Technology, Western Series, USDA Science and Education Administration, Beltsville, MD (1981).
6. Reynolds, J.F., and T. Murashige, *In vitro 15*:383 (1979).
7. Lioret, C., In *Oil Palm in Agriculture in the Eighties*, Kuala Lumpur, Incorporated Society of Planters (1981).
8. Giles, K.L., and W.M. Morgan, *Trends Biotechnol. 5*:35 (1987).
9. Hardon, J.J., R.H.V. Corley, and S.C. Ooi, *Euphytica 21*:257 (1972).
10. Jones, L.H., *J. Am. Oil Chem. Soc. 61*:1717 (1984).
11. Berger, K.G., and S.H. Ong, *Oleagineux 40*:613 (1985).
12. Corley, R.H.V., C.H. Lee, I.H. Law and E. Cundall, in *Proceedings of International Oil Palm Conference, Kuala Lumpur*, in press.
13. Corley, R.H.V., C.Y. Wong, K.C. Wooi and L.H. Jones, in *Oil Palm in Agriculture in the Eighties*, Incorporated Society of Planters, Kuala Lumpur, Malaysia (1981).
14. Corley, R.H.V., G.H. Lee, I.H. Law and C.Y. Wong, *The Planter 62*:233 (1986).

Biotechnology for Coconut Improvement

Yukio Sugimura[a], **Kazuya Otsuji**[a], **Shinta Ueda**[a], **Kikuhiko Okamoto**[a] and **Myrna J. Salvana**[b]
[a]Tochigi Research Laboratories, Kao Corporation Ichikaimachi, Haga, Tochigi, Japan, and [b]QC/R & D Department, Pilipinas Kao Incorporated, Cagayan de Oro City, Misamis Oriental, Philippines

Complete plantlets were regenerated from intact zygotic embryos by sequential culture from liquid to solid medium. When explants from zygotic embryo and immature rachilla tissues were cultured onto a solid medium supplemented with 2,4-dichlorophenoxy acetic acid (2,4-D), 6-benzylaminopurine (BAP), 2-isopentenyl adenine (2iP) and activated charcoal, various responses were observed. Nodular and root-like growths were initiated from embryo tissues while nodular and outgrowth projections were proliferated from rachilla tissues. Either root or shoot-like structures could be developed from nodular masses. Successful regeneration of plantlets was obtained from outgrowth projections induced from floral meristem of existing rachilla tissues. The gelling agent "Gelrite," as substitute for agar, gave better growth responses.

Coconut palm is at present propagated exclusively from seeds. Since this palm generally is cross-pollinated and heterozygous, the resulting genetic variation between seedlings is a serious problem for a perennial crop having a life span of 30 years or more. Lower productivity of coconut fruits may be due to a genetically variable population. The planting of high-yielding clones would enable average yields to be increased several-fold. In order to provide clone materials from proven, high-yielding and disease-resistant coconut, the development of rapid means of clonal propagation is of prime importance. The use of tissue culture techniques may make it possible to clone elite coconut palms.

The vegetative propagation of coconut by tissue culture has been attempted on many occasions. Significant progress in induction of callus and regeneration of plantlets in vitro was observed in the past few years. Current status of tissue culture study will be reviewed with special references including our recent results.

EXPERIMENTAL

For embryo culture, intact embryos were isolated aseptically from premature nuts of coconut palm (*Cocos nucifera* L. local tall variety) and then cultured in a liquid medium comprised of basal formulation of Y3 medium (1) until germination was initiated. To obtain complete plantlets, further culture was carried out onto the same medium containing 0.2% Gelrite and 0.25% activated charcoal. For callus induction, zygotic embryo and rachilla tissues were used as explant sources. The explants sliced transversely were inoculated onto different media tested. Subcultures were conducted at three to five week intervals.

RESULTS AND DISCUSSION

Embryo culture

One of the advantages in embryo culture is to rescue em-

bryos that cannot continue to develop normally within seeds for germination. Out of many varieties of coconut, the Makapuno coconut has no means of germination under natural environmental conditions. The endosperm of normal coconut has a layer about 10–15 mm thick lining the central cavity, which in a mature nut is almost filled up with coconut water. On the other hand, the anatomical observations of Makapuno seed showed that there are no layers of hard endosperm and no water inside the seed. The central cavity is occupied by jelly-like endosperm, instead of coconut water. This abnormality of the endosperm is believed to be governed by a single Mendelian recessive factor, and the tree that occasionally bears the Makapuno fruit is heterozygous for the character (2).

The Makapuno seed fails to germinate although its embryo morphologically is intact and indistinguishable from that of a normal coconut. Furthermore, microscopic examination showed that a well-organized plumule and radicle are present in the embryo. It was understood that a pure Makapuno plant may be obtained by taking the embryo out of the abnormal endosperm and culturing it in a suitable medium. Extensive studies on embryo culture of this seed were done to overcome nongermination in situ. De Guzman and her collaborators obtained plantlets from embryos by the modifications of the techniques and of the media used (3-7). The techniques developed for Makapuno could be applied to embryos of tall, dwarf and hybrid varieties (8,9).

The embryo isolated from a seed initially was cultured in liquid medium just before the start of germination and then transferred into solid medium several times until seedling reached a length of 10 cm, with two to three leaves and an adventitious root system (Fig. 1). Plantlets were removed from solid medium and transferred into vermiculite mixed with soil in pots. At this stage, the seedlings were exposed to different environmental conditions from that of the in vitro culture. In addition to drastic changes in growing conditions such as temperature and relative humidity, fungal and bacterial infection were very critical. These factors contribute to the low survival of the seedlings in pots. For increasing survival rate, various treatments have been recommended in date and oil palm (10,11): reculturing seedlings with primary root trimmed in auxin-containing medium for promoting adventitious roots, spraying fungicide before and after transplanting into pots, and maintaining high humidity.

In germplasm exchange, embryo culture could be a very practical tool. Collection of embryo instead of seednuts will facilitate the introduction of foreign varieties in a large scale. This technique could also be useful for the in vitro preservation of germplasm.

Tissue culture

Vegetative propagation by tissue culture is, in general, a difficult undertaking. Thus, there is relatively little information regarding callus initiation and its growth. In addition, differentiation from callus into plantlet is not

FIG. 1. Complete plantlet through embryo culture.

achieved consistently. Callus induction and subsequent development have been observed from various tissues such as zygotic embryo, stem (shoot tip), leaf, root, endosperm and immature inflorescence (Table 1).

Plant material

In general, greater success in callus induction has been obtained with younger tissue than mature one (12). Successful callus initiation occurred consistently with minimal browning when particular stages of rachilla tissue were used as explants. In addition, size of explant might be important. Smaller explants often had a better survival rate than larger ones.

Media components

The nutrient medium suitable for callus initiation and its growth was studied by Eeuwens (1). The medium, designated as Y3, gave good growth of coconut tissues. The Y3 medium contained both ammonium and nitrogen at lower level than in MS medium (13), and also contained high levels of potassium and iodine. Later, the medium components were modified by the inclusion of some amino acids and other organic substances (8,14,15).

A prominent problem of tissue browning has been noted in the culture of coconut (16,17). The incorporation of activated charcoal into media provided beneficial results in terms of preventing browning response during

TABLE 1

Tissue Culture of Coconut Palm

Tissue Source	Response	Reference
Zygotic embryo	callus	21
	callus/shoot–like structure	22
Leaf	callus/somatic embryogenesis	23
	root–like/embryo–like structure	8
	callus/somatic embryogenesis	25
Stem	callus	16
	callus	15
	callus/embryo–like structure	8
Rachilla	callus	1
	callus/embryo–like structure/ shoot and root–like growths	15
	callus/root and shoot–like structure	8
Endosperm	callus	17
	callus	26
Root	callus/plantlet regeneration	18

culture (8,10,18). The availability of other media components might be reduced due to the absorption by activated charcoal that was added into the medium.

Agar has been used widely as a gelling agent of the medium. Recently, "Gelrite," an agar substitute, was introduced for the study of plant tissue culture. In some cases, the growth of callus was enhanced more on the media solidified by Gelrite than by agar media (19). Our data showed that Gelrite gave better growth of coconut callus than agar. Possible explanations are some inhibitory substances were contaminants in the agar used, stimulative substances for cell growth were present in Gelrite, and/or some active substances, particularly oligosaccharins as elicitors, were formed during sterilization of the medium by autoclaving.

For callus induction, it clearly seems that a high level of auxin was required in the presence of activated charcoal. Out of auxin used, 2,4-D and 1-naphthaleneacetic acid were most effective in callus induction. Using rachilla explants, a wide range of 2,4-D concentration was examined for determination of optimum dose. The high frequency of callus induction was observed by the addition of 20-30 ppm of 2,4-D into the medium. For induction of callus from embryo tissue, 100 ppm of 2,4-D was required. It is likely that a level of auxin was somewhat different according to type of tissues and stages. Addition of cytokinin such as BAP and 2iP had a little effect on callus induction and growth, as reported previously (17).

Callus development and plantlet regeneration

Zygotic embryo. When the tissues of zygotic embryo were cultured in vitro, there were two types of proliferative growth induced: nodular growths initiated around the proximal end of embryo (Fig. 2A), during subculturing the nodules developed into discrete bodies consisting of small, prominently nucleated and compactly arranged cells; and root–like growths caused by the increasing thickness of procambial strands (Fig. 2B), this continued to produce many root–like structures without formation

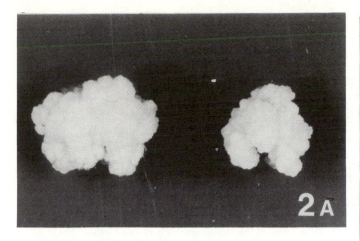

FIG. 2A. Nodular calluses initiated from procambial strands in embryo.

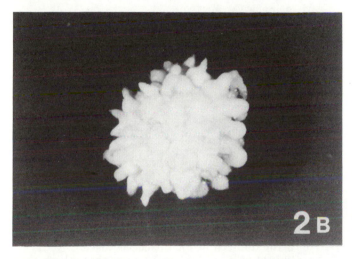

FIG. 2B. Spikelet projections initiated from procambial strands in embryo.

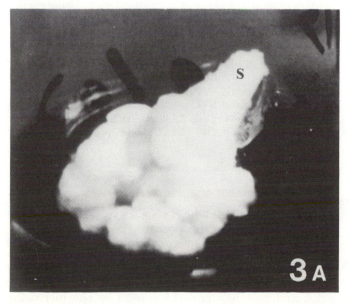

FIG. 3A. Shoot–like structure(s) in nodular masses induced from rachilla tissue.

FIG. 3B. Root–like growths proliferated from nodular masses.

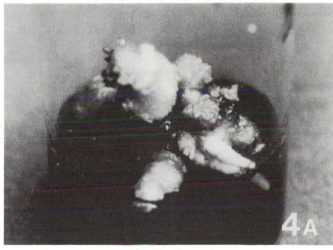

FIG. 4A Outgrowth projections induced from rachilla tissue.

of true shootlet. Similar responses were observed by De Guzman et al. (21) and D'Souza (22).

Rachilla. The developing inflorescence has been shown to be a promising source of explants for callus induction (12,15,20). The callus induced from existing floral meristems (male) could be subcultured and developed into nodular masses. When they were transferred into a lower level of auxin several times at three to five week intervals, some calluses produced shoot–like structure (Fig. 3A), and others proliferated to roots under illumination (Fig. 3B). Unsuccessful regeneration of complete plantlets, at present, was observed in our laboratories. A recent report has described production of plantlets through tissue culture (18). The normal development directed towards plantlet regeneration may fail to take place under our present conditions. It seems probable that hormonal balance in medium is critical at each stage of callus development. The first process for deriving plantlets from callus is to induce the organization in callus

FIG. 4B Plantlet regeneration from outgrowth projections.

leading to either the production of a shoot or a root meristem (organogenesis) or the development of a somatic embryo with both a shoot and a root meristem (embryogenesis) (12). Based on histological examination, Pannetier and Buffard-Morel (23) have observed structures identified as embryoids in the culture of a young leaf. Branton and Blake (15) also have observed meristematic areas in rachilla callus which have developed into embryoid-like structures. The growth pattern of these structures resembled that of a germinating zygotic embryo, suggesting that embryoids may have been present.

REFERENCES

1. Eeuwens, C.J., *Physiol. Plant. 36*:23 (1976).
2. Zuniga, L.C., *Philippine Agric. 36*:402 (1953).
3. De Guzman, E.V., and D.A. del Rosario, *Ibid. 48*:82 (1964).
4. De Guzman, E.V., *Ibid. 53*:65 (1970).
5. Balaga, H.Y., and E.V. de Guzman, *Ibid. 53*:551 (1971).
6. De Guzman, E.V., A.G. del Rosario, and E.C. Eusebio, *Ibid. 53*:566 (1971).
7. Del Rosario, A.G., and E.V. de Guzman, *Philippine J. Sci. 105*:215 (1976).
8. Gupta, P.K., S.V. Kendurkar, V.M. Kulkarni, M.V. Shirgurkar and A.F. Mascarenhas, *Plant Cell Reports. 3*:222 (1984).
9. Bah, B.A., *Oléagineux 41*:321 (1986).
10. Tisserat, B., in *Handbook of Plant Cell Culture*, Vol. 2, edited by W.R. Sharp et al., Macmillan Publishing Co., New York, NY, 1984, pp. 505–545.
11. Paranjothy, K., in *Handbook of Plant Cell Culture*, Vol. 3, edited by W.R. Sharp et al., Macmillan Publishing Co., New York, NY, 1984, pp. 591–605.
12. Blake, J., in *Tissue Culture of Trees*, edited by J. Dodds, The Avi Publishing Co., Westport, 1983, pp. 29–50.
13. Murashige, T., and F. Skoog, *Physiol. Plant. 15*:473 (1962).
14. Eeuwens, C.J., *Physiol. Plant. 42*:173 (1978).
15. Branton, R.L., and J. Blake, *Ann. Bot. 52*:673 (1983).
16. Apavatjrut, P., and J. Blake, *Oléagineux 32*:267 (1977).
17. Fisher, J.B., and J.H. Tsai, *In Vitro 14*:307 (1978).
18. Branton, R., and J. Blake, *New Scientist 98*:554 (1983).
19. Ichi, T., T. Kada, I. Asai, A. Hatanaka and J. Sekiya, *Agric. Biol. Chem. 50*:2397 (1986).
20. Blake, J., and C.J. Eewens, in *Proc. COSTED Symp. on Tissue Culture of Economically Important Plants*, Singapore, edited by A.N. Rao, 1981, pp. 145–148.
21. De Guzman, E.V., A.G. del Rosario and E.M. Ubalde, *Philippine J. Coconut Studies 3*:1 (1978).
22. D'Souza, L., in *Proc. 5th Intl. Cong. Plant Tissue & Culture*, edited by A. Fujiwara, 1982, pp. 179–180.
23. Pannerier, C., and J. Buffard-Morel, *Oléagineux 37*:349 (1982).
24. Eeuwens, C.J., and J. Blake, *Acta Horticulture 78*:227 (1977).
25. Raju, C.R., P.P. Kumar, M. Chandramohan, and R.D. Iyer, *J. Plantation Crops 12*:75 (1984).
26. Kumar, P.P., C.R. Raju, M. Chandramohan, and R.D. Iyer, *Plant Sci. 40*:203 (1985).

Genetic Modification of Polyunsaturated Fatty Acid Composition in Flax (*Linum Usitatissimum L.*)

Allan G. Green
C.S.I.R.O. Division of Plant Industry, Canberra, ACT, Australia

Linoleic and linolenic acid are the only two polyenoic fatty acids present in significant concentrations in the major economic seed oils. Linoleic acid (C18:2), the principal component of polyunsaturated oils, is an essential component of the human diet. Linolenic acid (C18:3), although also essential in human nutrition, is an undesirable component in edible oils due to flavor reversion problems associated with its oxidative instability.

Plant breeders have modified successfully the proportions of these two fatty acids in a number of oilseed crops in order to improve their quality as edible oils. These changes have been achieved through genetic modification of the activities of the desaturase enzymes that sequentially convert oleic acid (C18:1) to linoleic and linolenic acids. The conversion of linseed oil from an industrial oil into an edible oil is a striking example of such genetic modification.

Following EMS treatment of seeds of the high–linolenic (45–50%) linseed cultivar Glenelg, two mutants, M1589 and M1722, having reduced levels of linolenic acid (28–30%) were isolated. By recombining the M1589 and M1722 mutations into a single genotype, linolenic acid content was reduced further to around 1%. This reduction was associated with an equivalent increase in the highly desirable linoleic acid to 62% when grown in cool environments. Proportions of other fatty acids remain unaltered. The mutations thus appear to block the linoleic acid desaturation step of fatty acid biosynthesis. An additional 6% increase in linoleic acid content has been achieved by backcrossing the mutations into other genetic backgrounds.

The reduction in linolenic acid content has greatly improved the oxidative stability of the oil, making it suitable for use as a polyunsaturated edible oil.

The relatively minor chemical changes required to interconvert individual fatty acids combined with the apparent lack of functional constraints on seed triglyceride fatty acid composition have enabled this character to be manipulated readily by a variety of plant breeding techniques. The most notable example of this has been the development of rapeseed (*Brassica napus* and *B. campestris*) into a major source of edible oil through the selection in the 1960s and 1970s of genotypes lacking the nutritionally undesirable erucic acid (1,2).

More recently, attention has been focused on improving the level of polyunsaturated fatty acids in vegetable oils because of their reputed beneficial effects on cardiovascular health. The only two polyunsaturated fatty acids that occur in significant amounts in the economic oilseed species are linoleic acid (C18:2) and linolenic acid (C18:3), both of which are essential fatty acids in human nutrition. Linoleic acid is a dienoic fatty acid and is the principal component of polyunsaturated edible oils, such as sunflower, safflower, maize and soybean. Linolenic acid,

a trienoic fatty acid, is formed by the enzymatic desaturation of linoleic acid. It occurs in small proportions in some edible oils such as rapeseed and soybean but is regarded as an undesirable component in these oils because of flavor reversion problems associated with its oxidative instability. Consequently, in attempting to improve the polyunsaturated quality of seed oils, plant breeders have aimed to increase linoleic acid content and decrease linolenic acid content by genetically blocking the conversion of linoleic acid to linolenic acid that occurs during seed lipid synthesis. The successful breeding of an edible–quality linseed oil is a striking example of such genetic modification.

BACKGROUND

Linseed oil, obtained from the seed of the flax plant (*Linum usitatissimum* L.), is unique among the major vegetable oils in that it contains a very high level (45–65%) of linolenic acid. The susceptibility of this fatty acid to oxidation and polymerisation imparts the drying quality to the oil that has made it important in the production of paints, varnishes, inks and linoleum. However, the market for linseed oil has declined dramatically in recent years as synthetic compounds increasingly have replaced linseed oil in the manufacture of these products.

Since the market for edible vegetable oils is very much larger and has potential for further expansion, it was considered that the development of an edible-quality linseed oil would greatly improve the prospects for increased cultivation of this crop. In 1979, a plant breeding program was commenced at CSIRO Division of Plant Industry with the objective of developing an edible–oil linseed through the breeding of a low–linolenic genotype. The progress of that program is outlined in this paper.

NATURAL VARIATION

Initial studies demonstrated that natural variation for fatty acid composition within a germplasm collection was insufficient to reduce significantly linolenic acid content in *Linum* by intraspecific hybridization and selection (3). In contrast, wild *Linum* species varied widely in their fatty acid composition, with several species having very low levels (<3%) of linolenic acid (4). However, these species could not be exploited because of strong reproductive barriers preventing interspecific hybridization. Therefore, subsequent efforts concentrated on mutation breeding with the aim of inactivating the desaturase enzyme(s) involved in the synthesis of linolenic acid in the seed oil.

INDUCTION AND SELECTION OF MUTANTS

The availability of a rapid and sensitive chemical test specific for the presence of linolenic acid in vegetable oils enabled the screening of large populations of seeds. This assay, known as the TBA test, relies on the formation of a red color complex when thiobarbituric acid (TBA) reacts with the oxidation products of linolenic acid. When linolenic acid is absent, a yellow color is observed, grading through orange to red as linolenic acid content increases (5). Quantification can be achieved by comparison to standards of known linolenic acid content.

Seeds of the Australian linseed cultivar, *Glenelg*, were treated with various doses of the chemical mutagen ethyl methanesulphonate (EMS) using standard techniques (6). Ten M_2 seeds from each of 7072 field-grown M_1 plants were assayed nondestructively using the TBA test. Seeds appearing to have less than 40% linolenic acid were selected and progeny tested as M_2 plants. Two mutants subsequently were identified, M1589 and M1722, in which linolenic acid constituted 29% of the total fatty acids (mean for M_3 and M_4 plants) compared with 43% in seed oil from *Glenelg* control plants. The reduced level of linolenic acid was accompanied by an increase in the level of linoleic acid to 30%, compared with 18% in *Glenelg*, but there were no significant changes in the proportions of other fatty acids. Both mutants arose from the 0.4% EMS treatment but from different M_1 plants.

GENETIC ANALYSIS OF M1589 AND M1722

The inheritance patterns of the induced mutants were examined to determine whether they were allelic or in different genes that might be recombined to give even lower levels of linolenic acid. Analysis of the parental, F_1, and backcross generations of the crosses M1589 x *Glenelg* and *Glenelg* x M1722 indicated that both M1589 and M1722 carried a single gene mutation for which the normal and mutant alleles were codominant (7).

Analysis of 114 F_2 plants from the cross M1722 x M1589 revealed wide variation in linolenic acid content. Five distinct classes were apparent with means of 1.6%, 15.0%, 24.0%, 30.5% and 34.5%. The frequencies of F_2 plants in these classes fitted a 1:4:6:4:1 ratio expected for the segregation of two unlinked genes exhibiting additive gene action. The two lower linolenic acid classes arose by recombination of the M1589 and M1722 mutants; F_3 progeny tests confirmed that the seven F_2 plants having less than 2% linolenic acid were homozygous for the mutant alleles at both loci. This homozygous double-mutant genotype has been referred to as "Zero." The additional reduction in linolenic acid was accompanied by an equivalent increase in linoleic acid (Table 1). This further demonstrated that the M1589 and M1722 mutations specifically affect the linoleic desaturation step in fatty acid biosynthesis (7).

TABLE 1

Fatty Acid Composition of Seed Oil from Linseed cv. *Glenelg* and Induced Low-linolenic Mutants, Grown in a Glasshouse at 27/20 C (Day/Night)

	Fatty acid composition (%)				
	Palmitic	Stearic	Oleic	Linoleic	Linolenic
Glenelg	8.5	4.8	37.9	14.7	34.1
M1589	7.6	5.4	37.8	27.6	21.6
M1722	8.4	6.2	35.1	28.9	21.4
Zero	9.2	4.7	36.3	48.2	1.6

TEMPERATURE RESPONSE

The genetic analysis of the mutants outlined above was conducted on plants grown in a relatively high temperature glasshouse environment (27 C/20 C day/night). Because it was known that lower temperatures during seed maturation result in a more unsaturated fatty acid composition in several oilseeds, including high-linolenic linseed, a Phytotron study was conducted to determine the effect of temperature on the oil of the "Zero" genotypes (8). Plants were grown at 24/19 C until commencement of flowering and then were transferred into a range of day/night temperature regimes (Table 2). Thermoperiod control consisted of eight-hr day temperature and 16-hr night temperature. This study demonstrated that higher linoleic acid content results when seed development occurs at low temperatures. Linoleic acid content was above 62% when day/night temperatures were below 21/16 C, and in no case was linolenic acid content greater than 3%. Thus, this genotype should produce a high-quality polyunsaturated edible oil similar to sunflower and maize oils when grown in cool-temperate climates.

OXIDATIVE STABILITY

To determine if the large reduction in linolenic acid content had improved the oxidative stability of the oil, a 3 l sample of oil was refined and analyzed. The crude oil was

TABLE 2

Effect of Temperature During Seed Maturation on Fatty Acid Composition of the Low-linolenic Linseed Genotype "Zero"

Day/night temperature	Fatty acid composition (%)				
	Palmitic	Stearic	Oleic	Linoleic	Linolenic
15/10 C	6.2	3.2	17.3	70.2	3.1
18/13 C	6.4	3.3	25.1	62.7	2.3
21/16 C	7.4	4.4	29.0	57.3	1.8
24/19 C	8.1	6.0	28.0	55.9	1.8
27/22 C	8.5	6.3	32.1	50.9	1.9

TABLE 3

Comparison of Refined Low-linolenic Linseed Oil with Typical Sunflower Oil Specifications

Property	Sunflower*	Low-linolenic linseed
Fatty acid composition (%):		
Palmitic	6	7
Stearic	2	5
Oleic	26	24
Linoleic	66	63
Linolenic	—	1
Free fatty acid (%)	0.25 (max)	0.18
Iodine value	125-138	133
Refractive index	1.472-1.474	1.470
Relative density	0.914-0.920	0.917
Saponification value	190-196	185
Unsaponifiable matter (%)	1.5 (max)	0.9
Peroxide value	10 (max)	nil

*Specifications obtained from Australian Oilseeds Federation.

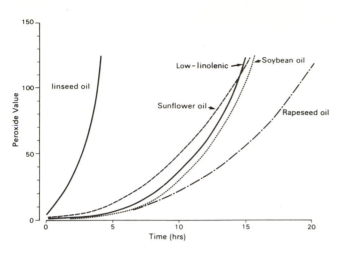

FIG. 1. Oxidative stability of low–linolenic linseed oil compared to that of sunflower oil, soybean oil, rapeseed oil and conventional linseed oil.

obtained by solvent extraction of a "Zero" seed harvested from a spring–sown outdoor nursery plot grown in Canberra. Processing of the oil consisted of degumming, alkali refining, bleaching and steam deodorizing. This combination produced a bland oil of light color with physical and chemical properties equivalent to those of typical sunflower oil (Table 3).

Oxidative stability was determined by the Active Oxygen Method using the IUPAC Method 11D21, in comparison with soybean, rapeseed, sunflower and high–linolenic linseed oils (with no antioxidants added). The results indicated that the low–linolenic linseed oil is suitable for use as an edible oil, having similar stability to sunflower oil and soybean oil and being far more resistant to oxidation than high–linolenic linseed oil (Figure 1).

TABLE 4

Fatty Acid Composition of Low-linolenic Backcross Lines Derived from Crosses Involving *Glenelg* and Croxton (Field-grown in Canberra, 1986)

Parent	Fatty acid composition (%)				
	Palmitic	Stearic	Oleic	Linoleic	Linolenic
Glenelg	8	5	21	64	2
Croxton	8	3	18	69	2

GENETIC IMPROVEMENT IN LINOLEIC ACID CONTENT

Future breeding will concentrate on further improvements in oil quality of the low–linolenic genotype. To ensure that polyunsaturated oil quality standards (>62% linoleic acid) consistently are met independently of temperature, it would be desirable to develop a genotype having a temperature–stable linoleic acid pathway similar to that of safflower. Alternatively, it might be possible to raise the level of linoleic acid genetically so that, although still temperature–sensitive, high levels of linoleic acid could be produced even under warm conditions. In this regard, it is encouraging to note that several linseed and flax genotypes have significantly higher oleic desaturation activity than "Zero." The low–linolenic acid genes have been backcrossed into some of these genotypes, and preliminary results indicate that their greater ability to desaturate oleic acid to linoleic acid can be combined with low linolenic acid content. Levels of linoleic acid about 6% higher than in "Zero" have been observed in low–linolenic derivatives from one of these crosses (Table 4). If the response to temperature of these lines is similar to that of "Zero," then linoleic acid levels greater than 62% could be achieved even in warm environments. This possibility currently is being investigated. The high–linoleic, low–linolenic linseed oil should be suitable for widespread use as an edible oil. At between 1–2%, the level of linolenic acid is below that of current commercial cultivars of soybean and rapeseed. Additionally, the large increase in linoleic acid content and the genetic potential for further increases have resulted in an oil that is more polyunsaturated than either rapeseed or soybean, the composition and physical properties being similar to sunflower and maize oils. The development of an edible–quality linseed oil greatly increases the market prospects for this crop and potentially provides expanded opportunity for diversification of cropping activities in many traditional cereal–growing regions because linseed is widely adapted to temperate climates throughout the world and has a long history of successful cultivation.

ACKNOWLEDGMENTS

This research was conducted with support from the Rural Credits Development Fund of the Reserve Bank of Australia and the Australian Oilseeds Research Committee. The author acknowledges the scientific collaboration of D.R. Marshall and M.L. Tonnet, and the technical assistance of R. Nurzynski, G. O'Niell, G. Smith, N. Hulse, M. Huber and T. Huber at various stages of the project. The laboratory scale refining and analysis of oil was performed by Vegetable Oils Pty Ltd. of Australia.

REFERENCES

1. Stefansson, B.R., F.W. Houghen and R.K. Downey, *Can. J. Plant Sci. 41*:218 (1961).
2. Downey, R.K., and B.L. Harvey, *Can. J. Plant Sci. 43*:271 (1963).
3. Green, A.G., and D.R. Marshall, *Aust. J. Agric. Res. 32*:599 (1981).
4. Green, A.G., *J. Am. Oil. Chem. Soc. 61*:939 (1984).
5. McGreggor, D.I., *Can. J. Plant Sci. 54*:211 (1974).
6. Green, A.G., and D.R. Marshall, *Euphytica 33*:321 (1984).
7. Green, A.G., *Theor. Appl. Genet. 72*:654 (1986).
8. Green, A.G., *Crop. Sci. 26*:961 (1986).

Biotechnology for Soybean Improvement

Niels C. Nielsen and **James R. Wilcox**

USDA/ARS and Department of Agronomy, Purdue University, West Lafayette, IN 47907

Nearly a quarter of the world supply of edible fats and oils is from soybeans. Though not as valuable as oil, soybean protein and carbohydrate coproducts also are of significant economic importance in the oilseed industry. Despite their obvious market value, soybeans suffer a number of constraints. Some of these problems might be eliminated by application of conventional or nonconventional genetic approaches and result in soybeans tailored for specific end uses. Topics to be discussed from this perspective during the presentation include strategies directed toward elimination of undesirable tastes, odors and antinutritional factors; manipulation of the oil composition to modify stability and physical properties; engineering of enzymes and other proteins to improve production efficiency or the quality of soy products; and extractability and bioavailability of seed carbohydrates to monogastric animals.

Genetic improvement of soybeans during the past 50 years has focused on increased production. These efforts have resulted in substantial increases in yield as well as resistance to certain diseases and pests that infest the crop. During this period, soybeans have become a major source of edible oils and fats in the western world and an important source of protein for livestock feed. However, for the most part equivalent improvement of the chemical composition of the seed has been neglected. That situation is changing due to increased competition for markets with other oilseed crops. The purpose of this communication will be to identify several problems relevant to improvement of soybean quality and approaches that are being directed toward their resolution.

TECHNOLOGICAL CONSIDERATIONS

In the past decade, technologies have become available that augment classical breeding techniques. Examples include cell transformation, plant regeneration and various methods for mutagenesis. Although these new technologies offer considerable potential, their application thus far has had only minimal economic impact on the soybean industry. Instead, major contributions have been made to a more detailed understanding about the structure of plant genes and their expression. The technologies have begun to be applied to selection of specific genetic traits but these are being used in conjunction with conventional breeding programs. However, it is reasonable to expect that more sophisticated applications of the biotechnologies to modification of the soybean genome will arise.

Cell transformation, combined with plant regeneration, probably will become the most direct and efficient approach for modification of quality components in seeds. Transformation is the process of introducing foreign genetic material into plant genomes. Both vector-mediated (for example, with the Ti plasmid of *Agrobacterium tumefaciens* [1,2]) and direct transfer techniques (for example, electroporation [3]) have been used to move foreign DNA into soybean cells. Unfortunately, the transformed cells were not regenerated into intact plants although other soybean cell lines have been regenerated but not transformed (4,5). These observations, together with the knowledge that cells from other legumes such as alfalfa have been transformed and regenerated suggest that a reproducible method for soybeans will become available.

Whole plant mutagenesis and somaclonal variation are two related technologies that can be applied to quality improvement. Somaclonal variation refers to the recovery of stable allelic variants when cells from homozygous plants undergo regeneration. Both whole plant mutagenesis and somaclonal variation are effective because the rate at which genetic variability can be recovered is increased many-fold over the rate of spontaneous mutation. The prevalence of somaclonal variants can be increased even further upon treatment of cells with mutagens prior to regeneration. Effective applications of both techniques depend on rapid and precise selection assays that permit the desired phenotypes to be distinguished from a large background of normal cells or organisms. Once genetic variability is identified, conventional breeding techniques can be used to introduce the traits into acceptable genetic backgrounds. The generation of somaclonal variants has the advantage that selection for specific characteristics can often be done in petri dishes prior to or during regeneration. As a consequence, selection of somaclonal variants does not have the space requirements encountered when whole plants must be propagated. On the other hand, whole plant mutagenesis requires less technical expertise, and mutants can be selected where only the desired gene product appears to be altered. The latter approach has been used by several groups to generate fatty acid mutants of soybeans, and some of these results will be summarized shortly.

Germplasm collections also provide a rich source of genetic variability. The key to recognition and utilization of this variability is again an effective assay for specific genes or gene products. As an example of this approach, results in which electrophoretic methodology was used to identify lipoxygenase null-alleles and then introduce them into a commonly grown soybean variety will be outlined. In retrospect, advantage could also have been taken of restriction fragment length polymorphisms (RFLP) now recognized to be associated with the lipoxygenase null-alleles (Nielsen, unpublished data). RFLP markers are useful for several reasons. They are inherited codominantly and provide complete genetic classification in segregating populations. Moreover, nondestructive assays that preserve the reproductive capability of the plants can be used.

FATS AND OILS

Soybean oil contains five predominant fatty acids: palmitate, stearate, oleate, linoleate and linolenate. Processors of soybean oil generally agree that decreasing the linolenic acid content would increase product stability although the widely publicized nutritional benefits of ω-3 oils raise questions about the degree to which it should be reduced. A decrease in the level of the saturated fatty acids palmitate and stearate could increase the nutritional value of the oil and would be beneficial in the salad oil industry. Conversely, an increase in the level of palmitate, stearate and/or oleate would change the melting point of the oil and could enhance the use of soy products in the soft margarine industry. It also may be possible to introduce genetic information for biosynthesis of rare and valuable fatty acids normally not present in the seed. Uses in the nonedible oil market also are being addressed. For example, soybean oil is being used in the manufacture of printers ink and as a suppressant of grain dust. Applications in processes that now utilize petroleum-based products may become feasible as mineral oil stocks are depleted or become more expensive. Hence, opportunities to tailor oil composition for specific end uses exist.

Porra and Stumpf (6,7) provided evidence nearly a decade ago that de novo synthesis of fatty acids in soybeans takes place by the general metabolic pathways discussed in Chapter 1. However, unequivocal identification of soybean genes in the pathways still is at a primitive stage. Whole plant mutation breeding programs at the Purdue and Iowa State Universities successfully have generated mutants with altered fatty acid compositions. A recurrent selection program has been used for the same purpose at North Carolina State University (8). Table 1 summarizes characteristics of mutants from the USDA/ARS program at Purdue. The mutants originated after treatment of seeds with ethylmethanesulfonate and were identified by chemical analysis of M_2-seeds (9). The commonly grown cultivar Century was used for mutation. The rationale for starting with a high-yielding variety was that only a single gene might be mutated and less effort would be required to incorporate the mutant alleles into commercial varieties.

One class of mutants, *fan*, typified by C1640 and C1725 results in a reduction of linolenate, in these cases from about 8% to less than 4% (9). The *fan* alleles have additive effects with normal *fan* alleles and segregate for low, intermediate and high levels of linolenic acid (10). A germplasm release of C1640 has been made (11). However, neither the molecular nature of the metabolic lesion

produced by the *fan* allele nor its location in the genome has been identified.

A second group of mutants have altered levels of palmitic acid (12). Line C1726 has a lower 16:0 content than the Century control while C1727 has a higher level. Genetic evidence indicates that the altered levels of 16:0 are controlled by genes at two loci that are inherited independently. Maternal effects appear not to be associated with these genotypes. Mutants with palmitate contents near 20% also have been identified in the Iowa State Program (W. Fehr, personal communication) but their relationships to the Purdue mutants have not been determined.

Mutants with increased stearic acid also have been obtained (12). In addition to C1728, the high stearate line described in Table 1, three mutants with similar phenotypes have been reported by the group at Iowa State University (13). They have released the mutant with the highest 18:0 content (30%) as a germplasm line. Although the three mutants from Iowa State are due to multiple alleles at the same locus, their genetic relationship to C1728 is unknown.

Another mutant, (C1729, Table 1), was discovered that produces elevated oleate and reduced linoleate contents (16). Unlike the mutant alleles discussed above, the maternal parent has some influence on the level of oleate and linoleate associated with this mutant. Knowles et al. (14) described two alleles at a single locus in the safflower genome that conditioned similar reciprocal changes in oleate and linoleate content. However, in the case of the safflower gene maternal effects were not observed.

A question frequently posed by breeders is to what extent alterations in seed fatty acid content affect yield. This question has been addressed in the case of the low 18:3 line C1640 (Table 2). Initial yield trials of C1640 revealed about a 25% depression in yield compared to Century, the variety from which it originated. As an extension of these experiments, segregants from a cross between C1640 and Century were selected for either high- or low-18:3 content. When lines derived from these selections were analyzed, the average yield of plants with the low linolenic acid levels did not differ significantly from those that had normal levels. Hence, genes other than the one that conditions low-18:3 content apparently were responsible for the low initial-yield values that were obtained. This result implies the possibility to generate high-yielding lines with both low-18:3 content and acceptable agronomic characteristics.

TABLE 1

Fatty Acid Composition of Soybean Mutants[1]

| Line | Percent Fatty Acids | | | | |
	16:0	18:0	18:1	18:2	18:3
Century	11.1	3.4	21.8	56.6	7.2
C1725	11.5	3.1	19.8	62.2	3.4
C1726	8.5	3.2	23.7	57.5	7.2
C1727	17.1	3.0	18.4	53.4	7.9
C1728	10.6	10.5	15.0	55.1	8.9
C1729	10.5	3.5	36.9	41.8	7.0

[1]Three-year mean, (1984-86). The mutants were generated by treatment of seeds from the variety Century with ethylmethylenesulfonic acid and identified as described by Wilcox et al. (9). Two low linolenic acid mutants (*fan*), C1640 and C1725, have been identified among the mutants that appear to be allelic.

TABLE 2

Mean Yield of Populations Derived from a Cross Between the Variety Century and Low Linolenic Acid Lines

Line	Percent 18:3	Yield (bu/a)
C1640	3.3	50.7
C1725	3.5	50.6
Low 18:3 lines (n-31)	3.2±0.05	54.6±0.55[a]
High 18:3 lines (n-22)	7.1±0.09	53.6±0.69[a]
Century	8.0	58.0

[a]Differences nonsignificant.

One cannot extrapolate the results with the low 18:3 mutant to situations that might be encountered for the other fatty acid mutants but similar experiments can be carried out to evaluate the effect these mutations have on yield. Before substantial additional effort goes into these mutants, however, the potential economic benefits that might be associated with them should be evaluated. The soybean industry needs to play an active role in such evaluations.

One might argue that the fatty acid mutants identified this far are unlikely to have direct economic impact on the soybean industry. However, the fact that such mutants can be obtained shows there is potential to alter oil composition of soybeans for specific end uses. It is possible that material generated in the future may be of more economic value. In this regard, the availability of mutants presently identified provides an opportunity to identify genes in the fatty acid biosynthetic pathway. Once purified, efforts can be made to engineer the genes to alter quality traits. Efforts along these lines await not only isolation of the genes but also development of effective assays to evaluate changes that are made.

FLAVOR ATTRIBUTES

Objectionable aroma and flavor attributes typically are associated with soy products. These have been described as being "greeny-beany," "grassy," "painty" or "fishy" and are considered to arise in part because of seed lipoxygenases (15,16). Lipoxygenase enzymes catalyze the hydroperoxidation of polyunsaturated fatty acids with *cis, cis*-pentadienes. The oxidation products from the reactions are metabolized further into short chain alde-

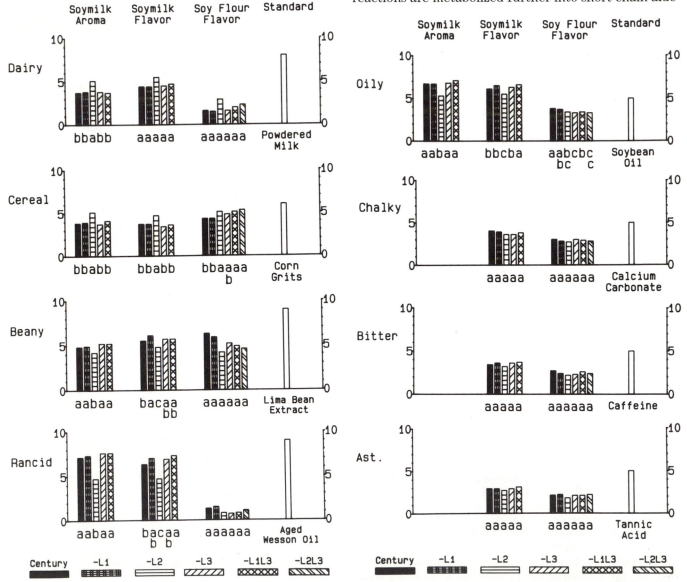

FIG. 1. Taste panel scores (y-axis) of soymilk preparations from seeds lacking one or two of the lipoxygenase isozymes. Preparations of Century seeds that contain all three isozymes were included as a control. The scores of 1-10 were subdivided into categories: 1, bland; 2-4, weak; 5-6, moderate; 7-8, strong; and 9-10, very strong intensity. Each bar represents the mean of 18 evaluations (six judges averaged over three days). Bars with the same letter are not different statistically at the 5% level of significance. Intensity values for the standards used to train the subjects, as well as additional experimental details and data, can be found in Davies et al. (22).

hydes, and it is the aldehydes that are thought to account for the undesirable flavor and aroma attributes. Genetic elimination of seed lipoxygenases, therefore, could influence the flavor profile of soy products. To approach this problem, the gene pool was searched to find null-alleles for lipoxygenases.

Axelrod and his colleagues showed that soybeans contain three lipoxygenase isozymes, denoted L1, L2 and L3 (17). The three isozymes can be visualized electrophoretically as a high molecular weight triplet at the top of SDS-polyacrylamide gels (18) or by immunological techniques (19). A screening of the USDA Northern Regional Research Center germplasm collection maintained at the University of Illinois resulted in identification of recessive null alleles for each of the isozymes (17–20). They are denoted $1x_1$, $1x_2$ and $1x_3$, respectively. Genetic evidence obtained using the null alleles is consistent with a model in which $1x_1$ and $1x_2$ are tightly linked but is separate from $1x_3$ (18). The tight linkage between $1x_1$ and lx_2 has precluded a generation of double null lines that lack L1 plus L2, although the double-mutants L1 plus L3 and L2 plus L3 have been produced.

Individual lines near-isogenic to the commonly grown variety Century have been developed for each null allele. These germplasm releases (21) were used to evaluate the flavor and aroma profiles of soy products made from seeds with reduced lipoxygenase. By comparison to Century controls, taste panel evaluations of products made from lines with the L2 null-allele had significantly lower scores for beany, rancid and oily flavors and aromas as well as higher scores for dairy and cereal flavors and aromas (Fig. 1) (22). Similar results were obtained with soy meal preparations but the differences were less pronounced. Lines from which either L1 or L3 were removed did not differ significantly from controls. The results indicated that removal of the L2 isozyme but not L1 or L3 may reduce undesirable flavors in soy products considerably. Either a reduction or removal of off–flavor compounds generated by lipoxygenases could improve consumer acceptability of soy products.

STORAGE PROTEINS

The most extensively studied soybean genes are those that encode subunits for the storage proteins glycinin and β-conglycinin. While not directly related to fats and oils, consideration of this work is important to this discussion for two reasons. First, improvement of protein quality will have a direct impact on the oilseed industry because it could change the relative economic value of seed components. Second, understanding how these genes are regulated and what approaches can be followed to evaluate changes in their coding regions illustrates some of the processes that will be followed when altering genes that affect fats and oils.

At least three genes for β-conglycinin subunits (23) and five for glycinin subunits (24,25), Nielsen, unpublished data), have been isolated and sequenced. These initial studies were important because they provided valuable structural information about the genes and the encoded proteins. The α' and β genes for β-conglycinin subunits have been used to transform both tobacco and petunia cells (26,27). When present in regenerated plants, the

genes are expressed under normal developmental and temporal control. For example, products from the α' and β-genes are located exclusively in seeds of transformed petunia plants. As in soybeans, products from the α'-gene are observed in transgenic petunia seeds several days before the ones from the β-subunit gene. Now, experiments underway in several laboratories are directed toward the identification of regulatory sequences upstream from coding regions in the genes. Sequences that enhance the level of expression of the α' and β-genes have been identified as have proteins that bind to this apparent element. Other factors that might influence the level of expression also are being evaluated: items such as gene copy number and orientation, introduction of antisense coding regions, and the consequence of the location in the genome into which genes are inserted. Although exciting progress has been made in understanding regulation of gene expression, one cannot ignore the fact that our understanding of basic principles is still at an early stage. However, development of routine protocols for genome manipulation looms in the future.

Efforts also are being made to modify quality characteristics of storage proteins. Research in my laboratory has focused on glycinin, the more sulfur-rich of the two prevalent storage proteins in soybean (25). Our prejudice is that changes introduced into these proteins should be conservative in nature and not perturb the normal assembly and packaging mechanisms that operate in the seed. This is considered necessary because the three–dimensional structure of glycinin subunits is unknown as are the structural requirements for packaging these proteins into protein bodies.

Figure 2 outlines our current understanding about the synthesis of glycinin subunits and their assembly in protein bodies. Polyadenylated glycinin messages are translated at the exterior of endoplasmic reticulum and the translation products (preproglycinins) are deposited vectorally within the lumen of these membranes. A signal sequence consisting of approximately 20 amino acids is removed cotranslationally during the process and then aggregation of proglycinin subunits into trimers takes place. The proglycinin trimers move from endoplasmic

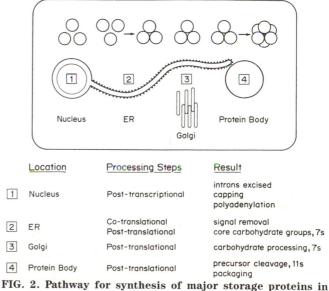

Location	Processing Steps	Result
[1] Nucleus	Post-transcriptional	introns excised capping polyadenylation
[2] ER	Co-translational Post-translational	signal removal core carbohydrate groups, 7s
[3] Golgi	Post-translational	carbohydrate processing, 7s
[4] Protein Body	Post-translational	precursor cleavage, 11s packaging

FIG. 2. Pathway for synthesis of major storage proteins in soybeans.

reticulum through golgi and into protein bodies. Proteolytic cleavages of proglycinin trimers occur within protein bodies to yield glycinin subunits composed of an acidic component linked to a basic one. Mature glycinin complexes isolated from seeds are hexamers. However, low angle scatter x-ray experiments (30) indicate that the storage proteins must be arranged in some higher order structure within protein bodies.

Synthesis and assembly of proglycinin trimers and glycinin hexamers have been carried out in vitro (28). For this purpose, nucleotides that encode the signal sequence of Gy4 preproglycinin were eliminated from a full–length cDNA clone by exonuclease digestion (Fig. 3). The remainder of the coding region was ligated to the SP6 promoter in pSP65 and produced a construction (pSP65/248) that encoded a Gy4 proglycinin with a methionine for leucine substitution at the NH$_2$-terminus. Run-off transcription by SP6 polymerase followed by translation in a rabbit reticulocyte extract yielded a product of the anticipated size. Under appropriate conditions, self-assembly of the proglycinin subunits into 9S trimers occurred. The trimers produced in vitro appeared equivalent to those formed within endoplasmic reticulum in vivo.

Although proglycinin trimers were formed in vitro, the trimers were unable to assemble into either hexamers or higher order aggregates similar to those found in vivo (Dickinson and Nielsen, unpublished data). To explore the basis for this inability, conditions that permitted reversible dissociation of native glycinin hexamers into trimers and monomers were used. When radioactive proglycinin trimers were mixed with dissociated native glycinin trimers and monomers, radioactivity was not recovered in hexamers after reassociation. Radioactivity was only associated with the trimer fractions from the sucrose gradients used for the analysis. However, when the same experiment was done with radioactive proglycinin monomers, radioactivity was recovered in the hexamer fraction. One explanation of the results is that structural differences exist between proglycinin and mature glycinin that preclude hexamer formation from trimers, but that are tolerated when proglycinin monomers become associated into hexamers.

Post-translational cleavage of proglycinin trimers within protein bodies is apparently required for assembly of glycinin hexamers (Dickinson and Nielsen, unpublished data). Evidence to support this idea was obtained by treating purified Gy4 proglycinin trimers with low levels of either papain or V8 endonuclease. The protease treatments caused trimers to assemble into hexamers in the reassociation assay. When the experiments were repeated using monomers instead of trimers, the monomers apparently were degraded so that neither trimers nor hexamers could be produced. The results are consistent with the idea that conformational changes occur in response to the post-translational cleavages, and the conformational changes in turn regulate assembly of the trimers into higher order structures. The lack of cleavage before arrival of trimers in protein bodies may

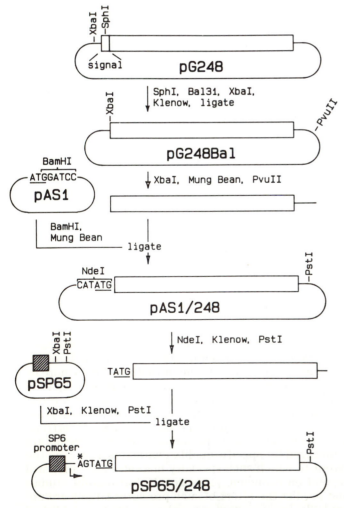

FIG. 3. Construction of pSP65/248. Open boxes indicate Gy4 coding sequence. An adenine (*) was engineered into the -3 position for efficient translation initiation (29). Data are from Dickinson et al. (28).

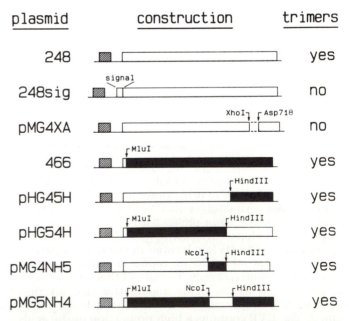

FIG. 4. Construction of coding regions for modified forms of proglycinin. Constructions were made using the same general approach outlined in Figure 3. Data are from Dickinson et al. (28). Cross-hatched boxes, SP6 promoter; open boxes, coding sequence from the Gy4 subunit gene; filled boxes, coding sequence from the Gy5 subunit gene.

prevent premature precipitation of the proteins as they traverse other subcellular locations.

We have used in vitro assembly to assay the consequence of changes engineered into glycinin subunits (28). This approach should permit identifications of changes in subunit structure that are detrimental to assembly of glycinin complexes without resorting to time–consuming transformation and regeneration techniques. To test this possibility, modified forms of pSP65/248 were constructed to see if trimer formation could be disrupted (Fig. 4). In one construction (pMG4XA), an 81-base-pair nucleotide sequence for a highly conserved 27 amino acid region in the basic polypeptide was deleted. The mutant proglycinin molecule produced from this plasmid migrated more rapidly in SDS-polyacrylamide gels than authentic proglycinin and did not self-assemble into trimers. The hydrophobic amino acids that were deleted earlier had been predicted to have β-sheet structural preferences and to be located in the interior of the subunits (33). Consistent with this hypothesis, disruption or elimination of the hydrophobic region was expected to destabilize subunit conformation and to interfere with assembly of trimers.

In another construction (pSP65/248sig, Fig. 4), nucleotides for the signal sequence at the NH_2-terminal of the coding region were left intact. Preproglycinins obtained from this plasmid were slightly larger than proglycinin from pSP65/248 and did not self-assemble into trimers. The signal sequence in the product from pSP65/248sig preceded an extensive region with predicted α-helix, coil and turn conformational preferences. Lack of assembly was considered due to the disruption of folding caused by the hydrophobic nature of the signal sequence.

Other constructions tested the extent to which alterations were tolerated by glycinin subunits. In one series (Fig. 3), use was made of a conserved MluI restriction site located at the beginning of Gy4 and Gy5. Proglycinin produced by pSP65/466 was identical to authentic Gy5 except for three amino acid substitutions at the NH_2-terminus; a methionine for isoleucine at position 1, a serine for threonine at position 2, and a phenylalanine for leucine at position 6. Plasmids pHG45H and pHG54H were constructed using a conserved HindIII site to trade sequences that encoded the basic polypeptides of Gy4 and Gy5. Plasmids pMG4NH5 and pMG5NH4 were made by using the NcoI and HindIII sites to trade sequences that encoded a naturally occurring hypervariable region in glycinin. The changes in all of these constructions were conservative because Gy4 and Gy5 are 80–90% homologous and were not expected to perturb assembly. As expected, each of these hybrid proglycinins self-assembled at a rate approximately equal to that observed with proglycinins made from pSP65/248.

Other efforts have concentrated on the hypervariable region (HVR). The HVR is located in the central part of the subunit, just on the NH_2-terminal side of the site for post-translational cleavage. The size of this region varies by nearly 100 amino acids among the five glycinin subunits whose primary structures have been determined. The HVR contains a high proportion of the acidic residues in the subunit, and these are predicted to be almost exclusively in either α-helical or turn conformations. The natural variability and physical properties of the HVR suggest it must be exposed to the surface of the subunits. Hence, structural changes might be tolerated in the HVR more easily than in other regions of the molecules.

The effect of both additions and deletions to the HVR have been evaluated. In one series of constructions, variable amounts of the HVR from Gy4 were deleted. The Gy4 subunit contains the largest HVR among the five genes, and the extent of the deletions in this region ranged between 18 and 90%. Consistent with the notion that the HVR might easily tolerate structural changes, proglycinins synthesized at the direction of constructions from the deletion series all self-assembled into both trimers and hexamers at rates similar to that for precursors made from the pSP65/248 control.

A second series of constructions involved insertions into the HVR of Gy4. For these, use was made of a unique AccI restriction site located in a region that encoded residues with predicted turn secondary structural preferences. The nature of the overhangs at the AccI site permitted multiple 6-mers with internal symmetry to be inserted into Gy4. These insertions encoded an alternating string of arginine and methionine residues. Proglycinins from constructions with one, two or three tandem insertions self- assembled to the same extent as precursors made from pSP65/248. However, assembly was reduced when five copies of the 6-mer were present. The latter result was anticipated since multiple arginine residues would tend to form a positively charged α-helix as the length of the insertion increased. The positively charged residues could interact with negatively charged glutamate and aspartate residues present in high concentration in the HVR and disrupt protein folding. However, despite the apparent disruption to folding, the results support the notion that extensive modification to structure should be tolerated in the HVR.

Although each of the three approaches to improve soybean quality that have been described had different objectives, they also had certain features in common. In each case, total yield, a general indicator often used by plant breeders to measure progress, was not an essential element. Rather, two other factors played initial key roles. These were the identification of a specific genetic or structural trait and the development of an effective assay for its measurement. Once these criteria were satisfied, yield became a consideration. Although total yield has played an important role in improvements for production agriculture, it seems likely that breeding schemes will be subjected to other criteria when quality factors are manipulated.

Some measure of productivity always will be required when economic considerations are involved, however new standards for measurement that are driven more by the content of specific quality components than by total yield may evolve. For example, there already is interest in including oil vs. protein composition in soybean grading factors in the U.S. It can be anticipated that this trend will become even more pronounced when soybeans are engineered for specific quality characters. However, the extent to which such specialty markets can evolve will depend on economic pressures. Unfortunately, studies that predict the effect of such pressures largely are unavailable. In this regard, priorities need to be established because the emerging biotechnology, although powerful, also is expensive both with respect to time and materials.

ACKNOWLEDGMENTS

Portions of the work described have been supported in part by grants from the American Soybean Association and by Central Soya.

REFERENCES

1. Facciotti, D., J.K. O'Neal, S. Lee and C.K. Shewmaker, *Bio/technology 3*:241 (1985).
2. Baldes, R., M. Moos and K. Geider, *Plant Mol. Biol. 9*:135 (1987).
3. Criston, P., J.E. Murphy and W.F. Swain, *Proc. Natl. Acad. Sci. U.S. 84*:3962 (1987).
4. Ranch, J.P., L. Oglesby and A.C. Zielinksi, *In vitro Cellular and Developmental Biology 21*:653 (1985).
5. Barwale, U.B., H.R. Kerns and J.M. Widholm, *Planta 167*:473 (1986).
6. Porra, R.J., and P.K. Stumpf, *Arch. Biochem. Biophys. 176*:53 (1976).
7. Porra, R.J., and P.K. Stumpf, *Arch. Biochem. Biophys. 176*:63 (1976).
8. Wilson, R.F., *Seed Metabolism in Soybeans: Improvement, Production and Uses*, edited by J.R. Wilcox, American Society of Agronomy, Madison, WI, 1986, p. 673.
9. Wilcox, J.R., J.F. Cavins and N.C. Nielsen, *J. Am. Oil Chem. Soc. 61*:97 (1983).
10. Wilcox, J.R., and J.F. Cavins, *Theor. Appl. Genet 71*:74 (1985).
11. Wilcox, J.R., and J.F. Cavins, *Crop Sci. 26*:209 (1985).
12. Frankenberger, E.M., *Studies Relating to the Inheritance of Altered Fatty Acid Content in* Glycine max., PhD Dissertation, Purdue University, West Lafayette, IN, 1986.
13. Graef, G.L., W.R. Fehr and E.G. Hammond, *Crop Sci. 25*:1075 (1985).
14. Knowles, P.F. and A.B. Hill, *Crop Sci. 4*:406 (1964).
15. Rackis, J.J., D.J. Sessa and D.H. Honig, *J. Am. Oil Chem. Soc. 56*:262 (1979).
16. Wolf, W.J., *J. Agr. Food Chem. 23*:136 (1975).
17. Axelrod, B., T.M. Cheesebrough and S. Laakso, *Meth. Enzymol. 71*:441 (1981).
18. Kitamura, K., C.S. Davies, N. Kaizuma and N.C. Nielsen, *Crop Sci. 23*:924 (1983).
19. Davies, C.S., and N.C. Nielsen, *Crop Sci. 26*:460 (1985).
20. Hildebrand, D.F., and T. Hymowitz, *Crop Sci. 22*:851 (1982).
21. Davies, C.S., and N.C. Nielsen, *Crop Sci. 27*:370 (1987).
22. Davies, C.S., S.N. Nielsen and N.C. Nielsen, *J. Am. Oil Chem. Soc. 64*:1428 (1987).
23. Doyle, J.J., M.A. Schuler, W.D. Godette, V. Zenger, R.N. Beachy and J.L. Slightom, *J. Biol. Chem. 261*:9228 (1986).
24. Nielsen, N.C. in *Molecular Biology of Seed Storage Proteins and Lectins*, edited by L.M. Shannon and M.J. Chispeels, American Society of Plant Physiologists, Waverly Press, Baltimore, MD, 1986, pp. 17-28.
25. Nielsen N.C., *J. Am. Oil Chem. 62*:1680 (1985).
26. Beachy, R.N., Z-L Chen, R.B. Horsch, S.G. Rodgers, N.J. Hoffmann, R.T. Fraley, *Eur. Mol. Biol. Org. 4*:3047 (1985).
27. Goldberg, R.B., *Phil. Trans. R. Soc. Lond. B 314*:343 (1986).
28. Dickinson, C.D., L.A. Floener, G.G. Lilley and N.C. Nielsen, *Proc. Natl. Acad. US*, in press.
29. Kozak, M., *Nucleic Acids Res. 12*:857 (1984).
30. Colman, P.M., E. Suzuki and A. van Donkelaar, *Eur. J. Biochem. 103*:585 (1980).

Biotechnology for *Brassica* and *Helianthus* Improvement

Raghav Ram, Terrence J. Andreasen, Alexi Miller and **David R. McGee**
Sungene Technologies Corp., 3330 Hillview Ave., Palo Alto, CA 94304

During the past 15 years, *Brassica* and *Helianthus* have enjoyed substantial growth in total production and new markets. In agronomic terms, this market expansion translates to improved total yield, environmental fitness and modified oil and meal components. Market opportunities continue to exist for new genotypes improved in these three areas. Of particular interest for *Brassica* are agronomic traits such as even seed maturity, shattering resistance and effective hybridization. Resistance to drought, disease and cold temperatures, and stability of oil profiles under variable growing conditions are important to *Brassica* and *Helianthus*. For both oilseeds, intensive interest exists for fatty acid profiles tailored to meet specific food and industrial market needs. Cell culture, plant regeneration and genetic transformation techniques are highly developed for *Brassica* species; they are less so for *Helianthus*. New phenotypes created by a few discreet mutations, especially a decrease in enzyme activity, readily are obtainable through mutagenesis and somaclonal variation regimes. Because these mutations occur randomly, effective selection of large-scale screens greatly enhances the chance of success. While such schemes are straghtforward for sunflower, the amphidiploid nature of *B. napus* and *B. juncea* makes rapeseed modification more complex. For the introduction of traits requiring new or enhanced enzyme activity, genetic transformation is the best approach. However, detailed molecular knowledge of key enzymes, structural genes and regulatory elements required for transformation is lacking for most useful phenotypic changes. The acquisition of this knowledge presents a major challenge to biochemists, geneticists and plant breeders.

Over the past 15 years, world production of plant oils has risen dramatically. The commercial application of these oils usually depends upon oil composition, price, availability and historical precedence. Oilseed crop improvement is driven by the needs of both industry users and the farmers who grow the crop.

Urbanization has taken a toll on the availability of quality agricultural land in many countries. In addition, increasing costs of transportation and other economic burdens have pressured farmers to reduce their crop-production costs. Improved agronomic traits such as cold tolerance, early maturity, increased seedling vigor and responsiveness to fertilizers would expand crops into less-than-optimal growing areas, thereby promoting overall cost reductions with a subsequent increase in the farmer's profit margin. Farmers would be encouraged to grow the crop, thus providing a predictable supply of raw materials to industry. The development of oils specifically tailored to meet consumer requirements would reduce industry's processing costs, provide higher quality products and expand product market potential.

Therefore, the goal of oilseed modification is to develop new oilseed varieties having consumer or processing benefits in addition to a high oil content or high yield. Biotechnology can provide an effective set of tools to accomplish such modifications. Rapeseed and sunflower present unique opportunities to apply biotechnology in changing these crops from producing a commodity product to a specialty product. Together, they account for nearly 30% of all major world edible oils produced and have higher oil content and higher oil value per hectare than soybean (Table 1). In addition, rapeseed and sunflower currently are among the crops most amenable to improvement through biotechnology.

TABLE 1

1985/86 Average World Yield and Value for Major Oil Crops

	Soybean	Sunflower	Rapeseed
Seed yield (MT/ha)	1.86	1.24	1.29
Oil yield (% oil)	18	40	37
(MT/ha)	0.34	0.50	0.48
Oil value ($/ha)	128	203	162
Meal yield (% meal)	79	46	61
(MT/ha)	1.47	0.57	0.79
Meal value ($/ha)	269	63	93

Source: USDA, FAS, Foreign Agricultural Circular, *Oilseeds and Products*, 1987.

Rapeseed belongs to the genus *Brassica*. Commonly grown species are *B. napus* and *B. campestris* in Europe, the U.S. and Canada, and *B. juncea* and *B. nigra* in the Orient. The erucic acid content of rapeseed oil determines its end use; oil from high-erucic rapeseed cultivars is used for lubricants in the steel and textile industries and in the manufacture of rubber additives, while low-erucic rapeseed (canola) is used primarily in salad oils, cooking oils, margarines and shortenings. During the past 10 years, rapeseed acreage has increased in Canada and several European countries, with much of this increase dedicated to the canola variety of rapeseed.

The commonly grown species of oilseed sunflower is *Helianthus annuus*. Most sunflower varieties produce oils that contain about 18% oleic acid and are used for margarines, salad and cooking oils, and mayonnaise products. High–oleic sunflower varieties, originating from the Soviet *Pervenets* strain, have been developed in the U.S. through conventional breeding (1,2). Two high–oleic varieties currently comprise about 2.5% of the total U.S. sunflower acreage. The high–oleic character imparts increased stability to oxidative degradation and high temperatures, making it especially useful in industrial frying operations and food processing and as a lubricant additive for auto engines and textile equipment. High–oleic sunflower oil also is recognized as having the positive nutritional qualities of monounsaturated oils such as olive oil.

The combination of modern plant breeding with the expanding capabilities of biotechnology already is formidable and will continue to grow. Cell biologists continue to develop systems for initiating and maintaining tissues, single cells or protoplasts in culture and regenerating whole plants. To date, most successes with plant transformation have been achieved with model systems in an effort to perfect the technology of gene packaging, transfer, integration and screening, and to understand basic gene expression and protein interactions. A well–integrated program for creating directed phenotypic modification requires input from many disciplines, from the field to the laboratory to the engineer's workbench.

Oilseed improvements addressable through biotechnology are of two general types: those that will increase the availability of the oil and lower the cost and those that will result in a premium quality oil. The first type includes improvements in agronomic traits affecting intrinsic yield (e.g., heterosis, delayed senescence, determinant flowering); environmental fitness (e.g., pest or disease resistance, cold, drought or herbicide tolerance, standibility, early flowering); and postharvest stability (e.g., improved storage, handling and processing).

Of these traits, the single most important is yield. Because yield cannot be attributed to a single gene or a small group of genes, high–yielding plants only can be selected through conventional plant breeding. Heterosis, or hybrid vigor, radically increases yield. In sunflower, commercial hybrids give yield increases of up to 60% over open–pollinated varieties (3). Although rapeseed hybrids are not yet available commercially, recent evidence shows that hybrid rapeseed gives up to a 40% yield increase over varieties (4). In addition to heterosis, agronomic traits such as disease resistance, cold tolerance and self–compatibility indirectly can affect yield.

Any additional improvements leading to enhanced complementary products such as improved seed meal quality can increase the value of the crop and may be a significant factor in inducing the grower to increase his acreage. Further, the development of subtropical and tropical varieties of sunflower and rapeseed could have a significant impact on developing countries.

The second type of improvement addresses only the oil fraction, including fatty acid composition and distribution, and lipid soluble components such as tocopherols and pigments.

The decision of which oil modifications to target is of course critical and involves several factors including: identification of superior oil sources and recognition of commercial markets and technical restraints. Markets for the improved product must be available either through enhanced product usage or through a premium price. Table 2 lists several fatty acid modifications of commercial importance. The fact that a superior oil source is identifiable or even attainable does not necessarily make it a good commercial research target. Perceptions of the ideal oil profile may change over time, especially in food markets.

Longevity of an opportunity to fill a market niche also is important. A new, transformed, field–tested oilseed may require 10 or more years to develop. Will the market still exist after 10 years? If so, will an alternate crop simultaneously have been developed? An example is the dual approach to producing medium chain-length fatty

TABLE 2

Goals for Modified Fatty Acid Composition in Triglycerides

Goal	Best target crop
100% oleic acid	H. annuus, B. napus
100% linoleic acid	H. annuus
100% erucic acid	B. napus
0% saturated acids	H. annuus, B. napus
High palmitic acid	H. annuus
High medium length acids	B. napus
Temperature stability of profile	H. annuus
Balanced fatty acid profile	H. annuus, B. napus

acids using cuphea. Does one transfer the biochemistry of cuphea to an existing oilseed, or does one develop cuphea agronomically into a useful crop itself?

OIL MODIFICATIONS

As modification of major crops has become feasible through biotechnology, attention has focused on oilseeds for three major reasons. 1) Inherent flexibility: The seed oil is essentially an inert reservoir of reduced carbon. As such, plants can tolerate wide fluctuations in triglyceride fatty acid structure without losing viability. 2) Seed specificity: Triglyceride biosynthesis in the seed involves a unique fatty acid biochemistry, perhaps due to seed–specific isozymes or enzyme regulation, or more subtle tissue– or developmental stage–specific gene expression. Specific biochemical targets may exist for selective modification of seed metabolism. 3) Single gene changes can have large effects: Research by many groups on soybean, sunflower, rapeseed and corn has shown that a single mutation can result in major shifts in fatty acid composition. Although a small number of modifier genes often play a minor role, a single mutation may result in an economically significant new phenotype.

Oils containing high levels of oleic or linoleic acid are desirable for both food and industrial uses, and a high erucic acid content is important for industrial purposes. Achievement of these goals would result in a premium oil that could fill a market niche devoid of equal competition. Although markets for oils containing high levels of palmitic acid and medium chain–length fatty acids already are served by palm oil, the growing demand for these oils and the desire for domestic sources make these feasible research targets also.

For food uses, the ideal oil profile may be a precise mixture of saturated and unsaturated fatty acids having a chain length of 16-20 carbons. Such mixtures provide the best organoleptic properties for margarine, salad oil, mayonnaise, chocolate and many other products. The exact fatty acid balance depends upon the end use, as well as companion technology and consumer preference. Decreased saturated fatty acids and increased polyunsaturates are both important targets. As a dietary requirement, low levels (10–15%) of linolenic acid (18:3) now are acceptable and even desirable, given new methods to stabilize against oxidation. A longer–range plant oil target is the production of stable, longer–chain polyunsaturated fatty acids in the ω-3 configuration, such as in fish oil. Theoretically, production of oils containing 100% of a single fatty acid may be possible; however, functional membrane/protein interactions necessary for plant viability may be destroyed.

Although sunflower has four major fatty acids in its seed oil, rapeseed may have seven or more. Both oilseeds show wide variability in individual fatty acid levels (Table 3), which seems to be augmented by tissue culture procedures. To date, no rapeseed variety has been found to contain more than about 65% erucic acid, perhaps because erucic acid has been found virtually to be absent from the acids esterified at the secondary hydroxyl on the glycerol moiety of the triglycerides. However, plants that do have the ability to attach erucic acid to this hydroxyl group have been identified (5). Therefore, the opportunity to purify the enzyme, find the appropriate gene and introduce it into *Brassica* exists.

TABLE 3

Extreme Fatty Acid Compositions Observed

Fatty Acid	*Helianthus annuus*		*Brassica napus*	
	Min.%	Max.%	Min.%	Max%
Palmitic (16:0)	1.9	33	1.0	20
Stearic (18:0)	0.5	12	0.6	24
Oleic (18:1)	8.0	95	3.8	78
Linoleic (18:2)	0.5	80	9.0	36
Linolenic (18:3)	0	0	0.5	27
Eicosenoic (20:1)	0	0	0	26
Erucic (22:1)	0	0	0	65

Temperature fluctuation during plant growth can greatly influence the fatty acid content in sunflower. Rochester & Silver (6) observed that the ratio of linoleic to oleic acid in seed oil inversely is proportional to the mean minimum temperature during cotyledon development. Because sunflower is grown in a wide range of latitudes, improvements to increase product uniformity during seasonal temperature fluctuations are desirable.

The elimination of antinutritional compounds from rapeseed and sunflower without causing a decrease in yield is of interest for both oil and meal markets. In *Brassica*, glucosinolates can be converted to toxic isothiocyanates upon hydrolysis by myrosinase, an endogenous rapeseed enzyme (7). Processing treatments to remove glucosinolates and other contaminants cause a loss of protein from the meal, and some of the breakdown products can be coextracted with the oil imparting an undesirable taste. To a lesser extent, the removal of chlorigenic acid from sunflower represents a desirable improvement.

AGRONOMIC MODIFICATIONS

Specific agronomic improvement targets in rapeseed include even seed maturity, resistance to pod shattering, earliness, drought resistance and decreased moisture and chlorophyll content in the seed. Short-statured plants with numerous branches that each have several pods provide an architecture better suited to harvesting.

Because sunflower is a hybrid crop, ideal sunflower parental lines show distinct morphological differences. Male inbreds should be multiheaded with one large head

maturing concurrently with the female lines and several smaller heads maturing later to allow complete pollination. Female lines should have a single, large, male-sterile head and a strong stalk. Hybrids should be unbranched, rather short, early maturing, resistant to environmental challenges, and should display marked heterosis. The oil produced should be 40% or more of seed mass with fatty acid profiles optimally arranged for end-use markets.

Increasing the number of seed per sunflower head is a direct method to increase yield. However, hybrids having traits conducive to mechanical harvesting may produce shorter, unbranched, stiff-stalked plants and fewer seeds. Increases in the seed number in a parental inbred is highly desirable. This can be accomplished through increased self-compatibility, larger head size or greater number of seeds per unit area on the inflorescence.

Sunflowers having heads positioned at a slight downward angle on the stalk are less prone to water accumulation and subsequent rotting, scorching from the sun and attack from birds. Concave-shaped heads are desirable for mechanical harvesting of the seed.

Disease damage is a regional problem having little impact in some areas, while in others farmers are prevented from planting the affected crop. A number of soil-born fungal diseases can cause overall crop losses of 50% or more. In sunflower, major fungal pathogens causing widespread problems include *Phoma* (black stem rot), *Sclerotinia* (root and stem rot), *Botrytis* (head rot) and *Alternaria* (leaf and stem rot) (8). Species of *Phoma* (black leg), *Alternaria* (black leaf spot) and *Sclerotinia* (stem rot) also affect rapeseed. In addition, *Peronospora* (downey mildew) can cause considerable damage (9). Although crop rotation, seed preparation and volunteer elimination effectively can combat the spread of infection, the use of resistant varieties of both rapeseed and sunflower would provide the simplest, most consistent solution. Resistance factors often can be found in wild species of a cultivated crop. Alternatively, disease-tolerant lines can be selected for tolerance to lethal concentrations of fungal toxins in culture. This approach has produced some promising results; however, toxin tolerance may not correlate with infection resistance.

BIOTECHNOLOGY APPROACHES

In vitro cell and tissue culture and mutagenesis programs have proven successful for producing a wide range of genetic variation in regenerated plants (10). These techniques can result in both single-gene and multi-gene changes. Elite genetic lines should be used, as they already are optimized for many desirable traits. Tissue culture-derived variations have been described as somaclonal variations. While some variations are transient, others are transmitted genetically in the normal Mendelian manner. Variations generally are expressed in the heterozygous condition, and selections must be made for one or two generations before the progeny genetically are stable. Reversal or disappearance of variations resulting from tissue culture is not uncommon.

It has been our experience that most genetic variations arising through in vitro mutagenesis are caused by recessive mutations. This approach appears to be more effective in decreasing rather than increasing enzyme activity.

In vitro mutagenesis programs are very promising for targets such as blocking desaturase activity for high oleic acid production and blocking elongase activity to eliminate erucic acid in *Brassica*. Amphidiploids such as *B. napus* and *B. juncea* may be more difficult to influence than diploids such as *B. campestris* or *H. annuus*, as more mutational events may be required to produce the new phenotype.

Gene amplification may be induced in culture under the correct in vitro selection pressure and may produce substantial enzyme activity increases (11). However, new phenotypes may not be stable when removed from the selection conditions, and the effect may be of transient duration. For increasing enzyme activities, transformation may be the best approach.

Interspecific hybridization provides an alternate method for transferring disease resistance and environmental fitness characteristics into an agronomic crop. It involves either the fusion of somatic cells (protoplasts) from different species (12) or wide crosses between genetically distant species to transfer organelles or genes across species barrier (13). In wide crosses of incompatible species, embryos usually abort after fertilization due to a lack of compatible endosperm. Embryos derived from wide crosses can be isolated (rescued) and cultured under in vitro conditions to give rise to whole plants. Usually, these plants must be back-crossed to a parental line to improve fertility and retain characteristics of the original crop. Such plants may express some degree of heterosis and the traits of interest.

Gene transformation technologies enable the specific selection of single genes or small groups of genes for introduction into the crop of interest. Effective transformation requires not only the construction of the desired gene package, but also stable insertion of the gene into the genome of a cell capable of regeneration.

Transformation with exogenous DNA offers the greatest potential for creating variation because of the wide range of available gene sources. Intra- and interspecific gene transfers, and even interkingdom gene transfers now are routine. Composite genes containing elements from several sources have been used and, where needed, totally synthetic genes eventually may be constructed to perform new tasks. In addition to the great potential for introducing new traits or increasing the copy number of desired genes, transformation also offers the possibility of controlled down-regulation of enzyme activity.

Modified genes that code for RNA ("antisense RNA") complementary to mRNA from a target gene can be inserted. These two RNAs can interact to prevent their translation into protein, effectively inactivating the target gene (14). By placing the modified gene under expression controls, a target gene could be inactivated selectively in a single tissue while still being expressed fully in the rest of the plant.

In theory, mixing and matching DNA sequences that specify mRNA production and the time, location and extent to which an enzyme is present will allow production of specifically designed crop phenotypes.

TECHNOLOGIES FOR RAPESEED IMPROVEMENT

Brassica species have been particularly amenable to several cell-biological techniques, including haploid and diploid tissue culture, genetic transformation and hybridization (12,15,16). Many phenotypic variations have been observed in regenerated rapeseed plants, including leaf shape and branching architecture, vigor differences, possible disease resistance to *Phoma* and *Alternaria*, and changes in glucosinolate levels and fatty acid profiles.

Haploid Tissue Culture

Haploid plants can be regenerated from anther (microspore) culture of several *Brassica* species. These sterile haploids then can be converted to homozygous diploids using chromosome–doubling techniques. Haploid culture technology allows a homozygous parental line to be produced in one–half to one–third the time required with conventional breeding methods using repeated self–pollination. With the increasing possibility of commercializing oil rapeseed hybrids in the near future, such a technology should be highly useful. Haploid protoplast systems also should play an important role.

Mutagenic treatments and the production of somaclonal variants via tissue culture methods would be more effective using anther culture. However, in amphidiploid species such as *B. napus*, *B. juncea* and *B. carinata* identification and homozygosity of an induced mutation could become complicated. For example, genes pertaining to fatty acid biosynthesis are present on all chromosome complements of an amphidiploid (AACC in *B. napus*, AABB in *B. juncea* and BBCC in *B. carinata*). In *B. napus*, effective fatty acid modifications may require mutations in both A and C complements. If a mutant AA can be distinguished from a mutant CC, then a plant homozygous for the mutation in all AACC can be synthesized by crossing a parent carrying a homozygous mutation in the AA complement with another parent carrying homozygous mutation in the CC complement.

Diploid Tissue Culture

Plants can be regenerated from a wide variety of somatic tissues in several *Brassica* species. Regeneration from leaf and hypocotyl tissues is important for transformation with *Agrobacterium* vectors. Single–cell suspension culture and protoplast culture technologies increasingly are becoming available as tools for genetic modification (15). Plants can be regenerated from protoplasts of *B. napus* hypocotyl and leaf mesophyll cells. Cell suspension cultures of *B. napus* and *B. nigra* yield protoplast populations that can produce somatic embryos in high frequency (15,17). *Brassica juncea* protoplasts are capable of undergoing direct embryogenesis.

Single–cell culture systems such as suspensions and protoplasts can contribute to in vitro selection studies such as herbicide and disease tolerance. Protoplast regeneration systems should be useful in producing novel hybrids between lines of interest.

Transformation

Advances in cell culture and gene transfer have immediate application in the genetic improvement of *Brassica*. Using various techniques, genes can be introduced into single cells, protoplasts or tissue slices from which whole plants can be regenerated. The most common method of

TABLE 4

Examples of *Brassica napus* Transformation to Achieve Oil Modification

Type of gene transfer	Trait	Gene product	Source
Intraspecific	Increased erucic acid	Elongase	*B. napus*
Interspecific	Increased erucic acid	Acyltransferase	*Tropaeolum majus*
Interkingdom	Medium chain-length fatty acids	Medium–chain thioesterase	Human
Composite	Modified fatty acid profile	ACP	*E. coli, B. napus*

gene transfer in dicots has been *Agrobacterium*–mediated transformation. *Agrobacterium* readily infects *Brassica* species (18).

Table 4 presents examples of *Brassica napus* transformations aimed at achieving specific oil modifications. Reports have appeared on transformation of *B. campestris* using a cauliflower mosaic virus vector (19,20), of *B. juncea* using *A. tumefaciens* (21) and of *B. napus* using *A. rhizogenes* (22,23) and *A. tumefaciens* (24,25). The *B. napus* transformations allow regeneration of transgenic plants.

Because regeneration of plants from protoplasts is highly efficient in *Brassica* (26), electroporation may become a preferred transformation method. Electroporation has been used for introducing genes into haploid protoplasts (27). All of these techniques allow for the introduction of multiple genes or gene copies into plant cells.

Reported successes with these directed DNA transfer techniques so far have been limited to nonagronomic genes such as antibiotic resistance. However, efforts are underway in several laboratories to isolate and identify genes of commercial importance in biochemical pathways and to modify their expression. Developmental– and tissue–specific expression using, for example, seed–specific promoters from storage proteins already has been achieved in other dicot systems (28). Some genes requiring constitutive expression throughout the development of a plant currently are being tested. These include genes for herbicide tolerance (e.g., sulfonyl, atrazine and glyphosate compounds) and insect tolerance (e.g., the *Bacillus thuringensis* toxin gene).

Hybridization

Rapeseed hybrids have not been available commercially, although considerable progress has been made towards this goal. Two basic requirements in the development of a commercial hybrid crop through conventional breeding are a clear demonstration of heterosis and the availability of male sterile and fertility restorer systems. Several cytoplasmic male sterile (cms) lines have been identified in *Brassica* (29). For all cms systems, maintainer and fertility restorer lines currently are available in either *B. napus* or *B. juncea*. In addition, nuclear male sterility has been reported in some *B. napus* lines, controlled by one or two recessive genes. F1 hybrids produced using nuclear male sterile lines have shown a 30–70% increase in seed yields over their male parents (29).

Interspecific and intergeneric hybridizations have been attempted in *Brassica* and other crucifers. Such crosses

are viable only in a few cases using conventional breeding methods, and recovery of hybrids is low. However, using cell biological techniques such as embryo culture hybrid rapeseed plants have been produced successfully (30). Several agronomic traits can be transferred from wild and noncrop species to crop species of interest using embryo rescue methods followed by repeated crossing with the agronomic crop. Examples include the transfer of *Plasimodiophora brassicae* (clubroot) resistance from *B. campestris* to *B. napus* and the transfer of *Erysiphe* (powder mildew) resistance from *B. oleraceae* to cultivated *Brassica* species.

Somatic hybridization using protoplasts of *Brassica* species has been attempted by several laboratories. Commercial application of this technique has not yet occurred due to poor ability of the fused protoplasts to regenerate. This technique is complicated by the difficulties in identifying fusion products at the cellular level. A major breakthrough has been the successful resynthesis of *B. napus* using protoplasts of *B. oleracea* and *B. campestris*. With this method, traits of importance can be transferred readily from the progenitors into resynthesized *B. napus* (31). Other successful reports of somatic hybrids are between *B. campestris* and *Arabidopsis* (32) and between *B. napus* and *B. hirta* (33). Three–way hybridizations also have been attempted in Canada and France with *B. napus, B. campestris* and *B. hirta*, and with *B. napus, B. campestris* and *Raphanus sativa* (15). Such crosses should prove useful for the development of commercially viable crops.

Somatic hybridization can be used for transferring specific cytoplasmic characteristics such as male sterility and herbicide tolerance. Although much research still needs to be done, prospects for quickly producing novel crop species are enormous. Genes not previously considered due to the biological limitations of conventional breeding programs could be utilized for crop improvement.

TECHNOLOGIES FOR SUNFLOWER IMPROVEMENT

Achievements in transformation methodologies with sunflower lag far behind those of *Brassica* species. In contrast, mutagenesis and culture systems in sunflower have produced more highly studied variant phenotypes than similar programs in *Brassica*.

Diploid Tissue Culture

Tissue culture and regeneration of sunflower has been successful using shoot tips (34) as well as undifferentiated callus tissue produced from stem pith tissue (35), hypocotyl segments (36) and immature embryos (37). Through tissue culture and mutagenesis programs in our laboratories, several genetically stable mutations have been produced in elite sunflower lines, including variations in plant height, flowering date, flower head size and morphology, branching characteristics, senescence, yield and fatty acid profile. Some of these variants are in advanced tests both as inbred lines and as hybrids.

Differences in plant height generally are caused by variations in plant vigor that when combined with other agronomic improvements may cause an overall increase in yield. Early maturing mutants that were shorter in height and flowered 20% earlier than controls may be useful for environments having a shorter growing season.

Sunflower head diameter and morphology both affect seed yield. One high–oil line showed a 69% increase in head diameter over the average control, nearly a tripling in head area. When characteristics such as increased vigor, head size and oil content are combined into one line, the resulting crop could be highly productive.

The development of an elite sunflower line having both branched (multi–headed) and unbranched (single–headed) types is useful for producing high–yielding hybrids. Branched variants of unbranched lines and unbranched variants of branched lines have been found in culture–derived plants.

Plants exhibiting delayed senescence without any change in flowering date undergo a longer seed–filling period during maturation, affecting the total yield of oil and seed. delayed senescence can be used as an indicator of general resistance to soil pathogens. Several variant lines from tissue culture have exhibited delayed senescence with up to 20% of a given line staying green when all control plants had turned brown.

Using elite parental lines of sunflower, progeny of tissue culture variants have shown up to a 100% yield increase in a California field nursery. Several other agronomic variants currently are under investigation, including those exhibiting improved self–compatibility and improved standability.

Fatty acid analysis of seed from regenerated plants revealed high–oleic variants that unlike the *Pervenets* high–oleic mutation appeared to segregate as a single gene recessive and responded to temperature changes. Crosses between these new high–oleic mutants and *Pervenets*–derived lines have produced hybrid seed containing up to 95% oleic acid and 0.5% linoleic acid. Seeds with greatly reduced palmitic acid and stearic acid have been identified as well as high–palmitic lines containing 12–33% palmitic acid.

Hybridization

Although sunflower hybrids of cultivated varieties commercially are available, hybrid crosses between *H. annuus* and many of its wild allies produce nonviable seeds. Embryo rescue methods have been developed to recover plants derived from such crosses (38). This technique has potential application in the transfer of genes of interest from wild species to agronomically useful sunflower lines.

Haploid Culture

Attempts have been made to produce haploid embryos from unfertilized ovules in sunflower varieties in China (39). Although no plants have been produced yet, this technology could become an important tool in sunflower improvement.

Transformation

While achievements in sunflower transformation have not been as extensive as in rapeseed, encouraging progress has been made. *Agrobacterium* vectors have been used to introduce genes for nopaline synthase, phaseolin, zein and antibiotic resistance (40–43). However, only one report of successful regeneration of transgenic plants exists (43). Electroporation of protoplasts with model genes has yielded transient expression in sunflower cells, although evidence of stable genomic incorporation is lacking. These protoplasts have produced callus and shoots but no plants were produced (27).

FUTURE

Market goals are the underlying driving forces for a crop improvement program. Once market goals are defined, they must be analyzed rigorously based upon available knowledge from several branches of science: molecular biology, cell biology, biochemistry and plant breeding. From this analysis may come a number of possible research targets and approaches. These must be screened for technical feasibility and for practical timelines. Approved approaches will move into the laboratory and field, eventually producing a product. This in turn may serve as a commercial impetus for still newer market goals.

The status of the ability to culture, transform and regenerate important plant species is continuing to advance both in terms of the breadth of cell types used and in the frequency of success. These methodologies are developed sufficiently for practical transformation approaches to produce new *Brassica* phenotypes. However, basic knowledge areas are lagging behind these technical achievements. When more knowledge of oilseed enzymology and directed gene expression becomes available, a proliferation of new, marketable phenotypes may occur. However, until then new traits with agronomic impact will be derived from cell culture and related technologies. These already have produced many variants in sunflower, which directly relate to goals such as production of high–yielding hybrids, disease resistance, movement into new growing regions and altered fatty acid profiles. Some of these are already in advanced field trials and most likely will provide the first upcoming products. The wave of new products that will come from transformation is several years away, at least. However, it inevitably is coming and will affect all phases of oilseed economics.

ACKNOWLEDGMENTS

The authors acknowledge the technical assistance of G. Cooley, J. Davies, E. Eikenberry, M. Ishizaki, A. McCann, S. Thorson and J. Van Dreser. This work was funded through research and development contracts with Lubrizol Enterprises, Inc.

REFERENCES

1. Soldatov, K.I., in *Proc. 7th Int. Sunflower Conf.*, 1976, p. 352.
2. Fick, G.N., in *Proc. of Sunflower Research Workshop*, 1984, p. 9.
3. Fick, G.N., in *Sunflower Science and Technology*, edited by J.F. Carter, Amer. Soc. Agronomy, Madison, WI, 1978, p. 279.
4. Grant, I., and W.D. Beversdorf, *Can. J. Genet. Cytol. 27*:472 (1985).
5. Tallent, W.H., *J. Am. Oil Chem. Soc. 49*:15 (1972).
6. Rochester, C.P., and J.G. Silver, *Plant Cell Reports 2*:229 (1983).
7. Fenwick, G.R., R.K. Heaney and W.J. Mullin, *CRC Critical Reviews in Food Science and Nutrition 18*:123 (1983).
8. Zimmer, D.E., and J.A. Hoes, in *Sunflower Science and Technology*, edited by J.F. Carter, Amer. Soc. of Agronomy, Madison, WI, 1978, p. 225.
9. Weiss, E.A., *Oilseed Crops*, Longman, New York, NY, 1983, p. 204.
10. Evans, D.A., and W.R. Sharp, in *Handbook of Plant Cell Culture*, Vol. 4, edited by D.A. Evans, W.R. Sharp and P.V. Ammirato, Macmillan, New York, NY, 1986, p. 97.
11. Shah, D.M., R.B. Horsch, H.J. Klee, G.N. Kishore, J.A. Winter, N.E. Tumer and C.M. Mironaka, *Science 233*:478 (1986).
12. Pelletier, G.R., C. Primard, F. Vedel and P. Chetrit in *Proceedings of World Conference on Emerging Technologies in the Fats and Oils Industry*, edited by A.R. Baldwin, AOCS, Champaign, IL, 1986, p. 337.
13. Hu, C., and P. Wang in *Handbook of Plant Cell Culture*, Vol. 4, edited by D.A. Evans, W.R. Sharp and P.V. Ammirato, Macmillan, New York, NY, 1986, p. 43.
14. Ecker, J.R., and R.W. Davis, *Proc. Nat'l. Acad. Sci. USA 83*:5372 (1986).
15. Downey, R.K., W.A. Keller and W.D. Beversdorf in *Proceedings of World Conference on Emerging Technologies in the Fats and Oils Industry*, edited by A.R. Baldwin, AOCS, Champaign, IL, 1986, p. 331.
16. Knauf, V.C., *Ibid.*, p. 340.
17. Klimaszewska, K., and W.A. Keller, *J. Plant Physiol. 122*:251 (1986).
18. Holbrook, L.A., and B.L. Miki, *Plant Cell Reports 4*:329 (1985).
19. Laliberte, J.F., and D.D. Lefebvre, *Crucifer Genetics Workshop*, 1986, p. 40.
20. Pazkowski, J., B. Pisan, R.D. Shillito, T. Hohn, B. Hohn and I. Potrykus, *Plant Mol. Biol. 6*:303 (1986).
21. Mathews, V.H., C.R. Bhatia, R. Mitra, T.G. Krishna and P.S. Rao, *Plant Sci. 39*:49 (1985).
22. Ooms, G., A. Bains, M. Burrell, A. Karp, D. Twell and E. Wilcox, *Theor. Appl. Genet. 71*:325 (1985).
23. Guerche, P., L. Jouanin, D. Tepfer and G. Pelletier, *Mol. Gen. Genet. 206*:382 (1987).
24. Radke, S., B. Andrews and M. Maloney, *Crucifer Genetics Workshop*, 1986, p. 42.
25. Pua, E., A. Mehra-Palta, F. Nagy and N. Chua, *Biotech. 5*:815 (1987).
26. Glimelius, K., *Physiol. Plant 61*:38 (1984).
27. Binding, H., R. Nehls, R. Kock, J. Finger and G. Mordhorst, *Z. Pflanzenphysiol. 101*:119 (1981).
28. Okamuro, J.K., K.D. Jofuku, R.B. Goldberg, *Proc. Nat'l. Acad. Sci. USA 83*:8240 (1986).
29. Shiga, T. in Brassica *Crops and Wild Allies*, edited by S. Tsunoda, K. Hinata and C. Gomezcampo, Japan Sci. Soc. Press, Tokyo, 1980, p. 205.
30. Downey, R.K., A.J. Klassen and G.R. Stringham in *Hybridization of Crop Plants*, edited by W.R. Fehr and H.H. Hadley, Amer. Soc. Agronomy, Madison, WI, 1980, p. 495.
31. Sundberg, E., and K. Glimelius, *Plant Sci. 43*:155 (1986).
32. Gleba, Y.Y., and F.Hoffmann, *Planta 149*:112 (1980).
33. Primard, C., *Cruciferae Newsletter 9*:37 (1984).
34. Paterson, K.E., *Amer. J. Bot. 71*:925 (1984).
35. Sadhu, M.J., *Ind. J. Exp. Biol. 12*:110 (1974).
36. Paterson, K.E., and N.P. Everett, *Plant Sci. 42*:125 (1985).
37. Cooley, G.L., and A.S. Wilcox, U.S. Patent No. 4,690,391 (1987).
38. Chandler, J.M., and B.H. Beard, *Crop Sci. 23*:1004 (1983).
39. Hongyuan, Y., Z. Chang, C. Detian, Y. Hua, W. Yan and C. Xiaoming in *Haploids of Higher Plants in vitro*, edited by H. Han and Y. Hongyuan, Springer-Verlag, Berlin, 1986, 1982.
40. Sutton, D.W., J.D. Kemp and E. Hack, *Plant Physiol. 62*:363 (1978).
41. Murai, N., D.W. Sutton, M.G. Murray, J.L. Slightom, D.J. Merlo, N.A. Reichart, C. Sengupta-Gopalan, C.A. Stock, R.F. Barker, J.D. Kemp and T.C. Hall, *Science 222*:476 (1983).
42. Matzke, M., M. Susani, A.N. Binns, E.D. Lewis, I. Rubenstein and A.J.M. Matzke, *Eur. Mol. Biol. Organ. 3*:1525 (1984).
43. *Ag. Biotechnology News*, May/June 1986, p. 4.

Discussion Session

In response to several questions, Colin Ratledge indicated γ-linolenic acid produced from microorganisms was more safe and contained no antinutritional properties such as those found in other natural oils containing this fatty acid. In addition, because such factors were not present there is no additional cost associated with their removal. Dr. Ratledge indicated that the work concerning atp-citrate lyase has been published in the *J. General Microbiology*. In response to a question regarding what we may learn from animal biochemistry in relation to plants, he indicated that although there were many different systems present in both species, many enzyme systems appear to have common goals that may be useful if studied from an evolutionary standpoint.

Tsuneo Yamane indicated that the term "microaqueous" referred to the water content of systems when describing the activity of enzymes that were active in organic solvents in the immobilized state, and that much of this depended upon the state of the water in such systems.

A.K. Stobart, in answer to a question, indicated that the PC-acyl CoA activity also was found in leaf organelles and leaf microsomes and that there was probably an equilibrium in the PC used for desaturation, although 60% was in the microsomal membrane. When queried on the selectivity of acylation of glycerol-3-phosphate, i.e., "Why do saturates end up in position-1 and unsaturated fatty acids in position-2?" he stated that this was due to substrate specificity. Stobart referred to older work of W. Lands that indicated in animal systems, as least, the presence of pi-electrons in fatty acids (the unsaturated fatty acids) caused placement into the 2-position, and the absence of such electrons (saturated fatty acids) caused placement into the primary hydroxyl positions. The mechanism could be the same in plants.

In repsonse to questions, P. Knowles indicated there is probably no possibility that a sunflower plant will be produced with a better resistance to "bird damage" before harvest because such crops always will attract birds. He indicated further that he was not aware if a high-stearic acid strain of soya was or could be used to produce a cocoa butter substitute.

Daniel J. Guerra responded to a question that transformation of *Brassica* may lead to altered content and structure of fatty acids in oils and indicated the Calgene Corporation was doing this in some of their high-oleic strains for sunflower oil.

A last question to L.H. Jones concerning the use of clones for palm was answered by the suggestion that 5-10 clones probably would be used per area, but they are not yet being used in large scale plantings.

Niels Nielson indicated in his responses that the function of lipoxygenases in the soybean is not known, however the L-2 isozyme largely was responsible for off-flavor generation in soy milk and meal produced from beans without this isozyme. He further indicated in response to a question that these beans are available from him for further planting. In answer to a query on the extractability and bioavailability of carbohydrates to monogastric animals, he indicated that those carbohydrates such as stachyose are being studied as to their pathways, but no work had been done to reduce their concentrations via crop improvement. When asked why protein engineering was done, he indicated protein was well-studied and could be used as a model soybean system to evaluate products made by protein engineering. One example given was the improvement of nutritional value by increasing methionine content.

Terrence J. Andreason indicated that up to 200 plants had to be screened to find lines with differing fatty acid content, and seed mutagens caused greater mutations when used.

Genetic Diversity of Lipids in Plant Germplasm

Robert Kleiman
Midwest Area Northern Regional Research Center, ARS/USDA, Peoria, IL 61604

To demonstrate the diversity and taxonomic utility of lipid structures found in seed oils, this paper addresses families with long chain fatty acids, those with short chain fatty acids and those with unusual double bond positions such as petroselinic acid and γ-linolenic acid. While dealing with these acids, other associated lipid structures (cyanolipids and acetotriglycerides, for example) and unusual fatty acids (hydroxy– and epoxy–containing acyl groups) are brought together and used as chemotaxonomic markers to relate lipid chemistry with phytotaxonomy.

Lipid chemists recognize the existence of over 500 different fatty acids (1), many from the plant kingdom. This number appears to be the ultimate in natural chemical diversity. However, considering that only about 10,000 of the approximately 250,000 species of plants have been studied, we realize that the list of different fatty acids probably represents only the tip of the proverbial iceberg. In general, we have little idea why one plant produces long chain fatty acids and another biosynthesizes short chain acyl groups in the same environment. In fact, we just are beginning to understand the role that fatty acids and other lipid structures play in the life cycle of the plant. However, we consistently find these compounds to be markers that point to a particular genus, family or order. In this paper, the multitude of lipid structures is indicated, and an attempt to identify compounds that are useful as chemotaxonomic markers is made.

This paper will not deal with the relationships of plants with seed oils with usual fatty acids (C_{16} and C_{18} chain length and double bonds only in the Δ-9, 12 or 15 position). It will be restricted to those plants that produce seed oils with unusual double bond locations, unusual chain length or oxygenated functional groups.

LONG CHAIN FATTY ACIDS

Sapindaceae

When examining analytical data of seed oils from plants of this family, the large levels of C_{20} fatty acids present are most striking (2). We analyzed 25 different species, and of these 19 had more than 10% C_{20} fatty acids. One, *Nephelium lappaceum*, had 33% 20:0, and 12 species had more than 30% 20:1. Two had 20:1 contents above 60%. This correlation is very high when talking about botanical classification and chemotaxonomy. However, this family has one additional chemical key, the cyanolipids (3). Although 28 out of 57 species examined contained cyanolipids, no evidence of cyanolipids has been found outside of the Sapindaceae. The basis of these cyanolipids is a five carbon nitrile skeleton with a double bond and either one or two hydroxyl groups. The hydroxyl groups usually are esterified with a fatty acid (3).

Cruciferae

When we consider long chain fatty acids, the Cruciferae certainly are the family that comes first to mind. Most species of the family examined have more than 10% erucic acid (22:1) in their seed oils. However, we have found 27 species in our collection alone that have more than 20% 20:1. The highest is in *Selenia grandis* (4) with 59% 20:1. The genus *Lunaria* is rich, up to 27%, in nervonic acid (24:1) and has more than 65% in combined 22:1 and 24:1. One member of this family, *Cardamine graeca* (5), has the most 24:1 found in a seed oil, 54%. Many other species in the family have up to 10% of this fatty acid.

The long chain character of the fatty acids produced by species in this family not only is expressed with simple unsaturation but also as hydroxy fatty acids. The latter are found principally in plants of the genus *Lesquerella* as lesquerolic (14-hydroxy-*cis*-11-eicosenoic) acid (6) and its dienoic analog auricolic (14-hydroxy-*cis*-11, *cis*-17-eicosadienoic) acid (7). We find lesquerolic acid at levels exceeding 70% in seed oils of this genus. Auricolic acid is present at the 32% level in *L. auriculata* seed oil. The C_{18} homologs of these two acids also are found in seed oils of this genus. Densipolic (12-hydroxy-*cis*-9, *cis*-15-octadecadienoic) acid (8) is found at the 51% level in *Lesquerella densipila* seed oil. All *Lesquerella* seed oils have small amounts of ricinoleic acid.

Lesquerolic acid and the related hydroxy acids are present in two other cruciferous species, *Physaria floribunda* and *Heliophila amplexicaulis* (9). The seed oil from the latter species had some additional unusual characteristics. All lesquerolic acid hydroxyl groups were esterified with other long chain fatty acids, thereby producing estolides. These same structures exist in the seed oil of *Lesquerella auriculata*, in which calculations showed that some glycerides had as many as six fatty acids per molecule.

The seed oil from another crucifer, *Cardamine impatiens*, also exhibited the presence of these types of compounds (10), that is, those having a hydroxyl group, chain length exceeding 18 carbons and the hydroxyl group being acylated. In this case, the fatty acids were *erythro*-13, 14-dihydroxydocosanoic acid and *erythro*-15, 16-dihydroxytetracosanoic acid with one of the hydroxyl groups per molecule acetylated. These unusual components comprise more than 25% of the seed oil.

Pittosporaceae

Our examination of seed oils from five different species of the Pittosporaceae showed a range of 23-36% 20:1 and 7-22% 22:1. Recently Stuhlfauth et al. (11) reported the composition of 10 species of this family and also found large quantities of C_{20} and C_{22} acids. No forthcoming analyses for this family have been showing oil composition without long chain fatty acids. Of interest is the finding that seeds of some species contain relatively large amounts of low molecular weight hydrocarbons.

Tropaeolaceae

This family, though sparsely analyzed, has a species, *Tropaeolum majus* (12), yielding seed oil with the largest content of erucic acid known (79 mol%). Another species, *T. speciosum* (13) has 17%-22:1, 42%-24:1, and 8%-26:1, all in the ω-9 series of fatty acids. The 24:1 content is second only to that found in the seed oil of *Cardamine graeca*, mentioned above in the Cruciferae section.

Limnanthaceae

As to seed oil compositions of species from Limnanthes, one of the only two genera in this small family has been determined. These seed oils contain long chain fatty acids, but they are different. Most of these long chain fatty acids have one double bond in the Δ-5 position. The major fatty acid is *cis*-5-eicosenoic acid (14), ranging from about 56–67% in *L. douglasii*. Two other Δ-5 fatty acids are present in these oils, *cis*-5-docosenoic (14) and *cis*-5, *cis*-13-docosadienoic (15) acids. These acids can make up 30% of the seed oil. Erucic acid also is present in these oils. Acids longer than C_{18} make up more than 95% of the acyl groups in the genus *Limnanthes*.

Proteaceae

While the Proteaceae are not exceptionally rich in long chain acids, some of their acids are unique to this family (16–18). These acids have double bonds in unusual locations. For example, *Hicksbeachia pinnatifolia* (16) seed oil contains monoenoic C_{20} acids with double bonds varying from Δ-9 through Δ-15. In *Grevillea decora* (18) seed oil, we found a series of ω-5 acids with chain lengths from C_{14} to C_{30}, with the greatest concentration (11%) of these as C_{26} acids. This species also yielded monohydroxy acids with ω-5 unsaturation and chain lengths of 22–30 carbon atoms.

Short and Medium Chain Fatty Acids

Up to this point in our survey, we have discussed the ability of plants to produce long chain fatty acids. I now would like to discuss the other end of the spectrum, the plants that yield acyl groups with less than 14 carbon atoms. Though species in many families such as Compositae, Convolvulaceae, Leguminosae and Meliaceae contain relatively small amounts of short chain acids, I report only on those families that have species producing seed oils with large quantities of these acyl groups.

Celastraceae

The plants from the family Celastraceae produce seed oils with about one third of the acyl groups as acetic acid (19). These acetyl groups are triglyceride substituents and esterified to the *sn*-3 position of the glyceride molecule, therefore, these molecules optically are active. The Celastraceae species examined contain 68–98% of their triglycerides as acetotriglycerides (19). Other families that also have species with seed oils rich in acetotriglycerides are Lardizabalaceae (*Akebia* ssp.) (19) and Balsiminaceae (*Impatiens edgeworthii*) (20). The Ranunculaceae (*Adonis aestivalis*) and Rosaceae (*Sorbus* ssp.) store these com-

pounds, but to a lesser degree (19). Two species from the Polygalaceae, *Polygala virgata* (21) and *Securidaca longipedunculata* (22), produce seed oils with acetotriglycerides as constituents. In these oils, the acetyl group is in the *sn*-2 position.

Lauraceae

Perhaps the richest source of medium chain fatty acids, particularly lauric acid, is the Lauraceae (2). The highest percentage of lauric acid that we found in this family was from *Litsea stocksii* (94%). The next highest concentration was in *Actinodaphne hookeri* with 90%. However, in terms of greatest amount of 12:0 in the seed, *A. hookeri* far exceeds *L. stocksii* because of the former's greater seed oil content (71% vs 32%). Many of the Lauraceae seed oils also contain large amounts of capric acid. In the Lauraceae, our analyses show that the largest concentrations of this acid are from *Sassafras albidum* (68%) and *Cinnamonum camphora* (56%).

All *Lindera* species reported contain lauric acid. However, a few species in this genus contain *cis*-4 analogs of this acid and also of capric and myristic acids (23). These acids are unique to this genus (24).

Palmae

The Palmae are prominent commercial sources of short chain fatty acids (25). *Elaeis guineensis* (coconut) is the major source of lauric acid for the production of a large variety of industrial products. The concentration of 12:0 in coconut oil varies from 45–52%, and the coconut probably is as good a source of this acid as any other species in this family.

Lythraceae

One genus in the family Lythraceae, *Cuphea*, is a source of short chain fatty acids (26–29). Some species in this genus have as much as 84% 12:0, others 95% 10:0 and still others 73% 8:0. These short chain acids are very useful as chemotaxonomic markers of this genus even though a few primitive species produce only C_{16} and C_{18} acids. We recently examined 25 genera, other than *Cuphea*, in this family and found none with major amounts of short chain fatty acids (30).

OTHER FAMILIES WITH SPECIES PRODUCING SHORT CHAIN ACIDS

A number of families have species, such as *Aglaia cordata* (Meliaceae) (31), *Otophora sp.* (Sapindaceae), and *Klainedoxa gabonensis* (Simaroubaceae) that produce short chain fatty acids. These species certainly are not representative of their family. There are other families that have not been surveyed extensively but those species analyzed give the impression that short chain acids are common within the family. This appears to be the situation with the Salvadoraceae. Only two species in the genus *Salvadora* (2) have been analyzed, but both have more than 40% 12:0 and 25% 14:0. Several Ulmaceae have more than 60% 10:0, but others have only the usual C_{16} and C_{18} acids (2,32).

UNUSUAL C$_{16}$ AND C$_{18}$ ACIDS

Umbelliferae

The seed oils of the Umbelliferae long have been known to contain petroselinic (cis-6-octadecenoic) acid. The average content of petroselinic acid in the 250 species we have analyzed is 57% of the acyl groups (33). The greatest concentration of this acid was 87% in Apium leptophyllum. Petroselinic acid is absent in only a few species in this family, thus this acid provides an excellent chemotaxonomic relationship, i.e., all species yielded petroselinic acid as a seed oil constituent (33). Indeed, the Umbelliferae, Umbelliflorae, display a strong chemotaxonomic relationship, i.e. all species yielded petroselinic acid as a seed oil constituent (33). Indeed, some of the richest sources of this oleic acid analog were found in these two families. However, three other families within this order, Alangiaceae, Nyssaceae and Davidiaceae, stored no petroselinic acid. We found one species in the Cornaceae, also in this order, that produced petroselinic acid. The remaining seven species analyzed from this family stored none of this unusual acid (33).

Labiatae

The most distinctive fatty acid found in the seed oils of the Labiatae is the allenic acid, laballenic (5,6-octadecadienoic) (34). This acid has been found only in this family and is associated primarily with plants in the subfamily Stachyoidae (35). The highest level found is 22% in Eremostachys speciosa (35). An analog of laballenic acid, lamenallenic (octadeca-5, 6-trans-16-trienoic) acid (36), has been found in two species of the genus Lamium, L. amplexicaule and L. purpureum. One other Lamium species, L. moschatum, showed only the simple 5,6-allenic acid.

Plants of the genus Teucrium also produced seed oils with unusual C$_{18}$ fatty acids. In this case, they are a mixture of all-cis-5,9,12-18:3 and trans-5,cis-9,cis-12-18:3 (37). We have analyzed 16 seed oils from this genus, and these components totaled from 4–15% of the acyl groups. Packed columns were used for the analysis, which did not separate these two fatty acids. However with Teucrium depressum, countercurrent distribution separation showed that the cis isomer was 6.7% and the trans was 2.0% of the total seed oil fatty acids (37).

This family also produced three α-hydroxy acids. α-Hydroxy linolenic was found in the seed oils of Thymus vulgaris (38) and Salvia nilotica (39), and both α-hydroxy oleic and α-hydroxy linoleic were detected in S. nilotica (39).

Ranunculaceae

The first discovery of trans-5,cis-9,cis-12-octadecatrienoic acid was from Thalictrum polycarpum (Ranunculaceae) (40). The seed oil of this species contains 35% of this unusual acid. Since that original find, we have examined a number of species in this family and found one seed oil, which is from Aquilegia alpina, with as much as 59% of this acid. Many of the species that produce the 5,9,12-18:3 also produce as much as 12% of the 18:2 analog, trans-

5,cis-12-octadecadienoic acid (41).

This family yields species that produce all-cis-5, 11,14-20:3 and all-cis-5,11,14,17-20:4 (42). Species in this family that bear seed with significant amounts of cis-5-20:1 and all-cis-6,9,12-18:3 (γ-linolenic acid) also are found. However, γ-linolenic acid has been found in only one species (Anemone cylindrica) (43) in this family and certainly is not a chemotaxonomic marker for this botanical unit.

Boraginaceae

γ-Linolenic acid might be considered as a chemotaxonomic indicator for the Boraginaceae family. Of the species on which we published seed oil fatty acid composition (44–46), only nine out of 64 did not contain γ-18:3. Borago officinalis had the largest amount of this acid with 21%. Along with the γ-18:3, many species produced the tetraene analog, the all-cis-6,9,12,15-octadecatetraenoic acid. Up to 17% of it was present in the seed oil of Lappa redowski.

Onagraceae

γ-Linolenic acid also is in Onagraceae seed oils (47). However, of the seven different genera of this family examined we found this triene only in the genus Oenothera, but not in every species within that genus. The greatest amount of γ-18:3 was in Oenothera biennis, Evening Primrose (10%), which is a commercial source of γ-linolenic acid.

Several other families include species with seed oils containing γ-linolenic acid. Examples are Aceraceae (Acer) (48) and Saxifragaceae (Ribes) (46).

OTHER CHEMICAL DIVERSITY

Compositae

No one fatty acid or lipid structure characterizes this complex family. However, several acids are prevalent in or unique to this botanical unit. One of these acids is dimorphecolic (9-hydroxy,trans-10,trans-12-octadecadienoic) acid (49). This hydroxy-conjugated dienoic compound is found at levels up to 75% only in the Compositae, mostly in the genera Dimorphetheca, Castalis and Osteospermum. Another fatty acid unique to this family and the genus Calendula is calendic (trans-8, trans-10, cis-12-octadecatrienoic) acid (50). This conjugated trienoic acid, an isomer of eleostearic acid, makes up about 64% of the fatty acids in the seed of C. tomentosa. With few exceptions, the trans-3 series of fatty acids (trans-3-16:1 [51], trans-3-18:1 [52], trans-3, cis-9-18:2 [53], and trans-3,cis-9,cis-12-18:3) (54) is found only in the seed oils of the Compositae. Crepis foetida seed was the first known source of crepenynic (cis-9-octadecen-12-ynoic) acid (55). This acid is present in all 12 species of Crepis analyzed. Crepis alpina seed oil contained the greatest concentration of this acetylenic fatty acid, 75%. Although all species of this genus contain crepenynic acid, half of them produce large amounts of the epoxymonoenoic acid, vernolic (cis-12,13-epoxy-cis-9-octadecenoic) (56). Vernolic acid certainly is not restricted to the Compositae; several species rich in this acid are found in this family. All old-world species of genus *Vernonia* contain significant amounts of this acid, the highest percentage being 80% in seed oil from *Vernonia galamensis*. Other Compositae that contain large amounts of vernolic acid in seed oils are *Erlangea tomentosa* (52%) (57), *Schlectendalia luzulaeflora* (46%) (58) and *Stokesia laevis* (74%) (58).

Euphorbiaceae

Plants from the Euphorbiaceae also produce seed oils rich in vernolic acid. From this family, the principal sources are *Cephalocroton joppica*, 62% (59); *C. peuschelii*, 72% (9); *C. cordofanus*, 61%; and *Euphorbia lagascae*, 70% (60). We also found the C_{20} homolog alchornoic (*cis* 14, 15-epoxy-*cis*-11-eicosenoic) at the 50% level in *Alchornea cordifolia* (61). This family is well-known for two commercial oils, castor (*Ricinus communis*) and tung (*Aleuites fordii*). Castor, the source of the hydroxy acid, ricinoleic (12-hydroxy-*cis*-9-octadecenoic), has no competitor in the Euphorbiaceae. In the 300 species analyzed in this family, we found only a few species that contained low levels of ricinoleic acid. Another hydroxy acid was reported from *Mallotus* ssp. Its structure, 18-hydroxy α-eleostearic (kamlolenic) acid, differs markedly from that of ricinoleic (62). The oil from *Mallotus philippinensis* also is unusual because it is made up primarily of estolides. In this case, the hydroxy acid is esterified with another kamlolenyl group. The seed oil of *Trewia nudiflora* has a similar composition (63).

Tung oil contains up to 82% α-eleostearic (*cis, trans, trans*-9,11,13-octadecatrienoic) acid. Although this acid is not found in many species in this family, we found that *Baliospermum montanum* seed oil contained 83% conjugated trienoic acids. Several other species in this family had much smaller amounts.

Two genera in the Euphorbiaceae, *Sapium* (64) and *Sebastiana* (65), contain seed oils that have a unique structure formed by unique acyl groups. These oils are a mixture of about two thirds usual triglycerides and one third estolide—containing glycerides. The estolide consists of two short chain acids, one an ω-hydroxy allenic C_8 acid (8-hydroxy-5,6-octadienoic acid) and the other a conjugated C_{10} acid (*trans*-2,*cis*-4-decadienoic acid).

Linaceae

The family Linaceae is notable for flax or linseed (*Linum usitatissimum*) which produces the highly unsaturated drying linseed oil. However, we found one species in the genus *Linum* with seed oil containing about 15% ricinoleic acid (66). This report led A.G. Green (67) to reinvestigate the composition of other *Linum* species. He found that all five *Linum* species in the section Syllinum produced ricinoleic acid. This hydroxy acid appears to be a chemotaxonomic indicator for this botanical group.

Apocynaceae

The 9-hydroxy-*cis*-12-ene isomer of ricinoleic acid (isoricinoleic) (68) is present in large quantities in the seed oils of several Apocynaceae species (69). Over 73% of the seed oil of *Holarrhena antidysenterica* is isoricinoleic

acid. *Wrightia tinctoria* also showed over 59% of its seed oil acyl groups to be this hydroxy acid. However, we and others have found that other samples of these same species have no isoricinoleic present. In our experience, species generally are consistent in their composition from place to place and from time to time. However, these species are either an exception or misidentified or environmental conditions affect these species more than others.

DISCUSSION

This paper is not intended to be an all-inclusive report on all fatty acids and lipid structures in seeds; it is intended to bring to mind the relationships between fatty acid composition and the botanical classification. Perhaps more importantly, I would like it to show that the plant kingdom is very diverse in the lipid compounds it biosynthesizes. Also, a particular structure may be found in an obscure plant with very little crop potential, but it still might be useful to man. Modern plant breeding, recombinant DNA procedures or immobilized enzyme techniques might be used to improve the species, to transfer this characteristic to another plant that can be grown as a commercially successful venture or to bypass the plant totally and produce the desired lipid in vitro.

REFERENCES

1. Gunstone, F.D., in *Comprehensive Organic Chemistry*, 1979, pp. 587–632.
2. Hilditch, T.P., and P.N. Williams, *The Chemical Constitution of Natural Fats*, 4th Edition, John Wiley & Sons, Inc., New York, NY, 1964.
3. Mikolajczak, K.L., *Prog. Chem. Fats Other Lipids 15*:97 (1977).
4. Mikolajczak, K.L., T.K. Miwa, F.R. Earle and I.A. Wolff, *J. Am. Oil. Chem. Soc. 38*:678 (1961).
5. Jart, A., *J. Am. Oil. Chem. Soc. 55*:873 (1978).
6. Smith, C.R., T.L. Wilson, T.K. Miwa, H. Zobel, R.L. Lomar and I.A. Wolff, *J. Org. Chem. 26*:2903 (1961).
7. Kleiman, R., G.F. Spencer, F.R. Earle, H.J. Nieschlag and A.S. Barclay, *Lipids 7*:660 (1972).
8. Smith, C.R., T.L. Wilson, R.W. Bates and C.R. Scholfield, *J. Org. Chem. 27*:3112 (1962).
9. Plattner, R.D., K. Payne-Wahl, L.W. Tjarks and R. Kleiman, *Lipids 14*:576 (1979).
10. Mikolajczak, K.L., C.R. Smith, Jr. and I.A. Wolff, *Lipids 1*:289 (1966).
11. Stuhlfauth, T., H. Fock, Huber and K. Klug, in *Biochemical Systematics and Ecology 13*:447 (1985).
12. Harlow, R.D., C. Litchfield and R. Reiser, *Lipids 1*:216 (1966).
13. Litchfield, C., *Lipids 5*:144 (1970).
14. Smith, C.R., M.O. Bagby, T.K. Miwa, R.L. Lohmar and I.A. Wolff, *J. Org. Chem. 25*:1770 (1960).
15. Bagby, M.O., C.R. Smith, T.K. Miwa, R.L. Lohmar and I.A. Wolff, *J. Org. Chem. 26*:1261 (1961).
16. Vickery, J.R., *Phytochemistry 10*:123 (1971).
17. Plattner, R.D., and R. Kleiman, *Phytochemistry 16*:255 (1977).
18. Kleiman, R., R.B. Wolf and R.D. Plattner, *Lipids 20*:373 (1985).
19. Kleiman, R., R.W. Miller, F.R. Earle and I.A. Wolff, *Lipids 2*:473 (1967).
20. Bagby, M.O., and C.R. Smith, Jr., *Biochim. Biophys. Acta 137*:475 (1967).
21. Smith, C.R., R.V. Madrigal, D. Weisleder and R.D. Plattner, *Lipids 12*:736 (1977).
22. Smith, C.R., Jr., R.V. Madrigal and R.D. Plattner, *Biochim. Biophys. Acta 572*:314 (1979).
23. Hopkins, C.Y., M.J. Chisholm and L. Prince, *Lipids 1*:118 (1966).
24. Wang, J., S. Meng and J. Li, *Zhiwu Xuebao 27*:117 (1985).
25. Idiem'opute, F., *J. Am. Oil Chem. Soc. 56*:528 (1979).

26. Miller, R.W., F.R. Earle, I.A. Wolff and Q. Jones, *J. Am. Oil Chem. Soc. 41*:279 (1964).
27. Graham, S.A., F. Hirsinger and G. Robbelen, *Am. J. Bot. 68*:908 (1981).
28. Wolf, R.B., S.A. Graham and R. Kleiman, *J. Am. Oil Chem. Soc. 60*:27 (1983).
29. Graham, S.A., and R. Kleiman, *J. Am. Oil Chem. Soc. 62*:81 (1985).
30. Graham, S.A., and R. Kleiman, *Biochem. System. Ecol.*, in press.
31. Kleiman, R., and K.L. Payne-Wahl, *J. Am. Oil Chem. Soc.61*:1836 (1984).
32. Ihara, S., and T. Tanaka, *J. Am. Oil Chem. Soc. 55*:471 (1978).
33. Kleiman, R., and G.F. Spencer, *J. Am. Oil Chem. Soc. 59*:29 (1982).
34. Bagby, M.O., C.R. Smith and I.A. Wolff, *J. Org. Chem. 30*:4227 (1965).
35. Hagemann, J.M., F.R. Earle and I.A. Wolff, *Lipids 2*:371 (1967).
36. Mikolajczak, K.L., M.F., Rogers, C.R. Smith, Jr. and I.A. Wolff, *Biochem. J. 105*:1245 (1967).
37. Smith, C.R., R.M. Freidinger, J.W. Hagemann, G.F. Spencer and I.A. Wolff, *Lipids 4*:462 (1969).
38. Smith, C.R., and I.A. Wolff, *Lipids 4*:9 (1969).
39. Bohannon, M.B., and R. Kleiman, *Lipids 10*:703 (1975).
40. Bagby, M.O., C.R. Smith, Jr., R.L. Mikolavczak and Z.A. Wolff, *Biochemistry 1*:632 (1962).
41. Bhatty, M.K., and B.M. Craig, *Can. J. Biochem. 44*:311 (1966).
42. Smith, C.R. Jr., R. Kleiman and I.A. Wolff, *Lipids 3*:37 (1968).
43. Spencer, G.F., R. Kleiman, F.R. Earle and I.A. Wolff, *Lipids 4*:99 (1969).
44. Kleiman, R., F.R. Earle and I.A. Wolff, *J. Am. Oil Chem. Soc. 41*:459 (1964).
45. Miller, Roger Wayne, F.R. Earle and I.A. Wolff, *Lipids 3*:43 (1968).
46. Wolf, R.B., R. Kleiman and R.E. England, *J. Am. Oil. Chem. Soc. 60*:1858 (1983).
47. Riley, J.P., *J. Chem. Soc.*, 1949, p. 2728.
48. Bohannon, M.B., and R. Kleiman, *Lipids 11*:157 (1976).
49. Smith, C.R. Jr., T.L. Wilson, E.H. Melvin and I.A. Wolff, *J. Am. Chem. Soc. 82*:1417 (1960).
50. Chisholm, M.J., and C.Y. Hopkins, *Can. J. Chem. 38*:2500 (1960).
51. Hopkins, C.Y., and M.J. Chisholm, *Can. J. Chem. 42*:2224 (1964).
52. Kleiman, R., F.R. Earle and I.A. Wolff, *Lipids 1*:301 (1966).
53. Morris, L.J., M.O. Marshall and E.W. Hammond, *Lipids 3*:91 (1968).
54. Bagby, M.O., W.O. Siegl and I.A. Wolff, *J. Am. Oil Chem. Soc. 42*:50 (1965).
55. Mikolajczak, K.L., C.R. Smith, M.O. Bagby and I.A. Wolff, *J. Org. Chem. 29*:318 (1964).
56. Earle, F.R., A.S. Barclay and I.A. Wolff, *Lipids 1*:325 (1966).
57. Phillips, B.E., C.R. Smith, Jr. and J.W. Hagemann, *Lipids 4*:473 (1969).
58. Earle, F.R., *J. Am. Oil. Chem. Soc. 47*:510 (1970).
59. Gunstone, F.D., and P.J. Sykes, *J. Sci. Food Agric. 12*:115 (1961).
60. Kleiman, R., C.R. Smith, S.G. Yates and Q. Jones, *J. Am. Oil Chem. Soc. 42*:169 (1965).
61. Kleiman, R., R.D. Plattner and G.F. Spencer, *Lipids 12*:610 (1977).
62. Chisholm, M.J., and C.Y. Hopkins, *J. Am. Oil Chem. Soc. 43*:390 (1966).
63. Madrigal, R.V., and C.R. Smith, *Lipids 17*:650 (1982).
64. Sprecher, H.W., R. Maier, M. Barber and R.T. Holman, *Biochemistry 4*:1856 (1965).
65. Heimermann, W.H., and R.T. Holman, *Phytochemistry 11*:799 (1972).
66. Kleiman, R., and G.F. Spencer, *Lipids 6*:962 (1971).
67. Green, A.G., *J. Am. Oil Chem. Soc. 61*:939 (1984).
68. Gunstone, F.D., *J. Chem. Soc.*, 1952, p. 1274.
69. Powell, R.G., R. Kleiman and C.R. Smith, *Lipids 4*:450 (1969).

Development of New Industrial Oil Crops

Gerhard Röbbelen
Institute of Agronomy and Plant Breeding, Georg-August-University, Göttingen, Federal Republic of Germany

Unusual fatty acids such as those present in the seed oils of rapeseed, coconut, palm kernel, tung and castor have been utilized as raw materials since the early days of the chemical industry. To date, these oils have formed a stable and valuable, although still relatively small, fraction of the world's oil market. Changes in agricultural production, diversification of industrial applications and the unreliable supplies of mineral oils as petrochemical raw materials prompted a search for new potential oilseed crops possessing sufficient agronomic performance for economical production of seed oils with unusual fatty acids. The first large screening program for new industrial plant oils was started in 1959 by the USDA, but it was not until the 1970s that interest increased in intensified agronomic investigations for the production of selected plant species.

This report summarizes over 12 years of research on the domestication of *Cuphea*, a herbaceous summer annual taxon native to Central America. Its seed oil contains large amounts of medium chain fatty acids with 8, 10 or 12 carbon atoms. In addition, the report deals with investigations to select new potential oilseed crops for European production of unusual fatty acids, i.e., $\Delta 6,7$-monoenoic, conjugated, acetylenic, hydroxy and epoxy fatty acids. In general, the existing knowledge supports the conclusion that today various options of biotechnology exist for breeding plant varieties with improved yield and performance and in particular for developing new industrial oil crops.

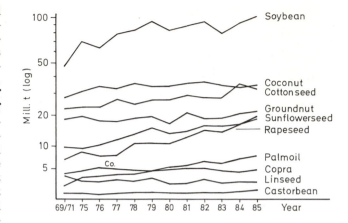

FIG. 1. World production of the most important oil crops (1).

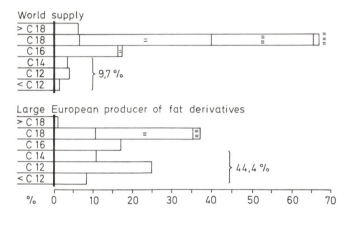

FIG. 2. Average fat composition: world supply of fatty acids and consumption by a large European producer of industrial fat derivatives (2).

CROPS FOR INDUSTRIAL OILS AND FATS PRODUCTION

More than 90% of the world's vegetable oil production is directed to food uses (Fig. 1). Many of these oils are composed of triglycerides with C_{18} fatty acids. Many of them are cultivated in tropical and subtropical countries in which they constitute the second most important export commodity after luxuries. Therefore, changes in their production can have serious consequences for the economy, development and even politics of many developing nations. But, a reliable supply of such raw materials can be of considerable significance for developed economies, too.

Less than 10% of all plant oils go into nonfood uses. In this case, industrial consumption often covers the same products that are used for food, sometimes by using tainted oils rejected for the food market. However, certain industrial demands are unique and irreplaceable (Fig. 2). This is determined by specific qualities of the respective oils and fats, mainly by higher contents of unusual fatty acids with either shorter or longer chain lengths, or by additional functional groups in the fatty acid molecules (2).

The classical example of an industrial oil crop is the castor bean, *Ricinus communis*. In the wild, this species forms a tree but now it is grown as an annual lush crop in tropical and subtropical regions. Nearly 60% of the present world market is exported from Brazil. Castor oil contains more than 90% ricinoleic acid (12 hydroxy-*cis*-9-octadecenoic acid), which this plant synthesizes from the common linoleic acid by oxidation to form a hydroxy group at the Δ-12 position. Ricinoleic acid easily can be converted to a dienoic fatty acid with conjugated double bonds. This rapidly polymerizes in fast drying paints, but darkens less than linolenic acid products such as those from linseed oil. At 275 C in alkaline medium, ricinoleic acid is split into sebacic acid for synthetic fiber and resins and into octanol useful as a solvent for the synthesis of perfumes. Many more such reactions are utilized worldwide, all depending on the reactivity of the Δ-12 hydroxy group (3).

Another traditional source of industrial oils is the tung tree, which yields elaeostearic acid (*cis, trans, trans*-9, 11,13-octadecatrienoic acid), i.e., a trienoic acid with three conjugated double bonds, serving similar purposes. Like castor bean, this genus *Aleurites* belongs to the *Euphorbiaceae* family. Its seed contains 40–58% oil. Tung trees are native to China but *A. fordii* has been under experimental cultivation in the southern states of the U.S. and in Mexico. Adult plantations yield up to 1.5 MT oil/ha/year (4), which is double an average rapeseed harvest, but the oilcake with 20–25% protein is toxic and can be used only as a fertilizer. The perennial nature also gives little production flexibility for changing demand. Therefore, soybean oil, linseed oil, castor oil and synthetic products are important competitors. They have caused stagnation in the production of traditional sources of industrial vegetable oils and fats during the last 50 years, regardless of steeply increasing demands. In many instances, it has been more economical to develop new processes for the chemical derivatization of a cheaply and widely available fatty raw material such as soybean oil than to develop a new oilseed crop for a specific nonfood use. Thus, it should be borne in mind that on the whole the recent developments towards renewable crop resources for industrial uses come more from the problems of agricultural surplus production in the developed countries than from urgent requirements of the chemical industry. However, a certain industrial interest for specific fatty raw materials does exist. In the Federal Republic of Germany, the last major production of industrial seed oil disappeared in 1974 with the total shift to zero erucic rapeseed varieties. At the same time the Henkel Company initiated a research and development program for the purpose of developing new plant sources for lauric acid (a C_{12} fatty acid).

CUPHEA, THE FIRST ANNUAL SOURCE OF MCT PLANT OILS

Proposition of the Development Program

For a long time, there has been a chronic deficiency in the world market of vegetable oils and fats containing medium chain triglycerides (MCT fats; Fig. 2). Because of a limited and unreliable supply of coconut and palm kernel oils, representing the only major sources of such "laurics," Stein (2) formulated the aims of breeding for industrial oil crops with greatest emphasis on the development of "crops with lauric acid in other species than coconut, in annual plant species and in higher contents than in the coconut."

In the early '60s, chemists of the USDA had discovered that the genus *Cuphea* of the *Lythraceae* family effectively stores MCT in its seeds (5). This genus includes a large number of annual, herbaceous species (6). It is native to Central America and the subtropics of Brazil but does not occur in Europe. It is the only genus of seed plants except palm trees that exhibits a wide array of fatty acids with medium chain fatty acids in seed oils harvested from annual plants. Therefore, in 1974 Hirsinger (7-9) started with growth studies of *Cuphea* in the greenhouse and fields in Göttingen. From more than 150 accessions of 45 wild *Cuphea* species, some that might have sufficient potential for agricultural production were found. These produced seeds with a 1,000-seed weight close to that of rapeseed, and they reached oil contents of 30% and more. Obviously, these *Cuphea* species met the requirements stated by Stein (2) for a new industrial oilseed crop producing lauric acid.

However, all the available *Cuphea* species exhibit wild plant characteristics that severely impede agricultrual production, such as seed dormancy conditioned by an impermeable seed coat, unique seed hairs that project with seed-moistening, a slow, nonuniform growth of the seedlings and thus tardy establishment of field stands, sticky glandular hairs on stems and flowers, a continuous flowering and unequal seed ripening and, in particular, early seed shattering. For this reason, breeding programs were started (10) with the aim of identifying useful genetic variation in the wild origins or after mutagenesis and to develop genotypes from these that might be adapted to agricultural production.

Variation of Existing Germplasm

The most remarkable characteristic of the genus *Cuphea* is the ability to produce in its seed oils unusually high quantities of single medium chain fatty acids that may range in length from the usual 18 down to 8 C atoms in different species (Table 1). By grouping the species according to their dominant fatty acid, an evolutionary trend from the C_{18} unsaturated linoleic acids as a major lipid component in the archaic species to the shorter-chained saturated capric and caprylic acids with C10:0 and C8:0 in the taxonomically most developed forms was recognized by following the taxonomic pedigree (Fig. 3) established earlier by Koehne (6). This finding leads us to assume that during evolution of the genus *Cuphea*, mutations had occurred in regulatory genes that caused the fatty acid production in the seeds to cease at progressively earlier stages, resulting in the accumulation of large amounts of single fatty acids of progressively shorter carbon chain lengths (11). Such variation provides most valuable raw materials not only for industrial chemistry but also for the cell scientists to study details of the enzymology for plant fatty acid biosynthesis or even to identify genes responsible for the production of the shorter chained fatty acids (12,13).

However, seed oil quality is not the only economic essential. An oil crop also needs good adaptability for agricultural production. As expected, *Cuphea* species differ widely in such performance traits (14,15). In Table 2, selected character traits associated with growth and seed yield are listed for a few *Cuphea* species that store lauric acid in their seeds. Large differences are evident in plant height and vigor, in flowering period and in flower distribution. Different seed number and 1,000-seed weight result in large differences of seed yield. From comparisions derived from such evaluations in Göttingen (14), some species, in particular those belonging to the section *Heterodon*, were determined to have a sufficiently high performance for further development.

In years with a late spring and/or a cool, rainy autumn, *Cuphea* development in the field was so poor in Göttingen

TABLE 1

Fatty Acid Composition in *Cuphea* Seed Lipids (11)[1]

Species	8:0	10:0	12:0	14:0	16:0	18:1	18:2
C. racemosa	-	-	0.1	0.2	15.3	17.4	58.8
C. palustris	19.7	1.4	2.0	63.7	6.7	3.0	2.9
C. tolucana	-	23.0	63.3	4.5	1.8	1.9	5.0
C. paucipetala	1.2	87.4	2.0	0.8	1.9	1.8	4.0
C. painteri	65.0	24.0	0.2	0.4	2.8	3.3	3.9

[1]Fatty acids in percent of total fatty acid content.

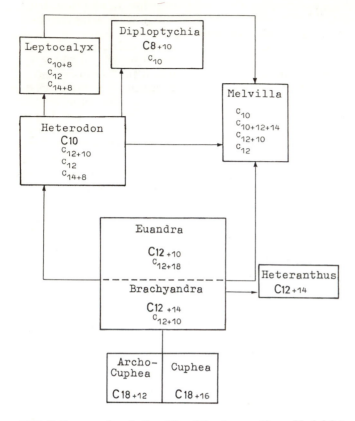

FIG. 3. Taxonomic relationship of *Cuphea* sections. Model fatty acid patterns in bold type; sectional boxes proportional to size of the section (11).

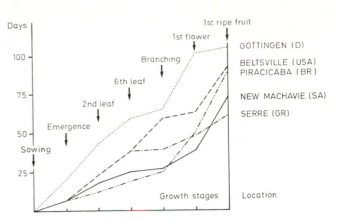

FIG. 4. Sequence of growth stages of *Cuphea wrightii* at five locations in 1980 (16).

(Serre), the first ripe seeds of *C. wrightii* could be harvested 61 days after sowing (Fig. 4). If sowing was done in late March, the ripening seed no longer was impaired by rainfall. Where sufficient water was available, seeds were formed continuously for four months, leading to seed yields of 2 MT/ha and more.

The main restriction for economic *Cuphea* production thus is not poor productivity or lack of ecological adaptability. It is the presence of the above–mentioned wild plant characteristics that prevents satisfying agronomic handling. In some cases, such as for the sticky hairiness of leaves and flowers, spontaneous mutants were found in accessions from wild habitats lacking such glandular hairs. But, for other characteristics, natural selection evidently had been so strong that spontaneous mutants could not survive in nature. In particular, no single species was detected with an indehiscent fruit. Seeds of all *Cuphea* species largely are shed to the ground before harvest. The peculiar shattering mechanism obviously is connected with the unique dorso-ventral flower symmetry

that virtually no seed could be harvested. In order to check for more suitable regions for the production of these subtropic species, cultivation experiments were initiated at 12 locations in eight countries throughout the Americas, Africa and Europe (16). Promising results were obtained from Mediterranean countries. In Greece

TABLE 2

Agronomic Traits of C12 *Cuphea* Species (14,15)

Cuphea species	H/W	Height cm	Early flower 0-6	Seeds/ flower	Seed weight mg	Seed/ plant g	Agron. potential
Sect. *Brachyandra*							
C. carthagenensis	W	41	6	7	0.69	6.7	+
C. parsonsia	W	20	3	6	0.51		-
C. calophylla	W	25	5	11	0.47	1.6	-
Sect. *Heterodon*							
C. glossostoma	H	92	6	7	1.95	9.2	+
C. laminuligera	H	90	6	9	2.13	10.1	+
C. lobophora	W	98	4	11	1.91	4.6	-
C. lutea	H	105	5	5	2.49	9.1	+
C. tolucana	H	69	6	6	1.36	2.3	+
C. wrightii	H	61	6	3	1.61	6.0	+
Sect. *Melvilla*							
C. melvilla	W	400	0	90	1.24	0	-
C. heterophylla	W	58	4	8	4.45	0	-

1) H, herbaceous; W, woody. 2) Greenhouse. 3) Rating 0-6; 0, none; 6, earliest flower.

ubiquitous in the *Cuphea* genus. Radial flowers such as those of the closely related species of *Lythrum* do develop indehiscent fruits.

Variation Induced by Mutagenesis

Experiments were performed to induce genetic variability by chemical mutagenesis and to select for radial flower symmetry. Because of seed availability and plant vigor, presoaked seed of the cross–fertilizing species *C. lanceolata* and *C. procumbens* were the first to be treated with EMS. Although some morphological variants appeared in the M_2 generation after open pollination of the M_1, no stable mutant with indehiscent fruits was detected (Table 3). The experiments were repeated on similar scales, i.e., 2,600 M_1 and 7,400 M_2 plants, with the autogamous *C. calophylla* (=*C. aperta*). This plant is tiny and has a short life cycle that is well-adapted to greenhouse screening. With this model plant, one mutant was found with a definitely improved shattering resistance that was caused by a changed flower symmetry. With this success attained, further mutation experiments were conducted with the productive, self-pollinating species *C. wrightii. C. tolucana* and *C. paucipetala*. More than 20,000 M_1 plants and 80,000 M_2 plants were grown during the past three years. In spite of this enormous effort, the desired indehiscent mutant with radial flowers was not found in these more productive *Cuphea* species. This negative result was quite unexpected because the mutagenic treatment had been effective, in general, if measured by the number of M_1 aberrations or other types of mutants in the M_2, e.g., those with nonviscid hairs, monoculm, dwarf growth.

In order to cope with this problem of early seed dispersal, an alternative solution was envisaged by technical means. A vacuum picking machine has been developed for non-damaging multiple harvests of *Cuphea* stands (9).

Our mutant screening still is being continued today. Fasciated mutants may have some promise in overcoming some of the restrictions of low seed retention not only because of their stiffer stem but even more because of their bunched and deformed flowers. Their tightly positioned fruits are hindered from early opening, and the seeds are kept in the fruits until ripe. In addition, fruit walls may be crooked, which inhibits their being slit open by the growing placenta.

In many cases the most severe drawback to these promising mutants is their reduced seed yield (Table 4). Moreover, as in the fasciata mutant performance also is more affected by adverse environmental conditions than the performance of the original line. The experienced plant breeder is well aware that every gene requires a suitable genetic background to be optimally expressed. A mutated gene cannot be expected to perform at its best within its original genotype. It requires an effective reshuffling of the rest of the genotype by adaptive recombination and selection. Consequently, plant breeders use recurrent selection schedules for improving induced mutants, particularly by inbreeding species to reach the necessary genetic fit of a new mutant locus within an adapted genotype (17).

TABLE 4

Seed Yield of *Cuphea* Lines and Mutants in a Field Test, 1986 (17)

Genotype	Seed yield*	
	abs. (g)	rel. (%)
C. tolucana 629	4.53	100.0
nonsticky hairs	1.72	38.0
red-shoot	0.65	14.3
rose petals	3.78	83.4
fasciated stem	0.59	13.0
C. wrightii 651	2.99	100.0
nonsticky hairs	2.08	69.6
red shoot	0.31	10.4
C. paucipetala C 80	3.63	100.0
nonsticky hairs	1.73	47.7
hairs in stripes	2.17	59.8
red shoot	0.88	24.2
semi-dwarf (rolled leaf)	1.19	32.8
dwarf (leaf deformation)	0.12	3.3
fasciated stem	2.43	66.9

*Based on three harvests of 12 plants/lm².

TABLE 3

Survey of Morphological Mutants of Eight *Cuphea* Species

Section	Brachyandra	Heterodon				
Species *Cuphea*	calophylla	lanceolata	procumbens	paucipetala	tolucana	wrightii
Main f.a.	12:0	10:0	10:0	10:0	12:0	12:0
Mutant 2n =	16	12	18	20	24	44
nonsticky hairs		+	+	+	+	+
no hairs	+			+		
monoculm	+	+				
shoot morphology	+	+		+	+	+
dwarf	+	+	+	+	+	+
fasciated stem	+	+		+	+	
double flower	+	+	+	+	+	
peloria					+	
radial flower	+					

DOMESTICATION FOR PRODUCTION OF OTHER INDUSTRIAL OILS

The abundance of problems that the breeder has to solve for the domestication of a wild or primitive species, as illustrated in our work with *Cuphea*, reappears with different severity whenever new plant species are developed for agricultural production. In its report (18), the U.S. Council for Agricultural Science and Technology published an *Optimistic Timetable for Domesticating a Wild Species*. Seven stages were designated in evaluating, developing and commercializing a new crop: collection and evaluation of germplasm, chemical and utilization studies, agronomic evaluation, breeding work, production and processing scale-up, and commercialization. Cost and time requirements depend on the level of earlier domestication of the species in question, on its ecological adaptation and on its seed yield potential.

We have gained more experience of this kind in a second research endeavor that we conducted during the recent decade in Göttingen to develop plant species that produce fatty acids with additional functional groups in their seed oil (19,20). Such C_{18} fatty acids containing unusual functional groups were investigated in 53 annual plant species, and seven of these were field-tested on larger scales. Table 5 provides examples of the biological variability available in fatty acid biosynthesis. There are fatty acids with unusual unsaturation, i.e., single unsaturation at the Δ-6,7 position, conjugated double bonds, or an acetylenic triple bond, and fatty acids with oxidative variations, i.e., hydroxy or epoxy groups at various positions.

Sources for Production of Petroselinic Acid

Petroselinic acid is an isomer of oleic acid in which the single double bond is shifted from the Δ-9 to the Δ-6 position. This results in an opportunity to generate lauric acid and adipic acid by oxidative ozonolysis. The latter is a valuable product, e.g., for nylon production. Petrose-

TABLE 5

Unusual Fatty Acids in Seed Oils of Annual Plant Species[1]

Elaeostearic acid				
Centranthus macrosiphon				COOH
Calendic acid				
Calendula officinalis				COOH
Parinaric acid				
Impatiens balsamina				COOH
Dimorphecolic acid				
Dimorphotheca pluvialis		OH		COOH
Densipolic acid				
Lesquerella lescurii		OH		COOH
Lesquerolic acid				
Lesquerella gracilis	OH			COOH
Vernolic acid				
Euphorbia lagascae	O			COOH
Crepenynic acid				
Crepis alpina				COOH
Petroselinic acid				
Coriandrum sativum				COOH

[1]Tested in Göttingen, 1983-1985.

TABLE 6

Climatic Adaptation and Agronomic Production Potential in Central Europe of the Most Prospective Oilseed Plant Species of the Present Investigation

Species Main fatty acid (%)	Oil %	Seed		Agron. potential
		yield kg/ha	1000 grain weight (g)	
Coriandrum sativum petroselinic acid (82.0)*	18.0	2800	6.6-18.9	+
Foeniculum vulgare petroselinic acid (71.3)*	14.5	7.1 g /plant	4.5	?
Calendula officinalis calendic acid (62.2)	19.4	1610	9.6	+
Crepis alpina crepenynic acid (74.0)	12.0	1362	1.2	-
Dimorphotheca pluvialis dimorphecolic acid (61.8)	15.8	826	2.7	-
Euphorbia lagascae vernolic acid (60.4)	46.8	685	12.1	?

*Includes oleic.

linic acid occurs in members of the families *Araliaceae* and *Umbelliferae* (21). Four species of the latter were included in the field test in Göttingen.

Coriandrum sativum showed the highest promise for the production of petroselinic acid under German conditions. In the four accessions tested, the agronomic potential appeared to be very high (20). Mechanical sowing and harvest gave no problems, and no special breeding efforts were necessary to initiate production. The oil content was 18.0%, and 82% of all the fatty acids consisted of oleic and petroselinic acid. In a 2 ha field trial with *C. sativum*, the seed yield was as high as 2.8 MT/ha (Table 6). This must reflect earlier selections for spice uses of *Coriandrum*. Further breeding work should be directed to increase the oil content of the fruit and to improve disease resistance, if necessary. In principle, this plant species is ready for immediate use as a new oil crop.

Foeniculum vulgare grows as a perennial or biennial species. All the four genotypes tested in Göttingen showed vigorous growth, but seeds matured too late after fall sowing. Perhaps flowering and harvest would be earlier if the plants were grown as perennials. Plant height was 155 cm; standing ability and seed retention were satisfactory for all genotypes. Mean plant yield averaged 7.1 g, and the weight of 1,000 seeds was 4.5 g. After spring planting in Nebraska, plot yield reached 0.6 MT/ha (22).

Ammi visnaga and *Petroselinum crispum* did not prove to be suitable for agronomic production of petroselinic acid because of their late maturity and their low plant yield and seed size.

Sources for Production of Fatty Acids with Conjugated Double Bonds

As a substitute for tung oil, *Calendula* contains trienoic seed oils, the double bonds that start with the even (8 *trans*, 10 *trans*, 12 *cis*) rather than the uneven positions (9 *cis*, 11 *trans*, 13 *trans*) found in *Aleurites*. In 1983-1985, 23 different accessions of *Calendula officina-*

lis and eight accessions of *Calendula arvenis* were tested in Göttingen. Both species exhibited wide variability in their vegetative growth and in their reproductive characteristics as a consequence of their earlier development as ornamental and medicinal plants. *C. officinalis* offered better agronomic potential than *C. arvensis*, particularly because of its better agronomical suitability and superior seed retention. The highest seed yield, 358 g/m², was harvested by hand from the most productive accessions of *C. officinalis*, corresponding to more than 3 MT/ha. The mean oil content of 12 genotypes tested in 1984 and 1985 was 19.4%; 62.2% of all fatty acids were (*cis, cis, trans*-8,10,12-octadecatrienoic) acid (Table 5). Earle et al. (23) also observed wide variations in the wild populations from South Africa that they tested. However, they concluded that none of the *Calendula* accessions was suitable for modern cultivation and harvest techniques, but our results of field tests in Göttingen prove that there is a good chance for *C. officinalis* to become a successful new crop (20).

All of the other species containing conjugated fatty acids in their seed oil did not show sufficient crop potential. They flowered too late, showed poor seed retention and/or reached a 1,000-seed weight of less than 1 g (19).

Sources for Production of Fatty Acids with an Acetylenic Bond

Of the group producing crepenynic (*cis*-9-octadecen-12-ynoic) acid, *Crepis*, *Lapsana* and *Picris* species were investigated in Göttingen. Only *C. alpina* had satisfactory crop potential with excellent growth habit and achene retention. However, seedling development of *C. alpina* was very slow, and the long and thin seed with its pappus caused difficulties for mechanical sowing and combine harvesting. The 1,000-seed weight was 1.2 g and the plot yield 136 g/m². Oil content was only 12%, but 74% of all fatty acids were crepenynic acid. In the field, 239 achenes were produced per seed head, and under insect–free conditions there were 179. This confirmed data of Babcock (24) that showed *C. alpina* to be highly self-compatible. According to White et al. (25), spring and fall plantings similarly are successful although fall may be the better planting date, resulting in earlier maturity. These authors reported plot yield equivalents of 1.8 MT/ha.

Sources for Production of Hydroxy Fatty Acids

Seed oils from the genus *Dimorphotheca* contain dimorphecolic (9-hydroxy-*trans, trans*-10,12-octadecadienoic) acid. The oil content of *D. pluvialis* was as low as 15.8%, of which about 62% was dimorphecolic acid. *Dimorphotheca* heads are composed of female fertile, male sterile ray flowers and hermaphroditic disk flowers. Achenes developing from the disk flowers are larger and winged; those produced by the ray flowers are small and unwinged (26). The percentage of oil, protein and pericarp, and the fatty acid composition did not differ in the two seed types (23). In Göttingen, the vegetative growth of *D. pluvialis* and *D. sinuata* was satisfactory. Both species hybridize readily when grown in close proximity (27). In 1984, harvest was accomplished by hand picking the ripe seeds once a week, but in 1965 a com-

bine was used successfully after desiccation of the plants with Deiquat five days before harvest (19). On the whole, crop potential and seed retention of the tested genotypes were not sufficient in either species although agronomic performance of *D. pluvialis* was better than that of *D. sinuata*. In Chico, California, Willingham and White (28) reached seed yields from 0.05 MT up to 1 MT per ha.

Lesquerella, a member of the *Cruciferae* that is native to the U.S. was found to contain densipolic (12-hydroxy-*cis, cis*-9,15-octadecadienoic) or lesquerolic (14-hydroxy-*cis*-11-eicosenoic) acid (29). Barclay et al. (30) observed large variations between and within species of the genus. *L. fendleri* prefers a cool and semi-arid climate, sandy soils and perfect drainage. It pioneers on disturbed sites along highways. *L. fendleri* has a strong tolerance of the cold. Gentry and Barclay (31) described seed retention, suitability for mechanical harvest and crop potential as sufficient for agricultural cultivation. But in Göttingen, only a few small seeds could be harvested from all the species tested. Seedlings developed very slowly; plants had poor vigor and not all of them flowered in the first year. 1,000-seed weight was not higher than 0.1-0.2 g in Göttingen, although Barclay et al. (30) reached 0.4–0.5 g. for *L. fendleri*, 0.7 g for *L. gracilis*, 1.0 g for *L. grandiflora*, and 0.8 g for *L. lescurii*.

Sources for Production of Epoxy Fatty Acids

Euphorbia lagascae seeds contain high oil and relatively high epoxy fatty acid contents, which assures great promise of economic uses. During the pioneering survey of the USDA Northern Regional Research Laboratory, Kleiman et al. (32) found only this one among 58 species within the *Euphorbiaceae* with such high contents of epoxy oleic acid. The seed of *E. lagascae* sown in Göttingen was collected from native habitats in Spain. There it flowers in early spring and sets fruits in April and May (33). Seed harvested in Göttingen had a 1000-seed weight of 12.1 g; they contained up to 46.8% oil with 60% vernolic (*cis*-12,13-epoxy-*cis*-9-octadecenoic) acid. Vegetative development of *E. lagascae* was adequate in Göttingen in all years but maturity was late, and growth was indeterminate. The most severe disadvantage is that the fruits burst upon ripening, but nonshattering mutants are known in other *Euphorbia* species. Therefore, it should be possible to find them in *E. lagascae* too, which then has good promise of becoming a new crop.

All other sources for the production of epoxy fatty acids proved to have little chance for crop development in Europe. *Vernonia* species, which are promising candidates in the U.S., require short days and did not flower in Göttingen in the first year. *Stokesia leavis* is a perennial *Compositae* native to the Southeastern U.S. Its seedling development is rather tardy, and it requires a long period of vernalization (34). Only 20% of the plants flowered in late September, and even in the second year after sowing flowers did not open earlier than September in Göttingen.

In 1983 and 1984, 12 different species of the genus *Crepis* were tested as sources of vernolic acid. Earle et al. (35) found one group of species high in vernolic acid, another high in crepenynic acid and a third group intermediate in the chemical composition of its seed oil (36). In Göttingen, all the species synthesizing vernolic acid produced light achenes with a long pappus in small heads. yielding a 1,000-seed weight mostly below 1 g. Yields per plot were low, and seed retention was poor. Flowering and fruiting were indeterminate, and the amount of oil also was low. For similar reasons, no crop potential was found in species of the genera *Erechtites, Cephalaria* and *Scabiosa*, the seed oils that all contain epoxy fatty acids. The results obtained in Göttingen during the 1983–1985 seasons on the agricultural performance of the six most promising plant species producing seed oils with unusual fatty acids are summarized in Table 6.

POTENTIAL OF BIOTECHNOLOGY IN OIL CROP DEVELOPMENT

As has been shown, certain species lend themselves better to domestication that others. A crucial requisite for success undoubtedly is a sufficient climatical and ecological adaptation of the crop candidate. Agronomic performance of a plant species cannot be anticipated if the conditions of cultivation are entirely different from its original habitat. To cite an extreme example, palm trees will never replace rapeseed at the shores of the Baltic Sea, but even those species that tolerate the conditions differ widely in suitability. It is most obvious that in order to obtain high yields of seed oils, seeds must reach a minimum size and oil content. In some cases plant species have been subject to earlier breeding for other uses, e.g., *Calendula* for medicinal and ornamental purposes, *Coriandrum* as a spice, and *Foeniculum* as a vegetable. In order to meet the new uses these species may require just a "second cycle of domestication." But in general, plant domestication and new crop introduction is a long and costly process.

As in evolution, any domestication is based on the mutation of genes. Plant populations often are considered domesticated with the acquisition of only one major gene mutation. For example, the nonbrittle rachis of the ear or its free-threshing habit separated the bread wheats in a "once only" event from their weedy, wild relatives. Seed dispersal is one of the most critical wild plant characteristics in many of these potential oil crops, too. However, there are other traits for which mutation appears to be an essential step toward final domestication. In addition to the previous examples, the removal of toxic compounds in particular deserves to be mentioned. Because of the high nutritional value of oilseeds with their high fat and protein contents, those species that succeeded in protecting their seed from animal consumption by repellent or toxic compounds obviously have been favored during evolution. On the other hand, the oilseed breeder searches for mutations to stop the synthesis of such compounds, which results in sweet, nutritious oilcakes after oil extraction. Low content of glucosinolates in rapeseed, of alkaloids in lupins, or of trypsin inhibitors in soybean are well-known examples of this kind. Whenever a mutational change was required during crop evolution, it happened spontaneously and was picked up and propagated by the primitive farmers.

Fifty years of experience in experimental induction of mutations has improved the chance of directed crop development considerably. Many mutations, even those with low natural survival such as those changing the natural system of seed dispersal, can be produced at will,

although some are more difficult to obtain than others. The rules that determine this different mutability are not at all clear, and they are not reflected by the spontaneous mutation spectrum. It could have been anticipated that the desired *Cuphea* mutants with radial flowers and with indehiscent fruits would be a rare occurrence because this shattering system is a taxonomic feature of all 200 *Cuphea* species (6). But, closely related genera like *Lythrum* and *Heimia* do produce the desired nonshattering capsules from radial flowers and so did *C. calophylla*. The latter species, as may be argued, belongs to the more primitive taxa in the genus *Cuphea* but *C. tolucana* and *C. wrightii* are regarded as being more developed in an evolutionary sense (17). However, the taxonomic value of a character is nothing more than the consequence of an extraordinarily low genetic variability.

Another characteristic of high specificity for the genus *Cuphea* is the unparalleled diversity of the fatty acid pattern in the seed oils. It has been assumed (11) that this variation evolved by an accumulation of genetically controlled blockings in the biosynthetic pathway of the fatty acid elongation. In such case, a reasonably high mutability of this process should be expected from experimental induction, too. Although the amount of gas chromatographic screening among progenies from our mutation experiments is limited, no mutant with changed fatty acid pattern has been discovered yet in the total of our mutagen-treated material. Thus, the spontaneous interspecific variability obviously is not paralleled by easy and high mutability through an otherwise effective mutagen treatment (10,17).

Because of all these genetic uncertainties and the many years that the domestication of a wild plant species takes, plant breeders always have been motivated to apply recent scientific progress at the earliest possible date. The historical pathways of Mendelian genetics, of quantitative genetics and heterosis, of cytogenetics and polyploidy as well as those of mutation genetics all exhibit early cornerstones of applications for crop improvement. No wonder modern biotechnologies and gene technology soon attracted active interests in the same directions. The leading idea is to escape from the lengthy and expensive procedures of traditional plant breeding and to obtain results in shorter times and by a more directed approach. It is appealing to take, for example, the genetic control of C_{12} fatty acid accumulation from *Cuphea* and to transfer such genes into rapeseed. This would enable us to produce palm oil quality at the shores of the Baltic Sea. It is not difficult to imagine other examples, but are such options realistic and what are their probabilities of success in terms of time?

It is beyond question that some plant breeding programs already are benefiting greatly from modern biotechnologies. Rapeseed is one of the most outstanding examples in which in vitro mass propagation, haploid production through microspore culture, or in vitro selection for disease resistance on pathotoxin-containing media at the cell level are now almost routine. Rapeseed culture in particular also lends itself to gene technological experiments. Protoplasts can be isolated effectively and plants regenerated thereof, even after fusion of widely distant species (37). DNA has been transferred with and without vectors and shown to be expressed in the donor cell. Experiments on gene identification using molecular techniques have been initiated and soon will enhance the body of knowledge, which finally may allow one to construct biochemical sequences of events needed for changing fatty acid production in a plant cell.

Notwithstanding all these challenges, the plant breeder has no other obligation than to produce a new and better variety for the farmer's field. In view of this task and of all the existing knowledge and experience, this paper will conclude with three comments on the possible uses of gene technology for the development of new industrial oil crops.

Presently, little is known about the molecular composition of the genetic factors involved in fatty acid and seed oil syntheses. The role of acyl carrier protein is beginning to be well understood, but whether it has any decisive role to play in determining the specific structure of the fatty acid produced still is open to debate. It would be difficult to explain the biosynthetic diversity in the *Cuphea* species by virtue of only this one factor. In particular, it still remains to be decided whether cell metabolism is completely keyed to synthesize C_{12} fatty acids exclusively or such activity is confined to the oleosome compartment, leaving the rest of the cell with a functional system of C_{18} fatty acid synthesis to supply all of the essential membrane lipids. A gene transferred for fatty acid synthesis, at any rate, must be expressed only in the mesophyll storage cells of the embryo cotyledons. It must not affect the synthesis of the usual C_{18} fatty acids, at least not in all other organs that otherwise could be lethal.

In principle, molecular gene transfer is not much different from mutagenesis. It creates a new allele in an otherwise unchanged genotype. Mutation breeders always have been burdened by the so-called pleiotropic effects of their mutants, as exemplified by the lower performance of fasciated *Cuphea* mutants under stress conditions, by a smaller number of flowers in the nonsticky forms or by a lower seed yield in a mutant with rose petals (17). No reason biologically is evident for these drawbacks, but such a situation does not surprise the breeder. He is accustomed to accounting for gene interactions or for background effects in the expression of single genes. Indeed, major attention is directed to improving gene interactions by skillful cross combinations and extensive selection of chance recombinants. By using the appropriate dimensions, this allowed the selection progress to speed up to the extent that made the "Green Revolution" possible. Directed gene transfer will provide valuable basic genotypes; this approach may have better specificity than traditional mutagenesis. But extensive recombination to adapt the genotypic background will require more effort as the new, transferred gene functions become more different. Mutants and transformants will come to their full right only when exploited in recombination breeding. If this were not so, sex would have been a luxury in evolution.

The last concern is directed to the topic of genetic erosion. Gene technological experimentation will be restricted to a few species. For mostly unknown reasons, some botanical taxa are especially suited for in vitro culture while others are clumsy materials. After a first success has been published, more attention usually is paid to that particular species and the number of scientific contributions multiplies with each new report. Also, crops are

more attractive the higher their given productivity and economic importance. These trends point in the same direction; a rather small number of oil crops throughout the world, e.g., rapeseed in the northern regions, soybean for the warmer, temperate zone and oil palm in the tropics will gain most of the attention. Each species will have to serve for food and feed, for technical applications and for chemical raw materials, and perhaps even for petroleum fuel substitutes. The biological diversity and the ecological stability of our flora may become endangered. But most certainly, the wealth of potential crop species, which could provide new industrial uses and could become new oilseed commodities, needs to be protected, maintained and evaluated before it is too late.

REFERENCES

1. FAO Production Yearbook, Vols. 31-39, Rome (1977-1985).
2. Stein, W., in *Improvement of Oil-Seed and Industrial Crops by Induced Mutations*, Int. Atomic Energy Agency, 1982, pp. 233–242.
3. Marter, A.D., in *Castor, Markets, Utilization and Prospects*, Trop. Prod. Inst., London, 1981, p. 152.
4. Rehm, S., and G. Espig, *Die Kulturpflanzen der Tropen und Subtropen*, 2nd ed., Eugen Ulmer, Stuttgart (1984).
5. Miller, R.W., F.R. Earle, J.A. Wolff, and Q. Jones, *J. Am. Oil Chem. Soc. 55*:25 (1964).
6. Koehne, E., in *Das Pflanzenreich. Regni vegetabilis conspectus*. Heft. 17 (edited by A. Engler), 1903, p. 216.
7. Röbbelen, G., and F. Hirsiner, in *Improvement of Oil-Seed and Industrial Crops by Induced Mutations*, Int. Atomic Energy Agency, Vienna, 1982, pp. 161–170.
8. Hirsinger, F., Fette, Seifen, *Anstrichmittel 82*:385 (1980).
9. Hirsinger, F., *J. Am. Oil Chem. Soc. 62*:76 (1985).
10. Hirsinger, F., and G. Röbbelen, *Z. Pflanzenzuchtg. 85*:275 (1980).
11. Graham, S.A., F. Hirsinger, and G. Röbbelen, *Am. J. Bot. 68*:908 (1981).
12. Singh, S.S., T.Y. Nee and M.R. Pollard, in *Structure, Function and Metabolism of Plant Lipids* (edited by P.-A. Siegenthaler and W. Eichenberger), Elsevier Science, Amsterdam, 1984, pp. 161–165.
13. Slabas, A.R., J. Harding, A. Hellyer, C. Sidebottom, H. Gwynne, R. Kessel and M.P. Tombs, *Ibid.*, pp. 3–11.
14. Hirsinger, F., *Angew. Bot. 54*:157 (1980).
15. Hirsinger, F., and P.F. Knowles, *Econ. Bot. 38*:439 (1984).
16. Lorey, W., *Anbauversuche mit Arten der Gattung* Cuphea (Lythraceae) *zur Nutzung als Quelle mittelkettiger Fettsauren (MCT)* Diss. Landw. Fak., University of Göttingen, Göttingen, FRG (1986).
17. Röbbelen, G., and S. von Witzke, in *Possible Use of Mutation Breeding for Rapid Domestication of New Crop Plants*, Int. Atomic Energy Agency, Vienna, in press.
18. *Development of New Crops: Needs, Procedures, Strategies, and Options*, Report No. 102, Council for Agricultural Science and Technology, Washington, D.C., (1984).
19. Meier zu Beerentrug, H., and G. Röbbelen, *Angew. Bot. 61*, in press.
20. Meier zu Beerentrup, H., and G. Röbbelen, *Fat Science Technology 89*:227 (1987).
21. Kleiman, R., *J. Am. Oil Chem. Soc. 59*:29 (1982).
22. Moreau, J.P., R.L. Holmes, T.L. Ward and J.H. Williams, *J. Am. Oil Chem. Soc. 43*:352 (1966).
23. Earle, F.R., K.L. Mikolajczak, I.A. Wolff and A.S. Barclay, *J. Am. Oil Chem. Soc. 41*:345 (1964).
24. Babcock, E.B., *Bot. 21/22*:1 (1947).
25. White, G.A., and W. Calhoun, *Econ. Bot. 27*:320 (1973).
26. Norlindh, T., in *The Biology and Chemistry of the Compositae*, Academic Press, London, 1977, pp. 961–987.
27. Barclay, A.S., and F.R. Earle, *Econ. Bot. 19*:33 (1965).
28. Willingham, B.C., and G.A. White, *Econ. Bot. 27*:323 (1973).
29. Mikolajczak, K.L., F.R. Earle and I.A. Wolff, *J. Am. Oil Chem. Soc. 39*:78 (1962).

30. Barclay, A.S., H.S. Gentry and Q. Jones, *Econ. Bot. 16*:206 (1962).
31. Gentry, H.S., and A.S. Barclay, *Econ. Bot. 16*:206 (1962).
32. Kleiman, R., C.R. Smith, Jr., S.G. Yates and Q. Jones, *J. Am. Oil Chem. Soc. 42*:169 (1965).
33. Krewson, C.F., and W.E. Scott, *J. Am. Oil Chem. Soc. 43*:171 (1966).
34. Campbell, T.A., in *New Sources of Fats and Oils* edited by E.H. Pryden, L.H. Princen and K.D. Mukherjee, AOCS, Champaign, IL, 1981, pp. 287–295.
35. Earle, F.R., A.S. Barclay and I.A. Wolff, *Lipids 1*:325 (1966).
36. Earle, F.R., *J. Am. Oil Chem. Soc. 47*:510 (1970).
37. Gleba, Y.Y., and F. Hoffmann, *Planta 149*:112 (1980).

Molecular Approaches to the Study and Modification of Oilseed Fatty Acid Synthesis

John B. Ohlrogge
Department of Botany and Plant Pathology, Michigan State University, East Lansing, MI 48824-1312

Studies of ACP and other soluble proteins involved in plant fatty acid metabolism have progressed to the stage in which their genes can be isolated and their levels manipulated in transgenic plants. These studies have and will continue to reveal the basic organization and control of fatty acid production in plants. In some cases, modification of the specificity or level of these soluble proteins may yield useful fatty acid chain length variations. More extensive modifications of acyl chain structure eventually will be possible. However, progress with the membrane-bound enzymes responsible for acyl chain modification and assembly into glycerolipids has been much slower. It is likely that application of recombinant DNA techniques to these enzymes will be delayed due to difficulties in their isolation and characterization at the protein level.

Development of the ability to isolate, modify and transfer genes has sparked interest in the possibility that such techniques can be used for the improvement of oilseeds. This paper will review some aspects of the biochemistry of fatty acid synthesis in seeds, discuss some prospects and limitations of oilseed genetic engineering and summarize studies on acyl carrier protein that provide insight into the organization and control of plant fatty acid production.

SOME ASPECTS OF FATTY ACID SYNTHESIS IN OILSEEDS

The fatty acid structures that occur in plant seeds are remarkably diverse; for example, acyl chain lengths range from 6 to 24 carbons, and they can contain double or triple bonds in almost all locations as well as many modifications in the acyl chain such as the addition of epoxy, cyclopropane, keto and hydroxy functions. In their classic 1964 compendium on natural fats, Hilditch and Williams recorded the occurrence of several hundred fatty acid structures from surveys of approximately 900 species (1). Subsequent studies have revealed many additional structures (2,3). Several of the ''unusual'' fatty acids produced in plants have commercial value as lubricants, soaps and plasticizers.

The rich diversity found in seed triacylglycerols contrasts markedly with the composition of other tissues and in particular with the composition of plant membrane acyl chains. Plant membrane lipids contain almost exclusively 16 and 18 carbon saturated and unsaturated (one to three double bands) fatty acids. This restricted range of structures of the so-called ''normal'' fatty acids is maintained even in the membranes of seed tissue. As shown below (Table 1), in Crambe the 20 and 22 carbon acids that characterize the seed oils almot completely are excluded from the seed phospholipids (4). Similarly, in *Cuphea*, the short chain length fatty acids primarily are restricted to the triacylglycerol fraction (5,6).

Why are seed triacylglycerols so diverse in composition whereas membrane composition is so uniform? It is likely that the relatively limited range of acyl structures found in most biological membranes reflects requirements for appropriate bilayer formation, protein binding and fluidity maintenance. Numerous studies have confirmed that major modifications in membrane fatty acid composition lead to decreased membrane integrity and disrupted functions. Thus, it is likely to be important that plant cells exclude from their membranes many of the unusual structures that can occur in seeds.

The source and function of the diversity of seed fatty acids is not well understood. Unusual fatty acids may serve as a defense against the digestive system of herbivores. Alternatively, it may be that unusual fatty acids are evolutionary artifacts whose structure has no advantageous or deleterious functions and thus is under no pressure for elimination. Regardless of their function, for the goals of plant biotechnology the important and encouraging point is that this diversity implies that seeds can tolerate widely different compositions. Thus, molecular genetic modifications of oilseed composition should be tolerable.

One of the goals of plant biotechnology is to modify the composition of oilseeds to provide new, improved or lower cost products. In order to maintain healthy plants, two potential requirements may apply to this goal: fatty acid modifications should be made only in seed tissue and not in other plant organs in which they might disrupt functions such as photosynthesis and fatty acid modifications should be made only in triacylglycerols (TAG) and not in membrane polar lipids. Information from several muta-

TABLE 1

Different Fatty Acid Compositions of Seed Polar Lipids and Triacylglycerols

Fatty acid chain length	*Cuphea lutea* (6)		*Crambe abyssinica* (4)	
	Triacylglycerol	Polar Lipids	Triacylglycerol	Polar Lipids
10	48	4	—	—
12	29	3	—	—
14	7	6	—	—
16	5	31	2	16
18	4	41	31	75
20	1	9	3	2
22	0	5	60	3

TABLE 2

Fatty Acid Composition of Seed and Leaf Lipids of High Stearic Acid Soybean (Variety A6)

Fatty acid	Weight % of total fatty acids	
	Seed	Leaf
16:0	9	17
18:0	25	5
18:1	17	4
18:2	42	25
18:3	7	47

TABLE 3

Fatty Acid Composition of Amsoy and High Stearic Acid Soybean (Variety A6)

Fatty acid	Weight % of total fatty acids			
	Triglyceride		Polar lipids	
	Amsoy	A6	Amsoy	A6
18:0	4	30	5	29
18:1	28	25	20	20
18:2	50	34	48	30

Fatty Acid Composition of High-linoleic (US-10) and High-oleic (UC-1) Safflower Varieties

Fatty acid	Weight % of total fatty acids			
	Triglyceride		Polar lipids	
	US-10	UC-1	US-10	UC-1
18:1	14	80	9	82
18:2	76	12	77	12

tion breeding experiments suggests that the first requirement may be achieved easily but that the second may present difficulties.

Table 2 indicates the fatty acid composition of the seed and leaf total lipids of a soybean variety selected for high seed stearic acid content (7). It is clear that the high stearic acid trait is not expressed in the leaf tissue. Similar data have been observed for several other fatty acid mutants including high oleic safflower varieties, low trienoic arabidopsis (8) and low linolenic linseed (9). These data demonstrate that the fatty acid composition of seeds and leaves is controlled by different genes. Therefore, it should be possible to modify seed composition without influencing other organs of the plant. (Although this also could be achieved using well-known seed specific promoters such as for storage proteins, these promoters may not offer ideal timing or levels of expression.)

The second requirement listed above may concern the health of the developing seed. A key question pertaining to the genetic engineering of seed oils is whether modifications that influence only the triacylglycerols and do not alter the properties of the cell membranes can be made. For example, if we manage to insert genes in oilseeds that lead to the seed specific production of short chain fatty acids, will these fatty acids be targeted to TAG as in coconut and *Cuphea* or will the short chain fatty acids enter into all lipid classes? If the latter case occurs, will the membranes be able to tolerate the introduction of perhaps substantial levels of unusual fatty acids?

A clue to the potential consequences of modifying oilseed fatty acid metabolism is again found by examining results of mutants obtained by conventional means. Table 3 shows the fatty acid composition of TAG and polar lipids in seeds of a high stearic acid soybean variety and a high oleic acid safflower variety. In both cases, the mutation that leads to altered TAG composition also is observed to influence the polar lipids. In these cases, it appears that the fatty acid composition of both classes of lipid in the seed are under the control of a common gene. If we extrapolate these data to future genetic engineering experiments, we might suggest that addition of a new fatty acid to seed tissue will result in its incorporation into all lipid classes, a situation that might have deleterious consequences to the developing seed if the alterations are more dramatic than the examples in Table 3.

However, mechanisms clearly exist in many oilseeds that assure unusual fatty acids are incorporated into tricyaglycerols and excluded from membrane phospholipids. It may be that such mechanisms exist only in seeds such as *Cuphea* or rapeseed in which unusual fatty acids are

produced. Plants such as soybean and safflower whose seeds produce "normal" fatty acids may not have mechanisms for dealing with unusual structures. Alternatively, all plants may have the ability to tailor their membrane composition such that unusual structures are excluded. Distinguishing between these various alternatives may have relevance to designing oilseed genetic engineering strategies.

Mechanism of this Specific Acyl Chain Targeting

Compartmentation. It is possible that reactions that form seed specific unusual fatty acids are compartmentalized such that the products of these reactions are not free to enter into membrane lipid synthesis. In this scenario, some subcellular fraction of the oilseed cell may contain enzymes for both fatty acid modification and esterification of unusual fatty acids to form triacylglycerol. Metabolite channeling may occur between these sets of reactions or physical barriers may prevent access of modified acyl chains to membrane biosynthetic reactions. Current evidence on the subcellular localization of fatty acid modification and TAG biosynthesis is inconclusive but implicates involvement of both endoplasmic reticulum and the oil body fraction (10,11). In some cases, fatty acid modification may occur only after acyl chains are esterified to TAG although this is not consistent with the mechanisms for producing short or long chain fatty acids.

Enzyme selectivity. For example, seeds may contain acyltransferases that specify the esterification of unusual acyl chains to TAG or which exclude their esterification to membrane phospholipids. The specificity of acyltransferases has been examined extensively in animals but only cursorily in plants. In general, the in vitro analysis of acyltransferases has not revealed the chain length specificity required to explain the data of Table 1. Recently, Cao and Huang examined the substrate preference of diacylglycerol acyltransferase from *Cuphea*, maize and *Brassica* (12). When microsome preparations from these species were provided mixtures of 12:0, 18:1 and 22:1-CoA, the shorter chain substrates were preferred in almost all cases. No strong selectivity for 12:0 was found in *Cuphea* or for 22:1 in *Brassica*. These data in vitro argue against TAG synthesizing acyltransferases as an explanation for the specific composition of TAG. Unfortunately, similar data on chain length specificity of plant phospholipid acyltransferases are not available. It should also be mentioned that in vitro examinations of acyltransferase specificity sometimes have been found insufficient to explain the in vivo distribution of acyl chains in phospholipids. This may be because the in vivo concentrations or availability of potential substrates at the active site of membrane enzymes generally are unknown and may contribute considerably to the observed patterns of lipid composition.

It is interesting to note that another unusual plant fatty acid, ricinoleic, when fed to animals is found esterified to TAG but is excluded completely from phospholipids (13). Barber et al. (14) demonstrated that this exclusion from phospholipids could be explained largely by the low activity of ricinoyl-CoA in assays of rat liver phospholipid acyltransferases. Thus, in this example, enzyme specificity appears to control the targeting of unusual fatty acids to TAG.

POTENTIAL FOR GENETIC ENGINEERING OF FATTY ACID COMPOSITION

Conventional plant breeding is more successful at eliminating undesirable fatty acids than in adding new acyl structures to existing oil crops. The advantages of applying recombinant DNA techniques to oilseeds is the potential for more rapid and dramatic alterations. It is clear from the results of several successful mutation breeding efforts that the oil composition of seeds can be altered substantially without damage to the agronomic properties of the species. These results offer encouragement that large and useful changes might be introduced through the use of recombinant DNA techniques to add or delete genes. However, two temporary limitations to this approach bear mentioning.

In most cases, we have not isolated either the proteins or genes that we might like to modify. Acyl carrier protein is the only plant FAS protein for which published amino acid or gene sequence data currently is available. Many of the proteins of greatest interest in plant fatty acid metabolism are of low abundance, membrane bound and difficult to purify. This makes the isolation and cloning of their genes difficult. Fortunately, the great increase in industrial research in this area is beginning to expand our knowledge about these proteins. In addition, genes isolated from other organisms may prove to be useful when combined with appropriate promoters and subcellular targeting sequences.

In addition to a lack of knowledge of individual FAS enzymes, we also have insufficient understanding of the organization and rate limiting steps in fatty acid metabolism. A sobering example of the unexpected results that can occur in genetic manipulation of lipid enzymes is provided by cloning of the phosphatidyl serine synthetase of *E. coli*. This enzyme is at a branch point in *E. coli* phospholipid synthesis and thus catalyzes the committed step in phosphatidylcholine (PE) and phosphatidylserine (PS) synthesis. Cloning of this enzyme and its 20–fold overproduction in *E. coli* led to almost no change in the phospholipid composition of the cells (15,16). This example from a very simple organism emphasizes how difficult it is to predict the outcome of genetic engineering experiments. Clearly other mechanisms in addition to the level of the biosynthetic enzyme are responsible for controlling cellular lipid composition. In a complex eukaryotic cell, we can expect several levels of control over lipid composition. The successful results of screening for oilseed composition mutants indicates that single gene changes can effect major lipid composition changes. However, as of yet the precise biochemical alteration has not been identified in any of these oilseed mutants.

Some strategies for oilseed modification (for example, introduction of mammalian medium chain acylhydrolase [17]) can be envisioned as leading to near–future genetic engineering of oilseeds. However, I hope the above discussion has demonstrated that substantial basic research in oilseed lipid biochemistry may be necessary before many desired modifications can be obtained predictably. One approach our lab has undertaken to obtain such basic information has been through the study of acyl carrier protein (ACP). ACP is central to many steps of plant lipid metabolism, both in the synthesis of fatty acids and of glycerolipids. As such, it serves as a powerful probe for investigating not only how lipid synthesis proceeds but also how it is regulated. It currently is very difficult to either identify the rate–determining steps or the key regulatory enzymes in plant lipid metabolism. Given the complexity of a higher plant, we can imagine that primary metabolism must be regulated by a wide range of signals, and that the pathway members must be coordinated in their expression and activity. Those factors that regulate ACP expression and activity also may be involved in regulation of the other FAS pathway members. The tools accumulated from studying ACP (amino acid sequence data, antibodies, synthetic and cDNA gene clones) are more complete than for any other plant FAS protein. Thus, we now are in a position to examine the regulation of ACP at both the protein and nucleic acid levels. Such information should not only provide insights into the functioning and regulation of the other members of the FAS pathway but also provide the framework for initial efforts directed toward modifying seed storage oils.

CELLULAR AND MOLECULAR ORGANIZATION OF PLANT FATTY ACID SYNTHESIS

ACP is a small, acidic protein that functions as a cofactor or cosubstrate for at least a dozen enzymes in plant lipid metabolism. In spinach, this protein has 82 amino acids with a phosphopantetine prosthetic group attached to a serine residue at position 38 of the protein sequence (18). ACP is the cofactor to which the growing acyl chain is attached during all steps of fatty acid assembly. In addition to the six enzymatic reactions of fatty acid biosynthesis, ACP also participates in the stearoyl-ACP desaturase, oleoyl-ACP hydrolase reactions, and serves as the acyl donor for two plastid acyltransferases (19).

Analysis of this relatively simple protein has yielded insights into the organization, localization and regulation of plant fatty acid metabolism. For example, in 1964 Overath and Stumpf found that plants contain a small heatstable protein similar to *E. coli* ACP (20). This provided the first evidence that plant FAS is organized in the type II nonassociated form rather than the type I multi-enzyme arrangement found in animals and yeasts. These earlier observations now have been extended in several labs to include purification and characterization of each of the component enzymes (21).

Recently, it has been observed that both monocots (22) and dicots (23) have isoforms of FAS proteins. In the case of ACP and malonyl-CoA:ACP transacylase, the different forms are expressed differently in leaves and seeds (23,24). Two major forms of ACP have been purified from barley and spinach leaves and based on amino acid sequencing appear to be the product of different genes. Our preliminary evidence indicates that the differential expression of acyl carrier protein isoforms may be one mechanism that plants use to regulate distribution of acyl chains within the plant cell (25). In spinach leaf tissue, we observed that oleate esterified to ACP I is hydrolyzed rapidly by the acyl-ACP hydrolase and thus becomes available for export from the plastid. In contrast, oleate attached to ACP II is transferred preferentially to glycerol-3-phosphate by a stromal acyltransferase and is retained in the plastid.

Cellular Organization of Plant FAS

In yeast and in animals it long has been known that fatty acid synthesis occurs primarily in the cytoplasm. In 1979, Ohlrogge, Kuhn and Stumpf (26) prepared antibodies against ACP and found that these inhibited all de novo FAS in spinach leaf homogenates. When protoplasts of spinach mesophyll cells were lysed gently and their organelles separated on sucrose gradients, essentially all of the ACP present could be attributed to the chloroplast fraction. This result indicated that plant mesophyll cells differ from other eukarotic cells in the absence of cytoplasmic FAS.

Although ACP is localized in the plastid, the genes for ACP are not in the plastid genome but rather ACP is coded in the nucleus. In vitro translation of poly A+ RNA has shown that ACP is synthesized first as a precursor that is approximately 5500 Da larger than the mature ACP (27). Thus, ACP is similar to most plastid proteins in its cytoplasmic synthesis with a transit peptide that is required for uptake into plastids.

Formation of active ACP requires attachment of the pantetheine prosthetic group. Recently, we determined that as in *E. coli* the pantetheine group can be donated from Coenzyme A by the enzyme holoACP synthase (28). The different subcellular sites of ACP synthesis and function raise the question of where the prosthetic group is attached, the cytoplasm or within the plastids. Assay of holoACP synthase in subcellular fractions of spinach leaves and developing castor oil seed endosperm cells indicated this enzyme largely is cytosolic (28). Thus, we propose that the prosthetic group is added to pre-apoACP in the cytoplasm, followed by uptake into the plastid and proteolytic processing to mature holoACP.

CLONING OF ACP

Acyl carrier proteins from spinach (18,29) and barley (30) were the first proteins of plant FAS for which amino acid sequence data became available. This has provided opportunities for cloning ACP genes through the use of oligonucleotide probes. The complete amino acid sequence of spinach ACP I also made possible the construction of a synthetic gene coding for this protein (31). We recently have succeeded in expressing such a gene in *E. coli*, and this has made available for the first time plant ACP in large quantities. This ACP then may be used for analysis of the protein and as a source of substrate for subsequent investigations of several other lipid biosynthetic enzymes. *E. coli* ACP also has been cloned recently by similar methods and at least two industry labs have succeeded in obtaining cDNA and/or genomic clones of ACP from spinach and *Brassica*. Efforts to transform plants with ACP genes currently are underway, and information on the effects of overproduction of ACP in plants should be available soon.

FATTY ACID SYNTHESIS IN SEEDS

The studies of Shimakata and Stumpf on spinach leaf and developing safflower seeds have established that the molecular organization of FAS in these two tissues essentially is similar (21). Although leaf tissue is the best characterized FAS system, studies from several labs have emphasized the role of plastids in nongreen tissues as a major site of FAS. Unfortunately, because of problems with subcellular fractionation of these tissues it has not been possible to rule out other subcellular cites. Some reports have indicated that FAS also may occur in oil bodies (4), and in vivo labeling data of Pollard led to the suggestion of two FAS systems in developing nasturtium seeds (32).

Despite uncertainties about the organization of lipid synthesis in oilseeds, a consensus hypothesis has emerged (10,11): Fatty acids first are synthesized in the plastid as ACP esters. Hydrolysis of the acyl chains by acyl-ACP hydrolase allows the export of fatty acids from the plastid. The fatty acids are esterified to CoA on the plastid envelope and the acyl-CoA's then are transported to the endoplasmic reticulum in which they serve as acyl donors for glycerolipid synthesis. In the ER, the acyl chains may be modified by elongation, desaturation, hydroxylation, etc., followed by their assembly into TAG. Part of the above scenario is based on extrapolations from leaf data. For example, whether acyl-CoA synthase also is localized in the envelopes of oilseed plastids is not known.

DEVELOPMENTAL REGULATION OF OILSEED ACP EXPRESSION

During the development of the soybean seed, fatty acid synthesis increases markedly, reaches a peak at 40–50 days after flowering (DAF), and then declines as the seed reaches maturity (Fig. 1A). The major increase in lipid synthesis occurs after cell division stops at 20 DAF. We have asked whether these developmental changes in fatty acid synthesis are coordinated with changes in levels of ACP (33).

The level of ACP per seed was measured enzymatically during soybean seed development (Fig. 1B). Comparison of Figures 1B and 1A shows that the increase in enzymatically active ACP occurs in close coordination with increased lipid synthesis. This result indicates that

MOLECULAR APPROACHES TO THE STUDY AND MODIFICATION OF OILSEED FATTY ACID SYNTHESIS

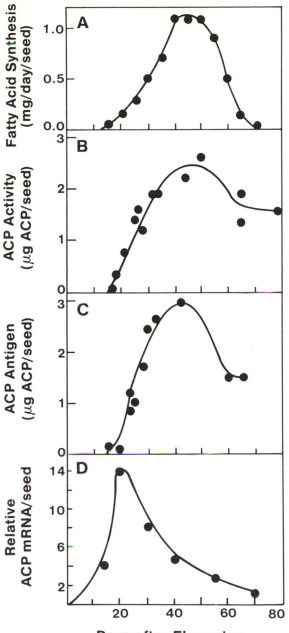

FIG. 1. Regulation of acyl carrier protein expression during soybean seed development. A, Rate of fatty acid synthesis per seed per day; B, ACP level per seed measured enzymatically; C, ACP level per seed measured immunochemically; D, ACP messenger RNA level per seed.

the soybean seed continues to increase the level of active ACP even after cell division ceases. Thus, the large increase in lipid synthesis seen between 25 and 50 DAF is not solely the result of increased photosynthate supply to a pre-existing fatty acid biosynthetic system. Furthermore, the close correlation between the level of active ACP and in vivo lipid biosynthesis indicates that the quantity of active fatty acid synthetase proteins present in the cell may be a rate–determining component of the cell's overall lipid biosynthetic capacity.

The increase in active ACP could be the result of several

processes, including de novo synthesis of ACP, activation of pre-existing ACP precursors, or transfer of the prosthetic group to ACP. Antibodies often are able to recognize both active and inactive forms of protein. Therefore, to distinguish between enzymatically active and inactive ACP we measured ACP immunochemically. In Figure 1C, the level of ACP per seed measured by radioimmunoassay is shown. ACP antigen also increases in parallel to lipid synthesis. In 1984, we were able to suggest that the most likely interpretation of these data is that the increase in ACP shown in Figure 1 is produced through de novo synthesis of ACP rather than through posttranslational activation of this protein.

We now have tested this hypothesis by examining the level of ACP mRNA during soybean seed development (D. Hannapel and J. Ohlrogge, unpublished data). This was accomplished by using RNA transcripts of the synthetic ACP gene as probe for the soybean polyA + (Fig. 1D). The abundance of ACP mRNA increases rapidly in the early stages of soybean seed development, peaks at approximately 20 DAF and declines to low levels as the seed reaches maturity. Thus, the level of ACP expression and in part the pattern of in vivo lipid synthesis can be seen to be a reflection of ACP mRNA levels. This is not to suggest that ACP levels determine rates of fatty acid synthesis. We expect that other members of the FAS pathway will be regulated coordinately in a fashion similar to that of ACP and that turning on lipid synthesis requires turning on many genes.

Although there is a close parallel between increases in lipid synthesis and ACP level, this correlation does not continue for the final stages of seed maturation. Between 50 and 70 DAF, in vivo lipid deposition (Fig. 1A) and acetate incorporation into lipids decrease to levels less than 10% of their maximum. In contrast, ACP levels decrease approximately one-half. This suggests that mechanisms such as desiccation or lack of substrates rather than decreased fatty acid synthetase proteins lead to the decrease in rate of soybean lipid accumulation. Alternatively, the activity of lipid biosynthetic proteins other than ACP may become rate limiting at the final stages of seed maturation.

REFERENCES

1. Hilditch, T.P., and P.N. Williams, *The Chemical Constitution of Natural Fats*, John Wiley and Sons, New York, NY (1964).
2. Princen, L.H., *Economic Botany 37*:478 (1983).
3. Kleiman, R., R.B. Wolf and R.D. Plattner, *Lipids 20*:373 (1985).
4. Gurr, M., J. Blades, R. Appleby, R. Appleby and C. Smith, *Eur. J. Bioch. 43*:281 (1974).
5. Slabas, A.R., P.A. Roberts, J. Ormesher and E.W. Hammond, *Biochim. Biophys. Acta 711*:411 (1982).
6. Singh, T., T. Nee and M. Pollard, *Lipids 21*:143 (1986).
7. Graef, G.L., L.A. Miller, W.R. Fehr and E.G. Hammond, *J. Am. Oil. Chem. Soc. 162*:773 (1985).
8. Browse, J., P. McCourt and C. Somerville, *Plant Physiol. 81*:859 (1986).
9. Tonnet, M.L., and A.G. Green, *Archives of Biochemistry and Biophysics 252*:646 (1987).
10. Slack, C.R., and J.A. Browse, in *Seed Physiology*, Vol. 1, edited by D.R. Murray, Academic Press, New York, NY, 1984, pp. 209-244.
11. Stymne, S., and A.K. Stobart, in *The Biochemistry of Plants*, Vol. 9, Academic Press, New York, NY, 1987, pp. 175-214.
12. Cao, Y-Z., and A. Huang, *Plant Physiol. 84*;762 (1987).

13. Watson, W.C., and E.S. Murray, *Biochim. Biophys. Acta 106*:311 (1964).
14. Barber, E.D., W.L. Smith and W.E.M. Lands, *Biochim. Biophys. Acta 248*:171 (1971).
15. Ohta, A., K. Waggoner, K. Louie and W. Dowhan, *J. Biol. Chem. 256*:2219 (1981).
16. Raetz, C.R., T.J. Larson and W. Dowhan, *Proc. Natl. Acad. Sci., USA 74*:1412 (1977).
17. Safford, R., J. de Silva, C. Lucas, J.H.C. Windust, J. Shedden, C.M. James, C.M. Sidebottom, A.R. Slabas, M.P. Tomba and S.G. Hughes, *Biochem. 26*:1358 (1987).
18. Kuo, T.M., and J.B. Ohlrogge, *Arch. Biochem. Biophys. 234*:290 (1984).
19. Frentzen, M., E. Heinz, T.A. Mckeon and P.K. Stumpf, *Eur. J. Bioch. 129*:629 (1983).
20. Overath, P., and P.K. Stumpf, *J. Biol. Chem. 239*:4103 (1964).
21. Stumpf, P.K. in *Fatty Acid Metabolism and Its Regulation*, edited by S.Numa, Elsevier Science Publishers B.V., Amsterdam, 1984, pp. 155–179.
22. Hoj, P.B., and I. Svendsen, *Carlsberg. Res. Commun. 49*:483 (1984).
23. Ohlrogge, J.B., and T.M. Kuo, *J. Biol. Chem. 260*:8032 (1985).
24. Guerra, D.J., and J.B. Ohlrogge, *Arch. Biochem. Biophys. 246*:274 (1986).
25. Guerra, D.J., J.B. Ohlrogge and M. Frentzen, *Plant Physiol. 82*:448 (1986).
26. Ohlrogge, J.B., D.N. Kuhn and P.K. Stumpf, *Proc. Natl. Acad. Sci., USA 76*:1194 (1979).
27. Ohlrogge, J.B., and T.M. Kuo, in *Structure, Function, and Metabolism of Plant Lipids*, edited by P.-A. Siegenthaler and W. Eichenberger, Elsevier Science Publishers B.V., Amsterdam, 1984, pp. 63–67.
28. Elhussein, S., J.A. Miernyk and J.B. Ohlrogge, in *The Metabolism, Structure, and Function of Plant Lipids*, edited by P.K. Stumpf, J.B. Mudd and W.D. Nes, Plenum Press, New York, NY, 1987, pp. 709–713.
29. Matsumura, S., and P.K. Stumpf, *Arch. Biochem. Biophys. 125*:932 (1968).
30. Hoj, P.B., and I. Svendsen, *Carlsberg Res. Commun. 48*:285 (1983).
31. Beremand, P., D.J. Hannapel, D.J. Guerra, D.M. Kuhn and J.B. Ohlrogge, *Archives of Biochemistry and Biophysics 256*:90 (1987).
32. Pollard, M., and P.K. Stumpf, *Plant Physiol. 66*:641 (1980).
33. Ohlrogge, J.B., and T.M. Kuo, *Plant Physiol. 74*:622 (1984).

Genetic Control of Fatty Acid Biosynthesis in Yeast

Eckhart Schweizer, Gerhard Müller, Lilian M. Roberts, Michael Schweizer, Johannes Rösch, Peter Wiesner, Joachim Beck, Dirk Stratmann and **Ira Zauner**
Lehrstuhl für Biochemie, Universität Erlangen-Nürnberg, Staudtstrabe 5, D-8520 Erlangen

In an attempt to investigate basic principles underlying the molecular architecture and functioning of fatty acid synthetase (FAS) multienzyme complexes, the FAS genes from selected organisms currently are studied in our laboratory. In this paper, we report the complete nucleotide sequence of the multifunctional FAS1 and FAS2 genes from *Saccharomyces cerevisiae*, of the FAS1 gene from *Yarrowia lipolytica* and the FAS2 gene from *Penicillium patulum*. In all three organisms, the FAS multienzyme complex is an $\alpha 6$ $\beta 6$ hexamer of the pentafunctional subunit β and of the trifunctional subunit α. Subunits α and β are encoded by FAS2 and FAS1, respectively. The obtained nucleotide sequences were used to deduce the coding region, amino acid sequence and size of the corresponding gene products together with other protein chemical parameters. All FAS genes studied exhibited a remarkably high degree of homology (60-80%) at the level of both DNA and protein structure. The order of catalytic domains in FAS1 is acetyl transferase (N-terminal), enoyl reductase, dehydratase and malonyl/palmityl transferase. In FAS2, the peripheral (cysteine) and central (pantetheine) SH groups were localized according to their known amino acid context, while the β-ketoacyl reductase domain was localized in the N-terminal part of the gene by deleting mapping. The *Penicillium* FAS2 gene contains two small (45 bp) introns, while the yeast genes were intron-free. From the presence of several nonhomologous deletions and insertions in the fungal FAS genes as well as from their size in comparison to that of the corresponding vertebrate gene, it is concluded that the DNA content of FAS1 and FAS2 in yeast may be considerably higher than is absolutely required for the catalytic process. By in vitro mutagenesis, it was intended to discriminate between functionally essential and nonessential parts of genes.

The enzyme system of de novo saturated fatty acid biosynthesis generally is considered as one of the so-called housekeeping functions of cellular metabolism. Besides their ubiquitous occurrence fatty acid synthetases (FAS) are characterized by their essentially uniform reaction mechanisms and product patterns. The general reaction scheme of saturated fatty acid biosynthesis is illustrated in Figure 1. Despite this uniformity, distinct variations of some functional and structural parameters are observed when FAS enzymes from different organisms are compared. For instance, fatty acid synthesis is constitutive in some yeasts (2), while being repressed by long chain fatty acids in others (3).

In vertebrates, the process is subject to hormonal (4–6) and dietary (7–9) control. In higher plants, cytoplasmic fatty acid synthesis appears to be repressed permanently in favor of a nuclear–encoded chloroplast enzyme system (10,11). In etiolated Euglena cells, fatty acid biosynthesis takes place in the cytoplasm. However, upon illumination the cytosolic fatty acid synthetase becomes repressed, while the chloroplast FAS system is induced (12). Some symbiotic or parasitic organisms such as mycoplasma (13), certain rumen bacteria (14) or the hair fungus *Pityrosporum ovale* (15) derive their fatty acids from the host and have no functional FAS system of their own. Furthermore, most *Archebacteria* apparently contain polyprenyl ethers rather than fatty acid esters in their cellular lipids and there is probably no FAS activity present in these organisms (16). Finally, some FAS systems synthesize multimethyl branched (17) or unsaturated (18) fatty acids in addition to saturated long chain fatty acids.

Besides these functional differences the FAS enzymes studied so far also may be differentiated according to their gross molecular structures. By this criteria, they are classified as so–called type I or type II fatty acid synthetases (for review see [19]).

Type II fatty acid synthetases are composed of eight structurally distinct and functionally different proteins. They are present in most bacteria and in plant chloroplasts (for review, see [20]). As evident from Figure 1, these constituents are the acyl carrier protein, the malonyl, acetyl and palmityl transacylases, the β-ketoacyl synthase, the β-ketoacyl reductase, the dehydratase and the enoyl reductase component enzymes. In contrast, type I fatty acid synthetases are multifunctional enzymes of high structural complexity and several or, at the extreme, all constituent functions combined within a single polypeptide chain (19,21). In these enzymes, the above–mentioned eight FAS functions are presumed to represent distinct domains within a multifunctional protein. So far, two different type I fatty acid synthetases of either α_2 or $\alpha_6\beta_6$ molecular structures have been identified. The α_2-FAS enzymes are found in vertebrates (19) and possibly also in some bacteria (20). The second, $_6\beta_6$ FAS complex, is found in lower fungi such as yeasts (22), *Neurospora crassa* (23) or *Penicillium patulum* (24,25). It consists of the trifunctional subunit α and the pentafuctional subunit β (22).

In the past, the multienzyme complex nature of type I fatty acid synthetases was one of the most fascinating aspects that attracted biochemists studying the enzymology of fatty acid biosynthesis. It is speculated but not yet proven experimentally that organized FAS enzymes are catalytically more efficient and, therefore, evolutionary advanced compared to nonorganized enzyme systems. Besides this, the possible correlation between enzyme architecture, catalytic efficiency and reaction mechanism represents a matter of central biochemical importance. The FAS multi-enzymes clearly represent one of the most complex but, at the same time, functionally best–characterized systems currently available to study this question. Reasonably, an appropriate topology of the various catalytic domains in the complex is necessary to control the sequence of more than 30 distinct reaction steps in the course of palmitate biosynthesis from acetyl- and malonyl-CoA. A detailed insight into this multistep reaction process only will be possible after elucidation of the exact three–D structure of the enzyme. Although yeast

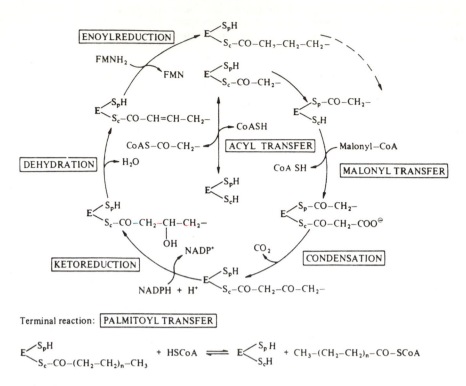

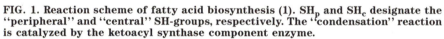

FIG. 1. Reaction scheme of fatty acid biosynthesis (1). SH$_p$ and SH$_c$ designate the "peripheral" and "central" SH-groups, respectively. The "condensation" reaction is catalyzed by the ketoacyl synthase component enzyme.

fatty acid synthetase has been crystallized by Lynen and coworkers (26), no x-ray crystallographic data for this protein as yet are available. Molecular genetic studies currently performed in our laboratory should be useful in a future approach to this question.

YEAST FATTY ACID SYNTHETASE

The multifunctional nature of the two yeast FAS subunits α and β became evident when fatty acid synthetase–deficient yeast mutants were shown to map to only two gene loci, FAS1 and FAS2 (22). FAS1 encodes subunit β and is located on chromosome XI (27,28) while FAS2, encoding subunit α recently was mapped on chromosome XVI (U. Siebenlist, unpublished data). Only limited amino acid sequence information is available for both yeast FAS subunits. Previously, peptide fragments specific for the active sites of the acyl carrier protein, the acetyl, malonyl and palmityl transacylases and the ketoacyl synthase, respectively, have been isolated and partially sequenced (1).

Specific mutations available for every catalytic domain of the yeast FAS complex have been used to attribute the acetyl, malonyl and palmityl transferase, the enoyl reductase and dehydratase functions to subunit β and the acyl carrier protein, ketoacyl reductase and ketoacyl synthetase domains to subunit α (22). The individual catalytic domains of subunits α and β apparently fold and function essentially independently of each other. This is concluded first from an extraordinarily regular and consistent pattern of intragenic complementation between allelic fas1 and fas2 mutants (22). Second, every FAS function may be specifically mutationally altered, leaving all others unaffected (22).

SEQUENTIAL ORDER OF FAS DOMAINS IN THE FAS1 AND FAS2 YEAST GENES

Recently, cloning of both yeast FAS genes has been achieved in our laboratory (29) and also was reported by Kuziora et al. (30). Using the isolated FAS-DNA we determined the sequential order of catalytic sites within both genes by deletion mapping (31). For this purpose, subcloned fragments of FAS1 and FAS2 were used to restore the function of defined fas1 and fas2 mutants. Figure 2 depicts the results of these studies indicating that in subunit α the ketoacyl reductase is located next to the N-terminus of the protein being followed by the ketoacyl synthase and the acyl carrier protein domains. In subunit β, the acetyl transferase is the N-terminal domain. It is followed by the dehydratase, the enoyl reductase and the malonyl/palmityl transferase active sites. In these studies, the acyl carrier protein function was mapped according to the ability of distinct DNA fragments to complement pantetheine-less yeast mutants. As will be demonstrated below, however, these data are not sufficient to locate the pantetheine binding site of the protein. By DNA sequencing, this binding site actually was located at a different position in FAS2 (Fig. 3).

NUCLEOTIDE SEQUENCE OF THE YEAST FATTY ACID SYNTHETASE GENES FAS1 AND FAS2

We have determined the complete nucleotide sequence of both S. cerevisiae FAS genes. The FAS1 coding sequence together with its flanking regions has been published (32). According to these data, the FAS1 encoded reading frame of subunit β comprises 5.535 base pairs, corresponding to a protein of 1,845 amino acids with a molecular weight of

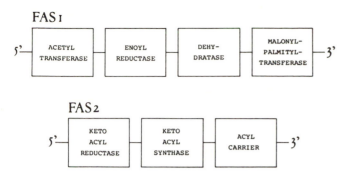

FIG. 2. Sequential order of FAS domains in FAS1 and FAS2 as determined by deletion mapping of respective mutants (31).

205, 130 daltons. The nucleotide sequence of FAS2 is depicted in Figure 3 together with the amino acid sequence of subunit α. Accordingly, the FAS2 coding sequence consists of 5661 nucleotides and thus subunit α contains 1887 amino acids and has a molecular weight of 206, 683 daltons. These findings agree with the observation that upon SDS polyacrylamide gel electrophoresis subunit α is slightly larger than subunit β (33). Nevertheless, the size difference between α and β as documented by the sequence data is smaller than that formerly derived from their electrophoretic mobilities (33,34). Therefore, either one or both subunits exhibit unusual electrophoretic mobilities or they are, to a limited extent, subject to posttranslational proteolysis.

Both *S. cerevisiae* FAS genes are free of introns. This is concluded from the absence in both FAS genes of the yeast–specific TACTAAC sequence present in every intron–containing yeast gene identified to date (35). Since the N- and C-terminal peptide sequences of subunits α and β are as yet unknown, the protein primary structures derived from the FAS1 and FAS2 open reading frames obviously represent maximal–length gene products. They invoke the first possible translational start and stop codons, respectively. Both start codons were confirmed by S1 mapping of the respective transcriptional initiation sites (32 and G. Müller, unpublished data), while the FAS2 termination codon was determined independently also by sequencing of the appropriate poly(A)–containing cDNA (R. Bauer-Hofmann, unpublished data).

The FAS1 and FAS2 coding sequences may be inspected for the occurrence of known subunit α and β active

sites.Thus, the acetyl-binding serine of acetyl transferase was identified at position 247, the malonyl-/palmityl-binding serine at position 1808 and the putative enoyl reductase-specific (36) sequence Gly-Ser-Ala at positions 772–774 in the β-chain. These positions are in perfect agreement with the functional subdivision of FAS1 as achieved earlier by deletion mapping (Fig. 2). Thus, the sequence data confirm the order of catalytic domains in subunit β as shown in Figure 2. Available data (22) suggest that these domains, being located in a distinct linear order along the chain, are connected by appropriate peptide spacers such that little or no protein-protein interaction occurs between them. This characteristic is by no means an expected property of all multidomain proteins. Even among type I fatty acid synthetases, it probably represents an exceptional rather than a general phenomenon. For instance, the *Candida lipolytica* fatty acid synthetase exhibiting essentially the same biochemical and gross protein-chemical properties as the *Saccharomyces cerevisiae* enzyme shows no interallelic complementation between differently affected fas1 mutants (R. Regler, unpublished data). This means that in this yeast, mutational alteration of one FAS domain concomitantly inactivates all others.

In vitro complementation studies suggested that there is one major interdomain region in FAS1 with a so-far unassigned function (32). This region, about 300 amino acids in length, is located between amino acids 850 and 1150, i.e., between the enoyl reductase and dehydratase domains. None of the fas1 mutants studied were reverted to FAS prototrophy by DNA fragments corresponding to this section of the FAS1 gene (32). In order to investigate whether this finding was of more than accidental significance, we introduced several in vitro constructed mutations into this region. The FAS1 DNAs thus modified were transformed into an intragenically noncomplementing fas1 mutant and analyzed for their ability to synthesize enzymatically active β subunits. As shown in Table 1, all amino acid replacements tested were functional and failed to produce a fas-negative phenotype. As it also is seen, the substituting amino acids deliberately were selected to exhibit different physicochemical characteristics compared to the genuine ones (Table 1). Therefore, these mutations would not be expected to be silent when located at critical positions. The only in vitro–produced mutation exhibiting fas-negative characteristics was the conversion of an arginine codon into a chain terminating TGA triplet (Table 1). It is concluded

TABLE 1

Replacement of Selected Amino Acids Within FAS Subunit β by In Vitro Mutagenesis

Amino acid position	Codon change	Amino acid replacement	fas1 Complementation
933	AGA→TGA	Arg→End	-
	→AGT	→Ile	+
957	CGT→AGC	Arg→Ser	+
972	CTA→AGA	Leu→Arg	+
985	AAT→AAA	Asn→Lys	+
1032	GAT→GCT	Asp→Ala	+
1108	CCT→GCT	Pro→Ala	+

E. SCHWEIZER ET AL.

```
   1  ATTCGATGGGTACGTACTCGCGGCCATTCCTTGTGCGTGCAGGAATCAAGAAAATAAACAAACCGGTACAGTACAATAAGCAGGCTGAGTAGGCGGAAA
 100  AAGTTTTATTCAGACGTGTACCGGAGCCTAAATCCTTTCTTACCCATAGCAACCACACTAAATATATTATTGTAACACGAGAAGATGTCACAAGGTCAG
 199  TCCAAAAAACTGGACGTAACTGTTGAGCAGCTTCGAATATATACCACCAGTTTCATGATATCTTGGAAGAAAAAACTGATTTGCATCTACCGAAGAAAG
 298  AATACGACGATGACGCTGTTAGGAGAGAGGTTCAGATACAGTTACAAGAATTTCTTTTGAGCGCTATGACGATGGCTTCGAAGTCACTAGAAGTTGTCA
 397  ACGCCGACACGGTAGGAAAGACGGTAAAGCAATTGATCATGGAATCACAAGAGAAGTACATGGAGACCTTGACCTTGACCTGAATGAGCAAGTTAGAA
 496  AGATGTACCAAGAGTGGGAAGACGAAACCGTTAAGGTGGCCCAGTTGAGGCAAACGGGGCCTGCAAAAATCAACGAAGTTTACAACAACTCAAAGGATG
 595  AGTATTTGGCACAATTGGATGGGAGAATCGGCGTTCTTCAAGCTAGAATGATGCAGCAACAATCTGCTGACCATGATGATAGTACCGACGATGCCGATG
 694  ATCACATCAACTGGGAGCACATCAAGCAGGATTACGTTGCCCTCACTCAATGAATTGTATCAAACACAGCAAGACCTACCCAAGGTAAGATATAACGTT
 793  GAAAAGGTCAAGCGCTTAATGGACTTCCTGGAGGAGGATTGAATAAATAACATATAAGACAGTCAACCTCGTGGTAAATGATAACTATATTCTGGGGCT
 892  TTATTCATTTTTTTACACTTTTTTATCTTTGTACGAAGAGCTGAAAATAAAAAGTATAGAGCCAGCCTTTTCCCACACACTAACGACAAAAGACGATA
 991  TGCAGGCTGTGCGCATGGGGGAGAATCCAGTTTTTTGCCCGTTAAAGAGAAAAAATTAGTGTCTGTGAAACGCGAATACAAAAAAAGAAAAGTAGCAGGA
1090  TCTTACTTCTCGTAAAGACGTCAAGAACCAAATCAAGTCAAATCGTGGAAGTTACAAGGGGAAAGACCAATAACTTTTAGTAAAGAACAAAGAAAGGTC
1189  TATCTCACGCAGTGACGGTCTTTGCGGTAAATCTGTGTATACTTGAAAGAAAACCCTTTTACAATTAAAAAAGGCAATTAAAAATAGAAACAAATCAAA
1288  TGAGTTATAATAATCCGTACCAGTTGGAAACCCCTTTTGAAGAGTCATACGAGTTGGACGAAGGTTCGAGCGCTATCGGTGCTGAAGGCCACGATTTCG
1387  TGGGCTTCATGAATAAGATCAGTCAAATCAATCGCGATCTCGATAAGTACGACCATACCATCAACCAGGTCGATTCTTTGCATAAGAGGCTACTGACCG
1486  AAGTTAATGAGGAGCAAGCAAGTCACTTAAGGCACTCCCTGGACAACTTCGTCGCACAAGCCACGGACTTGCAGTTCAAACTGAAAAATGAGATTAAAA
1585  GTGCCCAAAGGGATGGGATACATGACACCAACAAGCAAGCTCAGGCGGAAAACTCCAGACAAAGATTTTTGAAGCTTATCCAGGACTACAGAATTGTGG
1684  ATTCCAACTACAAGGAGGAGAATAAAGAGCAAGCCAAGAGGCAGTATATGATCATTCAACCAGAGGCCACCGAAGATGAAGTTGAAGCAGCCATAAGCG
1783  ATGTAGGGGGCCAGCAGATCTTCTCACAAGCATTGTTGAATGCTAACAGACGTGGGGAAGCCAAGACTGCTCCTTGCGGAAGTCCAGGCAAGGCACCAAG
1882  AGTTATTGGAAACTAGAAAAATCCATGGCAGAACTTACTCAATTGTTTAATGACATGGGAAGAACTGGTATTAGAACAACAAGAAAACGTAGACGTCATCG
1981  ACAAGAACGTTGAAGACGCTCAACTCGACGTAGAACAGGGTGTCGGTCATACCGGTAAGGGTGCCAGAAAAGCAAGAAAGAACAAGAATTA
2080  GATGTTGGTTGATTGTTATTCGCCATCATTGTAGTCGTTGTTGTTGTCGTTGTTGTCCCAGCCGTTGTCAAAACGCGTTAATTCCAACTATTTTCTATA
2179  TTTCTATTCTATCCGAACTCCCCTTTTGTATATCAATATATCTTAAATACTTTCGCCTATTATTTTCCTAAATTTTCTCTGGTTCTGCAGGCC
2278  AAAAACAACAACTTACTACTGAATCATGGACGTGTATTTAGTTTAGCCAAGCAATATTTAAATATCACTCTTCCTAAAAAATACATTGGGCATTACCCGC
2377  AATCTAACCCATCGCTTAGCAAAATCCAACCATTTTTTTTTTATCTCCCGCGTTTTCACATGCTACCTCATTCGCCTCGTAACGTTACGACCGAAATCT
2476  CACTAAGGCACGGTTTGTTGGGCAGTTTACAGATGTTGGATAACCAGTTGTTTCTAAACGGTTATGCCTC̲A̲T̲A̲T̲A̲T̲A̲A̲C̲TTGTTAACTGAAGGTTACAC

2575  AAGACCCACATCACCACTGTCGTGCTTTCTAATAACCGC̲T̲A̲T̲A̲T̲T̲A̲GACGTTTAAAGGGCTACAGCAACACCAATTGAAATACCATCATTATGAAGCCG      ]
                                                                                                  MetLysPro

2674  GAAGTTGAGCAAGAATTAGCTCATATTTTGCTAACTGAATTGTTAGCTTATCAATTTGCCTCTCCTGTGAGATGGATTGAAACTCAAGATGTTTTTTTG
      GluValGluGlnGluLeuAlaHisIleLeuLeuThrGluLeuLeuAlaTyrGlnPheAlaSerProValArgTrpIleGluThrGlnAspValPheLeu

2773  AAGGGATTTTAACACTGAAAGGGTTGTTGAAATCGGTCCTTCTCCAACTTTGGCTGGGATGGCTCAAAGAACCTTGAAGAATAAAATACGAATCTTACGAT
      LysAspPheAsnThrGluArgValValGluIleGlyProSerProThrLeuAlaGlyMetAlaGlnArgThrLeuLysAsnLysIleArgIleLeuArgAsp

2872  GCTGCTCTGTCTTTACATAGAGAAATCTTTATGCTATTCGAAGGATGCCAAAGAGATTTATTATACCCCAGATCCATCCGAACTAGCTGCAAAGGAAGAG
      AlaAlaLeuSerLeuHisArgGluIleLeuCysTyrSerLysAspAlaLysIleTyrTyrThrProAspProSerGluLeuAlaAlaLysGluGlu

2971  CCCGCTAAGGAAGAAGCTCCTGCTCCAACTCCAGCTGCTAGTGCTCCTGCTCCTGCAGCAGCAGCCCCAGCTCCCGTCGCGGCAGCAGCCCCAGCTGCA
      ProAlaLysGluGluAlaProAlaProThrProAlaAlaSerAlaProAlaProAlaAlaAlaAlaProAlaProValAlaAlaAlaAlaProAlaAla

3070  GCAGCTGCTGAGATTGCCGATGAACCTGTCAAGGCTTCCCTATTGTTGCACGTTTTGGTTGCTCACAAGTTGAAGAAGTCGTTAGATTCCATTCCAATG
      AlaAlaAlaGluIleAlaAspGluProValLysAlaSerLeuLeuLeuHisValLeuValAlaHisLysLeuLysLysSerLeuAspSerIleProMet

3169  TCCAAGACAATCAAAGACTTGGTCGGTGGTAAⒶTCTⒶCAGTCCAAAATGAAATTTTGGGTGATTTAGGTAAAGAATTTGGTACTACTCCTGAAAAACCA
      SerLysThrIleLysAspLeuValGlyGlyLysSerⓉhrValGlnAsnGluIleLeuGlyAspLeuGlyLysGluPheGlyThrThrProGluLysPro

3268  GAAGAAACTCCATTAGAAGAATTGGCAGAAACTTTCCAAGATACCTTCTCTGGAGCATTGGGTAAGCAATCTTCCTCGTTATTATCAAGATTAATCTCA
      GluGluThrProLeuGluGluLeuAlaGluThrPheGlnAspThrPheSerGlyAlaLeuGlyLysGlnSerSerSerLeuLeuSerArgLeuIleSer

3367  TCTAAGATGCCTGGTGGGTTTACTATTACTGTCGCTAGAAAAATACTTACAAACTCGCTGGGGACTACCATCTGGTAGACAAGATGGTGTCCTTTTGGTA
      SerLysMetProGlyGlyPheThrIleThrValAlaArgLysTyrLeuGlnThrArgTrpGlyLeuProSerGlyArgGlnAspGlyValLeuLeuVal

3466  GCTTTATCTAACGAGCCTGCTGCTCGTCTAGGTTCTGAAGCTGATGCCAAGGCTTTCTTGGACTCCATGGCTCAAAAATACGCTTCCATTGTTGGTGTT
      AlaLeuSerAsnGluProAlaAlaArgLeuGlySerGluAlaAspAlaLysAlaPheLeuAspSerMetAlaGlnLysTyrAlaSerIleValGlyVal

3565  GACTTATCATCAGCTGCTAGCGCTAGTGGTGCTGCCGGTGCAGGTGCTGCTGCCGGTGCAGCTATGATCGATGCTGGCGCTCTGGAAGAAATAACCAAA
      AspLeuSerSerAlaAlaSerAlaSerGlyAlaAlaGlyAlaGlyAlaAlaAlaGlyAlaAlaMetIleAspAlaGlyAlaLeuGluGluIleThrLys

3664  GACCACAAGGTTTTGGCGCGTCAACAACTGCAAGTATTGGCTCGTTATCTAAAAATGGACTTGGATAACGGTGAAAGAAAGTTCTTGAAAGAAAAGGAC
      AspHisLysValLeuAlaArgGlnGlnLeuGlnValLeuAlaArgTyrLeuLysMetAspLeuAspAsnGlyGluArgLysPheLeuLysGluLysAsp

3763  ACTGTTGCTGAACTTCAAGCTCAGTTGGATTACTTGAATGCCGAATTAGGTGAATTCTTTGTTAACGGTGTTGCTACTTCTTTCTCTAGAAAAAAGGCC
      ThrValAlaGluLeuGlnAlaGlnLeuAspTyrLeuAsnAlaGluLeuGlyGluPhePheValAsnGlyValAlaThrSerPheSerArgLysLysAla

3862  AGAACCTTCGATTCTTCCTGGAACTGGGCTAAACAATCTTTATTATCATTATACTTTGAGATAATTCATGGTGCTCTTGAAAAACGTTGATAGAGAGGTT
      ArgThrPheAspSerSerTrpAsnTrpAlaLysGlnSerLeuLeuSerLeuTyrPheGluIleIleHisGlyValLeuLysAsnValAspArgGluVal

3961  GTTAGTGAAGCTATCAATATCATGAACAGATCTAACGATGCTTTGATTAAATTCATGGAATACCCATATCTCTAACACTGATGAAACAAAAGGTGAAAAC
      ValSerGluAlaIleAsnIleMetAsnArgSerAsnAspAlaLeuIleLysPheMetGluTyrHisIleSerAsnThrAspGluThrLysGlyGluAsn

4060  TATCAATTGGTTAAAACTCTTGGTGAGCAGTTGATTGAAAACTGTAAACAAGTTTTGGATGTTGATCCAGTTTACAAAGATGTTGCTAAGCCTACCGGT
      TyrGlnLeuValLysThrLeuGlyGluGlnLeuIleGluAsnCysLysGlnValLeuAspValAspProValTyrLysAspValAlaLysProThrGly

4159  CCAAAAAACTGCTATTGACAGAACGGTAACATTACATACTCAGAAGAGCCAAGAGAAAAGGTTAGGAAATTATCTCAATACGTACAAGAAATGGCCCTT
      ProLysThrAlaIleAspLysAsnGlyAsnIleThrTyrSerGluGluProArgGluLysValArgLysLeuSerGlnTyrValGlnLeuMetAlaLeu

4258  GGTGGTCCAATCACCAAAGAATCTCAAACTACTATTGAAGAGGATTTGACTCGTGTTTACAAGGCAATCAGTGCTCAAGCTGATAAACAAGATATTTCC
      GlyGlyProIleThrLysGluSerGlnThrThrIleGluGluAspLeuThrArgValTyrLysAlaIleSerAlaGlnAlaAspLysGlnAspIleSer

4357  AGCTCCACCAGGGTTGAATTTGAAAAACTATATAGTTGATTTGATGAAGTTCTTGGAAAGCTCCAAAGAAATCGATCCTTCTCAAACAACCCAATTGGCC
      SerSerThrArgValGluPheGluLysLeuTyrSerAspLeuMetLysPheLeuGluSerSerLysGluIleAspProSerGlnThrThrGlnLeuAla

4456  GGTATGGATGTTGAGGATGCTCTTTGGACAAAGATTCCACCAAAGAAGTTGCTTCTTGCAAAAATCTACCATTTCTAAGACGGTATCTTCAACTATT
      GlyMetAspValGluAspAlaLeuAspLysAspSerThrLysGluValAlaSerLeuProAsnLysSerThrIleSerLysThrValSerSerThrIle

4555  CCAAGAGAAACTATTCCGTTCTTACATTTGAGAAAGAAGACTCCTGCCGGAGATTGAAATATGACCGCCAATTGTCTTCTCTTTTCTTAGATGGTTTA
      ProArgGluThrIleProPheLeuHisLeuArgLysLysThrProAlaGlyAspTrpLysTyrAspArgGlnLeuSerSerLeuPheLeuAspGlyLeu

4654  GAAAAGGCTGCCTTCAACGGTGTCACCTTCAAGGACAAATACGTCTTGATCACTGGTGCTGGTAAGGGTTCTATTGGTGCTGAAGTCTTGCAAGGTTTG
      GluLysAlaAlaPheAsnGlyValThrPheLysAspLysTyrValLeuIleThrGlyAlaGlyLysGlySerIleGlyAlaGluValLeuGlnGlyLeu

4753  TTACAAGGTGGTGCTAAGGTTGTTGTTACCACCTCTCGTTTCTCTAAGCAAGTTACAGACTACTACCAATCCATTTACGCCAAAATGGTGCTAAGGGT
      LeuGlnGlyGlyAlaLysValValValThrThrSerArgPheSerLysGlnValThrAspTyrTyrGlnSerIleTyrAlaLysMetValLeuLysGly
```

KETOACYL REDUCTASE

ACYL CARRIER

FIG. 3. Nucleotide sequence of yeast FAS2 and amino acid sequence of FAS subunit α. The pantetheine-binding serine and the "peripheral" SH-group of the enzyme are circled. The transcriptional initiation sites (•) and the polyadenylation site (*) are indicated. Putative transcriptional start and stop signals are underlined. The location of ketoacyl reductase and synthase domains was deduced from Fig. 2, that of the acyl carrier from the position of the pantetheine binding site.

GENETIC CONTROL OF FATTY ACID BIOSYNTHESIS IN YEAST

```
4852  TCTACTTTGATTGTTGTTCCATTCAACCAAGGTTCTAAGCAAGACGTTGAAGCTTTGATTGAATTTATCTACGACACTGAAAAGAATGGTGGTTTAGGT
      SerThrLeuIleValValProPheAsnGlnGlySerLysGlnAspValGluAlaLeuIleGluProIleTyrAspThrGluLysAsnGlyGlyLeuGly ┐

4951  TGGGATCTAGATGCTATTATTCCATTCGCGGCCATTCCAGAACAAGGTATTGAATTAGAACATATTGATTCTAAGTCTGAATTTGCTCATAGAATCATG
      TrpAspLeuAspAlaIleIleProPheAlaAlaIleProGluGlnGlyIleGluLeuGluHisIleAspSerLysSerGluPheAlaHisArgIleMet

5050  TTGACCAATATCTTAAGAATGATGGGTTGTGTCAAGAAGCAAAAATCTGCAAGAGGTATTGAAACAAGACCAGCTCAAGTCATTCTACCAATGTCTCCA
      LeuThrAsnIleLeuArgMetMetGlyCysValLysLysGlnLysSerAlaArgGlyIleGluThrArgProAlaGlnValIleLeuProMetSerPro

5149  AACCATGGTACTTTCGGTGGTGATGGTATGTATTCAGAATCCAAGTTGTCTCTTGGAAACTTTGTTCAACAGATGGCACTCTGAATCCTGGGCCAATCAA
      AsnHisGlyThrPheGlyGlyAspGlyMetTyrSerGluSerLysLeuSerLeuGlyThrLeuPheAsnArgTrpHisSerGluSerTrpAlaAsnGln

5248  TTAACCGTTTGCGGTGCTATTATTGGTTGGACTAGAGGTACTGGTTTAATGAGCGCTAATAACATCATTGCTGAAGGCATTGAAAAGATGGGTGTTCGT
      LeuThrValCysGlyAlaIleIleGlyTrpThrArgGlyThrGlyLeuMetSerAlaAsnAsnIleIleAlaGluGlyIleGluLysMetGlyValArg

5347  ACTTTCTCTCAAAAGGAAATGGCTTTCAACTTATTGGGTCTATTGACTCCAGAAGTCGTAGAATTGTGCCAAAAATCACCTGTTATGGCTGACTTGAAT
      ThrPheSerGlnLysGluMetAlaPheAsnLeuLeuGlyLeuLeuThrProGluValValGluLeuCysGlnLysSerProValMetAlaAspLeuAsn

5446  GGTGGTTTGCAATTTGTTCCTGAATTGAAGGAATTCACTGCTAAATTGCGTAAAGAGTTGGTTGAAACTTCTGAAGTTAGAAAGGCAGTTTCCATCGAA
      GlyGlyLeuGlnPheValProGluLeuLysGluPheThrAlaLysLeuArgLysGluLeuValGluThrSerGluValArgLysAlaValSerIleGlu

5545  ACTGCTTTTGGAGCATAAGGTTGTCAATGGCAATAGCGCTGATGCTGCATATGCTCAAGTCGAAATTCAACCAAGAGCTAACATTCAACTGGACTTCCCA
      ThrAlaLeuGluHisLysValValAsnGlyAsnSerAlaAspAlaAlaTyrAlaGlnValGluIleGlnProArgAlaAsnIleGlnLeuAspPhePro

5644  GAATTGAAACCATACAAACAGGTTAAACAAATTGCTCCCGCTGAGCTTGAAGGTTTGTTGGATTTGGAAAGAGTTATTGTAGTTACCGGTTTTGCTGAA
      GluLeuLysProTyrLysGlnValLysGlnIleAlaProAlaGluLeuGluGlyLeuLeuAspLeuGluArgValIleValValThrGlyPheAlaGlu

5743  GTCGGCCCATGGGGTTCGGCCAGAACAAGATGGGAAATGGAAGCTTTTGGTGAATTTTCGTTGGAAGGTTGCGTTGAAATGGCCTGGATTATGGGCTTC
      ValGlyProTrpGlySerAlaArgThrArgTrpGluMetGluAlaPheGlyGluPheSerLeuGluGlyCysValGluMetAlaTrpIleMetGlyPhe

5842  ATTTCATACCATAACGGTAATTTGAAGGGTCGTCCATACACTGGTTGGGTTGATTCAAAGACTCTCCCAAAACAAAGAACCAGTTGATGACAAGGACGTTAAGGCCAAG
      IleSerTyrHisAsnGlyAsnLeuLysGlyArgProTyrThrGlyTrpValAspSerLysThrLeuProValAspAspLysAspValLeuAlaLys

5941  TATGAAACATCAATCCTAGAACACAGTGGTATCAGATTGATCGAACCAGAGTTATTCAATGGTTACAACCCAGAAAAGAAGGAAATGATTCAAGAAGTC
      TyrGluThrSerIleLeuGluHisSerGlyIleArgLeuIleGluProGluLeuPheAsnGlyTyrAsnProGluLysLysGluMetIleGlnGluVal

6040  ATTGTCGAAGAACGACTTGGCAACCATTTGAGGCTTCGAAGGAAACTGCCGAACAATTTAAACACGCTGACAAAGGTTGCAAAGTGGATATCTTCGAAATCCCA
      IleValGluGluAspLeuGluProPheGluAlaSerLysGlnThrAlaGluGlnPheLysHisGlnHisGlyAspLysValAspIlePheGluIlePro

6139  GAAACAGGAGAGTACTCTGTTAAGTTACTAAAGGGTGCCACTTTATACATTCCAAAGGCTTTGAGATTTGACCGTTTGGTTGCAGGTCAAATTCCAACT
      GluThrGlyGluTyrSerValLysLeuLeuLysGlyAlaThrLeuTyrIleProLysAlaLeuArgPheAspArgLeuValAlaGlyGlnIleProThr

6238  GGTTGGAATGCTAAGACTTATGGTATCTCTGATGATATCATTTCTCAGGTTGACCCAATCACATTATTCGTTTTGGTCTCTGTTGTGGAAGCATTTATT
      GlyTrpAsnAlaLysThrTyrGlyIleSerAspAspIleIleSerGlnValAspProIleThrLeuPheValLeuValSerValValGluAlaPheIle

6337  GCATCTGGTATCACCGACCCATACGAAATGTACAAATACGTACATGTTTCTGAGGTTGGTAACTGTTCTGGTTCTGGTATGGGTGGTGTTTCTGCCTTA
      AlaSerGlyIleThrAspProTyrGluMetTyrLysTyrValHisValSerGluValGlyAsnCysSerGlySerGlyMetGlyGlyValSerAlaLeu

6436  CGTGGTATGTTTAAGGACCGTTTCAAGGATGAGCCTGTCCAAAATGATATTTTACAAGAATCATTTATCAACACCATGTCCGCTTGGGTTAATATGTTG
      ArgGlyMetPheLysAspArgPheLysAspGluProValGlnAsnAspIleLeuGlnGluSerPheIleAsnThrMetSerAlaTrpValAsnMetLeu

6535  TTGATTTCCTCATCTGGTCCAATCAAGACACCTGTTGGTGCCTGTGCCCACATCCGTGGAATCTGTTGACATTGGTGTAGAAACCATCTTGTCTGGTAAG
      LeuIleSerSerSerGlyProIleLysThrProValGlyAlaCysAlaThrSerValGluSerValAspIleGlyValGluThrIleLeuSerGlyLys

6634  GCTAGAATCTGTATTGTCGGTGGTGGATATGATTTCCAAGAAGAAGGCTCCTTTGAGTTCGGTAACATGAAGGCCACTTCCAACACTTTGGAAGAATTT
      AlaArgIleCysIleValGlyGlyGlyTyrAspPheGlnGluGluGlySerPheGluPheGlyAsnMetLysAlaThrSerAsnThrLeuGluGluPhe

6733  GAACATGGTCGTACCCCAGCGGAAATGTCCAGACCTGCCACCACTACCCGTAACGGTTTTATGGAAGCTCAAGGTGCTGGTATTCAAATCATCATGCAA
      GluHisGlyArgThrProAlaGluMetSerArgProAlaThrThrThrArgAsnGlyPheMetGluAlaGlnGlyAlaGlyIleGlnIleIleMetGln

6832  GCTGATTTAGCTTTGAAGATGGGTGTGCCAATTTACGGTATTGTTGCCATGGCTGCCACCGATAAGATTGGTAGAAGTGTGCCAGCTCCAGGT
      AlaAspLeuAlaLeuLysMetGlyValProIleTyrGlyIleValAlaMetAlaAlaThrAlaThrAspLysIleGlyArgSerValProAlaProGly

6931  AAGGGTATTTTAACCACTGCTCGTGAACACCACTCCAGTGTTAAGTATGCTTCACCAAACTTGAACATGAAGTACAGAAAGCGCCAATTGGTACTCGT
      LysGlyIleLeuThrThrAlaArgGluHisHisSerSerValLysTyrAlaSerProAsnLeuAsnMetLysTyrArgLysArgGlnLeuValThrArg

7030  GAAGCTCAGATTAAAGATTGGGTAGAAAACGAATTGGAAGCTTTGAAGTTGGAGGCCCAAGAGAAGACCCAAGCGAAGACCAAAACGAGTTCTTACTTGAA
      GluAlaGlnIleLysAspTrpValGluAsnGluLeuGluAlaLeuLysLeuGluAlaGluGlnThrProGlnAlaLysProGlnAlaGluAsnGluPheLeuLeuGlu

7129  CGTACCAGAGAAATCCACAACGAAGCTGAAAGTCAATTGAGAGCTGCACAACAACAATGGGGTAACGACTTCTACAAGAGGGACCCACGTATTGCTCCA
      ArgThrArgGluIleHisAsnGluAlaGluSerGlnLeuArgAlaAlaGlnGlnGlnTrpGlyAsnAspPheTyrLysArgAspProArgIleAlaPro

7228  TTGAGAGGACGACTGGCTACTTACGGTTTAACTATTGATGACTTGGGTGTCGCTTCAATCCACGGTACATCCACAAAGGCTAATGACAAGAACGAATCT
      LeuArgGlyAlaLeuAlaThrTyrGlyLeuThrIleAspAspLeuGlyValAlaSerPheHisGlyThrSerThrLysAlaAsnAspLysAsnGluSer

7327  GCCACAATTAATGAAATGATGAAGCATTTGGGTAGATCTGAAGGTAATCCCGTCATTGGTGTTTTTCCAAAAGTTCTTGACTGGTCATCCAAAGGGTGCT
      AlaThrIleAsnGluMetMetLysHisLeuGlyArgSerGluGlyAsnProValIleGlyValPheGlnLysPheLeuThrGlyHisProLysGlyAla

7426  GCTGGTGCATGGATGATGAATGGTGCTTTGCAAATTCTAAACAGTGGTATTATTCCAGGTAACCGTAACGCTGATAACGTGGATAAGATCTTGGAGCAA
      AlaGlyAlaTrpMetMetAsnGlyAlaLeuGlnIleLeuAsnSerGlyIleIleProGlyAsnArgAsnAlaAspAsnValAspLysIleLeuGluGlnGln

7525  TTTGAATACGTCTTGTACCCATCCAAGACTTTAAAGACCGACGGTGTCAGAGCCGTGTCCATCACTTCTTTCGGTTTTGGTCAAAAGGGTGGTCAAGCT
      PheGluTyrValLeuTyrProSerLysThrLeuLysThrAspGlyValArgAlaValSerIleThrSerPheGlyPheGlyGlnLysGlyGlyGlnAla

7624  ATTGTGGTTCATCCAGACTACTTATACGGTGCTATCACTGAAGACAGATACAACGAGTATGTCGCCAAGGTTAGTGCCAGAGAGAAAGTGCCTACAAA
      IleValValHisProAspTyrLeuTyrGlyAlaIleThrGluAspArgTyrAsnGluTyrValAlaLysValSerAlaArgGluLysSerAlaTyrLys

7723  TTCTTCCATAATGGTATGATCTACAACAAGTTGTTCGTAAGTAAAGAGCATGCTCCATACACTGATGAATTGGAAGAGGATGTTTACTTGGACCCATTA
      PhePheHisAsnGlyMetIleTyrAsnLysLeuPheValSerLysGluHisAlaProTyrThrAspGluLeuGluGluAspValTyrLeuAspProLeu

7822  GCCCGTGTATCTAAGGATAAGAAATCAGGCTCCTTGACTTTCAACTCTAAAAACATCCAAAGCAAGGACAGTTACATCAATGCTAACACCATTGAAACT
      AlaArgValSerLysAspLysLysSerGlySerLeuThrPheAsnSerLysAsnIleGlnSerLysAspSerTyrIleAsnAlaAsnThrIleGluThr

7921  GCCAAGATGATTGAAAACATGACCAAGGAGAAAGTCTCTAACGGTGGCGTCGGTGTAGATGTTGAATTAATCACTAGCATCAACGTTGAAAATGATACT
      AlaLysMetIleGluAsnMetThrLysGluLysValSerAsnGlyGlyValGlyValAspValGluLeuIleThrSerIleAsnValGluAsnAspThr

8020  TTTATCGAGCGCAATTTCACCCCGCAAGAAATAGAGTACTGCAGCGCGCAGCCTAGTGTGCAAAGCTCTTTCGCTGGGACATGGTCCGCCAAAGAGGCT
      PheIleGluArgAsnPheThrProGlnGluIleGluTyrCysSerAlaGlnProSerValGlnSerSerPheAlaGlyThrTrpSerAlaLysGluAla

8119  GTTTTCAAGTCCTTAGGCGTCAAGTCCTTAGGCGGTGGTGCTGCATTGAAAGACATCGAAATCGTACGCGTTAACAAAAACGCTCCAGCCGTTGAACTG
      ValPheLysSerLeuGlyValLysSerLeuGlyGlyGlyAlaAlaLeuLysAspIleGluIleValAlaArgValAsnLysAsnAlaProAlaValGluLeu

8218  CACGGTAACGCCAAAAAGGCTGCCGAAGAAGCTGGTGTTACCGATGTGAAGGTATCTATTTCTCACGATGACCTCCAAGCTGTCGCGGTCGCCGTTTCT
      HisGlyAsnAlaLysLysAlaAlaGluGluAlaGlyValThrAspValLysValSerIleSerHisAspAspLeuGlnAlaValAlaValAlaValSer

8317  ACTAAGAAATAGAGAGAGCACTACGTAGTCCCTCTTTTAATATGTAACGTGTCGCTTCTATTTATCAGACATAATAGTAATTACTTTGTTATTTTTCTA
      ThrLysLysEnd

8416  TCGTTTCCTTACTTTAGCCTCTGATTTCGATCTGGCCTAATCATTGTGTACGTCTATAATCATATGGCCCTGAGCGTACACACCGTTCAATTCTTCTAT
8515  TTGGTCGGCCCACACAAATTCGCCGCGGGGGCATATTGACGCGTGGAAGAATAGAAGTCCGCGTAGCAGCTCTGGGGTAATTAGGCTTTTACGATATCGG
8614  CTGGCCCAGCACCGGTTTTTATTCCTCCCGTGGTTGTTGTCTCTACCGTGAGGAGGGGAAAGGGTCAGGGACGGCCGGTAGTTATGTTGTCTCAACGAT
8713  TACTGCCATCTATTGTTTACACCAATCGCGTGGGGCAAGGGGCTGGAAGGCTGTGGTGTACAGAATAGGGTCTAATACCCGATGCGCGCTCCTGCG
8812  ACTGGTTGTGTGAGCCACCCCTTGATCGCCTCTGCCAAATACTCAGGGAAGAGAAAAATCGCGAAGGTGGACGAGTGCCCTTAAAGAACCAGCATCCTC
8911  GATCGCCGCTGCCGTTCGCCCTGGCTTAGCGGTGGGAAGCATCCTTTGGCTTTCGAATTCTCCGCCGTTGAGTCCCGCTAA  8991
```

KETOACYL SYNTHASE

FIG. 3. continued

from these results that some parts of FAS1 functionally are inert and that the yeast fatty acid synthetase genes probably contain more DNA than is required for encoding their catalytic domains. This may explain why the combined molecular weight of both yeast FAS subunits is more than 400,000 daltons while that of the octafunctional animal FAS subunit that accommodates the same set of active sites is only 250,000 daltons (19).

The FAS2 encoded α-subunit of yeast fatty acid synthetase contains the so-called ''central'' (SH_c) and ''peripheral'' (SH_p) SH-groups to which the malonyl and fatty acyl residues are attached during the overall catalytic process (1). Chemically, the ''central'' SH-group is part of acyl carrier protein–bound phosphopantetheine, while the ''peripheral'' thiol group is a distinct cysteinyl residue of the ketoacyl synthase active site (1). The amino acid sequences around both thiols are known, in yeast FAS, and may be used to identify the respective positions within the FAS2 gene. As indicated in Figure 3, the phosphopantetheine–binding serine of subunit α was identified at position 180 and the ''peripheral'' cysteine at position 1305 within the subunit α protein sequence.

The ''peripheral'' thiol group is located, as expected, from the complementation experiments depicted in Figure 2, within the ketoacyl synthase domain. On the other hand, the pantetheine-binding site was located at a position distal to that of the pantetheine-less mutations studied in Figure 2.

This latter area is indicated in Figure 3 by a broken line extending from the ketosynthase domain to the end of the gene. Thus, obviously there are two different parts of subunit α contributing to the attachment of phosphopantetheine to the fatty acid synthetase complex: one of them is located at the N-terminus and the other at the C-terminus of subunit α. Usually, the pantetheine-binding site of fatty acid synthetase is defined as the acyl carrier protein domain. In Figure 3, this domain is assigned to the N-terminus of FAS2 and, therefore conforms to this definition. On the other hand, in Figure 2 the C-terminal part of FAS2, mutations of which also prevent phosphopantetheine from binding to the enzyme, was designated as acyl carrier protein. Probably both parts of the protein are essential for this function and must interact in order to allow pantetheine binding to the enzyme. The acyl carrier protein function of subunit α thus may be divided into two half-domains located at the C- and N-terminal parts of the protein, respectively. As schematically depicted in Figure 4, intra- or intermolecular interaction between both sites may be necessary for activity.

SEQUENCE COMPARISON BETWEEN YEAST AND PENICILLIUM FAS

As part of a comparative study of the fatty acid synthetase and methylsalicylic acid synthetase complexes from *Penicillium patulum*, the FAS2 gene from this organism recently was isolated and sequenced in our laboratory (P. Wiesner, unpublished data). The two fungi, yeast and *Penicillium patulum*, contain fatty acid synthetases of very similar overall protein structure. Like the yeast complex, *Penicillium* FAS is composed of two subunits α and β, of approximately 200,000 daltons molecular weight (Fig. 5). Using specific antisera against the *Penicillium* enzyme, the corresponding FAS2 DNA was isolated from

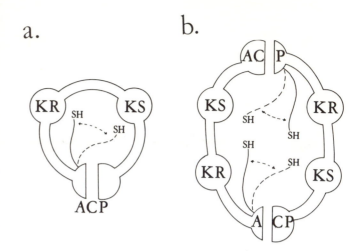

FIG. 4. Hypothetical formation of the acyl carrier (ACP) domain in subunit α by intra-(a) or intermolecular (b) interaction of two halfdomains.

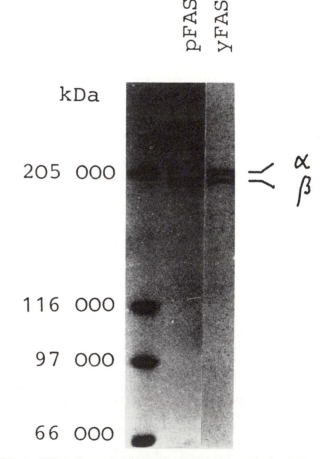

FIG. 5. SDS-polyacrylamide gel electrophoresis of purified *Penicillium* (pFAS) and yeast (yFAS) fatty acid synthetases. Experimental conditions were as described previously (33).

a genomic *P. patulum* λ gt11 expression library. The FAS2 open reading frame is 5571 nucleotides long, it is interrupted by two 54 base-pair introns and codes for a protein of 204,500 daltons molecular weight (P. Wiesner, unpublished data). The two introns are inserted into the

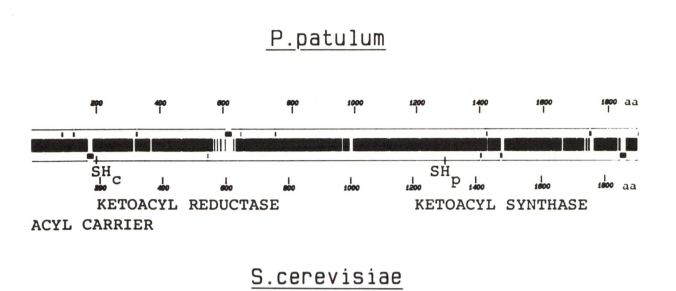

FIG. 6. Alignment of *P. patulum* and *S. cerevisiae* subunit α protein sequences according to the GAPSHOW algorithm of the UWGCG sequence analysis software package, Version 4 (39).

coding sequence at positions 514 and 5463. They were recognized and identified by several criteria. First, both contain a unique NNCTAAC internal consensus sequence as known from other fungal introns (37). Second, their exon-intron boundaries exhibit the sequence characteristics generally observed with nuclear eucaryotic splice sites (38). Furthermore, the high degree of sequence homology (50-70%) between the *S. cerevisiae* and *P. patulum* FAS2 genes abruptly is interrupted throughout the length of both introns (Fig. 6). Finally, sequencing of an appropriate cDNA clone revealed that it was lacking exactly the hypothetical intron II sequence (S. Ripka, unpublished data). These results agree with the general finding that in *Saccharomyces cerevisiae* nuclear introns occur less frequently than in other eucaryotic organisms.

In Figure 6, the nucleotide sequences of *S. cerevisiae* and *P. patulum* FAS2 were aligned using the GAPSHOW algorithm originally devised by Needleman and Wunsch (39). In this representation, both sequences are shown with appropriate gaps introduced in both genes in order to obtain maximal homology between them. Each pair of homologous nucleotides is connected by a vertical bar. Obviously, in addition to the two *Penicillium* FAS2 introns there are a number of mostly smaller insertions or deletions in both genes. From the otherwise high degree of homology of the two FAS genes, it is clear that the sequential order of the three α subunit catalytic sites must be the same in both organisms.

Nevertheless, the degree of sequence conservation obviously varies at different positions along the genes (Fig. 6). Though tempting, it certainly is premature to interpret Figure 6 in that areas of high sequence conservation correspond to individual FAS2 domains, while the less-conserved sequences represent redundant or interdomain regions. It is an open question too whether the intra- or the interdomain regions are more sensitive to amino acid sequence variation.

REGULATORY SIGNALS CONTROLLING FAS1 AND FAS2 TRANSCRIPTION

Since intact yeast fatty acid synthetase has an $\alpha_6\beta_6$ molecular structure, biosynthesis of this heterodimer should be most efficient if FAS1 and FAS2, though genetically unlinked, were coordinately expressed. In earlier experiments, we found that mutational loss of either FAS subunit resulted in the concomitant loss of the other (40). It was suggested that unassociated FAS subunits were proteolytically unstable and that any excess α- or β-subunits were degraded. The lack of free α- and β-subunits in wild type yeast and the absence of both free and aggregated FAS proteins in fas nonsense mutants was explained by this instability. As a result of this characteristic, the amount of α and β finally present in the cell would be stoichiometric even though they were not synthesized coordinately.

On the other hand, coordinate FAS1 and FAS2 transcription could be achieved by comparable strengths of their respective promotors. Polymerase II-promotor efficiencies are intrinsic properties of the 5′-nontranscribed DNA regions and depend on distinct sequence elements therein. In yeast, several gene-specific upstream regulatory sequences have been identified (41). However, besides the so-called TATA-box, little is known about signal sequences generally occurring in front of constitutively expressed yeast genes. To uncover such sequences possibly present in front of FAS1 and FAS2, we performed a computer-aided search for homologies in the upstream region of both genes. The result of this study is depicted in Figure 7. The 5′-flanking DNA-sequence of *Penicillium patulum* FAS2 was included in this study, too (Fig. 7). Transcription of both yeast FAS genes starts 36-38 nucleotides in front of the respective translational initiation codons (Fig. 3). Correspondingly, the TATA-sequences of FAS1 and FAS2 are at almost identical positions, i.e., 117 and 119 nucleotides upstream of ATG (Fig.

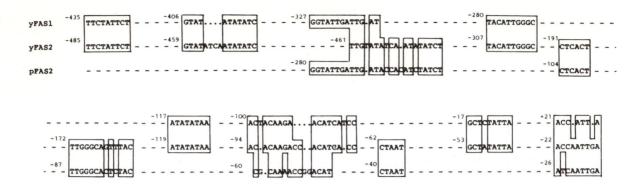

FIG. 7. Consensus elements within the 5'-flanking DNA of yeast fatty acid synthetase genes FAS1 (yFAS1) and FAS2 (yFAS2) and of the *Penicillium patulum* FAS2 gene (pFAS2). Indices refer to the first base of the tranlational start codon.

```
Yeast   L V A H K L K K S L D S I P M S K T I K D L V G G K S T V Q N E I L G D L G K E F G T T P E K P E E T P L E E L A E T F Q D
Rabbit  G E G Q R D L L K A V A H I L G I R D L A G I N L D S S L A D L G L D S L M G V E V R Q T L E R E H D L V L S M R E V R Q L
Barley  K E T V D K V ? M I V K K Q L A V P D G T P V T A E S K F S E L G A D S L D T V E I V M G L E E E F N I T V D E T S A Q D I
E.coli  S T I E E R V K K I I G E Q L G V K Q E E V T D N A S F V E D L G A D S L D T V E L V M A L E L E F D T E I P D E E A E K I
```

FIG. 8. Comparison of acyl carrier protein sequences from different sources around their respective panthetheine–binding serine group (circled). The yeast sequence was adapted from Fig. 3, the others from McCarthy and Hardie (42).

7). To concentrate on specific, nontrivial sequence homologies, homopolymeric stretches of mostly oligo d(T) or oligo d(A) have not been included in Figure 7. Despite this restriction, a considerable number and variety of limited sequence homologies were identified within the 400–450 nucleotides immediately preceding the transcriptional initiation sites. Further investigation is needed to show which of these sequences are nonstochastic and significantly influence FAS1 and FAS2 transcription. Among these, FAS-specific signals will have to be discriminated from those controlling yeast DNA transcription, in general.

Some sequence motifs obviously were common not only to yeast FAS1 and FAS2 but also to the *Penicillium* FAS2 promotor (Fig. 7). This possibly hints at the existence of controlling elements of general importance, as mentioned above. So far, it seems reasonable to interpret the similarities of FAS1 and FAS2 promotor sequences in terms of a coordinate transcription of both yeast genes.

The complete amino acid sequence of the yeast fatty acid synthetase complex now has been determined. In addition, the *Penicillium* FAS subunit α also has been sequenced. Information on the structure of other fatty acid synthetases, for instance that of certain vertebrates, probably will be available soon (M. Schweizer, personal communication). Available evidence suggests that not only the quaternary structures but also the sequential order of catalytic domains (19) and the primary structures of multifunctional FAS proteins differ considerably when members of different phyla are compared. Only between closely related organisms, as reported here for two lower fungi, is a high degree of FAS sequence homology encountered. For instance, the acyl carrier protein domains of animal, bacterial and yeast fatty acid synthetases have only very few amino acids conserved at homologous positions. When present, this conservation is observed mainly in close proximity to the pantetheine–bearing serine residue (Fig. 8). Almost identical sequences, such as those of *Saccharomyces cerevisiae* and *Penicillium patulum*

(Fig. 6) or those of rabbit and rat acyl carrier proteins (M. Schweizer, personal communication) are not compared in Figure 8. The homology between type II fatty acid synthetases like that of *E. coli* and barley chloroplasts is obviously higher than that between type I and type II enzymes. The yeast acyl carrier protein has the lowest, if any, sequence homology to the other FAS enzymes shown in Figure 8. Thus, the various FAS systems found in living cells differ not only in their gross enzyme architecture, their subunit composition, as well as to the extent and relative order of subunit fusion to multifunctional proteins, but also in the primary structures of their individual catalytic domains. Therefore, it is difficult to speculate from the limited data so far available on a possible common evolutionary origin of multifunctional FAS proteins. To what extent the various FAS systems have evolved different reaction mechanisms is a more important question that remains unanswered. Certainly, a definite answer to this quesiton only will come from an insight into the three-D enzyme structures. Obviously, primary structure divergencies do not preclude similarities at the respective secondary and tertiary structure levels. Future studies have to concentrate on these aspects.

ACKNOWLEDGMENTS

This work was supported by the Deutsche Forschungs-gemeinschaft and by the Fonds der Chemischen Industrie.

REFERENCES

1. Lynen, F., *Eur. J. Biochem 112*:431 (1980).
2. Meyer, K.H., and E. Schweizer, *Eur. J. Biochem.* 65:317 (1976).
3. Gill, C.O., and C. Ratledge, *J. Gen. Microbiol.* 78:337 (1973).
4. Joshi, V.C., and J.B. Sidbury, *Arch. Biochem. Biophys.* 173:403 (1976).
5. Watkins, P.A., D.M. Tarlow and M.D. Lane, *Proc. Natl. Acad. Sci. USA* 74:1497 (1977).

6. Dils, R.R., I.R. Cameron, C.J. Skidmore and A.F. Weir, *Biochem. Soc. Transact. 13*:830 (1985).
7. Alberts, A.W., A.W. Strauss, S. Hennessy and P.R. Vagelos, *Proc. Natl. Acad. Sci. USA 72*:3956 (1975).
8. Zehner, Z.E., V.C. Joshi and S.J. Wakil, *J. Biol. Chem. 252*:7015 (1977).
9. Back, D.W., M.J. Goldman, J.E. Fisch, R.S. Ochs and A.G. Goodridge, *J. Biol. Chem. 261*:4190 (1986).
10. Ohlrogge, J.B., *Trends Biochem. Sci. 7*:386 (1982).
11. Kekwick, R.G.O., *Biochem. Soc. Transact. 14*:570 (1986).
12. Worsham, L.M.S., Z.L.P. Jonak and M.L. Ernst-Fonberg, *Biochem. Biophys. Acta 876*:48 (1986).
13. Smith, P.F., in *The Biology of Mycoplasma*, Academic Press, New York, NY, 1971, pp. 109–158.
14. Hazlewood, G., and R.M.C. Dawson, *J. Gen. Microbiol. 112*:15 (1979).
15. Shifrine, M., and A.G. Marr, *J. Gen. Microbiol. 32*:263 (1970).
16. Langworthy, T.A., W.R. Mayberry and P.F. Smith, *J. Bacteriol. 119*:106 (1974).
17. Kim, Y.S., and P.E. Kolattukudy, *Arch. Biochem. Biophys. 190*:585 (1978).
18. Kawaguchi, A., Y. Seyam, T. Yamakawa and S. Okuda, *Methods Enzymol. 71*:120 (1981).
19. Wakil, S.J., J.K. Stoops and V.C. Joshi, *Ann. Rev. Biochem. 52*:537 (1983).
20. Bloch, K., and D. Vance, *Ann. Rev. Biochem. 46*:263 (1977).
21. Hardie, D.G., and A.D. McCarthy, *Biochem. Soc. Transact. 14*:568 (1986).
22. Schweizer, E., K. Werkmeister and M.K. Jain, *Mol. Cell. Biochem. 21*:95 (1978).
23. Elovson, J., *J. Bacteriol. 124*:524 (1975).
24. Holtermüller, K.H., E. Ringelmann and F. Lynen, *Hoppe-Seyler's Z. Physiol. Chem. 351*:1411 (1970).

25. Beck, J., *Diplomarbeit*, Universität Erlangen, 1986, pp. 13–23.
26. Oesterhelt, D., H. Bauer and F. Lynen, *Proc. Natl. Acad. Sci. USA 63*:1377 (1969).
27. Burkl, G., H. Castorph and E. Schweizer, *Molec. Gen. Genet. 119*:315 (1972).
28. Mortimer, R.K., and D. Schild, *Microbiol. Rev. 44*:519 (1980).
29. Schweizer, M., C. Lebert, J. Höltke, L.M. Roberts and E. Schweizer, *Mol. Gen. Genet. 194*:457 (1984).
30. Kuziora, M.A., J.H. Chalmers, Jr., M.G. Douglas, R.A. Hitzeman, J.S. Mattick and S.J. Wakil, *J. Biol. Chem. 258*:11, 648 (1983).
31. Schweizer, M., J. Höltke, K. Takabayashi, G. Müller, U. Hoja, E.Höllerer, B. Schuh and E. Schweizer, *Biochem. Soc. Transact. 14*:572 (1986).
32. Schweizer, M., L.M. Roberts, J. Höltke, K. Takabayashi, E. Höllerer, B. Hoffmann, G. Müller, H. Köttig and E. Schweizer, *Mol. Gen. Genet. 203*:479 (1986).
33. Schweizer, E., B. Kniep, H. Castorph and U. Holzner, *Eur. J. Biochem. 39*:353 (1973).
34. Stoops, J.K., E.S. Awad, M.J. Arslanian, S. Gunsberg, S.J. Wakil and R.M. Oliver, *J. Biol. Chem. 253*:4464 (1978).
35. Langford, C.J., F.J. Klinz, C. Donath and D. Gallwitz, *Cell 36*:645 (1984).
36. Poulouse, A.J., and P.E. Kolattukudy, *Arch. Biochem. Biophys. 220*:652 (1983).
37. Russell, P., and P. Nurse, *Cell 45*:781 (1986).
38. Parker, R., and C. Guthrie, *Cell 41*:107 (1985).
39. Needleman, S.B., and C.D. Wunsch, *J. Mol. Biol. 48*:443 (1970).
40. Dietlein, G., and E. Schweizer, *Eur. J. Biochem. 58*:177 (1975).
41. Brent, R., *Cell 42*:3 (1985).
42. McCarthy, A.D., and D.G. Hardie, *Trends Biochem. Sci. 9*:60 (1984).

Technical and Economic Aspects and Feasibility of Single Cell Oil Production Using Yeast Technology

R.S. Moreton
Cadbury-Schweppes Group Research, Lord Zuckerman Research Centre, University of Reading, Reading RG6 2LA, U.K.

There has been intermittent interest in the microbiological production of edible oils (Single Cell Oil, SCO) for at least the last 50 years as shown by the quite considerable number of research papers that have been published on this topic. Why have such processes been so slow to bring products to the market? The answer lies partially in the technical aspects but mainly with the economics of these processes. This paper will attempt to identify the important areas that are relevant to the commercial production of edible oils and fats by yeast fermentation processes.

It is thought that a microbial oil process has been operating in the U.S.S.R. for some years, probably a by-product of a Single Cell Protein process using a crude oil fraction as substrate. The microbial oil arises through a solvent extraction process to remove residual substrate, which also extracts the microbial lipids. This oil probably is not edible and therefore not a true SCO. Recent announcements by the British company John and E. Sturge Ltd. (1) that they are producing an oil rich in γ-linolenic acid (GLA (6,9,12-*cis,cis,cis*-octadecatrienoic acid) by fungal (*Mucor* sp.) fermentation is probably the first true commercial SCO process. The oil is intended to be ingested in the form of capsules claimed to be of benefit in the treatment of various stress-related disorders such as eczema. Similar oils from the seeds of evening primrose, borage and blackcurrant are already on the market in the U.K. and apparently are selling well.

This fungal oil is hardly an edible oil in the same class as soya oil, olive oil, etc., but an examination of the cost and price structure of the competitive products indicates why it is an attractive target for a fermentation process. Using the present exchange rate of £1 = $1.60, evening primrose seed currently sells in the U.K. for $3,600/ton, and borage seed sells for $4,480/ton. Evening primrose seeds contain 22-23% oil, of which 8-10% is GLA and 72-75% linoleic. Borage seeds contain 30% oil, of which 18-25% is GLA and 35% linoleic but also 2% erucic. Yields of evening primrose seeds vary from 0.25-0.50 tons/hectare, a better yield than borage that also is more difficult

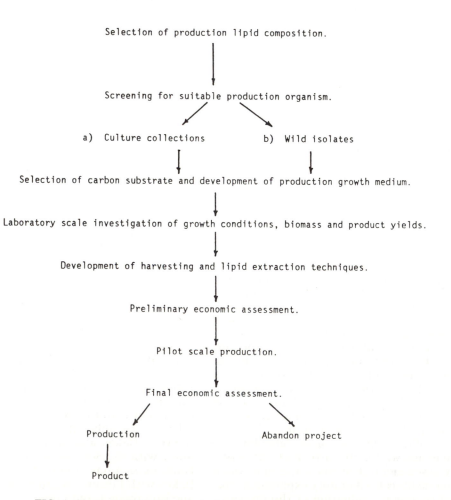

FIG. 1. Single Cell Production by Yeast Fermentation-Process Development Scheme.

to cultivate than evening primrose. Despite the lower oil and GLA content, evening primrose oil is arguably a better product than borage oil because of its higher yield, the absence of erucic and the high content of linoleic. Capsules of evening primrose oil containing 33 mg GLA sell for $6.38/50 capsules. From the above figures, evening primrose oil costs about $16,000/ton before extraction and packaging, or $100,000/ton of GLA, whereas the oil in capsule form sells for $3.868 x 10^6/ton GLA.

These figures place this product in an entirely different market from the traditional edible oils, most of which command about $400–800/ton. Sinden (1) quotes a price of $57,600/ton for evening primrose oil, which seems a large increase over the price of the oil in the seeds.

Descending from the dizzy heights of this exotic product to more realistic levels, what are the stages a potential producer of an edible yeast oil would go through to develop a product? An outline scheme is shown in Figure 1, the various stages of which could apply to most microbial products. If the oil is intended to substitute for an existing product, the chemical characteristics of this product will define the chemical composition of the required yeast lipid. Common plant oils, apart from the lauric group, consist of C_{16} and C_{18} fatty acids with varying degrees of unsaturation of the C_{18} acids. The exact composition of a particular oil can be found in reference works such as Gunstone, Harwood and Padley (2) and compared with analysis of yeast lipids in reviews such as Ratledge (3). Oils similar to palm, soya, sunflower, low erucic rapeseed, cottonseed, olive and groundnut oils have been reported for many lipid-accumulating yeasts. Alternatively, organisms from culture collections or wild isolates can be screened for the appropriate composition. The number of yeast species reported to be capable of accumulating worthwhile quantities of lipid (>40% of their dry biomass) is relatively small, 8 or so, so screening organisms from culture collections will not be too arduous.

After selection of the organism will come selection of the growth medium for the organism (one of the published media from previous workers will be adequate for the screening process, but it is unlikely that this will be an economic large-scale production medium). Growth media for lipid production usually are nitrogen-limited. That is, the medium contains less than a normal amount of the nitrogen source (an organic or inorganic N compound) with an excess of a carbon source usually a carbohydrate such as sucrose, glucose or lactose. Other carbon sources such as ethanol have been investigated, and the highest reported lipid yields are found with this substrate (4). Whatever the C source, the ratio of C to N in the medium is likely to be 30–40:1 for optimum lipid production. This must be determined for each organism. Biomass yields from common carbohydrates in N-limited cultures are usually about 0.4, that is, 2½ tons of carbon source will give one ton of cells. If 50% of the cell is lipid, then five tons of C-source are required for each ton of product. The economic implications of this will be discussed later, but low-cost carbon sources and wastes are particularly attractive.

A preliminary assessment of biomass and lipid yields can be made from small-scale laboratory experiments that will allow initial economic evaluation of the process.

The process so far contains no novel or difficult technology. All the steps described are common to many fermentation processes. Yeasts were probably the first microorganisms grown industrially on a large scale. The technology is well established and yeast products are aesthetically acceptable to consumers as food materials, unlike some other microbial products. Gill et al. (5) have shown that lipid-accumulating yeasts can be grown in continuous culture that has a considerable impact on the capital cost of a plant, as lower capacity fermenters can be used and downstream processing is more efficient. Yamauchi et al. (6) also have shown that fed-batch culture can be used to grow very high-density cultures (>150 g/l dry weight). One of the reasons this is possible is the low oxygen demand of lipid accumulating yeast, about half that of carbon limited cultures of the same organisms (3,7). These figures suggest that the maximum lipid productivity of a lipid fermentation could be >1 g/l/h. At this productivity a 1000 m³ fermenter would produce one ton of lipid per hour, about the same quantity of oil as one hectare of oilseed rape produces annually. For further details of the physiology of lipid-producing yeasts, see Ratledge (8) and Moreton (9).

At this point before proceeding to pilot scale fermentation and scale-up studies, the question of extraction of lipid from the microorganisms must be addressed. Small-scale methods for estimating the lipid content of yeasts are nearly all based on Bligh and Dyer's (10) modification of the Folch et al. (11) method. This method originally was devised to extract lipid from fresh, wet, animal tissues but it has been found by most authors that attempts to extract wet microbial cells using this technique result in intractable emulsions and inefficient extraction (12).

Freeze-dried or spray-dried cells extract well by this method, although the solvent:cell ratio needs to be in the order of 200:1 (v/w) for complete extraction. The original solvents used in this method, chloroform and methanol, are both unacceptable for use with potential food materials. Other solvent systems have been investigated such as hexane/isopropanol, hexane/ethanol and others; some of these have been successful partially.

On a commercial scale, however, there is one proprietary process for extracting lipids from yeast cells, whereby spray-dried yeast cells are partially rehydrated to between 10 and 20% moisture content, pelletized, flaked and the dried flakes solvent extracted (13). Results from a batch of *Trichosporon cutaneum* cells containing 50.05% lipid treated by this method are shown in Table 1. The cells were rehydrated to a moisture content of 10%. Higher moisture contents resulted in a glutinous paste that adhered to the flaking rolls, lower moisture contents produced pellets of low mechanical strength which disintegrated during flaking. After this mechanical treatment, the flakes were extracted either by the Folch cold extraction method or by a hot Soxhlet extraction. Soxhlet extractions were less efficient than Folch extraction with the same solvent (chloroform/methanol 2:1 v/v), and only extracted 80% of the lipid present in the yeast flakes, although the ratio of solvent:solids was lower with the Soxhlet extractions (20:1) than with the Folch extractions (100:1). Soxhlet extractions of yeast flakes with hexane, hexane/isopropanol 1:1 v/v, or hexane/isopropanol containing 1% acetic acid, 1:1 v/v, all at a solvent:solids ratio of 20:1 gave similar results, extracting about 50% of the total lipid. A continuously percolating extractor was even less efficient. This process was an attempt to pretreat yeast cells in such a way as to make them extractable using conventional oilseed extraction plant. Other workers have tried to turn yeast cells into oilseeds but generally this seems to be an inefficient process.

TABLE 1

Lipid Extraction from Pelleted and Flaked *Trichosporon cutaneum*. Spray-Dried Cells Rehydrated to 10% (w/w) Moisture Content

Sample	% Lipid (w/w)	Method	Solvent
Spray-dried powder	50.5	Folch	$CHCl_3$/MeOH 2:1
Dried flakes	49.6	Folch	$CHCl_3$/MeOH 2:1
10% Moisture flakes	40.9	Soxhlet	Hexane
10% Moisture flakes	24.6	Soxhlet	Hexane/IPA 1:1
10% Moisture flakes	23.4	Soxhlet	Hexane/IPA + 1%
10% Moisture flakes	24.2	Soxhlet	Hexane/IPA + 1% Acetic acid 1:1

Solvent: solids ratio 100:1 for Folch extractions, 20:1 for Soxhlet extractions.

A more radical approach was devised by Davies (14) who milled wet and dry yeast suspensions in solvents using a Dynomill KDL bead mill (Willy Bachofen AG, Basle, Switzerland). The Dynomill is a continuous-bead mill capable of working with very small beads of <0.5 mm diameter required to disrupt yeast cells. Attempts to mill cells well in hexane produced intractable emulsions, but a mixture of ethanol:hexane 1:1 was more successful. At a flow rate of 50 ml/min into the 600 ml disruption chamber of the mill about 80% extraction of lipid into the solvent was achieved. Centrifugation of the disrupted mixture produced three phases, hexane, containing ≈90% of the extracted lipid, an aqueous ethanolic phase containing ≈10% of the extracted lipid and a solid debris phase. Reextraction of the ethanol phase was necessary for optimum efficiency, but the process was simplified considerably by spray drying the cells before disruption of them in hexane.

An extraction efficiency of 100% was achieved with a 20% w/w suspension of spray-dried yeast in hexane at a feed rate of nearly 100 ml/min (Fig. 2). There were no problems separating the solvent and cell debris. The rate constant for extraction was 20 times higher than with the aqueous system and the energy input lower. This appears to be the first practical lipid extraction technique that could be scaled up to a commercial plant. Mills are available from the same manufacturer with chamber volumes of up to 275 liters that could process 687.5 kg dry yeast cells/hr per machine at the same relative flow rate as the KDL mill. Once a practical lipid extraction process has been developed, a final economic assessment of the process can be finalized from pilot plant data.

One of the main process costs already has been mentioned—the carbon source, five tons of which are required to produce one ton of lipid. The price of sugar on the world market in the recent past has been at its lowest value in real terms this century, less than $100/ton, although little trading has taken place at this price. One

of the causes of this has been the EEC support policy for European sugar beet growers, forcing the price of sugar within the EEC to nearly $500/ton. Not surprisingly, European farmers have been eager to produce sugar at high guaranteed prices that cannot be sold within the EEC, and therefore it has been dumped onto the world market at low, depressing prices for cane sugar. These artificial differences in the cost of the carbon source suggest that the EEC would not be the best place to build a yeast lipid plant.

Apart from the carbon source, the other main cost will be processing: fermentation and lipid extraction. Sinden (1) quotes manufacturing costs for a fungal lipid including extraction, as $4,000–4,800/ton, inclusive of $1,200 substrate cost, making the fermentation and extraction costs $2,800–3,600/ton. Lipid extraction from fungal biomass is likely to be easier and cheaper than from yeast cells. First, because fungal mycelium is easy to harvest by filtration and, second, because the mycelium is physically more suitable for an extraction process and will require no pretreatment such as pelleting and flaking or expensive technology such as the Dynomill. A wet extraction process is mentioned specifically.

Floetenmeyer et al. (15) described a process for continuous fermentation of the yeast *Candida curvata D* on cheese whey permeate, which they assumed to have zero cost. A 464 m³ fermenter would produce 2,773 tons of lipid and 3.273 tons of defatted biomass in a 20 hr period. The operating costs of the fermenter for this period were $2,000, making the cost of the oil $721/ton. Allowing for effluent disposal costs and revenue from the defatted biomass reduced the cost to $680/ton. These authors do not appear to have included a cost for lipid extraction from the yeast in their figures, so this figure of $680/ton of lipid presumably represents fermentation costs only. If we subtract this figure from that of Sinden (1) for the fermentation and extraction costs of the fungal oil process, we arrive at a figure of around $2,100–2,900/ton of lipid for the extraction process, say $2,500/ton, making the extraction process economically much more important than the fermentation process and equally as important as the substrate costs if the carbon source was $500/ton. Figure 3 shows the cost of the carbon source plotted against the total production cost of lipid for various fermentation and extraction costs using the equation: cost of lipid = (5 x price of carbon source) + production cost. For any given carbon substrate cost, the cost of lipid in $/ton at a given production cost can be read off the y–axis where the horizontal line from the carbon substrate cost meets the appropriate production cost line.

It immediately is apparent from the above formula that any oil that costs less than 5 x the price is the cheapest available carbon source is not a suitable target commodity, assuming no production costs whatsoever. Adding realistic production costs to the carbon source costs immediately rules out virtually all the common commodity oils. Sinden (1) predicts that prices for industrial carbohydrates within the EEC could fall to $240/ton in the near future. It seems unlikely that fermentation and processing costs could ever be less than $800/ton of lipid, making the lowest reasonable production cost of a yeast lipid $2,000/ton, with the figure more likely to be nearer to $2,500/ton.

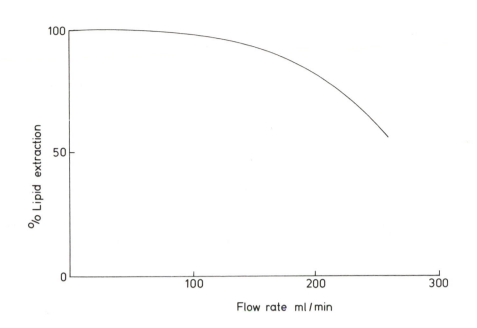

FIG. 2. Extraction of lipid from *Apiotrichum curvatum* ATCC 20509 with Dynomill KDL bead mill. Spray-dried yeast, 25% (w/w) suspension in hexane; bead diameter, 0.5–0.75 mm; bead volume, 300 ml; agitator speed, 2,000 rpm; rate constant for first order extraction, 2.1 min^{-1} (14).

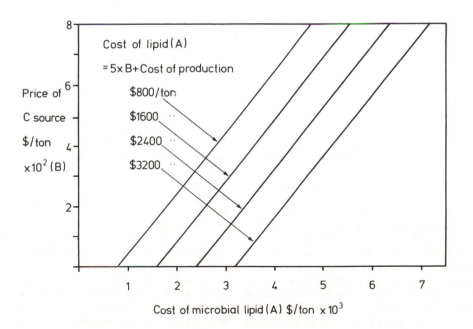

FIG. 3. Relationship between production cost of yeast lipid, cost of carbon source and processing costs.

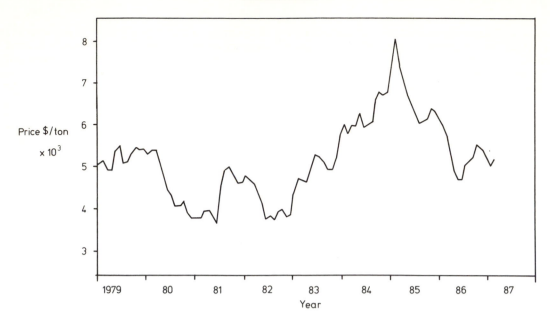

FIG. 4. Market price of cocoa butter. 1979-1987.

This logic would not apply if the carbon source had a zero or negative cost, in which the production cost would be equal to the fermentation and processing costs. Such a situation could occur in which large quantities of carbohydrate were being produced as a by–product of another process. Milk processing to produce cheese generates very large volumes of waste liquid containing 4–5% lactose, which is a suitable substrate for the growth of several lipid–accumulating yeasts such as *Candida curvata D* and *Trichosporon cutaneum*. Floetenmeyer et al. (15) and Davies (14) have published processes for the production of palm oil like lipids by fermentation of ultrafiltered cheese whey permeate. Moon and Hammond (16) quote the annual production of sweet cheese whey in the U.S. as 1.3×10^7 tons/annum that would produce 2×10^5 tons of yeast oil. Davies (14) quotes New Zealand production of whey as 2.5×10^7 tons/annum. New Zealand also has several sites that individually produce around 10×10^3 tons of whey solids annually, which could produce $1–2 \times 10^3$ tons of lipid/annum, probably about the smallest plant economically feasible. The quantities of lipid that could be produced from available sources of whey are not large in terms of world edible oil production but the process could reduce the cost of waste disposal and possibly be economically viable in countries such as New Zealand with no indigenous edible oil sources.

Davies (14) has produced a very well-documented study of a process for the production of lipid from *Candida curvata D* based on pilot plant data. Continuous cultivation at a dilution rate of $0.03–0.04$ h^{-1} when whey permeate containing 55 g/l total solids and 48–49 g/l lactose for periods up to 100 days gave an average oil yield of 20.8% of the incoming lactose. The fatty acid composition of the lipid virtually was unchanged over this period. The pH of the permeate was adjusted to 3.5, which had no effect on growth of the organism but reduced contamination. Because of this low pH, the permeate was not sterilized but given a partial heat treatment by holding

it at 95 C for 10 min. Whether this "pasteurization" treatment would be adequate from an operational point of view remains to be seen.

Davies' plant (14) was designed to produce 1,100 tons per annum (tpa) of SCO, but assuming a 10% refining loss actually would produce 1,000 tpa. Process equipment costs in 1985 were 1.3113×10^6, and total capital cost including land, buildings, design, installation and working capital for the first year was 2.77×10^6.

The fermenter used in this process was of 200 m³ capacity (2×10^5 liters). Using the approximation of Hacking (17), an estimate of installed fermentation equipment costs on a greenfield site can be made by multiplying the fermenter capacity in liters by a factor of 30-70, giving a range of $6-14$ times 10^6, suggesting that plant equipment costs, particularly of the fermenter itself, are very reasonable in New Zealand. This plant has an installed cost of $13.84/liter of fermenter capacity, the fermenter itself costed at $1/liter of capacity. Table 2 shows a breakdown of the capital equipment costs of this process. Moreton and Norris (18) quote a similar figure of $96–160 per installed liter of fermenter capacity for fermentation plants in general and our own work on a yeast lipid plant suggests a figure of around $60 per liter for a 5,000-ton-per-year plant using continuous fermentation and a wet extraction process. Fermentation vessels themselves are usually $6-10 per liter of capacity and represent only a small part of the total capital cost. Installed equipment costs include the site itself, buildings, roads, pipework, installation of services, canteens and trucks. Capital costs of this type of plant can be related to installed costs using Lang factors (17), which vary from three to five for fluid processing plants, a figure of 3.5 commonly being used for fermentation plants. In Davies' plant (14), fermentation and lipid extraction capital costs were about 50% each of the total capital cost. Our own work, costed at European prices, suggests that the split would be 4:1 rather than 1:1, the fermentation equipment being relatively more expensive than in New

Zealand. Davies (14) has calculated the Internal Rate of Return (IRR) of a yeast oil process using whey permeate in New Zealand. Assuming a selling price of $275/ton for the yeast cell debris/whey protein and a selling price of $700/ton for the oil, the IRR of the process would be 15.5% pre-tax and 10.6% after tax. To increase the IRR after tax to 20% would require a selling price for the oil of >$1,000/ton. Palm oil currently sells in the U.K. for around $370/ton. Even with very low capital and substrate costs this process does not look very attractive since most companies would want an IRR of 25–30% for a process of this type (18).

TABLE 2

Capital Costs of Equipment for Yeast Oil Process Using Whey Permeate as Substrate

Item	Cost U.S. $ x 10³	% of total
Substrate handling	87.25	6.65
Inoculum plant	33.00	2.51
Fermenter (air-lift)	200.00	15.25
Air compressor and filtration	110.00	8.39
Yeast recovery (centrifuges, pumps, etc.)	137.70	10.50
Drying	371.00	28.29
Disrupter (Dynomill)	55.00	4.19
Debris recovery (centrifuge)	33.00	2.51
Solids handling	55.00	4.19
Solvent evaporation	115.50	8.81
Oil refining	43.45	3.23
Services	66.00	5.03
Total	1,306.80	100.00

Total downstream costs, including drying, 51.22% of total (excluding yeast recovery).
Modified from Davies (14).

What yeast lipids could be made economically by a fermentation process? From what has been said, it seems unlikely that any of the common bulk edible oils would be appropriate targets. Higher-priced lipids include olive oil, castor oil, cocoa butter, jojoba oil and evening primrose oil or other oils high in GLA.

Olive Oil (ca $1,900/ton) is sold principally for its flavor characteristics. Chemically, it is quite similar to some other plant oils. A microbial oil of similar composition would be devoid of the desired flavor compounds and therefore of little value. Castor oil contains an unusual hydroxy acid not reported in substantial quantities in yeast lipids; jojoba is a volatile market; and evening primrose oil already has been discussed. This leaves cocoa butter, a commodity that has attracted the attention of American, Japanese and British research groups (9). The price of cocoa butter fluctuates as the price of cocoa beans fluctuates. Usually it is 2.2–2.4 times the price of cocoa beans, currently $2,000/ton, making cocoa butter worth $4,400–4,800/ton. The market price of cocoa butter from 1979 to 1987 is shown in Figure 4 (9). Present levels of less than $5,000/ton are low relative to prices in the recent past. Attempts currently are being made to stabilize the price of cocoa at a higher level than present to give the producers a fair return for their product.

Physically, yeast lipids have the same 1,3–saturated, 2–unsaturated composition as cocoa butter, making them

compatible with cocoa butter and important in the product type in which this commodity is used. However, cocoa butter contains an unusually high proportion of stearic acid ($\approx 35\%$), whereas very few yeast lipids described in the literature contain more than 10–12% stearic acid. The content of stearic acid in a yeast lipid can be increased to the level found in cocoa butter by use of $\Delta 9$ desaturase inhibitors such as the cyclopropene fatty acid sterculic acid found in Sterculia oil, the seed oil of *Sterculia foetida* (19). The effect of this inhibitor on the saturation index and fatty acid composition of the lipid from the yeast *Candida* sp. NCYC 911 treated with this compound is shown in Figure 5. A mass spectrum of this lipid, compared with that of cocoa butter, is shown in Figure 6, showing that they are very similar in composition.

Cocoa butter equivalents, such as this yeast lipid, can be used in chocolate in some parts of the world including Japan, Denmark, Ireland and the U.K., but not under current legislation in most of Europe or the U.S. If this situation were to change, the market for such materials would be substantial. The technology for the production of a yeast lipid product is relatively simple and

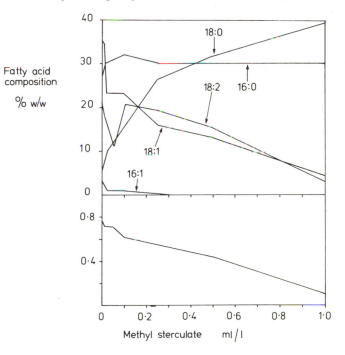

FIG. 5. Effect of methyl sterculate on the fatty acid composition and saturation index of lipid from the yeast *Candida* sp. NCYC 911 (19).

well-established, particularly now that a practical lipid extraction technique is available.

The reason there are so few processes in existence is mainly economic, i.e., there is a lack of suitably priced products that can be made by this technology. Products containing oils with the profitability potential of the GLA are few. Perhaps a few more will emerge with the increasing interest in the medical applications of the more unusual lipids. If single cell oil is to be economically produced, it will be to compete with products such as cocoa butter or other specialty materials with prices in excess of $5,000/ton.

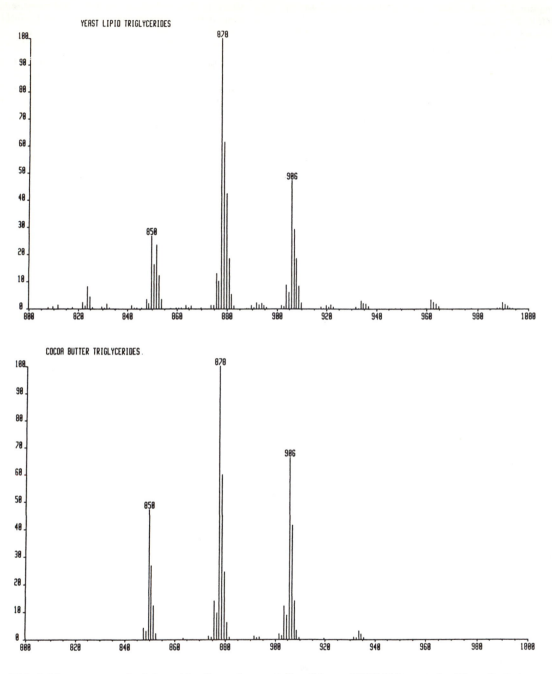

FIG. 6. Mass spectra of triglycerides from the yeast *Candida* sp. NCYC 911 treated with methyl sterculate compared with cocoa butter triglycerides.

REFERENCES

1. Sinden, K.W., *Enzyme and Microb. Technol. 9*:124 (1987).
2. Gunstone, F.D., J.L. Harwood and F.B. Padley, in *The Lipid Handbook*, Chapman and Hall, London, 1986.
3. Ratledge, C., and M.J. Hall, *Appl. and Environ. Microbiol. 34*:230 (1977).
4. Zhelifonova, V.P., N.I. Krylova, E.G. Dedyukhina and V.K. Eroshin, *Microbiol. 52*:219 (1983).
5. Gill, C.O., M.J. Hall and C. Ratledge, *Appl. and Environ. Microbiol. 32*:231 (1977).
6. Yamauchi, H., H. Mori, T. Kobayashi and S. Shimizu, *J. of Fermentation Technol. 61*:275 (1983).

7. Yoon, S.H., and J.S. Rhee, *Process Biochem. 18*:2 (1983).
8. Ratledge, C., *Progr. in Ind. Microbiol. 16*:119 (1982).
9. Moreton, R.S., in *Single Cell Oil*, edited by R.S. Moreton, Longmans, in press.
10. Bligh, E.G., and W.J. Dyer, *Can. J. Biochem. Physiol. 37*:911 (1959).
11. Folch, J., M. Lees and G.H. Sloane–Stanley, *J. Biol. Chem. 226*:497 (1957).
12. Hammond, E.G., and B.A. Glatz, in *Single Cell Oil*, edited by R.S. Moreton, Longmans, in press.
13. Alexander, D.G., A.Forster and D.W.Farmery, British Patent 1,466,853 (1977).
14. Davies, R.J., in *Single Cell Oil*, edited by R.S. Moreton, Longmans, in press.

15. Floetenmeyer, M.D., B.A. Glatz and E.G. Hammond, *J. Dairy Sci. 68*:633 (1985).
16. Moon, N.J., and E.G. Hammond, *J. Am. Oil Chem. Soc. 55*:683 (1978).
17. Hacking, A.J., in *Economic Aspects of Biotechnology*, Cambridge University Press, Cambridge, 1986.
18. Moreton, R.S., and J.R.Norris, in *Developments in Food Microbiology 3*, edited by R.K.Robinson, Elsevier Publishing Co., Amsterdam, in press.
19. Moreton, R.S., *Appl. Microbiol. and Biotechnology 22*:41 (1985).

Production of γ-Linolenic Acid by Fungi and Its Industrialization

Osamu Suzuki

National Chemical Laboratory for Industry, 1-1, Higashi, Yatabe-machi, Tsukuba, Ibaraki 305 Japan

A very efficient microbiological method has been developed for the production of a lipid rich in γ-linolenic acid or the preparation of γ-linolenic acid therefrom and also for the preparation of fungal body containing large amounts of lipid rich in γ-linolenic acid by the culture of strains of *Mortierella* genus fungi in a liquid culture medium. The growth of the fungal bodies of these strains is very rapid in a conventionally high concentration of a carbon source of 200 g/l or more to give a yield of the fungal bodies of 80-156 g/l on a dry basis containing 37-58 % by weight of the lipid corresponding to a yield of lipid of 30-83 g/l. The content of γ-linolenic acid in the fatty acids of the lipid obtained was as high as 4-11 % by weight. Moreover, we describe the extraction method for lipids from fungal bodies, the isolation of γ-linolenic acid from the lipids extracted and the industrial processes that have been developed in Japan.

It is well known that γ-linolenic acid, i.e., all *cis*-6,9,12-octadecatrienoic acid, is an intermediate in the transformation of linoleic acid into prostaglandins (PG), which play very important roles in the body. Thus, γ-linolenic acid is a precursor of and is converted into *bishomo-γ-linolenic* acid and further into arachidonic acid, which is converted into the PG 1 series and PG 2 series (1), respectively, as shown in Figure 1. On the other hand, diabetes, cancer, virus infections and aging are known to affect the activity of the enzyme Δ-6 desaturase that converts linoleic acid into γ-linolenic acid. This reaction is also the rate-determining step of the PG synthesis system, so that those who suffer with these diseases also suffer the possibility of the suppression of PG synthesis (2-5). The direct ingestion of γ-linolenic acid has been suggested as useful for the prevention of these situations. However, there is almost no opportunity for linolenic acid to be assimilated into the body from usual foods.

γ-Linolenic acid or a lipid containing it is obtained conventionally by extraction of the seeds of several plants such as evening primrose (*Oenothera biennis* L.) (6,7). The so-called "evening primrose" oil, which contains about 8% of γ-linolenic acid, is available commercially as a health food.

An alternative possibility for the industrial preparation of γ-linolenic acid, or a lipid rich in it could be based on a microbiological method when and if a microorganism efficiently producing such a lipid was discovered, and an industrially efficient method for the cultivation of such microorganisms was established. Because microbiological production usually can be performed without a supply of light energy, such productivity is not affected by the weather. Any large facility for the industrial production of γ-linolenic acid can be constructed on a relatively small land area in comparison with most plants, and the rate of production can be controlled as desired.

In Japan, the major sources of fat and oil such as soybean oil, palm oil, coconut oil and evening primrose oil are dependent upon other countries. For the purpose of insurance and preservation of the sources, we have studied the production of fats and oils by fungi. Recently, we have developed a microbiological method for the production of a lipid rich in γ-linolenic acid or the preparation of γ-linolenic acid therefrom. We also have developed a method for the preparation of fungal bodies containing a large amount of lipids rich in γ-linolenic acid by the culture of a particular fungal strain in a liquid culture medium with a markedly increased efficiency (8,9). Now, in Japan this production method has been successfully applied as the first industrial method for the production of fats and oils by a microorganism (10).

The following outlines the production method for a fat or oil rich in γ-linolenic acid, the preparation of γ-linolenic acid therefrom and also the industrial manufacturing process that has been developed in Japan.

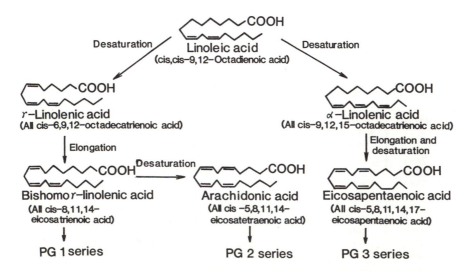

FIG. 1. Pathways of biochemical transformation of precursors of PG-TX series in vivo.

PRODUCTION METHOD FOR LIPIDS CONTAINING γ-LINOLENIC ACID

Various fungal strains obtained from culture collections of the Institute for Fermentation, Osaka, Japan, were screened for high productivity of a lipid rich in γ-linolenic acid or for the fatty acid. Some strains of *Mortierella* genus fungi were found to be suitable for this purpose (11).

The results of the cultures No. 1 to No. 9 under the optimum conditions for each strain are summarized in Table 1, including the yield of fungal mass (cell mass, DC) in g/l on the dry basis of the culture medium, the yield of the total lipid (TL) in g/l of the culture medium, the content of the total lipid in the fungal mass (TL/DC) in percentage, the content of the neutral lipid (NL) in the fungal mass (NL/DC) in percentage, which nearly corresponds with the content of the fat and oil, the content of the polar lipid (PL) in the fungal cells (PL/DC) in percentage, the content of γ-linolenic acid in the total lipid, neutral lipid and polar lipid in percentage, overall content of γ-linolenic acid in the fungal mass in percentage and the yield of γ-linolenic acid in g/l of the culture medium (12).

As shown in Table 1, the growth of the fungal mass of these strains was very rapid even in an unconventionally high concentration of the carbon source of 100 g/l or more and in some cultures, 200 g/l or more, yielding a fungal mass of 40–156 g/l on a dry basis containing 37–58% by weight of lipid corresponding to a yield of lipid of 13–83 g/l. The content of γ-linolenic acid in the fatty acids of the lipid obtained was as high as 4–11% by weight and is comparable with the content in the plant seeds used conventionally. Further, the content of γ-linolenic acid in the fungal bodies was 1.8–4.1% by weight on a dry basis, corresponding to a yield of γ-linolenic acid of 1.2–3.4 g/l of the culture medium.

The cultivation factors affecting the lipid and γ-linolenic acid production from glucose for the strain of *Mortierella ramanniana* var. *angulispora* IFO 8187 were examined in detail. It was found that the optimum culture conditions using a 30 l jar fermenter were at 30 C using a pH of 4.0 with the nutrient composition (M#1) shown in Table 2.

TABLE 2

Composition per Liter of Nutrient Solution in the Fermentation of the Fungal Mycelia of *Mortierella* Genus Fungi

Nutrient	P#1	M#1	F#1	F#2	F#3
Glucose	60 g	200 g	100 g	150 g	200 g
Urea	-	4.0		7.0	
NH$_4$SO$_4$	3.0	2.25		3.94	
KH$_2$PO$_4$	3.0	7.0		12.25	
MgSO$_4$	0.3	1.0		1.75	
NaCl	0.1	0.3		0.53	
Malt Extract	0.2	0.6		1.05	
Yeast Extract	0.2	0.6		1.05	
Peptone	0.1	0.3		0.53	
Minerals*	1.0 ml	3.0 ml		5.25 ml	
Glucose Feed (1F)			100	100	100
(2F)			50	50	50
(3F)			50	50	
(4F)			50		

*Minerals (mg/ml) FeSO$_4$•7H$_2$O 10.0, CaCl$_2$•2H$_2$O 1.2, CuSO$_4$•5H$_2$O 0.2, ZnSO$_4$•7H$_2$O 1.0, MnCl$_3$•7H$_2$O 1.0 P#1, Standard preculture medium. M#1, Standard growth medium with a 30 l jar fermenter. F#1, Culture medium by a batch-fed culture.

The cell growth, total lipids formed and glucose consumption under the above conditions are shown in Figure 2, and changes in the contents and amounts of γ-linolenic acid also are shown at progressive stages of incubation. As shown in this figure, the glucose was consumed continuously during culture growth process as the dry cell weight increased. However, the lipid formation started after a 48–hr incubation time. The cells proliferate on a logarithmic rate as the protein is accumulated, followed by an increase in the fungal lipid formation. The content of γ-linolenic acid in the total lipids was over 13%, after 44 hr of culture growth when the total lipid formation was low, then decreased continuously to 7% with the accumulation of total lipids. The γ-linolenic acid reached 2.2 g/l.

In addition, we explored the improvement of the productivity of the lipids and γ-linolenic acid by means of a batch–fed culture technique with this strain. As shown in Table 2, the nutrient compositions (F#1, F#2,

TABLE 1

Production of Lipids Containing γ-Linolenic Acid by *Mortierella* Genus Fungi (Jar fermenter, 30 l; Incubation temperature, 30 C)

No	IFO No	(C-1) source (g/l)	(N-2) source (g/l)	Cult. time (hrs)	DCW (g/l)	TL (g/l)	TL/ DCW (%)	GLA inTL (%)	GLA yield (g/l)
1	7884	390	17	168	156.4	83.1	53.1	4.5	3.4
2	7884	270	10	72	103.5	49.4	47.7	4.4	2.0
3	7884	280	10	108	104.2	53.1	51.0	3.9	1.9
4	7824	200	6.5	72	76.3	34.9	45.8	6.6	2.1
5	8309	140	3.5	57	43.2	15.9	36.7	10.0	1.4
6	6738	200	6.5	72	79.7	29.6	37.1	7.8	2.1
7	8794	200	6.5	96	72.5	33.1	45.6	7.5	2.2
8	6744	100	3.0	56	40.4	12.8	31.8	11.2	1.2
9	8187	190	6.5	66	72.5	33.1	45.6	6.4	1.7

C, carbon; N, nitrogen; DCW, dry cell weight; TL, total intracellular lipids; GLA, γ-linolenic acid. IFO No. 7884, 7824, 8309 = *M. isabellina*, 6738 = *M. vinacea*, 8794 = *M. nana*, 6744, 8187 = *M. ramanniana var. angulispora*. 1) The carbon source was glucose in all cultures except culture No. 3, in which molasses was used. 2) The nitrogen source was urea in all cultures except culture No. 8, in which ammonium sulfate was used.

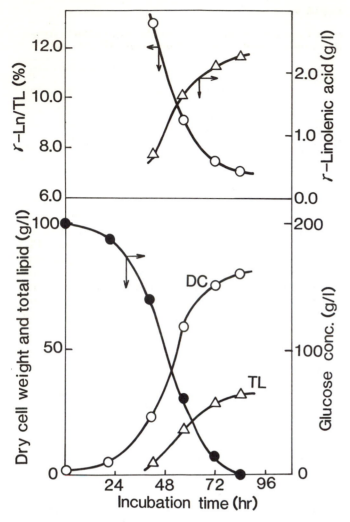

FIG. 2. Changes in the amounts of cell, total lipid and γ-linolenic acid at progressive stages of incubation at 30 C.

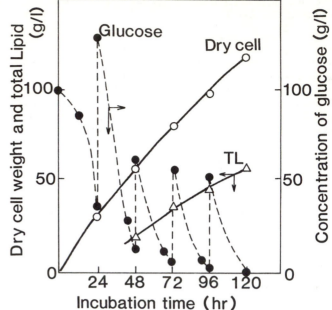

FIG. 3. Changes in the amounts of cell and total lipid and in the concentration of glucose at progressive stages of incubation by the repeated batch culture with a four-fold feed of glucose.

linolenic acid in the fatty acids of the lipid obtained was comparable with the content of that produced by the batch culture.

The cell growth and total lipids formed and changes of the concentration of glucose at progressive stages of incubation in the batch-fed culture No. 1 are shown in Figure 3. In this case, fungal bodies of high lipid content are observed as produced in the culture medium in high density. From the microscopic observation of these strains at the final stage of incubation, it is noted that the multiplication of the filamentous fungus commonly takes place with relatively suppressed extension of the mycelia in extremely minute discrete units, each containing only 1–10 cells, and also the form of fungal cell changes into spherical form during the incubation (12). This means the mature cells are easy to filter or dehydrate on a large scale.

EXTRACTION OF LIPIDS FROM FUNGAL BODIES AND ANALYSIS OF LIPID COMPOSITION

As mentioned above, the separation of the fungal mass from culture medium is greatly facilitated by use of a cen-

F#3) were increased in glucose concentration in comparison with that of the batch culture (M#1) by use of four, three and two times higher feeds, respectively, until they corresponded to a concentration of 350 g/l in the culture medium. The results of the batch-fed cultures No. 1 to No. 3 are summarized in Table 3.

As shown by the results in Table 3, the feeds of glucose to the culture medium during incubation increased the yield of lipids to almost twice that of the culture without the added glucose. This also affected the content of lipids in the fungal bodies. The content of γ-

TABLE 3

Lipid Production by *Mortierella ramanniana* var. *angulispora* in Batch-fed Cultures (Jar fermenter, 30 l; Incubation temp., 30 C)

No.	Nutrient solution	Glucose feed	Dry cell (DCW g/l)	Total lipids (TL g/l)	Lipid content (TL/DCW %)
Control	M#1		77.6	28.8	37.1
1	F#1	4F	120.3	56.0	46.6
2	F#2	3F	107.5	51.6	48.0
3	F#3	2F	115.7	50.2	43.3

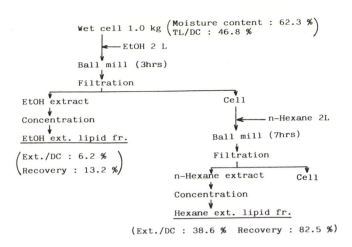

FIG. 4. Extraction procedure of lipids from fungal mass by the two-step extraction.

trifugal dehydrator to give a concentrate with water content as low as 60%. After sterilization of the wet fungus cell cake in an autoclave at 120 C for 10 min under pressure of 2 atmospheres, the extraction of the lipids from the wet cake was carried out with a stainless steel ball mill. Ethyl alcohol was used in the first extraction and hexane in the second extraction in the two-step process shown in Figure 4. An example of the data obtained from the two-step extraction process is shown in this figure (13).

The first and second extracts consist of neutral and polar lipids. From the results shown in Table 4, the ethanol extract contained a much larger amount (14.1%) of the polar lipids as compared with the hexane extract, with an amount of only 0.8%. These data revealed a large amount of polar lipid in the ethanol extract. On the other hand, the neutral lipids from the ethanol extract mainly consisted of diglycerides and free fatty acids. These results indicate that the extraction of lipids with polar or functional groups that have a strong affinity for water requires an appropriate solvent to achieve greater yields. For instance, in the first step, extraction hydrated ethyl alcohol was used. As for the composition of the hexane extract, the polar lipid amounts were extremely small, and the fraction was mainly neutral lipids, most of which are triglycerides, plus a small amount of diglycerides (9%). The rest are almost negligible and in trace amounts. These neutral lipid fractions can be a source of oleaginous product.

TABLE 4

Lipid Compositions of Ethanol and Hexane Extracts from *Mortierella ramanniana* var. *angulispora* IFO No. 8187

Solv	Polar lipids (%)	Neutral lipids (%)	Neutral lipid composition (%)					
			TG	DG	MG	FFA	FS	SE
E	14.1	85.9	4.4	66.6	tr	23.6	4.7	0.7
H	0.9	99.2	89.3	9.0	tr	0.3	0.4	tr

E, ethanol; H, hexane; TG, triglyceride; DG, diglyceride; MG, monoglyceride; FFA, free fatty acid; FS, free sterol; SE, sterol esters; tr, trace (<0.1).

The amounts of the individual polar lipids in the ethanol extract are listed in Table 5 (14). The glycolipids represent 38% of the total polar lipids, while phosphatidylcholine (PC) and phosphatidylethanolamine (PE) constitute 25%, phosphatidylinositol (PI), phosphatidylserine (PS), lysophosphatidylcholine (LPC), lysophosphatidylethanolamine (LPE), lysophosphatidylserine (LPS) and lysophosphatidylinositol (LPI) represent only a minor proportion. The results are consistent with those obtained for other fungi.

TABLE 5

Polar Lipid Composition of the Ethanol Extract from *Mortierella ramanniana* var. *angulispora* IFO No. 8187

Polar lipid	Composition (%)
Glycolipid	38.0
Unknown	7.1
Phosphatidylethanolamine	25.1
Phosphatidylcholine	25.4
Phosphatidylserine	2.4
Phosphatidylinositol	0.7
Lysophosphatidylcholine, lysophosphatidylethanolamine, lysophosphatidylinositol	1.3

Fatty acid composition of each fraction separated and isolated is shown in Table 6. The fatty acids in the lipid fractions extracted from the fungal mass of the *Mortierella* genus strain are mainly palmitic, stearic, oleic, linoleic and γ-linolenic acids with smaller proportions of myristic and palmitoleic acids. Differences in distribution were observed among the components. The glycolipids were

TABLE 6

Fatty Acid Composition of the Lipids of *Mortierella ramanniana* var. *angulispora* IFO No. 8187

Lipid	Fatty acid (% of total)						
	C14:0	C16:0	C16:1	C18:0	C18:1	C18:2	C18:3
Neutral lipids	1.0	34.1	0.3	5.5	39.3	11.3	7.9
Mixed polar lipids	0.6	21.2	0.2	2.5	37.9	18.2	17.9
Glycolipids	0.8	34.7	-	5.2	44.9	11.3	3.0
Phosphatidylcholine	0.6	11.2	0.3	1.0	18.3	41.0	27.5
Phosphatidylethanolamine	0.6	15.4	0.1	0.7	31.2	22.8	27.9

richest in oleic and palmitic acids and characteristically low in γ-linolenic acid. On the other hand, the high percentages of linoleic and γ-linolenic acids in PC and PE, especially the latter, was significant. The latter acid represented 27% of the total fatty acids in both phospholipids. γ-Linolenic acid content of the neutral lipid fraction obtained from hexane extract was 6.2%. This, therefore, can be comparable with the oil extracted from evening primrose seed but the fatty acid composition of this fraction is quite different from that of the evening primrose oil as it contains more than 70% linoleic acid. Thus, the fractional extraction with hexane can be a promising source for obtaining an oil containing γ-linolenic acid.

The extraction of lipids from the fungal mass also was carried out with the use of super–critical carbon dioxide alone or with addition of an organic solvent such as ethanol or hexane at the second step in the two-step extraction process (15). The results of the extraction of lipids from the fungal mass by using super-critical carbon dioxide prove this solvent is suitable for the extraction of lipids and especially neutral lipids.

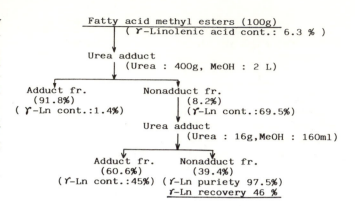

FIG. 5. Concentration of γ-linolenic acid methyl ester by the urea adduct method.

TABLE 7

Composition of the Dry Fungal Mass Containing γ-Linolenic Acid Produced by an Industrial Process

Component		Content
Moisture		5-6%
Protein		25-30%
Amino acid	Arginine	1.33 g/100 g fungal mass
	Lysine	1.83
	Histidine	0.61
	Phenylalanine	1.12
	Tyrosine	0.72
	Leucine	1.80
	Isoleucine	1.18
	Methionine	0.42
	Valine	1.31
	Alanine	1.55
	Glycine	0.98
	Proline	2.55
	Glutamic acid	2.38
	Serine	1.27
	Threonine	1.18
	Aspartic acid	2.38
	Tryptophan	0.33
	Cystine	0.28
Lipid		25–30%
Fiber		1–2%
Ash		6–7%
Carbohydrate		30–35%
Vitamin B	Thiamin (B₁)	0.4-0.5 mg/100 g fungal mass
	Riboflavin (B₂)	2.2-2.6

CONCENTRATION OF γ-LINOLENIC ACID

The concentration of γ-linolenic acid methyl ester was carried out by an urea adduct separation process; an example of the result is in Figure 5 (13).

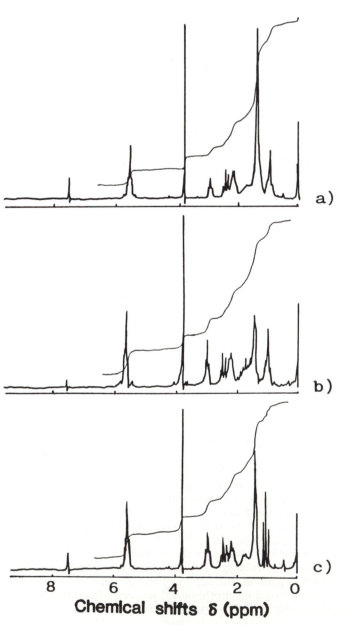

FIG. 6. PMR spectra of 10% esters of a) methyl linoleate, b) methyl γ-linolenate and c) methyl α-linolenate exhibiting differences in signals of the methylene groups.

The mixture of the fatty acid methyl esters obtained by the esterification of fungal lipids containing ca 7% of γ-linolenic acid was subjected to the urea adduct process to give a fatty acid mixture from which the γ-linolenic acid was concentrated up to 98% with recovery of about 50%.

The γ-linolenic acid methyl ester isolated was further analyzed and characterized as to its chemical structure by proton and carbon-13 nuclear magnetic resonance. The proton magnetic resonance spectra of 10% esters of methyl linoleate, methyl γ-linolenate and methyl α-linolenate exhibited a difference in their cis double bond numbers as indicated by the area of integration shown (Fig. 6). The proton magnetic resonance gave a disernible resolution for the terminal methyl group of α-linolenate different from linoleate and γ-linolenate, for the terminal methyl group of α-linolenate is adjacent to the methylene group that is next to the cis double bond. Further characterization was done by using carbon-13 NMR for more sensitive structural difference determination. These results confirmed the structure of γ-linolenate obtained from fungal lipids as all cis 6,9,12-octadecatrienoate.

INDUSTRIAL PROCESS FOR PRODUCTION OF OIL CONTAINING γ-LINOLENIC ACID

The industrialization of the technology for producing the oil containing γ-linolenic acid mentioned above was accomplished as the first case of the production of fat and oil by microorganism in Japan.

The process for the oil containing γ-linolenic acid by the fungus of Mortierella genus is mainly composed of a culture process and a crushing extraction/refining process shown in Figure 7 (10). The proliferation of fungal mass is carried out by using culture medium made with glucose as a carbon source in the culture process. Then a drying process is carried out after the centrifugation of fungal mass, yielding a dry fungus as a product. The composition of components in the dry fungus containing γ-linolenic acid is tabulated in Table 7. As shown by the results in Table 7, the dry fungus is an excellent material with high nutritive value that contains essential amino acids and the Vitamin–B group, not to mention γ-linolenic acid as a component of lipids.

In the process of crushing/extraction/refining, the extraction of oil from the fungal cells is carried out by the crushing of the cell walls of the fungal body in the presence of solvent. The extraction liquid (micella) is concentrated by evaporation under decreased pressure, yielding crude fungal oil. The refined oil containing γ-linolenic acid is obtained by carrying out deacidification, degumming, bleaching and deodorization of the crude oil. The analytical results on the refined fungal oil obtained are shown in Table 8. Although it is a fat originating from a microorganism, the safety of the product is confirmed as a food- or a cosmetic-based material from the physical, chemical and biological test.

TABLE 8

Analytical Data of Oil Containing γ-Linolenic Acid Produced by an Industrial Process

Items		Average value
General properties		
Appearance		Yellowish liquid
Water (ppm)		20
Specific gravity (25 C)		0.9103
Viscosity (30 C, Cst)		59.54
Cloud point (C)		5
Refractive index (25 C)		1.4664
Flash point (COC, C)		318
Quantity of phosphatide as stearoyloleoyl-lecithin (mg/100g)		15.0
Quantity of composition of fatty acid as each methyl ester	(wt%)	
Myristic acid	C14:0	0.7
Palmitic acid	C16:0	27.2
Palmititoleic acid	C16:1	0.9
Stearic acid	C18:0	5.7
Oleic acid	C18:1	43.9
Linoleic acid	C18:2	12.0
γ-linolenic acid	C18:3	8.3
Arachidic acid	C20:0	0.6
Eicosenoic acid	C20:1	0.4
Behenic acid	C22:0	0.1
Erucic acid	C22:1	0.2
Total		100.0

Items	Average value	Detectable limit
Properties of lipid		
Acid value	0.08	
Peroxide value (mg/kg)	0.10	
Iodine value (g/100g)	88	
Saponification value (mg/g)	192	
Quantity of heavy metals (ppm)		
As	N.D.	0.1
Pb	N.D.	0.05
Cd	N.D.	0.01
Total Hg	N.D.	0.01
Total Cr	N.D.	0.5
Sn	N.D.	5.0
Solvent Ethanol	N.D.	
n-Hexane	N.D.	
Quality of metal-elements by means of fluorescent x-ray analyzer		
(Na-U)	N.D.	
Biological test		
Acute toxicity	negative (55,200 mg/kg)	
Mutagenicity on gene mutation	negative	
Mutagenicity on chromosomal aberration	negative	

N.D., not detected.

(Culture process)

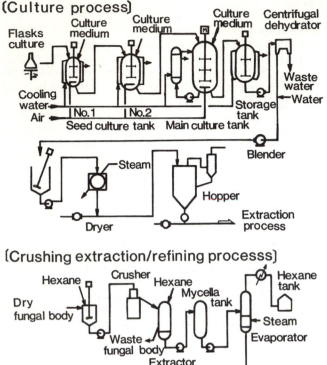

(Crushing extraction/refining processs)

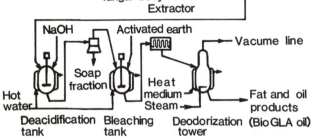

FIG. 7. Manufacturing flow sheet of oil containing γ-linolenic acid by *Mortierella* genus fungus.

REFERENCES

1. Hayashi, Y., and S. Yamamoto, *Kagaku to Seibutu 17*:684 (1979).
2. Peluggo, R.O., S. Ayala and R.R. Blenner, *Am. J. Physiol. 218*:669 (1970).
3. Horrobin, D.F., *Med. Hypothese 6*:929 (1980).
4. Brenner, R.R., *Drug Metab. Rev. 6*:155 (1977).
5. Horrobin, D.F., *Clinical Uses of Essential Fatty Acids*, Eden Press, Montreal (1982).
6. Hudson, B.F.J., *J. Am. Oil Chem. Soc. 61*:540 (1984).
7. Wolt, R.B., R. Kleiman and R.E. England, *J. Am. Oil Chem. Soc. 60*:1858 (1983).
8. Suzuki, O., *Kagaku to Seibutu 23*:11 (1985).
9. Suzuki, O., *Hakko to Kougyo 43*:1024 (1985).
10. Amano, K., *Food Chemicals*, 1987, p. 83.
11. Yokochi, T., and O. Suzuki, *Yukagaku 35*:929 (1986).
12. Yokochi, T., and O. Suzuki, *Proceedings of 23rd Symposium of Japan Oil Chemists' Society*, 1984, p. 128.
13. Suzuki, O., and T. Yokochi, *J. Am. Oil Chem. Soc. 63*:434 (1986).
14. Suzuki, O., T. Yokochi and M.T. Usita, *Proceedings of 24th Symposium of Japan Oil Chemists' Society*, 1985, p. 10.
15. Sako, T., T. Yokochi, T. Sugeta, N. Nakazawa, T. Hakuta, O. Suzuki, M. Sato and H. Yoshitome, *Yukagaku 35*:463 (1986).

Microalgae as a Source of EPA-containing Oils

D. Kyle, P. Behrens, S. Bingham, K. Arnett and **D. Lieberman**

Martek Corporation, 9115 Guilford Rd., Columbia, MD 21046

Recent epidemiological and clinical investigations have implicated eicosapentaenoic acid (EPA), a polyunsaturated fatty acid in certain fish oils, in reduced incidence of coronary heart disease and cancer. It is unclear at this point, however, whether the EPA in fish oil is synthesized directly by the fish or accumulated through the food chain from that produced by many species of marine phytoplankton (microalgae). Our recent work has focused on screening, selecting and improving the EPA yields of several oil-producing microalgal species as a prelude to determining if this source of EPA-containing oil may be of economic importance to the food or pharmaceutical industries. Our results from growing certain strains under a variety of culturing conditions have demonstrated the plasticity of lipid and EPA content of these microalgae. For example, a 10 C decrease in the culture temperature can double the EPA in some cases. Strain selected organisms in culture are now producing EPA to levels of one quarter of their total fatty acid complement. These organisms grow quickly (doubling time of six hr) and are amendable to controlled, fermentation-like culture by using newly developed photobioreactor technologies. As strain selection strategies continue to improve the oil content of these organisms while retaining their high EPA levels, this biotechnologically based source of EPA-containing oil is becoming more attractive as an alternative to fish oil for commercial use in the food industry.

INTRODUCTION

A major basic technology of the pharmaceutical industry involves the utilization of bacteria or fungi grown in tightly controlled fermentative culture conditions and highly selected for the optimal production of a finished pharmaceutical product (i.e., penicillin), or a precursor to a finished product (i.e., steroid biotransformations). Bacteria and fungi also are used in the food industry either as a source of specialty chemicals or in food processing itself (i.e., cheese making, yogurt production, etc.). Microalgae, however, have never been used in any important commercial application in either of these areas. The reasons for this include a lack of detailed knowledge about the physiology and biochemistry of these organisms and their capabilities and the absence of production scale-up technology for organisms requiring light.

Microalgae have characteristics in common with higher plants and are sometimes referred to as "microplants." They exhibit a tremendous diversity in ecological habitat and carry out the bulk of the photosynthesis on this planet (1). We are, perhaps, most familiar with phytoplankton, the microalgal primary producers in the oceanic food chain. In this paper, we will concentrate on one group of oleaginous phytoplankton, diatoms. These microalgae contain an abundance of lipid (30–50% of the dry weight in most cases), the principal component of which is the ω-3 polyunsaturated fatty acid, eicosapentaenoic acid (EPA). Although this fatty acid also is found in abundance in the oils of certain cold water fish, it is unclear whether the fish themselves can synthesize EPA, or whether EPA is accumulated as an essential fatty acid through the food chain. In any case, it is clear that most diatoms have the genetic information and biochemical ability to synthesize EPA de novo.

There is now an increasing demand for EPA-containing oils (exclusively fish oils at present) in both the pharmaceutical and food industries (2,3). Because the production capability of microorganisms in culture can usually be improved significantly by controlling nutrient composition and growth conditions and because selection and strain improvements have been so successful in other areas of industrial microbiology (4), we have embarked on a program to improve the productivity of selected diatoms with respect to EPA production. In order for diatoms to become competitive with fish oil as a source of EPA, both biological and engineering innovations are required. It is essential to improve the EPA productivity several-fold over levels produced endogenously by wild type species, and it is critical to develop a capacity for large-scale controlled culture of these organisms either through novel "photobioreactor" design (5) or through genetic alterations and/or selection of a microalgal species capable of growth with conventional fermentation technology.

EXPERIMENTAL

Growth Conditions

Throughout the course of this work, various types of photobioreactor geometries were utilized. Screening experiments were carried out using 250 ml shake-flask cultures with external illumination. Fluorescent lights provided illumination levels of ca 20 μE/m^2/s for autotrophic samples and 2-5 μE/m^2/s for mixotrophic samples. Heterotrophic cultures were grown in complete darkness. Cultures were maintained in a constant-temperature environmental room with elevated ambient CO_2, which averages about 2%.

To assess the effect of temperature, some cultures were maintained in a continuous culture mode using the cylindrical air-lift photobioreactor (ca 1.2 l working volume) described previously (6). This reactor consists of three concentric cylindrical chambers: an innermost chamber that houses a fluorescent light, a middle algal culture chamber and an outer chamber that serves as a water jacket for precise temperature control. At the constant cell density used in our experiments (regulated by turbidostat control), inorganic media nutrients did not limit culture growth. Rather, growth rates and productivities were set by the cell density and limited by the light intensity of the reactor (60 μE/m²/s).

Nutrient media limitation experiments were carried out using a flat-plate apparatus designed to allow the maximum photon flux through the culture. These temperature regulated air-lift reactors had an optical path of 3 cm, a working volume of 4 liters and a photon flux density of ca 300 μE/m^2/s on each side of the reactor. Biomass densities of up to 10 g/l could be obtained with this geometry under nutrient-sufficient conditions before light became severely limiting.

Lipid Analyses

Fatty acid analyses were carried out routinely on freeze-dried material. A 50 mg sample of freeze-dried material was resuspended in 2.0 ml of methanolic base reagent (Supelco, Inc., Bellefonte, PA), heated to 70 C for 15 min, diluted with 2.0 ml water and extracted with 2.0 ml hexane to remove the methyl esters of the fatty acids. Samples were then injected directly into a Shimadzu GC-9A gas chromatograph equipped with an SP-2330 packed column. Data were stored for subsequent analysis using a personal computer.

Separation of complex lipids involved repeated extraction of freeze-dried algal biomass with chloroform/methanol (1:1) at 60 C followed by drying of the extract under a stream of nitrogen. The crude lipid extract was resuspended in a minimal amount of hexane and applied to a disposable silica column (SEP-PAK, Waters Assoc., Inc., Millford, PA); various lipid fractions then were eluted with solvents of increasing polarity (i.e., hexane, benzene, chloroform, acetone and methanol). Triacylglycerides, the major nonpolar lipids, were recovered primarily in the hexane and benzene fractions, free fatty acids in the chloroform fraction and the polar membrane lipids (i.e., phospholipids and galactolipids) in the acetone and methanol washes. Greater than 95% of the lipid applied to the column could be recovered during this fractionation.

RESULTS AND DISCUSSION

Screening

Many species and strains of diatoms are available from culture collections and have been characterized with respect to their fatty acid profiles (7,8). A common feature in the fatty acid profiles of most diatoms is the abundance of palmitic acid, palmitoleic acid and EPA (Table 1). Unlike the oils of higher plants, most diatoms have only a small quantity of C_{18}-fatty acids, suggesting that fatty acyl chain elongation and/or desaturation may occur via a unique pathway. Little is known about the pathway of polyunsaturated fatty acid biosynthesis in diatoms, but one postulate suggests that palmitic acid is desaturated prior to chain elongation (9). Alternatively, palmitoleic acid may be elongated to C18:1, desaturated to C18:2 and further elongated to C20:2 prior to subsequent desaturation (10). Clearly, further details of the biosynthetic pathway need to be established before genetic engineering technologies can be utilized for strain improvement.

One of the principal objectives of the screening program was to identify diatoms with the capability of heterotrophic growth so that more conventional fermentation technologies could be utilized for large-scale production. Several species have been reported previously to exhibit heterotrophic growth; however, growth rates and lipid (specifically EPA) productivities are rarely mentioned (8,11,12). Indeed, *Chroomonas*, an alga of the class Cryptophyceae, has been reported to undergo a major physiological reorganization and change from a predominantly oil-producing organism to a starch producer as it converts from autotrophic to heterotrophic growth (13). The initial screen was set up to assess the extent of growth of various diatoms under both mixotrophic conditions (illumination is at or below the compensation point for photosynthesis) and truly heterotrophic conditions (complete darkness) using several different carbon sources. The results indicate that although most of the diatoms tested could use a supplementary carbon source, they were unable to grow to any great extent in complete darkness (Table 2). This finding is not uncommon among microalgae. It is believed that although light is not used as an energy source for growth under such mixotrophic conditions, it is necessary for energizing the uptake of the reduced carbon source (14). Although literature reports indicate that most of the species shown in Table 2 were capable of heterotrophic growth, only one unidentified species of *Nitzschia* (*Nitzschia* sp.) exhibited growth rates that were comparable to the same species grown in autotrophic conditions. However, in this case, heterotrophic growth did not occur on a single carbon source; both tryptone and glucose were required.

TABLE 1

Fatty Acid Profiles of Several Selected EPA-containing Diatoms

Fatty acid	Navicula Saprophilla	Navicula pelliculosa	Nitzschia angularis	Nitzschia sp.	Phaeodactylum tricornatum	Cylcotela cryptica	Skeletonema costatum
14:0	7	2	6	3	6	12	8
16:0	10	21	22	17	22	19	9
16:1	45	57	31	46	31	42	39
16:2	5	2	2	6	—	7	—
16:3	8	—	4	5	—	6	—
16:4	2	—	2	2	—	—	—
18:0	—	—	2	—	—	3	9
18:1	—	5	—	—	16	2	9
18:2	—	2	—	—	3	—	4
18:3	—	—	2	—	3	—	1
18:4	—	—	3	—	—	—	5
20:4	3	—	1	1	—	—	3
20:5 (EPA)	22	9	18	15	13	6	7

Cultures were grown in shake-flasks, then harvested and analyzed after at least one wk in stationary phase.

TABLE 2

Screen for growth of various EPA-producing diatoms under autotrophic, mixotrophic and heterotrophic conditions

Species	autotrophic	mixotrophic			heterotrophic		
	CO_2	gluc.	trypt.	glyc.	gluc.	trypt.	gluc.+trypt.
Cylindrotheca fusiformis	+++	++	++	++	-	-	-
Navicula pelliculosa	+++	+++	+++	+++	-	-	-
Navicula saprophilla	++	++	++	+++	n.d.*	n.d.	n.d.
Nitzschia angularis	+++	++	++	++	-	-	-
Nitzschia sp.	++	++	+++	++	-	-	++
Phaeodactylum tricornatum	+++	+++	n.d.	+++	n.d.	n.d.	n.d.

Carbon supplementation included 0.5% glucose (gluc), 0.1% tryptone (trypt), 0.5% glycerol (glyc), or 0.5% glucose plus 0.1% tryptone (gluc+trypt). Cultures were allowed to grow for 1–2 weeks prior to harvest and lipid analysis.
*Not determined.

Media Nutrient Limitation

It has been well established that nutrient limitation (especially nitrogen limitation) can cause major alterations in cellular physiology; in many cases nutrient limitation can trigger liponeogenesis in oil-producing algae (15–17). Diatoms have a silica-based cell wall, and several workers have shown that silica limitation also can induce lipid production (15). In order to determine the media concentrations of nitrogen and silica limiting for growth, continuous cultures of *Phaeodactylum tricornatum* and *Navicula saprophilla* were established using media with several-fold excess of these nutrients and the N and Si content of the biomass produced was analyzed. With the nutrient media adjusted to provide only enough of these nutrients to allow growth to 4 g/l and using photobioreactor geometry that would allow biomass densities of up to 10 g/l, the growth and lipid content of both species

were followed as a function of time. A diagramatic representation of the observed trends at the point of nutrient limitation is shown in Figure 1. As previously reported (18), when the cultures approach nutrient limitation, cell division stops, the amount of extractable fatty acid increases and biomass accumulation beyond this point is due almost exclusively to lipid accumulation.

When microalgae are triggered into oleogenesis by media limitation, the principal lipid fraction produced is neutral triglyceride (15). These triglycerides can be observed within the cells as oil droplets (Fig. 2) similar to those in oilseeds. The EPA content of these neutral fats is a critical factor that determines whether microalgal production of EPA will be feasible. When the complex lipids of an early log-phase culture (nutrient-sufficient) of *Navicula saprophilla* are fractionated on the basis of polarity, the major components are the membrain-derived polar lipids, i.e., phosphatides and galactolipids (Fig. 3). The lipids from the same culture in stationary phase (nutrient-deficient) exhibited a four-fold increase in the

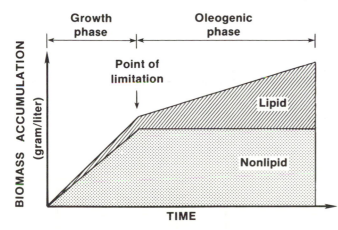

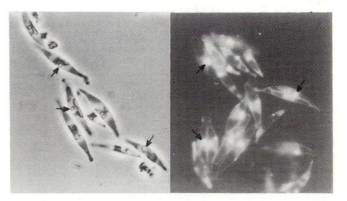

FIG. 1. The effect of nutrient limitation on growth and lipid content of an oil-producing diatom. Cell number increases during the growth phase and levels off at the point of nutrient limitation. Subsequent biomass accumulation in the oleogenic phase primarily represents the synthesis of storage lipid.

FIG. 2. *Phaeodactylum tricornatum* stained with the lipophilic fluorescent stain Nile Red observed using phase contrast (left) and fluorescence microscopy (right). Arrows mark the presence of oil droplets. Magnification is 1200X.

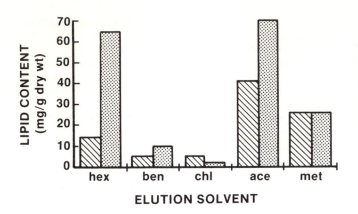

FIG. 3. Complex lipids from nitrogen-sufficient, early log-phase (hatched) and nitrogen-deficient, stationary phase (stippled) cultures of *Navicula saprophilla* fractionated on silica gel columns by successive elution with hexane (hex), benzene (ben), chloroform (chl), acetone (ace) and methanol (met). Fractions eluted with hexane and benzene represent neutral lipids; those separated with acetone and methanol represent polar lipids.

triglyceride component (eluted with hexane) and a modest increase in the polar lipid fraction eluted with acetone (phospholipid). We also have reported that the EPA content in the fatty acid profile exhibited little change or even increased slightly throughout all stages of growth of *N. saprophilla* even though the overall lipid content nearly doubled as the culture depleted its nutrients (18). Because this major increase in the triglyceride fraction had little effect on overall EPA content, we suggested that the triglyceride component must have a significant amount of EPA. This was confirmed by the fatty acid profiles of each of the classes of complex lipids (Table 3). Not only is there a relatively high level of EPA in the triglyceride fraction of this species but the fatty acid composition of the triglyceride does not change whether the culture is in a rapid growth or stationary phase. It is also interesting to note that a relatively high proportion of polyunsaturated C_{16} fatty acids is found in the membrain lipid fractions suggesting that acyl chain unsaturation and possibly elongation may take place in the membranes rather than in the oil bodies themselves.

Temperature Studies

Lowered growth temperatures have been correlated with increased levels of unsaturated fatty acids in the complex lipids of higher plants (19) as well as microalgae (7,15). As a method of improving EPA yields, we studied the effect of decreasing growth temperatures on lipid content and fatty acids of *P. tricornatum* and *N. saprophilla*. Although there was a general increase in the ratio of palmitoleic acid to palmitic acid with decreasing growth temperatures, we did not observe any major changes in the relative abundance of the polyunsaturated fatty acids and, specifically, no major change in the amount of EPA in these two species (Table 4). Furthermore, one must consider the decreased growth and biomass productivity associated with these reduced temperatures when considering its utility in commercial production. The productivities of *P. tricornatum* dropped 20% between 20 C and 15 C, while those of *N. saprohilla* dropped 45% between 20 C and 15 C. Thus, when one determines the EPA productivity (mg EPA/hr), there is clearly no advantage to lower growth temperatures for commercial batch production of an EPA-containing algal oil product.

TABLE 4

The effect of temperature on the fatty acid composition of nutrient-sufficient, log-phase continuous cultures of *Phaeodactylum tricornatum* and *Navicula saprophilla*

Fatty acid	Phaeodactylum			Navicula		
	10	15	20	10	15	20
14:0	6	5	6	3	3	4
16:0	16	22	23	17	18	18
16:1	36	28	19	31	28	26
16:2	7	5	4	10	10	5
16:3	—	—	—	8	11	7
16:4	—	—	—	7	4	3
18:1	4	3	5	—	—	—
18:2	10	13	9	—	—	—
18:3	3	7	9			
18:4	—	—	—	3	2	10
20:4	1	1	1	2	2	3
20:5	17	14	17	19	22	22

TABLE 3

Fatty acid profiles of the complex lipids of *Navicula saprophilla* extracted from early log-phase (nutrient-sufficient) and stationary phase (nutrient-deficient) cultures

Fatty acid	hexane		benzene		chloroform		acetone		methanol	
	N+	N-	N+	N-	N+	N-	N+	N-	N+	N-
14:0	6	7	3	4	6	4	4	5	8	8
16:0	20	22	9	22	14	17	10	14	25	24
16:1	29	27	34	35	15	43	33	45	26	23
16:2	8	5	3	3	5	1	14	4	—	1
16:3	7	3	5	3	5	3	13	6	2	6
16:4	3	3	3	5	11	2	3	3	5	4
18:4	2	9	12	9	23	8	5	2	10	10
20:4	3	5	4	2	6	3	3	2	6	6
20:5 (EPA)	19	18	19	12	6	12	15	20	14	14

Lipids were separated on the basis of polarity prior to fatty acid analysis. Values for each fatty acid are tabulated as its percentage in the lipid profile.

Strain Selection

If EPA is found in relatively large proportion in the neutral fats of diatoms, then it should be possible to select and/or optimize for elevated lipid levels, thereby improving EPA content. The storage oil droplets can be observed within the cells either by phase contrast microscopy or by visualization of a fluorescent dye such as Nile Red which partitions into the neutral lipid fraction of a cell (Fig. 2). Nile Red has been used previously to select populations of high lipid cells using cytofluorometry because it is nontoxic (15,20). When cells of *N. saprophilla* were stained with Nile Red and analyzed in a cytofluorometer, we observed a normal distribution pattern of fluorescence within a single population (Fig. 4). Although chlorophyll exhibits a strong fluorescence emission at 680 nm, the Nile Red-stained cells exhibited a clear difference in fluorescence intensity at 530–575 nm compared to unstained cells. The broad distribution in staining intensity gives us confidence that strain improvement using this procedure might be possible by repeated selection and subculture of the highest fluorescing fraction of the population. One must be aware of the potential limitations of this approach, however, which include selection of clumped (therefore highly fluorescent) cells or a physiologically distinct subpopulation (i.e., old cells) that may not be genetically dissimilar from the rest of the population.

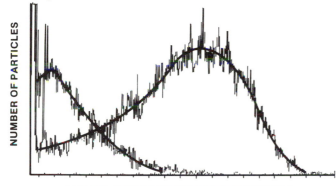

LOG RELATIVE FLUORESCENCE INTENSITY

FIG. 4. Distribution of fluorescence intensity in a population of *Navicula saprophilla* cells before staining (left curve) and after staining (right curve) with Nile Red. Cells were analyzed in a Coulter cytofluorometer (illumination at 488 nm) using a filter to detect fluorescence of the cells in the 530-560 nm range. 20,000 cells were analyzed in each experiment.

CONCLUSIONS AND FUTURE DIRECTION

It is clear that microalgae have the potential to become an economically important source of EPA for the food and/or pharmaceutical industry. They are primary EPA producers whose productivity can be manipulated by nutritional and cultural regulation. Even within natural populations, there appears to be a large variability with respect to lipid and EPA content that provides a basis for strain selection and the development of superproducing cell lines. A parallel can be drawn between the present state of EPA production by microalgae and that of penicillin production by *Penicillium chrysogenum* in the 1940s (4). Through strain improvement, tight cultural control of that organism's physiology and increases in biomass

densities, penicillin productivity has been increased several hundredfold over the last 40 years. We are now in the early stages of enhancing the EPA production of certain algal strains. In all likelihood, significant strain and cultural improvements that will make the microbiological production of EPA-containing oils a commercial reality will be made in the future.

The future of EPA-containing algal triglyceride will be determined by developments in two major areas: (1) the enhancement of EPA content on a unit biomass basis with a concomitant decrease in biomass production costs (i.e., by higher cell densities and faster growth rates) and (2) the isolation of genetic determinants associated with fatty acid chain elongation and unsaturation. If the genetic determinants for EPA biosynthesis can be isolated from these algae and transferred to higher plants, production costs would more closely approximate those for conventional vegetable oil. Success in the first realm may establish a new industry with algae as the primary producer of EPA; success in the second will require algae as the source of the relevant genes. In either case, an increased effort in the area of microalgal biotechnology will be required for the development of these EPA sources.

ACKNOWLEDGMENTS

The authors wish to thank M. Cole and N. Pai for their excellent technical help. This work was partly funded by an NIH grant (1-R43-HL38547-01).

REFERENCES

1. Lee, R.E., *Phycology*, Cambridge University Press, Cambridge, U.K., 1980, pp. 1–31.
2. Phillipson, B.E., D.W. Rothrock, W.E. Conner, W.S. Harris and D.R. Illingworth, *N. Engl. J. Med. 312*:1210–1216 (1985).
3. Bimbo, A.P., *J. Am. Oil Chem. Soc. 64*:706–715 (1987).
4. Prescott, S.C., and C.G. Dunn, *Industrial Microbiology*, McGraw-Hill Co. Inc., New York, NY, 1949.
5. Lee, Y.-K., *Trends in Biotechnol. 4*:186–189 (1986).
6. Radmer, R., P. Behrens and K. Arnett, *Biotechnol. Bioengineer. XXIX*:488–492 (1987).
7. Cohen, Z., in *Handbook of Microalgal Mass Culture*, edited by A. Richmond, CRC Press Inc., Boca Raton, FL, 1986, pp. 421–454.
8. Loeblich, A.R., and L.A. Loeblich, in *CRC Handbook of Microbiology*, edited by A.I. Laskin and H.A. Lechevalier, 2nd Ed., Vol. II, CRC Press Inc., West Palm Beach, FL, 1978, pp. 425–450.
9. Bloch, K., G. Constantopolous, C. Kenyong and J. Nagai, in *The Biochemistry of Chloroplasts*, edited by T.W. Goodwin, Vol. 2, Academic Press, NY, 1967, pp. 197–211.
10. Nichols, B.W., and R.S. Appleby, *Phytochemistry 8*:1907–1913 (1969).
11. Lewin, J.C., *J. Gen. Microbiol. 9*:305–313 (1959).
12. Lewin, J.C., and R.A. Lewin, *Can. J. Microbiol. 6*:127–134 (1960).
13. Antia, N.J., J.P. Kalley, J. McDonald and T. Bisalputra, *J. Protozool 20*:377–383 (1973).
14. Droop, M.R., in *Algal Physiology and Biochemistry*, edited by W.D.P. Stewart, University of California Press, Berkeley, CA, 1974, pp. 530–559.
15. FY 1986 Aquatic Species Program Annual Report of the Solar Energy Research Institute, Golden, CO.
16. Shifrin, N.S., and S.W. Chisholm, in *Algae Biomass*, edited by G. Shelef and C.J. Soeder, Elsevier, Amsterdam, 1980, pp. 627–645.
17. Metzger, P., N. Descouls and E. Casadevall, *Com. Eur. Communities Rep. EUR 8245*:339–343 (1983).
18. Kyle, D.J., and P. Behrens, *Dev. Industr. Microbiol*, in press.
19. Lyons, J.M., *Ann. Rev. Plant Physiol. 24*:445–466 (1973).
20. Greenspan, P., E.P. Mayer and S.D. Fowler, *J. Cell Biol. 100*:965–973 (1985).

Discussion Session

Robert Kleiman said he was aware that there were no species in whcih galactosyl diglycerides were esterified with fatty acids on the sugar moiety. The question, "Why is there still interest in developing *Cuphea* as a source of lauric and myristic acids instead of relying on conventional sources?" was asked of G. Röbbelen. L.H. Princen answered that alternate sources are needed for industry in case problems arise with current suppliers or if price and availability changed. R.S. Moreton indicated, in response to a question, that cocoa butter from microbial production would be economical only if the price of cocoa butter was high enough to make this production method attractive.

Colin Ratledge agreed that, contrary to his previous statement, γ-linolenic acid is not essential to humans because man has the ability to desaturate linoleic to this isomer. However, he felt that it may be useful or become essential when the person is in a "stressed state" and in some medical instances. In reference to his paper, David J. Kyle indicated there is a temperature-dependent effect on growth rate and degree of unsaturation produced. He further indicated that the final EPA and oil content after N-starvation was from 0.1–6% in the biomass and decreased from 25% to 16–20% in the lipid.

Immobilized Lipases in Organic Solvents

Atsuo Tanaka, Takuo Kawamoto, Masaya Kawase, Toshiki Nanko and **Kenji Sonomoto**
Department of Industrial Chemistry, Faculty of Engineering, Kyoto University, Yoshida, Sakyo-ku, Kyoto 606, Japan

Interesterification of triglycerides (olive oil) with stearic acid was mediated successfully in water-saturated *n*-hexane by *Rhizopus delemar* lipase entrapped in a hydrophobic photo-crosslinkable resin prepolymer or adsorbed on celite. The entrapped enzyme preparation was found to be more stable than the celite-adsorbed enzyme. Esterification of terpenoids could be catalyzed by various kinds of lipases. A primary alcohol, citronellol, served as a good substrate for lipases, although stereoselective reaction was not observed. Secondary alcohols, such as menthol and borneol, were stereoselectively esterified by entrapped lipase from *Candida cylindracea* in water-saturated isooctane or cyclohexane. Long chain fatty acids were good substrates for esterification but not for stereoselective reactions. Short chain fatty acids, especially 5-phenylvaleric acid, were found to be satisfactory acyl donors from both points of view. Effects of water content in the gels and kinds of organic solvents on the esterification activity also are discussed together with the effect of chemical modification of the enzyme on the activities of diverse reactions.

Lipases from various sources are able to catalyze hydrolysis, synthesis and interesterification of different classes of esters. To apply lipases to ester synthesis or ester exchange, it is essential to carry out the reactions in organic solvents to avoid hydrolytic reactions. Immobilization on or in suitable supporting materials renders the enzymes resistant to denaturation caused by organic solvents. We have developed novel methods to immobilize biocatalysts, including lipases, by entrapping them in synthetic polymer gels prepared from photo-crosslinkable resin prepolymers and urethane prepolymers. These methods are very useful for the bioconversions of lipophilic compounds, such as substrates for lipases, because of an easy choice of hydrophilicity-hydrophobicity balance of gels, which seriously affects the diffusion of lipophilic compounds inside gel matrices.

This report deals with several examples of applications of lipases for different reactions together with the effects of hydrophilicity-hydrophobicity balance of gel materials and polarity of solvents on the activities of gel-entrapped lipases.

SYNTHETIC RESIN PREPOLYMERS

Two types of synthetic resin prepolymers (1) were employed to entrap lipases. Photo-crosslinkable resin prepolymers containing polyethylene glycol (ENT) or polypropylene glycol (ENTP) as the main chain (Fig. 1) can be polymerized by illumination with near-UV light for several minutes in the presence of a sensitizer such as benzoin ethyl ether. ENT is water-soluble and gives a hydrophilic gel, but ENTP is water-insoluble and forms a hydrophobic gel. Urethane prepolymers having polyether diols of different ratios of polyethylene glycol/polypropylene glycol can be polymerized in the presence of water. PU-3 of which 57% is polyethylene glycol moiety and PU-6, 91% is polyethylene glycol moiety, are both water-miscible. PU-6 gives a hydrophilic gel, but PU-3 gives a hydrophobic gel.

IMMOBILIZATION OF LIPASES

To carry out the enzymatic reactions in a water-saturated organic solvent system, lipases usually were adsorbed on celite or entrapped with photo-crosslinkable resin prepolymers or urethane prepolymers in the presence of a controlled amount of water. In some cases, celite-ad-

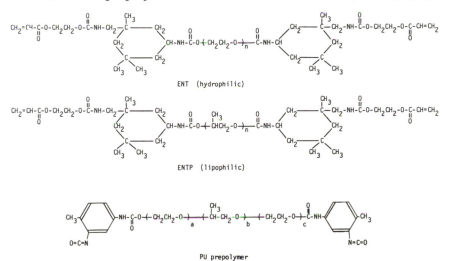

FIG. 1. Structures of synthetic resin prepolymers. ENT, Hydrophilic photo-crosslinkable resin prepolymer; ENTP, hydrophobic photo-crosslinkable resin prepolymer; PU, urethane resin prepolymer.

A. TANAKA ET AL.

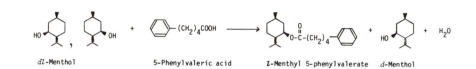

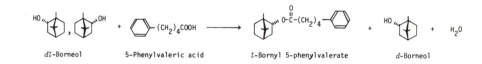

FIG. 2. Esterification of alcohols catalyzed by *C. cylindracea* lipase in organic solvents.

TABLE 1

Enzymes Having a High Activity of Citronellyl 5-Phenylvalerate Formation in Organic Solvent (2)

Enzyme	Source	Maker	Conversion (%)
Lipase OF 360	*Candida cylindracea*	Meito Sangyo	100
Lipase 643 335	*Candida cylindracea*	Boehringer	100
Lipase CE	*Humicola langinosa*	Amano	100
Lipase SP225	*Mucor miehei*	Novo	99
Lipase Saiken 100	*Rhizopus japonicus*	Osaka Saiken	100
Lipase P-1	*Phycomyces nitens*	Takeda	91
Lipase T-01	*Chromobacterium viscosum*	Toyo Jozo	100
Lipase Blend	Fungus & Bacterium	Scripps Lab.	99
Lipase (Steapsin)	Hog pancreas	Tokyo Kasei	99
Lipase Type II	Hog pancreas	Sigma Chemical Co.	99
Lipase Type II	Porcine pancreas	Sigma Chemical Co.	97
Lipase 644072	Porcine pancreas	Boehringer	92
Lipase II	Porcine pancreas	Cooper Biomedical	100
Lipase	Pancreas	Wako	91
Lipoprotein lipase Type A	*Pseudomonas* sp.	Toyobo	98
Lipoprotein lipase LPL	*Pseudomonas* sp.	Amano	100
Lipoprotein lipase	*Pseudomonas* sp.	Sigma Chemical Co.	100
Lipoprotein lipase	*Pseudomonas* sp.	ICN Nutritional	100
Cholesterol esterase Type A	*Pseudomonas* sp.	Toyobo	100
Cholesterol esterase CHE Amano II	*Pseudomonas* sp.	Amano	100
Cholesterol esterase CE	Pancreas	Oriental Yeast	100
Cholesterol esterase T-18	Microorganism	Toyo Jozo	100

The reaction was carried out for 48 hr in water-saturated cyclohexane with celite-adsorbed enzymes.

sorbed lipases were further entrapped with resin prepolymers.

ESTERIFICATION OF PRIMARY ALCOHOL

Activities of lipases and esterases for esterification in organic solvent systems have been examined with citronellol and 5-phenylvaleric acid as substrates to explore organic solvent-tolerant enzymes (Fig. 2) (2). Of 50 enzyme preparations adsorbed on celite, 22 preparations showed high esterification activity in water-saturated cyclohexane (Table 1), although none of the enzyme preparations exhibited stereoselectivity. When free or celite-adsorbed enzymes were subjected to entrapment, *Candida cylindracea* lipase OF 360 entrapped with hydrophobic urethane prepolymer (PU-3) or photo-crosslinkable resin prepolymer (ENTP) exhibited a higher activity than the enzyme entrapped with hydrophilic prepolymers (ENT and PU-6) did. As organic solvents, isooctane and cyclohexane were found to be most suitable and carbon tetrachloride and n-hexane gave a moderate effect, but more polar solvents inactivated the enzyme (Table 2). It is very interesting that the inactivation of the enzyme by polar solvents, such as acetone and chloroform, was reversible. The enzyme recovered activity in isooctane after treatment in these solvents. However, methanol caused irreversible inactivation under the conditions employed (Table 3). Long chain fatty acids, especially oleic acid, served as excellent acyl donors (Table 4). Different from the esterification of secondary alcohols, stereoselective reaction was not observed with any acyl donors.

Not only is water a product of esterification, but water in the gels used for immobilization of enzymes also hinders the diffusion of hydrophobic substrates. Therefore, it is necessary to remove water from the reaction

TABLE 2

Effect of Organic Solvents on Citronellyl 5-Phenylvalerate Formation by Polyurethane Gel-entrapped *C. cylindracea* Lipase (2)

Organic Solvent	Relative activity (%)
Acetone	0
Benzene	2
Carbon tetrachloride	25
Chloroform	0
Cyclohexane	89
Dioxane	0
n-Hexane	40
Isooctane	100
Methanol	0
Methyl isobutyl ketone	0
Tetrahydrofuran	0

TABLE 4

Effect of Acyl Donors on Citronellyl 5-Phenylvalerate Formation by Polyurethane Gel-entrapped *C. cylindracea* Lipase (2)

Acid	Relative activity (%)
Acetic	1
Pivalic	0
n-Valeric	18
Enanthic	32
3-Phenylpropionic	4
5-Phenylvaleric	15
Myristic	29
Stearic	17
Oleic	100

TABLE 3

Effect of Treatment with Polar Solvents on Citronellyl 5-Phenylvalerate Formation by Polyurethane Gel-entrapped *C. cylindracea* Lipase (2)

Treatment (24 hr) with addition of 300 μl water (+ or -)	Enzyme reaction in	Ester formation μmol.h^{-1}. mg lipase^{-1})
Acetone (+)	Isooctane	2.02
Acetone (-)	Isooctane	2.02
Methanol (+)	Isooctane	0
Methanol (-)	Isooctane	0
Chloroform (+)	Isooctane	1.06
Chloroform (-)	Isooctane	1.66
Water-saturated isooctane (-)	Water-saturated isooctane	2.19
Water-saturated isooctane (-)	Isoctane	2.41

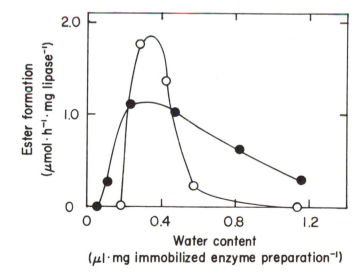

FIG. 3. Effect of water volume on citronellyl 5-phenylvalerate formation by *C. cylindracea* lipase (2). (○), celite-adsorbed lipase; (●), hydrophobic polyurethane gel-entrapped celite-adsorbed lipase.

TABLE 5

Effect of Acyl Donors on Menthyl Ester Formation by Poly-urethane Gel-entrapped *C. cylindracea* Lipase (3)

Acid	Relative activity (%)	E.E.
Acetic	nil	—
n-Valeric	34	100
n-Heptanoic	37	84
n-Nonanoic	49	88
Undecanoic	68	60
Myristic	100	58
Palmitic	58	16
Stearic	86	2
Oleic	100	18
5-Phenylvaleric	59	100
11-Phenylundecanoic	4	100

system for carrying out the reactions efficiently. However, the presence of a certain amount of water in the vicinity of the enzyme was found essential for the maximum esterification activity (Fig. 3). A small amount of water seems to be indispensable for the initiation of the reaction and also for the maintenance of the active conformation of the enzyme molecules. The entrapped enzyme was more stable than the celite-adsorbed enzyme during repeated use in water-saturated isooctane.

ESTERIFICATION OF SECONDARY ALCOHOLS

Secondary alcohols such as menthol and borneol were esterified stereoselectively by gel-entrapped lipases. These reactions are very important for resolution of chemically synthesized terpene alcohols whose L-isomers are useful as ingredients for perfumes, foods and medicines.

Optical resolution of DL-menthol was performed successfully by polyurethane gel-entrapped *C. cylindracea* lipase OF 360 in water-saturated isooctane (Fig. 2) (3). Fatty acids having long chains such as myristic, stearic and oleic acids served as excellent acyl donors for the esterification of DL-menthol, but shorter chain fatty acids

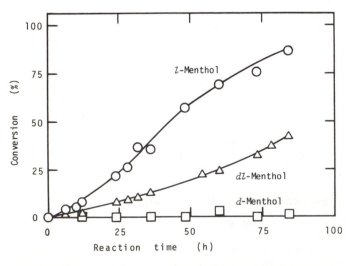

FIG. 4. Stereoselectivity of menthol 5-phenylvalerate formation catalyzed by hydrophobic polyurethane gel-entrapped *C. cylindracea* lipase (3). (○), L-menthol; (△), DL-menthol; (□), D-menthol.

were not good substrates. Moderate activity was observed with medium chain fatty acids (Table 5). However, stereoselectivity was not obtained with longer chain and medium chain fatty acids in contrast to the high stereoselectivity with shorter chain acids. Of the acids examined, 5-phenylvaleric acid was selected as the acyl donor based on the yield of menthyl ester and optical purity of the product. As shown in Figure 4, D-menthol barely was converted to an ester even though more than 80% of L-menthol was esterified with 5-phenylvaleric acid by gel-entrapped *C. cylindracea* lipase. The yield of the ester with DL-menthol was about half that obtained with L-isomer.

For this reaction, yeast and porcine pancreas lipases were found to be active, but fungal lipases did not work under the conditions employed. Cyclohexane and isooctane were suitable solvents for the yeast enzyme. On the contrary, this enzyme did not show the activity in polar solvents such as acetone, chloroform, dioxane, methanol and methyl isobutyl ketone (Table 6). In the esterification of menthol with 5-phenylvaleric acid, the activity of entrapped lipase was almost independent of the hydrophilicity-hydrophobicity balance of urethane gels differing from the results observed on steroid conversions (4). In fact, partition coefficients (ratios of concentration in gel to that in external solvent) of 5-phenylvaleric acid were very high with the hydrophilic and hydrophobic gels, and those of menthol also were high enough with both types of gels. These results indicate that the esterification reaction will not be limited by the diffusion of the substrates inside gel matrices under the conditions employed.

Effect of entrapment on the stability of *C. cylindracea* lipase was studied with the celite-adsorbed enzyme and the polyurethane gel (PU-3)-entrapped celite-adsorbed enzyme because the free enzyme did not disperse in the organic solvent. Although the enzyme was not very stable in both cases, the entrapped preparation showed a higher productivity of ester from DL-menthol than did the adsorbed preparation through 10 batches of reaction (operational period, 40 days) (Table 7). The optical purity of the product was over 96%, even after 40 days of reaction.

As in the case of menthol, borneol also was esterified with 5-phenylvaleric acid by *C. cylindracea* lipase OF 360 in water-saturated cyclohexane (Fig. 2), but not by lipases from fungi after entrapment with a hydrophobic

TABLE 6

Effect of Organic Solvents on Menthyl 5-Phenylvalerate Formation by Polyurethane Gel-entrapped *C. cylindracea* Lipase (3)

Organic solvent	Relative activity (%)
Acetone	0
Benzene	2
Carbon tetrachloride	14
Chloroform	0
Cyclohexane	100
Dioxane	0
n-Hexane	32
Isooctane	77
Methanol	0
Methyl isobutyl ketone	0

TABLE 7

Production of *l*-Menthyl 5-Phenylvalerate by Lipase Preparations in Repeated Reactions (3)

Reaction temperature (C)	Total amount of ester formed (mg per 10 batches)	
	Celite-adsorbed lipase	Polyurethane-entrapped celite-absorbed lipase
30	1,380	2,080
35	1,220	1,880
40	1,100	1,690

Each batch was carried out for 93 hr.

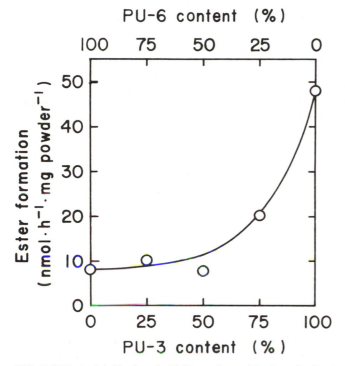

FIG. 5. Effect of gel hydrophobicity on bornyl 5-phenylvalerate formation by polyurethane gel-entrapped *C. cylindracea* lipase. PU-3, hydrophobic urethane prepolymer; PU-6, hydrophilic urethane prepolymer.

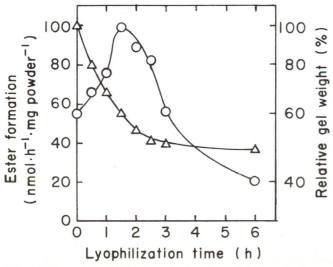

FIG. 6. Effect of lyophilization of gel on bornyl 5-phenylvalerate formation by hydrophobic urethane gel-entrapped *C. cylindracea* lipase. (○), Activity; (△), relative gel weight.

urethane prepolymer (PU-3). As an acyl donor, *n*-valeric acid and 5-phenylvaleric acid were found to be suitable from the standpoint of stereoselectivity and yield of ester although the optical purity of the product was lower (E.E., about 70%) than that of the menthyl ester (E.E., about 100%). Oleic acid again was an excellent substrate for the esterification, but not for the optical resolution of **DL**-borneol.

The hydrophilicity-hydrophobicity balance of polyurethane gels had a remarkable effect on the activity of the entrapped enzyme with the hydrophobic gel-entrapped enzyme exhibiting a higher activity (Fig. 5). These results are quite different from the case of menthol esterification. In addition to the gel hydrophobicity, water content of the gels was supposed to have a critical effect on the activity of the gel-entrapped enzyme. When the hydrophobic gel (PU-3) entrapping the enzyme was lyophilized for the indicated periods, the activity of esterification increased along with the rapid decrease in the gel weight, that is with removal of water from the gel, and then decreased with the slow decrease in the gel weight (Fig. 6). The enzyme in the gel lyophilized for six hr was not found to be denatured because the addition of water to the gel restored the original enzyme activity. A similar phenomenon also was observed with the enzyme entrapped in a hydrophilic polyurethane gel (PU-6), although the maximum activity obtained was lower than that of the hydrophobic gel (PU-3)-entrapped enzyme (Fig. 7). These results again indicate that at least a small amount of water in the vicinity of the enzyme molecules is essential to keep the enzyme active and that a large part of water in the gel hinders the diffusion of the substrates, lowering the apparent enzyme activity.

As described above, *C. cylindracea* lipase OF 360 showed a high activity for a primary alcohol, citronellol, and a moderate activity for secondary alcohols, menthol and borneol, while tertiary alcohols, such as linalool, nerolidol and α-terpineol, did not serve as substrates for the enzyme. Different enzymes should be screened for the esterification of these tertiary alcohols.

ESTER EXCHANGE OF TRIGLYCERIDES

Production of fat with desired physical and chemical properties by replacing the fatty acid moieties of triglycerides with other fatty acids is of great importance and interest from an industrial viewpoint. Enzymes such as lipases, which produce new types of triglycerides depending on their substrate and position specificities, are very useful for this purpose.

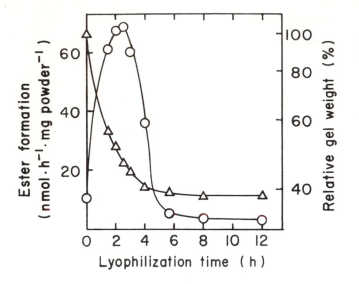

FIG. 7. Effect of lyophilization of gel on bornyl 5-phenylvalerate formation by hydrophilic urethane gel-entrapped *C. cylindracea* lipase. (O), Activity; (△), relative gel weight.

The formation of cocoa butter-like fat from olive oil and stearic acid or palmitic acid by enzymatic interesterification in an organic solvent system was attempted with celite-adsorbed 1- and 3-positional specific lipase from *Rhizopus delemar* (Fig. 8); this system has been proven to be superior to the aqueous reaction system (5). To prepare a more stable enzyme preparation, lipase was immobilized by different methods such as covalent binding, ionic binding and adsorption on porous silica beads and entrapment with hydrophilic or hydrophobic resin prepolymers with or without celite adsorption. Of the methods employed, entrapment with a hydrophobic photo-crosslinkable resin prepolymer (ENTP) after adsorption on celite was found to be most suitable for the

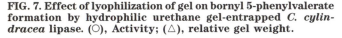

FIG. 8. 1- and 3-positional specific interesterification of triglyceride catalyzed by *R. delemar* lipase.

TABLE 8

Ester Exchange Activity of *R. delemar* Lipase Immobilized by Different Methods (6)

Method	Support	Relative activity (%)
—	None	0
Adsorption	Celite	100
	Silica beads	16
Ionic binding	Silica beads	28
Covalent binding	Silica beads	9
Entrapment*	Hydrophilic PC prepolymers	14–22
	Hydrophobic PC prepolymer	75
	Hydrophilic PU prepolymers	14–17
	Hydrophobic PU prepolymer	21

*Entrapped after adsorbed on Celite.
PC, photo-crosslinkable resin; PU, urethane resin.

immobilization of *R. delemar* lipase (Table 8). The activity of the ENTP-entrapped celite-adsorbed enzyme was about 75% of that of the enzyme merely adsorbed on celite. Direct entrapment of the enzyme with the hydrophobic prepolymer (ENTP) did not give reproducible results, probably due to the difficulty of homogeneous distribution of lipase in the gel during the entrapment procedure (6).

Water-saturated *n*-hexane was selected as the reaction solvent based on the solubility of the substrates and products and on the stability of the enzyme. The concentration of water in the reaction system affects the yield of the transformed fat with an excess of water tending to favor hydrolysis of triglyceride rather than interesterification. When the amount of water was controlled in the preparation of celite-adsorbed lipase, there was an optimal condition, as seen in Figure 9. Entrapment of the celite-adsorbed enzyme with the prepolymer (ENTP) did not shift the optimal range of the water volume. Incorporation of stearic acid into olive oil was linear with time up to four hr and, thereafter, increased gradually to reach the maximum of about 40%. This value is rather small compared with the theoretical value of 65%, suggesting the existence of an equilibrium or competition between stearic acid added as the substrate and oleic acid released from olive oil.

Celite-adsorbed *R. delemar* lipase lost about half of its original activity after five batches of reaction (operational period, 5 days), while the gel-entrapped celite-adsorbed enzyme was far more stable. More than 90% of the original activity remained after 12 batches (operational period, 12 days) (Fig. 10). Maintenance of the enzyme activity by entrapment may be ascribable to the protection of the enzyme from denaturation by *n*-hexane, prevention of leakage of the enzyme protein from celite or to both.

Not only *R. delemar* lipase but also fungal lipases from *Aspergillus niger*, *Geotrichum candidum* and *Rhizopus niveus* were applied to the interesterification of different triglycerides (5,7).

CHEMICAL MODIFICATION

Chemical modification often gives some important information concerning the roles of amino acid residues and the reaction mechanisms of enzymes. Furthermore, this technique sometimes is effective in improving the catalytic activties of enzymes and, therefore, the results will be useful for the genetic modification of enzyme proteins to produce industrially applicable enzymes.

C. cylindracea lipase, purified from a commercial enzyme preparation by gel filtration chromatography, was subjected to chemical modification; and the activities of the modified enzyme, i.e., hydrolysis, esterification and ester exchange, were examined. The latter two reactions were carried out in organic solvent systems with the celite-adsorbed enzyme (Table 9).

When arginine residues were modified with phenylglyoxal, the enzyme lost all the activities mentioned above. Modification of aspartic acid and glutamic acid residues with 1-ethyl-3-(3-dimethylaminopropyl) carbodimide (EDC) also resulted in the significant decreases in three activities of the enzyme. These results indicated that the arginine residues and carboxyl groups in the ac-

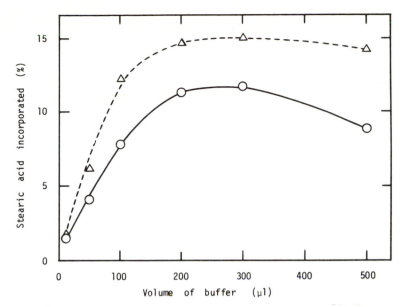

FIG. 9. Effect of water volume on interesterification of olive oil by *R. delemar* lipase (6). (○), Hydrophobic photo-crosslinked gel-entrapped celite-adsorbed lipase; (△), celite-adsorbed lipase.

TABLE 9

Effect of Chemical Modification on *C. cylindracea* Lipase

Target amino acid residue	Modifier	Relative activity (%)[1]	
		Hydrolysis[2]	Esterification[3]
Arg	Phenylglyoxal	0–9	nil
Asp, Glu	EDC	17–53	23–37
His	EFA	71–77	80–98
Cys	NEM	72–83	135–140
Cys-Cys	DTT	67–97	225–467
Lys	PLP	83–86	161–200
Ser	PMSF	84–90	146–152
Tyr	TNM	113–127	89–95

[1]Expressed as relative value to the respective control runs.
[2]Hydrolysis of olive oil. [3]Esterification of citronellol with 5-phenylvaleric acid.
DTT, dithiothreitol; EDC, 1-ethyl-3-(3-dimethylaminopropyl) carbodiimide; EFA, ethoxyformic anhydride; NEM, *N*-ethylmaleimide; PLP, pyridoxal 5'-phosphate; PMSF, phenylmethylsulfonyl fluoride; TNM, tetranitromethane.

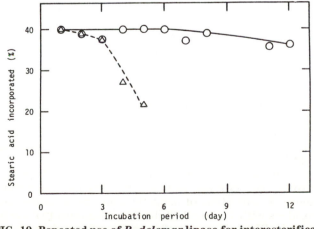

FIG. 10. Repeated use of *R. delemar* lipase for interesterification of olive oil (6). Each batch was carried out for 24 hr. (○), Hydrophobic photo-crosslinked gel-entrapped celite-adsorbed lipase; (△), celite-adsorbed lipase.

tive site or substrate binding site of this enzyme participate in these reactions. On the other hand, the esterification activity and the ester exchange activity were enhanced markedly without an appreciable loss of the hydrolytic activity by the modification of lysine residues with pyridoxal 5'-phosphate (PLP) or by the reduction of disulfide bonds with dithiothreitol (DTT). These results might be useful for the genetic improvement of *C. cylindracea* lipase in order to render the enzyme more active in organic solvent systems.

DISCUSSION

In addition to our work mentioned here, lipases from different sources have been used in organic solvents to produce useful compounds. For example, Klibanov and his coworkers (8,9) reported wide applications of enzymes to esterification and interesterification in preparing optically active alcohols and acids. Inada et al. (10) synthesized polyethylene glycol-modified lipase, which is soluble in organic solvents and active for ester formation. These results reveal that lipases are very useful enzymes to catalyze different types of reactions with rather wide substrate specificities. However, it will be essential to screen or prepare lipases of high tolerance to organic solvents for extensive applications of the enzymes in industrial synthesis of useful lipophilic compounds. Fundamental studies on proteinaceous properties of lipases also will be necessary for the further modification of the enzymes by chemical and biological methods and for the production of enzymes having excellent activities and stabilities in organic solvents.

REFERENCES

1. Fukui, S., and A. Tanaka, *Adv. Biochem. Eng./Biotechnol. 29*:1 (1984).

2. Kawamoto, T., K. Sonomoto and A. Tanaka, *Biocatalysis*, in press.

3. Koshiro, S., K. Sonomoto, A. Tanaka and S. Fukui, *J. Biotechnol. 2*:47 (1985).

4. Fukui, S., and A. Tanaka, *Endeavour 9*:10 (1985).

5. Tanaka, T., E. Ono, M. Ishihara, S. Yamanaka and K. Takinami, *Agric. Biol. Chem. 45*:2387 (1981).

6. Yokozeki, K., S. Yamanaka, K. Takinami, Y. Hirose, A. Tanaka, K. Sonomoto and S. Fukui, *Eur. J. Appl. Microbiol. Biotechnol. 14*:1 (1982).

7. Macrae, A.R., *J. Am. Oil Chem. Soc. 60*:291 (1983).

8. Kirchner, G., M.P. Scollar and A.M. Klibanov, *J. Am. Chem. Soc. 107*:7072 (1985).

9. Zaks, A., and A.M. Klibanov, *Proc. Natl. Acad. Sci. USA 82*:3192 (1985).

10. Inada, Y., K. Takahashi, T. Yoshimoto, A. Ajima, A. Matsushima and Y. Saito, *Trends Biotechnol. 4*:190 (1986).

Enzymatic Fat Splitting

Warner M. Linfield

Eastern Regional Research Center, USDA/ARS, Philadelphia, PA 19118; National Renderers Association, Des Plaines, IL 60018

A brief review of the more recent research and technical literature on enzymatic fat splitting is given. It was found that the rate of enzymatic hydrolysis is directly proportional to the logarithm of reaction time and to the logarithm of enzyme concentration. The lipase from *C. rugosa* is available commercially and appears to be most suitable for fat splitting. Different fats or oils will hydrolyze at different rates, probably due to differences in chemical structure. Temperature has little influence on reaction rate. Addition of electrolytes, albumin or other additives either has no effect or adversely affects hydrolysis. A novel assay method based on the above logarithmic relationship is described. The most favorable aspects of the enzymatic approach are low-energy input, high-quality and relatively low capital cost. Its practicality has been demonstrated by the existence of industrial batch processes as well as continuous processes described in the patent literature.

There are three routes towards hydrolysis of triglycerides: high-pressure steam splitting, alkaline hydrolysis and enzymatic hydrolysis. Of these, the high-pressure steam splitting at 50 bars and 250 C or more for a contact period of about two hr involves the highest capital investment and energy cost. Alkaline hydrolysis (saponification), whether carried out batchwise (kettle process) or continuously, is also energy intensive, and the resulting soap has to be acidified to yield fatty acid—a cumbersome process. Enzymatic hydrolysis appears to be the cleanest process, using the least energy and, apart from the cost of the lipase, involving the lowest chemical cost.

Most of the fatty acid producers in industrialized countries have used the high-pressure steam splitting technique because this was the only process available at the time. The commercial availability of lipases at present should make enzymatic fat splitting very attractive for new producers just entering the fatty acid market.

Until recently, the scientific literature on enzymatic lipolysis was confined almost entirely to rather academic biochemical studies in which pancreatic lipase typically was the enzyme studied. Unfortunately much of this information has little practical utility and some of the published data appear to be erroneous. At present, lipases from *Candida rugosa*, *Rhizopus arrhizus* and *Aspergillus niger* have become commercially available. Although these are still high-priced, increased usage undoubtedly will result in lower prices.

A 1965 patent (1) discloses the purification of a highly active lipase obtained from the yeast *Candida rugosa (cylindracea)*, and it was shown in another patent (2) that this lipase catalyzed the hydrolysis of olive, soybean and linseed oils. We chose this lipase as the catalyst of choice for most of our research. We determined the kinetics of hydrolysis of olive oil, coconut oil and tallow. Because Na and Ca ions as well as albumin had been cited in the biochemical literature as greatly enhancing hydrolysis, the effects of these were studied as were those of the addition of a nonionic surfactant and hydrocarbon solvent (3).

Initially, the results from the hydrolysis of olive oil were quite erratic. This apparently was due to impurities in the oil; treatment with bleaching clay removed the enzyme inhibitors. The tallow used already had been bleached, and no inhibition problems were observed. Therefore, it would stand to reason that the substrate fat or oil always should be bleached to remove such impurities, which could range from polyvalent metal cations to chlorophyll.

The major reaction parameters studied were temperature, pH, enzyme concentration and reaction time. Although the experimental procedure used was quite simple, a careful technique must be used. The reaction vessel obviously has to be clean. It should not be oversized as excessive stirring speed could bring about a centrifuge effect whereby the oil and water phases are separated. Intimate contact between oil and water phases is essential, and this is achieved most easily by initial homogenization or short-term ultrasonic agitation.

In a typical hydrolysis run, the buffer is dissolved in water, then the enzyme is added, and the water phase is stirred until all of the enzyme is dissolved. The oil or melted fat then is added and the mixture stirred at a predetermined constant temperature. Samples are withdrawn at fixed time intervals usually until the substrate is completely hydrolyzed. Alcohol is added to deactivate quickly the enzyme in each, and the sample is titrated for free fatty acid content. As most fats and oils contain some free fatty acid, a blank determination of the initial fatty acid content has to be made. In this fashion, the percentage of enzymatic hydrolysis is determined. Further experimental details can be found (3-5).

It was found that the percentage of hydrolysis is directly proportional to the logarithm of reaction time and, in another series of experiments, it could be shown that the percentage of hydrolysis also is directly proportional to the logarithm of the enzyme concentration. Such a linear relationship cannot be explained readily on the basis of conventional kinetics. This linearity held for all three substrates studied as well as for the lipases from *C. rugosa* and *A. niger*. These findings were confirmed by an independent study on palm oil and other oils carried out by Khor and coworkers (6). It also was observed that the Ca ion slowed down hydrolysis catalyzed by *C. rugosa* lipase, which again was confirmed by Khor et al. (6). Addition of sodium ion or albumin had essentially no effect. Addition of small amounts of nonionic surfactant had little effect, but larger concentrations of surfactant inhibited hydrolysis.

We found the hydrolysis rate to be optimal in a pH range of 4.8–7.2 (3), whereas Khor (6) established the optimal pH at 7.5 in studies with palm oil. Addition of a hydrocarbon solvent to semisolid fats such as tallow or palm oil eliminates the need for keeping the hydrolysis mixture above the melting point of the fat. A ratio of 0.5 ml of hexane to 1 g of lipid appeared to give optimum results (6), whereas a ratio of 1 ml of hexane to 1 g of fat gave rise to substantial slowing of the hydrolysis (3).

With respect to the effect of temperature on hydrolysis rate there are some obvious limitations. At 25 C, tallow and palm oil are solids and parts of coconut and other oils also begin to crystallize, while above 50 C the enzymes become denatured. in the 26–46 C range, no statistically significant difference in hydrolysis rate could be observed (3). Tomizuka et al. (7) reported that the optimal temperature for the lipase from *C. rugosa* was 45 C, but Khor et al. (6) conducted their studies at 37 C. Because a catalyst's function is to overcome the energy barrier, it would stand to reason that temperature does not play an important part in enzymatic hydrolysis.

Not all fats and oils are hydrolyzed at the same rate. We found that olive oil hydrolyzed more rapidly than tallow or coconut oil (3), and Khor et al. found that palm, soybean and corn oil were hydrolyzed at the same rate, whereas peanut oil was hydrolyzed more slowly. These differences may be due to the physical structure rather than to the presence of inhibiting impurities (6).

As mentioned earlier, the lipases from *R. arrhizus* and *A. niger* were not considered suitable for the purpose of total hydrolysis of fats and oils because these two enzymes exhibit specificity towards hydrolysis of the α-glyceride linkage. Because the β-monoglyceride will slowly rearrange to the α-isomer, this rearrangement then becomes the rate-determining step in the hydrolysis rather than the catalytic effectiveness of the enzyme.

The determination of the activity of the enzyme is another important consideration. All lipases will slowly degrade with age, particularly if they are kept warm during shipment. Thus, their activity is seldom as high as claimed by the manufacturer. Because the Worthington procedure (8) is rather awkward and not too reliable, we developed a simple assay method (6) base on the above research findings and applicable to a variety of lipases. It used bleached olive oil as substrate instead of pure triolein, and a linear plot of mmols of fatty acid formed after a one-hr reaction time vs logarithm of enzyme concentration was made. The activity was calculated at that point when 24% of the oil had been hydrolyzed to fatty acid.

It should be mentioned here that because hydrolysis is a reversible reaction, the lipase also can be used to catalyze esterification. In this case, instead of running the reaction in a fairly large volume of water, the enzyme is dissolved in a minimal amount of water, and as the reaction proceeds, water periodically is withdrawn by vacuum distillation (5).

Some practical aspects of enzymatic fat splitting follow: In a batch operation, the fat is mixed with an equal or slightly lesser weight of water in which the enzyme has been dissolved. The amount of enzyme used will depend on the reaction time desired for complete hydrolysis. As mentioned above, the reaction rate is a linear function of the logarithm of enzyme concentration. Intimate mixing of the two phases either by homogenization or by ultrasound is needed initially, thereafter only gentle agitation is required. The enzyme is denatured even at room temperature by Cu, Fe and Ni ions, therefore, stainless steel or glass-lined equipment must be used. If samples of the mixture are withdrawn at regular time intervals and percentage hydrolysis is plotted against logarithm of time, the reaction time required for 97-98% hydrolysis can be estimated fairly accurately by extrapolation. Table 1 gives a summary of data for complete hydrolysis of various substrates.

TABLE 1

Minimum Lipase Requirement for Complete Hydrolysis of 100 g Substrate after 72-hr Reaction Time

Substrate	mg *C. rugosa* lipase (30,000 u/g)	Temp. (C)
Tallow	166	43–46
Coconut oil	198	23–26
Olive oil	66	23–26

There are several advantages of enzymatic fat splitting over the conventional high-pressure steam splitting or saponification processes. The considerably lower energy requirement and lower chemical cost already have been mentioned. The color of the crude fatty acid obtained enzymatically is the same as that of the starting fat. Thus, if a light-colored fat is used, distillation may not be needed. By contrast, the acid obtained via the steam-splitting method is pitch black and always must be distilled. The aqueous layer, which typically contains 8–10% glycerine, likewise is light-colored and, unlike the saponification lyes, contains no salt, and the glycerine obtained often can be decolorized with carbon black or other bleaching agents, thus avoiding distillation. Currently, a major concern is the relatively high cost of the lipase, which could be expected to drop with increased usage. At present, the cost of enzymatic fat splitting appears to be higher than that of steam splitting. Recycling of an immobilized enzyme would lower the splitting cost substantially. In addition, the higher initial capital investment for high-pressure equipment would favor the enzymatic process. The mild enzymatic hydrolysis conditions would be especially favorable for the splitting of heat-sensitive oils such as castor oil.

The major drawback of enzymatic hydrolysis lies in the slowness of the reaction. Speeding up the reaction time will increase the enzyme cost logarithmically, but a continuous process involving immobilized enzyme would circumvent this. Furthermore, conversion rates of greater than 98% are required for fatty acid production. This again involves longer reaction time and/or an increased amount of water in the reaction mixture in order to shift the equilibrium to the right.

The practicality of enzymatic fat splitting on an industrial scale has been shown in two Japanese patents on enzymatic fat splitting. In one of these (9), molten fat was dispersed in cold aqueous lipase solution to give a stable dispersion and hydrolyzed. Thus, tallow heated to 45 C was homogenized at 10 C with lipase solution, and the semisolid dispersion was kept at that temperature for 48 hr. After heating the mixture to 70 C, the dispersion broke, and the oily layer was 98% hydrolyzed. Another patent (10) describes the lipolysis of fats and oils with lipase immobilized on gels of alkylated, phenylated or triethylated polysaccharides such as ocytylagarose gel. The immobilized enzyme could be reused five consecutive times and still produce 98.5% hydrolysis. The activity of commercial lipase from *C. rugosa* has been increased significantly from 30,000 units/g to 360,000 units/g. The current price of this enzyme in Japan is about 30,000 yen ($200)/kg. A batch process has been in use in Japan for some time (11). Here, the *C. rugosa* lipase is dissolved in water, then mixed with the molten tallow and immediately chilled to room temperature. After it stands for 24 hr, the reaction mixture is heated and the glycerine layer separated from the fatty acid layer. A small interface layer containing some enzyme is removed and recycled. The hydrolysis is 95–97% complete. The fatty acid thus produced is neutralized and made into soap. Some continuous processes also have been used industrially, at least on a pilot plant stage.

REFERENCES

1. Yamada, K., and H. Machida, Meito Sangyo Kabushiki Kaisha, U.S. Patent 3,189,529 (1965).
2. Shinota, A., H. Machida and T. Azuma, Meito Sangyo Co., Japan Patent 71:16,509 (1971).
3. Linfield, W.M., D.J. O'Brien, S. Serota and R.A. Barauskas, *J. Am. Oil Chem. Soc. 61*:1067 (1984).
4. Linfield, W.M., R.A. Barauskas, L. Sivieri, S. Serota and R.W. Stevenson, *J. Am. Oil Chem. Soc. 61*:191 (1984).
5. Linfield, W.M., S. Serota and L. Sivieri, *J. Am. Oil Chem. Soc. 62*:1152 (1985).
6. Khor, H.T., N.H. Tan and C.L. Chua, *J. Am. Oil Chem. Soc. 63*:538 (1986).
7. Tomizuka, N., Y. Otta and K. Yamada, *Agric. Biol. Chem. 30*:576 (1966).
8. Worthington Enzymes and Related Biochemcial Co. Cat. #WC790, Cooper Biomedical, Palo Alto, CA, 1982, pp. 117–118.
9. Nippon Oils and Fats Co. Ltd., Japan Kokai Tokkyo Koho JP 82 57,799; *Chem. Abstr. 97*:111593t (1982).
10. Nippon Oils and Fats Co. Ltd., Japan Kokai Tokkyo Koho JP 53,146,284 (83,146,284); *Chem. Abstr. 99*:190630v (1983).
11. Sato, S., *Proc. China Internatl. Soap and Detergent Symposium*, 1987, p. 89.

Ester Synthesis with Immobilized Lipases

Peter Eigtved[a], **Tomas T. Hansen**[a] and **Carl A. Miller**[b]

[a]Novo Industri A/A, DK-2880 Bagsvaerd, Denmark, and [b]Novo Laboratories, Inc., Danbury, CT

The use of immobilized lipases for the preparation of a variety of esters is reviewed here. Fatty acid ester products can be waxes, emulsifiers and flavors. Biochemical synthesis offers a number of advantages compared with chemically catalyzed processes such as selectivity, gentle reaction conditions leading to pure products and no use of toxic catalysts. Examples of thermostable microbial lipases useful for the synthesis of primary and secondary alcohol esters are given. Also, it is illustrated that lipases may accept a number of acid and alcohol components apart from their natural fatty acid/glyceride substrates. Lipase stereoselectivity may be used in the preparation of optically pure substances. The benefit of immobilization in terms of enzyme activity, stability and handling will be demonstrated. Lipase development also has included process technology such as a simple and efficient vacuum system for making wax esters. Furthermore, the use of substrate derivatives or solvents does not exclude the use of lipases in ester synthesis reactions.

A large number of esters are of importance in our daily life (Table 1). Natural products like triglycerides, phospholipids, galactolipids, cutin and many waxes, steroids, flavors and fragrances are all esters. These substances have many different functions. They are energy sources, membrane constituents, emulsifiers, viscosity builders, protective coatings, and they affect taste and pleasure. Although very diverse even in terms of chemical structure, the ester bonds are common to all of them.

It has been realized that lipases and esterases, a class of enzymes used by living matter to digest triglycerides, also will hydrolyze many esters other than their natural substrates. Further, synthesis of such esters can be made with lipases, given the appropriate reaction conditions and enzyme preparations. New product opportunities and biochemical synthesis routes will be possible, and this has led us into the development of immobilized, microbial lipases and ester synthesis processes.

BIOCHEMICAL SYNTHESIS OF ESTERS

Although many esters may be produced in high yields and good purity by existing chemical methods, the lipase-catalyzed reactions have a number of advantages. Improvements can be made in terms of investment and energy cost, safety, and purity. Quantitative yields have been obtained. More importantly, pure ester products based on natural raw materials and biocatalysis can be expected to substitute certain scarce natural products and create new ones. One of the examples is the various attempts to make glycolipid emulsifiers. Many organic chemists also are aware of the potential lipase stereoselectivity in hydrolysis and synthesis reactions.

This paper deals only with simple esterifications between alcohols and acids. Esters also may be produced by alcoholysis and transesterification reactions. We have developed an immobilized lipase that can work efficiently in all three reaction types with small amounts of water present (1,2). The basic esterification reactions are shown in Scheme 1.

$$R_1 - OH + HOOC - R_2 \rightleftarrows R_1 - O - CO - R_2 + H_2O$$

Alcohol   Acid   Ester   Water

SCHEME 1

Some lipases may have a narrow substrate specificity, a property that can be used to make products not easily obtained by conventional chemistry. They have regiospecificity toward primary and secondary hydroxyl groups and stereospecificity with which lipases may react preferentially with one optical isomer.

The increasing number of publications and patent applications reflects a worldwide interest in the use of lipases for ester synthesis. An extensive study covering many lipases and reactions has been published by Lazar and coworkers (3). Baratti et al. have investigated enantioselective synthesis (4). Gatfield (5) has focused on short and medium chain length fatty acid flavor esters, and Seino et al. focused on sugar esters (6). Klibanov and coworkers have broadly exploited the use of lipases in organic chemistry (7). At the 78th annual meeting of the American Oil Chemists' Society in May 1987, we presented some characteristics of Lipozyme® in ester synthesis (8).

PRODUCT AND PROCESS POSSIBILITIES

We have analyzed a number of model systems to illustrate the kind of products that can be made, necessary lipase properties and possible process configurations and conditions.

TABLE 1

Esters: Typical Substances and Uses

Emulsifiers	Emulsification of
Monoglycerides	Foods
Glycolipids	Feeds
Phospholipids	Pharmaceuticals
Alkyl esters	Personal hygiene products
Flavors and fragrances	
Fatty acid/alcohol esters	Food and feed flavoring
Polyesters (cutin)	Perfumes
Specialty chemicals	
Fatty acid/alcohol esters	Lubricants, visc. builders
Polyesters (cutin)	Protective coatings
Specialty Chemicals	
Optical active esters	Pesticides, pheromones

Wax Esters

One of the fundamental problems in esterifications is to shift the equilibrium efficiently towards synthesis. This can be accomplished by maintaining the concentration of water low, e.g., by keeping the reaction mixture under slight vacuum if the acid and alcohol components are less volatile than the water formed. This is the case in the synthesis of wax esters from fatty acids and fatty alcohols. Some natural wax esters are high-value products being either difficult to isolate from natural sources or difficult to make by chemical synthesis due to chain length and unsaturation. Examples are spermaceti wax and jojoba oil.

As a model compound, we have prepared myristyl myristate under vacuum with reuse of the enzyme. In a 3-liter stirred reactor at 60 C at .05 atm for 20 hr, we used 1 kg reactants (equimolar) with a 10 g enzyme charge. We also used 12 cycles of catalyst at 96[+]% yield leading to productivity of at least 1.2 tons of product per kg catalyst.

The immobilization method used for our enzyme solves at least two problems with water-soluble lipase powders. The water-binding capacity of the carrier assures that sufficient water is kept around the enzyme even at low pressure. The particulate preparation makes multiple re-uses and easy separations possible because sticky aggregates of lipase and water with low surface areas and activities can be avoided. Therefore, the increased cost of producing an immobilized lipase preparation may be more than compensated for by the increase in productivity.

Table 2 shows the effect of pressure and time on the yield. Without vacuum, an equilibrium conversion dependent on the composition of the reaction mixture can be expected, in this case, around 85%. When the pressure is decreased, nearly quantitative yields can be obtained. Please also note the influence of pressure on the reaction rate; high vacuum results in a high rate.

Emulsifiers

The selective monoesterification of polyols is one of the potential uses of regiospecific lipases. Cesti et al. (9) showed that pancreatic lipase could be used with a number of diols in transesterification reactions. We have investigated the formation of esters from 1,2-propyleneglycol and myristic acid using a 1,3-regiospecific enzyme.

In contrast to many other polyol/fatty acid (triglyceride) mixtures, this substrate is a single phase (at 60%). However, as can be seen in Figure 1, it turned out to be impossible to obtain high degrees of selectivity and conversion at the same time. Significant amounts of diester were formed in spite of the low reactivity seen with secondary hydroxyls in other substrates. Other lipase preparations or reaction conditions will be needed.

Ninety percent monoglycerides with mixed fatty acid composition are made industrially from glycerol and fats of around 200 C with chemical catalysis followed by molecular distillation.

Pure 1-monoglycerides from glycerol and free fatty acids using regiospecific lipase would be another attractive synthesis. However, acyl migration leads to substantial amounts of di- and triglycerides. Furthermore, fatty acids and glycerol are not miscible and have a poor solubility in each other at 60-80 C.

Godtfredsen (10) has described a route in which the problems with both solubility and selectivity are solved. The route to a very pure monoglyceride follows: glycerol is reacted with acetone at mild conditions to form isopropylidene glycerol (a ketal). This is miscible with fatty acid and has only one primary hydroxyl group available for esterification. The derivatized glycerol is a good substrate for our enzyme, and the ester product easily can be deblocked with mild acid hydrolysis. Another example of this approach is the synthesis of glycolipids. A diketal was made of galactose, and this was monoesterified with myristic acid using our enzyme at 60 C at atmospheric pressure.

We believe surface-active compounds based on natur-

TABLE 2

Synthesis of Myristyl Myristate, Influence of Pressure

Reaction time hours	Pressure, bar[a]		
	1.0	0.2	0.04
1	77	90	98
2	85	99	99

[a]Percentage of yield.
Conditions: 0.05 mol acid, 0.05 mol alcohol, 60 C, 1 g enzyme.

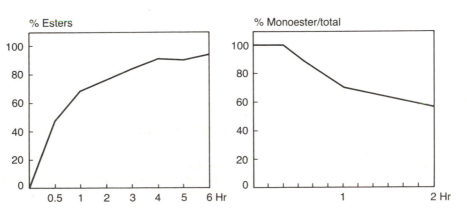

FIG. 1. Synthesis of propyleneglycol esters.

al raw materials like sugars and lipids that are produced by biochemical synthesis will be very important in the future. Pure, food-grade and biodegradable surfactants with specific properties are in demand to substitute some large, existing products such as sorbitan esters and LAS.

Flavors and Fragrances

Soluble lipases are used in situ for flavor enhancement in some foods, typically those based on milk fat. The mechanism may be liberation of short chain fatty acids, further oxidation to aldehydes, etc., but re-esterification with ethanol or other alcohols produced by fermentation also is involved. Other esters of natural occurrence are important as fragrances, for instance those based on geraniol. The concept of a natural ester made by enzymatic synthesis with lipase and natural substrate components is therefore an interesting alternative to extraction and organic chemical synthesis. Gatfield (5) has investigated *Mucor miehei* lipase in this light. In the future, production of more flavors and fragrances in this way as well as by fermentation can be expected.

Optically Active Compounds

Lipases and esterases now very often appear as tools for resolution of optical isomers either by selective ester hydrolysis or by synthesis. Although some of these enzymes are available in sufficient amounts for industrial purposes, two problems limit application outside the laboratory scale: their stereospecificity is not yet characterized in detail to allow selection of the proper lipase for a given task; and the productivity often is very low because the lipases are not immobilized or otherwise stabilized to resist inactivation by the organic chemical substrates and solvents.

Yet, a number of promising and inspiring results have been published. In a recent work by Sonnet (11), Lipozyme® was used to make a pheromone ester by selective esterification of a (S)-2-alcohol component as shown in Scheme 2.

Another example of stereoselective synthesis appears below, where *Candida cylindracea* lipase was used to esterify a (R)-2-chloropropionic acid with butanol, leaving the desired (S) acid unreacted for separation (12). This

and the corresponding bromo compound are important in herbicide synthesis (Scheme 3).

Lipase Specificity

It is our experience that lipases are difficult to classify in terms of specificity because the apparent specificity may depend very much on the preparation and reaction conditions. From the characterization of Lipozyme® (8), we know that many "unnatural" alcohols and acids are quite good substrates (Tables 3 and 4) and that a number of water-immiscible solvents can be used in esterification reactions if needed (Fig. 2).

As our enzyme reacts much better with primary than secondary alcohols, we have screened for a nonspecific, thermostable lipase with a high reactivity towards both primary and secondary hydroxyl groups. Recently, an experimental preparation of such a lipase has been developed (Novo SP 344). In Figure 3, synthesis of isopropyl- and n-propylmyristate with Lipozyme® and immobilized SP 344 is shown. Only SP 344 is effective with the isopropanol, but it also is preferred in terms of reaction rate with n-propanol. Water was not removed, and equimolar substrate components were used, which explains the less than quantitative yields.

The development, characterization and application technology of lipases now have reached the stage at

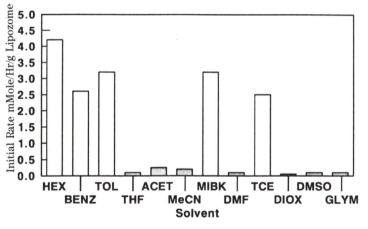

FIG. 2. Synthesis of propyl myristate in solvent.

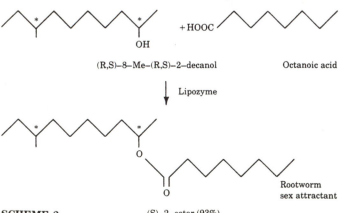

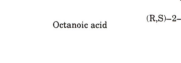

SCHEME 2   (S)–2–ester (93%)

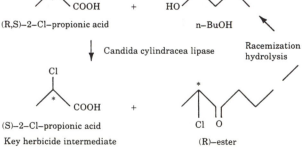

SCHEME 3

TABLE 3

Esterification Rates of Cyclohexyl vs Phenyl Substituted Carboxylic Acids with Octanol

	mM Ester/HR/G Lipozyme®		
◯–COOH	1.41	.15	◯–COOH
◯–COOH	.02	.29	◯–COOH
◯–COOH	4.88	7.95	◯–COOH
◯–COOH	1.42	—	◯–COOH

TABLE 4

Synthesis of Myristate Esters in Hexane

Alcohol	Structure	Initial Rate mM/hr/g enzyme
n-propanol	⌣ OH	10.3
isopropanol	Y OH	0.42
neopentanol	⋋ OH	4.45
cyclohexanemethanol	◯⌣ OH	12.8
benzyl alcohol	◯⌣ OH	3.34
cyclohexanol	◯ OH	0.70

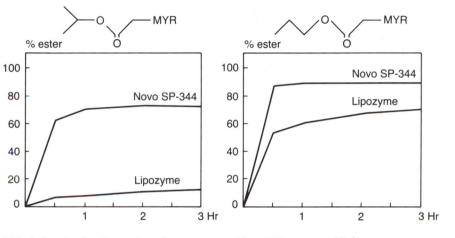

FIG. 3. Synthesis of propyl myristate esters effect of lipase specificity.

which some products made by enzymatic synthesis can begin to move from the laboratory to a larger scale. Both high-value specialty chemicals and commodity esters may be future lipase-based products. The examples presented here should give an idea of the scope and state-of-the-art. It is our hope that Lipozyme® and other immobilized lipases will solve today's problems in ester synthesis and make a broader use of enzymatic synthesis attractive to the industry.

REFERENCES

1. Hansen, T.T., and P. Eigtved, in *Proc. World Conf. Emerging Technol. Fats and Oils Ind.*, edited by A.R. Baldwin, American Oil Chemists' Society, Champaign, IL, 1986, pp. 365–369.
2. Eigtved, P., T.T. Hansen and H. Sakaguchi, *J. Am. Oil Chem. Soc.* 63:463 (1986).
3. Lazar, G.A. Weiss and R.D. Schmid, in *Proc. World Conf. Emerging Technol. Fats and Oils Ind.*, edited by A.R. Baldwin, American Oil Chemists' Society, Champaign, IL, 1986, pp. 346–354.
4. Baratti, J., G. Buono, H. Deleuze, G. Lagrand, M. Secchi and C. Triantaphylides, in *Proc. World Conf. Emerging Technol. Fats and Oils Ind.*, edited by A.R. Baldwin, American Oil Chemists' Society, Champaign, IL, 1986, pp. 355–358.
5. Gatfield, I., *Lebensm.-Wiss. u. Technol. 19*:87 (1986).
6. Seino, H., U.S. Patent 4,614,718 (1986).
7. Klibanov, A.M., in *Biocatalysis in Organic Media*, edited by Laane, Tramper and Lilly, Elsevier Publishing Co., Amsterdam, 1987, pp. 115–116.
8. Miller, C.A., H.F. Austin, L.H. Posorske and J. Gonzales, *J. Am. Oil Chem. Soc. 64*:653 (1987).
9. Cesti, P., A. Zaks and A.M. Klibanov, *Appl. Biochem. Biotechnol. 11*:401 (1985).
10. Godtfredsen, S.E., European Patent Application EP 215,038.
11. Sonnet, P.E., and M.W. Baillargeon, *J. Chem. Ecology 13*:1279 (1987).
12. Krichner, G., M.P. Scollar and A.M. Klibanov, *J. Am. Oil Chem. Soc. 107*:7072 (1985).

Enzymatic Conversion of Diglycerides to Triglycerides in Palm Oils

J. Kurashige
Oils and Fats Research Laboratories, Central Research Laboratories, Ajinomoto Co., Inc.

The development of a bioreactor system that converts diglycerides and fatty acids into triglycerides in crude palm olein has been studied. Some of lipases showed lipase activities at below 300 ppm of moisture in palm olein when immobilized onto diatomatious earth (Celite). Using these immobilized lipases, the equilibrium moisture concentration of palm olein was found to be around 150 ppm. Among active lipases at such microaqueous conditions, Lipase P (*Pseudomonas fluorescence*) was most active and was able to increase TG from 87% to 95% in palm olein at 100 to 20 ppm of moisture level within three days. However, Lipase P lost all of this esterification activity when purified to around 98% (50-fold purification). Finally, lecithin was found to restore the esterification ability of pure Lipase P under ultra microaqueous conditions. With these results, an ultra Micro Aqua Bioreactor will be constructed.

World production of palm oil is now at about eight million tons, and according to the forecast by Oil World 1985 the production will reach 20 million tons in 2003. This means that palm oil, together with soybean oil, will be one of the most abundant sources of edible oils in the near future. Now, one of the targets of the oil industries in Japan and, perhaps, in the world is how to make the most use of palm oil. On the other hand, it is well-known that triglycerides (TG) in palm oil are partially hydrolyzed by lipases included in the palm tissue itself during the lag time after the harvest up to the extraction producing partial diglycerides and free fatty acids in the crude oil (1). These free fatty acids are removed during purification processes, resulting in lower yields than with other oils. Another product of hydrolysis, diglycerides (DG), cannot be removed by the present purification processes and thus remain in refined palm oil (1).

The several percentage of remaining DG is reported to cause the following phenomena setting limits to the utilization of palm oils: they lower the crystallization speed of TG by preventing the formation of crystalline nuclei; they restrain the *trans*-crystallization of TG; they decrease solid fat contents, increasing difficulties in the fractionation of fats and oils; and they promote the increase of acid value in frying oils. Thus, reducing DG in palm oils could be very useful in widening the use of palm oils.

EXISTING WAYS TO REMOVE DG FROM PALM OIL

As removal methods for DG from palm oils, the following ways are known (2): partition chromatography by silica gel columns, solvent partition chromatography, and hydrolysis of DG to fatty acids and glycerol using partial glycerides lipases.

These methods are effective. However, none of these methods can satisfy conditions for industrial use because of low yield and high costs. We have been studying a bioreactor system to convert DG to TG in crude palm oils using lipases as a way to solve the problems.

MATERIALS AND METHODS

No solvent system was used to permit easier application to the existing processes. Palm olein was selected as the substrate because the melting point is low enough (15 C) allowing it to be handled as a liquid phase at ordinary temperature.

Lipases

Eight kinds of lipases, which were provided by Amonao Pharmaceutical Co. Ltd., were used. Lipase P and Pure-P (purified Lipase P) from *Pseudomonas fluorescence* were mainly used (3).

Immobilization of Enzymes

Celite No. 535 (diatomatious earth of Johns Manville, New York, NY) was used as the carrier. Lipase solution and Celite were mixed fully, and the mixture was dried to provide the immobilized lipase.

Reaction Vessels

Three types of closed reactors such as vials (100 ml), conical flasks (200 ml) or four-necked flasks (Fig. 1) were used.

ANALYTICAL METHODS

Fatty acids (FA), monoglycerides (MG), diglycerides (DG) and triglycerides (TG) were determined by gas-liquid chromatography. Moisture of the enzymes and immobilized enzymes were determined by the reduction of weight under the conditions of 105 C for three hr. Moisture of raw oils and reacted ones were determined by the Karl Fischer method.

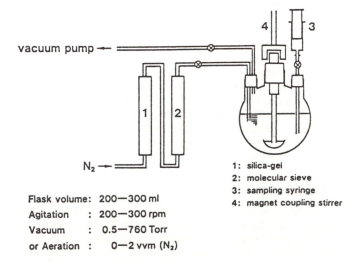

Flask volume:	200—300 ml
Agitation :	200—300 rpm
Vacuum :	0.5—760 Torr
or Aeration :	0—2 vvm (N₂)

1: silica-gel
2: molecular sieve
3: sampling syringe
4: magnet coupling stirrer

FIG. 1. Batch Esterification Reactor.

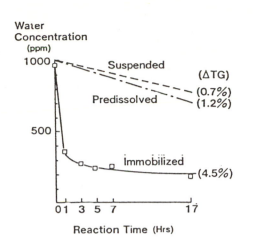

FIG. 2. Effects of Enzyme Treatments on Hydrolysis Reaction.

---- 4 mg of powdered Crude-P directly suspended
—·— 4 mg of Crude-P preliminary dissolved in 20c1+l of water
—— 4 mg of Crude-P preliminary immobilized onto 200 mg of celite

IMMOBILIZATION OF LIPASES

Lipase activities in ultra microaqueous oil and the moisture conditions to promote esterification in palm olein when lipase is immobilized onto Celite are as follows: the immobilization enabled some of lipases to manifest lipase activities significantly in ultra microaqueous oil below 300 ppm water (Fig. 2); and thermostability of Lipase P was remarkably improved; half-life at 60 C was prolonged from 72 hr to above 1800 hr. Surface of Celite seems to play an important role for the manifestation of lipase activity under ultra microaqueous condition.

EQUILIBRIUM MOISTURE CONCENTRATION (EMC) IN PALM OLEIN

EMCs were measured while changing initial moisture concentration (IMC) from 1000 ppm to 100 ppm (Fig. 3, Table 1). As shown in Figure 3, in spite of the difference of IMC, EMC ended at around 150 ppm. Table 3 shows that as the change of monoglycerides (MG) is quantitatively very slight, the hydrolysis reaction and accordingly the synthetic reaction (esterification) also can be discussed based on this following equation: TG + H$_2$O $\leftrightarrows$ DG + FA. As a result of the triglyceride synthesis reaction, water comes out. So a dehydration system is necessary to maintain moisture concentration below 100 ppm, and to continue the synthetic reaction.

BATCH REACTOR SYSTEMS

Enzymatic esterification results under dehydration by vacuum are shown in Figure 4. Esterification occurs at 100 down to 30 ppm of moisture yielding low DG palm olein with 95% TG within three days. However, this reaction rate is not high enough to construct a continuous reactor system (Bioreactor).

TABLE 1

Results of Hydrolysis in Response to Initial Water Concentration

Water conc.		Oil composition				Equilibrium constant (K*)
Initial	End	TG	DG	MG	FA	
ppm	ppm					
1000	218	80.8	12.3	0.20	6.7	0.22
500	189	81.9	11.6	0.18	6.3	0.22
200	147	83.5	10.5	0.13	5.9	0.20
96	143	84.0	10.1	0.13	5.8	0.21

(Palm olein: 2% of lipase and 1.5% of lecithin were immobilized onto celite)

$$DG + FA \leftrightarrows TG + H_2O$$
$$K^* = \frac{(TG)(H_2O)}{(DG)(FA)}$$

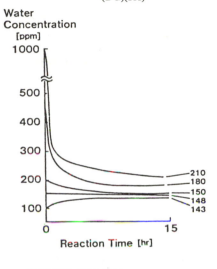

Substrate: Palm olein

Lipase: *Pseudomonas fluorescence*

Enzyme immobilized / Celite: 2%

FIG. 3. Equilibrium Concentration of Water in Response to Initial Water Concentration.

To improve reaction rates, the adsorption of Lipase P on Celite was studied. As shown at Figure 5, adsorption of more than 12.5% of Lipase P onto Celite does not increase the esterification rate. Actually, adsorbed enzyme specific activity reaches a peak at ca. 7.5% of adsorbed enzyme.

PURIFICATION OF LIPASE P

The reason why a limit of effective adsorption existed was considered to result from the 98% of impure compounds included in Lipase P itself (Table 2). Thus, Lipase P was purified up to pure P of almost 98% purity (around 50-fold purification) (Fig. 6). Lipase activity of pure P is shown in Table 3. Pure P lost only the synthetic activity at 50 ppm of moisture level.

ACTIVATOR FOR LIPASE UNDER ULTRA MICROAQUEOUS CONDITIONS (UMC)

Fraction A, which was the first precipitate of the purification process of Lipase P (Fig. 6), restored the esterification activity at UMC of pure P as shown at Figure 7. This showed that there are activators for lipase under

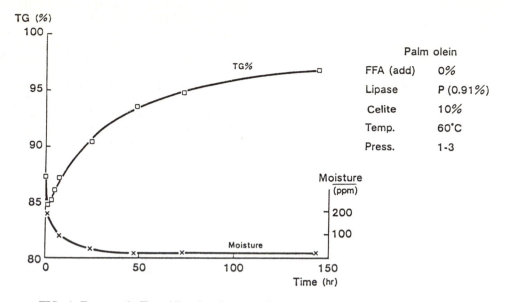

FIG. 4. Enzymatic Esterification Pattern (Batch) by the Vacuum System.

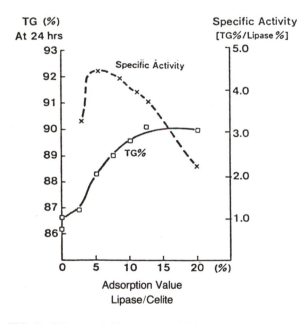

FIG. 5. Effects of Adsorption of Lipase on Celite on TG Synthesis.

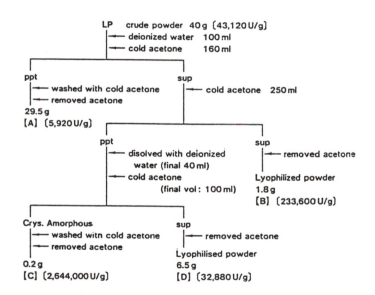

FIG. 6. Purification Flow Chart of Lipase P.

TABLE 2

Purity of Lipase P and Effective Utilization of Surface Area of Celite

	Purity	Impurity	Relative Ratio of SA*/Enzyme
Lipase P	Ca 2%	Ca 98%	1
Purified P	98%<	<2%	49

*Surface Area of Celite No. 535; 1.4 m^2/g. By BET method.

TABLE 3

Activity of Purified Lipase P (Pure-P)

	Hydrolysis activity	Synthesis activity	
Moisture	10%	10%	50 ppm
(A) crude-P	43$^{U/mg}$	113$^{SU/g}$	4.0*
(B) pure-P	2,644$^{U/mg}$	2,064$^{SU/g}$	0*

*: (ΔTG)g/day/lipase g.

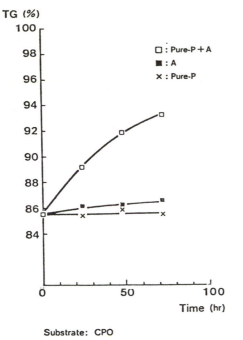

Substrate: CPO

10% of Immobilized enzyme added

Enzyme / Celite : 10%

Fraction-A / Celite: 0.33%

FIG. 7. Effects of Fraction-A (1st PPT.) as an Activator for Pure-P in Ultra Microaqueous Condition.

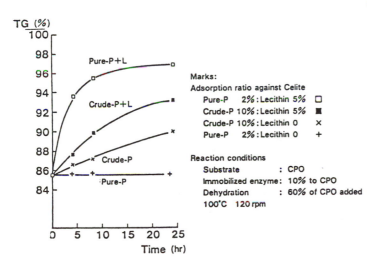

Marks:

Adsorption ratio against Celite

Pure-P	2% : Lecithin 5%	□
Crude-P	10% : Lecithin 5%	■
Crude-P	10% : Lecithin 0	×
Pure-P	2% : Lecithin 0	+

Reaction conditions

Substrate : CPO

Immobilized enzyme: 10% to CPO

Dehydration : 60% of CPO added

100°C 120 rpm

FIG. 8. Comparison of TG Synthesis Speed.

UMC. As one of these activators, lecithin was shown to enhance the lipase activity under UMC. Figure 8 shows some of the effects of lecithin on TG synthesis. When 2% of pure P together with 1.5% of lecithin was immobilized on Celite, the synthesis of TG from 86% to 95% in palm olein is accomplished within six hr. Two percent of Pure P means (pure P vs pure P + Celite) is 2%. This means 99% of crude P could be effectively absorbed onto Celite without losing efficient specific lipase activity. Thus, to convert diglycerides and fatty acids to triglycerides in palm olein, the following were necessary: immobilization of a lipase onto carrier such as Celite, maintenance of moisture levels of oil at below 150 ppm, and selection of a lipase active at below 150 ppm of moisture. It also was found that an enzymatic reaction can go on even at below 30 ppm of moisture concentration, and that there exist such activators as lecithin that enhance lipase activities at ultra microaqueous conditions below 100 ppm of moisture.

With these results, an Ultra Micro Aqua Bioreactor System to produce low DG palm olein will be constructed.

ACKNOWLEDGMENTS

This study has been made in cooperation with AMANO Pharmaceutical Co. Ltd., and JGC Corporation, as a part of activities of the Japanese Research and Development Association for Bioreactor-System under the auspices of the Ministry of Agriculture, Forestry and Fisheries of Japan.

REFERENCES

1. Goh, E.M., *J. Am. Oil Chem. Soc. 62*:730 (1985).
2. Mase, T., Japanese published unexamined PA, (1987).
3. Inukai, T., Japanese published PA, 58-37835 (1983).

Omega-Hydroxylations

Franz Meussdoerffer
Biochemical laboratories, Henkel KGaA, Düsseldorf, Federal Republic of Germany

Hydroxycarboxylic acids constitute a chemically very interesting class of compounds because they carry two different functional groups within the same molecule. Thus, hydroxycarboxylic acids can serve as precursors for polymers or cyclic lactones, which are used in fragrance and antibiotics. Moreover, hydroxy acids are constituents of some biological surfactants. Although it is feasible to synthesize hydroxy fatty acids by chemical means, the procedures involved are complicated, and extensive purification is needed. On the other hand, hydroxy fatty acids occur widely in vegetable oils. However, ω-hydroxy fatty acids are rather exceptional. It is known that alkane-degrading microorganisms can hydroxylate alkanes or fatty acids to the respective alcohol or hydroxy fatty acid. However, although several patents have been filed for microbial hydroxy acid formation, no such products are sold to a large extent. The reason might be that fermentation yields are too poor to sustain a rentable process. For the selection of strains suitable for ω-hydroxy fatty acid production, a detailed knowledge of the biochemical reaction involved in fatty acid metabolism is required. Besides ω-hydroxylation, OH groups also might be introduced at other places in the molecule. Thus, β-hydroxy fatty acids are a common byproduct of fermentations leading to dioic or hydroxy acids. This can be explained by the biochemical reactions involved. For hydroxy-acid production, alkane-degrading microorganisms as yeasts of the Candida-type or Corynebacteria are more suited. To prevent further degradation of the product formed, these microorganisms are blocked in the alkane/fatty acid metabolism. A block within the β-oxydation system might result in the formation of 3-OH acids. The terminal hydroxylation is accomplished by monoxygenase systems that most commonly contain a cytochrome P450 component. The further oxidation of the alcohol formed in this reaction proceeds through dehydrogenases of different specificity. These reactions are usually much faster than the hydroxylation and have to be slowed down to obtain hydroxy-acid formation. Alternatively, enhanced levels of the monoxygenase system must be obtained.

Omega-hydroxy fatty acids, HO-$(CH_2)_n$-COOH, and dicarboxylic acids, HOOC-$(CH_2-)_n$-COOH, comprise bifunctional compounds of considerable interest for a variety of applications. In particular, dicarboxylic acids like adipic acid, azelaic acid and dodecanedioc acid are produced in large amounts as ingredients for polyesters and polyamides. However, these compounds are manufactured in polymer grade quality by petrochemical processes at prices that cannot be matched by biotechnological products at this time.

On the other hand, longer chain (C_{12}-C_{18}) hydroxy and dioic acids could also be considered as valuable chemicals, in particular as precursors for polymers, macrolides for antibiotics and fragrances and adhesives. There are three principal routes to obtain such compounds, namely:

Extraction from plant material: some plant oils contain ω-hydroxycarboxylic acids in the form of cyclic lactones. Thus Kerschbaum (1) demonstrated the presence of 15-hydroxypentadecanoic acid in angelica root oil and

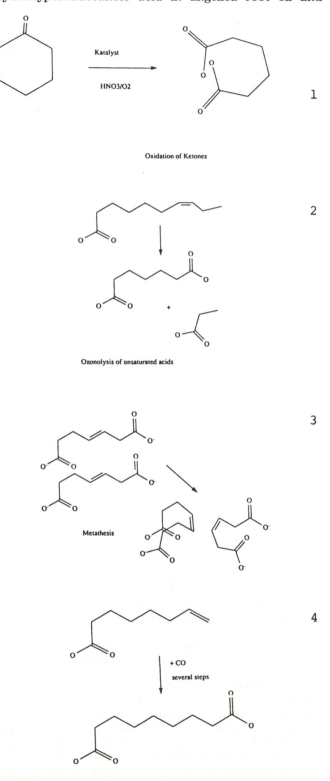

FIG. 1. Chemical routes to ω-hydroxy and dioic acids.

TABLE 1

Some Patents on Microbial Dicarboxylic Acid Production

Products	Educts	Biocatalysts	Documents
C_{14}-dioic acid C_{12}-dioic acid C_8-dioic acid C_6-dioic acid	N-C_x	*Pichia carboniferus*	Dainippon Ink Chemicals, Inc., Japan. Kokai Tokkyo Koho 82 129 694)1982)
C_{13}-dioic acid	n-C_{13}	*Brettanomyces petrophilum* ATCC 20 224	Mitsubishi Petrochemical Co., Ltd., Jpn. Kokai Tokkyo Koho 8 279 889
C_{13}-dioic acid	n-C_{13}	*Brettanomyces petrophilum* ATCC 20 224	Mitsui Petrochemical Ind. Ltd., Japan Pat. 7 079 889 (1982)
C_{12}-dioic acid C_{12}-dioic acid	n-C_{11}-n-C_{18}	*Candida tropicalis* 1098 (FERM-P 3291	Bio-Research Center Co. Ltd., Jpn. Kokai Tokkyo Koho 81 154 993 (1981)
C_{18}-dioic-di-Me C_{16}-dioic Me C_{18}-oic Me	n-C_{18}	*Torulopsis bombicola* PRL 319-67	Phillips Petroleum Co., US Pat. 3 796 630 (1974)
C_{16}-dioic acid	n-C_{16}	*Candida* sp.	Ajinomoto Co. Inc., DE 2 140 133 (1973)
C_{12}-C_{22}-dioic dioic acids	n-C_{12}-C_{22}	*Corynebacterium* sp. ATCC 21 745	DuPont, US Pat. 3 773 621 (1973)

hydroxymyristic acid in angelica fruit oil (2). These acids are used for fragrances.

Chemical syntheses: there are several chemical routes to bifunctional fatty acids. For instance, dicarboxylic acids might be obtained by oxidation of the respective hydroxy acids, ozonolysis of unsaturated fatty acids (3), metathesis and subsequent hydrogenation, CO-addition to α, ω-olefins derived by the SHOP (Shell Higher Olefin Process) or ethenolysis of unsaturated fatty acids, alkaline cleavage of unsaturated hydroxy acids (ricinoleic acid).

ω-hydroxy acids can be obtained by the same reaction as above or Bayer-Villiger oxidation of the respective cyclic ketones. Some examples are summarized in Figure 1.

Microbial fermentation: there is a vast scientific and patent literature on the production of ω-hydroxy and dioic acids by microbial fermentation. Some patents are summarized in Table 1. Typical substrates are alkanes, fatty acids or esters. Both bacteria and yeasts have been found suitable for the production of these compounds.

Although long chain dicarboxylic and ω-hydroxy acids are not readily produced chemically, to date chemical synthesis is still the preferred route to these bifunctional acids. One exception is brassylic acid, which is manufactured on a technical scale by fermentation of alkanes with *Candida tropicalis* (5). One reason for preferring chemical syntheses over fermentations might be cheaper chemical production with existing equipment. Certain purification steps are necessary in both production systems. However, with changing environmental standards, improved fermentation techniques and better strains, microbial functionalisation of fatty acids and alkanes might come of age. Prerequisites are: a high specificity of the microbial oxidation (homogeneous products), high conversion rates and new fermentation techniques (con-

tinuous, cell recycling, etc.).

This paper will deal with the biochemical aspects of the formation of dioic- and ω-hydroxy fatty acids with respect to strain selection and improvement. We will focus on alkane metabolizing yeasts, which are the most representative organisms for the fermentations envisaged. The topic has been reviewed previously (4,6,12).

As shown in Figure 2, three basic steps in the conversion of alkanes or fatty acids by *Candida* yeasts

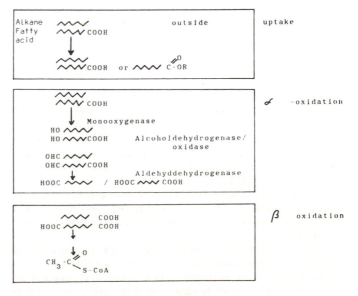

FIG. 2. Basic metabolic steps in the oxidation and utilization of alkanes and fatty acids by microorganisms.

can be distinguished: uptake, microsomal oxidation and peroxisomal oxidation.

Uptake. Uptake of both fatty acids and alkanes appears to be an active process requiring energy (7,8). At least 16 different inducible genetic loci have been found to be involved in tetradecane uptake by *C. tropicalis* (7). This indicates that alkane uptake is a rather complicated process that might require the presence of biosurfactants plus proteinaceous components (9). Active transport also has been found in the case of fatty acid uptake in *S. lipolytica* (8). Here, at least two different fatty acid carrier systems can be distinguished—one being specific for fatty acids with 12 and 14 carbon atoms and another one for unsaturated and saturated fatty acids with 16- and 18-carbon atoms.

Microsomal oxidation. There are three different types of oxidations involved in alkane/fatty acid metabolism in yeasts: α- oxidation of alkanes to alcohols, ω-oxidation of

The first component of the terminal oxidase, the cytochrome P450, belongs to the wide class of cytochromes P450 that occur in almost all microbial and animal systems. Typically, in alkane assimilating yeasts two cytochrome P450 systems are found with distinct substrate specificities. One system catalyses the demethylation of lanosterol (one step in the biosynthesis of the membrane component ergosterol). The other is the system involved in alkane and fatty acid hydroxylation. It is induced by the respective substrate. Alkane-hydroxylating cytochromes P450 from several yeasts have been extensively characterized (reviewed in (11)). They are b-type hemoproteins with a molecular weight of approximately 55 Kd. The gene encoding the alkane-hydroxylating cytochrome P450 from *Candida tropicalis* has been isolated and characterized (13). It exhibits homology to other cytochrome P450 genes, in particular to the phenobarbital-inducible protein from rat liver. Another

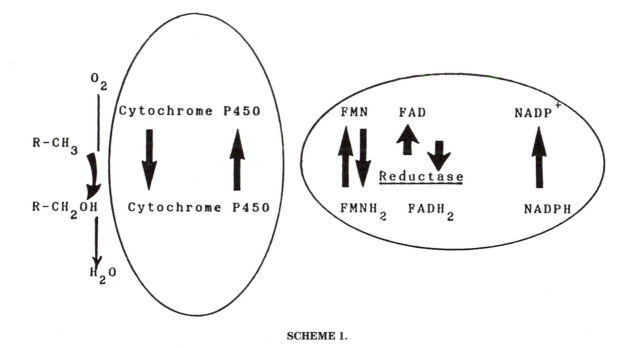

SCHEME 1.

fatty acids to ω-hydroxy acids and in β- oxidation as degradation of fatty acids. The first two types of oxidations are catalyzed by microsomal enzymes while the last type is compartmentalized in the peroxisomes.

The series of oxidations of the terminal methyl group to the carboxyl group are the essential reactions for the production of ω-hydroxy and dioic acids. The three components involved in these steps are indicated in Figure 2. The predominant enzyme system is the terminal hydroxylase complex consisting of a cytochrome P450 and a reductase enzyme in yeasts. It is noteworthy that in bacteria the corresponding systems are cytochrome P450 with the reductase and an additional Fe-protein channeling the electrons from the cytochrome to the reductase system. Alternatively, systems consisting of a rubredoxin and a reductase are found in bacteria (10).

In yeasts, terminal oxidation of alkanes follows scheme 1.

cytochrome P450-dependent monoxygenase that oxygenates fatty acids has been cloned from *Bacillus megaterium* (14). This gene encodes a single protein of 119 Kd which contains heme, FMN and FAD. In contrast to the monoxygenase systems from yeasts that are localized in the membrane of the endoplasmic reticulum, the enzyme from *Bacillus megaterium* is soluble. By limited proteolysis this protein can be split into one cytochrome P450 domain of 55 Kd and one 66 Kd domain with reductase activity. This indicates a gene fusion between a cytochrome P450 and a reductase gene. In contrast to the 119 Kd protein the combined domains are no longer capable of hydroxylating fatty acids. This shows that a close, specific contact between both domains is required for monoxygenase activity. Although the contact of reductase and cytochrome is usually defined by their orientations in the membrane, in the case of the soluble enzyme from *Bacillus* the correct orientation seems to be brought about by the folding of the protein chain.

It is not clear yet whether fatty acids are hydroxylated by the same monoxygenase system as alkanes. Thus, two distinct cytochromes P450, one (M_r=51 Kd) which hydroxylates alkanes and one (M_r=69 Kd) that hydroxylates fatty acids preferentially, have been described (15) for *Candida* yeasts. On the other hand, fatty acids are superior to alkanes as substrates for the monoxygenase from *Candida tropicalis* and *Saccharomycopsis lipolytica* (16, 17; Table 2). Of the fatty acids tested, laurate proved to be the optimal substrate and is used routinely to test the alkane hydroxylase activity in yeasts.

TABLE 2

Substrate Specificity of the Monoxygenase from *Candida Tropicalis* According to Lebeault et al. (16)

Compound tested	Concentration	Specific activity[a]
	M	
Laurate	5.0×10^{-4}	3.1
Myristate	5.0×10^{-4}	2.8
Undecanoate	5.0×10^{-4}	2.3
Decanoate	5.0×10^{-4}	2.2
Palmitate	5.0×10^{-4}	1.2
Stearate	5.0×10^{-4}	0.9
Octanoate	5.0×10^{-4}	0.8
Hexanoate	5.0×10^{-4}	0.5
Hexadecane	5.0×10^{-4}	0.8
Decane	5.0×10^{-4}	0.2
Octane	5.0×10^{-4}	0.2

[a])nmol/min. mg protein.

The following oxidations leading from the alcohol to the aldehyde and subsequently to the acid are catalyzed by microsomal alcohol and aldehyde dehydrogenases (18). They are nucleotide coenzyme-dependent enzymes. In addition, a long chain alcohol oxidase has been described for *Candida maltosa* (19). Interestingly, conversion rates of the alcohol and aldehyde dehydrogenases measured so far are orders of magnitude greater than those determined for the terminal monoxygenase (Table 3). This might explain why the hydroxy and aldehyde intermediates are normally undetectable. Moreover, as all enzyme systems involved in the oxidation of the terminal methyl group to the carboxylic function are localized in the same compartment (microsomes), tunneling of the substrate may take place. Thus, several different enzyme systems quickly convert the alcohols produced by the terminal oxidation to the corresponding acids.

β-Oxidation. This is the series of enzymatic reactions leading from fatty acids or dioic acids to acetyl CoA. A prerequisite for this degradation is the esterification of the fatty (ω-hydroxy dioic) acid with coenzyme A. The order of events taking place during β-oxidation is summarized in Figure 3. It has been shown at least for liver, that long chain monocarboxylic acids, ω-hydroxy fatty acids and dicarboxylic acids are good substrates for acyl-CoA activation and β-oxidation (20). All three substrates are esterified with CoA at a comparable rate. Thus, to

TABLE 3

Rates of Alcohol and Aldehyde Oxidation Reported in Studies on Alkane Grown *Candida* Yeasts

Alcohol Dehydrogenase[a]	Aldehyde Dehydrogenase[a]	Source
350	300 CT (cell free)	Gallo et al. (1971)
270	320 Loddermyc.	Mauersberger et al. (1984)
130	- Ct	Sanglard (1984)
160	140 Ct	Gallo et al. (1973)
133	133	Krauzova et al. (1985)
187	88	Krauzova et al. (1986)
220 (162)	150	G.D. Kemp. Pers. comm. (1/4/87-30/6/78)

[a])nmol/min. mg protein

avoid degradation of any of the products formed in the microsomal oxidation, β-oxidation must be blocked. After esterification with CoA the fatty acids or their derivatives are converted to the β-keto acids as a prerequisite for the cleavage of the C-C bond between the α- and β-carbons. This conversion takes place in a three-enzyme sequence involving oxidation, hydration of the resulting olefin and subsequent nicotine-amide-dependent oxidation to the keto group. Any block in these reactions results in the formation of unsaturated or 3-hydroxy-mono- or dicarboxylic acids. These are often unwanted products in the dicarboxylic acid production. The keto acid subsequently is split into acetyl CoA and a fatty acid-CoA ester shortened by two carbon atoms by thiolysis with coenzyme A. This degradation by subsequent splitting off acetyl groups from the acid is the reason for the occurrence of shorter products with some yeast mutants.

DISCUSSION

In recent years, a clearer picture of the biochemical reactions involved in ω-hydroxy and dicarboxylic acid formation by yeasts has evolved. In parallel, the means to construct yeasts strains with defined properties have been developed. Thus, cloning of *Candida* genes (21,22) has become possible and protocols for genetic engineering manipulations have become available (23–26). This might boost the construction of strains which effectively and selectively convert fatty acids or alkanes to hydroxy or dioic acids.

By conventional mutagenesis, genes occurring in multiple copies or encoding large proteins are impaired preferentially. A widely preferred screening strategy for dioic acid producers aims at mutants that do not degrade fatty acids with the intention of blocking β- oxidation (27–29). As a result of conventional mutagenesis and this screening rationale, low yields or unwanted by-products often had to be accepted. For instance, three enzymic activities of the peroxisomal β- oxidation system of *Candida tropicalis* are combined in a single protein chain and thus encoded by a single gene (30). Any mutation (deletions, frameshift, stopcodons) affecting transcription or translation of this gene or the protein (correct folding or turnover rate) might abolish all three activities. Because of the large size of the gene, a higher mutation frequency of this

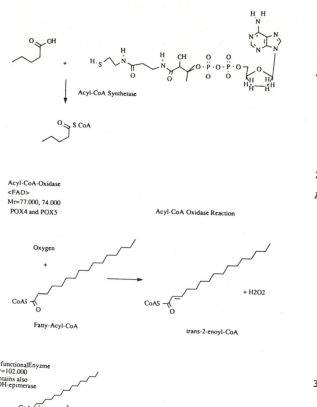

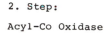

1. Step:

Acyl-CoA Synthetase

2. Step:

Acyl-Co Oxidase

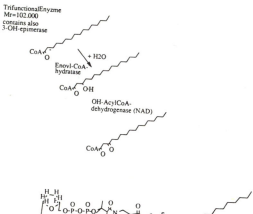

3. Step:

Enoyl-Hydratase

Hydroxiacyl-dehydrogenase

4. Step:

Thiolase

FIG. 3. Peroxisomal β-oxidation in yeasts.

protein might be expected. A mutation impairing the trifunctional protein must result in the formation of unsaturated or hydroxylated byproducts (see above) that are often found in alkane or fatty acid fermentations. A suitable alternative to the conventional techniques is offered by genetic engineering. Thus, the copy number of desired genes (e.g., encoding the hydroxylase) might be enhanced while other genes (β-oxidation) might be selectively deleted. Although such manipulations require haploid yeasts, most industrial yeasts are aneuploid or diploid.

However, new strain development and improved cloning protocols will yield yeasts with the capacities required for economical production of hydroxy and dioic acids.

REFERENCES

1. Kerschbaum, M., *Ber.* **60**:902 (1927).
2. Kerschbaum, M., *Chem. Abstracts* **21**:2118 (1927).
3. Witthaus, M., in *Fats & Oils*, Fonds der chem. Industrie, 1986, p. 19.

4. Rehm, H.-J., and I. Reiff, *Adv. Biochem. Eng. 19*:175 (1981).
5. Okino et al., *Nippon Mining*, Biobusiness Department.
6. Bühler, M., and J. Schindler, in *Biotechnology*, edited by H.-J. Rehm and G. Reed, Vol. 6a, 329–385 (1984).
7. Bassel, J.B., and R.K. Mortimer, *Curr. Genet. 9*:579 (1985).
8. Kohlwein, S., and F. Paltauf, *Biochem. Biophys. Acta 792*:310 (1984).
9. Ito et al., *Appl. Environ. Microbiol. 43*:1278 (1982).
10. Ratledge, C., *J. Am. Oil Chem. Soc. 61*:447 (1984).
11. Käppeli, O., *Microbiol. Review 50*:244 (1986).
12. Fukui, S., and A. Tanaka, *Adv. Biochem. Eng. 19*:217 (1971).
13. Sanglard et al., *Biochem. Biophys. Res. Comm. 144*:251 (1987).
14. Nahri, L.O. and A.J. Fulco, *J. Biol. Chem. 262*:6683 (1987).
15. Sokolov, Y.I. et al., *Doklady Akademii Nauk SSSR 286*:1509 (1986).
16. Lebeault, J.-M. et al., *Biochem. Biophys. Res. Comm. 42*:413 (1971).
17. Marchal, R. et al., *J. Gen. Microbiol. 128*:1125 (1982).
18. Yamada, T. et al., *Arch.Microbiol. 128*:145 (1980).
19. Patent DD-239005; 87-007940/02; Akad. Wiss. DDR.
20. Vamecq, J., and J.-P. Draye, *J. Biochem. 102*:225 (1987).
21. Kamiryo, T., and K. Okazaki, *Mol. Cell. Biol. 4*:2136 (1984).
22. Rachubinski et al., *Proc. Natl. Acad. Sci. USA 82*:3973 (1985).
23. Gaillardin, C., and A.-M. Ribet, *Curr. Genet. 11*:369 (1987).
24. Kelly, R. et al., *Mol. Cell. Biol. 7*:199 (1987).
25. Takagi, M. et al., *J. Bacteriol. 167*:551 (1986).
26. Kunze, G. et al., *J. Basic Microbiol. 25*:141 (1985).
27. Uchio, R., and I. Shiio, *Agric.Bio. Chem. 36*:426 (1972).
28. Uchio, R., and I. Shiio, *Agric. Bio. Chem. 36*:1169 (1972).
29. Uchio, R., and I. Shiio, *Agric. Bio.Chem. 36*:1389 (1972).
30. Moreno de la Garza, M. et al., *Eur. J. Biochem. 148*:285 (1985).

Production of Dicarboxylic Acids by Fermentation

Namio Uemura[a], **Akira Taoka**[b] and **Motoyoshi Takagi**[a]

[a]Nippon Mining Company, Limited Bioscience Research Laboratories, 3-17-35 Niizo-minami, Toda City, Saitama, 335 Japan, and
[b]Biobusiness Development Department, 1-34 Akasaka 4-chome, Minatoku, Tokyo, 105 Japan

Out of 1,600 strains isolated from natural sources as n-alkane-assimilating microorganisms, *Candida tropicalis* 1098 showed the greatest and the steadiest productivity of dicarboxylic acids (DCAs), was selected and mutated to increase the productivity of DCAs. M2030, a mutant obtained after several mutations, showed a very high productivity of DCAs and a low-assimilating activity of monocarboxylic acids (MCAs), DCAs and n-alkanes. This mutant also produced DCAs from n-alkenes, fatty acids and their esters and fats and oils with various carbon chain lengths. Optimization of production conditions of DCAs was carried out using 500 ml flasks and 3l fermenters. The optimum conditions established in small-scale experiments were successfully applied to 30l fermenters and a 2 kl pilot scale fermenter with a slight modification. On a commercial scale, a DCA manufacturing process with a 20 kl fermenter achieved production of 140 g/l of brassylic acid (DC-13). DC-13 now is being commercially produced at a rate of 150 t a year by this fermentation process.

Dicarboxylic acids (DCAs) are important chemical intermediates. DCAs with long carbon chain length are used mainly as raw materials for the preparation of plasticizers, lubricants, polyurethanes, polyamides and perfumes.

It is well known that the production of long-chain DCA by chemical synthesis or chemical oxidation is not easy and that most of the DCAs produced by these methods exhibit relatively short chains. On the other hand, many research papers have reported that the DCAs could be produced from n-alkanes by using microorganisms (1,2). Also, several chemical companies have claimed inventions in the field of fermentative production of DCAs (2). However, a commercial process for producing DCAs by fermentation has not been available until recently.

We have developed the fermentation process for producing long-chain DCAs from n-alkanes using *Candida tropicalis* (3). This report describes the process for manufacturing the long-chain DCAs from n-alkanes by fermentation.

SCREENING OF DICARBOXYLIC ACID-PRODUCING MICROORGANISMS

In order to obtain microorganisms that produce DCAs from n-alkanes, screening of microorganisms was carried out first. Many samples of soil were collected from Japanese oil fields and petroleum refineries. Then, 1,600 strains that could grow on n-dodecane (n-C_{12}) as the sole source of carbon were isolated from the soils. By analyzing the culture broths of these microorganisms, 113 strains of microorganisms were found to produce DCA with the same carbon chain length as that of n-alkane. Considering the amounts of DCA and the stability in DCA production, No. 1098 strain finally was selected and identified as *Candida tropicalis* 1098. Studies on cultivation conditions of this strain were made, and as a result

1.6 g/l of DC-12 (dodecane dioic acid) was obtained from n-C_{12}.

IMPROVEMENT OF THE MICROORGANISMS BY MUTATION

To improve the DCA-producing activity, UV irradiation and/or NTG (N-methyl-N'-nitro-N-nitrosoguanidine) treatment were used on *Candida tropicalis* 1098 to cause mutations. After several mutations, a mutant *Candida tropicalis* M2030 was obtained. This mutant showed very high productivity of DCAs and low assimilating activity of monocarboxylic acids (MCAs), DCAs and n-alkanes.

Figure 1 shows the improvement of DCA productivity by multistage mutations of microorganisms. Strain M2030 was finally obtained and found to provide productivity 100 times greater than that of the original wild strain. As the growth rate of strain M2030 on n-alkane markedly decreased in comparison with that of wild strain, the desired amount of DCA was not produced by using strain M2030 when the conventional cultivation method was applied. Upon further investigation, a new cultivation method called DSF (Dual Substrates Fermentation) was developed, wherein another carbon source such as sugar, e.g., glucose, sucrose or molasses, which supported the growth of strain M2030, was fed in addition to culture medium with n-alkane. This method made it possible to obtain enough biomass yield and to convert a large amount of n-alkane to DCA in a comparatively short period of time. The optimum culture medium compositions could be obtained by determining a proper balance among the amount of charged n-alkanes, the concentration of a nitrogen source and the concentration of sugar that was used as a cosubstrate. Under optimum

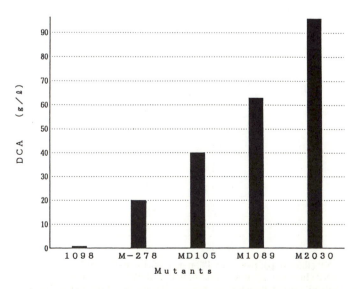

FIG. 1. Increase of production yield by the improvement of DCA-producing yeast.

conditions, the highest level of DC-13 (brassylic acid), i.e., 140 g/l, was obtained from n-C$_{13}$ after 120 hr cultivation.

DICARBOXYLIC ACIDS PRODUCTION FROM VARIOUS SUBSTRATES

Strain M2030 can produce dicarboxylic acids from various kinds of substrates such as n-alkenes, fatty acids and their esters and fats and oils other than n-alkanes.

Dicarboxylic Acid Production from Various n-Alkanes

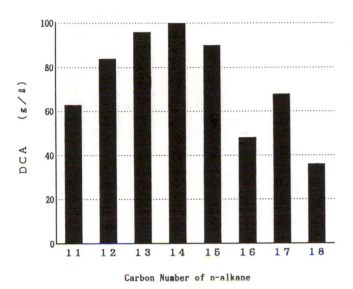

FIG. 2. DCA Production from n-alkanes. Three l jar cultivation for 96~113 hr substrate concentration, 30v/v%.

TABLE 1

DCA Production from Mixed n-alkanes

Substrate	DCA	(g/l)	Total DCA (g/ℓ)
LNP	DC-10	7.0	
	DC-11	31.0	60.0
	DC-12	22.0	
HNP	DC-12	7.5	
	DC-13	9.3	20.2
	DC-14	3.4	
SHyNP	DC-14	1.9	
	DC-15	20.0	28.6
	DC-16	4.7	
	DC-17	2.0	

LNP (light n-p): n-C$_{16}$, 18.4%; n-C$_{11}$, 55.1%; n-C$_{12}$, 26.5%.
HNP (heavy n-p): n-C$_{12}$, 26.5%; n-C$_{13}$, 46.9%; n-C$_{14}$, 26.5%.
SHyNP (super-heavy n-p): n-C$_{14}$, 2.9%; n-C$_{15}$, 54.6%; n-C$_{16}$, 19.5%; n-C$_{17}$, 6.9%; n-C$_{18}$, 2.3%.
Three l jar cultivation for 96 hr. Substrate concentration, 30 v/v%.

DCAs production from various n-alkanes was examined by using strain M2030. The results are shown in Figure 2. It was found that a remarkable amount of DCAs with the same carbon chain length as that of n-alkanes used was produced by using strain M2030. Among DCAs produced, those with carbon chain length from 12 to 15 were more abundantly accumulated, while the number of DCAs produced from n-alkanes with carbon chain length either greater than 15 or smaller than 12 was fewer.

When a mixture of various n-alkanes was used as the substrate, DCAs corresponding to the respective n-alkanes were produced as a mixture (Table 1).

Dicarboxylic Acid Production from Various n-Alkenes

TABLE 2

DCA Production from 1-alkenes

Substrate	Total product	Frac. No	%
1-Dodecene	30.0g/l	1	7.7
		2	9.0
		3	4.5
		4	4.2
		5	74.6
1-Tetradecene	24.1g/l	1	14.0
		2	7.0
		3	3.1
		4	18.1
		5	57.8
1-Hexadecene	12.8g/l	1	18.8
		2	7.8
		3	2.7
		4	17.6
		5	53.1
1-Octadecene	4.1g/l	1	15.3
		2	6.5
		3	2.8
		4	38.7
		5	36.7

Three l Jar cultivation for 89 hr with 15 v/v% of substrate.

	Frac.
$C_nH_{2n} \rightarrow HOOC\text{-}(CH_2)_{n-2}\text{-}COOH$	1
$HOOC\text{-}(CH_2)_{n-3}\text{-}CHCOOH$ $\quad\quad\quad\quad\quad\quad\quad OH$	2
$HOOC\text{-}(CH_2)_{n-3}\text{-}CHCH_2OCH_3$ $\quad\quad\quad\quad\quad\quad\quad OH$	3
$HOOC\text{-}(CH_2)_{n-2}\text{-}CH_2OH$	4
$HOOC\text{-}(CH_2)_{n-3}\text{-}CHCH_2OH$ $\quad\quad\quad\quad\quad\quad\quad OH$	5

Strain M2030 can utilize n-alkenes for DCA production. In the case of l-alkenes used as substrates, ω, ω-1 dihydroxymonocarboxylic acid as a main product and other oxidative compounds as minor products were produced as a mixture (Table 2).

When inner-alkenes were used as substrates, unsaturated DCAs with the same carbon chain length as that of substrates were produced. In this case, DCAs of trans(E) type were more abundantly accumulated than that of *cis*(Z) type (Table 3).

TABLE 3

DCA Production from Inner-alkenes

	Substrate	Product (g/l)
(Z)-alkene	Δ^5-nC$_{10}$	0
	Δ^3-nC$_{11}$	0.8
	Δ^5-nC$_{12}$	0.1
	Δ^5-nC$_{12}$	0.1
	Δ^7-nC$_{14}$	1.1
	Δ^8-nC$_{16}$	3.8
(E)-alkene	Δ^5-nC$_{10}$	1.0
	Δ^3-nC$_{11}$	2.3
	Δ^5-nC$_{12}$	10.2
	Δ^5-nC$_{12}$	12.1
	Δ^7-nC$_{14}$	4.5
	Δ^8-nC$_{16}$	7.5

Flask cultivation for 48 hr. Substrate concentration, 4.3 v/v%.

Dicarboxylic Acids Production from Fatty Acids and their Esters and Fats and Oils

TABLE 4

DCA Production from Unsaturated Fatty Acids

	Substrate	Product (g/l)
(Z)-fatty acid	Δ^6-MC18	53.8
	Δ^9-MC18	48.1
	$\Delta^{9,12}$-MC18	0
	$\Delta^{9,12,15}$-MC18	0
(E)-fatty acid	Δ^9-MC18	13.9
	Δ^{11}-MC18	0
α, β-unsat. Fatty acid	Δ^2-MC10	18.4
	Δ^2-MC11	41.2
	Δ^2-MC13	4.7
	Δ^2-MC16	13.6

Flask cultivation for 96 hr. Substrate concentration, 9.5 v/v%.

The production of DCAs from saturated fatty acids was examined by using strain M2030. As shown in Figure 3, large amounts of DCAs were obtained from saturated

fatty acids with 12 to 16 carbons. Production of DCAs from unsaturated fatty acids was possible. Remarkable amounts of DCAs were produced from petroselinic acid (*cis*-(Z)- Δ^6-C-18) and oleic acid (*cis*-(Z)- Δ^9-C-18) (Table 4).

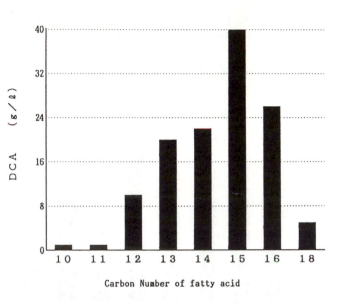

FIG. 3. DCA production from fatty acids. Flask cultivation for 96 hr. Substrate concentration, 9.5 v/v%.

The production of DCAs from palmitic acid esters was examined by using strain M2030. As shown in Table 5, large amounts of DC-16 (hexadecanedioic acid) were obtained from all kinds of palmitic acid esters used as substrates and the highest production of DC-16 was obtained from n-butyl palmitate. When the fats and oils, e.g., coconut oil, palm oil and extra-fancy tallow, were used as substrates, DCAs corresponding to the fatty acids contained in the substrates were produced as a mixture (Table 6).

TABLE 5

DCA Production from MC16 Esters

Substrate	Product (g/l)
nC$_{16}$	72.5
MC$_{16}$Me	34.1
MC$_{16}$Et	32.6
MC$_{16}$nPr	43.6
MC$_{16}$isoPr	51.1
MC$_{16}$nBu	57.5
MC$_{16}$isoBu	48.5

Five l Jar cultivation for 96 hr. Substrate concentration, nC$_{16}$: 40v/v%; fatty acid ester : 30v/v%.

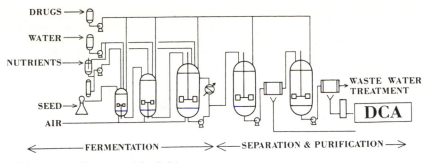

FIG. 4. Flow diagram of the DCA process.

TABLE 6

DCA Production from Fats and Oils

Fats and oils	Total DC (g/l)
Crude coconut oil	15.0
RBD Palm oil	14.7
Extra-fancy Tallow	6.9

Flask cultivation for 96 hr. Fats and fatty oil concentration, 15 w/v%.

SCALE-UP OF DICARBOXYLIC ACID PRODUCTION

Experiments to optimize fermentation conditions were carried out using 500 ml flasks and 3 l jar fermenters. The variables of fermentation conditions examined were: culture medium compositions, pH, fermentation temperature, concentration of n-alkane in the culture medium and so on. The optimum fermentation conditions were established in smaller-scale experiments. Aeration and agitation conditions then were examined using 30 l jar fermenters. From these experiments, basic data for scale-up were obtained. On the basis of the scale-up data, a pilot plant with a 2 kl fermenter was constructed.

For purposes of the establishment of fermentation process conditions, the countermeasure of contamination and foaming of fermentation media, and the establishment of separation and purification process conditions, studies were carried out with a pilot plant. These studies were successfully pursued for about one year, and then a plant for the commercial production of DCA was constructed.

DICARBOXYLIC ACID MANUFACTURING PROCESS

The commercial-scale dicarboxylic acid manufacturing process, or DCA process, is shown in Figure 4. This process consists of the fermentation section and the DCA separation and recovery section. The main fermenter has a 20 kl capacity. It is equipped with a powerful agitator to mix culture broth and oil uniformly and to supply a large volume of oxygen into the culture broth. To remove the heat generated during fermentation, part of the culture broth can be circulated and cooled through the heat exchanger outside the fermenter. The DCA separation and recovery section is extremely simplified because n-alkane is almost entirely consumed and no by-product remains at the end of the fermentation process. Figure 5

shows the time course of DC-13 (brassylic acid) production from n-C_{13} in a 20 kl fermenter. The high level of DC-13 production, i.e., 140 g/l achieved in the small-scale experiment, also was obtained in the commercial-scale fermenter.

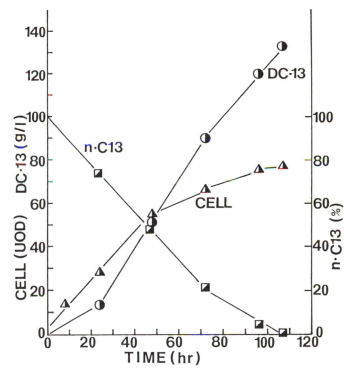

FIG. 5. DC-13 Production from n-C_{13} by fermentation using 20 kl fermenter.

FEATURES OF THE DCA PROCESS AND ITS PRODUCTS

This fermentation or DCA process, has several remarkable features. For example, dicarboxylic acids with various carbon chain lengths can be manufactured by simply changing the fermentation substrates. We have commercially manufactured 150 tons of brassylic acid (DC-13) from n-tridecane (n-C_{13}) in a year. This is almost the entire quantity required to produce ethylene brassylate (a macrocyclic musk compound) for the world. Our work represents the first petroleum fermentation process for manufacturing chemicals on a commercial scale. The product quality profile of brassylic acid (1,13-tridecane-

TABLE 7

Profile of 1,13-Tridecanedioic Acid Produced by Fermentation of n-Tridecane

1. Brassylic acid	
1,13-Tridecanedioic acid	
1,11-Undecane dicarboxylic acid	
1,11-Henedecane dicarboxylic acid	
Brasileic acid, Brasilic acid, Brazilic acid, etc.	
2. Molecular formula	$HOOC(CH_2)_{11}COOH$
3. Molecular weight	244
4. Appearance	white pellet
5. Melting point	112-113 (C)
6. Density	1.12 (g/cm^3)
7. Purity (GC Method)	94 (%)
8. Water content	0.15 (%)
9. Acid value	440 (mg KOH/g)
10. Solubility	
Water (24 C)	insoluble
Methanol (20 C)	13 (g/100g)
Toluene (24 C)	1.3 (g/100g)

dioic acid) from the DCA process is summarized in Table 7. It is obvious that the chemical purity of brassylic acid from the DCA process is markedly high in comparison with the product resulting from the ozonization of rapeseed oil.

REFERENCES

1. Fukui, S., and A. Tanaka, in *Japn. Adv. Biochem. Eng. 17*:1 (1980).
2. Buhler, M., and J. Schindler, in *Biotechnology*, Vol. 6, edited by H.J. Rehm and G. Reed, Verlag Chemie,Weinheim, 1984, pp. 229–385.
3. Uemura, N., *Hakko to Kogyo 43*:436 (1985).

Enzymic Modification at the Mid-chain of Fatty Acids

Roger C. Hammond

Unilever Research, Colworth Laboratory, Sharnbrook, Bedford, MK44 1LQ, U.K.

Natural fatty acids commonly show a range of mid-chain modifications including mono- to polyunsaturation, hydroxyl substituents and methyl branches. Enzymes responsible for these modifications and others involved in degradation are found in most enzyme classes, with oxidoreductases predominating. Unsaturated fatty acids are the most common substrates giving hydroxy-, epoxy- and hydroperoxy-products after oxygenase attack. Further enzyme reactions can lead to keto- and dihydroxy-fatty acids, and lipoxygenase can oxidize arachidonic acid to a prostaglandin precursor. Unsaturated fatty acids also can be reduced by enzymes found in rumen microorganisms. A mixed bacterial culture can fully reduce α-linolenic acid to stearic acid by a series of reactions mimicking the hardening process currently used by the oils and fats industry. Most individual chemically catalyzed mid-chain modification processes are efficient and controllable, but their enzymic counterparts display regiospecificity and, perhaps more significantly, stereospecificity that could be the key to novel products. Some chemical processes such as ozonolysis have no identified enzymic parallel although there are alternative biotechnological routes to the same products. Consideration of the differences between chemical catalysis and possible biocatalytic routes highlights the need to solve the problem of reductant supply that is common to most oxidoreductases, or to find alternative routes. Progress is being made in using whole cells for these types of biotransformations where the sequential and spatial requirements for complex reaction sequences and recycling of cofactors have not been disrupted by isolating the enzymes. Derivatization or immobilization of cofactors such as NAD+ shows promise for the repeated use of isolated redox enzymes although further development is needed. A limited range of mid-chain products may be made using enzymes that do not require cofactors. Potential processes using whole cells or isolated enzymes must be operated intensively with essentially nonaqueous lipid feedstreams if they are to compete with chemical catalysis.

Biological sources account for by far the largest proportion of fatty acids used in foods and as oleochemicals. Evolution has led to the development of biological enzyme systems that can catalyze a variety of reactions at the mid-chain of fatty acids or esterified fatty acyl residues. Such enzyme systems are responsible for the range of naturally occurring fatty acids with other than saturated alkyl mid-chain structures. Chemical reactions of fatty acids, particularly at the carboxyl group and at the double bonds of mono-, di- and polyunsaturated acids, have been carried out for many years in synthetic and analytical chemistry laboratories and on a large scale by the oils and fats industry. It may be appropriate to consider using some enzymes reacting at the mid-chain to modify fatty acids on a commercial scale.

Aliphatic acids potentially can be modified by enzymes at virtually any carbon of the chain. Reactions at the carboxyl and methyl termini are, of course, well known in biosynthetic and degradative pathways. Fatty acids are intermediates in degradation of alkanes and other hydrocarbons (1) and some reactions considered here are part of biodegradation pathways. Apart from reactions at the termini, enzymes may catalyze reactions close to the termini ("subterminal" reactions) or further removed from either terminus ("mid-chain" reactions). The occurrence of subterminal products arising from essentially terminal reaction processes is noted here but not further discussed. For example, fatty acid degradation processes may lead to the formation of subterminal by-products: β-oxidation commences with formation of the coenzyme A (CoA) thioester and proceeds with desaturation to give the α, β-enoyl CoA followed by hydration to the β-hydroxyacyl CoA and oxidation to the β-oxoacyl CoA ester. Under some circumstances, such intermediates may accumulate. ω-oxidation of fatty acids leading to bifunctional products commences with a mono-oxygenase-catalyzed hydroxylation of the terminal methyl group but ω-1 hydroxy fatty acid accumulates as a result of incomplete specificity of the enzyme. In contrast, the mid-chain reactions considered here take place in regions of the fatty acid removed from either terminus by a considerable number of carbon-carbon bonds. This is not to say that these reactions take place at random along the length of the fatty acid. On the contrary, many are highly specific for the "mid-chain" region. This regiospecificity is due to accurate positioning of the fatty acyl substrate at the active site of the enzyme. This is accomplished by the binding of the carboxyl group and probably, in some instances, the saturated ω-alkyl chain of the substrate to the enzyme. Some enzymes are less regiospecific, carrying out reactions at hydroxyl substituents or olefinic linkages occurring at various positions along the substrate chain.

Although enzyme regiospecificity may be coupled with stereospecificity, both major advantages of enzyme-catalyzed reactions, the use of biological systems for the synthesis of desirable products from cheap, plentiful starting materials such as fatty acids is often less attractive than would appear at first sight. The severe problems of complex, unstable catalysts operating at near-ambient conditions of temperature and pressure in predominantly aqueous suspension or solution at high dilution has limited the adoption of bioprocessing routes. Avoidance of these problems and successful development of suitable reaction systems exploiting lipase enzymes has led to the adoption by the oils and fats industry of at least one biocatalytic process for the preparation of cocoa butter-type triacylglycerols (2). It may be possible in the future to build on our current expertise to carry out mid-chain modifications of fatty acids by isolated enzymes or whole cells. In this paper, I shall discuss the types of mid-chain reactions, the aspects of several examples that render them particularly suitable or unsuitable for industrial exploitation and the advances being made in biocatalyst and bioreactor technology that could lead to the commercialization of such processes by the industry.

ROGER C. HAMMOND

TYPES OF MID-CHAIN REACTIONS

Enzymes catalyzing reactions at the mid-chain of commonly occurring fatty acids of importance to the oils and fats industry have been known for many years. Examples are found in most of the major enzyme classes. Table 1 shows a range of enzymes that catalyze reactions of potential interest. Most of the enzymes in Table 1 are in the oxidoreductase class. Of particular significance are the desaturases (EC 1.14.99.5), which act on the saturated fatty acyl products of the fatty acid synthetase complex to yield mono-, di- and polyunsaturated fatty acids. The unsaturated fatty acids are the substrates for nearly all other mid-chain reactive enzymes and enzyme sequences.

TABLE 1

Enzymes That Catalyze Mid-chain Reactions of Potential Interest to the Oils and Fats Industry

Class	Enzyme	EC Number	Reaction Catalyzed	Comments
Oxidoreductases	Lipoxygenase	1.13.11.12	Linoleate + O_2 $\longrightarrow$ 13-hydroperoxy-octadeca-9, 11-dienoate	Also related arachidonate enzymes: 1.13.11.31 1.13.11.33 1.13.11.34 Further rx. gives wide variety of products. Plant enzyme. Commercially available.
	Oleate Δ^{12}-hydroxylase	1.14.13.26	1-Acyl-2-oleoylphospholipid + NADH + O_2 + H^+ $\longrightarrow$ 1-Acyl-2-[(S)-12-hydroxyoleoyl] phospholipid + NAD^+ + H_2O	Plant enzyme, ricinoleic acid synthase.
	P-450 (Flavoprotein-linked mono-oxygenase)	1.14.14.1	e.g. Oleate + O_2 $\longrightarrow$ 9, 10-epoxystearate	Group of enzymes with different specificities and other activities.
	Acyl-CoA desaturase	1.14.99.5	Stearoyl-CoA + AH_2 + O_2 $\longrightarrow$ oleoyl-CoA + A + $2H_2O$	Liver, microorganisms. Plant enzyme (1.14.99.6) is ACP-linked. Other desaturases, e.g., Δ^6, Δ^{12}, Δ^{15}.
	Secondary alcohol dehydrogenase	1.1.1.1 (1.1.1.2.)	2° Hydroxyfatty acid + NAD^+ $(NADP^+)$ $\longrightarrow$ 2° ketofatty acid + NADH (NADPH)	Few enzymes specific for 2° alcohols.
	Saturase		9-cis-11-*trans*-Octadecadienoate + NADH + H^+ $\longrightarrow$ 11-*trans*-octadecenoate + NAD^+	Other microbial enzymes reduce polyunsaturated C_{18} fatty acids to stearate.
	Unsaturated phospholipid methyltransferase	2.1.1.16	*S*-Adenosyl-L-Methionine + Δ^9-olefinic fatty acyl phospholipid $\longrightarrow$ *S*-adenosyl-L-homocysteine + methylene-acylphospholipid	*Mycobacterium phlei* alkylenation at 10-carbon, enzymically reduced to 10-methyl-product. Cyclopropane fatty acids formed similarly.
	Epoxide hydrolase	3.3.2.3	Epoxy fatty acid + H_2O $\longrightarrow$ dihydroxy fatty acid	Further reaction of EC 1.14.14.1 product.
	Oleate hydratase	4.2.1.53	Oleate + H_2O $\longrightarrow$ (*R*)-10-hydroxystearate	Microorganisms.
	Linoleate isomerase	5.2.1.5	9-*cis*-12-*cis*-Octadecadienoate $\longrightarrow$ 9-*cis*-11-*trans*-octadecadienoate	Initial enzyme in biohydrogenation sequence.
	Hydroperoxide isomerase	5.3.99.1	13-Hydroperoxyoctadeca-9, 11-dienoate $\longrightarrow$ 12-oxo-13-hydroxy-octadeca-9-enoate	Acts on product of lipoxygenase.

Most mid-chain reactive enzymes have been studied extensively and have received Enzyme Commission (EC) numbers. These enzymes have been purified, their kinetic parameters measured, and in some cases, a knowledge of their active site chemistry gleaned. Other reactions, themselves well documented, are catalyzed by enzymes or sequences of enzymes less well understood. Similar enzymes may be found in animals, plants and microbes, the latter being the source of choice for most biocatalytic applications because of their frequently stated advantages of controllable monoculture, high enzyme titres by genetic modification and so on. In some instances, intact cells may be suitable for carrying out a reaction. In other cases, isolated enzymes are more suitable. The factors influencing source and type of enzyme catalyst and its use can be illustrated by several examples.

Oxidoreductase Reactions

Biohydrogenation. The saturation of unsaturated fatty acids in the guts of some animals has been long recognized (3,4). Ruminants ingest linoleic and α-linolenic acids derived from dietary herbage galactolipids (5) or from triacylglycerols in supplementary feed grain (6), but the fatty acids assimilated by the animal are more saturated (4). This "biohydrogenation" is brought about by the mixed culture of rumen bacteria and involves several enzyme catalyzed steps. Figure 1 indicates the sequence of reactions in hydrogenation of unsaturated dietary C_{18} fatty acids to stearic acid. The extent of reaction depends on a number of factors, but the use of a stable mixed rumen population can lead to complete hydrogenation of linoleic and α-linolenic acids to stearic acid (7) although a minimum of two types of organism only is sufficient for the reaction (8).

Several of the enzyme steps in biohydrogenation of linoleic acids have been studied by Tove and coworkers. This work used the rumen bacterium *Butyrivibrio fibrisolvens*, which is able to hydrogenate linoleic acid to the 11-*trans*-octadecenoic acid level. The initial isomerization leading to a conjugated diene is catalyzed by linoleate Δ^{12}-*cis*-Δ^{11}-*trans*-isomerase (EC 5.2.1.5) (9,10). This enzyme has marked substrate specificity requirements (11). A *cis*-9, *cis*-12-diene substrate with a free carboxyl and an ω-chain length of six carbon atoms is required, hence the preferred chain length is C_{18}.

The *cis*-9-*trans*-11-octadecadienoic acid reductase catalyzing the hydrogenation of the conjugated acid to *trans*-11-octadecenoic acid is apparently an iron protein that interacts with a rare endogenous electron donor, α-tocopherolquinol (2-[3-hydroxy-3, 7, 11, 15-tetramethylhexadecyl]-3, 5, 6-trimethylbenzoquinol), which is involved as the semiquinone (12,13). The reducing power derives originally from NADH and passes along a short chain (Fig. 2) to the enzyme (14). The hydrogens involved

in the reduction derive from water and hydrogenation occurs stereospecifically by *cis* addition to the D side of the double bond between carbons 9 and 10 of the substrate (15). Examination of the substrate specificity of the reductase indicated that while the carboxyl group was not necessary, the conjugated diene structure was required, probably to orient the substrate correctly at the active site (15). The biohydrogenation of oleic and vaccenic acids that occurs in other organisms of the mixed culture probably is catalyzed by a different enzyme because the reaction involves *cis* addition to the L-side (16). This explains why complete reduction of unsaturated fatty acids to stearic acid requires a mixed culture with representatives of two distinct groups of microorganisms—those carrying out the reduction of di- and trienoic acids to monoenoic acids and others reducing the monoenoic acids (8).

Comparison of Biological and Chemical Catalysis for Hydrogenation

The observation of biohydrogenation in rumen microorganisms has suggested the investigation of this route as an alternative to chemical catalysis. A general consideration of the existing chemical process and the attributes of a putative microbial route suggests (Table 2) that the products of the two reactions would be little different although free fatty acid would be the product of the microbial route even if an ester substrate was supplied, because hydrolysis of esters is a necessary prerequisite for biohydrogenation (5,6,17). Partial hydrogenation by either process would lead to *trans* acids with double-bond migration. Undesirable cyclic by-products of chemical hydrogenation (18) are not likely with the biohydrogenation route, but the nature of other potential by-products of the latter route is unknown.

The major difference between conventional hardening processes and a biohydrogenation route is in the nature of the catalyst and process conditions. The use of high temperatures and pressures (Table 2) would not be necessary for an enzyme-catalyzed route, but the relatively low physiological temperature (37 C) might present the problem of solid products. The major technical advances that would be necessary to allow consideration of a biohydrogenation route as a rival to current chemical catalysis

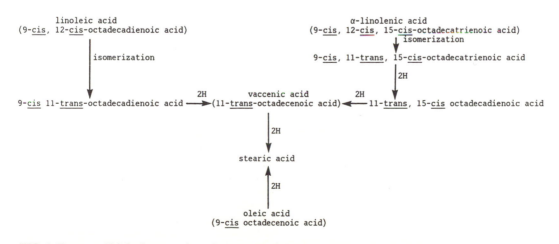

FIG. 1. Routes of biohydrogenation of unsaturated C_{18} fatty acids in the rumen.

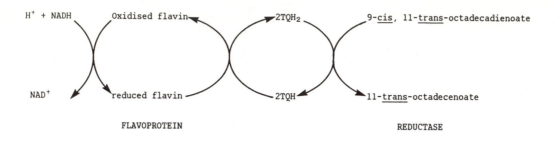

TQH: α-tocopherolsemiquinone.

FIG. 2. Pathway of reducing equivalents during biohydrogenation.

TABLE 2

Comparison of Biohydrogenation with Existing Chemical Process

	Biohydrogenation	Chemical catalysis
Reaction	H_2O linolenic → stearic	H_2 linolenate → stearate
Partial reactions	yes	yes
Reactants	acids	esters
DB positions	restricted	wide
By-products	trans acids DB migration	trans acids DB migration cyclic compounds
Catalyst	mixed bacterial culture	Ni, Cu
Re-use	yes (continuous?)	single or multi-use
Poisoning	?	yes
Process Temperature Pressure	aqueous/fatty acid 37 C atmospheric anaerobic	dry liquid oil up to ~ 200°C up to ~ 4 bar anaerobic

DB, double bond.

concern the nature of the biocatalyst that would have to be of a reusable and recoverable form and the process stream that would need to be liquid, with substrate dissolved in a solvent as necessary, and of minimal water content. Progress toward these goals is being made and is further discussed below.

Oxidation at the double bond. A number of oxidoreductases acting on unsaturated fatty acids generates oxidized products (Table 1). Some transformations of oleic and linoleic acids are shown in Figures 3 and 4. The enzymes catalyzing these reactions also can act on polyunsaturated acids. An appreciation of the range of products potentially possible requires examination of the substrate specificity of the individual enzymes.

Oxygenases

Lipoxygenase (EC.1.13.11.12) is a dioxygenase catalyzing

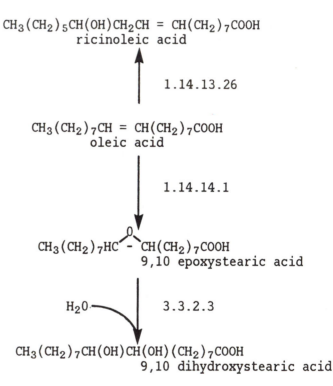

FIG. 3. Mid-chain oxidative reactions of oleic acid.

the conversion of polyunsaturated fatty acids having a 1,4-*cis*,*cis*-pentadiene structure to conjugated hydroperoxides (Fig. 4). The animal lipoxygenases are heme-containing, while the plant enzymes have nonheme iron and function via a free radical mechanism. Elucidation of the mechanism of action of soybean lipoxygenase I (19,20) partly explains the range of products obtained. The enzyme normally shows regio- and stereospecificity, but leakage of alkyl radicals from the enzyme active site may lead to nonenzymic production of racemic products. Under conditions of restricted O_2 supply, the reactive alkyl radicals may form other by-products including dimers and disproportionation products.

A further complication in considering the exploitation

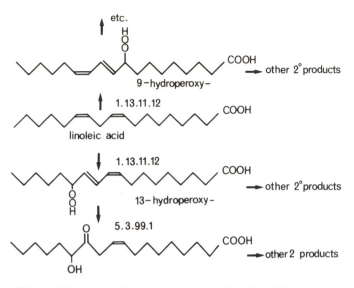

FIG. 4. Mid-chain oxidative reactions of linoleic acid.

of lipoxygenase is the presence in the highly active, most readily available soybean preparations of isozymes giving different products (21,22). The isozymes differ in their pH optima, regio- and stereospecificity, giving wide variations in the ratio of 9:13 hydroperoxides (2.5:97.5 for soybean I to 75:25 for soybean II lipoxygenases) and similarly large differences in the proportions of S and R enantiomers (23,24). The hydroperoxide products are the substrates for further reactions, including reduction to the hydroxydienoic acid, isomerization and hydroperoxide cleavage (25). Despite the problems of by-product and secondary product formation, the use of these relatively cheap, stable, highly active oxygenases that do not require an external reductant and redox carrier chain is an attractive route to high- and medium-value products.

The reaction carried out by lipoxygenase is analogous to that of the initial enzyme in prostaglandin synthesis in that a stereospecific proton abstraction is the initial step (24). Bild et al. (26) showed that prostaglandin $F_{2\alpha}$ could

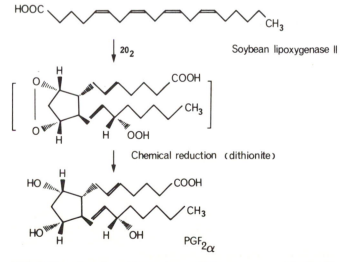

FIG. 5. Use of soybean lipoxygenase II to prepare prostaglandin $F_{2\alpha}$.

be prepared by the action of soybean lipoxygenase II on arachidonic acid (5, 8, 11, 14-eicosatetraenoic acid) followed by a simple chemical reduction (Fig. 5). The synthesis of low-volume, high-added value products exemplified by this example is probably the most attractive area of use of enzymes.

Another group of oxygenases catalyzing mid-chain oxidations of fatty acids is the P-450 group (EC 1.14.14.1). These heme monooxygenases are widely distributed throughout nature and catalyze a wide variety of oxidative transformations. Several isozymes may be present in any one organism or tissue, and this may account for the breadth of substrates attacked. For consideration in the context of fatty acid modification, microbial sources are the most likely candidates. Our main knowledge of the mechanism of microbial P-450s has come from the work of Gunsalus and coworkers on the camphor P-450 of *Pseudomonas putida* (27). Reducing power for the reaction derives from NADH or NADPH and is channeled via a reductase (FAD-linked) and a redoxin (with a [2 Fe.S*] center in the case of the camphor system). This redox chain is analagous to that proposed to operate in biohydrogenation (Fig. 2) and commonly is present in P-450 systems.

The product of mid-chain reaction of unsaturated fatty acids with P-450 enzymes is the epoxy fatty acid, and this is probably the route to naturally occurring plant epoxy fatty acids (28). The soluble P-450 system from *Bacillus megaterium* can carry out both subterminal and mid-chain reaction (Fig. 6) (29). The proportion of the epoxy product was greatest (35%) with palmitoleic acid as substrate and only 11% with oleic acid. Both *cis*- and *trans*-double bonds were attacked. The organism also contains an epoxide hydrolase (EC 3.3.2.3) that gives rise to 9, 10 dihydroxypalmitate from palmitoleic acid (Fig. 3) (30,31). In view of the relatively substantial commercial use of epoxidized oils prepared by reaction with peroxyacids (10^4-10^5 tpa), it may be worthwhile considering the possibility of a biocatalytic route to these compounds.

Since most P-450 enzymes are membrane-bound, particulate, cofactor-requiring multiprotein complexes, it is unlikely that stable cell-free preparations could be obtained easily or that appropriate strategies and reactors for supplying the necessary reducing equivalents will be designed in the near future. However, it is reasonable to consider the use of whole microbial cells with the appropriate enzyme complement for biotransformations. This approach suggests itself for the modification of fatty acids and alkanes (32) and has been applied commercially for the oxidation of alkanes to dicarboxylic acids by *Candida* yeast in conventional aqueous submerged culture in a fermenter (33). Some advantages of whole cells for biotransformation of lipophilic substrates are discussed below.

The castor bean monooxygenase responsible for the oxidation of oleic to ricinoleic acid is unusual in that its substrate is not the free acid or CoA ester but a 2-oleoylphospholipid. Oleate Δ^{12}-hydroxylase (EC 1.14.13.26) oxidizes 1-acyl-2-oleoyl-*sn*-glycero-3-phosphocholine regiospecifically to give the Δ^{12}-hydroxyoleoyl product (34). The enzyme uses NADH as electron donor and appears similar in some respects to the oleate desaturase

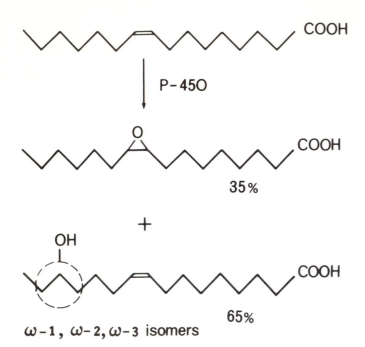

FIG. 6. Reaction of *Bacillus megaterim* P-450 with palmitoleic acid.

(EC 1.3.1.35) in some seeds that require a similar substrate (35). The activity against an oleoylphospholipid also is analogous to the unsaturated phospholipid methyltransferase (EC 2.1.1.16) involved in the bacterial synthesis of methyl- and cyclopropane fatty acids (see below).

A group of enzymes other than the oxygenases that has been investigated for oxidative transformations is the peroxidases (EC 1.11.1.7). These heme enzymes use hydrogen peroxide as oxidant, do not require NAD(P)H or intermediate redox proteins, are readily available from plant sources, and are relatively stable. Unfortunately, fatty acids are not substrates although short-chain, water-soluble acids are attacked (36).

Further Oxidation

Preliminary oxidative mid-chain transformations of fatty acids are likely to be restricted to the products of oxygenase attack, with lipoxygenase being the enzyme of choice for appropriate substrates because it is effective as a free enzyme without cofactor requirements. Further oxidation is potentially possible with a number of other enzymes acting on mid-chain substituents, including alcohol dehydrogenases (EC 1.1.1.1 and 1.1.1.2) able to act on secondary alcohols and secondary alcohol oxidase (EC 1.1.3.18) (37). The oxidases (e.g., EC 1.1.3.18) are possibly of more interest since the electron acceptor (O_2) can be easily supplied, while dehydrogenases need adenine nucleotide, although, in some cases, a redox dye may suffice (38) (Fig. 7). Of some interest is the stereospecific broad substrate specificity secondary oxoacid reductase from *Proteus* described by Neumann and Simon (39), which reacts in vitro with reduced viologen dyes and not with NADH or NADPH. Unfortunately, the enzyme is specific for 2-oxo-acids and apparently is not reactive at the

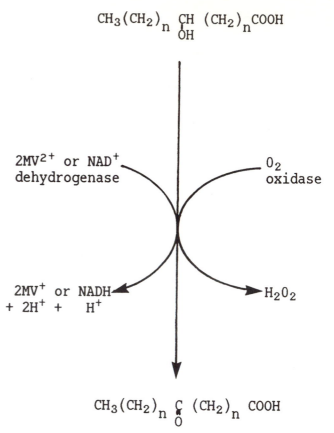

FIG. 7. Secondary alcohol dehydrogenase and oxidase MV: methylviologen.

mid-chain. The NAD^+-dependent *Pseudomonas* secondary alcohol dehydrogenase (see below) is active against a range of secondary hydroxy-fatty acids and secondary fatty alcohols.

Other Reactions

Transferases. Enzymes other than oxidoreductases can lead to potentially interesting products via mid-chain reactions. The enzyme system giving rise to 10-methyl fatty acids has been studied in *Mycobacterium phlei*. An unsaturated phospholipid methyltransferase (EC 2.1.1.16) catalyzes the methylenation of a Δ^9-unsaturated fatty acylphospholipid (40). The methyl group is derived from methionine and S-adenosylmethionine serves as the donor. The 10-methylene product is subsequently enzymically reduced to the 10-methyl fatty acyl residue. When an oleoyl residue is the substrate, 10-methylstearate (tuberculostearate) is produced (Fig. 8). The *Mycobacterium* enzyme appears very similar to the cyclopropane synthase system found in *Clostridium* (41) in requiring olefinic acyl residues of phospholipids as substrates and using S-adenosylmethionine as a methylene donor, although the *Clostridium* enzyme is somewhat more fastidious in substrate requirement than the *Mycobacterium* one.

The requirement of enzymes such as those responsible for tuberculostearic or ricinoleic acid synthesis (see

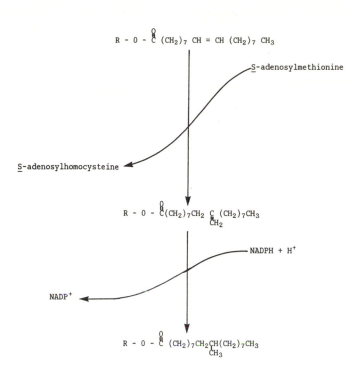

$$R - O - \overset{O}{\overset{||}{C}} (CH_2)_7 \; CH = CH \; (CH_2)_7 \; CH_3$$

S-adenosylmethionine

S-adenosylhomocysteine

$$R - O - \overset{O}{\overset{||}{C}}(CH_2)_7 CH_2 \; \underset{CH_2}{C} \; (CH_2)_7 CH_3$$

NADPH + H⁺

NADP⁺

$$R - O - \overset{O}{\overset{||}{C}} (CH_2)_7 CH_2 \underset{CH_3}{CH} (CH_2)_7 CH_3$$

R = phospholipid.

FIG. 8. Enzymic synthesis of tuberculostearate. R: phospholipid.

above) for acyl phospholipid substrates is of interest because there appears to be no a priori reason why the free acid or another ester could not serve as substrate. In the case of tuberculostearic and cyclopropane fatty acid residues, these are predominantly found in structural membrane phospholipids; it seems probable that their synthesis takes place in situ after deposition by the phospholipid synthases. The synthesis of ricinoleic acid for castor seed storage lipids may well take place at a membrane (42), and this might account for the substrate requirement. In these instances, the nature of the substrate may be dictated by the physical properties and/or subcellular localization of the product.

Lyases. A potentially useful group of enzymes for mid-chain modification is the hydro-lyase group. These are hydratase and dehydratase enzymes generally catalyzing the reaction of water with double bonds. The hydration is generally regiospecific. For example, the ergot fungus *Claviceps purpurea* contains an hydratase that converts linoleic acid to 12-hydroxy-*cis*-Δ^9-octadecenoic acid (ricinoleic acid) (43). The best studied example of a mid-chain hydratase is oleate hydratase (EC 4.2.1.53), which catalyzes the stereospecific addition of water across the double bond of oleic acid to give (*R*)-10-hydroxystearic acid (Fig. 9). This activity is widely distributed among both anaerobic and aerobic intestinal bacteria (44,45), soil microorganisms (46–48) and others (49). The intracellular hydratase from *Pseudomonas* has been purified and shown to be a soluble enzyme that requires no cofac-

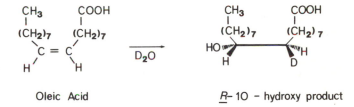

Oleic Acid   *R*- 10 - hydroxy product

FIG. 9. Stereochemical hydration of oleic acid in D₂O enriched medium.

tors (50). The substrate specificity has been studied in detail (50–53). A *cis* Δ^9 double bond is required and a free carboxyl group is necessary although other double bonds and substituents can be tolerated. For example, linoleic acid was converted to (*R*)-10-hydroxy-*cis*-Δ^{12}-octadecenoic acid (51). In addition to *cis*-Δ^9 double bonds, both *cis*- and *trans*-9,10 epoxystearic acids were substrates, giving *threo*- and *erythro*-9,10 dihydroxystearic acids respectively (52). In this case, the enzyme functions as an epoxide hydrolase.

The activity also could be observed in whole cells (47) although two other activities were responsible for side reactions. An isomerase activity was responsible for the production of *trans*-Δ^{10}-octadecenoic acid from oleic acid (54) and, under aerobic conditions, further oxidation of the hydrated product to the oxoacid occurred. This was due to the induction of a secondary alcohol dehydrogenase (55). This enzyme is specific for hydrophobic secondary alcohols with several methylene groups on each side of the carbinol and is inhibited by primary alcohols.

The inducible nature and specificity of the hydratase and secondary alcohol dehydrogenase suggest a specific role for the hydration/oxidation route in unsaturated fatty acid metabolism in these microorganisms, but the full pathway has not been elucidated yet. Although the enzyme in *Corynebacterium* appears to have the same specificity for oleic acid hydration as the *Pseudomonas* enzyme (48), the *Corynebacterium* is able to hydrate the polyunsaturated alkene squalene at all double bonds. In this case, the hydration step may be part of a degradation pathway. The use of microorganisms for hydration reactions has been considered (56,57) because hydroxyl-substituted fatty acids can be produced without the need to supply or recycle cofactors as would be the case with the oxidative transformations discussed above.

Mid-chain fission reactions. Cleavage of unsaturated fatty acids at the double bond is important analytically and industrially. Oxidative ozonolysis of oleic and other unsaturated fatty acids is the common chemical route to C_9 and $C_{13}\alpha$, ω-dicarboxylic acids and other reactions of ozonides give alcohols, aldehydes or amines. Although mid-chain and subterminal fission of alkanes (32) and alkenes such as squalene (58) has been reported in microorganisms, the mid-chain fission of fatty acids does not appear to be a major pathway in nature. It is not surprising that under aerobic conditions, fatty acids are preferentially degraded from the carboxyl terminus by α- or β-oxidation since these latter processes generate one or two carbon units respectively which are immediately available for fixation into cell biomass and/or oxidative energy metabolism. Whether mid-chain and subterminal processes are of major importance in anaerobes has not been established.

PROSPECTS FOR COMMERCIAL MID-CHAIN MODIFICATION PROCESSES

I will not attempt to predict whether any particular enzymic mid-chain reaction of a fatty acid will become the basis of a successful commercial process. That will depend not only on the successful exploitation of an enzyme or enzymes but also on the advantages (or disadvantages) of the product over existing or potentially available materials. I will use some examples of mid-chain reactions to highlight advances in our concepts of biocatalyst preparation and application and to focus on areas needing more work. As some areas are covered elsewhere in these proceedings, this is by no means a comprehensive survey.

Whole Cells or Isolated Enzymes?

The mid-chain reactions discussed above, in general, have been known for years. Current developments in biotechnology are pointing the way toward exploitation. To appreciate these possibilities, it is useful to divide the types of reactions not into classes of enzymes involved, but into those reactions most effectively accomplished using whole cells or those possible with isolated enzymes. For example, the biohydrogenation reaction would be

best performed in whole cells because the complete pathway involves at least two bacteria, multistep enzymic reactions and the need to supply cofactors to carry out the reaction. On the other hand, lipoxygenase is highly effective as an isolated enzyme requiring no cofactors.

Table 3 shows the mid-chain enzymes discussed above divided into these groups. It may be possible to use some enzymes, e.g., isomerases and hydrolyases, as either whole cells or isolated enzymes. The choice between the two will depend on other factors such as enzyme isolation cost weighed against potentially higher volumetric productivity with enzyme catalyst.

TABLE 3

Likely Preferred Form of Mid-chain Enzymes as Whole Cell Catalysts or Isolated Enzymes

Whole cells	Either	Isolated Enzyme
Biohydrogenation system	hydrolyase	lipoxygenase
Monooxygenases-P-450, etc.	epoxide hydrolase	
Desaturases	Isomerases	
Transferase system		

Whole Cells

Whole cells may have positive advantages over catalysts based on isolated enzymes (Table 4). In addition to cofactor and reductant supply essential to many of the oxidoreductases, a major potential advantage of whole-cell catalysts for transformation of hydrophobic substrates such as fatty acids is that cells metabolizing such compounds have efficient mechanisms for uptake of the substrate and, often, excretion of the product (59). This is an important attribute because our current expertise in cell cultivation is based on submerged culture in aqueous media. While it is possible to operate such processes for transformation of lipophilic substrates at commercial intensity (33), much development work is directed toward the use of immobilized cells for continuous or repeated operation. Full exploitation of immobilized biocatalysts will require the minimizing of the aqueous phase and maximizing of the organic substrate/product stream to avoid expensive downstream extraction and water removal processes.

The area of cell immobilization has been reviewed frequently and recently (60). The major effort is devoted to entrapment within a polymeric matrix to prevent cell loss. Most matrices are hydrophilic polysaccharide gels that have been successfully applied to yeast alcohol fermentation systems (61) but that are less suitable for biotransformation of hydrophobic substrates (62). Hydrophobic entrapment matrices may be more suitable for transformation of lipophilic compounds (63) because mass transfer of the substrate to the biocatalyst may be promoted. Use of support materials to which the cells may adhere is commonly practiced in waste water treatment and has been applied occasionally in the food industry (e.g., vinegar production using a film of *Acetobacter* on

TABLE 4

Advantages of Whole Cells for Biomodification of Lipophlic Substrates

Process problem	Advantage of whole cells
Supply of lipophilic substrate to enzyme	1. Secrete emulsifiers. 2. Have specific uptake systems. 3. Intracellular enzymes are membrane-bound/compartmentalized for efficiency. 4. Substrate activated intracellularly as required.
Supply of other reactants	1. Redox chain for reductant supply. 2. Other reactants and cofactors present at appropriate concentrations. 3. Cofactor recycle.
Presence of inhibitors	1. Inhibitors may not penetrate cell or reach enzyme. 2. Inhibitor may be metabolized to harmless product.
Product recovery	1. Product may be excreted from cell. 2. Product toxicity/inhibition may be avoided. 3. Cell catalyst may survive product extraction.

wood chips in a trickle reactor). Mass transfer limitations may be lessened by using organisms adhering to a support compared with a gel matrix because the distance over which the substrate/product has to diffuse between the bulk phase and biocatalyst may be shorter. Mass transfer limitations remain a major hurdle to be overcome in the use of immobilized cells as compared with conventional submerged culture in an STR. This problem is exacerbated in considering the transformation of fatty acids added as a discrete liquid phase or particulate suspension and probably constitutes a major barrier to exploitation.

The type of reactor suitable for use with immobilized cell biocatalyst also deserves mention since the impeller-mixed STR has high shear characteristics important for gas transfer (necessary for many of the O_2-requiring transformations) but that is likely to result in attrition of supported biocatalyst. Gas-mixed reactors, fluidized beds and packed beds (64) all may be suitable for certain reactions but no one reactor configuration is likely to be universally effective. The hollow fiber membrane reactor, which has been successfully applied to lipid transformations such as lipase catalyzed triacylglycerol hydrolysis (65) and to whole-cell reactions (60,61), may be worthy of further investigation for fatty acid modifications.

Isolated Enzymes

Major areas of improvement in our understanding and exploitation of free-enzyme biocatalysts are in the nature of the biocatalyst microenvironment and in how to supply cofactors for redox enzymes.

Maintenance of an appropriate enzyme microenvironment is essential for the activity and stability of isolated enzymes. This is particularly true of most enzymes involved in mid chain fatty acid reactions because they are particulate and/or membrane-bound enzymes that

can be isolated from cells only by substantially altering their normal in vivo microenvironment. This normally would favor application of whole-cell catalysts, but some enzymes of interest can be successfully used in isolated form. In considering the use of enzymes in predominantly organic media, the importance of the essential water layer around the enzyme has been recognized (66) and from this has come an understanding of the interaction of this water layer with the continuous phase. To avoid removal of the essential water by partition into a continuous organic phase, the organic bulk phase must be selected with care. One predictive parameter that has been used to select organic solvents is their log P (P = partition coefficient of bulk phase in a standard ocatanol-water two-phase system (67,68). Bulk phases with log P > 4 are suitable for use with enzymes because the essential water layer is not removed or distorted. Liquid fatty acid substrates have high log P and are probably quite suitable in this respect as bulk continuous phases for enzyme catalysis.

The different relationships of biocatalyst with the substrate phase are summarized in Figure 10. In some cases (e.g., whole cells), emulsified substrate may be the necessary substrate for reaction. In other cases, mainly with isolated enzymes, close contact of the bulk substrate phase with an immobilized biocatalyst or accommodation of the biocatalyst within the substrate phase are the most effective options.

Although whole cells are undoubtedly suitable for most cofactor-requiring reactions, the generally low enzyme concentrations within cells, particularly of interesting redox enzymes, has led to continued work on such enzymes as isolated biocatalysts. A significant advantage would be greater volumetric biocatalyst loading in the reactor. The predominant hurdle to exploitation is the difficulty of providing any necessary oxidized or reduced cofactors.

The redox cycle of whole cells may be maintained to support biotransformation by the use of a cosubstrate (69) to supply (or consume) the energy required for the exploited reaction. A similar approach is possible with free enzyme reactions when a reduction is linked with an oxidation to recycle pyridine nucleotide cofactors. For NADH regeneration from NAD^+, the use of formate dehydrogenase (EC 1.2.1.2) often has been proposed. This has the advantage that the enzyme is readily available; the substrate is cheap; and the product is gaseous (70). Unfortunately, it may be less than suitable for hydrophobic transformations.

It may be possible to provide reducing equivalents from an electrode to a redox enzyme (71,72) using low molecular weight redox mediators such as viologens. This approach also is possible with whole cells (73). The coupled enzyme and electrochemical resupply of reductant approaches are probably more suitable to reactions carried out in aqueous media because the mediators and cofactors are water-soluble. However, it is possible to include the aqueous phase in an organic environment by forming reversed micelles using a surfactant (74). In this environment it still is possible to regenerate NADH electrochemically to drive steroid reduction (75) or to use lipoxygenase to oxidize linoleic acid (76). Another significant advance into redox biocatalysis in organic media is

Types of microenvironment of biocatalyst and hydrophobic substrate

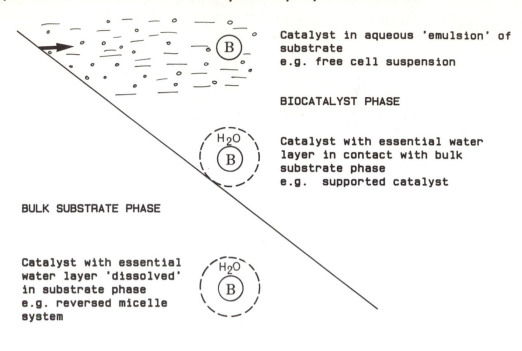

B: biocatalyst

FIG. 10. Types of microenvironment of biocatalysts and hydrophobic substrate.

in the use of PEG-modified enzymes and cofactors to allow their use in low-water, organic environments (70,77). The PEG chains allow the enzyme or cofactor to "dissolve" in the organic phase. The nature of this dissolution is not fully clear, and some water is required to maintain enzyme activity under these circumstances. The pyridine nucleotide cost element in appropriate biotransformations can be lowered significantly due to efficient recycling of PEG-NADH over prolonged reaction times. Productivities of 100,000 mol product per mol of NAD^+ consumed (by dimerisation, denaturation or other loss) are achievable.

REFERENCES

1. Britton, L.N., in *Microbial Degradation of Organic Compounds*, edited by D.T. Gibson, Marcel Dekker, New York, 1984, pp. 89–129.
2. Macrae, A.R., and R.C. Hammond, *Biotechnol. Genet. Eng. Rev. 3*:193 (1985).
3. Reiser, R., *Fed. Proc. Fed. Am. Soc. Exp. Biol. 10*:236 (1951).
4. Dawson, R.M.C., and P. Kemp, in *Physiology of Digestion and Metabolism in the Ruminant*, edited by A.T. Phillipson, Oriel Press, Newcastle-upon-Tyne, 1970, pp. 504–518.
5. Dawson, R.M.C., N. Hemington, D. Grime, D. Lander and P. Kemp, *Biochem. J. 144*:169 (1974).
6. Hawke, J.C., and W.R. Silcock, *Ibid. 112*:131 (1969).
7. Kellens, M.J., H.L. Goderis and P.P. Tobback, *Biotechnol. Bioeng. 28*:1268 (1986).
8. Kemp, P., and D.J. Lander, *J. Gen. Microbiol. 130*:527 (1984).
9. Kepler, C.R., K.P. Hirons, J.J., McNeill and S.B. Tove, *J. Biol. Chem. 241*:1350 (1966).
10. Kepler, C.R., and S.B. Tove, *Ibid. 242*:5686 (1967).
11. Kepler, C.R., W.P. Tucker and S.B. Tove, *Ibid. 246*:2765 (1971).
12. Hughes, P.E., W.J. Hunter and S.B. Tove, *Ibid. 257*:3643 (1982).
13. Hughes, P.E., and S.B. Tove, *Ibid. 255*:4447 (1980).
14. Anon., *Nut. Rev. 38*:284 (1980).
15. Rosenfeld, I.S., and S.B. Tove, *J. Biol. Chem. 246*:5025 (1971).
16. Morris, L.J., *Biochem. J. 118*:681 (1970).
17. Body, D.R., *Ibid. 157*:741 (1976).
18. Patterson, H.B.W., *Hydrogenation of Fats and Oils*, Applied Science Publishers, London, 1983.
19. de Groot, J.J.M.E., G.A. Veldink, J.F.G. Vliegenthart, J. Boldingh, R. Wever and B.F. van Gelder, *Biochim. Biophys. Acta 377*:71 (1975).
20. Slappendel, S., G.A. Veldink, J.F.G. Vliegenthart, R. Aasa and B. Malmstrom, *Ibid. 667*:77 (1981).
21. Alexrod, B., *Adv. Chem. Ser. 136*:324 (1974).
22. Alexrod, B., T.M. Cheesebrough and S. Laasko, *Meth.Enzymol. 71*:441 (1981).
23. Van Os, C.P.A., M. Vente and J.F.G. Vliegenthart, *Biochim. Biophys. Acta 574*:103 (1979).
24. VanOs, C.P.A., G.P.M. Rijke-Schilder and J.F.G. Vliegenthart, *Ibid. 575*:479 (1979).
25. Galliard, T., and H.W.S. Chan, in *The Biochemistry of Plants*, edited by P.K. Stumpf and E.E. Conn, Academic Press, New York, Vol. 4, 1980, pp. 131–161.
26. Bild, G.S., S.G. Bhat, C.S. Ramadoss and B. Axelrod, *J. Biol. Chem. 253*:21 (1978).
27. Gunsalus, I.C., and S.G. Sligar, *Adv. Enzymol. 47*:1 (1978).
28. Croteau, R., and P.E. Kolattukudy, *Arch. Biochem. Biophys. 170*:61 (1975).
29. Ruettinger, R.T., and A.J. Fulco, *J. Biol. Chem. 256*:5728 (1981).
30. Buchanan, J.F., and A.J. Fulco, *Biochem. Biophys. Res. Commun. 85*:1254 (1978).
31. Michaels, B.C., R.T. Ruettinger and A.J. Fulco, *Ibid. 92*:1189 (1980).
32. Ratledge, C., *J. Am. Oil Chem. Soc. 61*:447 (1984).
33. Uemura, N., *Hakko to Kogyo 43*:436 (1985).
34. Moreau, R.A., and P.K. Stumpf, *Plant Physiol. 67*:672 (1981).
35. Stumpf, P.K., in *The Biochemistry of Plants*, edited by P.K. Stumpf and E.E. Conn, Academic Press, New York, Vol. 4, 1980, pp. 177–204.

36. Mason, H.S., *Adv. Enzymol. 19*:79 (1957).
37. Suzuki, T., *Agric. Biol.Chem. 40*:497 (1976).
38. Mahler, H.R., and E.H. Cordes, in *Biological Chemistry*, Harper and Row, New York, 1966.
39. Neumann, S., and H. Simon, *FEBS Lett. 167*:29 (1984).
40. Akamatsu, Y., and J.H. Law, *J. Biol. Chem. 245*:701 (1970).
41. Chung, A.E., and J.H. Law, *Biochemistry 3*:967 (1964).
42. Kolattukudy, P.E., in *The Biochemistry of Plants*, edited by P.K. Stumpf and E.E. Conn, Academic Press, New York, Vol. 4, 1980, pp. 571–645.
43. Morris, L.J., S.W. Hall and A.T. James, *Biochem. J. 100*:29c (1966).
44. Thomas, P.J., *Gastroenterology 62*:430 (1972).
45. Pearson, J.R., H.S. Wiggins and B.S. Drasar, *J. Med. Microbiol. 7*:265 (1974).
46. Wallen, L.L., R.G. Benedict and R.W. Jackson, *Arch. Biochem. Biophys. 99*:249 (1962).
47. Davis, E.N., L.L. Wallen, J.C. Goodwin, W.K. Rohwedder and R.A. Rhodes, *Lipids 4*:356 (1969).
48. Seo, C.W., Y. Yamada, N. Takada and H. Okada, *Agric. Biol. Chem. 45*:2025 (1981).
49. Takatori, T., N. Ishiguro, H. Tarao and H. Matsumiya, *Forensic Sci.Int. 32*:5 (1986).
50. Niehaus, W.G., A. Kisic, A. Torkelson, D.J. Bednarczyk and G.J. Schroepfer, *J. Biol. Chem. 245*:3790 (1970).
51. Schroepfer, G.J., W.G. Nichaus and J.A. McCloskey, *Ibid. 245*:3798 (1970).
52. Niehaus, W.G., A. Kisic, A. Torkelson, D.J. Bednarczyk and G.J. Schroepfer, *Ibid. 245*:3802 (1970).
53. Kisic, A., Y. Miura and G.J. Schroepfer, *Lipids 6*:541 (1971).
54. Mortimer, C.E., and W.G. Niehaus, *Biochem. Biophys. Res. Commun. 49*:1650 (1972).
55. Niehaus, W.G., T. Frielle and E.A. Kingsley, *J. Bacteriol. 134*:177 (1978).
56. Wallen, L.L., U.S. Patent 3,115,442 (1963).
57. Litchfield, J.H., and G.E. Pierce, U.S. Patent 4,582,804 (1986).
58. Yamada, Y., H. Motoi, S. Kinoshita, N. Takada and H. Okada, *Appl. Microbiol. 29*:400 (1975).
59. Cooper, D.G., and J.E. Zajic, *Adv. Appl. Microbiol. 26*:229 (1980).
60. Scott, C.D., *Enzyme Microb. Technol. 9*:66 (1987).
61. Godia, F., C. Casas and C. Sola, *Process Biochem. 22*:43 (1987).
62. Yi, Z-H., and H.J. Rehm, *Eur. J. Appl. Microbiol. Biotechnol. 16*:1 (1982).
63. Fukui, S., and A. Tanaka, *Adv. Biochem. Eng. Biotechnol. 29*:1 (1983).
64. Bungay, H.R., in *Biotechnology for the Oils and Fats Industry*, edited by C. Ratledge, P.S.S. Dawson and J. Rattray, American Oil Chemists' Society, Champaign, IL, 1984, pp. 45–54.
65. Kloosterman, J., in *Proceedings of the World Conference on Biotechnology for the Fats and Oils Industry*, edited by T.H. Applewhite, American Oil Chemists' Society, Champaign, IL, 1988, pp. xxx–xxx.
66. Laane, C., *Biocatalysis 1*:17 (1987).
67. Hilhorst, R., R. Spruijt, C. Laane and C. Veeger, *Eur.J. Biochem. 144*:459 (1984).
68. Laane, C., S. Boeren and K. Vox, *Trends Biotechnol. 3*:251 (1985).
69. Perry, J.J., *Microbiol. Revs. 43*:59 (1979).
70. Kula, M-R., in *Proceedings Third European Congress Biotechnology 4*:475 (1985).
71. Higgins, I.J., R.C. Hammond, E. Plotkin, H.A.O. Hill, K. Uosaki, M.J. Eddowes and A.E.G. Cass, in *Hydrocarbons in Biotechnology*, edited by D.E.F. Harrison, I.J. Higgins and R.J. Watkinson, Institute of Petroleum, London, 1980, pp. 181–193.
72. Gunther, H., and H. Simon, *Appl. Microbiol. Biotechnol. 26*:9 (1987).
73. Bader, J.H. Gunther, S. Nagata, H.J. Schentz, M.L. Link and H. Simon, *J. Biotechnol. 1*:95 (1984).
74. Martinek, K., A.V. Levashov, N.L. Klyachko and I.V. Berezin, *Dokl. Akad. Nauk SSSR (Eng. Trans.) 236*:951 (1978).
75. Laane, C., R. Hilhorst, R. Spruijt and C. Veeger, *Abstr. Third Eur. Congr. Biotechnol. 1*:357 (1984).
76. Luisi, P.L., Angew, *Chem. Int. Ed. Eng. 24*:439 (1985).
77. Takahashi, K., H. Nishimura, T. Yoshimoto, M. Okada, A. Ajima, A. Matsushima, Y. Tamaura, Y. Saito and Y. Inada, *Biotechnol. Lett. 6*:765 (1984).

A Simple Test for the Determination of Lipase Fatty Acid Specificity

J. Baratti[a], **M-S Rangheard**[b], **G. Langrand**[b] and **C. Triantaphylides**[b]

[a]CNRS, Laboratoire de Chimie Bactérienne, BP 71, 13277 Marseille Cedex 9, France, [b]Ecole Sùperieure de Chimie de Marseille, rue H. Poincaré, 13397 Marseille Cedex 13, France

A lipase-catalyzed reaction in organic solvent was used to quantitatively determine the fatty acid chain length specificity of some commercial lipase preparations. A mixture of the ethyl esters of fatty acids from acetic to stearic was submitted to an alcoholysis reaction in heptane with n- propanol. The reaction rate was determined by gas chromatography. The kinetic analysis of two substrates in competition for one enzyme active site described earlier in our laboratory was extended to multiple substrates. The ratio of the catalytic powers (V/K), the competitive factor α, was used to analyze the results. A scale of reactivity for acyl donor that corresponds to the fatty acid specificity of the lipase tested was established. The results were not greatly dependent upon the conditions of reaction. Comparison with previous reports on lipase specificity showed little differences in the results. The new test is simple and can be used to construct a bank of specificity spectra for commercial lipases.

Because lipases (EC 3.1.1.3) are enzymes that split ester bonds of triglycerides, the knowledge of the fatty acid specificity is of great importance for the understanding of their biochemical role in substrate hydrolysis, for instance, during digestion in mammals. Early studies with porcine pancreatic lipase have demonstrated that triglycerides with short chain fatty acids (butyric acid) are hydrolyzed faster than any other triesters containing acids from C_2 to C_{18} (1). Several other authors also have studied the specificity of microbial lipases using the same approach (2-5), but full details on the specificity of commercial lipases presently are not available.

More recently, lipases have been used in organic solvents to catalyze ester synthesis or transesterification reactions (6-11). The latter reaction is of special interest in the fats and oils industries for modification of the fatty acid content of a mixture of triglycerides to change its physical-chemical properties (12-15). For this purpose, the knowledge of the fatty acids specificity is necessary to select the suitable lipase from commercially available enzymes. In addition, the screening for new microbial lipases will be much easier with a simple test for the determination of fatty acids specificity. It is the purpose of this work to define such a test.

When acting on insoluble substrates, the kinetics of lipase hydrolysis are rather complex (16, 17). By contrast, Michaelis-Menten kinetics have been demonstrated for lipase-catalyzed reactions in organic media (18). Furthermore, a "ping pong" mechanism has been proposed including the formation of an acyl enzyme intermediate. In previous work, we have extended this analysis to a reaction involving two substrates in competition for the enzyme active site. The use of the competition factor α_D defined as the ratio of the catalytic powers (V/K) for each substrate was very useful for the prediction of reaction rates (19).

Here we report the use of the competition factor to describe the fatty acid specificity of lipases quantitatively, and a simple test is proposed for the determination of this specificity.

MATERIALS AND METHODS

Enzymes

The following commercial lipase preparations were used: *Mucor miehi* (Gist-Brocades), *Mucor miehi* (Lipozyme, Novo), *Candida rugosa* (Meito Sangyo, also named *C. cylindracea*), porcine pancreatic lipase (Koch-Light Laboratories).

Reaction Conditions

A mixture containing the ethyl esters of fatty acids from C_2 to C_{18} at a concentration of 0.1 M each, with 2 M propanol, 0.1% (w/v) water and 1% (w/v) of the lipase preparation in 10 ml of pentane was incubated under agitation at 38 C. At regular time intervals, aliquots of the reaction mixture were analyzed for their ethyl and propyl esters content using gas liquid chromatography.

RESULTS AND DISCUSSION

Competition with Two Substrates

According to experimental conditions and choice of substrates, several reactions are catalyzed by lipases in organic solvent: ester hydrolysis, ester formation and transesterification, which include ester acidolysis and ester solvolysis (8,11). In this study, the alcoholysis with n-propanol of a fatty acid ethyl ester was used according to the reaction: $RCOOC_2H_5 + C_3H_7OH \rightleftarrows RCOOC_3H_7 + C_2H_5OH$ (1) with an excess of n-propanol the reaction equilibrium is shifted to the right side of the equation. The reaction rate can be determined easily from analysis of the variations of the ethyl and propyl ester concentrations with time.

In the presence of the ethyl esters of two different fatty acids in the reaction mixture, a competition for the enzyme active site will occur according to reaction 2: $R_1COOC_2H_5 + C_3H_7OH \rightleftarrows R_1COOC_3H_7 + C_2H_5OH \quad R_2COOC_2H_5 + C_3H_7OH \rightleftarrows R_2COOC_3H_7 + C_2H_5OH$ (2). The kinetic analysis of this system (19) led to the conclusion that the ratio of the reaction rates for substrates 1 and 2 (v1/v2) is given by equation 3:

$$\frac{v_1}{v_2} = \frac{(V/K)_1 (S_1)}{(V/K)_2 (S_2)}$$ (3), in which V, maximal reaction rate; K, Michaelis constant; S_1 and S_2, concentrations of ester 1 and 2. Introducing the competition factor α_D leads to equation (4):

$$\frac{v_1}{v_2} = \alpha_D \left(\frac{S_1}{S_2}\right)$$ (4), which can be integrated to

give: $\text{Log } S_1 / (S_1)_0 = \alpha_D \text{ Log } S_2 / (S_2)_0$ (5), in which $(S_1)_0$ and $(S_2)_0$ = initial concentrations of substrates 1 and 2.

From kinetic data on the variations of S_1 and S_2 with

time and the initial substrates concentrations S_{10} and S_{20}, the competition factor α_D can be determined using the integrated equation 5. When the initial concentrations of the two substrates are identical, the competition factor is the ratio of the two initial reaction rates. Then, a value of one for α_D means identical initial reaction rates and a value of 10 means an initial reaction rate 10 times higher for substrate 1. Because the competition factor is the ratio of the catalytic powers V/K (maximal velocity/Michaelis constant) it is a quantitative measure of the enzyme specificity for the two substrates. In the special case studied here, (ethyl esters of two fatty acids) α_D is a quantitative measure of the fatty acid specificity.

Competition Between n Substrates

The above analysis can be readily extended to the case of n substrates in the reaction mixture:

$$\frac{\begin{array}{l} R_1COOC_2H_5 + C_3H_7OH \rightleftarrows R_1COOC_3H_7 + C_2H_5OH \\ R_2COOC_2H_5 + C_3H_7OH \rightleftarrows R_2COOC_3H_7 + C_2H_5OH \end{array}}{R_nCOOC_2H_5 + C_3H_7OH \rightleftarrows R_nCOOC_3H_7 + C_2H_5OH} \quad (6)$$

The reaction rate for each substrate (v_n) is obtained from the variations of the substrate (S_n) and product (P_n) concentrations with time. One of the substrates (usually the best one) is taken as a reference, and a competition factor (α_{Dn}) is calculated for each substrate using equation 5. Each substrate reactivity toward lipase then is characterized by its competition factor, and the whole values give a scale of substrate specificity. In the special case of ethyl esters of different fatty acids, the resulting scale describes the chain length specificity of the lipase tested.

Application to the Obtention of a Lipase Specificity Spectrum

The previous kinetic analysis was applied to the alcoholysis of a mixture of the ethyl esters of fatty acids with chain length from C_2 (acetic) to C_{18} (stearic) with n- propanol under the conditions described in Materials and Methods. A typical gas chromatographic separation of the ethyl and propyl esters is given in Figure 1. The degree of conversion was estimated from the areas of the peaks.

The resulting reaction kinetics are given in Figure 2 for *Mucor miehi* lipase (lipozyme, Novo). It clearly appeared that the ethyl esters of fatty acids from C_8 to C_{18} were reacting more rapidly than those from C_2 to C_6. From these curves, the competition factors were calculated using the C_8 acid as reference by the use of equation 5 as shown in Figure 3. This representation was linear as soon as initial reaction rates were measured (less than 30% conversion). At higher degree of conversion, the reverse reaction cannot be neglected, and equation 5 no longer is valid. From the slope of the straight lines, the competition factors were calculated; the results are given in Table 1. The *Mucor miehi* lipase showed a pronounced specificity for fatty acids containing 8 or 12–18 carbon atoms. Short chain fatty acids (C_6 and C_4) were poor substrates; no reaction was observed with acetic acid as donor in the conditions used. It can be concluded that this lipase mainly is specific for long chain fatty acids (n > 8) with little differences among them. Furthermore, the total reaction rate was 4.92 μmol/min/g of enzymne, and the

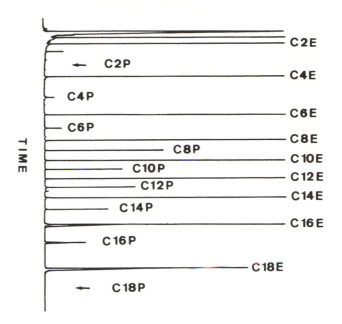

DETECTOR SIGNAL

FIG. 1. Gas chromatographic separation of ethyl (CnE) and propyl (CnP) esters of fatty acids containing n-carbon atoms.

reaction rates for the C_8 ester (the best substrate) was 1.16 μmol/min/g. As the lipase activity was very different on a weight basis for the enzymes tested, (due to large variations in the enzyme content of the crude preparations) the reaction rates are important process parameters for industrial applications.

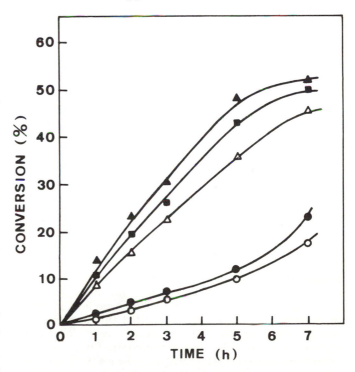

FIG. 2. Kinetics of propyl esters formation by *Mucor miehi* lipase (Lipozyme, Novo). Experimental conditions are given in Materials and Methods. ▲, C_8 and C_{14}; ■, C_{12}–C_{16}, and C_{18}; △, C_{10}; ●, C_6; ○, C_4.

J. BARATTI ET AL.

TABLE 1

Competition Factors of Some Lipases

	C_2	C_4	C_6	C_8	C_{10}	C_{12}	C_{14}	C_{16}	C_{18}	v_t	v_m
Mucor miehi (Gist Brocades)	—	4.8	6.7	1.0	1.8	1.5	1.6	2.0	2.3	2.88	10.77
Mucor miehi (Lipozyme, Novo)	—	6.3	5.9	1.0	1.6	1.2	1.1	1.2	1.2	1.16	4.92
Candida rugosa	—	1.0	16	4.0	7.5	10.8	15.3	16.1	—	0.087	0.126
Porcine pancreas	2.2	1.0	1.9	1.9	3.2	3.3	3.6	5.2	5.0	0.004	0.012

v_t = total reaction rate; v_m = reaction rate for the best substrate. Both rates are expressed in μmol/min/g of enzyme.

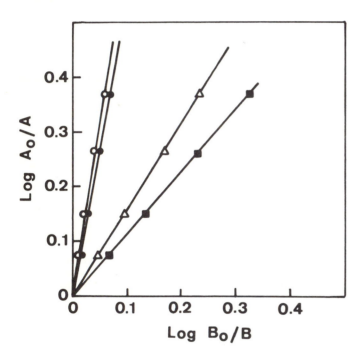

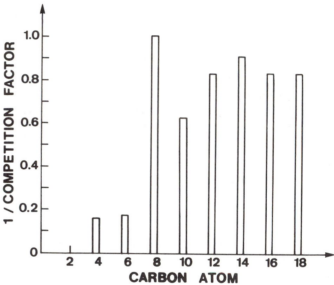

FIG. 4. *Mucor miehi* (Lipozyme, Novo) fatty acid specificity represented as $1/\alpha_D$ vs the number of carbon atoms.

FIG. 3. Estimation of competition factors α_D from the data in Figure 2. For the signification of symbols, see Figure 2.

A different representation of the lipase specificity is given in Figure 4 where $1/\alpha_D$ is plotted against the chain length of the fatty acid. It is clear from these results that the *Mucor miehi* lipase (lipozyme, Novo) was specific for long chain fatty acids containing 8–10 carbon atoms and showed a very low specificity for lower fatty acids (from C_2 to C_6).

Checking the Validity of the Test

The influence of experimental conditions on the quantitative values of the competitive factors was examined. The analytical error of a standard measure was in the range of 5-10%. The effect of number of esters in the mixture, nature of the alcohol used in the reaction, substrate concentration, nature of organic solvent, temperature and water content was checked. These validity experiments were done with the *Mucor miehi* lipase (lipozyme, Novo) as a model system for lipase and with two pairs of

ester butyric-caprylic and caprylic-lauric ethyl esters. In all cases, the variations observed on the experimental results were within the range of the experimental error. It then was concluded that the assay conditions did not influence the results of the specificity test, which can be considered as independent of reactions conditions. It was of special interest to demonstrate that the competition factor did not vary with the alcohol used in the reaction. This point further confirms the two steps reactions (19): formation of an acyl enzyme from the fatty acid ethyl ester and nucleophilic attack of the acyl enzyme by n-propanol. With this mechanism in mind, one should expect that the reaction rate will be determined by the rate of the acylation reaction, which involved different acyl groups, rather than acyl enzymes alcoholysis, which involved the same alcohol.

Comparison with Previous Work

Several attempts to describe the fatty acid specificity of lipases are reported in the literature. Usually, the authors measured the individual rate of hydrolysis of a triglyceride with the same fatty acid in all three positions of glycerol as a function of the fatty acid (triacetin, tributyrin . . .). In some cases, the methyl esters of fatty acids were used. The results using this approach are compared with ours for porcine pancreatic lipase (1) and *Candida rugosa* (ex. *cylindracea*) lipase (2) in Figure 5. For pancreatic lipase, the results are quite similar with butyric acid being the best acyl donor and then a slow decrease being observed by increasing the chain length. This decrease is more pronounced in our test conditions. In addition, acetic acid is a better substrate in our conditions than in hydrolysis. For *C. rugosa* lipase, we found a pronounced specificity for butyric acid. The comparison with previous results (2) is difficult because the authors only have studied fatty acids with a chain length higher than C_{12}. However, the two kinds of results are in agreement for both lipases tested, which demonstrated that the fatty acid specificity is not very different when measured in organic solvent or in a two-phase system. This was unexpected because the experimental conditions were quite different as well as the chemical nature of the substrates and the kind of reaction involved. In one case (hydrolysis of triglycerides), reactions rates are measured while in the other one competition factors (ratio of V/K) are determined. Furthermore, some authors (5) have postulated an influence of physical properties of substrate on specificity in addition to the well-known chemical specificity based on enzyme-substrate interactions.

Specificity Spectrum of Commercial Lipases

The test was applied to several commercial lipase preparations and the results are shown for four of them in Table 1. The two lipases from *Mucor miehi* (Gist-Brocades and lipozyme Novo) showed a very similar specificity for long chain fatty acids. In contrast, *Candida rugosa* and porcine pancreatic lipases were strongly specific for butyric acid. On a weight basis, the observed reaction rates varied from 0.004 to 2.88 μmol/min/g of enzyme. It appeared clearly that under the conditions used, the activities of these preparations were very different. This factor is of great practical influence, for instance, in industrial applications.

We currently are investigating the specificity of some other commercial lipases using the simple test defined during this work to compare the fatty acid specificity spectrum of these lipases.

ACKNOWLEDGMENTS

This work was supported by a grant from Société Nationale Elf Aquitaine. We thank Seris and Gancet (SNEA, Groupe de Recherche de Lacq) for discussions.

REFERENCES

1. Desnuelle, P., *Adv. Enzymol. 23*:129 (1961).
2. Benzonana, G., and S. Esposito, *Biochim. Biophys. Acta 231*:15 (1971).
3. Iwai, M., S. Okumura and Y. Tsujisaka, *Agric. Biol. Chem. 39*:1063 (1975).
4. Sugiara, M., and M. Isobe, *Chem. Pharm. Bull. 23*:681 (1975).
5. Sugiara, M., and M. Isobe, *Chem. Pharm. Bull. 23*:1226 (1975).
6. Cambou, B., and A.M. Klibanov, *J. Am. Chem. Soc. 106*:2687 (1984).
7. Cambou, B., and A.M. Klibanov, *Biotechnol. Bioeng. 26*:1449 (1984).
8. Marlot, C., G. Langrand, C. Triantaphylides and J. Baratti, *Biotechnol. Letters 7*:647 (1985).
9. Baratti, J., G. Buono, H. Deleuze, G. Langrand, M. Secchi and C. Triantaphylides, *Proceedings of the World Conference on Emerging Technologies in the Fats and Oils Industries*, edited by A.R. Baldwin, American Oil Chemists' Society, Champaign, IL, 1986, p. 355.
10. Langrand, G., M. Secchi, G. Buono, J. Baratti and C. Triantaphylides, *Tetrahedron Lett. 26*:1857 (1985).
11. Langrand, G., J. Baratti, G. Buono and C. Triantaphylides, *Tetrahedron Lett. 27*:29 (1986).
12. Wisdom, R., P. Dunnil, M. Lilly and A. Macrae, *Enzyme Micro. Technol. 6*:443 (1984).
13. Wisdom, R., P. Dunnil and M. Lilly, *Enzyme Microb. Technol. 7*:567 (1985).
14. Tanaka, T., E. Ono, M. Ishihara, S. Yamanaka and K. Takinami, *Agric. Biol. Chem. 45*:2387 (1981).
15. Tjujisaka, Y., S. Okumura and M. Iwai, *Biochim. Biophys. Acta 489*:415 (1977).
16. Verger, R., and G. de Haas, *Ann. Rev. Biophys. Bioeng. 5*:77 (1976).
17. Verger, R., *Methods Enzymol. 64*:340 (1980).
18. Zaks, A., and A.M. Klibanov, *Proc. Natl. Acad. Sci.USA 82*:3192 (1985).
19. Deleuze, H., G. Langrand, H. Millet, J. Baratti, G. Buono and C.Triantaphylides, *Biochim. Biophys. Acta 911*:117 (1987).

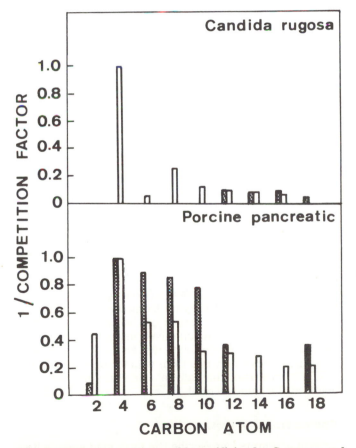

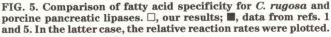

FIG. 5. Comparison of fatty acid specificity for *C. rugosa* and porcine pancreatic lipases. □, our results; ■, data from refs. 1 and 5. In the latter case, the relative reaction rates were plotted.

Discussion Session

Atsuo Tanaka indicated there is a relationship between water content and lipase activity; he added that the structure of the lipases was very difficult to change, but one could produce changes by using either highly hydrophilic or hydrophobic gels that would change enzyme substrate interactions. He also indicated that about 80% of the enzyme activity could be recovered; this was a measure of the activity while the enzyme was immobilized.

A question for Peter Eigtved regarding the reason for using a derivative in the esterification of galactolipids was answered by indicating that polyols were not soluble or miscible with triglycerides, and the reaction rates were very slow at the interfaces of immiscible materials. In addition, the use of a derivative allows the formation of monoester when other groups are blocked. In a similar question, J. Kurashige indicated it was not possible to esterify phospholipids as diglycerides as this was probably a two-step reaction. The last questions directed at F. Meussdoerffer concerned the enzymes involved in ω-oxidation of alkanes. He indicated it may be possible to eliminate alcohol dehydrogenase activity and still have alkane hydroxylation. Meussdoerffer also noted that other enzymes also can oxidize alkanes.

Namio Uemura indicated that the growth of *Candida* on sucrose, glucose or molasses did not repress alkane oxidation, and yields of DCA were about 65%. He also said some of the mutants were strain-specific for fatty acids and others for alkanes. D. Estell responded to his question by stating to his knowledge, no one has determined yet the crystal structure of a lipase since it is very difficult to crystallize the lipases.

Unusual Fatty Acids and Their Scope in Biotechnology

Rolf D. Schmid

Gesellschaft fuer Biotechnologische Forschung mbH (GBF), D-3400 Braunschweig, West Germany

Whereas as little as seven types of fatty acids account for more than 90% of the fatty acid-derived lipids occurring in nature, an astonishing variety of unusual fats and fatty acids have been reported from many genera of animals, plants and microorganisms. In fact, more than 1,000 different fatty acid structures easily were compiled from publications concerning chemotaxonomical markers. In most cases, the biological significance of such unusual fats and fatty acids is unknown. In a few cases, however, their biological role has become clearer. It is inferred that from a more systematic investigation into the relationship between their structure and biological function, industrial applications of such compounds and their derivatives might be derived.

Most forms of life rely on membranes to both separate themselves from and communicate with the environment. With a few exceptions, such membranes are composed of acyl phosphoglycerol esters, whose acyl moieties are composed of as few as six varieties of fatty acids. As a consequence, six types of fatty acids make up for more than 95% of the fatty acids used as membrane building blocks. The same is true for the fatty acids found as triglycerides in the seeds of plants (1). If one starts asking about the structures of the remaining few percentages, however, one rapidly becomes intrigued by the permutations on fatty acid structure that can be found in nature. In fact, in an attempt to set up a catalogue of unusual fatty acids (R.D. Schmid and V. Fischer, unpublished data), it turned out to be quite simple to identify more than 1,000 different fatty acid varieties (Fig. 1).

Unusual fatty acids
– chain-length distribution –

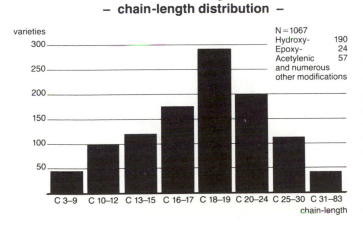

FIG. 1. Statistical Composition of Unusual Fatty Acids.

Unusual fatty acid structures are encountered in most, if not all, families of living beings. In some cases, they constitute but a small fraction of total fatty acids; in other cases, their derivatives represent the major class of lipids in a membrane or a compartment. Although such observations date back to the early days of structure elucidation

of natural compounds, and because quite a few reviews exist on special aspects of this topic (R.D. Schmid and V. Fischer, unpublished data, 2–5), practical applications of unusual fatty acids have remained scarce. As major reasons for this, it might be stipulated that: availability of naturally occurring unusual fatty acids was, and still is, below the quantities required for assessment of industrial application; and in some cases, the particular properties of a natural unusual fatty acid can be mimicked by the appropriate formulation of mixtures of standard fatty acids, fatty alcohols and their derivatives available in commercial quantities.

It should be kept in mind, however, that recent years have seen considerable progress in the application of plant breeding and biotechnology to fatty acid production and modification (reviews, 6–10). Thus, the manufacture of unusual fatty acids by modern breeding methods is presently being investigated in both the U.S. and W. Germany; in addition, a search for novel metabolic pathways in microorganisms has been started (11), which eventually might be used to transform standard fatty acids into value-added specialties useful for a variety of industrial sectors. To select the right targets for such endeavors, it may be helpful to ask the following questions: What is the biological function of an unusual fatty acid in the organism by which it is synthesized and does such biological function teach us how we might apply unusual fatty acids and their derivatives in research and technology?

OCCURRENCE AND FUNCTION OF UNUSUAL FATTY ACIDS

Though the field in general seems to be little explored, a detailed analysis of the published information is beyond the scope of this paper. Thus, a few examples shall suffice to outline structure-function relationships of unusual fatty acids and their naturally occurring derivatives.

Plants

The occurrence of unusual fatty acids in plant seeds, e.g., ricinoleic acid in castor, has long been known. A systematic investigation, however, was started only in 1957 when the Northern Regional Research Center of the U.S. Department of Agriculture began a screening program on unusual plant lipids with a goal to alleviate overproduction of agricultural commodities and find new raw materials for industry. As has been published (12, 13), by the late 1970s more than 7,000 species of plants were screened and more than 75 new fatty acids were discovered. Some of the plants identified during this screening now are being studied by breeding stations for commercialization. A particularly important outcome of this work is experimentation on various *Cuphea* species in the U.S. (14) and Europe (15,16) with the target to introduce a new crop for the production of laurics during the next decade.

In the meantime, a project of similar ambition has been initiated in W. Germany under the title of "Renewable Raw

Materials for Industry." Several university groups that consult with four companies will attempt over a five-year period to develop improved oil crops and apply microbial and enzymatic conversion to the upgrading of triglycerides and fatty acids (17).

Information as to why a plant synthesizes an unusual fatty acid so far is scarce. However, in the case of seed lipids of unusual fatty acid structure, it generally is maintained that unusual structures provide stability against premature enzymatic hydrolysis or spoilage from microbial attack. This argument must be reconciled with the fact that the seeds of most plants contain normal triglycerides. Furthermore, little if any work has been undertaken to clarify the specificity of plant and microbial lipases with respect to triglycerides composed of unusual fatty acids.

In a few cases, the biological role of unusual fatty acids in plants has been elucidated. Thus, various epoxy and hydroxy fatty acids identified in the rice plant were found to possess biocidal activity toward the rice plant fungus, *Pyricularia oryzae* (18).

Birds

Whereas minor amounts of branched-chain fatty acids have been observed in many mammals including man (5), preen glands of most birds secrete waxes that predominantly consist of branched-chain fatty acids and alcohols (19). These waxes probably assist in protecting the bird from the cold aqueous environment. As indicated in Figures 2 and 3, branched-chain triglycerides show enhanced spreading power, and branched-chain waxes and triglycerides facilitate perspiration through the skin (20). Synthetic analogues have been studied in cosmetic applications (20,21).

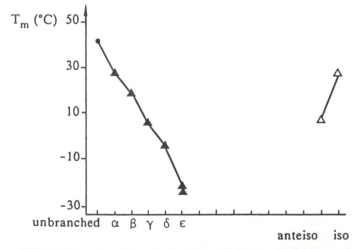

FIG. 2. Phase-inversion temperature (T_m) of methyl-branched fatty acid derivatives (1,2-Diacylglycero-3-phosphocholines). Modified after ref. 6.

Fish

As the ultimate predator in the aquatic food chain, fish accumulate various kinds of unusual fatty acids pro-

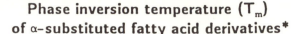

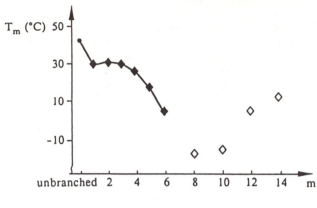

(m = chain-length of α-branch)
* 1,2-Diacylglycero-3-phosphocholines

FIG. 3. Phase Inversion temperature (T_m) of α-substituted fatty acid derivatives (m=chain-length of α-branch). 1,2- Diacylglycero-3-phosphocholines.

duced by the plankton. In addition, fish are able to desaturate fatty acids resulting in levels of 65–80% of highly unsaturated fatty acids in fish oil (22). From the fact that the proportion of such fatty acids is higher during the cold season in fish adapted to cold habitats, it may be inferred that unsaturation plays a crucial role in membrane fluidity. Diets rich in fish oil ("eicosapentenoic acid") are claimed to result in reduced occurrences of heart ischemia (23).

Mammals

Branched-chain fatty acids have been invoked as components of mammalian pheromones, "copulins" and "ecomones." Thus, estrous female goats are interested in 4-ethyl C-8, C-10, C-12, C-14 and C-16 acids secreted by mature male goats (24). Primates, including the human female, secrete branched-chain volatile fatty acids in their vaginas that seem to possess sex-attractant properties (25). And, branched-chain volatile fatty acids such as 2- or 3-methyl C- 4, 2-methyl C-6 and 4-methyl C-9 are major components of the subcaudal gland secretion of badgers, beavers, raccoons and other mustelides that might act as chemical markers ("ecomones") (26,27).

Insects

Ants of the subfamily *myrimcinae* secrete β-hydroxy decanoic acid, a compound exhibiting antimicrobial properties (28,29). The same compound, enriched by other β-hydroxy fatty acids is produced in the abdominal tips of the water beetle *Laccophilus minutus* and is released for defense; the hydroxy fatty acid mixture is assumed to act as both a repellent and an antimicrobial agent (30). The antimicrobial activity of various unusual fatty acids has been studied in detail. Thus, compounds such as 13-methyltetradecanoic acid and 12-methyltridecanoic acid act as potent promoters of chemical biocides such as tetramethylthiuramdisulfide (TMTD) (31).

Microorganisms

The microorganisms are especially versatile in synthesizing phosphoglycerides with unusual acyl residues. In fact, this observation has been used as a taxonomic marker tool for identifying bacterial strains by means of gas chromatography of their fatty acids (32).

In Table 1, the phase transition temperatures of various 1,2-diacylglycero-3-phosphocholines are given. It is clear from this list that microorganisms can "choose" from a considerable variety of phosphoglycerides to optimize their membrane composition for a given environment (33). Thus, an appropriate fluidity of the membrane can be achieved even under hostile environmental conditions such as high or low pH-values (alkalophilic or acidophilic bacteria existing at pH 11 or 2), high or low temperatures (thermophilic or psychrophilic bacteria growing around 90 or 0 C) (34,35) or in the presence of detrimental chemicals such as high concentrations of ethanol (36).

TABLE 1

Transition temperatures (T_m) of Fatty Acid Derivatives* (46–48)

Acyl Residue	Trivial name	T_m (°C)
$C_{18:0}$	stearic	58
$C_{16:0}$	palmitic	41
iso-$C_{17:0}$		28
$C_{14:0}$	myristic	23
C_{12}-ω-cyclohexyl		16
anteiso-$C_{17:0}$		9
iso-$C_{15:0}$		7
$C_{12:0}$	lauric	0
C_{19}-cp, Δ9	dihydrosterulic	-7
C_{19}-cp, Δ11	lactobacillic	-7
anteiso-$C_{15:0}$		-14
$C_{18:1}$, Δ9	oleic	-22
$C_{18:1}$, Δ11	cis-vaccenic	-23
$C_{16:1}$' Δ9	palmitoleic	-40
C_{17}-cp, Δ9		-40

*1,2-Diacylglycero-3-phosphocholines.

POSSIBILITIES FOR THE PREPARATION OF UNUSUAL FATTY ACIDS AND THEIR DERIVATIVES

To date, only a few unusual fatty acids have become available in larger quantities. They are mostly derived from the seed oil of plants presently under agronomical investigation (e.g., Crambe) and from fish oil. In addition, chemical synthesis of unusual fatty acids, e.g., of acetylenic acids (37), has been explored, and in some cases phosphoglyceride derivatives of such fatty acids have been prepared and their properties studied (38,39). However, chemical synthesis often lacks the stereospecificity of biosynthetic pathways and that is often the prerequisite for biological activity.

Biotechnology can contribute to this task in several ways. Thus, eukaryotic microorganisms produce and sometimes store a number of unusual lipids that can be prepared through fermentation as "single-cell oil." In fact,

yields exceeding 100 g/l of γ-linolenic acid have been reported from the fermentation of fungi (e.g., from the *Mortierella* group) with sugars as substrates (40). A second possibility is the use of unusual substrates for fatty acid synthesis. Microorganisms use branched-chain amino acids as precursors for the synthesis of iso- or anteiso-branched fatty acids. And, some strains of *Escherichia coli* are capable of incorporating alicyclic substrates ranging from cyclopropane- to cycloheptane carboxylic acids for the synthesis of the corresponding C_{18} acid derivatives (41). In the case of mid-chain branched or cyclopropyl fatty acids such as tuberculostearic acid, the methyl group of S-adenosyl- methionine is attached to an activated precursor (42). Perhaps the greatest potential lies in the screening of microorganisms capable of functionalizing fatty acid acyl chains. A large project has been initiated on this subject (11), and first results for the position-specific hydroxylation or desaturation of fatty acids by microorganisms have been reported (43, 44). Finally, lipases and phospholipases are prime candidates for transforming unusual fatty acids into the corresponding triglycerides, phophoglycerides or wax esters.

Unusual fatty acids, a very large class of natural compounds, are important "specialties" occurring in many organisms as phospholipid components of membranes, as storage lipids or as waxes with functional properties. In some cases, relationships between the structure and the biological function of such lipids have been discovered; in the vast majority of cases, however, such correlations are as yet unknown. It is probable that from a study of such biological functions, in some cases technical possibilities for the application of unusual fatty acids and their derivatives could be derived. From that point, economic considerations would dictate the use of chemistry or biotechnology for the manufacture of such lipids by extraction, fermentation, biotransformation or bioorganic synthesis.

ACKNOWLEDGMENTS

Special thanks to U. Fischer, U. Griesbach, R. Kindervater and K. Hellberg. This work was supported by Henkel KGaA, Düsseldorf, and by the Ministry of Research and Technology.

REFERENCES

1. Stumpf, P.K., in *The Biochemistry of Plants*, edited by P.K. Stumpf and E.E. Conn, Academic Press, New York, 1980, p. 177.
2. SenGupta, A.K., *Fette Seifen Anstrichm.* 74:693 (1972).
3. Pohl, P., and H. Wagner, *Fette Seifen Anstrichm.* 74:424 (1972).
4. Pohl, P., and H. Wagner, *Fette Seifen Anstrichm.* 74:541 (1972).
5. Nuhn, P., M. Gutheil and B. Dobner, *Fette Seifen Anstrichm.* 87:135 (1985).
6. Werdelmann, B.W., and R.D. Schmid, *Fette Seifen Anstrichm.* 84:436 (1982).
7. Schmid, R.D., *Fette Seifen Anstrichm.* 88:555 (1986).
8. Schmid, R.D., *J. Am. Oil Chem. Soc.* 64:563 (1987).
9. Ratledge, C., *Fette Seifen Anstrichm.* 86:379 (1984).
10. Tsujisaka Y., and M. Iwai, *Yukagaku 31*:772 (1984).
11. Koritala, S., C.W. Hesselteine, E.H. Pryde and T.L. Mounts, *J. Am. Oil Chem. Soc.* 64:509 (1987).
12. Princen, L.H., *J. Am. Oil Chem. Soc.* 61:281 (1984).
13. Princen, L.H., *Economic Botany 37*:478 (1983).
14. Anon., *J. Am. Oil Chem. Soc.* 62:7 (1985).
15. Roebbelen,G., *Fette Seifen Anstrichm.* 86:373 (1984).
16. Hirsinger, F., *J. Am. Oil Chem. Soc.* 62:76 (1985).
17. Biehl,H.M., and E. Sanders, in *Nachwachsende Rohstoffe*, WeKa Druck Linnich,1986.

18. Kato,T., Y. Yamaguchi, T. Uyehara and T. Yokoyama, *Tetrahedron Letters 24*:4715 (1983).
19. Jacob, J., in *Fortschritte in der Chemie organischer Naturstoffe*, Vol. 30, edited by W. Herz, H. Grisebach and G.W. Kirby, Springer-Verlag Heidelberg, 1972, p. 374.
20. Huettinger, R., *Parfumerie und Kosmetik 63*:326 (1982).
21. Berg, A., *Dragoco report*, p. 159 (1976).
22. Stoffel, W., and E.H. Ahrens, Jr., *J. Am. Oil Chem. Soc. 80*:6604 (1958).
23. Tamura, Y., A. Hirai, T. Terano, M. Takenaga, H. Saitoh, K. Tahara and S. Yoshida, *Progr. Lipid. Res. 25*:461 (1986).
24. Sugiyama, T., *Kagaku to Seibutsu 21*:428 (1983).
25. Michael, R.P., *J. Steroid Biochem. 6*:161 (1975).
26. Schildknecht, H., and H. Hiller, *Chemiker-Zeitung 108*:1 (1984).
27. Schildknecht, H., and J. Ubl, *Chemiker-Zeitung 109*:135 (1985).
28. Schildknecht, H., and K. Koob, *Angew. Chem. 82*:181 (1970).
29. Schildknecht, H., and K. Koob, *Angew. Chem. 83*:110 (1971).
30. Schildknecht, H., H. Winkler and U. Maschwitz, *Z. Naturforsch. 23b*:637 (1968).
31. Larsson, K., B. Noren and G. Odham, *Biochem. Pharmacol. 21*:947 (1972).
32. Hewlett Packard Co., *HP 5898A Microbial Identification System*, (company brochure).
33. Thompson, G.A., in *Low Temperature Stress Crop Plants: Role of Membranes*, edited by J.M. Lyons, D. Graham and J.K. Raison, Academic Press, New York, 1979, p. 347.
34. Heinen W., *Forum Städte-Hygiene 35*:167 (1984).
35. Oshima, M., and A. Miyagawa, *Lipids 9*:476 (1974).
36. Bringer, S., T. Härtner, K. Poralla and H. Sahm, *Arch. Microbiol. 140*:312 (1985).
37. Nakatani, M., Y. Fukunaga, H. Haraguchi, M. Taniguchi and T. Hase, *Bull. Chem. Soc. Jpn. 59*:3535 (1986).
38. Blume A., R. Dreher and K. Poralla, *Biochim. Biophys. Acta 512*:489 (1978).
39. Eibl, H., *Angew. Chem. 96*:247 (1984).
40. Suzuki, O., *Hakko to Kogyo 43*:1024 (1985).
41. Jaureguiberry, G., M. Lenfant, R. Toubiana, R. Azerad and E. Lederer, *Chem. Commun. 23*:855 (1966).
42. Dreher, R., K. Poralla and W.A. Koenig, *J. Bacteriol. 127*:1136 (1976).
43. Soda, K., in *Proceedings of the World Conference on Biotechnology for the Fats and Oils Industry*, edited by T.H. Applewhite, American Oil Chemists' Society, Champaign, IL, 1988, pp. 178-179.
44. Yamada, H., in *Proceedings of the World Conference on Biotechnology for the Fats and Oils Industry*, edited by T.H. Applewhite, American Oil Chemists' Society, Champaign, IL, 1988, pp. 178-179.
45. McGarrity, J.T., and J.B. Armstrong, *Biochim. Biophys. Acta 640*:544 (1981).
46. Lewis, R.N.A.H., and R.N. McElhaney, *Biochem. 24*:2431 (1985).
47. Lewis, R.N.A.H., and R.N. McElhaney, *Biochem. 24*:4903 (1985).

Production of Arachidonic Acid and Eicosapentaenoic Acid by Microorganisms

Hideaki Yamada, Sakayu Shimizu, Yoshifumi Shinmen, Hiroshi Kawashima and **Kengo Akimoto**

Department of Agricultural Chemistry, Kyoto University, Sakyo-ku, Kyoto 606, Japan

Arachidonic acid, bishomo γ-linolenic acid and eicosapentaenoic acid (EPA) are C-20 fatty acids that are important as precursors of prostaglandins. The dietary effects of these C-20 fatty acids recently have attracted increasing interest. Currently, several protozoal and fish oil products are available as sources of these fatty acids. From an economic point of view, however, these sources are not so suitable for large-scale preparation. Recently, we have found that fungal mycelia are rich sources of these fatty acids, which would make preparation of arachidonic acid, bishomo γ-linolenic acid and EPA far simpler.

A soil isolate, *Mortierella alpina* 1S-4, was found to show high productivity of arachidonic acid through our screening of a wide variety of microorganisms. The production of arachidonic acid reached 3.6 g/l (147 mg/g dry mycelia) when the fungus was grown in a medium containing glucose and yeast extract as the main carbon and nitrogen sources, respectively. The fungus produced EPA together with arachidonic acid when grown at low temperature. Therefore, we assayed the productivity of EPA in various fungal strains under growth conditions at low temperature and found that *M. alpina* 20-17 grows well at low temperature (6–24 C) and accumulates more than 0.5 g/l of EPA (27 mg/g dry mycelia). Bishomo γ-linolenic acid also was accumulated in mycelia of *M. alpina* 1S-4 in a high concentration (0.6 g/l, 28 mg/g dry mycelia). These fatty acids were found to accumulate especially in polar lipids in mycelial membrane. The fatty acids were isolated as either oils or highly pure fatty acid esters from the mycelia in high yields. High productivity of the fatty acids shows their practical significance.

The C-20 fatty acids were desaturated on incubation with microbial acyl CoA synthetase and acyl CoA oxidase. The former enzyme catalyzed the thioester formation of the fatty acids with CoASH to yield the corresponding acyl CoAs, which were in turn desaturated by the latter enzyme. The double bond was specifically introduced between the C_2 and C_3 positions of the fatty acyl CoAs, yielding 2,3-*trans*-fatty enoyl-CoAs.

Arachidonic acid (all *cis*-5,8,11,14-eicosatetraenoic acid) and eicosapentaenoic acid (all *cis*-5,8,11,14,17-eicosapentaenoic acid, EPA) are polyunsaturated fatty acids (PUFAs) with 20 carbon atoms and 4 and 5 double bonds, respectively, and have been shown to exhibit several unique biological activities. In particular, the dietary effects of these PUFAs recently have attracted increasing interest (1–4). In addition, they are natural precursors of a large family of structurally related C-20 compounds that include prostaglandins, thromboxanes, leucotrienes and prostacyclins, all of which are potent biological regulators (5,6).

The available lipid sources relatively rich in these C-20 PUFAs are animal tissues and algal cells (see Table 1). For practical purposes, however, these conventional sources are not satisfactory, either in their lipid contents or the PUFA contents of the resultant lipids. To obtain more suitable sources for large-scale preparation of these

TABLE 1

Novel Sources of PUFAs

Source	PUFA type	FA content		Potential PUFA productivity	
		Total (mg/g)	PUFA (% in total FA)	(mg/g)	(g/l)
Fungi					
Mortierella isabelina (7)	C18:3	553	10	55	5.0
Penicillium cyaneum (8)	C20:4	90	3-11	1	—
M. elongata 1S-5 (9)	C20:4	88	25	22	1.0
M. alpina 1S-4	C20:3	456	6.6	28	0.6
M. alpina 1S-4	C20:4	420	35	147	3.6
M. alpina 20-17	C20:5	243	10.9	27	0.5
Others					
Chlorella minutissima (10)	C20:5	—[c]	35-40	—	—
Porphyridium cruentum (11)	C20:4	30-220	5-36	20-80	0.07
Euglena gracilis (12)	C20:4	—	5-15	—	0.1-0.5
Seed oil[a]	C18:3	—	6-10	—	—
Porcine liver	C20:4	65	2-8	0.5	—
Fish oil[b]	C20:5	50-150	10-15	20	—

[a]*Oenothera biennis.*
[b]*Scomber scrombrus.*
[c]Not reported

PUFAs, we have screened for microorganisms capable of accumulating lipids containing PUFAs (9). Very little effort has been made so far to produce C-20 PUFAs using microorganisms, in spite of their extremely high growth rates in simple media and the simplicity of their manipulation. We have now found that several fungal strains produce large amounts of lipids rich in arachidonic acid or EPA, or both. Here we summarize the results of our recent work, which included the selection of potent C-20 PUFA producers, determination of the culture conditions for PUFA production, preparation of PUFA-rich lipids and enzymatic desaturation of the resultant PUFAs to yield novel PUFAs.

SCREENING OF PUFA-PRODUCING MICROORGANISMS

As the first step of the screening, about 300 fungal microorganisms capable of growing on an agar plate containing 1% stearic or elaidic acid as the main carbon and energy source for growth were isolated from natural sources for evaluation. Stock cultures in our laboratory (AKU culture collection) also were subjected to the screening. Each strain was inoculated into a test tube (16 x 165 mm) containing 2 ml of YM medium, comprised of 1% glucose, 0.5% peptone, 0.3% yeast extract and 0.3% malt extract, pH 5.0, or P medium, composed of 0.1% glucose, 1% peptone, 0.5% yeast extract and 0.5% NaCl, pH 7.2. Following 4–7 days of incubation at 28 C with shaking (240 stroke/min), the cells were harvested, washed with water and dried in a centrifugal evaporator at 50–60 C. Fatty acids in the cells were subjected to methanolysis, extracted with n-hexane and then analyzed by gas-liquid chromatography (9).

We tested 610 stock strains comprising 40 strains of bacteria (23 genera), 50 strains of actinomycetes (4 genera), 120 strains of yeasts (20 genera), 40 strains of basidiomycetes (11 genera) and 350 strains of filamentous

fungi (45 genera) and 324 fungal isolates. Bacteria, actinomycetes, yeasts and basidiomycetes, in general, did not produce detectable amounts of PUFAs intracellularly. Most PUFA producers were found to be filamentous fungi belonging to the orders *Mucorales* and *Entomophthorales*. Through this screening, we found that 60 strains of *Mortierella* and 45 isolates produce relatively large amounts of C-20 PUFAs (arachidonic acid and bishomo γ-linolenic acid) together with C-18 PUFAs (mainly γ-linolenic acid). Most of the PUFA-producing isolates were found to belong to the genus *Mortierella*. It should be noted that all of these *Mortierella* strains found as C-20 PUFA producers belong to the subgenus *Mortierella*. Neither the stock cultures nor isolates belonging to the subgenus *Micromucor* showed any detectable accumulation of C-20 PUFAs, although they produced C-18 PUFAs such as γ-linolenic acid well. Isolates 1S-4, 1S-5 and 20-17 were taxonomically identified as *Mortierella alpina*, *M. elongata* and *M. alpina*, respectively. The arachidonic acid contents of most of these strains accounted for more than 15% of the total extractable fatty acids. These values represent more than 50% of the PUFAs and are particularly high when compared with those in the case of C-18 PUFAs (Table 2). No detectable amount of free fatty acids was found in lipid fractions extracted with chloroform-methanol, suggesting that the PUFAs produced are present as esters (9,13,14).

PRODUCTION OF ARACHIDONIC ACID

First, the culture conditions for the production of arachidonic acid were optimized. The results with *M. elongata* 1S-5 are summarized below.

Effect of the culture time.

M. elongata 1S-5 grew rapidly in YM medium; after four

TABLE 2

Arachidonic Acid Production by Selected Fungi and Their Cellular Fatty Acid Compositions

Strain	Medium	Arachidonic acid productivity		Fatty acid composition (%)						
		(mg/g)	(mg/ml)	20:4	16:0	18:0	18:1	18:2	18:3[a]	Others[b]
M. elongata IFO 8570	YM	nd	0.10	16.5	9.4	3.5	50.9	8.2	3.5	8.0
	GY[c]	26.9	0.23	13.9	25.2	4.3	49.0	3.5	2.6	1.5
M. exigua IFO 8571	YM	nd	0.01	22.5	15.1	4.3	36.7	10.1	3.6	7.7
M. hyalina IFO 5941	YM	nd	0.04	18.3	21.1	5.6	32.4	5.6	4.2	12.8
	GY	32.8	0.23	17.9	24.7	2.9	37.4	9.5	5.5	2.1
Isolate 1S-4	YM	nd	0.10	25.6	14.1	9.0	32.1	7.7	5.1	6.4
Isolate 1S-5	YM	nd	0.16	15.2	14.2	4.3	51.1	4.0	3.8	7.4
	GY	17.6	0.13	19.4	20.4	12.0	37.6	4.3	4.4	1.9
Isolate 2S-13	YM	nd	0.12	19.8	5.5	2.2	48.4	5.5	14.3	4.3
Isolate 2O-17	YM	nd	0.005	34.9	4.7	2.3	18.6	7.0	23.2	9.3

nd, not determined.
[a]γ-linolenic acid.
[b]Mainly bishomo γ-linolenic acid.
[c]Medium GY contained 2% glucose and 1% yeast extract, pH 6.0.

days of cultivation, the dry mycelial mass reached 20 mg/ml. The arachidonic acid content of the mycelia increased in parallel with growth, reaching the maximum on the fourth day (0.2 mg/ml, 12.4 mg/g dry mycelia).

Effects of growth temperature and pH.

The amount of arachidonic acid produced increased in parallel with mycelial growth, reaching the maximum at 24 C on incubation for 4-7 days (0.25 mg/ml, 12.1 mg/g dry mycelia). Cultivation at higher temperatures resulted in decreased accumulation of total lipids in mycelia, although mycelial growth was rapid and dense. The PUFA content of the lipids was also markedly decreased at these temperatures. The suitable initial pH for the growth and production was found to be between 5 and 7.

Effects of carbon sources.

Several carbon compounds were tested as carbon sources for growth and arachidonic acid production. Glucose, maltose, glycerol, n-hexadecane and n-octadecane were found to be effective carbon sources. Among them, glucose (2-10%) was most effective as to both mycelial growth and arachidonic acid production. The maximum production of arachidonic acid amounted to 0.62 mg/ml (19.1 mg/g dry mycelia) under the conditions of a glucose concentration of 10% and seven days of cultivation.

From the results of studies on individual and combined factors affecting arachidonic acid production previously described, the optimal culture conditions were determined to be as follows: 50 ml of a medium containing 10% glucose, 0.5% peptone and 0.3% yeast extract, pH 6.0, in a 500 ml shaking flask inoculated with M. elongata 1S-5 and shaken at 120 strokes/min. After four days at 24 C, the intracellular production of arachidonic acid reached 0.99 mg/ml (22 mg/g dry mycelia). This value accounts for about 25% of the total extractable fatty acids.

The arachidonic acid methyl ester (37.5 mg) was obtained from the lipids extracted from 236.3 g of wet mycelia of M. elongata 1S-5 grown under the optimal culture conditions. Analysis of the isolated arachidonic acid methyl ester by PMR and gas-liquid chromatography-mass spectrometry showed it was identical to authentic methyl arachidonate.

Fractionation of the extracted lipids by thin layer chromatography demonstrated that triglycerides (73.8%) and phospholipids (21.6%) were their major components, and 50.0% of the arachidonic acid extracted was present in the phospholipid fraction.

In a similar manner, the culture conditions for arachidonic acid production with other strains were optimized. As shown in Table 3, all of the strains produced more than 1 g/l of arachidonic acid under 5 liter bench scale fermentor conditions. Based on these results, we selected M. alpina 1S-4 as the most promising producer. At present, using a 2,000 liter fermentor, 22.5 g/l mycelia (dry wt) containing 44.0% by weight of the lipids can be produced under the conditions of continuous feeding of glucose and 10 days of cultivation at 28 C. The major fatty acids in the resultant lipids are palmitic (13.5%), stearic (9.0%), oleic (20.5%), linoleic (12.0%) and arachidonic (31.0%).

PRODUCTION OF EPA

Because we could not find any strains capable of accumulating a detectable amount of EPA under the above screening conditions, we investigated the growth conditions causing compositional changes in cellular fatty acids. For this purpose, the 105 arachidonic acid producers obtained through the aforementioned screening were used. It has been suggested that differences in the viscosities of fatty acids are exploited by several biological systems to control cellular membrane fluidity in response to changes in environmental temperature. This can be done through adjustments in the proportions of unsaturated and saturated fatty acids within cells (15). Therefore, we investigated fungal growth and changes in the cellular PUFA at low temperature (14). Lowering of the cultivation temperature led to the additional accumulation of a PUFA with five double bonds, EPA, by all the arachidonic acid producers tested. Results of M. alpina 20-17 are shown in Figure 1. It should be noted that most of these arachidonic acid producers grew well at low temperature (6-12C) and yielded enough mycelia (10-20 mg/ml) after five to seven days of cultivation in YM or GY medium, although the growth rates were somewhat lower than those at higher temperature (24-30 C). Among these strains, M. alpina 20-17 was found to accumulate 0.5 g/l (26.6 mg/g dry mycelia) of EPA on cultivation for seven days at 12 C. This value accounted for 11% of the total fatty acids in the extracted lipids. Other major fatty acids in the lipids were palmitic (6.4%), stearic (4.8%), oleic (3.2%), linoleic (3.1%), γ-linolenic (4.5%) and arachidonic

TABLE 3

Production of Arachidonic Acid Under Optimal Culture Conditions

| Strain | Culture conditions | | | | Productivity | | | | |
	Total glucose used (g/l)	Temp/ (°C)	pH	Culture period (days)	Mycelial yield (g/l)	Total FA (mg/g)	C20:4 content (% in total FA)	C20:4 yield (mg/g)	(g/l)
M. elongata 1S-5	100	24	5.0	4	22	88	25	22.0	1.0
M. alpina 1S-4	50	24	6.0	7	24.5	420	35	147.0	3.6
M. alpina 1S-4	104	28	6.0	10	22	440	31	136.4	3.0
M. alpina 20-17	20	20	5.5	5	10	485	52	252.5	2.5

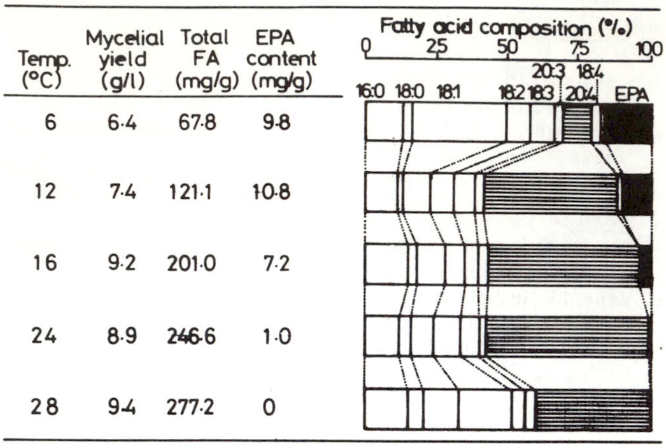

Temp. (°C)	Mycelial yield (g/l)	Total FA (mg/g)	EPA content (mg/g)	Fatty acid composition (%)
6	6.4	67.8	9.8	
12	7.4	121.1	10.8	
16	9.2	201.0	7.2	
24	8.9	246.6	1.0	
28	9.4	277.2	0	

FIG. 1. Effect of growth temperature on the accumulation of EPA. *M. alpina* 20-17 was grown in medium GY (See Table 2) for 6 days at the indicated temperatures.

(63.8%). Cultivation at low temperature significantly increased the mycelial phospholipid content. Over 65% of the EPA was found in the phospholipid fraction.

A concentrate containing 98% EPA methyl ester (5.4 mg) was obtained from the lipids extracted from 39 g wet mycelia of *M. alpina* 20-17.

PRODUCTION OF BISHOMO γ-LINOLENIC ACID

Bishomo γ-linolenic acid (all *cis*-8,11,14-eicosatrienoic acid) is another biologically interesting C-20 PUFA. A practical method for the preparation of this fatty acid has not been reported. We found that *M. alpina* 1S-4 accumulates relatively large amounts of this fatty acid (16). The maximum production of 0.6 g/l was attained under similar 500 ml shaking flask scale conditions as to those given above.

ENXYMATIC TRANSFORMATION OF PUFA

The aforementioned PUFAs may be further transformed enzymatically. We attempted to introduce a C-C double bond into the PUFA molecules on the basis of the following enzymatic reactions:

$$RCH_2CH_2COOH + CoASH + ATP$$
$$\rightarrow RCH_2CH_2COSCoA + AMP + PPi$$
$$RCH_2CH_2COSCoA + O_2$$
$$\rightarrow RCH=CHCOSCoA + H_2O_2$$

Acyl-CoA synthetase (EC 6.2.1.3) catalyzes the thioester formation of a fatty acid with CoASH to yield fatty acyl CoA, which is in turn desaturated by acyl CoA oxidase (EC 1.1.3.-). The double bond is specifically introduced between the C_2 and C_3 positions of the fatty acyl CoA, yielding 2,3-*trans*-fatty enoyl CoA (17).

Previously, we reported that these enzymes are abundantly produced by microorganisms (17–19). Both enzymes have been isolated as crystalline forms and characterized in some detail in our laboratory (17,20), and it has been shown that they are very useful for free fatty acid determination in biological fluids (18,21–23). To use these enzymes for the present purpose, we first tested their reactivities toward fatty acids with more than 18 carbon atoms. As shown in Table 4, hydrogen peroxide production based on the above reactions was observed with each of the tested fatty acids, suggesting that these enzymes can be used for double bond introduction.

Preparative scale synthesis was carried out, at 37 C for eight hr, by incubating arachidonic acid (10 mM) with CoASH (30 mM), ATP (40 mM), MgCl₂ (5 mM), KCl (50 mM), acyl CoA synthetase (*Pseudomonas aeruginosa*, 10 units/ml) and acyl CoA oxidase (*Candida tropicalis*, 5 units/ml) in 100 mM Tris-HCl buffer containing 0.25% Triton X-100. About 80% of the added arachidonic acid was converted to the corresponding 2,3-*trans*-enoyl CoA

TABLE 4

Activation and Desaturation of Long Chain Fatty Acids by Microbial Acyl-CoA Synthetase and Acyl-CoA Oxidase[a]

Fatty acid	Formation of 2,3-*trans* enoyl-CoA (%)[b]	Fatty acid	Formation of 2,3-*trans* enoyl-CoA (%)[b]
18:0	100	20:1 (*cis*-11)	100
18:1 (*cis*-9)	100	20:2 (*cis*-11,14)	100
18:1 (*cis*-6)	100	20:3 (*cis*-8,11,14)	72.6
18:1 (*trans*-9)	100	20:3 (*cis*-11,14,17)	58.8
18:1 (*cis*-11)	100	20:4 (*cis*-5,8,11,14)	79.0
18:1 (*trans*-11)	92.1	20:5 (*cis*-5,8,11,14,17)	75.9
18:2 (*cis*-9,12)	41.6	22:0	43.1
18:2 (*trans*-9,12)	100	21:1 (*cis*-13)	86.4
18:3 (*cis*-9,12,15)	100	22:6 (*cis*-4,7,10,	
18:3 (*cis*-6,9,12)	100	13,16,19)	35.5
18:4 (*cis*-9,15)		24:0	16.3
trans-11,13)	47.9	24:1 (*cis*-15)	34.6
20:0	85.4	28:0	5.6

[a]Each fatty acid (0.1 mM) was incubated under the conditions described previously (23).
[b]100% represents complete conversion of the added fatty acid to the corresponding 2,3-*trans*-enoyl-CoA.

ester. After methanolysis of the reaction mixture and purification by high performance liquid chromatography on an ODS column, the product was isolated as the methyl ester. In a similar manner, γ-linolenic acid, bishomo γ-linolenic acid and EPA were transformed to the corresponding 2,3-*trans*-enoyl fatty acids. Testing of these novel PUFAs as to their biological activities may be interesting.

DISCUSSION

The results reported here show that several dietetically important PUFAs are efficiently produced in the mycelia of the selected fungi. In particular, *M. alpina* 1S-4 was found to accumulate a large amount of arachidonic acid (147 mg/g dry mycelia, 3.6 g/l). The value as to the mycelial content attained is more than 100 times higher than that reported for *Penicillium cyaneum* (8). Other advantageous characteristics of this fungus as an arachidonic acid producer are that the mycelia contain no PUFAs other than arachidonic acid in high concentrations and that contents of undesirable saturated and monounsaturated fatty acids also are low. These characteristics would make the use of *M. alpina* 1S-4 as a source of arachidonic acid very promising when compared with previously reported sources. As to EPA, no reports prior to ours have shown the possibility that microorganisms are practical sources of it.

As shown in Table 2, the mycelia of most of the arachidonic acid producers found in the present study contained small amounts of γ-linolenic acid and bishomo γ-linolenic acid in addition to arachidonic acid. Fungi belonging to the order *Mucorales* are known to contain γ-linolenic acid as a characteristic lipid component (24). Some of them were reported to accumulate it in their mycelia in large quantities (7). These observations suggest that the arachidonic acid producing *Mortierella* strains found in the present study may have the ability to elongate the carbon chain of γ-linolenic acid to yield bishomo γ-linolenic acid, which is followed by its desaturation toward the carboxyl end of the molecule to yield arachidonic acid.

All the arachidonic acid producers tested here accumulated EPA in their mycelia only when they were grown at low temperature. This suggests that arachidonic acid may be a precursor of EPA. If this is the case, cold adaptation may induce activation of enzymes that catalyze the methyl-end directed desaturation of arachidonic acid to yield EPA (25). The resultant EPA may be necessary for proper membrane fluidity in a low-temperature environment. The high proportion of EPA in the phospholipid fraction of the mycelia supports this assumption.

REFERENCES

1. Kromhout, D., E.B. Bosschieter and D. de Lezenne Coulander, *N. Engl. J. Med. 312*:1205 (1985).
2. Phillipson, B.E., D.W. Rothrock, W.E. Connor, W.S. Harris and D.R. Illingworth, *N. Engl. J. Med. 312*:1210 (1985).
3. Wong, S.H., P.J. Nestel, R.P. Trimble, G.B. Storer, R.J. Illman and D.L. Topping, *Biochim. Biophys. Acta 792*:103 (1984).
4. Glomset, J.A., *N. Engl. J. Med. 312*:1253 (1985).
5. Samuelsson, B., in *Les Prix Nobel*, Almqvist and Wiksell International, Stockholm, 1982, p. 153.
6. Vane, J.R., in *Les Prix Nobel*, Almqvist and Wiksell International, Stockholm, 1982, p. 176.
7. Suzuki, O., *Hakko to Kogyo 43*:1024 (1985).
8. Iizuka, H., T. Ohtommo and K. Yoshida, *Eur. J. Microbiol. Biotechnol. 7*:173 (1979).
9. Yamada, H., S. Shimizu and Y. Shinmen, *Agric. Biol. Chem. 51*:785 (1987).
10. Seto, A., *BioIndustry 3*:20 (1986).
11. Ahern, T.J., S. Katoh and E. Sada, *Biotechnol. Bioeng. 25*:1057 (1983).
12. Anonymous, *Nikkei Sangyo 5* (1986).
13. Yamada, H., S. Shimizu and Y. Shinmen, *Abstracts of Annual Meeting of Agricultural Chemical Soc. Jpn.*, Kyoto, 1986, p. 502.
14. Shimizu, S., Y. Shinmen and H. Yamada, *Abstracts of Annual Meeting of Soc. Fermentation Technology, Jpn.*, Osaka, 1986, p. 91.
15. Watanabe, T., H. Fukushima and Y. Nozawa, *Biochem. Biophys. Acta 575*:365 (1979).
16. Shinmen, Y., H. Kawashima, S. Shimizu and H. Yamada, *Abstracts of Annual Meeting of Agricultural Chemical Soc. Jpn.*, Tokyo, 1987, p. 673.
17. Shimizu, S., K. Yasui, Y. Tani and H. Yamada, *Biochem. Biophys. Res. Commun. 91*:108 (1979).
18. Shimizu, S., K. Inoue, Y. Tani and H. Yamada, *Anal. Biochem. 98*:341 (1979).
19. Shimizu, S., H. Morioka, K. Inoue, K. Yasui, Y. Tani and H. Yamada, *Agric. Biol. Chem. 44*:2659 (1980).
20. Yamada, H., and S. Shimizu, *Abstracts of 2nd European Congress of Biotechnology*, Eastbourne, 1981, p. 225.
21. Shimizu, S., Y. Tani, H. Yamada, M. Tabata and T. Murachi, *Anal. Biochem. 107*:193 (1980).
22. Shimizu, S., Y. Tani and H. Yamada, in *Enzyme Engineering*, Vol. 6, edited by I. Chibata, S. Fukui and L.B. Wingard, Jr., Plenum Publishing, New York, 1982, p. 467.
23. Shimizu, S., and H. Yamada, in *Methods of Enzymatic Analysis*, Vol. 8, edited by H.U. Bergmeyer, VCH Verlagsgesellschaft, Weinheim, 1985, p. 19.
24. Show, R., *Biochim. Biophys. Acta 98*:230 (1965).
25. Gellerman, J.L., and H. Schlenk, *Biochim. Biophys. Acta 573*:23 (1979).

Biotransformation of Oleic Acid to Ricinoleic Acid

Kenji Soda

Laboratory of Microbial Biochemistry, Institute for Chemical Research, Kyoto University, UJI, Kyoto-Fu 611, Japan

Various microorganisms were tested for their ability to transform oleic (cis-9-octadecenoic) acid to ricinoleic (12-hydroxy-cis-9-octadecenoic) acid. Ricinoleic acid formed was analyzed routinely by thin layer chromatography (TLC) and reversed-phase high-performance liquid chromatography (HPLC) and identified by gas-liquid chromatography (GLC). We selected three fungal strains, four yeast strains and six bacterial strains as producers of ricinoleic acid. A strain of soil bacterium, BMD I 20, which was incapable of utilizing oleic acid as a sole carbon source, formed the highest amount of ricinoleic acid from oleic acid in the culture fluid. Approximately 3.6 mg/ml of ricinoleic acid were produced when oleic acid was added to the fermentation broth at the stationary phase. Accumulation of ricinoleic acid began about 48 hr after the addition of oleic acid. Resting cells grown in the medium containing oleic acid catalyzed conversion of oleic acid to ricinoleic acid.

A variety of fatty acids is produced directly by microorganisms and secondarily from fats. However, several useful fatty acids are not produced microbially or only in small amounts. It is not easy to modify fatty acids specifically and effectively by chemical procedures; therefore, the biotransformation of abundant fatty acids to more useful ones is very important. Ricinoleic acid is industrially important as a raw material from which coatings, synthetic lubricants and other chemicals are made but it is produced only by higher plants such as castor. This study was undertaken to investigate the microbial transformation of readily available and inexpensive oleic acid to ricinoleic acid by selective 12-hydroxylation.

EXPERIMENTAL PROCEDURES

Media

Medium A contained 6.0% glycerol, 1.0% $(NH_4)_2HPO_4$, 0.2% K_2HPO_4, 0.05% yeast extract, 0.1% $MgSO_4 \cdot 7H_2O$, 0.001% $FeSO_4 \cdot 7H_2O$, 0.0008% $MnSO_4 \cdot H_2O$, 0.0008% $ZnSO_4$ and 0.01% nicotinic acid. The pH of the medium was adjusted to 7.2 for bacteria with 3N NaOH and to 5.5 for yeasts and fungi with phosphoric acid. Medium B was composed of 1.5% polypeptone, 0.2% K_2HPO_4, 0.2% KH_2PO_4, 0.5% NaCl, 0.01% $MgSO_4 \cdot 7H_2O$, 0.01% yeast extract, 0.01% meat extract and 0.1% glycerol (pH, 7.2).

Analytical Methods

The reaction products from oleic acid were detected by TLC on a silica gel plate (Merck Type 5721) with toluene/1,4-dioxane/acetic acid = 18/3.2/1.6 as a solvent mixture and on a RP-C_{18F} plate (Merck Type 15423) with acetonitrile/tetrahydrofuran/0.05% trifluoroacetic acid = 60/25/20 as a solvent mixture. After development, the plates were treated with 2,7-dichlorofluorescein, vanillin and p-anisaldehyde for analyses of carboxylic acids, hydroxy fatty acids and unsaturated fatty acids, respectively. HPLC was carried out with a Toyo Soda CCPD and 8000

High Performance Liquid Chromatograph. Oleic acid and the reaction products were analyzed at 35 C by reversed-phase HPLC with a column of Ultron S-C_{18} (Shinwakako, Japan), and eluted with a system, acetonitrile/tetrahydrofuran/0.05% trifluoroacetic acid = 60/25/20. The eluates were monitored by absorption at 215 nm.

RESULTS AND DISCUSSION

Various microorganisms mainly isolated from soil were examined for their ability to transform oleic acid to hydroxy unsaturated fatty acids. Organisms were grown at 30 C on a reciprocating shaker in 12 ml test tubes containing 3.0 ml of medium A. After one to two days of organism growth, oleic acid was added to the fermentation broth to make the concentration 0.5%. After incubation for three to five days and depending on oleic acid consumption, the culture was acidified with 1N HCl and extracted with 4.0 ml of ethyl acetate. The ethyl acetate layers were analyzed by TLC and reversed-phase HPLC. Among the organisms examined, three fungal strains (FH 31, FH 33 and FH 23), four yeast strains (YHH II 21, YHH I 16, YME I 22, and YMA I 2) and six bacterial strains [BK7 II 14, BMF I 10, BMD I 20, BHB I 6, BD7 I 17 and BKδ (OHL$^+$)] produced hydroxy unsaturated fatty acids that reacted with 2,7-dichlorofluorescein, vanillin and p-anisaldehyde reagents. These were examined in more detail for hydroxy unsaturated fatty acid production. These microorganisms were grown at 30 C for one or two days with shaking in 500 ml flasks containing 100 ml of medium A or B supplemented with 0.5% oleic acid. The hydroxy unsaturated fatty acids produced were determined by reversed-phase HPLC after incubation for 3–5 days with oleic acid consumption. The cells of a bacterial strain, BMD I 20, which were incapable of utilizing oleic acid as a sole carbon source formed the highest amount of

TABLE 1

Production of Ricinoleic Acid

Strains	Pheno-type	Growth of cells or mycelia (wet wt)	Ricinoleic acid formed
BK7 II 14	S^{+a}0^{-b}	0.91 $^{g/100\ ml}$	0.56 $^{mg/ml}$
BMF I 10	S$^+$ 0$^+$	1.4	0.54
BMD I 20	S$^+$ 0$^-$	1.1	2.6
BHB I 6	S$^+$ 0$^-$	1.8	0.55
BD7 I 17	S$^+$ 0$^-$	0.94	0.57
BKδ (OHL$^+$)	OHL$^+$	2.2	0.35
YHH II 21	S$^+$ 0$^-$	3.9	0.1
YHH II 6	S$^+$ 0$^-$	1.6	1.5
YME I 22	S$^+$ 0$^+$	1.0	0.43
YMA I 2	S$^+$ 0$^+$	3.5	0.73
FH 31	S$^-$ 0$^+$	3.5	0.37
FH 33	S$^+$ 0$^+$	0.84	0.20
FH 23	S$^+$ 0$^-$	6.1	0.26

S, Stearic acid; 0, Oleic acid; OHL, β-Hydroxy-β-methyl lauric acid.
[a]Utilized a fatty acid as a sole carbon source.
[b]Not utilized a fatty acid as a sole carbon source.

TABLE 2

Effect of Co-carbon Sources on Ricinoleic Acid Formation in BMD I 20

Co-carbon source	Growth of cells (wet wt) (g/100 ml)	Ricinoleic acid formed (mg/ml)
Glucose	1.1	2.8
Glycerol	1.1	2.4
Sucrose	0.96	1.9
Acetate	1.3	1.4

TABLE 3

Production of Ricinoleic Acid with BMD I 20 Cells

Cells	Addition	Oleic acid remained (mg/ml)	Ricinoleic acid formed (mg/ml)
Grown in the glucose-oleic acid medium	None	0.5	5.8
	Oleic acid	6.3	22.2
Grown in the glycerol-oleic acid medium	None	1.3	2.9
	Oleic acid	21.2	6.0
Grown in the glucose medium	None	0	0
	Oleic acid	20.8	4.3

Incubation conditions: 100 mg of oleic acid, 100 mg of cells, 4.0 ml of potassium phosphate buffer (0.05 M, pH 7.2), 24 hrs, 30 C, on a rotary shaker.

an hydroxy unsaturated fatty acid from oleic acid in the culture medium (Table 1). A hydroxy unsaturated fatty acid produced was identified as ricinoleic acid by GLC of its methyl ester and silylized derivatives. The extract of cells of BMD I 20 grown in the oleic acid medium contained ricinoleic acid although it was not determined quantitatively. When BMD I 20 cells were grown in medium B in which 0.1% glucose was substituted for 0.1% glycerol as a co-carbon source, ricinoleic acid was produced most abundantly (Table 2). Approximately 3.6 mg/ml of ricinoleic acid were produced when oleic acid was added to the BMD I 20 fermentation broth containing glucose as a co-carbon source at the stationary phase (Fig. 1). Accumulation of ricinoleic acid began about 48 hr after addition of oleic acid. Resting cells of BMD I 20 catalyzed the hydroxylation of oleic acid to ricinoleic acid. Those grown in the glucose-oleic acid medium were the most active (Table 3).

Niehaus et al. (1) reported that the cells of a pseudomonad catalyze the interconversion of oleic acid and 10-hydroxystearic acid through an hydration reaction. A hydroxylation reaction of unsaturated fatty acids has been characterized in β- and ω-oxidation systems of fatty acid degradations (2,3). Cells of BMD I 20 probably add molecular oxygen to C_{12} of oleic acid although further study is currently in progress to elucidate the enzymology of the hydroxylation reaction.

REFERENCES

1. Niehaus Jr., W.G., A. Torkelson, A. Kisic, D.J. Bednarczyk and G.J. Schroepfer Jr., *J. Biol. Chem. 245*:3790 (1970).
2. Tahara, S., and J. Mizutani, *Agric. Biol. Chem. 42*:879 (1978).
3. Peterson, J.A., M. Kusunose, E. Kusunose and M.J. Coon, *J. Biol. Chem. 242*:4334 (1967).

Temperature Effects in the Biosynthesis of Unique Fats and Fatty Acids

Saul L. Neidleman
Cetus Corporation, 1400 53rd St., Emeryville, CA 94608

The effects of temperature on the structure of lipids in living systems are illustrated. The dominant response is to increase lipid unsaturation upon exposure to low temperature and to decrease lipid unsaturation upon exposure to high temperature. This relationship is universal, and data for microorganisms, plants and animals are given to support this point. In addition to alterations in lipid unsaturation, other chemical changes in lipid structure in response to temperature stress are noted. These include modifications in chain length, branching, hydroxylation and cyclization in fatty acids. From a commercial point of view, knowledge of the relationship between lipid structure and temperature will suggest approaches for the improved production of desirable fats and oils and the reduction of crop losses due to temperature stress.

Temperature, directly or indirectly, is one determinant in altering the biosynthesis and distribution of fats and fatty acids. Both in nature and in the laboratory, a decrease in temperature may cause liquids to solidify while an increase in temperature may cause solids to liquefy. These gel/liquid crystalline transformations are an integral part of the phase changes by which naturally occurring lipids may respond to an environmental stress such as temperature. In many instances, these phase modifications may be detrimental to the routine performance of life functions and, perhaps, even to life itself. Living systems have devised response networks to retain suitable membranous liquidity or phase relationships and function within their tissues by homeoviscous or homeophasic adaptations (1). One of these biochemical mechanisms in times of thermal stress is an adjustment of lipid unsaturation: low-temperature survival favoring an increase in lipid unsaturation and high-temperature survival favoring a decrease in lipid unsaturation. Lipid changes in response to temperature alteration transcend effects on the degree of unsaturation. They may include, for example, changes in chain lengths of fatty acids, levels of fatty acid branching, hydroxylation and cyclization, and the distribution and relative proportions of members of the glycolipid and phospholipid families (2). At least in some cases, there still remains the question of whether effects of temperature on lipid unsaturation and other structural responses act directly on genetic or biochemical expression, or whether the temperature effect is indirect, with the direct mediator being oxygen availability or cell growth rate, for example.

These reactions to synthesize modified structures and proportions of fats and fatty acids often occur in microorganisms, plants and animals in a universal response to thermal stress. While this affords a survival advantage to living systems, in many instances, such as in production of polyunsaturated fatty acids, unsaturated cooking oils and unsaturated wax esters, there also is the promise of commercial relevance.

TEMPERATURE EFFECTS ON MICROBIAL LIPIDS

Microorganisms have been the primary experimental material for studying the effects of temperature on the chemistry of fatty acids (2,3-6). Some representative data are shown in Tables 1-4. Table 1 illustrates that decreasing temperature increased the amounts of linoleic and linolenic acids at the expense of less unsaturated fatty acids, primarily oleic, in the triglycerides of the fungus *Emericellopsis salmosynnemata* (7). Although the general effect of increasing unsaturation in response to decreasing temperature is prevalent, it should be emphasized that this response may be tempered by species-to-species variation. Table 2 shows that in the fungus *Neurospora crassa* the major effect of low temperature on phospholipid fatty acids was an increase in linolenic acid mainly at the expense of linoleic acid (8), but in Table 3 in the total lipids of the yeast *Hansenula polymorpha*, linolenic acid increased at the expense of a wide variety of less unsaturated fatty acids (9).

TABLE 1

Effects of Temperature on Major Fatty Acids of Triglycerides of the Fungus *Emericellopsis salmosynnemata*

Fatty acid		Relative %		% 20 C - % 36 C
		20 C	36 C	
Palmitic	(16:0)	26.7	29.4	- 2.7
Stearic	(18:0)	9.5	14.3	- 4.8
Oleic	(18:1)	24.6	44.2	-19.6
Linoleic	(18:2)	26.0	12.2	+13.8
Linolenic	(18:3)	13.3	Trace	+13.3

TABLE 2

Effects of Temperature on Major Fatty Acids of Whole Cell Phospholipids of the Fungus *Neurospora crassa*

Fatty acid		Relative %		% 15 C - % 37 C
		15 C	37 C	
Palmitic	(16:0)	13.2	19.0	- 5.8
Oleic	(18:1)	4.7	8.8	- 4.1
Linoleic	(18:2)	39.7	60.6	-20.9
Linolenic	(18:3)	42.5	11.6	+30.9

A more dramatic species difference is shown in Table 4 with the bacterium *Pseudomonas aeruginosa* (10). In this case, decreasing temperature increased not only the more unsaturated palmitoleic and oleic acids at the expense of stearic, palmitic and dodecanoic acid, but also resulted in an increase in hydroxylated fatty acids in the total lipid fatty acids. The insertion of hydroxyl groups, as is the case with double bonds, contributes positively to preserving membrane liquidity at low temperatures.

TABLE 3

Effect of Temperature on the Total Lipid Fatty Acids of the Thermotolerant Yeast *Hansenula polymorpha*

Fatty acid		Relative %		% 20 C - % 50 C
		20 C	50 C	
Myristic	(14:0)	0.3	1.7	− 1.4
Palmitic	(16:0)	16.3	24.4	− 8.1
Palmitoleic	(16:1)	2.3	3.5	− 1.2
Stearic	(18:0)	3.3	7.4	− 4.1
Oleic	(18:1)	24.9	29.8	− 5.1
Linoleic	(18:2)	29.5	33.1	− 3.6
Linolenic	(18:3)	23.4	0.1	+23.3

TABLE 4

Temperature Effects on the Total Fatty Acids of the Bacterium *Pseudomonas aeruginosa*

Fatty acid	nmol fatty acid/mg dry wt of cells		
	15 C	45 C	15 C-45 C
Dodecanoic (C 12:0)	5.0	14.0	− 9.0
Palmitic (C 16:0)	60.0	96.2	−36.2
Stearic (C 18:0)	Trace	4.4	− 4.4
Palmitoleic (C 16:1)	69.7	16.6	+53.1
Oleic (C 18:1)	109.8	78.4	+31.4
3-Hydroxydecanoic (C 10:0, 3-OH)	17.9	12.7	+ 5.2
2-Hydroxydodecanoic (C 12:0, 2-OH)	24.3	13.4	+10.9
3-Hydroxydecanoic (C 12:0, 3-OH)	16.7	16.8	− 0.1

Another example of this response is indicated in Table 5. The effect of decreasing temperature on the bacterium *Thermomicrobium roseum* is to cause an increase in branched chain fatty acids at the expense of straight chain compounds (11). The branched chain acids have lowered melting points. Microorganisms have several strategies, therefore, to produce fatty acids with lower melting points in response to decreasing temperature. These include unsaturation, branching, hydroxylation, cyclization and chain shortening (2,12–14).

TABLE 5

Effect of Temperature on the Straight and Branched Chain Fatty Acids of the Thermophilic Eubacterium *Thermomicrobium roseum*

Fatty acids	Relative %		% 60 C - % 75 C
	60 C	75 C	
Total straight chain	24.1	46.3	−22.2
Total branched chain	75.6	52.4	+23.2
Straight chain C 18:0	15.1	22.0	− 6.9
12-Methyl-C 18:0	66.9	46.0	+20.9

A final and clear example of the inverse relationship between temperature and lipid unsaturation is illustrated in Table 6 in the instance of wax ester biosynthesis from ethanol by *Acinetobacter* sp. H01-N (15). It can be

TABLE 6

Effect of Temperature on Unsaturation of Wax Esters Derived From Ethanol Using *Acinetobacter* sp. HO1-N

Wax ester fraction	% Total		
	17 C	25 C	30 C
Dienes* (32:2, 34:2, 36:2)	72	29	9
Monoenes** (32:1, 34:1, 36:1)	18	40	25
Saturated*** (32:0, 34:0, 36:0)	10	25	66

$*CH_3(CH_2)_7CH=CH(CH_2)_{5-7}CO_2(CH_2)_{6-8}CH=CH(CH_2)_7CH_3$
$**CH_3(CH_2)_7CH=CH(CH_2)_{5-7}CO_2(CH_2)_{15-17}CH_3$ and
$CH_3(CH_2)_{14-16}CO_2(CH_2)_{6-8}CH=CH(CH_2)_7CH_3$
$***CH_3(CH_2)_{14-16}CO_2(CH_2)_{15-17}CH_3$

seen that the percentages of the three classes of wax esters as obtained at 30 C were reversed at 17 C; in the former, the saturated esters dominated, while in the latter, the diunsaturated esters were most prevalent. The application of temperature stress made possible the production of unusual derivatives.

It would be misleading to claim that the inverse relationship between temperature and unsaturation is without exception. Studies with psychrophilic marine *Pseudomonas* sp. showed little increase in fatty acid unsaturation as a result of temperature reduction (16). In the fatty acids of polar lipids of the fungus *Rhizopus arrhizus*, the major effect of lowering the growth temperature was a decrease in unsaturation and an increase in short chain compounds (12).

In addition, the effect of temperature may be mediated through or modulated by a number of other factors that are part of the fermentation process. These include oxygen availability, nature of nutrients, dilution rate and growth stage (2). Two other aspects of the experimental protocol also can be involved: species-to-species variability in response and the lipid material under investigation, whether it be total lipid or a particular subset such as phospholipids, glycolipids, triglycerides or a specific membrane preparation (2). However, given that exceptions and alternatives exist, it is still reasonable to expect that reducing temperature will increase lipid unsaturation.

Temperature Effects on Plant Lipids

In most of the microbiological examples of the effect of temperature on lipid composition, the data are more academic than commercial in value. Certainly, the production of unusual fatty acids by a microorganism under temperature stress is a strong possibility, but much work and many decisions are required. In the case of plants, and in particular crop plants, however, the commercial significance of temperature stress on lipid unsaturation is more obvious.

Table 7 illustrates the effect of temperature stress during the last 10 days of sunflower seed maturation on the percentages of oleic and linoleic acid in the oil: the lower the temperature, the higher the level of linoleic acid (17).

TABLE 7

Effect of Late Maturation Temperature on Fatty Acid Unsaturation in Sunflower Seeds (*Helianthus annuis* Tainan #1)

Late maturation temperature C	% Fatty acid	
(Final 10 days)	Oleic (18:1)	Linoleic (18:2)
25.4	62.2	28.6
24.8	37.1	53.1
23.3	34.1	55.6
19.3	19.9	69.6
18.7	15.2	74.2

TABLE 8

Effect of Temperature on Fatty Acid Unsaturation in American Peanut (*Arachis hypogea*) Oil Triacylglycerol

Peanut Variety	Temperature C*	mol %		
		Oleic acid (18:1)	Linoleic acid (18:2)	mol % 18:1- mol % 18:2
Florigiant	26.6	54.8	26.1	28.7
Florigiant	14.7	44.7	36.7	8.0
Florunner	26.6	52.7	29.7	23.0
Florunner	14.7	44.5	37.4	7.1

*Temperature six wk before harvest.

Similarly, Table 8 indicates that a decrease in temperature six wk prior to the harvest of American peanuts also increased the percentage of linoleic acid with a reduction in oleic acid as in the sunflower example (18). In addition, Table 9 shows effects of varying the day and night temperatures during soybean seed maturation (18); lower temperatures favor an increase in linoleic and linolenic acid at the expense of oleic acid. Therefore, in these three examples of important commercial crop oils, it is clear that temperature during seed maturation has a definite effect on the unsaturation level in the product.

A final example of the effect of temperature on plant lipid unsaturation is the seasonal response of the aquatic fern *Azolla caroliniana* harboring the cyanobacterium *Anabaena azollae*, utilized as a green manure for rice cultivation in the Orient and under study for use in Europe (20). The data in Table 10 illustrate that the colder season (B) caused an increase in linolenic acid and a decrease in palmitic acid. Interestingly, there also is an increase in arachidonic acid.

TABLE 9

Effect of Maturation Temperature on Fatty Acid Unsaturation in Soybean Seeds (*Glycine max* Fiskeby V)

Fatty acid		% Total		
		18/13*	33/28*	% 18/13 - % 33/28
Palmitic	(16:0)	11	12	- 1
Stearic	(18:0)	3	4	- 1
Oleic	(18:1)	13	39	-26
Linoleic	(18:2)	56	40	+16
Linolenic	(18:3)	17	5	+12

*Temperature C day/night.

TABLE 10

Effect of Seasons on Fatty Acid Unsaturation in the Aquatic Fern *Azolla caroliniana*

Major fatty acids	% Total		
	July 15- August 15 (A)	November 15- December 15 (B)	% B-%A
Palmitic (16:0)	51.6	36.6	-15.0
Stearic (18:0)	2.4	0.9	- 1.5
Oleic (18:1)	6.3	8.2	+ 1.9
Linoleic (18:2)	6.8	7.7	+ 0.9
Linolenic (18:3)	12.7	22.4	+ 9.5
Arachidonic (20:4)	1.1	4.0	+ 2.9

*The fern harbors the cyanobacterium, *Anabaena azollae*.

The inverse relationship between temperature and lipid unsaturation is significant for two major reasons in the plant field. The first is that the lipid composition of agricultural products is influenced by the growth and seed maturation temperatures. This strongly supports the existence of a geographical and climatic impact on both product quality and consumer health as affected by the unsaturation level. Lipid unsaturation generally is accepted as desirable from a health standpoint, but, from a product stability point of view, it may be detrimental. The second reason is that high levels of lipid unsaturation are believed by some researchers to favor low-temperature hardiness, resulting in decreased crop losses. While this relationship is still the subject of controversy, considerable work continues to be devoted to genetic and chemical means of increasing lipid unsaturation in crop plants (2).

Temperature Effects on Animal Lipids

Less data are available from the animal kingdom than from the microorganism and plant kingdoms to support the inverse relationship between temperature and lipid unsaturation. However, there is sufficient information to support the universality of the response. The most studied animal is the protozoa *Tetrahymena pyriformis* NT-1. Data on the effect of temperature on total lipid fatty acids are shown in Table 11 (21). It can be seen that palmitoleic and linoleic acids increased at the expense of the less unsaturated palmitic, stearic and oleic acids at the lower temperature. It is of further interest that the shorter-chain myristic acid also increased at the lower temperature, an alternative response also noted in microbial systems.

Another response of an animal system's lipids to temperature is that of female rats shifted to five days at 15 C after 20–25 days at 30 C (22). The data are shown in Table 12 for serum fatty acids. The major increase at the lower temperature is that of arachidonic acid (20:4) with the major decreases shown by palmitic acid (16:0) and oleic acid (18:1).

A final example is among fish in which it has been reported that tissue lipid unsaturation was increased by decreasing environmental temperatures. Representative fish from tropical, deep water and Antarctic locations were compared (23). One unanswered question is

TABLE 11

Effect of Temperature on Fatty Acid Unsaturation in Total Lipids of the Protozoa Tetrahymena pyriformis NT-1

Major	% Total		
fatty acids	20 C	38 C	% 20 C – % 38 C
Myristic (14:0)	22.0	15.5	+6.5
Palmitic (16:0)	19.7	22.8	-3.1
Palmitoleic (16:1)	9.0	2.1	+6.9
Stearic (18:0)	2.4	8.3	-5.9
Oleic (18:1)	5.1	10.3	-5.2
Linoleic (18:2)	21.7	16.2	+5.5
Linolenic (18:3)	18.2	18.1	+0.1

TABLE 12

Effect of Temperature on Unsaturation of Serum Fatty Acids in Female Rats

Fatty acid	mol %		
	15 C	30 C	mol % 15 C – mol % 30 C
16:0	18.3	23.6	-5.3
16:1	1.9	2.9	-1.0
18:0	13.8	13.7	+0.1
18:1	11.3	15.8	-4.5
18:2	26.1	25.9	+0.2
20:3	1.0	0.8	+0.2
20:4	24.1	15.0	+9.1
20:5	1.1	0.7	+0.4
22:6	2.4	1.6	+0.8

whether the effects were directly on the fish or whether they were mediated by the fatty acid composition of the algal diet consumed by the fish (2). A simpler indication of the effect of temperature on lipid unsaturation was a report on the ability of liver microsomes of *Pimelodus maculatus* to unsaturate and elongate 18:1, 18:2 and γ-18:3 fatty acids. Fish kept at 14–15 C had higher activity in both regards than those at 29–30 C (24).

Three regulatory mechanisms have been suggested to occur in eucaryotic cells, such as *Tetrahymena* and mammalian cells. These are modulation of existing desaturase enzyme molecules by membrane fluidity, synthesis of new desaturase enzyme molecules induced by low temperatures and desaturase inhibition by polyenoic fatty acids. In procaryotic cells, other regulatory mechanisms have been offered. In *Escherichia coli* a temperature-labile enzyme, β-ketoacyl-ACP-synthase II (EC 2.3.1.41), specifically increases the rate of unsaturated fatty acid synthesis at low temperatures. In *Bacillus megatherium* desaturase induction is mediated by a temperature-sensitive modulator, temperature labile at 35 C and most active, therefore, at low temperatures. This microorganism has a second mechanism in which the desaturase is high-temperature labile. It has a lower turnover rate at low temperatures, so that when the growth temperature is lowered, more enzyme activity is available and this activity only slowly declines in response to the new growth conditions. This mechanism, using pre-existing desaturase molecules, provides the cell with a rapid response device.

DISCUSSION

The primary objective of this paper is to support the idea that the lipids of living systems can respond in several ways to temperature stress to yield a variety of fatty acids. In particular, the production of unsaturated fatty acids is favored at low temperatures. Therefore, to produce unsaturated lipids, the biosynthetic steps should be performed at the lowest temperature compatible with the demands of technology and economics. The influence of other variables that may modulate the inverse relationship between temperature and unsaturation should not be neglected. These include species variability, nutrient supply, growth rate, age, and oxygen availability, for example. The value of the prospective product will, to a great degree, determine the magnitude and cost of the research that can be tolerated en route to a commercial process.

REFERENCES

1. McElhaney, R.N., in *Biomembranes, Membrane Fluidity*, edited by M. Kates and L.A. Manson, Plenum Press, New York, NY, 1984, p. 249.
2. Neidleman, S.L., *Biotechnol. Genetic Eng. 5*:245 (1987).
3. Herbert, R.A., in *Effects of Low Temperatures on Biological Membranes*, edited by G.J. Morris and A. Clarke, Academic Press, New York, NY, 1981, p. 41.
4. Langworthy, T.A., *Current Topics in Membranes and Transport 17*:45 (1982).
5. Ratledge, C., *Prog. Ind. Microbiol. 16*:119 (1982).
6. Russell, N.J., *Trends Biol. Sci. 9*:108 (1984).
7. Parmegiani, R.M., and M.A. Pisano, *Dev. Ind. Microbiol. 15*:318 (1974).
8. Aaronson, L.R., A.M. Johnston and C.E. Martin, *Biochem. Biophys. Acta 713*:456 (1982).
9. Wijeyaratne, S.C., K. Ohta, S. Chavanich, V. Mahamontri, N. Nilubol and S. Hayashida, *Agric. Biol. Chem. 50*:827 (1986).
10. Kropinski, A.M.B., V. Lewis and D. Berry, *J. Bact. 169*:1960 (1987).
11. Pond, J.L., and T.A. Langworthy, *J. Bact. 169*:1328 (1987).
12. Gunasekaran, M., and D.J. Weber, *Trans. Brit. Mycol. Soc. 65*:529 (1975).
13. Reizer, J., N. Grossowicz and Y. Barenholz, *Biochem. Biophys. Acta 815*:268 (1985).
14. Monteoliva-Sanchez, M., and A. Ramos-Cormenzana, *FEMS Microbiol. Lett. 33*:51 (1986).
15. Neidleman, S.L., and J. Geigert, *J. Am. Oil Chem. Soc. 61*:290 (1984).
16. Bhakoo, M., and R.A. Herbert, *Arch. Microbiol. 126*:5 (1980).
17. Nagao, A., and M. Yamazaki, *Agric. Biol. Chem. 48*:553 (1984).
18. Sanders, T.H., *J. Am. Oil Chem. Soc. 59*:346 (1982).
19. Wolf, R.B., J.F. Cavins, R. Kleinman and L.T. Black, *J. Am. Oil Chem. Soc. 59*:230 (1982).
20. Paoletti, C., F. Bocci, G.Lercker, P. Capella and R. Materassi, *Phytochem. 26*:1045 (1987).
21. Connolly, J.G., I.D. Brown, A.G. Lee and G.A. Kerkut, *Comp. Biochem. Physiol 81A*:287 (1985).
22. Gonzalez, S., A.M. Nervi and R.O. Peluffo, *Lipids 21*:440 (1986).
23. Patton, J.S., *Comp. Biochem. Physiol. 52B*:105 (1975).
24. De Torrengo, M.P., and R.R. Brenner, *Biochem. Biophys. Acta 424*:36 (1976).
25. Cossins, A.R., in *Cellular Acclimisation to Environmental Change*, edited by A.R. Cossins and P. Sheterline, Cambridge University Press, Cambridge, 1983, p. 2.
26. Thompson, G.A., Jr., *Ibid*, p. 33.
27. Brenner, R., *Prog. Lipid Res. 23*:69 (1984).
28. Kasai, R., T. Yamada, I. Hasegawa,Y. Muto, S. Yoshioka, T. Nakamaru and Y. Nogawa, *Biochem. Biophys. Acta 836*:397 (1985).
29. Thompson, G.A., Jr., in *Frontiers of Membrane Research in Agriculture*, edited by J.B. St. John, E. Berline and P.C. Jackson, Rowman and Allenheld, Totowa, 1985, p. 347.

Lipids of *Acinetobacter*

W.R. Finnerty

Department of Microbiology, University of Georgia, Athens, GA 30602

The alkane-utilizing microorganism, *Acinetobacter* species HO1-N, synthesizes a number of unique lipids when grown at the expense of long chain alkanes, fatty alcohols, fatty acids and symmetrical alkoxyalkanes. The metabolism of n-alkanes and fatty alcohols yields wax esters as primary end products. The characterization of peroxypalmitate in lipid extracts of hexadecane-grown cells plus an enzyme that converts peroxypalmitate to palmitaldehyde has led to the possible existence of a new pathway in alkane dissimilation by *Acinetobacter*. Alkane-inducible fatty alcohol dehydrogenase is absent in *Acinetobacter*; whereas, a NADP-dependent fatty aldehyde dehydrogenase and a fatty acyl-CoA reductase are induced by growth on n-alkane, fatty alcohol and fatty aldehyde. Isotope dilution experiments support fatty aldehyde as a central, branch-point intermediate with fatty alcohol a metabolic endproduct rather than a metabolic intermediate. A double mutant of *Acinetobacter* converts fatty acid to short chain alkane and decanedioic acid. These products result from an internal carbon-carbon scission of the endoperoxide, 10-hydroperoxy-*trans*-8-hexadecenoic acid. Further, *trans*-9-hexadecenoic acid and *trans*-9-octadecenoic acid accumulate in the culture broth. Extension of these studies to the metabolism of long chain, symmetrical alkoxyalkanes (C_{14}-C_{20}) demonstrated the oxidation of a homologous series of alkoxyalkanes to alkoxyacetic acids and dibasic acids. The reaction mechanism involves an internal carbon-carbon scission two carbons removed from the ether oxygen. The alkoxyacetic acids accumulate in the culture broth as nonmetabolizable endproducts. Similar bioconversions result from the oxidation of phenoxy-substituted fatty acids to dibasic acids and phenoxy-substituted short chain acids.

The study of microbial hydrocarbon utilization has documented many unique lipid bioconversions in diverse microorganisms (1,2). *Acinetobacter* species HO1-N represents a particularly novel alkane-utilizing microorganism that has received extensive study with respect to its lipid biochemistry (3). This report documents some of the more unusual facets of lipid metabolism in *Acinetobacter* species HO1-N in relationship to alkane oxidation and specialized biotransformations of hydrophobic molecules.

METHODS AND MATERIALS

Acinetobacter species HO1-N was grown at pH 7.8 in a mineral salts medium containing (in g/l): $(NH_4)_2SO_4$, 2; KH_2PO_4, 4; $Na_2HPO_4.7H_2O$, 4; $MgSO_4.7H_2O$, 0.2; $CaCl_2.2H_2O$, 0.001; $FeSO_4.7H_2O$, 0.001. The mineral salts medium was supplemented with 0.3% (vol/vol) n-hexadecane or other hydrocarbon analogues as indicated. Complex growth medium consisted of 0.8% nutrient broth-0.5% yeast extract (NBYE). All cultures were grown on a gyratory shaker and harvested during the late exponential growth phase.

Extraction and Fractionation of Lipids

Cells and spent culture media were extracted and fractionated as previously described (4,5). Peroxy acids were extracted from intact cells with cold concentrated H_2SO_4 for 12 hr. The acid extract was diluted with chipped ice and the peroxy acids extracted into hexane. Thin layer chromatography (TLC) of lipids was as described (4). Thin-layer plates containing silica gel were prewashed with 0.1 M HCl and reactivated before analysis of peroxy acids.

Quantification of Lipids

The procedures for analyses and quantification of various lipids has been described (4-6). Peroxy acids were synthesized as described by Swern (7).

RESULTS

The total cellular and extracellular lipid composition of hexadecane- and hexadecanol-grown *Acinetobacter* sp. HO1-N is shown in Table 1.

Qualitative and quantitative changes in the lipid composition of NBYE-, hexadecane- and hexadecanol-grown cells relate to the growth substrate. A two- to three-fold

TABLE 1

Cellular and Extracellular Lipids of Acinetobacter sp. H01-N

Lipid class	Cellular lipid (μmol/g dry cell weight) NBYE,-hexadecane-, hexadecanol-grown cells			Extracellular lipid μmol/l Culture medium NBYE-, hexadecane-, hexadecanol-grown cells		
Phospholipid	46.00	129.0	100.0	0	0	0
Mono- and diacylglyceride	0.38	6.8	20.4	0	410.0	0
Triacylglycerol	1.78	2.5	8.4	2.4	25.6	30
Free fatty acid	7.50	8.2	0.3	4.0	60.0	0
Free fatty alcohol	trace	2.6	3.5	0	0.5	not determined
Wax ester	11.50	18.0	306.0	0	280.0	280.0
Hexadecane	0	360.0	0	0	not determined	—

increase in cellular phospholipid occurs following growth on hexadecane or hexadecanol in comparison to NBYE-grown cells. Mono- and diacylglycerols increase 18- and 54-fold in hexadecane- and hexadecanol-grown cells, respectively, with a corresponding large accumulation in the culture medium of hexadecane-grown cells. Notably, hexadecane-grown cells produce large amounts of extracellular wax ester (cetylpalmitate); whereas, hexadecanol-grown cells accumulate significantly large amounts of intracellular wax ester (cetylpalmitate) and extracellular wax ester (cetylpalmitate). Interestingly, the intracellular hexadecane and wax ester present in hexadecane- and hexadecanol-grown cells, respectively, are membrane-limited cytoplasmic inclusions (8,9).

Indentification of Peroxypalmitate

Hexadecane-grown cells (10 g) were extracted with cold, concentrated H_2SO_4 to stabilize cellular peroxy acids. The final hexane extract was treated with diazomethane to form the fatty acid methyl esters. Gas chromatographic analyses on polar and nonpolar columns established the presence of a component with the retention time of authentic peroxypalmitate. Treatment of the hexane extract with methanol quantitatively eliminated this putative peroxy acid from the fatty acid profile. Methanol causes the destruction of peroxy acids (7), with authentic peroxy acids reacting identically following methanol treatment. Hexadecane-grown cells were exposed to 2 μcuries of carboxyl-labeled C^{14}-peroxypalmitate. This 1-C^{14}- peroxypalmitate was incorporated into the cellular phospholipids (25% of total radioactivity) as 1-C^{14}-palmitate and into the wax ester cetylpalmitate as 1-^{14}C-palmitate and 1-^{14}C-hexadecanol (44% of total radioactivity). NBYE-grown cells incorporated less than 0.001% of the total radioactivity into cellular lipid.

A cell-free supernatant fraction was prepared by centrifugation of hexadecane-grown cells at 100,000 x g to analyze for an enzyme activity that converted peroxypalmitate to some product. Product formation was determined by GC on a polar 5% DEGS column operating at 150 C. Reaction mixtures (1 ml) were incubated at 30 C for 15 min, stopped by the addition of 2 ml of acidic hexane (1% concentrated HCl in hexane), the hexane phase was recovered and evaporated to dryness and the residue reconstituted in 50 μl of hexane. The chromatographic conditions were optimized for separation of fatty alcohol, fatty aldehyde and free fatty acid. Palmitaldehyde was identified as the product formed in the presence of peroxypalmitate. The formation of palmitaldehyde was dependent on active enzyme because boiled extracts did not catalyze this product's formation.

Fatty Alcohol Dehydrogenase, Fatty Aldehyde Dehydrogenase and Fatty Acid Reductase

Multiple alcohol and aldehyde dehydrogenase activities have been described in Acinetobacter sp. HO1-N (10,11). Hexadecanol dehydrogenase (HDH) was present in extracts of hexadecane- and hexadecanol-grown cells as both a soluble and membrane-bound activity. The specific activity of HDH varied considerably with the growth phase, reaching a maximum level during late-exponential growth and declining in the stationary growth phase (10).

An approximate five-fold induction of HDH activity was observed in hexadecane- and hexadecanol-grown cells in addition to a high constitutive level of hexadecanol oxidation in intact cells. The isolation of mutants lacking HDH have been unsuccessful to date reflecting the possible existence of multiple fatty alcohol dehydrogenase activities with broad substrate specificity.

In contrast with nonspecific fatty alcohol dehydrogenase activity in Acinetobacter, two distinct fatty aldehyde dehydrogenase (FALDH) activities were demonstrated in Acinetobacter sp. HO1-N: a membrane-bound NADP-dependent FALDH activity induced 5-, 9- and 15-fold by growth on hexadecanol, hexadecanal and hexadecane; and a constitutive, NAD-dependent, membrane-localized FALDH. FALDH isoenzymes, ald-a, ald-b, and ald-c were demonstrated in extracts of hexadecane- and hexadecanol-grown cells. Fatty aldehyde mutants were isolated and grouped into two phenotypic classes based on growth. Class 1 mutants were hexadecane and hexadecanol negative and class 2 mutants were hexadecane and hexadecanol positive. The specific activity of the NADP-dependent FALDH in class 1 mutants was 85% lower than that of wild-type FALDH, while the specific activity of class 2 mutants was 55% higher than that of wild-type FALDH.

Fatty aldehyde reductase has been demonstrated in Acinetobacter sp. HO1-N (M.E. Singer, unpublished data). In this reaction, palmitaldehyde is reduced to hexadecanol in the presence of NAD. The enzyme has not been characterized in this system, although its physiological importance has been studied in intact cells. An isotope dilution study of 1-^{14}C-hexadecane oxidation in intact cells demonstrated the existence of an alternative pathway for hexadecane dissimilation in Acinetobacter (M.E. Singer, unpublished data). Hexadecane-grown cells, which were oxidizing 1,2-^{3}H-hexadecane to radioactive oxidation products, were challenged with nonradioactive substrates considered as intermediates in the pathway of hexadecane dissimilation, i.e., hexadecanol, hexadecanal and palmitic acid. The cell suspension was extracted for total lipid after 30 min incubation, saponified and the specific activity of total ^{3}H-fatty acid and ^{3}H-fatty alcohol was determined. The addition of a nonradioactive challenging substrate that is an intermediate in the pathway would reduce the specific activity of subsequently formed intermediates. Addition of a nonradioactive challenging substrate that is not an intermediate in the pathway would not reduce the specific activity of radioactive hexadecane oxidation products. Table 2 shows the specific activities of total ^{3}H-fatty acid and ^{3}H-fatty alcohol derived from control cells and from cells challenged with the nonradioactive intermediates.

Addition of fatty acid reduced the specific activity of ^{3}H-fatty acid 17-fold, but did not significantly reduce the specific activity of ^{3}H-fatty alcohol. Addition of fatty aldehyde reduced the specific activity of ^{3}H-fatty acid and ^{3}H-fatty alcohol 50-fold and 31-fold, respectively. Addition of fatty alcohol reduced the specific activity of ^{3}H-fatty acid only two- to three-fold, regardless of whether exogenous fatty alcohol concentration was high (41.2 mM) or low (8.2 mM). However, exogenously supplied fatty alcohol reduced the specific activity of ^{3}H-fatty alcohol 666- and 1,000-fold at low and high concentrations, respectively.

TABLE 2

Isotope Dilution with Intermediates in the Pathway of Hexadecane Dissimilation in *Acinetobacter* sp. H01-N

Nonradioactive Challenging Substrate	Concentration mM	Specific activity			
		^{3}H-Fatty acid μCuries/mg		^{3}H-fatty alcohol μCuries/mg	
		Control	Experimental	Control	Experimental
Hexadecanoic acid	1.96	10.66	0.64	76.0	47.00
Hexadecanal	9.20	31.50	0.63	1140.0	37.00
Hexadecanol	8.20	8.65	4.19	73.0	0.11
Hexadecanol	41.20	7.11	2.47	21.0	0.02

Fatty Acid Metabolism in Acinetobacter

A study of long-chain fatty metabolism in *Acinetobacter* sp. HO1-N demonstrated repression of fatty acid biosynthesis in alkane-grown cells (12) and the formation of short chain alkanes. Table 3 shows the products resulting from the oxidation of alkanes, fatty acids and phenoxy-substituted fatty acids by *Acinetobacter* sp. HO1-N. The nature of products arising from the oxidation of alkanes, fatty acids and phenoxy-substituted fatty acids suggests a mechanism involving the scission of internal carbon-carbon bonds. Two facts of importance emerge from these results: a short chain alkane and a phenoxy-substituted short chain fatty acid result from the oxidation of alkanes, long chain fatty acids and phenoxy-substituted long chain fatty acids, and alkanedioic acids are produced as oxidation products and appear to provide the principal source of carbon and energy to the cell. The phenoxy-short chain fatty acids appear refractory to further metabolism in that they do not support growth. The short chain alkanes are oxidized to the corresponding short chain acid although these acids do not support cell growth probably as the result of substrate limitation due to alkane volatilization.

TABLE 3

Bacterial Oxidation of Alkanes, Fatty Acids and Phenoxy-Substituted Fatty Acids

Substrate	Products
Pentadecane	Pentane; decanedioic acid
Hexadecane	Hexane; decanedioic acid
Heptadecane	Heptane; decanedioic acid
Pentadecanoic acid	Pentane; decanedioic acid
Hexadecanoic acid	Hexane; decanedioic acid
11-Phenoxy-undecanoic acid	3-Phenoxy-propionic acid; octanedioic acid
12-Phenoxy-dodecanoic acid	4-Phenoxy-butyric acid; octanedioic acid

Genetic and biochemical analyses of hexadecane-grown *Acinetobacter* sp. HO1-N were initiated to provide further understanding into the mechanism(s) of product formation, namely, short chain alkane and alkanedioic acid. A series of mutants that were unable to grow on hexadecane as a sole source of carbon and energy were isolated. One mutant, B2074, was selected for further study due to its inability to use C_{10} through C_{18} alkanes, C_6 through C_{16} alkanedioic acids and C_{12} through C_{18} fatty acids. B2074 was grown on 0.2% acetate and challenged with hexadecane. The culture broths of both acetate-grown and hexadecane-challenged cells were examined qualitatively and quantitatively for fatty acid compositions (Table 4).

TABLE 4

Extracellular Fatty Acids of B2074

Fatty acid	Acetate-grown cells	Hexadecane-challenged cells
	(mg total fatty acid)	
Hexadecanoic acid	61	70
cis-9 hexadecenoic acid	14	16
trans-9 hexadecenoic acid	5	10
Octanedioic acid	not detected	20
Decanedioic acid	8	230
Octadecanoic acid	37	47
cis-9 octadecenoic acid	39	35
trans-9 octadecenoic acid	14	25
Total	178	453

B2074 was characterized as a double mutant (Dca[-], Lfa[-]); one phenotype being dicarboxylic acid-negative (Dca[-]) and long chain fatty acid-positive (Lfa[+]) and the other phenotype being dicarboxylic acid positive (Dca[+]) and long chain fatty acid negative (Lfa[-]). The production of extracellular dodecanedioic acid requires both Dca[-], Lfa[-] loci in that Dca[-], Lfa[+] phenotypes grew normally on alkanes and long chain fatty acid. The accumulation of *trans*-fatty acids in the culture broth is unusual, not having been shown previously in prokaryotic microorganisms.

The derivation of decanedioic acid from uniformly labeled ^{14}C-palmitate by B2074 was demonstrated by the recovery of radioactive decanedioic acid (53%), octanedioic acid (25%) and hexanedioic acid (22%). The major dioic acid cleavage product of fatty acid was decanedioic acid (87%), octanedioic acid (10%) and hexanedioic acid (3%). The minor amounts of C_6 and C_8 dioic acids are attributed to the leaky nature of the Dca[-] locus.

The spent culture broth of hexadecane- or palmitate-challenged B2074 cells yielded two unknown fatty acid components, designated X and Y. Compounds X and Y were highly labile; loss of X and Y followed treatment with bromine, iodine, sulfite, potassium permanganate, sodium hydroxide or zinc. X and Y were separated by TLC and co-chromatographed with 13-hydroperoxy linoleic methyl ester standard.

Thermal decomposition of purified X and Y resulted in the formation of hexane. Permanganate-periodate treatment of X and Y yielded octanedioic acid. These analyses suggested that compound Y is a thermal decomposition product with a structure corresponding to 10-8-oxo-decenoic acid methyl ester.

Oxidation of Alkoxyalkanes by Acinetobacter sp. HO1-N

The oxidation of a homologous series of symmetrical alkoxyalkanes ranging in carbon number from 14- to 20-carbon atoms has been studied in *Acinetobacter* sp. HO1-N (13,14). These compounds were selected due to their structural similarity to the corresponding alkane, differing only by the presence of an oxygen atom symmetrically located in the carbon chain. This study resulted in the description of a new metabolic pathway involving an internal carbon-carbon scission reaction, producing a homologous series of alkoxyacetic acids that accumulated in the culture broth in direct stoichiometric proportions to the disappearance of the substrate plus alkanedioic acids (Table 5). The alkoxyacetic acids are not further metabolized by the cells, whereas the dioic acids serve as the sole source of carbon and energy for growth of the bacteria. The pathway of alkoxyalkane oxidation by *Acinetobacter* is:

$$CH_3(CH_2)_n\text{-}O\text{-}(CH_2)_nCH_3 \rightarrow$$
$$CH_3(CH_2)_n\text{-}O\text{-}(CH_2)_n\text{-}COOH \rightarrow$$
$$CH_3(CH_2)_n\text{-}O\text{-}CH_2\text{-}COOH$$
$$+$$
$$HOOC\text{-}(CH_2)_n\text{-}COOH$$

The production of alkoxyacetic acids as nonmetabolizable endproducts represents a novel bioconversion.

TABLE 5

Products of Alkoxyalkane Metabolism

Substrate	Products
Heptoxyheptane	2-Heptoxyacetic acid; glutaric acid
Octoxyoctane	2-Octoxyacetic acid; adipic acid
Nonoxynonane	2-Nonoxyacetic acid; pimelic acid
Decoxydecane	2-Decoxyacetic acid; suberic acid

Extension of this metabolic reaction to phenoxy substituted long chain fatty acids resulted in the production of 3-phenoxy-propionic acid and suberic acid from 11-phenoxy-undecanoic acid and 4-phenoxy-butyric acid and suberic acid from 12-phenoxy-dodecanoic acid (Table 3).

DISCUSSION

The cellular and extracellular lipids of *Acinetobacter* sp. HO1-N reflect dramatic qualitative and quantitative changes following growth on long chain alkanes. The two-to three-fold increase in cellular phospholipids correlates with the induction of intracytoplasmic membranes in alkane-grown cells (8,15). Such increased phospholipid content also has been observed in hexadecane-grown *A. lwoffi (16)*, in alkane-grown *Candida tropicalis* (17,18) and in *Candida* 107 (19). The increase in cellular neutral lipids documented in *Acinetobacter* have been described for *Mycobacterium vaccae* (20), *Nocardia* (21), *Mycobacterium convolution* (22) and *Candida parapsilosis* (23).

The extracellular lipids of hexadecane-grown *Acinetobacter* contain mono- and diglyceride, wax ester and triglyceride as contrasted with extracellular and intracellular accumulation of wax ester in hexadecanol-grown cells.

The oxidation of alkanes is considered to represent an obligatory oxygen-dependent reaction in which molecular O_2 reacts with alkane to form a n-alkyl-hydroperoxide, the reaction being mediated by a dioxygenase. The alkyl-hydroperoxide is further considered to be reduced to fatty alcohol which, in turn, is oxidized to fatty aldehyde and fatty acid by alcohol dehydrogenase and aldehyde dehydrogenase, respectively. The pathway for carbon transformation is represented by the reaction sequence:

$$R\text{-}CH_3 \xrightarrow{O_2} RCH_2OOH \longrightarrow RCH_2OH \longrightarrow RCHO \longrightarrow RCOOH$$

Alkylhydroperoxide has been implicated as a potential metabolic intermediate by indirect means, having never been isolated from metabolizing cells. The isolation and identification of a possible new intermediate in alkane oxidation, peroxy acid, suggests the possibility of an alternative pathway in bacterial alkane oxidations. The finding of an enzyme activity in hexadecane-grown cells that converts peroxypalmitate to palmitaldehyde suggests the possible oxidation of n-hexadecylhydroperoxide to peroxypalmitic acid, which is converted to palmitaldehyde, eliminating the reduction of alkylhydroperoxide to fatty alcohol. Enzymological and genetic analyses of hexadecane-grown *Acinetobacter* demonstrated multiple alcohol dehydrogenase and aldehyde dehydrogenase activities with the induction of a specific NADP-dependent fatty aldehyde dehydrogenase. Fatty aldehyde mutants with significantly reduced FALDH activity were unable to grow at the expense of either hexadecane, hexadecanol or hexadecanal. We also have described a fatty aldehyde reductase activity that reduces fatty aldehyde to fatty alcohol. Multiple alcohol dehydrogenases have been documented in a number of hydrocarbon-utilizing microorganisms (2). *Pseudomonas aeruginosa* 473 has several pyridine-nucleotide dependent alcohol dehydrogenases of broad substrate specificity and two NAD(P)-independent alcohol dehydrogenases induced by growth on alkanes, alcohols and α, ω-diols (24). In addition, pyridine-nucleotide independent alcohol dehydrogenases have been implicated in bacterial alkane metabolism (25-27). The question of the direct involvement of a hexadecanol dehydrogenase as a principal enzyme in hexadecane oxidation remains unresolved. Currently, it is believed that fatty alcohol (hexadecanol) represents an endproduct rather than a metabolic intermediate in alkane oxidation by *Acinetobacter*. Circumstantial evidence to support this interpretation relates to the large amount of fatty alcohol (hexadecanol) bound in wax ester as well as the isotope dilution experiments that demonstrated a large dilution factor (666- to 1000-fold) effected by nonradioactive hexadecanol.

Aldehyde dehydrogenases have been described in a number of hydrocarbon-utilizing microorganisms, including *Candida tropicalis* (28), *Pseudomonas aeruginosa* (29), *Pseudomonas putida* (30), and *Acinetobacter calcoaceticus* 69V (31). In hexadecane-challenged cells, the prolonged lag period before NADP-dependent FALDH induction corresponds to the extended lag period preceding hexadecane oxidation (10). Thus, hexadecane does not induce FALDH but rather a product of hexadecane oxidation acts as the inducer molecule. Evidence has shown that NADP-dependent FALDH was not induced in hexadecane-exposed, alkane-negative (alk⁻) mutants, but the enzyme was induced in fatty aldehyde exposed cells. The alk⁻ mutants were unable to oxidize hexadecane but were capable of growth on fatty aldehyde, fatty alcohol and fatty acid. These mutants appear to be blocked in one of the initial steps of hexadecane oxidation.

A current working model for hexadecane oxidation in *Acinetobacter* sp. HO1-N is:

$$RCH_3 \xrightarrow[1]{O_2} RCH_2OOH \xrightarrow{2} \underset{RCOOH}{\overset{O}{\underset{\|}{}}} \xrightarrow{3} RCHO \xrightarrow{4}{\underset{5}{}} \begin{matrix} RCH_2OH \\ RCOOH \end{matrix}$$

This model is supported conceptually by the identification of enzyme activities catalyzing steps 3, 4 and 5. Peroxy acids previously have not been described from biological systems, however, most lipid studies employ chloroform/methanol as the extractive solvent system. Peroxy acids are destroyed spontaneously by lower alcohols, essentially eliminating the possibility of recovering such oxygenated lipids.

The study of fatty acid metabolism in *Acinetobacter* has revealed a new fatty acid degradative pathway in prokaryotic microorganisms. This pathway involves an internal carbon-carbon scission reaction of long chain unsaturated fatty acids yielding a short chain alkane and alkanedioic acid. The mechanism resembles a lipoxygenase type reaction. Analyses of an unknown compound isolated from the spent culture broth indicated the structure, 10-hydroperoxy-*trans*-8-hexadecenoic acid. The formation of an endoperoxide from a *cis* double bond is always associated with a *cis* to *trans* and positional isomerization in both autoxidation and lipoxygenase reactions (32). The thermal decomposition of 10-hydroperoxy-*trans*-8-hexadecenoic acid would result in the formation of hexane and 10-oxo-*trans*-8-decenoic acid. The oxidation of this compound would yield *trans*-2-decendioic acid, which upon hydrogeneration forms decanedioic acid. These reactions are not specific to alkane-grown bacteria, occurring with nonhydrocarbon growth substrates such as alkoxyalkanes, fatty acids, succinate and complex media.

The ability of *Acinetobacter* to oxidize alkanes and derivatives thereof to specific lipids is unique. A number of specialty chemicals are produced by *Acinetobacter* through the metabolism of simple substrates. The application of recombinant DNA technology offers the means for significant improvements in product yields. A gene-cloning system has been developed for *Acinetobacter* genes with expression of these genes in *Pseudomonas*. A potential remains untouched for the development of specific microbial processes for new and old products of commercial value and utility.

REFERENCES

1. Finnerty, W.R., in *Biotechnology for the Oils and Fat Industry*, edited by C. Ratledge, P. Dawson and J. Rattray, American Oil Chemists' Society, Champaign, IL, 1984, pp. 199–216.
2. Singer, M.E., and W.R. Finnerty, in *Petroleum Microbiology*, edited by R. Atlas, MacMillan Publishing Co., New York, 1984, pp. 1–59.
3. Finnerty, W. R., and M.E. Singer, in *Organization of Procaryotic Membranes*, Vol. III, edited by B.J.Ghosh, CRC Press, Inc., Boca Raton, FL, 1985, pp. 1–40.
4. Makula, R.A., P.J. Lockwood and W.R. Finnerty, *J. Bacteriol.* 121:250 (1974).
5. Modzrakowski, M.C., R.A. Makula and W.R. Finnerty, *Ibid.* 131:92 (1977).
6. Modzrakowski, M.C., and W.R. Finnerty, *Arch. Microbiol.* 126:285 (1984).
7. Swern, D., in *Organic Peroxides*, Vol. III, edited by D. Swern, Wiley-Interscience, New York, 1970, pp. 313–516.
8. Scott, C.L., and W.R. Finnerty, *J. Bacteriol.* 127:481 (1976).
9. Singer, M.E., S.M. Tyler and W.R. Finnerty, *Ibid.* 162:162 (1985).
10. Singer, M.E., and W.R. Finnerty, *Ibid.* 164:1017 (1985).
11. Singer, M.E., and W.R. Finnerty, *Ibid.* 164:1011 (1985).
12. Sampson, K., and W.R. Finnerty, *Arch. Microbiol.* 99:203 (1974).
13. Modzrakowski, M., R.A. Makula and W.R. Finnerty, *J. Bacteriol.* 131:92 (1977).
14. Modzrakowski, M., and W.R. Finnerty, *Arch. Microbiol.* 126:285 (1980).
15. Kennedy, R.S., and W.R. Finnerty, *Ibid.* 102:83 (1975).
16. Vachon, V., J.T. McGarrity, C. Brevil, J.B. Armstrong and D.J. Kushner, *Can. J. Microbiol.* 28:660 (1982).
17. Hug, H., H. Blanch and A. Fiechter, *Biotechnol. Bioeng.* 16:965 (1974).
18. Mishina, M., M. Isugi, A. Tanaka and S. Fukui, *Agric. Biol. Chem.* 41:517 (1977).
19. Thorpe, R.F., and C. Ratledge, *J. Gen. Microbiol.* 72:151 (1972).
20. Vestal, J.R., and J.J.Perry, *Can. J. Microbiol.* 17:445 (1971).
21. Raymond, R.L., and J.B. Davis, *Appl. Microbiol.* 8:329 (1960).
22. Hallas, L.E., and Vestal, J.R., *Can. J. Microbiol.* 24:1197 (1978).
23. Omar, S.H., and H.J.Rehm, *Eur. J. Appl. Microbiol. Biotechnol.* 11:42 (1980).
24. Van der Linden, A.C., and R. Huybregtse, *Antonie van Leeuwenhoek, J. Microbiol. Serol.* 35:344 (1969).
25. Tauchert, H., M. Grunow and H.Aurich, *Z. Allg. Mikrobiol.* 18:675 (1978).
26. Tauchert, H., M. Roy, W. Schopp and H. Aurich, *Ibid.* 15:457 (1975).
27. Benson, S., and J. Shapiro, *J. Bacteriol.* 126:794 (1976).
28. Yamada, T., H. Nawa, A. Kawamoto, A. Tanaka and S. Fukui, *Arch. Microbiol.* 128:145 (1980).
29. Guerrillot, L., and J.-P. Vendecasteele, *Eur.J. Biochem.* 81:185 (1977).
30. Baptist, J.N., R.K. Gholson and M.J. Coon, *Biochim. Biophys. Acta* 69:40 (1963).
31. Aurich, H., and G. Eitner, *Z. Allg. Microbiol.* 17:263 (1977).
32. Privett, O.S., C. Nickell, W.O. Lundberg and P.D. Boyer, *J. Am. Oil Chem. Soc.* 32:505 (1955).

Strategies for Biosurfactant Production

Fritz Wagner

Institute of Biochemistry and Biotechnology, Technical University, D 3300 Braunschweig, Federal Republic of Germany

Extracellular amphiphilic microbial metabolites are called biosurfactants. Biosynthesis of biosurfactants is often associated with microbial growth on lipophilic substrates. Depending on the microbe used, the synthesis of a biosurfactant can be growth- or nongrowth-associated and enhanced by growth-limiting conditions or in the presence of a precursor. After a short account on the general mechanisms of biosynthesis in the field, special strategies are presented for the overproduction of glycolipids as the main group among biosurfactants. We also discuss screening methods for glycolipid-producing microorganisms in situ extraction of the growth correlated and cell wall associated formation of α, α-trehalose-6,6'-dicorynomycolates and tailoring the surfactant properties in the case of the anionic trehalose-tetraester produced under growth-limiting conditions. Further strategies take into account the overproduction of rhamnolipids with resting cells, the stabilization of the biocatalyst by immobilization techniques of viable cells and a process design for the continuous production of rhanolipids with immobilized viable cells. An integral part of the process is the continuous on-line product recovery by foam flotation.

Biosurfactants are produced by a variety of microorganisms including bacteria, yeast and fungi. Basically, there are five major classes of biosurfactants: Glycolipids, Lipopeptides/lipoproteins, Lipopolysaccharides, Substituted fatty acids and Phospholipids.

The topic of these extracellular amphiphilic metabolites—their production, properties, and utilization for various applications—was presented recently in a comprehensive treatise by Kosacic et al. (1). Biosynthesis of biosurfactants is often associated with microbial growth on hydrocarbons or other lipophilic substrates. The phys-

iological role of biosurfactants in hydrocarbon metabolism is connected with the mechanism for the initial interaction of the lipophilic substrate and the microbial cell (2,3). Besides hydrocarbons, nonhydrocarbon substrates such as fats, oils, glycerol or carbohydrates can serve as carbon substrates for the microbial synthesis of biosurfactants (4). Depending on the microorganism used, the biosynthesis of a biosurfactant can be growth or nongrowth associated and enhanced by growth-limiting conditions or by the addition of a precursor.

Microbial surfactants are generally extracellular amphiphilic metabolites. The hydrophilic part of the molecule can be a mono-, oligo- or polysaccharide, an amino acid, cyclic peptide, protein or a phosphoester. The lipophilic portion of the molecule usually consists of one or more long chain fatty acids, hydroxy fatty acids or α-alkyl-β-hydroxy-fatty acids with a carbon number in the range of 30–40 that are esterified via the acyl group or glycosidically linked to the hydrophilic portion (5). In Table 1, a few examples of chemical-characterized biosurfactants are presented.

In many cases, the biosynthesis of the hydrophilic as well as hydrophobic portion of a biosurfactant derives directly from the microbes' primary metabolism. This is the reason that there are some common rules concerning the metabolic pathways and regulation in the multistep synthesis of these metabolites (6,7). The chief regulatory mechanisms that control the synthesis of biosurfactants are induction, nitrogen catabolite regulation and feedback regulation. Thus, microbes have an amazing ability to change their metabolic activities leading to overproduction of metabolites in response to changes in their environment.

Another aspect concerning the physical properties of biosurfactants is the alteration of the product composition from a certain microbe by variation of the carbon substrate or the culture conditions. With these methods we have in hand the abilities to change the structure of a

TABLE 1

Example of Chemical Characterized Biosurfactants (4,5)

Biosurfactant	Fatty acid portion	Type of linkage	Charge	Location
Cellobioselipids	15,16-dihydroxy-hexadecanoic + β-hydroxy-C-6 and -C-8	glycosidic + ester	anionic	extracellular
Rhamnolipids	β-hydroxy-decanoic	glycosidic + ester	anionic	extracellular
Sophoroselipids	17-hydroxy-octadecanoic	glycosidic + Lacton	nonionic	extracellular
Trehalose-diester	Corynomycolic[a]	ester	nonionic	cell-wall
Trehalose-tetra-ester	C-7→C-18 + Succinate	ester	anionic	extracellular
Mono-, di-, tri-saccharides	Corynomycolic	ester	nonionic	cell-wall
Surfactin	β-hydroxy-C-15 + peptide	ester + amide	anionic	extracellular

[a] β-Alkyl-β-hydroxy-fatty acid C-32 to C-40.

certain biosurfactant and to tailor the surfactant properties. Glycolipids are the greatest group among biosurfactants. This presentation is primarily concerned with different strategies for the production of glycolipids which were worked out mainly in our laboratory.

RESULTS AND DISCUSSION

General screening of glycolipid-producing microorganisms

Classical enrichment of techniques designed to isolate microbes that grow at the expense of hydrophobic carbon substrates were applied with success to those from natural sources such as oil-contaminated and normal soils, refinery effluent waste, sewage sludge, sea water and plant material. It is fortunate that the last two transfer steps during this type of enrichment procedure are carried out under medium conditions with a carbon to nitrogen ratio of 20:1. This ratio takes into account the positive influence of nitrogen limitation on the regulation of the biosynthesis of biosurfactants (7). Under such growth conditions, more than 60% of the pure cultures selected are normally positive for glycolipid formation. The glycolipid production is detected after solvent extraction of the whole broth, with dichloromethane/methanol (2:1, v/v) evaporation of the organic solvent and thin layer chromatography (TLC) of the residue. The sugar moiety of glycolipids is detected by spraying the plates with anisaldehyde/sulfuric acid reagent. The lipid portion at the same spots is detected by spraying them with 2', 7'-dichlorofluorescin (8). With these general procedures, both nonionic and ionic glycolipids are detected independent of their location. The glycolipids can be cell wall associated or in solution in the culture suspension.

Direct screening for water-soluble glycolipids

One approach to detect water-soluble and diffusible glycolipids was worked out by Singer et al. (9). The rapid assay is based on the ability of surfactants to lyse red blood cells. A red blood cell agar plate is inoculated with microbial isolates from enrichment cultures. Incubated isolates exhibit clear zones surrounding the colonies due to the red blood cells lysis during growth on the plate indicating the formation of a surface active compound. Besides detecting water-soluble, mostly ionic glycolipids, the test also is positive if lipopeptides are produced. Microorganisms that form cell-wall associated glycolipids that are not diffusible or which produce surfactants only during growth on hydrocarbons are not detected by this assay.

Direct screening for cell-wall associated glycolipids

This approach is adopted from a procedure published by Matsuyama et al. (10). Agar nutrient plates, for example, with different carbon sources, are inoculated with microbial isolates from enriched cultures and incubated for a certain time to generate robust colonies on the agar plate. Each colony is transferred directly on a silica gel TLC plate and predeveloped with n-hexane or a mixture of n-hexane/dichloromethane. After drying this predeveloped TLC plate, the cell mass is removed and the plate

is developed with dichloromethane/methanol (2:1, v/v). Sequential examination with reagents for glycolipids is carried out to characterize the spots developed. For further separation and characterization of glycolipids, a two-dimensional TLC can be utilized.

With these simple and rapid screening methods, a higher yield of microbes producing biosurfactants, especially low molecular weight compounds, will be found with a variation of the nutrient composition because these microbes synthesize biosurfactants depending on the culture media.

Production of microbial glycolipids

In this section, some general aspects of the practical applicability for the production of glycolipids are considered and a few processes are selected for illustration of different strategies. Mainly, two topics for a process design have to be taken into consideration. First, from a physiological view, the product formation can be growth-correlated, uncoupled from the growth phase or under resting cell conditions. Second, the product is located on the outer cell wall or is secreted in the surrounding medium. It is obvious that different strategies must be followed to optimize the glycolipid production.

Growth-correlated glycolipid formation

A typical example of a growth-correlated and cell-wall associated glycolipid production is the formation of α,α'-trehalose-6,6'- dicorynomycolate (Fig. 1) by *Rhodococcus*

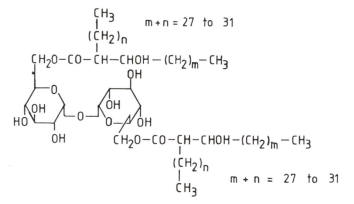

FIG. 1. Trehalose-6,6'-dicorynomycolate from *Rhodococcus erythropolis* (11).

erythropolis in a mineral salt medium on an n-alkane fraction C-13 to C-16 (2,11). The growth and product kinetics for the batch cultivation shown in Figure 2 can be divided into four phases. In the first phase, the exponential growth takes place in the given alkane droplets with a specific growth rate of 0.22 h^{-1}. This phase is characterized by a strong orientation of the cell toward the hydrocarbon droplets. At a certain proportion between the concentrations of hydrocarbons, cell mass and glycolipid, an oil/water phase inversion is obtained and large clumps of bacteria are built up. At this stage the growth rate is nearly zero due to nutrient limitation while the production rate is still increasing. After further decreases of the

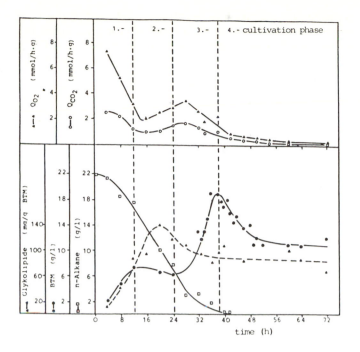

FIG. 2. Trehalose-6,6'-dicorynomycolate formation by *Rhodococcus erythropolis* during a 50 l batch cultivation on 2.2% n-alkanes C-13 to C-18 (2).

hydrocarbon substrate, the large clumps are disintegrated into much smaller cell aggregates following a second exponential growth rate of 0.15 h^{-1}. These kinetics are in accordance with the oxygen uptake rate and the carbon dioxide production rate. More than 95% of the glycolipid produced during the batch process starting with 2.2% hydrocarbons are cell wall associated. At the end of the production phase a ratio of glycolipid to cell dry mass of 0.1 and a volumetric productivity of 50 mg trehaloselipid per liter per hour is obtained. These values could not be markedly influenced by variation of the culture conditions at a constant C/N ratio of 10:1, for example, by an increased oxygen tension resulting from oxygen enrichment of the air at constant aeration rate.

To overcome this disadvantage, an in situ extraction method was developed. To prevent the aggregation of the hydrocarbon substrate and the cells during the cultivation process and to extract the cell-wall associated glycolipid continuously, the addition of solvents which cannot metabolize by the microorganisms in the presence of n-alkanes was studied. Production of trehalose- dicorynomycolates under the aforementioned conditions in the presence of 1% xylene or 4% kerosene increased the volumetric productivity four- to five-fold without aggregation of the cell mass. Less than 5% of the glycolipid is associated with the cells. It seems that in situ extraction is an effective strategy for the overproduction of cell-wall associated glycolipids. However, no general rule for solvents can be given. It depends on the microorganism and on the structure of the cell-wall associated glycolipid.

Glycolipid formation under growth-limiting conditions

The synthesis of glycolipids under growth-limiting conditions is mainly influenced by changing the metabolic pathway or the metabolic activity leading to an overproduction and/or an alteration of the structure (7). This is demonstrated on two examples, the formation of trehalose-tetraester and rhamnolipids.

Rhodococcus eytropolis grown on hydrocarbon substrates as a carbon source without limitation of nitrogen synthesize only nonionic trehalose-dicorynomycolates. However, under N-limitation a dramatic change in the structure of the glycolipid is observed (12). The main compound is an anionic trehalose-tetraester (Fig. 3) and

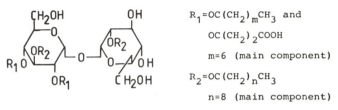

FIG. 3. Trehalose-2,3,4,2'-tetraester from *Rhodococcus erythropolis* (12).

the nonionic trehalose-dicorynomycolat is formed only in small amounts. A similar result is obtained by limitation of multivalent cations by addition of EDTA to the growing culture. In addition to N-limitation, a further increase of trehalose-tetraester formation could be obtained with a temperature shift from 30 to 21 C, which decreases the growth rate of the organism (Fig. 4). Not only is the

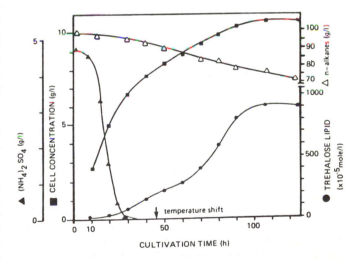

FIG. 4. Trehalose-2,3,4,2'-tetraester formation by *Rhodococcus erythropolis* during a 50 l batch cultivation. Conditions: mineral salt medium, 0.1% yeast extract and 9.8% n-alkanes C-14 to C-15, T = 30 C at the beginning, T = 21 C after 45 hr, N-limitation (13).

change in the structure remarkable, but the strategy to put in a limitation step also increases the efficiency expressed as conversion yield of the carbon substrate to the product (Table 2).

The formation of trehalose- tetraesters by *Rhodococcus erythropolis* is a typical example of the possibility of altering the product composition in the lipophilic moiety by variation of the hydrocarbon substrate. The results in Table 3 show that the quantitative composition of the

FRITZ WAGNER

TABLE 2

Trehalose Lipid Formation by _Rhodococcus erythropolis_ on n-alkane (13)

Conditions	Conversion-yield (g Product/g substrate)	Product composition (%)	
		Trehalose-dicorynomycolate	Trehalose-tetraester
Growing cells			
-No limitation	0.15	100	—
-N-limitation	0.28	10	90
Resting cells	0.65	8	92

TABLE 3

Alteration of the Fatty Acid Composition in Trehalose-tetraester Dependent from the Composition of the Carbon Substrate

Fatty acid	Carbon substrate			
	n-Undecane %	n-Octadecane %	n-Paraffin[a] %	Kerosene[b] %
C-7	—	—	3	—
C-8	—	—	24	17
C-9	18	3	9	2
C-10	4	7	59	35
C-10α-Methyl	—	—	—	11
C-11	72	11	2	6
C-12	—	—	—	5
C-14	—	22	—	6
C-15	—	—	—	3
C-16	5	27	—	6
C-18	1	30	—	2

[a]89 % n-Tetradecan, 9 % n-Pentadecan

[b]85 % n-paraffins + iso-paraffins, 15 % aromatic compounds

trehalose-tetraesters can be influenced greatly by these variations. In addition, with this strategy, tailoring of the surfactant properties is a possibility.

Concerning the biosynthesis of trehalose-lipids by _Rhodococcus erythropolis_, it is obvious that the structure of the sugar moiety is not influenced by different carbon substrates. Irrespective of the hydrocarbon substrate, the sugar moiety of the biosurfactant will be principally lipolytic and gluconogenic. The lipophilic moiety of the trehalose-lipids may be readily synthesized from hydrocarbons rather than the corresponding fatty acid and may then be modified and incorporated into surfactants (6,14). Similar pathways were observed by the formation of mannosyl erythritol-lipids (15) and sophorose lipids from hydrocarbons by growing cultures (16).

The metabolic pathways involved in the synthesis of rhamnolipids are diverse. The carbon substrate such as carbohydrates, glycerol or alkanes with different chain lengths has no influence on the structure of the lipophilic part of the surfactant (7). Both the sugar and lipophilic part in the surfactant are synthesized de novo independent of the carbon substrate. De novo syntheses of biosurfactants also were observed by the formation of sophorose lipids under resting cell conditions (17) and for cellobiose lipids with growing or resting cells (18).

The anionic rhamnolipids are real extracellular metabolites and produced under growth-limiting conditions (13,19). Experiments on medium optimization for the growth of _Pseudomonas_ spec. indicated that a limitation of nitrogen (20,21) or multivalent cations (19) caused an

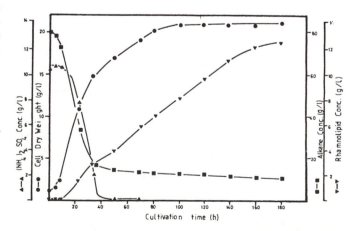

FIG. 5. Rhamnolipid production by _Pseudomonas_ sp. DSM 2874 during a 30 l batch cultivation. Conditions: mineral salt medium with 8% n-alkanes C-14 to C-15, T = 30 C, N-limitation (22).

overproduction of rhamnolipids. The growth and product kinetics showed that in the presence of ammonium ions rhamnolipids are not formed in substantial amounts (Fig. 5). However, when the ammonium concentration decreased to zero, the n-alkane consumption and the rhamnolipid formation is linear up to 140 hr. This interrelation also is affected by oxygen limitation. A further increase in rhamnolipid production was obtained by limitation of iron, magnesium or calcium, which all are essential for the growth of *Pseudomonas* sp. (19,22).

From these results, we have worked out a general strategy for biosurfactant production with resting cells. In all studies to date, this method was successful when the production phase was clearly separated from the growth phase. For this purpose, the microbes are grown under optimal growth conditions, separated from the culture broth and resuspended in a reaction medium for optimal production conditions. Such a procedure has the advantage that possible byproducts in the culture broth are eliminated. Also, single parameters that influence the synthesis of the biosurfactants can be studied under chemically well-defined conditions.

The production of rhamnolipids with resting cells suspended in 100 mM sodium chloride solution with n-alkanes as the carbon substrate is demonstrated in Figure 6. An apparent increase in the cell biocatalysts' stability and higher specific rhamnolipid production resulted in comparison with that of growing cells under nitrogen-limitation. Although growing cells formed the two rhamnolipids RI and RIII, two new rhamnolipids RII and RIV (23) were obtained under resting cell conditions (Fig. 7). The influence of different carbon substrates and reaction conditions on the productivity and product composition is summarized in Table 4. As has been shown for the production of rhamnolipids and trehalose-tetraesters (Table 2), with resting cells this strategy is even successful for the overproduction of sophorose lipids (17) and cellobiose lipids (18).

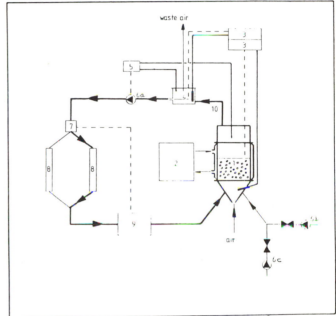

FIG. 8. Continuous production of rhamnolipid in a fluidized bed reactor and on-line isolation of rhamnolipid by foam flotation. 1, fluidized bed reactor; 2, thermostate; 3, pH-adjustment; 4, product precipitation; 6, pump for a)medium circulation, b)substrate feeding, c)regeneration of biocatalysts; 7, magnetic valve; 8, XAD-2 adsorption column; 9, photometer; 10, continuous product foam flotation.

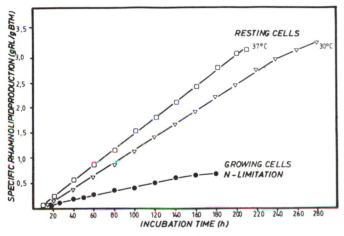

FIG. 6. Specific rhamnolipid formation with resting cells in 100 mM sodium chloride solution at T = 30 C and 37 C in comparison with N-limited growing cells at T = 30 C by *Pseudomonas* spc. DSM 2874 (22).

A primary technical problem in the production of extracellular biosurfactants under aerobic conditions with growing cells or resting cells is the extensive formation of foam. The rhamnolipids are one of the most effective surfactants known today. After reaching a concentration of about 2 g per liter in an aerated, mechanical stirred reactor the whole reactor system is filled with foam and there is no chance to keep the cells in the reactor systems. Therefore, we have developed a strategy to optimize the production phase with immobilized viable cells in which the immobilized cell biocatalysts can be easily retained in the fluid reaction phase. Several cell entrapment methods were tested (24) in polymers such as carrageenan, chitosan, alginate or poly-acryl amide. The highest rhamnolipid production was obtained in the

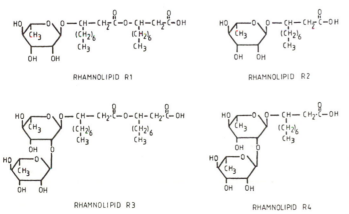

FIG. 7. Rhamnolipids RI to RIV from *Pseudomonas* sp. DSM 2874 (23).

TABLE 4

Rhamnolipid Formation by *Pseudomonas* spec. DSM 2874. Influence of different reaction conditions on the conversion yield (g product/g carbon substrate) and product composition.

Reaction conditions	Conversion yield (g/g)	Product composition (%)			
		RI	RII	RIII	RIV
[a]Growing cells					
—No-limitation	0.04	50	—	50	—
—N-limitation	0.18	65	—	35	—
[b]Resting cells					
—C-source: n-alkanes, 30 C	0.23	42	15	41	2
—C-source: glycerol, 30 C	0.10	22	15	62	1
—C-source: n-alkanes, 37 C	0.23	57	—	43	—
[c]Immobilized cells					
—C-source: glycerol, 30 C	0.11	22	15	62	1

ionotrophic network of calcium-alginate (22). With the active cell biocatalysts entrained in calcium-alginate, we have built a successful unit operation for the continuous production of rhamnolipids in a fluidized bed reactor.

The operational stability under continuous process conditions is at practically a constant level over 10 days, and during this period no leakage of cells could be detected. After this time, it is necessary to interrupt the production phase to regenerate the cell biocatalysts by incubating them with a nutrient medium for a period of about seven hours. Then, another production phase is run for 10 days.

In such a continuous process with immobilized cell biocatalysts for the production of a biosurfactant we have focused on another strategy. As demonstrated in Figure 8, an integral part of the process is the continuous on-line isolation of rhamnolipids by foam flotation. The foam contains more than 95% of the total rhamnolipid concentration in the reaction system. The flotated product in the foam is precipitated in a second reactor. From this, the supernatent is passed through an adsorber column to complete the isolation of rhamnolipids from the fluid phase. After the adsorber column, the liquid phase is recycled to the fluidized bed reactor.

In the course of this paper some general strategies for the practical applicability of various systems for the overproduction of biosurfactants have been selected for detailed illustration. It is safe to predict that biosurfactants can be produced economically. However, depending on the functional characteristics of such biosurfactants, strong competition from chemically synthesized surfactants must be expected.

REFERENCES

1. Kosaric, N., W.C. Cairus and N.C.C. Gray, *Surfactant Science Series*, Vol. 25, Marcel Dekker Incorp., New York, NY, 1987.
2. Wagner, F., H. Bock and A. Kretschmer, in *Fermentation*, edited by R.M. Lafferty, Springer Verlag, Wien, 1981, pp. 181-192.
3. Singer, M.E. and W.R. Finnerty, in *Petroleum Microbiology*, edited by R.M. Atlas, MacMillan Publishing Co., New York, NY, 1984, pp. 1-59.
4. Haferburg, D., R. Hommel, R. Claus and H.P. Kleber, *Adv. Biochem. Eng. 33*:53-93 (1986).
5. Lang, S., and F. Wagner, in *Biosurfactants and Biotechnology*, edited by N. Kosaric, W.L. Cairus and N.C.C. Gray, Marcel Dekker, Inc., New York, NY, 1987, Vol. 25:21-45.
6. Boulton, C.A., and C. Ratledge, *Ibid. 25*:47-87 (1987).
7. Syldatk, C., and F. Wagner, *Ibid. 25*:89-120 (1987).
8. Kretschmer, A., S. Lang, G. Marwede, E. Ristau and F. Wagner, *Advances in Biotechnology*, edited by Cl. Vezina and K. Singh, Pergamon Press, Canada, Vol. 3:475-479 (1981).
9. Singer, M.E., W.R. Finnerty, P. Bolden and A.D. King, *Am. Chem. Soc. Symp. 28*:785-788 (1983).
10. Matsuyama, T., M. Sagowa and I. Yano, *Appl. Environ. Microb. 53*:1186-1188 (1987).
11. Rapp, P., H. Bock, V. Wray and F. Wagner, *J. Gen. Microbiol. 115*:495-503 (1979).
12. Ristau, E., and F. Wagner, *Biotechnol. Letters 5*:95-100 (1983).
13. Wagner, F., J.S. Kim, Z.Y. Li, G. Marwede, U. Matulovic, E. Ristau and C. Syldatk, in *3rd Eur. Congr. Biotechnol.*, Verlag Chemie, Weinheim, Vol. 1:3-8 (1984).
14. Kretschmer, A., and F. Wagner, *Biochim. Biophys. Acta 753*:306-313 (1983).
15. Kawashima, H., T. Nakahara, M. Oogaki and T. Tabuchi, *J. Ferment. Technol. 61*:143-149 (1983).
16. Tulloch, A.P., J.F.T. Spencer and A.J. Gorin, *Can. J. Chem. 40*:1326-1338 (1962).
17. Gobbert, U., S. Lang and F. Wagner, *Biotechnol. Letters 6*:225-230 (1984).
18. Frautz, B., S. Lang and F. Wagner, *Ibid. 11*:757-762 (1986).
19. Syldatk, C., S. Lang, U. Matulovic and F. Wagner, *Z. Naturforsch. 40c*:61-67 (1985).
20. Wagner, F., U. Behrendt, H. Bock, A. Kretschmer, S. Lang and C. Syldatk, in *Microbial Enhanced Oil Recovery*, edited by J.E. Zajic, D.G. Cooper, T.R. Jack and N. Kosaric, Pennwell Books, Oklahoma, 1983, pp. 55-60.
21. Guerra-Santos, L., O. Käppeli and A. Fiechter, *Appl. Environ. Microbiol. 48*:301-305 (1984).
22. Syldatk, Ch., U. Matulovic and F. Wagner, *Biotechforum 1*:58-66 (1984).
23. Syldatk, Ch., S. Lang, F. Wagner, V. Wray and L.Witte, *Z. Naturforsch. 40c*:51-60 (1985).
24. Klein, J. and F. Wagner, *Appl. Biochem. Bioeng. 4*:11-51 (1983).

Biosurfactants as Food Additives

Satoshi Kudo
Yakult Central Institute for Microbiological Research, 1796 Yaho, Kunitachi-shi, Tokyo 186, Japan

This paper gives a general review of the use of biosurfactants in food featuring recent work on transphosphatidylation of soy lecithin. A biosurfactant is defined as a natural substance that lowers interfacial tension and stabilizes emulsions. The structures and properties of proteins and lecithins conforming to this criterion are reviewed in relation to their emulsifying abilities. Milk proteins, which despite their relatively moderate surface activities are referred to as components of a refined emulsion system, also are discussed. Many attempts at chemical modification of both proteins and lecithins, which cause a marked change in surface activities, are introduced. With respect to lecithins, their naturally emulsifying power, production, market, structure, composition, analysis (including a new and improved method), processing and modification are reviewed. In addition, a new biosurfactant, transphosphatidylated lecithin, in which most of the phosphatidylcholine and phosphatidylethanolamine has been converted to phosphatidylglycerol by phospholipase D, is introduced. The properties and screening of the enzymes, the conditions of transphosphatidylation in an aqueous system or an inorganic solvent system and the emulsifying properties of the transphosphatidylated products are detailed. Although the safety of adding chemically modified products to food is questionable, transphosphatidylation, which is an enzymatic modification, may be acceptable. A possible future application of this reaction also is described.

Biosurfactants may be defined as natural substances that are produced biologically and which lower interfacial surface tension and also stabilize emulsions. Biosurfactants are distinguished from synthetic surfactants only on the basis of their origin. According to this definition, biosurfactants include a wide range of natural surfactants and emulsifiers, such as glycerophospholipids, sphingophospholipids, glycoglycerolipids, glycosphingolipids, saponins, bile acids, proteins, polysaccharides, etc. A number of microorganisms that are economical sources of biosurfactants is known. Some biosurfactants produced by microorganisms are shown in Table 1.

To improve its emulsifying property, an intact biosurfactant molecule is often subjected to modification involving fractionation and enzymatic or chemical alteration (Table 2). Enzymes can construct monoacylglycerol, sucrose esters (20) and N-acyl amino acids (15), which previously was believed to be possible solely by chemical synthesis. Whether these enzymatic derivatives of natural sources fall within the category of biosurfactant has not yet been decided.

The use of biosurfactants in food, however, is more restricted by questions of safety, sensory quality, cost, functional properties, custom and law. The safety of biosurfactants weighs most heavily and requires that the biosurfactants be biologically degraded into a form non-toxic not only for the human body but also for the environment, which has become a social "requirement" for chemical compounds. Biosurfactants for use as food additives are classified as lipids, proteins and polysaccharides (Fig. 1). The lecithins (phospholipids) can substantially reduce surface tension and are virtually a single entry in the lists of food emulsifiers in many countries

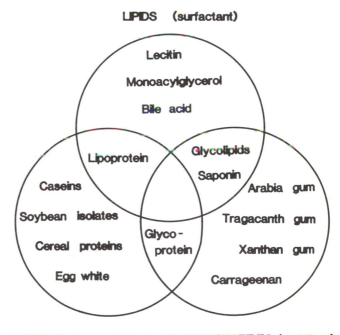

FIG. 1. Biosurfactants as food additives.

TABLE 1

Microbial Biosurfactants

Product	Microorganisms	Hydrophilic residue	Max. yield
			g / L
Spiculisporic acid (1)	*Penicillium spiculisporum*	Carboxylic acid	110
N-acyl amino acids (2)	*Pseudomonas* sp.	Amino acid	—
Surfactin (3)	*Bacillus subtilis*	Heptapeptide	0.8
Emulsan (4)	*Acinetobacter calcoaceticus*	Polysaccharide	5
Bioemulsifier (5)	*Corynebacterium hydrocarboclastus*	Protein, sugar	6
Rhamnolipid (6)	*Pseudomonas aeruginosa*	Rhamnose	15
Sophorolipid (7)	*Toluropsis bombicola*	Sugar	120

SATOSHI KUDO

TABLE 2

Production and Modification of Biosurfactants

Methods		Biosurfactants
Bacterial	Fermentation	Xanthan gum, polysaccharide AX
		Cardiolipin (8)
		Cyclodextrin (9)
Enzymatic	Hydrolysis	Protein hydrolysate (10)
		Lysolecithin (11)
		Lysophosphatidic acid (12)
	Transfer	Transphosphatidylated lecithin (13)
	Synthesis	Sugar ester (14)
		N-acyl amino acid (15)
		Leucine SDS-gelatin (16)
		Mono or diacylglycerol
Solvent	Fractionation	Proteose-peptone (17)
		Fractionated lecithin (18)
		Phytoglycolipid (19)
Chemical	Hydrolysis	Protein hydrolysate, lysolecithin
	Ester exchange	Mono or diacylglycerol
	Synthesis	Soap, sugar ester
	Modification	Acetylated lecithin (18)
		Acetylated protein (16)
		Hydroxylated lecithin (11)

(21). Proteins function as both surfactants and emulsion stabilizers through adsorption at interfacial membranes, although their surface activities are relatively moderate. Although they will not be discussed in this review, polysaccharides, which have little or no surface activity themselves, play a role as stabilizers by increasing viscosity, hydration or electric charge interactions at an interface. This review will describe proteins and lecithins as biosurfactants. Special attention will be paid to transphosphatidylation with phospholipase D (PLD. EC 3.1.4.4), which produces a new and powerful biosurfactant, i.e., transphosphatidylated lecithin or phosphatidylglycerol (PG).

PROTEINS

Surface Action of Proteins

A protein consists of both hydrophilic and hydrophobic amino acids. Uneven distribution of these amino acids can result in a surface active nature in a protein molecule. An example is milk, which is the most sophisticated natural emulsion system containing two types of particles, i.e., casein micelles and milk fat globules. The amphiphilic nature of κ-casein in casein micelles was elaborated by Hill and Wake (22) who found that the N-terminal two-thirds of κ-casein are hydrophobic, and the C-terminal third and its carbohydrate moiety are hydrophilic. This property is the most important factor in retaining the size distribution of casein micelles (23). Other caseins display, more or less, these same amphiphilic properties, which partly may explain why edible caseins can participate in surface active reactions with the aid of other surfactants. In addition to the uneven distribution of amino acids, the surface activity of a protein, which also is a function of its flexibility (24), can be predicted from the ratio of polar to nonpolar amino acids, with a ratio of 1.3 or less indicating

more flexibility (25). Consequently, β-casein, with a ratio of 0.82, is highly flexible and therefore can easily diffuse to the interface and cause a rapid attainment of equilibrium interfacial tension. Nevertheless, the surface activity of a protein is still moderate when compared with other surfactants. In contrast to surfactant properties, the rheological stabilization of an interfacial membrane is more attributable to the rigidity, large molecular size and reduced charge of proteins adsorbed at the interface. Although such molecules require an input of energy to partition them into the interface (26), the emulsions, thus made, are generally more stable than those formed solely with a surfactant despite the moderate surfactant properties of proteins. In addition to proteins, milk fat globules also contain other natural surfactants such as polar lipids, lipoproteins and glycoproteins (27). Identification of proteins with these incompatible properties is still more an art than a science. In addition to milk proteins, egg protein, serum albumin, soy protein isolates and many other plant proteins display these properties.

Modification of Proteins

Proteins, however, are tailored naturally not for industrial purposes but for functionality in cells, etc. Hence their emulsifying properties do not always satisfy industrial needs. Many attempts have been made to correct the imperfections of natural surfactants by modification (28). The covalent attachment of fatty acid residues to proteins can change their surface activity. Acylation of soy glicinin (29) or casein (30) causes an increase in their emulsification activity. The emulsifying power of glicinin increased as much as 2.7-fold when one acyl group was attached to one subunit of glicinin. When casein was modified in the same manner, the resulting product showed increased emulsifying properties depending on the numbers of bound fatty acids (30). A great deal of research has been published on acylation, alkylation, phosphorylation and other chemical modifications of edible proteins to improve their emulsifying properties. However, none of these modified proteins has yet been released to the market because of questions as to their safety to the human body. L-alanyl myristate was attached enzymatically to the C-terminal end of a very hydrophilic protein, gelatin, and the resulting derivative, which has strong interfacial activity, is used in cosmetics (31). However, to my knowledge, this product has not been adopted for food use despite the fact that the modification was made by enzymatic treatment. In contrast, degradation of edible proteins is generally recognized as safe. Therefore, a variety of protein hydrolysates from less useful sources is on the market (32). Hydrolysis can bring about changes in functional properties including an improvement in emulsifying activity (33). Enzymatic hydrolysis of proteins is preferable to alkali or acid hydrolysis from a nutritional standpoint. Alkaline protease, trypsin, chymotrypsin and many other proteolytic enzymes are used to degrade soy protein, caseins, egg white, gluten, collagen and so on.

LECITHINS

Lecithin is a term commonly used for a mixture of phosphatides that are derived from plants and animals. Another ambiguous use of lecithin is as the chemical structural name of phosphatidylcholine (PC), which comprises ca. 80% of egg yolk phosphatides. (The word "lecithin" is derived from the Greek "lekitos" meaning egg yolk.) The latter mode of expression is not recommended. Major oilseed crops, such as soybeans, cottonseed, peanuts, sunflower seed or rapeseed, and egg yolk are the main sources of industrial lecithins. However, soybeans are almost the only source of food-grade lecithins because of their continuous availability and function, while egg yolk, though its industrial output is comparatively small, provides lecithins for medical use with a good market price. Oilseed and egg yolk lecithins differ considerably in their phosphatide (Table 3) (16) and fatty acid (Table 4) (17) composition.

Industry is estimated to use 130,000 out of the 380,000 tons of lecithins available with 20,000 tons being used in Asia, 15,000 tons in South America, 45,000 tons in the U.S. and 50,000 tons in Western Europe.

TABLE 3

Composition of Soy and Egg Yolk Lecithin

Phosphatide	Soybean (34)	Egg Yolk (35)
	% (w/w)	
Phosphatidylcholine	31.7	78.8
Phosphatidylethanolamine	20.0	17.1
Phosphatidylserine	3.0	0.0
Phosphatidylinositol	17.5	0.6
Phosphatidic acid	2.0	0.0
Plasmalogen	—	1.0
Sphingomyelin	—	2.5
Phytoglycolipid	14.8	0.0
Others	10.2	—

TABLE 4

Composition of Fatty Acids in Soy and Egg Yolk Lecithins (36)

Fatty acid	Soy lecithin	Egg yolk lecithin
	% (w/w)	
Palmitic acid	15.9	30.4
Palmitoleic acid	1.5	trace
Stearic acid	3.7	15.2
Oleic acid	16.6	27.7
Linoleic acid	55.6	15.5
Linolenic acid	6.8	trace

Structure and Composition of Soy Lecithin

The molecular structures of glycerophospholipids found in soy lecithin are shown in Figure 2. In the case of x=choline in Figure 2, some confusion may occur as to the nomenclature of this compound. Lecithin, phosphatidylcholine, L-α-phosphatidylcholine, 3-sn-phosphatidylcholine and 1,2-diacyl-sn-glycero-(3)-phosphocholine are all names for the same substance provided that the compositional heterogeneity of the fatty acids is neglected. Phosphatidylcholine is frequently used for convenience. Phospholipids in which an acyl group in position-1 is replaced by an O-alkyl (i.e., plasmanylcholine) (37) or O-(1-alkenyl) (i.e., plasmanylcholine) are included in the family of glycerophospholipids. These ether bond-containing phospholipids are known to be widely distributed in the animal kingdom (e.g., egg yolk), and to a lesser extent in plants (e.g., soy lecithin) and bacteria.

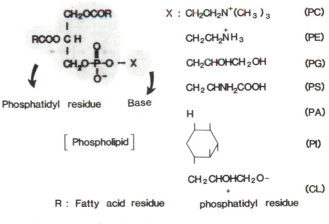

FIG. 2. Structure of glycerophospholipids.

In addition to glycerophospholipids, soy lecithin also contains neutral lipids, glycolipids and carbohydrates. Egg yolk lecithin also contains sphingophospholipids (Table 3).

Analysis of Phospholipids in Lecithin

To establish a standard method for phospholipid analysis, the appropriateness of thin layer chromatography (TLC), high pressure liquid chromatography (HPLC) and TLC-FID (flame ionization detection) has been examined with two-dimensional TLC on a silica plate (38) followed by phosphorus determination proposed as the first choice (39). It is believed that TLC on silica plates gives the best resolution provided careful calibration is followed (40). The second choice is HPLC and whether it could replace TLC is still under examination. HPLC, in conjunction with an ultraviolet detector at 206 nm, gives quantitatively reliable data on discrete peaks because the molecular species found in soy lecithins are fairly constant (40). A mixture of hexane, 2-propanol and water as the mobile phase on silica is recommended for relatively good resolution (40,41). However, there are still some difficulties in simultaneously separating the phospholipids, glycolipids and neutral lipids in commercial lecithin even if a gradient of water, which brings about drift of the base line (35), is adopted to achieve better resolution than an isocratic solvent system. Both two-dimensional TLC and HPLC with a gradient are laborious and expensive. These shortcomings can be overcome to some extent by TLC-FID although resolution is inferior to that of TLC or HPLC. Consequently, an analytical method that satisfies both qualitative and quantitative analysis requirements has not yet been developed for commercial lecithins. We have developed an improved method of one-dimensional TLC (unpublished data), which obviously is advantageous to two-dimensional TLC. More than 13 components in 20 commercial lecithins are separable on a silica plate (10 x 10 cm) within 25 min (Fig. 3) when ethanol is added to the developing solvent as a ratio of chloroform: ethanol: methanol: acetic acid (13:3:2:2 v/v/v/v).

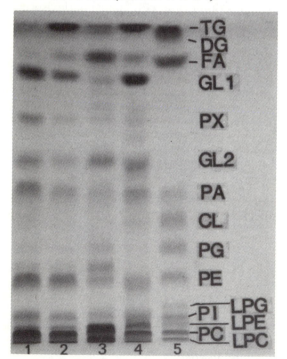

FIG. 3. Silica-gel TLC of powdered lecithin (1), crude lecithin paste (2), rapeseed lecithin (3), corn lecithin (4), and the standard mixture (5) containing triacylglycerol (TG), diacylglycerol (DG), fatty acid (FA, oleate) and various phospholipids (for abbreviations, please refer text). Other abbreviations: Glycolipids (GL1 and GL2); Unknown lipid positive with Dittmer and Lester reagent (42) (PX); Lyso form phospholipids (LPG, LPE and LPC). Developing solvents; (chloroform:ethanol:methanol: acetic acid = 13:3:2:2 v/v/v/v). Detection; 50% sulfuric acid.

Processes

Soy lecithins are separated from crude soybean oil by a traditional degumming process. An excellent review on progress in the recovery and processing of plant lecithins has been published (43). In addition, new technological processes have been developed that isolate phospholipids using supercritical carbon dioxide (44) or ultrafiltration (45). The latter may be competitive with the traditional degumming process whereas the former will be so when a large-scale plant becomes available accompanied by an increase in the cost of hexane. Phospholipids can be synthesized chemically (46) or extracted from *Rhodococcus* (47) and *Corynebacterium* (8). However, neither chemical nor bacterial production is much more economical than extraction of phospholipids from oilseeds.

Modification of Lecithins

The use of lecithins often is narrowed by their shortcomings as emulsifiers under conditions of low pH, increased ionic strength, heating, oxidation and solvation and in the presence of divalent cations. To overcome these shortcomings, a lecithin is often modified by either enzymatic or chemical procedures.

Alcohol fractionation based on the better solubility of PC than cephalin in ethanol can concentrate this lipid by more than 80% (48). The concentrated phosphatidylcholine works as an antispattering agent in saltless margarine at less cost than egg yolk lecithin of the same grade.

Lysophosphatides were developed in 1969 (49) by the partial hydrolysis of lecithins with pancreatin until 2–15% lysophosphatides formed. Later on, at least 60% hydrolysis was claimed to be necessary to obtain sufficient efficiency when mixed with oil (50), although the more hydrolysis the more liberation of free fatty acids which cause a bitter taste. Combined use of pancreatin and PLD removes both fatty acids and bases respectively from phospholipids with the resulting lysophosphatidic acids showing improved emulsifying properties (12).

Reports of chemical modifications such as acetylation (18), hydroxylation (51) and hydrogenation (52) can be found elsewhere. However, their use in food has been legally restricted in most countries except the U.S. where an hydroxy lecithin is allowed as a food additive. Therefore, the market value of chemically modified lecithins is higher in the cosmetics, pharmaceutical and high-technology industries than in the food industries.

TRANSPHOSPHATIDYLATED LECITHIN— A NEW BIOSURFACTANT

It was reported that an emulsion containing PG as the emulsifier underwent flocculation and coalescence under conditions of low pH and/or increased calcium concentration (13). PG is ubiquitously distributed in plant and bacterial cells, and is present in the mitochondria of animal cells. A commercially available source of PG, however, has not yet been found (soy lecithins contain little or none [53]). On the other hand, the enzymatic formation of a small amount of PG first was documented by Dawson and Hemington (54). This enzymatic conversion of phospholipids is identical to the transphosphatidylase activity of PLD, which was reported by Benson et al. in 1965 (55). They observed an artificial phosphatidylethanol in eth-

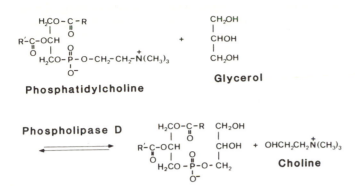

FIG. 4. Conversion of phosphatidylcholine into phosphatidylglycerol by phospholipase D.

anol extracts of plant tissue containing PLD activity. Further investigations (56,57) confirmed that the transfer of the phosphatidyl residue from PC to primary alcohols in general was catalyzed by PLD. Hence, the author and coworkers began research on transphosphatidylation and its use in industrial production of PG from lesspotent phospholipids, i.e., soy lecithins (Fig.4).

Phospholipase D

An enzymatic entity in carrot roots and cabbage leaves, which cleaves PC into phosphatidic acid (PA) and choline, was described by Hanahan and Chaikoff (58). Numerous investigations followed to elucidate the properties and mechanism of this enzymatic reaction (59). It has been postulated also that such activity exists in microorganisms as well as in mammalia. The first demonstration of transphosphatidylation (55), plant PLD, and above all cabbage PLD, has been used most frequently to study or prepare transphosphatidylated phospholipids but only on a laboratory scale for delivering reagent-grade products. In addition, the potency, substrate specificity, supply and cost of plant PLD are not always satisfactory for industrial purposes. With respect to the different sources of PLD, attention has been paid mainly to the metabolism and biosynthesis of phospholipids in tissues.

On the other hand, several reports on microbial PLD from actinomycetes and its use in transphosphatidylation recently have been released. PLD (PLD-M) from *Nocardiopsis* (60), *Actinomadura* (61) and *Nocardia* (62) were characterized as acting on a wide range of substrates. However, organisms that belong to the rare genus of actinomycetes usually display laborious fermentation properties and poor productivities. In addition, both *Actinomadura* and *Nocardia* are pathogenic, and the PLD enzymes from all three species are relatively unstable on heating and expensive because of poor recoveries in culture broth. Consequently, little information is available on the use of PLD for producing transphosphatidylated lecithin in the food industry.

To date, we have screened fungi, bacteria and actinomycetes by directly measuring transphosphatidylation activity and have found several strains of actinomycetes that produce enzymes having powerful activities in transphosphatidylation and acceptable heat stabilities and productivity. One strain, which liberates large quantities of PLD (PLD-Y1) into culture broth (80 unit/ml), was identified as *Streptomyces prunicolor* (63). The properties of PLD-Y1 were somewhat similar to the enzyme from *Streptomyces* sp. (PLD-P, Toyo Jozo Co.) although the heat stability and productivity of PLD-Y1 are superior to PLD-P. Some properties of PLD-Y1 are shown in Table 5.

Donors and Acceptors

PLD-Y1 displays a broader range of both donors and acceptors than plant PLD. The PLD-Y1 donors are PC, PG, phosphatidylethnolamine (PE), phosphatidylserine (PS), or their lyso-type phospholipids, and sphingomyelin. Compounds containing primary, secondary or phenolic hydroxyl groups are acceptors, such as pentose, hexose, glycosides, amino sugars, alcoholic sugars, di- or oligosaccharides, amino acids, nucleic acids, dinitrophenol, primary and secondary alcohols. Similar properties of PLD-M were reported (68). PLD-M can act on sphingophospholipids and transfer phosphoryl ceramide to not only primary but also secondary alcohols or their derivatives (68,69). This enzyme will be valuable for industries other than the food industry for the reasons previously mentioned.

Activators

It has been believed that both calcium and organic sol-

TABLE 5

Properties of Phospholipase D from Actinomycetes

Actinomycetes	Transphosphatidylation	Optimum		Activity of hydrolysis		Molecular weight
		pH	C	Broth	After 50 C - 30 min	
				—unit/ml—	— % —	
Streptomyces hachijoensis (64)	No	7.0	50	0.05	100	19,000
Streptomyces chromofuscus (65)	No	7.5	37	0.50	—	50,000
Micromonospora chalsea (66)	No	8.0	50	0.10	95	—
Nocardiopsis sp. (60)	Yes	7.0	70	0.54	81	—
Actinomadura sp. (61)	Yes	7.2	70	1.70	63	—
Nocardia sp. (62)	Yes	9.0	45	0.17	24	45,000
Streptomyces prunicolor (63)	Yes	6.0	55	80.00	100	52,000
Kitasatosporia sp. (67)	Yes	5.5	75	2.00	100	—

TABLE 6

Influence of Ethyl Acetate and Calcium on the Velocity of Phosphatidylglycerol formation[a] from Phosphatidylcholine by Transphosphatidylation

Sample mixture[b]	Ethyl acetate (μL)					
	0	10	30	50	75	100
0 mM calcium	0	0	35	124	143	101
10 mM calcium	185	333	255	177	249	250
	(56)[c]	(29)	(21)	(6)		

[a]Increased phosphatidylglycerol (μg/unit, min).
[b]Phosphatidylcholine (5 mg), glycerol (100 μL), 12.5 mM acetate buffer pH 5.5 (400 μL), phospholipase D-Y1 (1 unit).
[c]Increased cardiolipin (μg/unit, min).

vents such as ether or hexane are essential for PLD activity (70). During examination of the effects of various organic solvents on PLD activity, both esters and ketones, derived from combinations of alcohols and carboxylic acids containing 2-10 carbons, were found to have remarkable stimulating effects. Ethyl acetate was selected finally as the most suitable activator for use in the food industry based on considerations of safety, sensory quality and economy, and was used in all subsequent experiments. Either calcium or ethyl acetate, however, is not necessary for PLD activity, which is at variance with what was previously believed. In the absence of ethyl acetate, calcium is indispensable for PLD-Y1 to exhibit the transphosphatidylation activity and vice-versa (Table 6). In this case, however, two kinds of side reactions inevitably occur: hydrolysis and the formation of cardiolipin (CL) by two-step transphosphatidylation. In contrast, PG is the predominant product of normal or one-step transphosphatidylation when PLD-Y1 is activated solely by ethyl acetate.

Water Content

When transphosphatidylation is carried out in an aqueous reaction system containing PLD-Y1, PC (donor), glycerol (acceptor), calcium, ethyl acetate and buffer solution, several products including PG, PA, the residual PC and surprisingly CL, can be detected in the resulting mixture. The yields of such products vary considerably with the concentration of glycerol. In the absence of glycerol, phosphatidyl residues are transferred to hydroxyl groups of water, thus PA is the sole product (i.e., hydrolysis). When glycerol is added to the reaction mixture, the yields of PG and CL increase proportionally to the concentration of added glycerol (i.e., transphosphatidylation) up to a maximum value above which additional glycerol becomes inhibitory. Thus, there is a maximal production rate (Fig. 5). The initial acceleration of transphosphatidylation reflects the successful competition of the primary hydroxyl groups of glycerol against water as an acceptor, while the latter inhibitory effect of glycerol can be explained partly as follows: alcohols can interact with water through hydrogen bonds (71); therefore, glycerol, when present in excess, robs the water bound to PLD-Y1, which is essential for enzyme activity. This explanation could be extended to other compounds possessing high

dielectric constants or polarity such as dimethylsulfoxide, dioxane, methanol and so on. Consequently, to activate PLD and complete the transphosphatidylation reaction, a water content somewhere between that needed for hydrolysis and that for esterification, as a rule, is required in the reaction mixture.

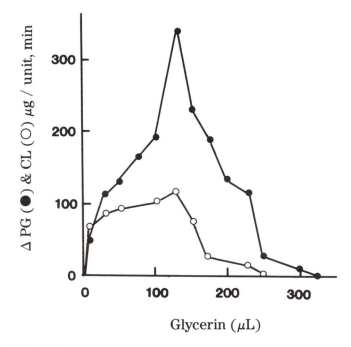

FIG. 5. Effect of glycerol concentration on the velocity of producing phosphatidylglycerol and cardiolipin at 50 C in an aqueous reaction mixture.

Reaction Phase

Addition of an organic solvent to the reaction mixture is essential for performance of one-step transphosphatidylation. However, if excess ethyl acetate, which dissolves in 10 parts of water, is added, the reaction mixture becomes biphasic. This mixture must be shaken vigorously in order for the enzyme to act at the interface. This is not practical on a large scale. Alternatively, transphosphatidylation in a uniform phase system was developed (72) in which appropriate amounts of phospholipids function to mix water with water-immiscible solvents. Although water-miscible solvents also can aid such mixing, they tend to inactivate the enzymes, limiting their use. Employing the uniform phase system for transphosphatidylation facilitates the adoption of immobilized enzymes. An emulsion system for transphosphatidylation, which consists of phospholipids, primary alcohol, and water [w/w ratios of phospholipids + alcohols:water not less than 1:1 also is known] (73).

Cardiolipin

It has been documented elsewhere (74) that two molecules of PG serve as the donor and the acceptor of the transphosphatidylation moiety in a reaction leading to the formation of one molecule of CL. However, tedious methods of extracting CL from animal brain or heart, and bacterial cells (8) were employed in the past because an abundant source of PG had not been identified. The surface activity of CL is not as strong as PG. However, it is a very important material in the medical and pharmaceutical fields (75). In this context, the direct formation of CL from PC during transphosphatidylation in an aqueous system was described in the previous section. The mechanism of CL synthesis is revealed by the structure of CL itself (Fig. 2). Obviously, CL is a composite of two phosphatidyl and one glyceryl residue which suggests that it can be formed when the product PG rather than glycerol is transphosphatidylated using either PC (76) or a second PG molecule as the phosphatidyl donor (74). This new principle for preparing CL will make it possible to exploit lecithin as a resource for the more valuable phospholipids by very simple but efficient procedures.

Outlook on Future Development

The search for new biosurfactants will be continued. However, it is unlikely that new biosurfactants suitable for use as food additives will be found readily. Enzymatic modification of known biosurfactants may be an alternative and must be extended further. Devising new techniques is also a potent and promising way of developing new emulsifiers. By studying the detailed nature of surfactants, new emulsion techniques such as liposomes (77), w/o/w emulsion (78), liquid crystal (79), gel base (80) and so on will be developed further. In addition, it is possible to control natural emulsion systems by a manner other than adding biosurfactants. For instance, the acid precipitation of a certain fermented milk beverage (81,82) or the collapse of dough (12) can be prevented by enzymatically decomposing phospholipids. Finally, application of transphosphatidylation to synthesizing novel phospholipids also is promising.

REFERENCES

1. Tabuchi, T., et al., *J. Ferment. Technol.* 55:37 (1977).
2. Wilkinson, S.G., *Biochim. Biophys. Acta* 270:1 (1972).
3. Arima, K., *Biochim. Biophys. Res. Com.* 31:488 (1968).
4. Rosenberg, E., et al., *Appl. Env. Microbiol.* 37:402 (1979).
5. Zajic, J.E., et al., *Biotech. Bioeng.* 19:1285 (1977).
6. Jarvis, F.G., and M.J. Johnson, *J. Am. Chem. Soc.* 2:4124 (1949).
7. Spencer, J.F.T., et al., *Can. J.Chem.* 39:849 (1961).
8. Ajinomoto Inc., Japan Patent 71013677 (1971).
9. Kobayashi, S., et al., *Carbohyd. Res.* 61:229 (1978).
10. Kaminogawa, S., et al., *J. Am. Oil Chem. Soc.* 63:445 (1986).
11. Unilever N.V., DE Patent 19 00 959.7 (1969).
12. Kyowa Hakko Kogyo Co. Ltd., Japan Patent Kokai 58-51853 (1983).
13. Kabushiki Kaisha Yakult Honsha, Japan Patent Kokai 61-199749 (1986).
14. Uchibori, T., et al., *J. Am. Oil Chem. Soc.* 63:464 (1986).
15. Ajinomoto Inc., Japan Patent Kokai 57-129696 (1982).
16. Ball Jr., H.R., *J. Am. Oil Chem. Soc.* 63:447 (1986).
17. Aschaffenburg, R., *J.Dairy Res.* 15:316 (1946).
18. Aneja, R., and J.S. Chadha, *Fette Seifen Anstrichm.* 73:643 (1971).
19. Cater, H.E., et al., *J. Am. Oil Chem. Soc.* 35:335 (1958).
20. Seino, H., et al., *J. Am. Oil Chem. Soc.* 61:1761 (1984).
21. European Food Emulsifere Manufactures Association, Monographs for emulsiferes for food (1970).
22. Hill, R.J., and R.G. Wake, *Nature* 221:635 (1969).
23. Slattery, C.W., and R. Evard, *Biochim. Biophys. Acta* 317:529 (1973).
24. Bull, H., *J.Colloid Interfacial Sci.* 41:305 (1972).
25. Birdi, K.S., *Ibid.* 43:545 (1973).
26. Inklaar, P., and J. Fortuin, *Food Technol.* 23:103 (1969).
27. Anderson, M., and T.E. Cawston, *J. Dairy Res.* 42:459 (1975).
28. Feeney, R.E., and J.R. Whitaker, in *Modification of Proteins: Food, Nutritional, and Pharmacological Aspects*, Am. Chem. Soc. (1982).
29. Haque, Z., et al., *J. Agric. Food Chem.* 30:481 (1982).
30. Haque, Z., and M. Kito, *Ibid.* 31:1231 (1983).
31. Morita, K., et al., *Bio Industry* 4:280 (1987).
32. Gunther, R.C., *J. Am. Oil Chem. Soc.* 56:345 (1979).
33. Das, K., et al., *Indian J. Food Sci. Technol.* 16:58 (1979).
34. Wagner, H., and P. Wolff, *Fette Seifen Anstrichm.* 66:425 (1964).
35. Rhodes, D.N., and C.H. Lea, *Biochem. J.* 65:526 (1957).
36. Chigira, J., *New Food Industry* 26:22 (1984).
37. IUPAC-IUB, Hoppe-Seyler s Z., *Physiol. Chem.* 358:599 (1977).
38. Erdohl, W.L., et al., *J. Am. Oil Chem. Soc.* 50:513 (1973).
39. Watanabe, M., et al., *J. Jap. Oil Chem. Soc.* 35:1018 (1986).
40. Nasner, A., and Lj. Kraus., *Appl. Environ. Microbio.* 44:25 (1982).
41. Rivnay, B., *J. Chromatography* 294:303 (1984).
42. Dittmer, J.C., and R.L. Lester, *J. Lipid Res.* 5:126 (1964).
43. Pardum, H., in Soya Lecithin Proceed. 2nd Inter. Colloquium Soya Lecithin, Semmelweis-Verlag, England, 1982, p. 37.
44. Kali-Chemie Pharma, DE Patent 3011185 (1981).
45. Iwama, A., and Y. Kazuse, *J. Membrane Sci.* 11:297 (1982).
46. Rosenthal, F.A., *Methods in Enzymology XXXV*:429 (1975).
47. Kretschmer, A., et al., *Appl. Environ. Microbiol.* 44:864 (1982).
48. Unilever N.V., DE Patent 1492925 (1973).
49. Unilever N.V., DE Patent 1900961.1 (1969).
50. Egi, M., *Food Chemical* 1:51 (1987).
51. Julian, P.L., and H.T.Iverson, U.S. Pat. 2629662 (1953).
52. Eichberg, J., U.S. Patent 3,359,201 (1967).
53. Mangold, H.K., in Soya Lecithin Proceed. 2nd Inter. Colloquium Soya Lecithin, Semmelweis-Verlag, England, 1982, p. 9.
54. Dawson, R.M.C., and N. Hemington, *Biochem. J.* 102:76 (1967).
55. Benson, A.A., et al., The 9th Inter. Confer. Biochem. Lipids, Noordwijk Aan Zee, The Netherlands, Sept. 5-10, 1965.
56. Yang, S.F., et. al., *J. Biol. Chem.* 242:477 (1967).
57. Dawson, R.M.C., *Biochem. J.* 102:205 (1967).
58. Hanahan, D.J., and I.L. Chaikoff, *J. Biol. Chem.* 168:233 (1947).
59. Heller, M., *Adv. Lipid Res.* 16:267 (1978).
60. Meito Sangyo Co. Ltd., Japan Patent Kokai 58-63388 (1983).
61. Meito Sangyo Co. Ltd., Japan Patent Kokai 58-67183 (1983).
62. Meito Sangyo Co. Ltd., Japan Patent Kokai 60-164483 (1985).
63. Kabushiki Kaisha Yakult Honsha, Japan Patent Appl. 62-52990 (1987).
64. Okawa, Y., and T. Yamaguchi, *J. Biochem.* 78:363 (1975).
65. Imamura, S., and Y. Horiuti, *J. Biochem.* 85:79 (1979).
66. Kyowa Hakko Kogyo Co. Ltd., Japan Patent Kokoku 58-52633 (1983).
67. Kabushiki Kaisha Yakult Honsha, Japan Patent Appl. (1987).
68. Meito Sangyo Co. Ltd., Japan Patent Kokai 59-187786 (1984).
69. Meito Sangyo Co. Ltd., Japan Patent Kokai 59-187787 (1984).
70. Saito, M., et al., *Arch. Biochem. Biophys.* 164:420 (1974).
71. Reichardt, C., in *Solvent Effects in Organic Chemistry*, Verlag Chemie, Weinheim, 1979, p. 12.
72. Kabushiki Kaisha Yakult Honsha, Japan Patent Appl. 62-75908 (1987).
73. The Nisshin Oil Mils Ltd., Japan Patent Kokai 62-36195 (1987).
74. Stanacev, N.Z., *Can. J. Biochem.* 51:747 (1973).
75. Teijin Ltd., Japan Patent Kokai 58-222022 (1983).
76. Kabushiki Kaisha Yakult Honsha, Japan Patent Appl. 62-96243 (1987).
77. Ostro, M.J., *Scientific American*, January:90 (1987).
78. Matsumoto, S., and P. Sherman, *J. Texture Studies* 12:243 (1981).
79. Rydrag, L., and I. Wilton, *J. Am. Oil Chem. Soc.* 58:830 (1981).
80. Kumano, Y., et al., *J. Soc. Cosmet. Chem.* 28:285 (1977).
81. Kudo, S., et al., *Jap. J. Zootech. Sci.* 49:753 (1978).
82. Kabushiki Kaisha Yakult Honsha, Japan Patent Kokai 57-189637 (1982).

Biosurfactants for Petroleum Recovery

Melanie J. Brown[a] and **V. Moses**[b]

[a]QMC Industrial Research Ltd., Mile End Road, London, [b]School of Biological Sciences, Queen Mary College, London

Chemically synthesized surfactants currently are being evaluated by the oil industry for enhanced oil recovery (EOR) applications. The primary purpose of a surfactant flood is to lower the interfacial tension between the injection fluid and the oil to displace crude oil that cannot be mobilized by water alone. The petroleum sulphonates have been the favored group of surfactants for this application due to their ability to lower interfacial tension against crude oil from 30 to less than 10^{-3} mNm^{-1}. These compounds, however, show optimal activity over a narrow range of temperatures and salinities and can therefore rapidly lose their effectiveness as the surfactant "slug" moves through a reservoir.

A wide variety of surfactants is produced by microorganisms. Many of these compounds have been identified and characterized, but only recently have quantitative assessments of biosurfactant performance for EOR been described. Certain biosurfactants have been reported to display some very promising properties for oil recovery. Crude bacterial culture broths can reduce interfacial tension against oil to 10^{-2} mNm^{-1}, and in the presence of an alcohol cosurfactant, interfacial tensions as low as 10^{-5} mNm^{-1} have been measured. Biosurfactant performance is improved by the addition of sodium chloride, and several biosurfactants have been shown to aid recovery of residual oil from laboratory sandpacks. Some biosurfactants are effective at low concentrations and most can be made from cheap carbon sources. In the future they may find applications as emulsifiers or wetting agents in many other industries.

Only one third of the earth's oil reserves, currently estimated to be close to 7×10^{11} barrels (1), are recoverable by conventional oil extraction methods. Primary production of crude oil is driven by the release of natural reservoir pressure. Secondary oil recovery results from the injection of water or gas into the reservoir to sweep oil toward production wells. After waterflooding, as much as two-thirds of the original oil-in-place remains in the reservoir rock pores as residual oil; this is the target for enhanced oil recovery (EOR).

Methods for recovery of residual oil have been under investigation for many years and currently are undergoing extensive field testing. EOR processes may be divided into four categories: thermal, miscible, chemical and microbial. Few processes have found widespread application, and the economics frequently are marginal. Thermal processes have been successful at recovering heavy oils; however, there is still a need for a viable, economic process to recover light oils.

Chemical oil recovery methods include injection of surfactants and polymers. Polymers increase the viscosity of the injected waterflood and hence improve sweep efficiency. The primary aim of surfactant flooding is to lower the interfacial tension between oil and injected water to displace the oil that is trapped after waterflooding. Surfactant injection normally is followed by a polymer "slug" to control the mobility of the flood. Microbial processes for EOR are based on the ability of microbes to produce chemicals with similar properties to the conventional EOR chemicals that may be generated either in fermenters at the surface or within an oil reservoir from an injected carbon source. Microbial polymers, such as xanthan, already are produced on a commercial scale and have found application in the oil industry as EOR polymers.

CONVENTIONAL SURFACTANTS FOR EOR

Residual oil is trapped in the pore structure of reservoir rock by capillary forces. These can be overcome by increasing the capillary number by 10,000-fold; this can be achieved under reservoir conditions by reducing the interfacial tension of the injection fluid from 30 to 10^{-3} mNm^{-1} by adding an appropriate surfactant (2). An increase in the velocity and viscosity of the injection water also could increase the capillary number; however, an increase of this magnitude would cause fracturing of the reservoir rock and equipment failure (2).

Petroleum sulphonates are the major class of chemically synthesized surfactants currently under evaluation for EOR applications. Surfactant formulations are developed for individual reservoirs depending on the temperature, salinity and rock type. They normally consist of oil, water, surfactant, cosurfactant (usually an alcohol) and salts (principally sodium chloride) (3). The surfactant may be injected as a large volume at a dilute concentration or as a small volume of concentrated surfactant (4). The former produces an immiscible displacement of oil; the latter is initially miscible with the oil, although it becomes immiscible as the "slug" moves through the reservoir and surfactant is lost by adsorption on rock. The primary purpose of both types of flood is to lower the interfacial tension between the oil and the injected fluid and hence to recover trapped residual oil.

Petroleum sulphonates can be effective at recovering oil. However, they show optimal activity over a very narrow range of temperatures and salinities (5,6). Thus a small alteration in reservoir conditions can greatly affect their performance. Loss of surfactant due to adsorption on rock is also a problem. Since petroleum sulphonates are synthesized from crude oil distillates, their cost is linked to the price of oil; blending them with alcohols as cosurfactants also increases cost. After reviewing the field trials of surfactant flooding, Kuuskraa and Hammershaimb (7) concluded that oil prices in excess of $30 per barrel would be required to achieve substantial implementation of surfactant-based EOR. In order to compete with the chemically-synthesized surfactants, a biosurfactant must be one of the following: more tolerant of reservoir conditions; effective at lower concentrations; effective without alcohol cosurfactants.

DIVERSITY OF BIOSURFACTANTS

The production of surface active compounds by microorganisms has been known to occur for more than 50 years

(8). All the biosurfactants described contain a lipid component, usually the hydrocarbon chain of a fatty acid. Three principal classes of biosurfactant have been identified: glycolipids, lipopeptides and phospholipids. Phospholipids, although present in every microorganism, are rarely extracellular products. Glycolipids and lipopeptides are produced by a number of different organisms and released into the culture broth in significant quantities.

A list of some of the microorganisms reported to produce biosurfactants is presented in Table 1. The properties of certain of these biosurfactants have been evaluated in some detail.

TABLE 1

Biosurfactants Produced by Some Microorganisms

Organism	Biosurfactants
Bacillus subtilis	crystalline peptidelipid (9)
Bacillus licheniformis	lipopeptide (10)
Bacillus circulans	peptidelipid (9)
Serratia marcescens	aminolipids (11)
Corynebacterium lepus	lipopeptide, phospholipids (12)
Corynebacterium fasciens (13)	—
Pseudomonas aeruginosa	rhamnolipids (14, 15)
Arthrobacter sp.	glycolipid (16, 17)
Ustilago maydis	cellobiose lipids (18)
Rhodococcus erythropolis	trehalose-tetraester (19)
Myxococcus xanthus (20)	—

Lipopeptides

Bacillus subtilis produces a crystalline, lipopeptide surfactant that possesses cytolytic properties against other organisms, such as *Pseudomonas aeruginosa* (9).

A strain of *Bacillus licheniformis*, designated JF2, produces an ionic biosurfactant containing a free amino group and a lipid moiety when grown on a glucose carbon source under both aerobic and anaerobic conditions (10). The biosurfactant is produced during the exponential growth phase. A peptide biosurfactant produced by a species of *Bacillus*, designated 1165, also was produced during the exponential phase when grown on glucose (21,22). Three different aminolipid wetting agents are produced in extracellular vesicles by *Serratia marcescens* (11).

Glycolipids

A species of *Arthrobacter* isolated from soil, designated H-13A, produces a variety of glycolipid surfactants when grown on a hydrocarbon substrate (16,17). Glycolipid is not produced in significant quantities on nonhydrocarbon substrates and is produced maximally on tridecane with a yield of 1.6 gl^{-1}. The major glycolipid contains a disaccharide (trehalose), glycerol, an amino sugar and fatty acids. The exact structure of the glycolipid varies with the carbon number of the hydrocarbon substrate for growth. Glycolipid is synthesized by *Arthrobacter* sp. throughout the growth cycle, mostly occurring in the stationary phase (17).

Pseudomonas aeruginosa produces rhamnolipid biosurfactants as secondary metabolites when grown on hydrocarbons or sugars. They consist of one or two rhamnose units attached to a β-hydroxy decanoic acid lipid component (14,15). Rhamnolipid production by "resting cells" of *P. aeruginosa* was observed when the organism was incubated without a nitrogen source (23).

FUNCTION OF MICROBIAL SURFACTANTS

Many theories have been proposed to explain the production of biosurfactants by bacteria; however, their exact physiological function still remains uncertain in most cases (24). Some biosurfactants are involved in bacterial gliding (20). Phospholipids excreted by *Thiobacillus* play a role in the wetting of sulphur (24). Other biosurfactants (for example, subtilysin produced by *Bacillus subtilin*) display biocidal activities (9). The most common explanation for the production of biosurfactants has been the emulsification of hydrocarbon substrates to facilitate uptake by microorganisms.

BIOSURFACTANT PROPERTIES FOR EOR

The most important property in a biosurfactant for EOR is the ability to reduce the interfacial tension between oil and water by several orders of magnitude. It must achieve this under reservoir conditions, remaining stable at reservoir temperatures and salinities. Finally, it should show minimal adsorption to reservoir rock.

Interfacial Tension

Many biosurfactants have been reported to lower the surface tension of an aqueous medium from 72 to approximately 30 mNm^{-1}. Interfacial tension against hydrocarbon frequently has not been measured or has been measured with a de Nuoy ring tensiometer, which reaches the limits of detection at about 0.5 mNm^{-1}. A spinning drop tensiometer is the instrument used by the oil industry for measuring the ultralow interfacial tensions of petroleum sulphonates for EOR (5); it also has been used to evaluate the performance of some biosurfactants.

Interfacial tensions of some biosurfactants that have been evaluated for EOR are shown in Table 2. Interfacial tension values of $10^{-2} mNm^{-1}$ have been recorded for three different biosurfactants against hydrocarbon: the lipopeptide produced by *Bacillus Licheniformis* JF2 (25), the glycolipid produced by *Arthrobacter* H-13A (16) and the peptide surfactant produced by *Bacillus* 1165 (21). These values were measured in the crude, cell-free culture broth for the *Bacillus* biosurfactants without the addition of cosurfactants although salts were present in the culture growth media and the JF2 medium contained 5% sodium chloride. The addition of 0.5% pentanol to the H-13A glycolipid as a cosurfactant reduced interfacial tension by a further three orders of magnitude to 6 x 10^{-5}, a value well below that required to mobilize residual oil.

Interfacial tension values vary depending on the hydrocarbon against which they are measured. Values obtained for H-13A biosurfactant ranged from 0.57 mNm^{-1} against hexane to 0.02 mNm^{-1} against decane (18).

TABLE 2

Interfacial Tension Values for Some Biosurfactants

Bacterium	Carbon Substrate	Biosurfactant	Minimum Interfacial Tension		
			mNm^{-1}	Salinity (% NaCl)	Hydrocarbon
Pseudomonas aeruginosa	glucose	rhamnolipid	0.5	9	Mixture hydrocarbons (15)
Bacillus licheniformis	sucrose	lipopeptide	0.012	5	Crude oil (10,25)
Bacillus sp. 1165	glucose	peptide surfactant	0.032	0	Hexadecane (21,22)
Arthrobacter sp. H-13A	hexadecane	glycolipid	0.02	1.7	Decane (16)
		glycolipid + 0.5% pentanol	6×10^{-5}	1.7	Undecane (16)

Sodium Chloride Tolerance

Sodium chloride concentration greatly influences the interfacial tension of petroleum sulphonate solutions. The biosurfactants of *Bacillus* 1165 (21) and *B. licheniformis* JF2 (25) are more effective at lowering interfacial tension in the presence of sodium chloride. Strain JF2 was isolated from water injection brine and grows at sodium chloride concentrations up to 10%. Surface tension decreased as sodium chloride was added to the biosurfactant from 31 mNm^{-1} to a minimum of 27 mNm^{-1} at 5% NaCl. The critical micelle concentration (CMC) also decreased from 1500 mgl^{-1} to 1.5 mgl^{-1} as sodium chloride concentration increased to 5%. Interfacial tension measurements on extracted *Bacillus* 1165 biosurfactant were reduced from 0.09 to 0.07 on the addition of 5% NaCl and a much lower concentration of biosurfactant was required to reach the CMC value.

Temperature Stability

In only a few cases has the temperature tolerance of biosurfactants been explored. The *Bacillus* strains JF2 and 1165 both retained surface activity following heating to 100 C for short periods (21,26). Temperatures above 100 C caused a gradual decrease in JF2 biosurfactant surface activity. The JF2 strain grows and produces biosurfactant at temperatures up to 50 C.

Oil Mobilization from Sandpacks

Bacillus 1165 biosurfactant has been tested for its ability to mobilize residual oil from laboratory sandpacks (21). Sandpacks with permeabilities of approximately 25 Darcies were saturated with oil and then waterflooded until no further oil was released from the pack. A biosurfactant flood injected into the pack at a rate of 24 ftd^{-1}, either as an extract or in the culture supernatant, recovered up to 87% of the residual oil. The flow velocity and permeability are considerably higher than would be encountered in the field. However, these results indicate the considerable potential of biosurfactants for EOR. The biosurfactants produced by *B. licheniformis* JF2 and *P. aeruginosa* also have been reported to aid recovery of oil from sandpacks (26,27).

Biosurfactant Adsorption

Little information is available on the adsorption of biosurfactants on sand or reservoir rock. Sandpack studies on *Bacillus* 1165 biosurfactant have revealed that extracted, redissolved biosurfactant passed straight through the pack, appearing in the second pore volume of the eluate, whereas biosurfactant in the culture supernatant was delayed until the fourth pore volume. This may have been due to the adsorption of biosurfactant in the supernatant on to the sand delaying its appearance until all the active sites were saturated (21).

PRODUCTION AND EXTRACTION OF BIOSURFACTANTS

Pseudomonas aeruginosa rhamnolipids have been produced in a 50 l continuous pilot bioreactor operated at a dilution rate of 0.065 h^{-1} using a medium containing 3% glucose and growth-limiting concentrations of nitrogen and iron (15). Higher glucose concentrations were not practical as they caused excessive foaming. These conditions produced a yield of biosurfactant in the culture broth of 2.5 gl^{-1}. The rhamnolipid was extracted by adsorption on to a polystyrene resin and was subsequently eluted with methanol. It was further purified by anion exchange chromatography. The final product was 90% pure and the overall recovery of active material was above 60%. Semicontinuous rhamnolipid production was obtained by immobilizing *P. aeruginosa* with Ca-alginate (23).

B. licheniformis JF2 has been cultivated in a 50 l fermentor from which 0.1 to 0.7 gl^{-1} of lipopeptide was extracted (10). The extraction process involved the precipitation of the surfactant by adding hydrochloric acid to the culture broth supernatant to decrease the pH to 2.

Biosurfactants have been extracted on a smaller scale using solvents. *Bacillus* 1165 surfactant was successfully extracted with butanol (21) and the *Arthrobacter* glycolipid was extracted with methanol (16).

BIOSURFACTANT PRODUCTION WITHIN THE RESERVOIR

If an effective biosurfactant could be synthesized by an anaerobic bacterium growing within an oil reservoir, all production and extraction costs would be eliminated.

The economics of such a process could be very attractive. However, biosurfactants have yet to be field tested and adopted as viable EOR chemicals as the microbial polymer, xanthan gum, has been for polymer flooding.

Bacillus licheniformis JF2 produces biosurfactant under anaerobic conditions using nitrate at salinities up to 10% NaCl and at temperatures up to 50 C in the presence of crude oil. This organism is thus a potential candidate for in situ biosurfactant production.

Certain microorganisms have been shown to penetrate Berea sandstone cores with permeabilities as low as 200 mD (25). A computer search of reservoir data in the 10 major oil-producing states in the U.S. has shown that 27% of reservoirs have environmental conditions readily compatible with the growth of microorganisms (28).

DISCUSSION

Although the production of biosurfactants by microorganisms has been known for many years, very little quantitative assessment of biosurfactant performance has been undertaken. Biosurfactants with the ability to reduce interfacial tension against oil to values close to those required for EOR, without the addition of cosurfactants, have been described. Sodium chloride improves their performance, and exposure to temperatures up to 100 C does not appear to reduce their activity. The ability of biosurfactants to recover crude oil from sandpacks also has been demonstrated. These conditions require further testing and optimization.

The greatest economic benefit to EOR probably lies in the production of biosurfactant from cheap sugar sources by organisms growing within an oil reservoir, thus eliminating production and extraction costs. Biosurfactants also may find applications as emulsifiers, wetting agents and dispersing agents in a number of other industries in the future.

REFERENCES

1. Institute of Petroleum, Petroleum Statistics, 61 New Cavendish St., London W1M 8AR (1985).
2. Shah, D.O., Introduction, in *Surface Phenomena in Enhanced Oil Recovery*, edited by D.O. Shah, Plenum Press, New York, 1981, p. 1.
3. Putz, A., J.P. Chevalier, G. Stock and J. Philippot, *France. J. Petrol. Technol. 33*:710 (1981).
4. Oh, S.G., and J.C. Slattery, *Proc. 2nd ERDA Symp. Enhanced Oil and Gas Recovery*, Tulsa, OK, D-2/1 (1976).
5. Burkowsky, M., and C. Marx, *Tenside Detergents 15*:247 (1978).
6. Chan, K.S., and D.O. Shah, *J. Dispersion Sci. Technol. 1*:55 (1980).
7. Kuuskraa, V.A., and E.C. Hammershaimb, *Int. Bioresources J. 1*:85 (1985).
8. Wolf, C.G.L., *Biochem. J. 17*:813 (1923).
9. Takahara, Y., Y. Hirose, N. Yasuda, K. Mitsugi and S. Muroo, *Agric. Biol. Chem. 40*:1901 (1976).
10. Javaheri, M., G.E. Jenneman, M.J. McInerney and R.M. Knapp, *Appl. Environ. Microbiol. 50*:698 (1985).
11. Matsuyama, T., T. Murakami, M. Fujita, S. Fujita and I. Yano, *J. Gen. Microbiol. 132*:865 (1986).
12. Cooper, D.G., J.E. Zajic and D.E. Gerson, *Appl. Environ. Microbiol. 37*:4 (1979).
13. Akit, J., and J.E. Zajic, in *Dev. Ind. Microbiol.*, p. 445 (1981).
14. Edwards, J.R., and J.A. Hayashi, *Arch. Biochem. Biophys. 111*:415 (1965).
15. Reiling, H.E., U. Thanei-Wyss, L.H. Guerra-Santos, R. Hirt, O. Kappeli and A. Fiechter, *Appl. Environ. Microbiol. 51*:985 (1986).
16. Singer, M.E., W.R. Finnerty, P. Bolden and A.D. King, in *Symp. Biol. Pressures Related to Petrol Rec.*, Am. Chem. Soc. Div. Petrol. Chem., Seattle, WA, 1983, p. 785.
17. Finnerty, W.R., in *Contracts for Field Projects and Supporting Research in Enhanced Oil Recovery*, DOE/BC-85/4, Progress Review no. 44, U.S. Department of Energy, Bartlesville, OK, 1986.
18. Frautz, B., S. Lang and F. Wagner, in *Proc. III Eur. Cong. Biotechnol.*, Verlag Chemie GmbH, Weinheim, 1-79, 1984.
19. Kretschmer, A., H. Bock and F. Wagner, *Appl. Environ. Microbiol. 44*:864 (1982).
20. Dworkin, M., K.H. Keller and D. Weisberg, *J. Bact. 155*:1367 (1983).
21. Brown, M.J., M. Foster, V. Moses, J.P. Robinson, S.W. Shales and D.G. Springham, *3rd Eur. Meet. on Improved Oil Recovery*, Rome, Agip S.p.A., 1985, p. 241.
22. Brown, M.J. and V. Moses, British Patent Specification 8508233, (1985).
23. Wagner, F., J-S Kim, S. Lang, Z-Y Li, G. Marwede, U. Matulovic, E. Ristau and C. Syldatk, *Proc. II Eur. Cong. Biotechnol.*, Verlag Chemie GmbH, Weinheim, 1-3 (1984).
24. Haferburg, D., R. Hommel, R. Claus and H-P Kleber, *Adv. Biochem. Eng. Biotechnol. 33*:53 (1986).
25. Jenneman, G.E., R.M. Knapp, M.J. McInerney, D.E. Menzie and D.E. Revus, *3rd Joint SPE/DOE Symp. on Enhanced Oil Recovery*, Tulsa, OK, SPE/DOE 10789, 1982, p. 921.
26. Jenneman, G.E., M.J. McInerney, R.M. Knapp, J.B. Clark, J.M. Feero, D.E. Revus and D.E. Menzie, *Dev. Ind. Microbiol. 24*:485 (1983).
27. Guerra-Santos, L.H., U. Thanei-Wyss, L. Katterer, O. Kappeli, A. Fiechter and E. Puhar, *Symp. on Biol. Pressures Related to Petrol. Rec.*, Am. Chem. Soc. Meet., Seattle, WA, 1983, p. 814.
28. Clark, J.B., D.M. Munnecke and G.E. Jenneman, *Dev. Ind. Microbiol. 22*:695 (1981).

Biosurfactants in Cosmetic Applications

Shigeo Inoue
Kashima plant, Kao corporation, 20, Higashi Fukashiba, Kamisu-machi, Kashima-gun, Ibaragi, 314-02, Japan

As the first step for the development of biosurfactants, we focused on glycolipid surfactants, the development of which has been delayed due to some limitation in raw materials and difficulties in production technology in the field of industrial synthetic surfactants. We started our research by finding a variety of oligo-glycolipid biosurfactant-producing microorganisms from nature. We first collected soil samples, separated hydrocarbon-utilizing microorganisms and then screened oligo-glycolipid producing microorganisms. As a result, we obtained 23 oligo-glycolipid-producing strains. One of these strains, a yeast, *Torulopsis bombicola* KSM-36, produced 10 g or more of glycolipids per liter. As a result of an analytical study on the structures of these glycolipids, it has been found that they are a series of derivatives containing as their backbones sophorolipid (SL) structures in each of which ω- or ω-1-hydroxy fatty acid has been β-glucosylated with a sophorose group obtained by bonding two moles of glucose together. We determined production conditions under which sophorolipid can be produced at a stable rate of about 100 g to 150 g per liter of a combined carbon source of palm oil and glucose. We prepared alkyl-SL esters through esterification between a group of long-chain fatty alcohols and SL. On the other hand, being attracted by the hydroxyl group of sophorose group of SL, we prepared P-SL by subjecting a propylene oxide and SL to addition polymerization. We found that oleyl-SL and P-SL, the latter being an addition-polymerized product of one mole of SL and about 12 mol of propylene oxide, have specific compatibility to the skin. They have found commercial utility as skin moisturizers.

Owing to the remarkable achievement in the study of hydrocarbon fermentation over the last 20 years, some hydrocarbon-utilizing microorganisms have been found to exude and produce emulsifying materials, namely biosurfactants, upon incorporation of hydrocarbons into their cells. Biosurfactants produced by these microorganisms have certain structural characteristics that are not seen in industrial synthetic surfactants. These biosurfactants thus are believed to bring about some new possibilities in the current field of industrial surfactants provided that they could be developed as industrial surfactants.

Accordingly, we focused our primary attention on glycolipid surfactants as the first step in the development of biosurfactants. Their development has been delayed due to some limitations in raw materials and difficulties in production technology in the field of industrial synthetic surfactants.

CURRENT SITUATION IN BIOSURFACTANT RESEARCH

Microbial biosurfactants which have been reported to date may be classified, depending on the structures of their hydrophilic moieties, into five types as shown in Table 1.

TABLE 1

Type of Biosurfactants

Hydrophilic group	Biosurfactant
Sugar	Glycolipid
Oligo-peptide	Acylpeptide
Phosphate	Phospholipid
Carboxylate	Fatty acid
Protein, polysaccharide	Biopolymer

Application of biosurfactants has been attempted in a wide variety of fields, including tertiary petroleum recovery from exhausted oil wells, pollution-free agents for removing petroleum pollution, flotation reagents for metals, hydrocarbon fermentation, industrial surfactants (with which we are concerned) and so on (Table 2). The usefulness of biosurfactants in the technical field of tertiary petroleum recovery in particular has been widely known. It is well known that active development and research work in Europe and the U.S. currently is under way.

SCREENING OF GLYCOLIPID-PRODUCING MICROORGANISMS

We collected samples from natural sources, separated hydrocarbon-utilizing microorganisms from these samples, and then screened for oligo-glycolipid-producing microorganisms. As a result, we obtained 23 oligo-glycolipid-producing strains. One of these strains, which is a yeast, produced 10 g or more glycolipid per liter. Thus, we decided to proceed with a further study on the glycolipid of this yeast strain, namely KSM-36 (Table 3).

From identification tests conducted on the yeast in the usual manner, it has been found that this glycolipid-producing yeast, KSM-36, is very similar to *Torulopsis bombicola* ATCC 22214, which had been found earlier by Spencer and Tulloch (1). ATCC 22214 is known to produce sophorolipid, similar to our yeast, but this particular ATCC strain was isolated from a Canadian wildflower by means of a high concentration glucose medium. On the other hand, our yeast was isolated from cabbage leaves, using a hydrocarbon medium (2,3).

STRUCTURE OF GLYCOLIPID

The glycolipid produced by the KSM-36 strain is not a single substance, but, as shown in Figure 1, is a mixture of many glycolipid compounds. From an analytical study of the structures of these glycolipids, it was found that they are a series of sophorolipid (SL) derivatives containing as their backbone Acid-SL (type IV) in each of which ω- or ω-1-hydroxy fatty acid has been β-glucosylated with a sophorose group obtained by bonding two moles of glucose.

TABLE 2

Application of Biosurfactants

Application	Source
Enhance of oil recovery (Tertiary oil recovery)	
(1) Injection of biosurfactants	Whole fermentation broth
(2) Microbially assisted	*Desulfovibrio* sp.
	Clostridium sp.
Removal of oil from tar sand and oil shale	*Pseudomonas* sp., *Clostridium* sp., *Coryne.* sp., Corynomycolic acid
Oil spill clean-up	Emulsan, Bioemulsifier, Rhamnolipid
Flotation separation of minerals	*Pseudomonas* sp., *Alcaligenes* sp.
Stimulant of cells-growth in hydrocarbon fermentation	Trehalose lipid-I, II Rhamnolipid-I, II, III, IV Sophorolipid-I Mannosyl erythitol lipid Hydrocarbon-emulsifying factor Protein-like activator
Industrial surfactant	Trehalose lipid-I, II Sophorolipid I Spiculosporic acid
Other uses	Surfactin (Subtilysin) Surfactin analog

TABLE 3

Screening of Glycolipid-Producing Microorganisms from Natural Sources

1. Natural sources (Total 91 Places)
   Soil 20
   Oily waste water 9
   Plants 46
   Honey 16

2. Enrichment culture (n-Hexadecane Medium)

3. Single colony isolation
   Bacteria 78 Strains
   Yeast 134 Strains

4. Glycolipid-producing microorganisms
   Glycolipid (g/l)
   0-1.0 21 Strains
   1.0-10.0 1 Strains
   10.0- 1 Strains (KSM-36)

5. Identification

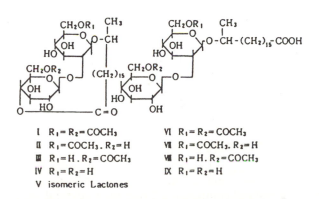

I $R_1 = R_2 = COCH_3$ VI $R_1 = R_2 = COCH_3$
II $R_1 = COCH_3. R_2 = H$ VII $R_1 = COCH_3. R_2 = H$
III $R_1 = H. R_2 = COCH_3$ VIII $R_1 = H. R_2 = COCH_3$
IV $R_1 = R_2 = H$ IX $R_1 = R_2 = H$
V   isomeric Lactones

FIG. 1. Structure of sophorolipids produced by *Torulopsis*.

FERMENTATION PRODUCTION OF SOPHOROLIPID

When a single carbon source was employed, an oil or wax was found to bring about higher sophorolipid production than that generated by n-paraffins. In view of the structure of sophorolipid, there seemed to be a chance to produce sophorolipid from glucose and a fatty acid through the salvage biosynthetic pathway. Accordingly, further investigation was carried out on carbon sources, each making combined use of aliphatic substances. As a result, high sophorolipid productivity was observed, as estimated, in systems containing two or more carbon sources in combination (Fig. 2). A radioisotopic experiment has shown that the sophorolipid was produced from glucose and palm oil through the salvage biosynthesis pathway.

The scale-up of sophorolipid fermentation and production was investigated using optimum conditions as to the medium composition and culturing environment that we had investigated. As a result, we succeeded in determining conditions under which sophorolipid can be produced with a source of glucose and palm oil (Fig. 3). Although more sophorolipid can be produced by increasing the concentrations of glucose and palm oil, such high glucose and palm oil concentrations develop a foaming problem of the liquid medium due to sophorolipid which is disadvantageous from the viewpoint of production engineering. Recovery of the sophorolipid was carried out with a heated-phase separation of the liquid medium.

APPLICATION OF SOPHOROLIPID

It is feasible to esterify SL with a fatty alcohol through the carboxyl group. Accordingly, we prepared alkyl-chain fatty alcohols and SL. On the sophorose group of SL, we prepared a series of polyoxyalklene-sophorolipid compounds by reaction of SL with an alkylene oxide, specifically propylene oxide or P-SL (Fig. 4).

The basic surface activity levels of a group of long-chain alkyl-SLs prepared from sophorolipid are shown in Table 4.

It was found that P-SLs complement the natural moisturizing factor (NMF) (Fig. 5).

It has been found through application tests that these sophorolipid derivatives display some unique properties unseen in conventional synthetic surfactants. Oleyl-SL

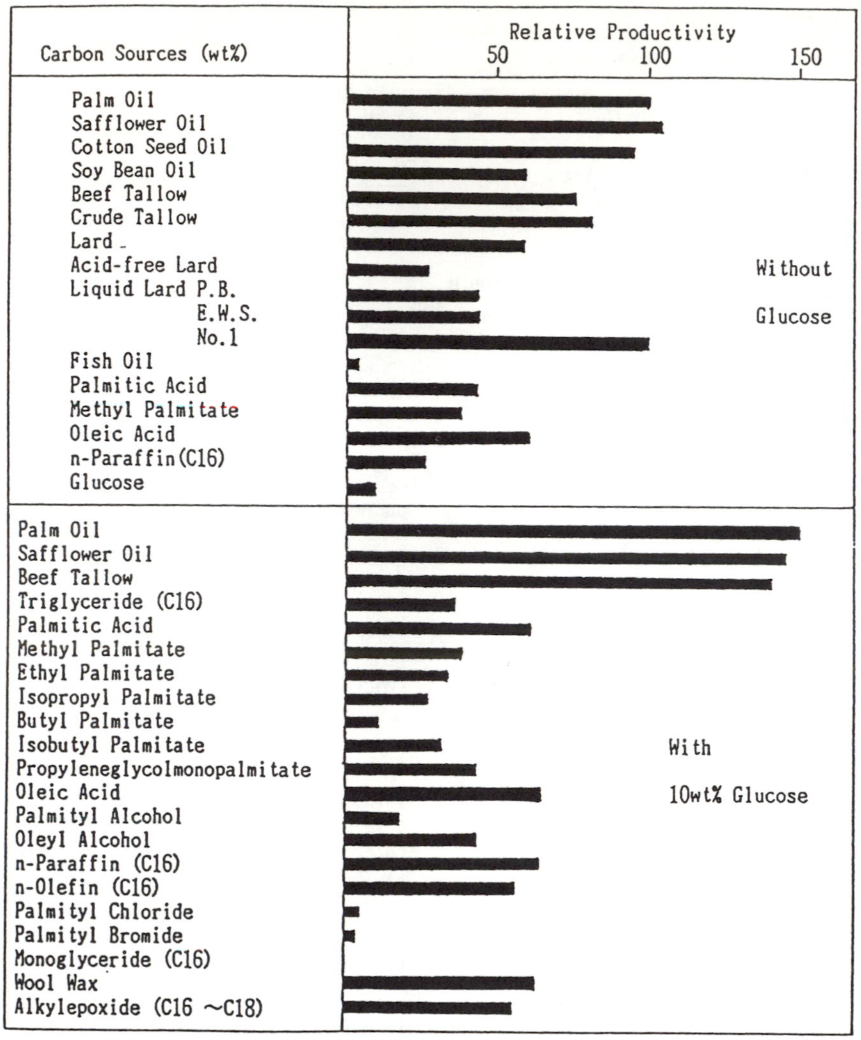

Carbon Sources (wt%)	Relative Productivity

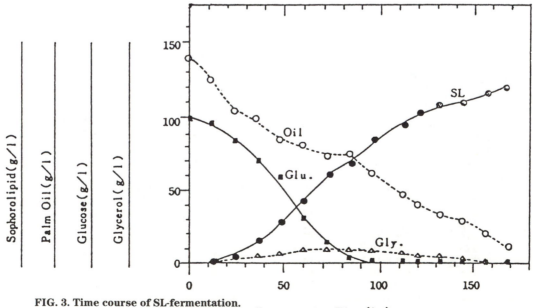

FIG. 2. Effect of carbon sources on the SL production by *T. bombicola* KSM-36. Basal medium: 0.5 wt% yeast extract in a total volume of 50 ml. Cultivation was carried out using shaking flasks.

FIG. 3. Time course of SL-fermentation.

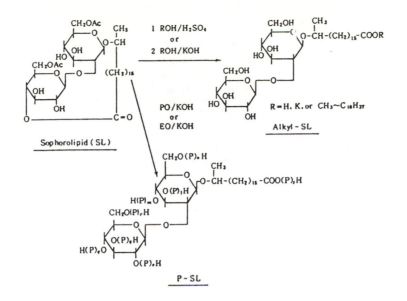

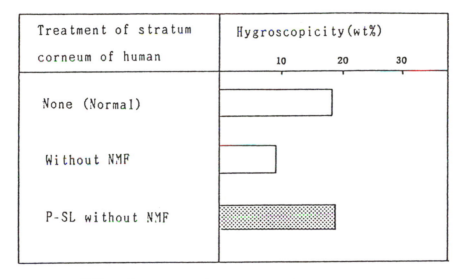

FIG. 4. SL and their derivatives.

NMF : Natural moisturizing factor

FIG. 5. Effect of P-SL to hygroscopicity of stratum corneum.

TABLE 4

Surface Activity of Alkyl-SL

Alkyl-SL	Surface Tension dyne. cm^{-1}	CMC wt %	HLB
Acid-SL	34.8	3.2×10^{-2}	$> 25 \sim 45$
Methyl-SL	38.0	5.4×10^{-3}	$> 25 \sim 45$
Ethyl-SL	40.0	2.9×10^{-3}	$\geqq 20 \sim 45$
Propyl-SL	39.5	1.4×10^{-3}	19.2
Butyl-SL	37.5	7.6×10^{-4}	18.5
Amyl-SL	36.5	4.6×10^{-4}	17.7
Hexyl-SL	36.5	3.2×10^{-4}	17.0
Octyl-SL	35.8	2.8×10^{-4}	15.5
Decyl-SL	38.5	5.3×10^{-4}	14.0
Myristyl-SL	48.8	6.9×10^{-3}	11.0
Oleyl-SL	— —	— —	$7 \sim 8$

and P-SL in particular have specific compatibility to the skin. Thus, they have found commercial utility as skin moisturizers.

REFERENCES

1. Tulloch, A. P., J. F. T. Spencer and M. H. Deinema, *Can. J. Chem. 46*:345 (1968).
2. Inoue, S., S. Itoh, *Biotechnol. Lett. 4*:3 (1982).
3. Itoh, S., and S. Inoue, *Appl. Environ. Microbiol., 43*:1278 (1982).

Discussion Session

In a question directed to Saul L. Neidleman, Colin Ratledge commented that all of his examples quoted had used cell systems growing at uncontrolled growth rates. When experiments were carried out with microorganisms growing at constant growth rates (in chemostats), the changes were much less obvious. He said there might be some changes in the degree of unsaturation of membrane fatty acids, but fatty acids of nonmembrane lipids largely are unaffected. Further, he indicated that the results could be explained by the decreased temperature inducing a decreased growth rate. The speaker generally agreed and indicated that because all work could not be done in chemostats, lowering temperature caused the desired effect of changing unsaturation. W.R. Finnerty indicated that the isolation procedure he employed was to treat the biomass with sulfuric acid, remove the cells by filtration, and extract the product with hexane. In answer to a second question regarding the mechanism of conversion of alkane to acids and alcohols, he indicated there probably were two pathways involved: one for production of carboxylic acids and one for the alcohols.

This session demonstrated the many ways microorganisms can be used to prepare a variety of lipid materials without the use of advanced biogenetic engineering processes.

Bioreactors for Hydrolysis of Fatty Oils

Jiro Hirano
Tsukuba Research Laboratory, Nippon Oil & Fats Co. Ltd., Tsukuba, Ibaraki, Japan

Lipases that can be mass produced have been developed through studies on microbial lipase that have been conducted since the 1960s. In particular, the highly active lipase from *Candida cylindracea* has been widely used in studies for the commercialization of the hydrolysis of oils and fats. The hydrolytic reaction has been investigated mainly using the two-phase reaction system. Various bioreactors, including stirred-tank, packed-bed and membrane systems, have been progressively developed. The membrane-type bioreactor equipped with many disperser membranes has been evaluated as a highly efficient and compact reactor. For high-melting fats, the solid-phase static process has been evaluated.

The oleochemical industry is the chemical industry that utilizes fatty acids obtained by hydrolyzing oils and fats as a basic material. In Japan, about 240,000 tons of fatty acids are produced by this industry annually.

For the industrial processes involved in hydrolysis of fatty oils, those of caustic hydrolysis, Twitchell, autoclave and the like have been adopted to date (1). Presently, the Colgate-Emery process, which was established by modifying the Mills' continuous and countercurrent hydrolytic method (1935), is widely employed (2). In this process, the fatty oil material and water continuously are fed into a fat splitter (height, about 20m) from the bottom and top inlets respectively at 250–260 C and 55 kg/cm^2. This process is very efficient, allowing the degree of hydrolysis to reach 98–99% for a residence time of 2–3 hours (3).

However, following the 1973 oil crisis, production systems aimed at saving resources and energy were sought in most industries. Since then, in the fat processing industry the conversion of processes based on chemical reactions to processes based on biochemical reactions has been seriously investigated.

PROCESSES OF ENZYMATIC HYDROLYSIS

With regard to methods to hydrolyze fatty oils using lipases, there exists an historical report by Connstein in 1902 on the lipase from castor beans. Studies (4) on microbial lipases, aggressively conducted from the 1960s through the 1970s, have made it possible to mass produce and appropriately apply these lipases, leading to the extensive development of enzymatic hydrolysis technology as an industrial process. In particular, the lipase from *Candida cylindracea*, which exhibits high hydrolytic activity, has been used successfully in industrial studies (5). Furthermore, heat-resistant lipases have been developed for hydrolysis of high-melting fats (6).

On the other hand, the repeated use of costly lipases for a long period is one of the most important problems of industrialization. Hence, the techniques for immobilizing lipase have been investigated, and as a result the method for immobilizing the enzyme by entrapping it with a lipophilic gel was developed using a polypropylene glycol-type photo-crosslinking prepolymer (7).

In the biochemical hydrolysis of fatty oils, both the substrate and the product are almost water-insoluble, requiring the use of reaction systems in which the lipid is dissolved in a water-saturated organic solvent at a high concentration (homogeneous), or the use of a biphasic reaction system that is composed of a water phase and an oil phase (or an organic solvent phase containing the lipid at a high concentration).

BIOREACTORS FOR HYDROLYSIS

A biphasic reaction system that is composed of aqueous and organic phases is used mainly in the hydrolytic reaction of fatty oils. The organic phase is composed of the lipid itself or of an organic solvent phase in which the lipid is dissolved at a high concentration.

The primary advantages of this biphasic reaction system follow:
1. Because concentrations of the substrate and product in the reactor are high, the productivity of the apparatus is improved, and the product is easily recovered.
2. As the product is transferred into the organic phase, inhibition against the enzyme is less.
3. The immobilization of the enzyme is not always essential, but the enzyme can be easily separated from the product.

In such a biphasic reaction system, the lipase exhibits its activity by being adsorbed on the interface between the organic and aqueous phases. Thus, it is necessary to provide an effective interfacial area. To achieve this, the following systems have been used: stirred-tank reactors in which stirring is conducted efficiently, packed-bed reactors in which immobilized enzymes are packed and membrane-type reactors with porous thin films.

In addition, the hydrolysis of triglyceride by a homogeneous reaction system has been investigated, and there has been a report on the continuous hydrolysis of 2.5% olive oil in diisopropyl ether by the lipase from *Rhizopus arrhizus* (8).

Stirred-tank Reactor

Hydrolysis of solid fats (9). The lipase from *C. cylindracea* was dissolved in M/15 phosphate buffer, and solid fat (tallow) was dissolved in a nonpolar solvent (isooctane). The biphasic reaction mixture composed of these aqueous and organic phases was submitted to a hydrolytic reaction at 37 C in a stirred tank with a six-wing impeller turbine. The degree of hydrolysis was 62% after 96 hr in a system without the addition of isooctane, and 100% after 48 hr in a system with the addition of isooctane. The reaction rate attained a maximum when isooctane was added at 20% of the aqueous phase and was lowered when added at more than 20%. The reaction rate on isooctane addition of 60% reached a level similar to that without isooctane. The reaction rate was increased with the increase in the rotational rate of the impeller up to a maximum level at 500 rpm.

Hydrolysis of olive oil (10). Olive oil was submitted to an hydrolytic reaction at room temperature in a phosphate buffer solution by stirring it with the lipase from *Candida rugosa*. Linear relationships were found between the degree of hydrolysis (in %) and the logarithm of the reaction time or the enzyme concentration in the reaction medium. The reaction temperature did not affect the degree of hydrolysis in a range from 26–46 C. Olive oil was hydrolyzed more rapidly than tallow and coconut oil.

Packed Bed Bioreactor

Hydrolysis of olive oil (7). An olive oil/water emulsion (1:1,v/v) was prepared in a service tank. A column reactor was packed with *C. cylindracea* entrapped in a hydrophobic polypropylene glycol resin prepolymer. The emulsion was recycled into the column reactor. Each batch was recycled at 30 C for 48 hr. A stable degree of hydrolysis was obtained at the third batch after 48 hours and after that the degree of hydrolysis was almost constant (about 75%) through the tenth batch.

Membrane Bioreactor

Hydrolysis of olive oil (11–13). A flow of oil and a flow of buffer (containing 17–20% glycerol) made interfacial contact with each other continuously through a hydrophobic membrane made of microporous polypropylene (maximum pore size, 0.4 μm; void fraction, 55%; thickness, 25 μm; total effective area of membrane, 11.6 cm x 31.3 cm x 2) to be submitted to the hydrolytic reaction. *C. cylindracea* lipase was adsorbed on the membrane and stabilized with added glycerol beforehand. The half-life of the adsorbed enzyme was 15 days at 40 C. The fatty acids produced were distributed in the oil phase and collected. The degree of hydrolysis was maintained at around 80% for 50 hr.

As a result of the investigation on the dynamics of the continuous hydrolytic reaction with plate and hollow-fiber membrane reactors, it was found that the ratio of flow rate of oil (F)/total membrane area (A) was important, and that a first-order reversible reaction formula was found over a wide range of the ratio F/A for the countercurrent plate membrane reactor. However, such a reversible formula was found only for a higher range of F/A for the hollow-fiber reactor.

In this hydrophobic membrane reactor, higher degrees of hydrolysis were attained with use of the countercurrent system than with the cocurrent system. It was supposed that the reason for this was that oleic acid, the main product of hydrolysis of olive oil, was solubilized in the buffer-glycerol phase more homogeneously and more concentratedly, and that it affected adsorbed lipase.

Sandwich-type membrane bioreactor (14). This reactor was composed of water phase, hydrophilic cellulose acetate membrane, enzyme chamber, hydrophobic porous polyethylene membrane and oil phase in sequence. The hydrolytic reaction was conducted in an enzyme chamber (2 mm thick) sandwiched between the two membranes. An enzyme solution was made to flow through the chamber, and the water and oil entered into the chamber through the respective membranes to form a W/O-type emulsion. The *C. cylindracea* enzyme that exists on the surface of this emulsion catalyzed the hydrolytic reaction. In a stationary state, the concentration of glycerol in the chamber was 15–20%. This glycerol stabilized the lipase and maintained the degree of hydrolysis at a level of 80% without any further supply of lipase.

Hydrolysis of tallow with microporous membrane bioreactor (15). A heat-resistant lipase obtained from *Thermomyces lanuginosus* was contained in a microporous membrane made of acrylic resin (thickness, 0.1 mm; pore size, 0.2 μm). The activity yield was 0.4–0.9%. Tallow heated at 50 C was fed into a bioreactor equipped with a pleated filter element and submitted to the hydrolytic reaction.

When buffer was made to flow, and pH of the product in the aqueous phase was held at higher than 5.0, the half-life of the enzymatic activity was raised to at least more than 20 days.

Solid-phase Static Hydrolysis (16)

For high-melting fats, a concentrated W/O-type suspension was submitted to the hydrolytic reaction under static state (no mixing). The aqueous solution (0.7 parts) of *C. cylindracea* at 20 C, and one part of molten tallow at 40 C was rapidly mixed together using a homogenizer to prepare a homogeneous W/O dispersion system in which the mean size of water particles was less than 0.03 mm (see Fig. 1). The dispersion obtained was maintained at temperatures lower than the melting point of the fat and submitted to the hydrolytic reaction in a solid state without any stirring.

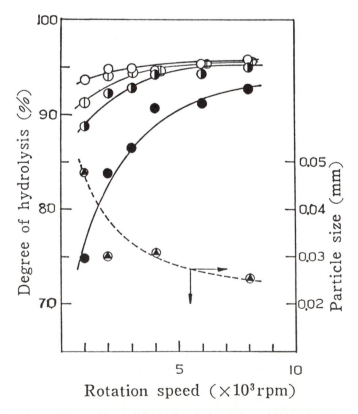

FIG. 1. Effect of the rotation speed of a homomixer on the degree of hydrolysis after one day (●); two days, (◐); three days, (◉); four days, (○); and on water particle size in W/O emulsion (⊕).

In this system, a 95% degree of hydrolysis of the tallow was achieved by the solid-phase static hydrolytic method without use of any surfactant, organic solvent and long-period mechanical agitation (Fig. 2).

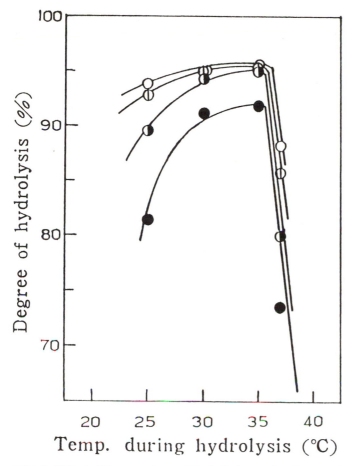

FIG. 2. Effect of temperature of hydrolysis on the degree of hydrolysis (%) after one day, (●); two days, (◐); three days, (◑); four days, (○).

MULTI-STAGE HETEROGENEOUS BIOREACTOR (17)

A hydrolytic reaction of fatty oils was conducted using a continuous, multistage, heterogeneous bioreactor in which oil particles were dispersed into the aqueous phase through a dispersing membrane made of high polymer to increase the area of the interface between the oil and aqueous phases (Fig. 3). A polymer membrane that has a large hydrogen bond component of the critical surface tension and a large adhesive tension between the membrane and the water phase is effective in such cases. The homogeneous dispersion of oil droplets (diameter 0.1–2 mm) was obtained continuously by passing the oil through the dispersing membrane using the gravity difference between the oil and water phase. In addition, the efficiency of the interfacial reaction was high because the oil/water interface was renewed and oil droplets were redispersed at every stage by utilizing many dispersing membranes in the multistage column system.

When, using *C. cylindracea* lipase (500 U/ml), oil droplets were passed through such a multistage heterogeneous bioreactor under conditions of oil droplet size of 0.2 mm and an ascension rate of 0.5 g/min at 37 C, the

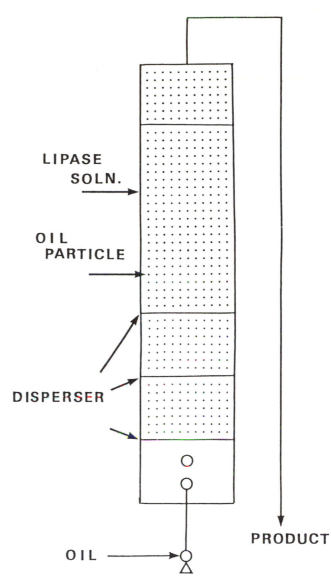

FIG. 3. The scheme of the FN-2 bioreactor. The bioreactor is consisted of connecting sets. A set is 30 mm diameter and 70 mm height and has a disperser.

one-pass degree of hydrolysis was 90% in case of a five-stage reactor and 96% in case of an eight-stage reactor (Fig. 4).

DISCUSSION

Currently, in the oil and fat chemical industry, research and development has been promoted on various lipid products of high added value including useful products from unsaturated fatty acids and phospholipids.

These useful lipid products are synthesized efficiently by making use of biochemical reactions; therefore, the research and development of various bioreactors is being actively pushed forward for industrial production. Because intermediates and products of the oil and fat industry are, in general, relatively low-priced, their production by bioreactors rather than conventional chemical pro-

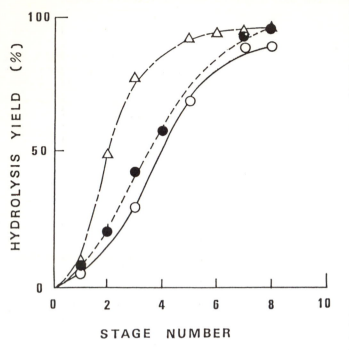

FIG. 4. Hydrolysis yield at the top of each stage of the FN-2 bioreactor after one hour of feeding the substrate at 37 C. Particle size was 0.2 mm diameter. Enzymatic concentrations were ○, 100 U/ml; ●, 300 U/ml; △, 500 U/ml.

cesses is economically less competitive. Thus, bioprocessing is not yet established as a large-scale industrial process in the oil and fat chemical industry. Therefore, it should be expected that the industrialization of bioreactors will be attained for the production of lipid products which cannot be synthesized by conventional chemical processes, or for highly functional products of high added value.

REFERENCES

1. Rattray, J.B.M., *J. Am. Oil Chem. Soc. 61*:1701 (1984).
2. *Kirk-Othmer Encyclopedia of Chemical Technology*, 2nd edn., Interscience Publishers, New York, 1967, p. 826.
3. Inaba, K. and J. Hirano, *Shibosan Kagaku (Chemistry of Fatty Acids)*, Saiwai Shobo, Tokyo, Japan, 1981, p. 17.
4. Iwai, M., *Bulletin of Osaka Municipal Research Institute, 39*:1 (1965).
5. Machida, H., *Fragrance Journal 60*:44 (1983).
6. Funada, T., *Papers on Advanced Technology of JITA*, Oct. 1984, Japan Industrial Technology Association.
7. Kimura, K., et al., *Eur. J. Microbiol. Biotech. 17*:107 (1983).
8. Bell, G., et al., *Biotechnol. Bioengrg. 23*:1703 (1981).
9. Kobayashi, T., et al., *Hakko Kogaku 63*:439 (1985).
10. Linfield, W.M., et al., *J. Am. Oil Chem. Soc. 61*:191, 1067 (1984).
11. Hoq, H.M., et al., *Ibid. 62*:1016 (1985).
12. Yamane, T., et al., *Yukagaku 35*:10 (1986).
13. Hoq, H.M., et al., *Enzyme Microb. Technol. 8*:236 (1986).
14. Taniguchi, M., et al., *Rikagaku Kenkyujo Symposium*, p. 45 (1987).
15. Taylor, E., et al., *Biotechnol. Bioengrg. 28*:1318 (1986).
16. Sato, T., et al., *J. Chem. Soc. Japan*, 1358 (1983).
17. Hirano, J., et al., *Bio Industry 4*:237 (1987).

Biosensors for Lipids

Isao Karube and **Koji Sode**

Research Laboratory of Resources Utilization, Tokyo Institute of Technology, 4259 Nagastuta, Midori-ku, Yokohama, Japan

The determination of lipids for the precise control of industrial processes is very important. Quantitative determination of lipids conventionally is carried out colorimetrically. However, complicated pretreatments are required before the measurement. Therefore, the development of rapid and simple determination of lipids is required. In this study, novel biosensors for lipids such as neutral lipids, phosphatidylcholine, cholesterol and free fatty acids are examined. These sensors are based on electrochemical methods using enzyme membranes. Neutral lipids were determined by lipoprotein lipase immobilized membrane and a pH electrode. The measurement was based on sensing pH change caused by hydrolysis of neutral lipids. Phosphatidylcholine was determined amperometrically, measuring oxygen consumption by sequential enzyme reaction (phospholipase D and choline oxidase). Cholesterol and free fatty acid also were determined amperometrically by using an enzyme immobilized membrane and an oxygen electrode. These methods require no special pretreatment of the sample. Furthermore, measurements were completed within 10 min. These sensors appear to be promising and attractive methods for the routine measurement of lipids.

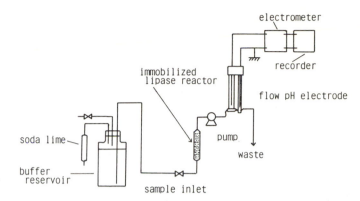

FIG. 1. Schematic diagram of neutral lipid sensor using flow-through pH electrode.

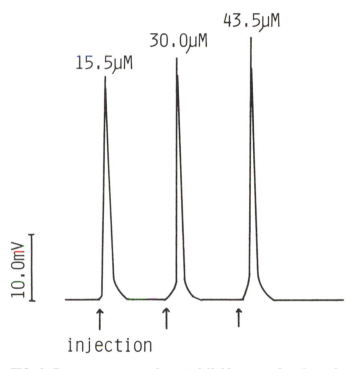

FIG. 2. Response curves of neutral lipid sensor for glyceryl trioleate.

NEUTRAL LIPID SENSOR

Lipoprotein lipase hydrolyzes neutral lipids to glycerol and fatty acids. The fatty acids can be isolated with a pH electrode so that neutral lipids can be determined indirectly by lipoprotein lipase and a pH electrode. A neutral lipid sensor utilizing a lipase-immobilized collagen membrane and a glass electrode has been developed by the authors (1). However, each determination required 20 min, because of the low activity of the immobilized lipase. Therefore, an improved neutral lipid sensor was developed (2). Lipoprotein lipase was covalently bound to polystyrene sheets coated with γ-aminopropyltriethoxysilane, and a new flow-through pH electrode was employed for the sensor. The flow system used for the determination of neutral lipids is illustrated in Figure 1. The electrode system consisted of a tubular glass-responsive membrane electrode and a saturated calomel electrode connected via a ceramic junction (2mm diameter) at the bottom of the tubular electrode. The resistance of the tubular electrode was 10 MΩ, and the total internal volume of the electrode was 110 mm³.

Figure 2 shows the response of the sensor to neutral lipid. The potential of the glass electrode increased with time until a maximum was reached. The time required to reach the maximum was one min; the potential returned to its initial level within three min. The relationship between the logarithm of the concentration of trioleate and the potential difference was linear, changing by 8 mV over the range of 5–50 μM.

The effect of flow rate on the sensor response was examined. If the flow rate was 70 cm³h⁻¹, the immobilized lipase had sufficient time to measurably hydrolyze cholesterol esters. Therefore, a flow rate of 72 cm³h⁻¹ is recommended for lipid determination if cholesterol esters are present.

The reusability of the sensor was examined with various concentrations of neutral lipid. Lipid determinations were carried out 20–25 times per day, and no significant decrease in the response was observed over a 10-day period. The reproducibility of the potential difference was within 5%.

In addition, an amperometric neutral lipid sensor has been developed by the authors. The sensor consisted of a membrane containing lipoprotein lipase and glycerol oxidase and an oxygen probe. A linear relationship was observed between the neutral lipid concentration below 500 mg·dl⁻¹ and the current increase. This sensor can be used for a long time in neutral lipid determination.

PHOSPHOLIPID SENSOR

An enzymatic method for the determination of phospholipids has been developed and used for clinical analysis. The reactions were as follows:

$$\text{Phosphatidyl choline} \xrightarrow{\text{phospholipase D}} \text{phosphatydic acid} + \text{choline}$$

$$\text{Choline} + 2O_2 + H_2O \xrightarrow{\text{choline oxidase}} \text{betaine} + 2H_2O$$

To achieve a rapid and simple phospholipid assay, electrochemical monitoring of these reactions may have definite advantages. The use of immobilized enzymes linked to the direct amperometric measurement of the hydrogen peroxide liberated appeared to be the best approach. Therefore, the authors have developed an enzyme sensor system utilizing phospholipase D and choline oxidase (3).

Phospholipase D and choline oxidase were immobilized together on octyl-Sepharose, porous glass, polystyrene and in a collagen membrane. The enzyme immobilized on octyl-Sepharose showed the highest activity. Therefore, the hydrophobicity of octyl-Sepharose may play an important role.

The measurement system for phosphatidyl choline consisted of an immobilized enzyme reactor with the sensing electrodes positioned close to the reactor (Fig. 3).

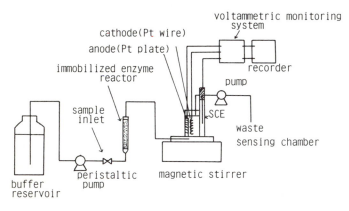

FIG. 3. Schematic diagram of phospholipid sensor.

Figure 4 shows typical responses obtained when various concentrations of phosphatidyl choline were injected. An assay was completed within four min if the phosphatidyl choline concentration was lower than 3 g·dm⁻³. If at least 0.3 I.U. of phospholipase was used, a reliable assay of phosphatidyl choline was obtained. The proportion of the two immobilized enzymes is an important factor in obtaining maximum reaction rate and complete reaction. When the phospholipase/choline oxidase weight ratio after immobilization was 0.9, the reaction was completed within four min.

A flow system was used for more rapid phosphatidyl choline assay and the calibration graph was linear up to 3 g·dm⁻³ for phosphatidyl choline (Fig. 5). More concentrated sample solutions should be diluted with the

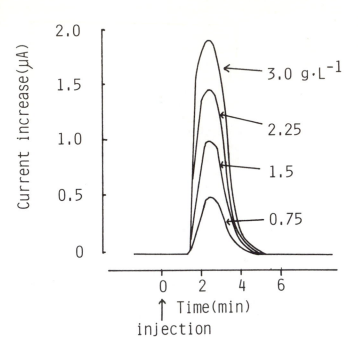

FIG. 4. Response curves of phospholipid sensor for phosphatidyl choline.

buffer. The standard deviation for the determination of 3 g·dm⁻³ of phosphatidyl choline was 0.15 g·dm⁻³ (50 experiments).

The immobilized enzymes were stable for two months when stored at 4 C. About 4% of the activity of the immobilized enzymes was lost after 50 assays.

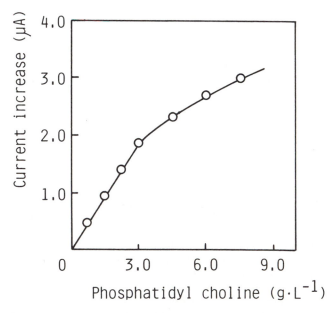

FIG. 5. Calibration curve for flow analysis of phosphatidyl choline.

CHOLESTEROL SENSOR

An enzyme sensor for free cholesterol was developed by the authors (4). This measurement was based on moni-

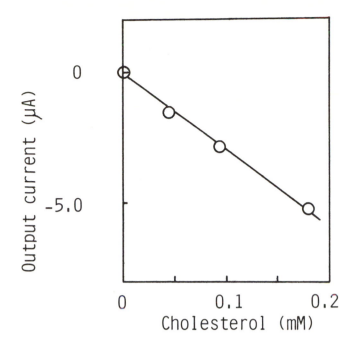

FIG. 6. Calibration curve for cholesterol.

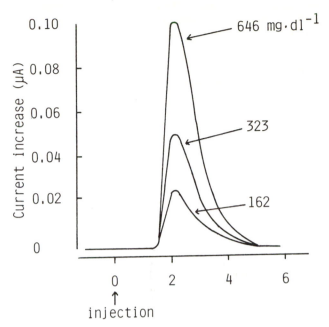

FIG. 7. Response curves of total cholesterol sensor for cholesterol palmitate.

toring the decrease in dissolved oxygen consumed by cholesterol oxidase. Figure 6 shows the calibration curve for cholesterol. A good linear relationship was obtained up to 0.2 mM cholesterol.

The determination of total cholesterol is important in clinical analysis. For this purpose, an immobilized enzyme reactor containing cholesterol esterase and cholesterol oxidase was coupled to an amperometric detector system (5). Both enzymes were covalently immobilized on octyl-Sepharose activated with cyanogen bromide. The hydrogen peroxide liberated by the enzyme reaction was monitored with an amperometric system based on a platinum electrode at 0.60 V vs a saturated calomel electrode. The counter electrode was platinum. The signal obtained was displayed on a recorder. The peak current increased with increasing cholesterol hexadecanoate concentration below 10 g·dm^{-3}. One sample could be assayed in five min (Fig. 7).

Human serum samples were diluted and examined for the concentration of cholesterol with the cholesterol test kit. A linear relationship was obtained between the peak current and the cholesterol concentration in the range 1000–4000 mg·dm^{-3}; a 2000 mg·dm^{-3} sample gave a current of 0.05 μA. The standard deviation for the determination of 3000 mg·dm^{-3} cholesterol was 60 mg·dm^{-3} (that is 2% over 50 experiments). Since human serum contains less than 4000 mg·dm^{-3} of cholesterol, the flow system can be used for such determinations.

The activity of the immobilized enzymes was retained for at least one month at 4 C and no appreciable decrease of their activity was observed after 300 successive assays.

FREE FATTY ACIDS SENSOR

The measurements of free fatty acids in food processing are now of great interest. The conventional methods for free fatty acids analysis are based mostly on spectrometry. Yamada et al. have reported an enzymatic determination of free fatty acids utilizing acyl CoA synthetase as a specific enzyme for free fatty acids having a carbon chain length of 8–20 (6). This enzyme catalyzes the following reaction, subsequently producing AMP, in the presence of ATP, CoA and magnesium:

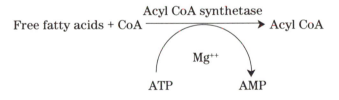

An AMP sensor was developed in our laboratory based on four sequential enzyme reactions leading to dissolved oxygen consumption (7). Combining acyl CoA synthetase with an AMP sensor, the free fatty acids' sensor was constructed (unpublished data).

Acyl CoA synthetase, xanthine oxidase, nucleosid-phosphorylase, 5′-nucleotidase and AMP-deaminase were immobilized in a porous cellulose nitrate membrane by absorption. The free fatty acids' sensor consisted of the enzyme membrane, a Teflon membrane, a dialysis membrane and an oxygen electrode.

In this study, free fatty acids were well-emulsified in phosphate buffer solution, by sonication (200w, 60KHz) for one min. The emulsion thus formed was stable for more than one day. The enzyme sensor was immersed in a 30 ml reaction vessel containing phosphate buffer with 3.2 mM of ATP and 6 mM of magnesium chloride. After the output current became steady, the enzyme sensor was taken out of the reaction vessel. Free fatty acids were then injected into this solution and emulsified by the method described above. The enzyme sensor was again immersed in this sample solution and current decrease was recorded. Within one min after the free fatty acids' sensor was immersed in the sample solution, the output current gradually decreased, and in three min the minimum current was obtained. This assay was completed within five min. Figure 8 shows the typical response curve for oleic acids.

Figures 9 and 10 show the calibration curves of the sensor toward oleic acids and palmitic acids. In both free fatty acids, good linearity was observed between the current decrease and the concentration of free fatty acids below 15 mM. The minimum concentration for the determination was 1.77 mM.

These results suggested that this enzyme sensor appears to be a promising and attractive method for the routine measurements of free fatty acids.

REFERENCES

1. Satoh, I., I. Karube and S. Suzuki, *J. Solid-Phase Biochem.* *2*:1-7 (1977).
2. Satoh, I., I. Karube, S. Suzuki and K. Aikawa, *Anal. Chem. Acta* *106*:369-372 (1979).
3. Karube, I., K. Hara, I. Satoh and S. Suzuki, *Anal. Chem. Acta* *106*:243-250 (1979).
4. Satoh, I., I. Karube and S. Suzuki, *Biotechnol. Bioeng.* *19*:1095-1099 (1977).
5. Karube, I., K. Hara, H. Matsuoka and S. Suzuki, *Anal. Chem. Acta* *139*:127-132 (1982).
6. Shimizu, S., K. Inoue, Y. Tani and H. Yamada, *Anal. Biochem.* *98*:341-345 (1979).
7. Watanabe, E., K. Toyama, I. Karube, H. Matsuoka and S. Suzuki, *J. Food Sci.* *49*:114-116 (1984).

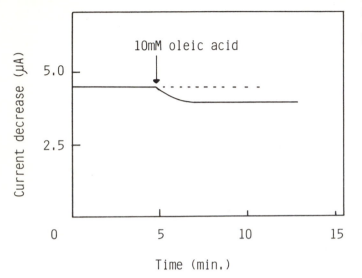

FIG. 8. Response curve of free fatty acids sensor for oleic acid.

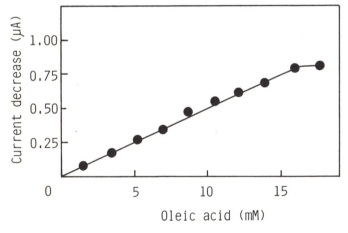

FIG. 9. Calibration curve of free fatty acids sensor for oleic acid.

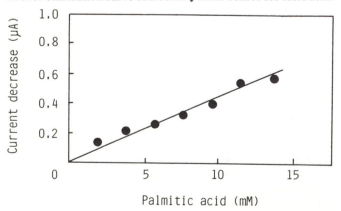

FIG. 10. Calibration curve of free fatty acids sensor for palmitic acid.

Modification of Fats and Oils in Membrane Bioreactors

J. Kloosterman IV*, P.D. van Wassenaar and **W.J. Bel**
Unilever Research Laboratorium Vlaardingen, The Netherlands

Membrane bioreactors (MBRs) offer promising possibilities for an efficient integration of bioconversion and separation processes. The possibility of performing biocatalytic conversions between two immiscible phases separated by a membrane makes this technology particularly interesting for the oils and fats industry in processing emulsions. Examples are the hydrolysis of oils and the esterification of fatty acids and glycerol using lipases. For the esterification of fatty acids, a controlled water activity in the vicinity of the enzymes, which is roughly attainable in MBRs, is important. Several approaches for operating membrane bioreactors are discussed.

Optimization of a membrane bioreactor (network) in terms of biocatalytic and economical performance requires more research. A simplified general costing procedure is presented which may be used to identify factors greatly influencing the economic aspects and help to establish research targets.

Conventionally, lipase-catalyzed reactions are carried out in emulsion systems where reactions take place at the interface of oil droplets. Such an emulsion system has certain drawbacks for industrial processes. Enzyme recovery often is troublesome and the emulsion sometimes is difficult to break.

In general, one may distinguish four types of lipase-catalyzed reactions: lipolysis, esterification, interesterification and glycerolysis (Fig. 1). The composition of the

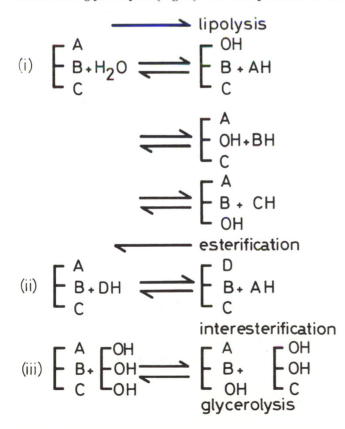

FIG. 1. Examples of (i) lipolysis and esterification, (ii) interesterification and (iii) glycerolysis reactions. AH, BH, CH and DH represent fatty acids (additional reactions giving various isomers are omitted in reactions [ii] and [iii]).

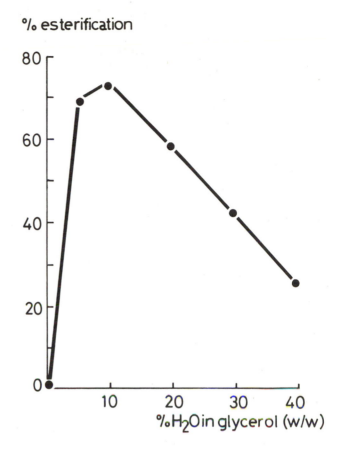

FIG. 2. Effect of percentage of water in glycerol on the degree of esterification of oleic acid using lipase from *Chromobacter viscosum* at 40 C. The weight ratio of the glycerol and oleic acid phases was 1. The percentage esterification was calculated on the acid value after reaction for 44 hr.

product mix resulting from such reactions is dependent on the water activity of the system (Fig. 2). For the production of acylglycerols by direct esterification of glycerol, low water activity is required and during interesterification no water should be produced so that the reactor can operate at optimal water activity. We have developed a process for the interesterification of oils and fats using a packed bed of immobilized lipase. The feedstock to the reactor is dissolved in hexane saturated with water. When free fatty acids are esterified with, for instance, glycerol, water is produced so that one is then faced with increasing water activity as the reaction proceeds.

Membrane bioreactors (MBRs) are an alternative to the more conventional stirred-tank and packed-bed immobilized biocatalyst reactors. In membrane bioreactors, two fluids are separated by a permeable membrane such that the reaction and separation of substrates and/or products can take place in one unit. Potential advantages of MBRs over other reactor configurations are several. The enzyme is retained in the reactor (either by immobilization or by entrapment by the membrane) and may be reused. This is not possible with conventional stirred-tank reactors. In general, pressure drops are lower in

MBRs than in packed beds. Fluid channeling should then not occur. In the case of two immiscible fluids, oil-water interfaces occur only at the membrane surface and no emulsion is formed. In practice, the specific interfacial area available for reaction will be lower than in a stirred-tank reactor. However, problems commonly encountered with the formation of stable emulsions, which must be broken for product and enzyme recovery, thus are avoided. The water activity in the vicinity of the enzyme can be controlled to some extent in a membrane bioreactor. MBRs are reported to have been used for lipolysis, esterification and glycerolysis.

LIPOLYSIS

Membrane bioreactors can be operated in different ways. For the hydrolysis of fats and oils, mainly microporous (0.1–0.4 μm) hydrophobic membranes have been used (1–4). The lipase is adsorbed onto polypropylene membranes and the interfacial enzyme concentration is dependent on the purity of the enzyme preparation used. The maximum value determined was ca. 15 Lu/cm^2 (2). Using hydrophobic membranes, the enzyme must be immobilized on the membrane side that will be exposed to the aqueous phase. The membrane will be saturated with oil, which results in a very poor diffusion of the substrate water and of the product glycerol through the membrane and consequently in low reaction rates if the enzyme is immobilized on the oil-phase side of the membrane. By properly balancing the pressure on either side of the membrane, it is possible to keep the two phases completely separated. Separation of the resulting glycerol solution and free fatty acids is realized in the membrane unit. Using a thermostable lipase immobilized on a hydrophobic membrane, Taylor et al. (4) succeeded in passing tallow through the membrane. The two phases were separated by settling the mixture after which the aqueous phase could be recycled.

Alternatively, the enzyme can be immobilized (via ultrafiltration) onto a hydrophilic membrane on the oil side of the reactor (5,6). For such a configuration, the following mechanism is required for the lipolysis reaction. First, water must diffuse through the membrane to the enzyme active site. Lipolysis then takes place and the resulting glycerol molecule must diffuse back through the membrane to the aqueous effluent stream. Failure of these two diffusion processes to take place would result in a rapid build-up of the glycerol concentration in the vicinity of the enzyme and consequently decrease the reaction rate. Pronk et al. (5) demonstrated a quantitative diffusion of glycerol through the membrane during the total recycle lipolysis of olive oil. Whether the reaction rate is transport- or kinetically controlled is not yet known, but in either case it would be dependent on specific enzyme activity, membrane thickness and morphology.

In most cases, the stability of the immobilized lipases was high. Half-lives in the order of several months have been reported. It generally is accepted that lipolysis can be described by Michaelis-Menten kinetics if the substrate concentration is replaced by the interfacial surface area (7). In an MBR, this area is determined by the total membrane area. Therefore, we added enzyme dissolved in a

small amount of water to the oil phase, so that the co-substrate water and interfacial area (emulsion) become available to the enzyme. As expected, this approach resulted in high rates of hydrolysis (Fig. 3). In the membrane unit, glycerol moves to the aqueous phase and water to the oil phase. Using this approach, only 5–10% (v/v) water had to be added to the oil phase to obtain a maximum rate of hydrolysis at 40 C (Fig. 4). Whether this route offers better opportunities for commercialization in relation to the immobilized enzyme system will depend on the stability/reusability of the enzyme and on the reaction rates of both systems. The influence of these parameters is discussed later in this paper.

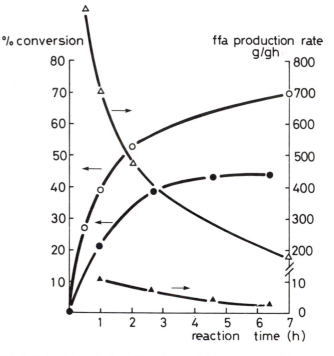

FIG. 3. Hydrolysis of olive oil in a hydrophilic membrane bioreactor (cellulose hollow fiber, 0.8 m^2) at 40 C.

Closed symbols: 1 g of *Candida cylindraceae* (Meito Sangyo, 360 OF) lipase was physically immobilized onto the cellulose membrane. The total fiber volume was 54 ml through which olive oil was pumped at different flow-rates. Water was circulated through the shell side (95 ml) at a flow-rate of 5.5 1/h. The FFA production rate (g FFA/g lipase·h) was calculated as follows:

$$\text{rate} = \frac{54 \times \text{reaction time} \times \text{degree of hydrolysis} \times \text{density of oil}}{\text{reaction time} \times 1}$$

Open symbols: 0.45 g lipase was dissolved in 45 ml water, which was added to 900 ml olive oil. Circulation of the oil-phase through the shell side and the aqueous phase (850 ml) through the fibers of the reactor was 5.5 1/h. The oil-phase was agitated at 325 rev/min using a paddle agitator (vessel diameter 0.1 m, impeller diameter 0.075 m).

The average FFA-production rate (g FFA/g lipase·h) was calculated according to:

$$\text{rate} = \frac{900 \times \text{density} \times \text{degree of hydrolysis}}{\text{reaction time} \times 0.45}$$

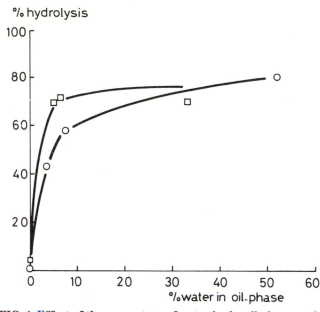

FIG. 4. Effect of the percentage of water in the oil-phase on the hydrolysis of olive oil after reaction for 7 hr. *Candida cylindraceae* **lipase (0.55 mg/g oil) was dissolved in the amount of water, which was added to the oil phase. The reaction was started by adding the enzyme solution to 900 ml olive oil in an agitated vessel. The oil was circulated through the shell side of a 0.8 m² cellulose hollow fiber unit at 5.5 l/hr □ : 40 C; 0: 35 C.**

ESTERIFICATION

During esterification, water is produced while glycerol is consumed, so that ever-increasing water activity results if such emulsion reactions are conducted in processes. Using an MBR, there are several approaches to control the water activity. Hoq et al. (8,9) advocated the use of a microporous hydrophobic membrane onto which lipases could be adsorbed. The adsorbed lipases desorbed relatively easily in glycerol solutions (97%). Therefore, lipase was added to the glycerol solution, which was pumped across the membrane surface. In this way, an equilibrium was established between surface-adsorbed enzyme and enzyme in free solution. On the opposite side, free fatty acid (oleic acid) was circulated through the reactor. The water activity in this system was controlled by circulating the glycerol solution through a continuous dehydrator, a molecular sieve, in a manner similar to that employed by Yamane et al. (10).

Hydrophilic membranes also can be used in cases where the enzyme is physically immobilized via ultrafiltration on the fatty acid/acylglycerol side of the membrane (personal communication, A. van der Padt, Wageningen Agricultural University, The Netherlands). A glycerol solution with a given water content is pumped along the opposite side of the membrane. The water activity of this phase can be controlled either by a molecular sieve or by evaporation. We are engaged in a joint project with the Wageningen Agricultural University (Wageningen, The Netherlands) studying the esterification of capric acid with glycerol in a hydrophilic membrane reactor as described above as a model system. In view of the generally high melting points of other fatty acids and acylglycerols, organic solvent (hexadecane) is used to solubilize the fatty acid, making lower operating temperatures possible. Figure 5 shows an example of such a conversion (11).

As for lipolysis in the previous examples, the reaction is limited to the oil-water interface at the membrane surface. Direct addition of the lipase dissolved in a small

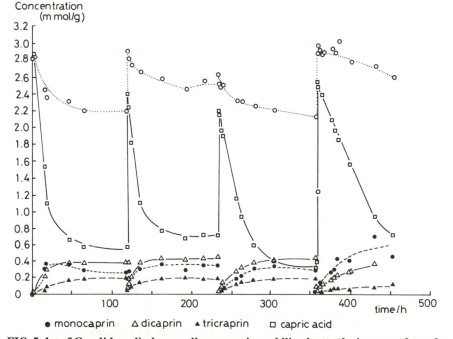

● monocaprin   △ dicaprin   ▲ tricaprin   □ capric acid

FIG. 5. 1 g of *Candida cylindraceae* **lipase was immobilized onto the inner surface of a cellulose hollow fiber reactor (0.8 m²). Per batch, 100 g of 50% capric acid in hexadecane was circulated through the fibers at 25 C and 30 ml/min. The aqueous phase was circulated through the shell side at 47.5 ml/min. The glycerol concentration of this phase varied between 88 and 25% (Courtesy: Wageningen Agricultural University, The Netherlands).**

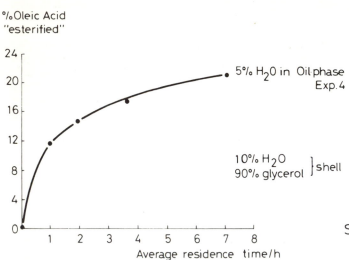

FIG. 6. Esterification of oleic acid with glycerol in a 0.8 m² cellulose hollow fiber reactor.

Lipase (*Chromobacter viscosum*) was dissolved in 50 ml water and added to 950 ml oleic acid agitated at 40 C. This phase was pumped at various flow-rates through the hollow fibers (total volume 54 ml) of a 0.8 m² membrane unit. A 90% glycerol solution was circulated through the shell side.

amount of water to the oil phase also may be considered. This water phase will equilibrate with the glycerol solution on the opposite side of the membrane. Figure 6 shows that even when starting with lipase in pure water added to the oleic acid, a mean residence time of seven hr in the hollow fiber membrane unit resulted in an esterification of 21% using a 90% glycerol solution on the opposite side of the membrane. For this to occur, the glycerol and water activities in the droplets containing enzymes must equilibrate with those in the glycerol feed solution. Thus, by this approach an emulsion reaction is combined with a membrane process to control the water activity in the vicinity of the enzymes.

GENERAL ECONOMIC EVALUATION

There are several modes for operating membrane bioreactors. The industrial potential of such bioreactors (networks) will mostly depend on their economics as compared with other systems. Factors such as kinetics and reusability of the biocatalyst, degree of substrate conversion, scale of operation, size of the membrane unit and the lifetime of the membranes are important for the overall economics of membrane bioreactors. In turn, these factors are influenced by the type of biocatalyst and membrane used, the hydrodynamics of the membrane unit, the purity of the substrates, etc. Therefore, optimization of a membrane bioreactor (network) in terms of biocatalytic and economic performance requires further research.

GENERAL PROCESS DESCRIPTION

For a general membrane bioreactor scheme (Fig. 7), the following reactions are considered:

$$kS + 1A \xrightleftharpoons{\text{biocatalyst}} mP + nB$$

where k, l, m and n are stoichiometric parameters. S and A are (co-)substrates and P and B (by-)products. The mode of operation can either be batchwise (total recycle), continuous or continuous combined with recycle. In this paper, it is assumed that k, l, m and n are equal to l, the recovery of product P is 90% and l year is 8,000 hr. Substrate costs and therefore the degree of substrate conversion are not taken into account.

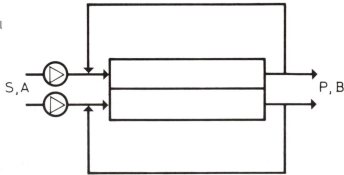

FIG. 7. General scheme of a membrane bioreactor.

EQUIPMENT SIZING AND COSTING

Starting from the annual production requirements for product P (90% recovery), the membrane surface area required can be calculated according to:

$$A_m = \frac{1\,000\,Pr}{0.9\,V_r t_{op}} \qquad [1]$$

where: A_m is the installed membrane surface area (m²); Pr the amount of P required (kmol); V_r the average reaction rate of the biocatalyst (mol S/m²h) and t_{op} the operational time (hr).

The capital investment for membrane units (excluding membranes) is calculated according to Peters and Timmerhaus (12).

$$CI = UC\,(A_m)^{CE} \qquad [2]$$

where CI = capital investment (Dfl), UC = unit costs (Dfl/m²) and CE = cost exponent. Membrane filtration equipment prices of various manufacturers were compared by us and from the data obtained a cost exponent of 0.6 was estimated for manually operated ultrafiltration equipment including pumps, feed vessels, piping and CIP facility. A fair estimate for the unit costs is Dfl 10,000 so that the capital investment is

$$CI = 10,000\,A_m^{0.6} \qquad [3].$$

Fixed capital investment is calculated by adding 2.5% and 15% of the capital investment for plant erection (existing site) and automation (13), respectively. The electricity used by membrane systems is dependent on reactor configuration, operational conditions, operational time and membrane area installed. Based on the power consumption for milk concentration by ultrafiltration, which ranges between 0.2 and 0.5 kW/m², and the fact that the reactor unit is not operated as a filtration unit, a power consumption of 0.1 kW/m² is estimated.

Consumption of substrates S and A is not considered

here although, of course, their percentage conversion and losses influence the total cost of the product. Calculation of biocatalyst consumption depends on the reaction rate and the reusability or lifetime of the biocatalyst.

$$E_c = \frac{V_r\, A_m\, t_{op}}{E_r} \qquad [4]$$

where E_c is the amount of biocatalyst required (kg) and E_r the amount of substrate effectively converted per unit of enzyme used (mol S/kg). Thus, E_r is a measure of the stability and reusability of the biocatalyst in the membrane reactor system. The cost of the biocatalyst is a process variable in this paper. The operational lifetime of the membranes used is assumed to be 8,000 hr. Costs of membrane cleaning are calculated as 10% of the annual membrane costs. The price of membranes is very much dependent on type and supplier. The cost contribution of materials and utilities is calculated from their consumption and prices. The unit is operated by one operator. Annual production costs are calculated as indicated in Table 1.

SENSITIVITY ANALYSIS

Unless otherwise stated, the following values have been used: P_r, 10^3 kmol; V_r, $5 \cdot 10^{-2}$ mol/m^2 h; t_{op}, 8,000 hr; E_r, 50 mol/kg; enzyme costs, Dfl 300/kg; membrane costs, Dfl 400/m^2. Table 1 also gives the cost calculation data for the process variables stated below. The installed membrane area for this "base case" is 2,778 m^2. Figure 8 shows the breakdown of product costs into depreciation (D), enzyme cost (E), membrane cost (M), labor and supervision (L) and others (O).

TABLE 1

Major Cost of Operating a Membrane Bioreactor.

	kDfl
Capital investment	932
Cost of erection @ 2.5% of capital inv.	23
Automation, electrical @ 15% of cap. inv.	140
Fixed capital investment	1,095
Direct Production Costs	
Cost of enzyme	6,670
Cost of membrane	1,110
Membrane cleaning @ 10% of membrane cost	111
Electricity @ 0.25/kWh	556
Maintenance @ 6% FCI	66
Labour @ 60,000/year	60
Supervision + laboratory 20% L	12
	8,585
Fixed Charges	
Depreciation @ 7% FCI	77
Insurance @ 1% FCI	11
Overheads @ 40% TLSM	55
	143
Annual production costs	8,728

Cost of product P Dfl 8.72/mol.
P_r, 1,000 kmol; V_r, 0.05 mol/m^{2}h; t_{op}, 8000 h; E_r, 50 mol/kg; enzyme cost, Dfl 300/kg; cost of membrane, Dfl 400/m^2.

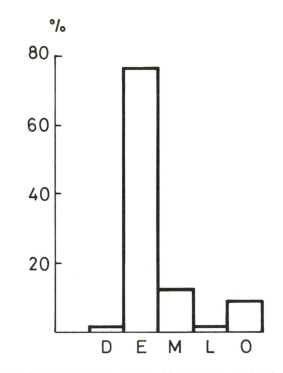

FIG. 8. Breakdown of membrane bioreactor costs for the "base case" data of Table 1: D, depreciation; E, enzyme cost; M, membrane cost; L, labor and supervision; O, others.

Enzyme costs are a major item of the overall economic process. Enzyme reusability is, therefore, an important process parameter to be optimized. Figure 9 shows the result. The benefits of increased enzyme reusability/stability are more pronounced at higher enzyme prices. For a relatively cheap enzyme (Dfl 25/kg), optimization of its reusability to more than 200 mol/kg only results in a small price reduction. The opposite is true of an enzyme of say Dfl 1,000/kg (Fig. 9).

The reaction rate V_r and the annual production determine the size of the membrane reactor unit. Figure 10 shows that the reaction rate V_r also is an important parameter to be optimized. At the given process parameters, optimization to values higher than 0.05 mol/m^2 h does not yield any major benefits. The price of the membranes used is only significant at relatively low reaction rates.

Setting research targets for process optimization clearly depends on the relative ratios between several process parameters. Optimizing the reusability of a cheap enzyme is not very research sensitive, whereas for an expensive biocatalyst it could be the prime research target. Whatever the cost of a biocatalyst, it always is possible to define a target beyond which further improvement of, for instance, enzyme reusability is no longer very effective. In most cases, the choice of biocatalyst and type of membrane will be decided on in a relatively early stage of process development and their prices virtually will be fixed entities. Major variables are then the reaction rate and the reusability or stability of the biocatalyst. If market conditions are more or less known, a maximum product price can be estimated. Once the practical and/or theoretical capabilities of the biocatalyst are roughly

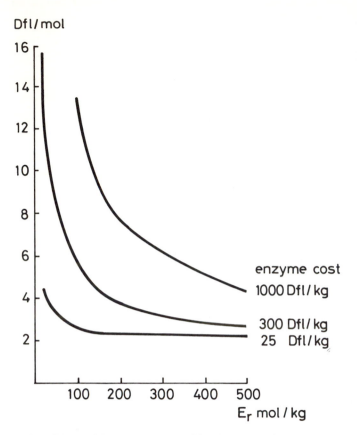

FIG. 9. Effect of the enzyme reusability E_r on the price of product at different enzyme cost.

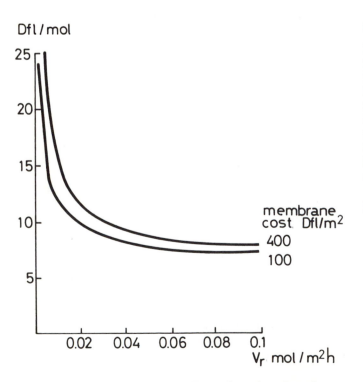

FIG. 10. Effect of reaction rate V_r on the price of product at different membrane prices.

known, it can be decided whether the estimated price is realistic against a certain set of process conditions. For example, on the basis of a maximum acceptance cost of Dfl 4/mol for membrane bioreactor operation, the reaction rate definitely must be higher than 0.01 mol/m² h (Fig. 11). Various combinations of biocatalyst reusability and reaction rate may, however, result in the same cost.

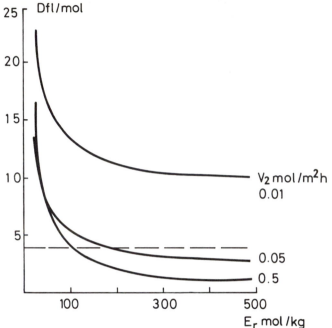

FIG. 11. Effect of various combinations of reaction rate (V_r) and biocatalyst reusability (E_r) on the cost of product produced in a membrane bioreactor.

CONCLUSION

These are general and simplified procedures for calculating the cost of operating a membrane bioreactor. Even in the absence of hard quantitative data, this procedure can be used to assess whether for a particular bioconversion, membrane reactors are a feasible process option if realistic parameter estimates are used. Furthermore, this approach will identify research-sensitive areas. The reliability of the conclusions drawn from such a study will improve with the availability of quantitative data obtained through experimentation or experience (6,12).

REFERENCES

1. Hoq, M.M., M. Koike, T. Yamane, and S. Shimizu, *Agric. Biol. Chem. 49*:3171 (1985).
2. Hoq, M.M., T. Yamane, S. Shimizu, T. Funada, and S. Ishida, *J. Am. Oil Chem. Soc. 62*:1016 (1985).
3. Hoq, M.M., T. Yamane, and S. Shimizu, *Enz. Microb. Technol. 8*:236 (1986).
4. Taylor, T., C.C. Panzer, J.C. Craig, and D.J. O'Brien, *Biotech. and Bioeng. 28*:1318 (1986).
5. Pronk, W., C. van Helden, and K. van 't Riet, Poster at the Membrane Symposium of the Wageningen Agricultural University, Wageningen, The Netherlands (1986).
6. Kerkhof, P.J.A.M., and K. van 't Riet, *I²-procestechnologie 1*:17 (1987).
7. Macrae, A.R., *Microbiol Enzymes and Biotechnology*, Applied Science Publ. Ltd, 1983, p. 225.

8. Hoq, M.M., H. Tagami, T. Yamane, and S. Shimizu, *Agric. Biol. Chem. 49*:335 (1985).

9. Hoq, M.M., T. Yamane, and S. Shimizu, *J. Am. Oil Chem. Soc. 61*:776 (1984).

10. Yamane, T., M.M. Hoq, S. Itok, and S. Shimizu, *J. Jap. Oil Chem. Soc. 35*:632 (1986).

11. van der Padt, A., L. van Dorp, K. van 't Riet, World Conference on Biotechnology for the Fats and Oils Industry, 1987, Hamburg.

12. Peters, H.S., and K.D. Timmerhaus, in *Plant Design and Economics for Chemical Engineers*, 3rd Ed., McGraw-Hill Book Company, New York, 1980.

13. Kerkhof, P.J.A.M., and K. van 't Riet, *I²-procestechnologie 1*:17 (1987).

Mass Transfer In Bioreactors

George Abraham

Southern Regional Research Center, Agricultural Research Service, USDA, P.O. Box 19687, New Orleans, LA 70179

Mass transfer or, more precisely, mass transfer limitations, occur at interfaces. In reaction systems containing immiscible liquids mass transfer limitations take place at the liquid-liquid interfaces. If an enzyme is immobilized on a solid support, there will be a liquid-solid interface through which mass transfer must occur. If the solid support is smooth, mass transfer occurs by convective flow. If the support is a porous solid or membrane then, depending on pore size, internal diffusion would control mass transfer.

The effect of mass transfer limitations is to alter the observed or overall reaction rate. This rate is what is measured during a particular experiment with a given set of experimental conditions. It is the superposition of mass transfer effects on intrinsic kinetics. Changing an experimental parameter such as temperature changes the observed rate by affecting both intrinsic and mass transfer rates. However, changing such parameters as stirring rates, vessel size or enzyme support pore size does not affect the intrinsic kinetic rates but could affect the mass transfer rates, consequently changing the observed rate. The variability of the mass transfer rates complicates the scale-up of reaction systems. To properly size a batch, continuous stirred tank or packed-bed reactor from laboratory bench scale data, the intrinsic kinetic rates must be determined, and the mass transfer rates must be cast in dimensionless form.

In a heterogeneous system, the conversion rate of a chemical reaction determined from bulk fluid properties is a superposition of chemical and physical rate processes. If the physical processes are much faster than the chemical processes, then the observed rates are said to be intrinsic, and the reaction is chemically limited. If the opposite occurs, then the physical rates dominate, and the reaction is physically, or more specifically, mass-transfer limited. Under mass transfer limitations, the observed rates are functions of mixing speed and the vessel size and configuration. Consequently, under these conditions, the rates and rate constants determined for an enzymatic reaction in two different experiments could be different even if substrate concentration, enzyme level and reaction temperature are identical. The rates certainly will be different for a pilot plant or commercial reaction system as compared with rates obtained from bench-top data. For these reasons, it is important that mass transfer limitations be identified in a reaction system and minimized if possible.

Mass transfer limitations occur at interfaces. In reaction systems containing immiscible liquids, mass transfer limitations can occur at the liquid-liquid interfaces. If an enzyme is immobilized on a solid support, then there is a liquid-solid interface through which mass transfer must occur. If the enzyme is located inside a porous solid support, then diffusive mass transfer must occur through the pores. To understand these processes, a model from classical heterogeneous catalysis will be used (1). If this model is applied to substrates being catalyzed to prod-ucts by an enzyme immobilized on a porous support, either a solid particle or a membrane, then the following steps occur (Fig. 1): mass transfer of substrates through the fluid boundary layer surrounding the immobilized enzyme; molecular diffusion of substrates from exterior surface of the support into the interior pore structure; attachment of substrates to enzyme, reaction and detachment; diffusive transfer of products from the interior of the porous structure to the external surface; and mass transfer of products from the exterior surface of the support through the boundary layer to the bulk fluid.

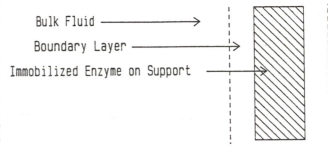

Bulk Fluid ——————→

Boundary Layer ——————→

Immobilized Enzyme on Support ——————→

FIG. 1. Model of interface in heterogeneous catalysis.

Step 3 is chemical in nature while the others are physical. Steps 1 and 5 are dependent on the bulk fluid motion of the system. Far from the support surface, the substrate concentration has a value characteristic of the bulk fluid. This value is what is measured by typical analytical methods. Near the support surface, the actual concentrations of the substrates decrease due to gradients formed between the bulk fluid, where no substrate is being consumed, and the support where the reaction is occurring. For products in which the opposite occurs, their concentrations are higher at the enzyme sites than in the bulk mixture. Fluid velocities, mixing rates, support size and diffusional characteristics of the substrates are parameters directly affecting these concentration profiles and consequently affecting the rates of steps 1 and 5. When fluid velocities or mixing rates are high relative to the motion of the support, external mass transfer limitations are diminished. However, while increased mixing decreases these limitations, it also may present a problem because high fluid turbulence leads to shear forces that can denature the enzyme. If the support is porous, then the external surface area is only a small fraction of the total area to which enzymes can be attached. In this case, steps 2 and 4 play a key role in the observed reaction rate. Here the substrate must travel through the pores to the enzyme; consequently, the substrate concentration at the support surface is greater than at the enzyme. This gradient can be reduced by decreasing the size of the support. However, in a packed-bed reactor, for example, the smaller support particles will cause an increase in pressure drop, which can degrade reactor performance.

EVALUATION OF MASS TRANSFER LIMITATIONS

There is a large body of information that treats simultaneous mass transfer and chemical reaction. Many similar approaches have been reported. They tend to be complex, and in many cases it is difficult for the nonspecialist to extract the needed information. Consequently, we will present only the essential techniques, equations and methods for calculating the necessary parameters. Some excellent detailed reviews are given in the references (2,3). External mass transfer will be considered first, then internal mass transfer will be considered. Finally, relationships treating each phenomenon will be presented.

External Mass Transfer

One method for testing external or convective mass transfer resistance in a packed-bed or membrane reactor is to measure the conversion of reactants for different bed heights or membrane surface areas while maintaining constant residence times. If film resistance is significant, the increased velocity will cause a change in the conversion rate even for constant residence times. In a continuous-stirred tank reactor, a similar approach is to increase stirring rate while keeping residence times constant. Figure 2 shows this method applied to the hydrolysis of DL-N-benzoyl-arginine-p-nitroanilide by trypsin immobilized in a packed-bed reactor (4). The reaction rate obtained at each flow rate is v_o while v^* is the rate measured at high flow rates. As is seen in Figure 2, v_o equals v^* at flow rates greater than about 0.015 l/min. At this or greater flow rate, the reactor is operating in the reaction-limited regime. Kinetic constants determined under these conditions will be intrinsic and can be used in design equations.

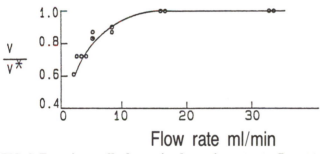

FIG. 2. Experimentally determined reaction rates vs flow rates (4).

Because enzymes can be denatured or dislodged from their supports by high shear forces, it would be more desirable to use a method that allows determination of kinetic constants at flow rates less than those needed in the method described above. Such a technique has been presented (5). For a packed tubular reactor at steady state with Michaelis-Menten kinetics, the following equation can be derived:

$$s_o x = K_m^{app} \ln (1 - x) + v_{max}^{app} F/V \qquad [1]$$

in which s_o is the observed or bulk substrate concentration, x is the fractional conversion, F is the volumetric flow rate of substrate, V is the reactor void volume, and K_m^{app} and v_{max}^{app} are the apparent Michaelis constant and maximum reaction velocity. The values of these last two

quantities change as mass transfer conditions change and reach their intrinsic values, K_m and v_{max}, in the absence of mass transfer limitations. To obtain these intrinsic values, a series of experiments is performed in which x is measured for several different values of s_o and F. A plot, as shown in Figure 3(a), then is made with each line representing a different flow rate. As can be seen from Equation 1, the slope of each line equals the K_m^{app} for that flow rate. Figure 3(b) shows a plot of K_m^{app} vs flow rate. It can be seen that as the flow rate increases, corresponding to a reduction in mass transfer limitations, K_m^{app} approaches its limiting value of K_m. The advantage of this method over the previous one is that the reactor does not have to be run under conditions free of mass transfer limitations. The limitations of this method and variations that can be tried if this method fails have been discussed (6-8).

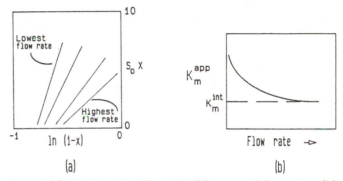

(a)   (b)

FIG. 3. (a) Typical plot of Equation 3 for several flow rates. (b) Apparent Michaelis-Menten constants, from (a), vs flow rates.

Internal Diffusive Mass Transfer

When the enzyme is immobilized inside the pores of a solid support or in the cells of a membrane, the substrate must diffuse from the outside surface of the support to the enzyme for reaction to occur. When this rate is slower than the chemical kinetic rate, the reaction is said to be diffusion-limited. Methods used to extract intrinsic constants for these cases must take into account the substrate concentration profile along the diffusion path. To do this, a differential mass balance is written across a thin shell inside the support (9) with the boundary condition that at the surface of the support the substrate concentration is s_s. If this equation is made dimensionless and integrated across the support, the following functional relationship is found:

$$\eta = \underset{f}{|} (\Phi, s_s/K_m \qquad [2]$$

in which η, the effectiveness factor, is defined as the observed rate divided by the rate that would be obtained with no concentration gradients in the pellet and Φ, the observable modulus, is defined as:

$$\Phi = (v_s/D_{eff} s_s) (V/A)^2 \qquad [3]$$

in which D_{eff} is the effective diffusivity of the substrate in the support, v_s is the reaction rate at the surface, and V and A are the support volume and area. This modulus is called observable because the parameters in it can be experimentally measured. This is true, however, only if external mass transfer limitations do not exist, since under that condition, v_s and s_s can be replaced by the observable quantities v_o and s_o. Correlations that allow calculation of D_{eff} exist (10, 11); however, the uncertainties involved usually make it necessary to experimentally measure D_{eff}. A method for this is given below.

The functional relationship for Equation (2) cannot be found analytically for reactions following Michaelis-Menten kinetics; it must be solved numerically. It also will have different solutions for different support geometries (12,13). Useful information can be obtained, however, by looking at an analytical form of Equation (2) for a limiting case. When first-order kinetics are assumed, valid if s_o is much smaller than K_m, and diffusional limitations are significant, the following equation is obtained (9,14):

$$v_o/s_o = A/V \, (D_{eff} \, v_{max}/K_m)^{1/2} \qquad [4]$$

If experimental measurements can be made on a system in which the above assumptions are valid, then Equation (4) can be used to find D_{eff} if v_{max} and K_m are known. These last two quantities can be found by designing a second experiment in which diffusional limitation is absent. In the first experiment, v_o is measured for several values of s_o in a system with enzyme immobilized on large particle supports; hence, diffusion limited the conditions. In the second experiment, the same measurements are made except that the enzyme is immobilized on small particles, small enough so that there is no diffusional limitation. Both of these sets of data are then plotted as v_o/s_o vs v_o. These coordinates come from a rearranged form of the Michaelis-Menten equation

$$v_o/s_o = v_{max}/K_m - v_o/K_m \qquad [5]$$

and the graph is called the Eadie-Hofstee plot. If the small particle system is not diffusion limited, then its plot will be a straight line. At the y-axis intercept the value of v_{max}/K_m is found, and at the x-axis intercept the value of v_{max} is found. If diffusional limitation exists for the large particle system, its plot will be curved. At the y-intercept of this line, where s_o and v_o go to zero, which corresponds to first-order kinetics, the value for the righthand side of Equation 4 is found, and D_{eff} can be calculated.

Figure 4 shows the solution of Equation 2 for three values of s_s/K_m and for spherical geometry, e.g., the enzyme supports are spherical pellets, and slab geometry, as a membrane might be shaped. It is evident from Figure 4 that for Φ less than about 0.3, η is about 1 which implies that chemical reaction is the rate-limiting process. Further, if Φ is greater than about 3, there are mass transfer limitations and η is proportional to $1/\Phi$.

Simultaneous External and Internal Mass Transfer Limitations

The two previous sections provide methods for identifying the existence of mass transfer limitations that were either external or internal but not when they existed simultaneously. This latter case now will be considered. A differential mass balance is made as described before for internal diffusion except that the boundary condition at the surface of the pellet is $k_s(s_s-s_o)$; k_s is the external mass transfer coefficient. If this equation is solved for first-order kinetics and made dimensionless, the following is obtained:

$$\eta/\eta_s \, 1 + \Phi/Bi \qquad [6]$$

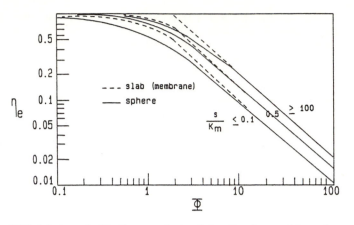

FIG. 4. Internal effectiveness factor (η) vs the observable modulus (Φ) for values of s/K_m and two geometries (15).

in which η_s is an overall effectiveness factor and Bi, the Biot number, is $k_s V/AD_{eff}$. Since correlations are available for estimation of k_s and D_{eff} (10,11), Φ/Bi can be calculated for a given system. If this value is much less than 1, internal diffusion limitations will dominate. If it is much greater than 1, external convective limitations will dominate.

Engasser (15) provides a quick method for estimating simultaneous external and internal mass transfer limitations. The method requires knowledge of V, A, k_s, D_{eff} and K_m for the system under study. He defines an external effectiveness factor, η_e, so that $\eta_s \, \eta_e \eta$ [7]. η and η_e are obtained from Figures 4 and 5 respectively. To use these figures, s_s is calculated from $s_s = s_o - v_o/k_s$ [8], and K_m is assumed to be equal to the K_m of the nonimmobilized soluble enzyme.

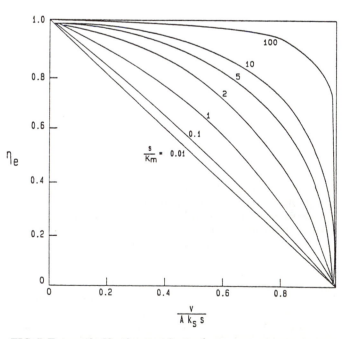

FIG. 5. External effectiveness factor (η_e) vs the observed rate of reaction divided by the catalyst area (A), mass transfer coefficient (k_s) and observed substrate concentration (s_o) for values of s_o/K_m (15).

The analyses so far presented have considered only single substrate diffusion. It has been shown (16) that in some cases, product diffusion also can considerably modify the kinetic properties of the reaction. In the case of two substrates, it has been shown (17) that the effects of diffusional resistances on the overall reaction rate are related mainly to the transport of the limiting substrate.

MASS TRANSFER IN LIPASE SYSTEMS

In comparison with other immobilized enzyme systems, little information about mass transfer effects on lipase-catalyzed reactions is available in the literature. Probably the most important reason for this lack is that lipases are somewhat more complex in their action as compared with other enzyme systems and consequently are more complex in their analyses. For example, Sarda and Desnuelle (18) showed that the rate of triacetin hydrolysis using a pancreatic lipase increased significantly when the substrate concentration in the mixture was greater than its water solubility limit, thus producing two phases. Brockman (19) points out that enzyme orientation and substrate diffusion to the interfacial site are critical components of these reactions. Whether this diffusion is rate limiting is difficult to analyze in these systems because an increase in mixing, for example, increases surface area and hence increases reaction rate.

Two studies have been reported on the investigation of mass transfer effects in lipase systems. One study (20), using pancreatic lipase immobilized onto stainless steel beads, compares the observed rate of tributyrin hydrolysis to rate of external mass transfer calculated from the collection theory. This theory takes into account Brownian diffusion, bulk motion and gravitational settling. Because the calculated external mass transfer rates were over an order of magnitude greater than the observed rates, they concluded that the system was not external mass transfer limited. Internal diffusion was not investigated. In the second study (21), olive oil was hydrolyzed using *Rhizopus arrhizus* mycelia as a source of insolubilized lipase. Using a packed-bed reactor, they found a decrease in conversion only at low flow rates. They concluded that this decrease was due to channeling in the reactor and not to an increase in mass transfer resistance. As in the previous study, internal diffusion was not studied.

DISCUSSION

Knowledge of mass transfer coefficients and diffusivities along with intrinsic rate constants allows estimation of apparent rates. In principle, this will allow reactor scale-up from laboratory data. In practice, however, it has been found that some reactor configurations scale-up better than others. Until the knowledge base of bioreactor design becomes sufficiently broad, it will be necessary to generate design data under conditions approaching production scale for optimum efficiencies to be assured.

Much of the discussion so far has assumed a reduction in overall reaction conversion due to mass transfer limitations. While this usually is true, in some instances desirable effects could result. In sequential reactions, for example, the production of a wanted intermediate species could be increased if conversion to the final species were mass-transfer limited (22). Work has been done using the shape and size selectivity characteristics of the pores in zeolite catalysts to preferentially react molecular species (23). Such concepts might have potential applications in immobilized enzyme systems. Much of the theoretical base of classical heterogeneous catalysis can be applied directly to immobilized enzyme and whole-cell reactors. However, because interest in these systems is relatively recent, little application work has been done. Just as the rise in importance of the petrochemical industry spurred research in the mass transfer and reactor design of inorganic catalytic systems, the biotechnical industry will spur interest in the design of bioreactors. The last decade has seen a great deal of effort put into the study of enzyme immobilization with many successful techniques being developed. The work now involves using these techniques to develop large-scale reaction systems of optimal efficiency.

REFERENCES

1. Hill, G.H., *Chemical Engineering Kinetics and Reactor Design*, John Wiley and Sons, Inc., New York, 1977, pp. 178–179.
2. Goldstein, L., in *Methods in Enzymology*, edited by K. Mosbach, Vol. 44, Academic Press, New York, 1976, pp. 397–443.
3. Engasser, J.M., and C. Horvath, in *Applied Biochemistry and Bioengineering*, edited by L.B. Wingard, E. Katchalski and L. Goldstein, Vol. 1, Academic Press, New York, 1976, pp. 127–220.
4. Ford, J.R., in *Enzyme Engineering*, edited by L.B. Wingard, Wiley-Interscience, New York, 1972.
5. Lilly, M.D., and W.E. Hornby, *Biochem. J. 100*:718 (1966).
6. Lee, S.B., and D.D.Y. Ryu, *Biotech. and Bioeng. XXI*:1,499 (1979).
7. Patwardhan, V.S., and N.G. Karanth, *Ibid. XXIV*:763 (1982).
8. Karanth, N.G., and V.S. Patwardhan, *Ibid. XXIV*:2269 (1982).
9. Bailey, J.E., and D.F. Ollis, *Biochemical Engineering Fundamentals*, 2nd edn., McGraw-Hill, New York, 1986, pp. 208–220.
10. Satterfield, C.N., *Heterogeneous Catalysis in Practice*, McGraw-Hill, New York, 1980.
11. Reid, R.C., J.M. Prausnitz and T.K. Sherwood, *The Properties of Gasses Liquids*, 3rd edn., McGraw-Hill, New York, 1977.
12. Ghim, Y.S., and H.N. Chang, *J. Theor. Biol. 105*:91 (1983).
13. Hamilton, B.K., C.R. Gardner and C.K. Colton, *Am. Inst. Chem. Eng. J. 20*:503 (1974).
14. Clark, D.S., and J.E. Bailey, *Biotech. and Bioeng. XXV*:1027 (1983).
15. Engasser, J., *Biochim. Biophys. Acta 526*:301 (1978).
16. Marc, A., and J.M. Engasser, *J. Theor. Biol. 94*:179 (1982).
17. Leypoldt, J.K., and D.A. Gough, *Biotech. and Bioeng. XXIV*:2705 (1982).
18. Sarda, L., and P. Desnuelle, *Biochim. Biophys. Acta 30*:513 (1958).
19. Brockman, H.L., in *Lipases*, edited by B. Borgstrom and H.L. Brockman, Elsevier, New York, 1984, pp. 4–46.
20. Lieberman, R.B., and D.F. Ollis, *Biotech. and Bioeng. XVII*:1401 (1975).
21. Bell, G., and J.R. Todd, *Ibid., XXIII*:1703 (1981).
22. Barker, S.A., and P.J. Somers, in *Advances in Biochemical Engineering*, Vol. 10, edited by T.K. Ghose, A. Fiechter and N. Blakebrough, Springer-Verlag, New York, 1978, pp. 27–50.
23. Weisz, P.B., *Pure Appl. Chem. 52*:2091 (1980).

Continuous Use of Lipases in Fat Hydrolysis

Matthias Bühler[a] and **Christian Wandrey**[b]
[a]Biotechnological Laboratories of Henkel KGaA, D-4000 Düsseldorf, Federal Republic of Germany,
and [b]Institute for Biotechnology of the Nuclear Research Center Jülich, D-5170 Jülich, Federal Republic of Germany

Compared with other enzymatically catalyzed reactions, especially in the pharmaceutical field, the product-added value to be derived from the fat-splitting process is low. On the other hand, fatty acids have a remarkable market volume. Continuous processing, with reuse of the enzyme, seems to be the biotechnological method of choice. To achieve high space-time yields, carrier fixation of a lipase may be unfavorable due to mass transfer limitations, especially in the case of two-phase reaction systems. The aim of this investigation was to develop a method for the continuous use of lipases without carrier fixation. Because the enzyme is enriched ("immobilized") at the phase boundary (fat/water) where the reaction takes place, a microemulsion is desirable, and phase separation is necessary for product isolation and for recovery of the enzyme. This was accomplished by the use of two stirred tank reactors and two continuously operating centrifuges. The conditions were selected so that about 90% of the pure aqueous phase containing glycerol (first stage) and about 90% of the pure fat phase containing fatty acids (second stage) were separated. By this series of two incomplete separations, it was possible to recycle about 90% of the enzyme together with the interfacial layer. This procedure allows a kinetically and thermodynamically desirable counter-current flow of the fat and the aqueous phases. The feasibility of such a process was demonstrated with a continuous hydrolysis of soybean oil in two-stage mixer-settler setups with a total volume of up to 15 liters. Thereby, a degree of hydrolysis of up to 98% and a space-time yield of 11.4 kg fatty acid/l.d were achieved.

The continuous hydrolysis of fats to produce fatty acids and glycerol has been a standard large-scale process for several decades (1). An economically high reaction rate is achieved by high temperatures and pressures as well as by acidic or alkaline catalysis (2). It has been known since the beginning of this century that natural fats and oils can be split by enzymes (lipases) (3–5). Over the last few years, enzymatic fat-splitting has attracted increasing attention because of the expectation that biotechnical processes will better meet increasingly strict ecological regulations. Additionally, renewable resources such as fats are becoming more and more important as raw materials in the chemical industry (6,7). Thus, studies on the development of processes for the enzymatic splitting of fat are important to evaluate the feasibility and the limitations of the biotechnical alternative.

The enzymatic hydrolysis of fat takes place under mild, so-called physiological conditions (temperatures in the range 30-60 C at normal pressure), thus reducing energy consumption. In this paper, the expression "fat" generally is used for triglycerides independent of their state of aggregation at room temperature. Furthermore, product selectivity is increased, and thermal damage to the reactants is reduced, resulting in lower concentrations of undesirable by-products. However, in applying lipases to hydrolyze fats completely, the typical advantages of enzymatic catalysis such as regio- and stereo-specificity or high product selectivity are lost. However, if enzymatic fat-splitting proves to be technically feasible, the presently successful industrial application of other hydrolases such as proteases or amylases leads us to hope that cheap and stable lipases also will become available.

The present state-of-the-art in the conventional splitting-process by high-pressure steam is a continuous procedure. This makes sense because of the high production volume and the low product-added value. For the same reasons, a continuous procedure also has to be considered for the enzymatic process. Contrary to conventional fat-splitting, in which the reaction takes place in a one-phase system because of the high temperature and pressure, the enzymatic conversion takes place at the phase boundary between fat and water (8). An intensive mixing of the phases is thus essential. Various reactor systems for a continuous process of enzymatic fat-splitting have been described (9–11). Most of them, however, use one-stage reactor setups in which the space-time yield and the degree of hydrolysis are in principle lower than in a two-stage system. This is because in a one-stage setup, no counter-current flow performance with a sufficiently large phase boundary can be realized. Such a performance, however, is favorable due to thermodynamic and kinetic properties of the reaction.

In the enzymatic process, the high cost of the enzyme itself is one of the most significant factors. To reduce these costs, fixation of the enzyme to a carrier has been widely suggested (12–16). The immobilized enzyme has to be equally accessible to both the water and the fat phase. By the use of carrier particles, an enrichment of one or other of the phases can take place on the carrier surface depending on its properties. To circumvent this problem, the use of very small particles (17,18) or the immobilization of lipases on membranes (11,19) have been described. The first approach leads to additional costs involved in retaining the catalyst in the reactor. In the case of a membrane reactor, the volume-specific interface might not be so high.

In this paper, we describe a method for the recycling of lipases in continuous fat hydrolysis without using carrier fixation or membranes. It exploits the fact that lipases accumulate at the phase boundary (10,20,21). The process meets the thermodynamic and kinetic requirements of fat hydrolysis by using a mixer-settler setup. Furthermore, the conditions for the reaction in the mixer and for the separation of the enzyme in the settler can be adjusted independently.

EXPERIMENTAL

Materials and Equipment

All the experiments described in this paper were done with a commercial lipase preparation from *Candida cylindracea* (type OF-360), which was purchased from

Meito Sangyo (Tokyo, Japan) with an activity of 188 U/mg enzyme and was used without further purification. The enzyme was kept at 4 C in a sealed container under a dry atmosphere. There was no loss of activity under these conditions for more than one year. More detailed information on storage and operational stability is given in the literature (20,22). The lipase selected was the most suitable, among a great variety tested (20), for the complete hydrolysis of various fats and oils. Its thermal stability, however, probably is not yet sufficient for an economic enzymatic fat-splitting process.

Olive oil of pharmaceutical-grade (average acid value 1.5, average saponification value 186) was purchased from G. Hess, D-7000 Stuttgart. It was used primarily as a reference substance in the determination of lipase activity because much of the published literature deals with olive oil exclusively. Semi-refined soybean oil (average acid value 0.3, average saponification value 191) was purchased from O.L. Sels, D-4040 Neuss. Soybean oil is one of the main raw materials in commercial fat processing, and it exhibits reaction kinetics in enzymatic fat splitting very similar to those of olive oil. Both oils were used without further purification. All other materials were purchased from E. Merck GmbH, D-6100 Darmstadt or from Riedel-de Haen, D-3016 Seelze.

As a mixer (and reactor) for batchwise and continuous reactions, Biostate E-type fermenters (total volume 3 liters) from B. Braun AG, D-3508 Melsungen, were used with a three-stage turbine impeller (D/d 1.9). Alternatively, a polyethylene tube (i.d. 0.6 cm, length 106 m) was employed as a plug-flow reactor. To minimize the phase separation and the pressure drop within the tube, a group of three static mixers (type SMX DN 4, Sulzer AG, CH-8401 Winterthur) was positioned every 20 m along the tube. For the continuous supply of oil, enzyme or buffer solution peristaltic pumps (types 202 U and 501 U, Watson-Marlow Ltd., Falmouth, U.K.) at low-pressure drops, or gear pumps (type V 150.12, Verder GmbH, D-4000 Düsseldorf) at higher-pressure drops (plug-flow reactor with static mixers) were used.

As settlers, separators of the type TA1 (Westfalia Separator AG, D-4740 Oelde) were used. To recycle the lipase with the interfacial layer, an incomplete separation of the phases was necessary. This was achieved by adjusting the appropriate machine parameters, such as rotation speed, number of discs, type and diameter of the centripetal pump, back pressure at the discharge outlets of the two phases and the position of the rising channels.

For the determination of glycerol online, a differential refractometer equipped with flow-through cuvettes (Knauer KG, D-6380 Homburg) was used. All measurements of pH to determine the degree of hydrolysis or the activity of the lipase were carried out by an automatic titration device (Titroprocessor P 0100, Metrohm AG, CH-9100 Herisau) equipped with an autosampler for 96 samples.

Analytical Procedures

To evaluate the activity of the lipase or the degree of hydrolysis, free fatty acid was titrated with 0.025 M or 0.05 M ethanolic potassium hydroxide. To determine the degree of hydrolysis, an aliquot (1-2 g) was taken from the reaction mixture and heated at 95 C for 10 min. From the separated lipid phase, 0.1-0.2 g were titrated as described above. The degree of hydrolysis was calculated on the basis of the saponification value. In this procedure, a small, systematic error towards higher degrees of hydrolysis may occur. This is due to the transfer of glycerol from the lipid to the water phase, thereby increasing the acid value especially at high conversion levels. The deviation is maximally 5%.

For the determination of the lipase activity, 1.0 g olive oil and 9.0 ml 0.05 M acetate buffer pH 5.6 containing 1-10 units of lipase were mixed by stirring them (400 rpm) at 30 C for one hr. One unit is equivalent to the number of μmol of fatty acids produced per minute. The reaction was stopped by adding 30 ml of ethanol; the reaction mixture was then titrated in the same vessel. All the data presented are averages of at least two experiments or independent analyses. The deviations from the mean values were less than 3% for the determination of the degree of hydrolysis and up to 10% for the lipase activity test.

To determine enzyme activity during continuous hydrolysis, an aliquot of the reaction mixture (0.025-0.5 ml) was directly transferred into the assay instead of a sample from the original lipase preparation. The amount of enzyme was calculated from calibration curves. Blank values, which were obtained by titrating the same mixture with ethanol added at time zero (to deactivate the enzyme), were subtracted.

To follow conversion during continuous operation, glycerol concentrations in the water phase were measured by refractometry after the separation of the phases, removal of the residual emulsion by a filter, and dilution (1:100) with water in a mixing chamber. In Figure 1, the correlation between the degree of hydrolysis measured by titration and the glycerol concentration determined by refractometry is shown. This method is very suitable for determining the steady state. After this is reached, the precise degree of hydrolysis can be measured by discontinuous sampling as described above.

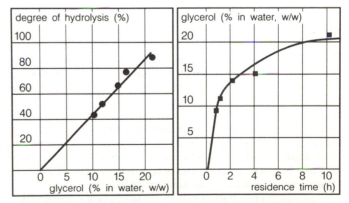

FIG. 1. (A) Correlation between the degree of hydrolysis (titrimetric) and the glycerol concentration (refractometric) in the water phase. (B) Correlation between residence time and glycerol concentration. For continuous olive oil hydrolysis, a stirred laboratory reactor (total volume 0.5 liters) was used at 30 C and 400 rpm; residence times of .7, 1, 2, 4 and 10 hr were chosen. Reaction mixture: 70% (w/w) olive oil, 30% (w/w) 0.05 M acetate buffer pH 5.6; 0.5 g lipase/kg oil.

RESULTS AND DISCUSSION

Kinetics

The kinetics of enzymatically catalyzed fat splitting under technically relevant conditions have been described recently (20,23). In a stirred tank reactor, the time course for the splitting of soybean oil with lipase OF from *Candida cylindracea* was followed in a batch experiment (Fig. 2). From the slope of the plot showing the degree of hydrolysis vs time, the reaction rate was determined and depicted as a function of the degree of hydrolysis. The reaction rate decreases rapidly with an increasing degree of hydrolysis, indicating severe product inhibition. This can be interpreted as resulting from an enrichment of the products (fatty acids, mono- and diglycerides) at the interface, leading to a dilution there of the triglyceride concentration. Therefore, an evaluation of the inhibition constants for a kinetic model of the Michaelis-Menten type is not possible simply by adding products to the reaction mixture. Furthermore, the definition of a bulk concentration for the substrate and the determination of a K_M-value according to Michaelis-Menten turns out not to be very useful because the surface concentrations of the substrate, the products and the enzyme (which cannot be directly measured) are determining the rate. Thus, the only practicable approach seems to be a direct use of the measured values in a numeric model.

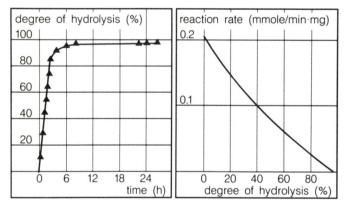

FIG. 2. Discontinuous hydrolysis of soybean oil with lipase. The reaction was carried out at 30 C and 1,600 rpm (see Experimental). Reaction mixture: 70% (w/w) soybean oil; 30% (w/w) 0.05 M acetate buffer pH 5.6; 0.5 g lipase/kg oil. (A) Time-course of the hydrolysis. (B) Reaction rate as a function of the degree of hydrolysis. The reaction rate is correlated to mg lipase preparation in the reaction mixture.

Due to the equilibrium, a technically relevant degree of hydrolysis can be achieved only when the aqueous phase is in high excess. However, a high-water content reduces the space-time yield and extracts the enzyme from the interface, increasing the amount of lipase in the bulk water phase (20). From these considerations, the following guidelines for the design of a reactor for a continuous fat-splitting process can be drawn up: (a) Countercurrent extraction of glycerol is necessary, because of the enrichment of glycerol in the water phase, which favors the reverse reaction. (b) The devices of choice are a tubular reactor, or a cascade of stirred tank reactors, because of the drastically decreased reaction rate at high degrees

of hydrolysis. (c) Recycling the enzyme with the interfacial layer is possible, due to the enrichment of the lipase at the interface. (d) Residual lipase in the aqueous phase may be recovered by means of ultrafiltration because of the solubility of lipase in water.

Enzyme Recycling

A process scheme of the two-stage mixer-settler set up for continuous enzymatic fat hydrolysis is shown in Figure 3. Oil is added to reactor (mixer) 1. About 90% of the aqueous phase is separated in separator (settler) 1. Oil and the remaining part of the aqueous phase (together with the enzyme) is pumped to reactor (mixer) 2. About 90% of the organic phase is separated from separator (settler) 2, while the rest, consisting of the aqueous phase and the interfacial layer containing the enzyme, is recycled to reactor (mixer) 1. Part of the enzyme is lost with the aqueous phase from separator (settler) 1. This may be recovered and recycled by ultrafiltration.

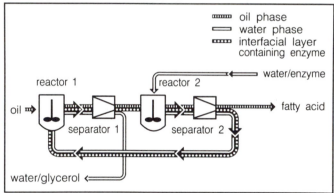

FIG. 3. Process scheme of a continuous two-stage mixer-settler device for oil hydrolysis with enzyme recycling.

As the process scheme (Fig. 3) shows, the enzyme can be transferred with the interfacial, incompletely separated emulsion layer either together with the oil phase (first reaction stage), or together with the aqueous phase (second reaction stage). The technical feasibility of these two approaches had to be established experimentally.

The transfer of the enzyme-containing interfacial layer together with the oil phase corresponds to the first stage of the proposed reaction scheme (Fig. 3). To simulate this part of the process, soybean oil (70% w/w) and a buffered solution (30% w/w) containing the lipase were continuously added to a stirred tank reactor (mixer). A continuously working centrifugal disc separator (settler) was adjusted to give incomplete separation, generating a pure water phase, and an oil phase that contained 10-20% of the water phase as a microemulsion within the interfacial layer. This was achieved by using a separator configuration with a centripetal pump for the outlet of the water phase and a free outlet with a valve for applying back pressure for the oil phase (Fig. 4A; Fig. 5, curve 1).

After the steady state was reached, the oil phase together with the interfacial layer was recycled into the reactor. To supplement the separated water phase, fresh buffer without enzyme was added. This process results in

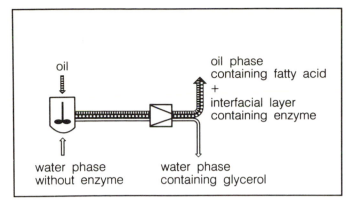

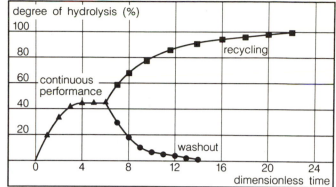

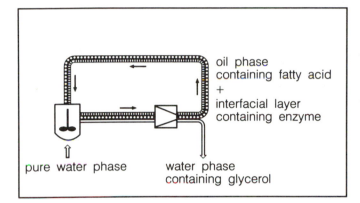

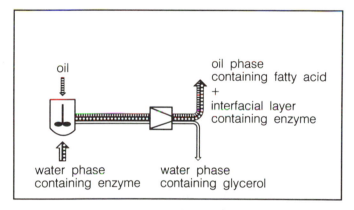

FIG. 4. One-stage mixer-settler setup for recycling the lipase with the oil phase during oil hydrolysis (first stage of the process). (A) Continuous performance with continuous supply of lipase. The enzyme-containing interfacial layer is separated together with the oil phase. (B) Semicontinuous performance. Recycling of the lipase-containing interfacial layer together with the oil phase during continuous supply of pure buffer. (C) Continuous performance. Washout of the enzyme by addition of pure buffer instead of enzyme solution to the steady state (according performance A).

FIG. 5. Recycling of lipase with the oil phase. Degree of hydrolysis vs dimensionless time (i.e., time referred to the mean residence time) for the various mixer-settler setups (Fig. 4). Reaction mixture: 70% (w/w) soybean oil; 30% (w/w) 0.05 M acetate buffer pH 5.6; 0.5 g lipase/kg oil. Reaction conditions: 30 C; impeller speed 1,600 rpm; 0.25 hr residence time in the mixer (total volume 3.0 liters); volumetric flows: oil phase 8.8 l/hr, water phase: 3.5 l/hr. (Curve 1) continuous performance (Fig. 4A). (Curve 2) recycling: semicontinuous performance (Fig. 4B). (Curve 3) washout: continuous performance with washout of the enzyme (Fig. 4C).

a semicontinuous system, which is batchwise with respect to the oil phase and continuous with respect to the water phase (Fig. 4B). Under these semicontinuous reaction conditions, the degree of hydrolysis of the oil phase increases as its reaction time is prolonged (Fig. 5, curve 2). The glycerol formed is extracted continuously with the water phase, so that nearly 100% hydrolysis is achieved with no further addition of enzyme to the mixture.

Furthermore, if pure buffer is added instead of enzyme solution under continuous reaction conditions (Fig. 4C), a decrease in the degree of hydrolysis was found resembling a normal washout function (Fig. 5, curve 3).

These results show that the lipase can be recycled in an active form together with the oil phase (Fig. 5, curve 2) and that the reactor has nearly no capacity for storing the enzyme (Fig. 5, curve 3).

To evaluate the separation conditions necessary for recycling the enzyme-containing interfacial layer together with the water phase, the analogous experimental procedure was performed. A suitable configuration for the outlets of the separator turned out to be a centripetal pump for the oil phase and a ring dam for the water phase. After having reached the steady state under continuous reaction conditions (Fig. 6A; Fig. 7, curve 1), semi-continuous performance was established (Fig. 6B; Fig. 7, curve 2). The water phase was recycled together with 10-20% of the oil phase within the lipase-containing interfacial layer, while fresh olive oil was added continuously to the reactor. Thus, the process is a continuous one with respect to the lipid phase and a batch process with respect to the water phase. However, this means that glycerol accumulates in the water phase. The results of such an experiment are shown in Figure 7, curve 2. The decrease in the degree of hydrolysis during the recycling of the enzyme might be due to the accumulation of glycerol, as well as to deactivation and loss of the enzyme. The effect of enzyme-recycling is shown by a comparison of curves 2 and 3 in Figure 7. The plot for the degree of hydrolysis vs dimensionless time (Fig. 7, curve 3) during a continuous reaction without adding or recycling the enzyme (Fig. 6C) again shows a normal washout function. Similar results were obtained using unreacted or hydrolyzed soybean oil as a substrate. The latter case simulated conditions in the second reaction stage.

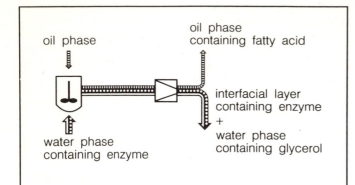

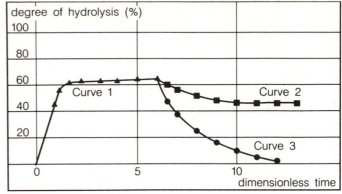

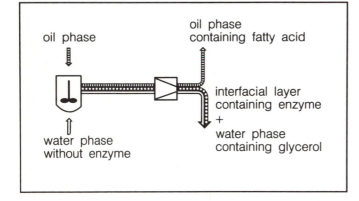

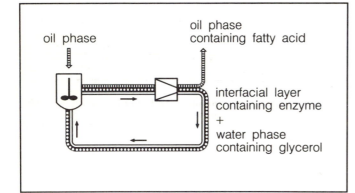

FIG. 6. One-stage mixer-settler setup for recycling the lipase with the water phase during oil hydrolysis (second stage of the process). (A) Continuous performance with continuous supply of lipase: the enzyme-containing interfacial layer is separated together with the water phase. (B) Semicontinuous performance: recycling of the lipase-containing interfacial layer together with the water phase during continuous supply of fresh oil. (C) Continuous performance: washout of the enzyme by addition of pure buffer instead of enzyme solution to the steady state (according performance A).

Two-stage Reactor System

The results shown in the Figures 4–7 confirm that the enzyme-containing interfacial layer can be transferred together with the aqueous phase or with the oil phase by incomplete separation procedures achieved by adjusting the appropriate mechanical parameters. On the basis of these results, a two-stage mixer-settler setup was established according to Figure 3. Due to the rapid decrease of

FIG. 7. Recycling of lipase with the water phase. Degree of hydrolysis vs dimensionless time (i.e., time referred to the mean residence time) for the various mixer-settler setups (Fig. 6). Reaction mixture: 70% (w/w) olive oil; 30% (w/w) 0.05 M acetate buffer pH 5.6; 2.0 g lipase/kg oil. Reaction conditions: see Figure 5. (Curve 1) Continuous performance (Fig. 6A). (Curve 2) Semicontinuous performance (Fig. 6B). (Curve 3) Continuous performance with washout of the enzyme (Fig. 6C).

TABLE 1

Characteristic Data of the Two-stage Mixer-settler Setup (also Fig. 10)

Parameter	Reactor 1[a]	Reactor 2[a]	Complete setup
Degree of hydrolysis (%)	65	93	93
Lipase activity (U/mg enzyme)[b]	52.9	7.2	14.3
Lipase concentration (g enzyme/l)[c]	1.87	1.99	1.98
Productivity (kg fatty acid/d)	119.5	51.5	171
Space-time yield (kg fatty acid/l.d)	39.8	5.7	11.4
Specific enzyme consumption			
(g enzyme/kg fatty acid)	0.08[e]	0.19[e]	0.62[d]
(kg fatty acid/g enzyme)	12.45	5.35	1.62

Volumes: Reactor 1: 3.0 liters; reactor 2: 9.0 liters; separator 1 and 2: 1.2 liter each; tubing: 0.6 liter; total system: 15.0 liters. Reaction mixture: 70% (w/w) soybean oil; 30% (w/w) 0.05 M acetate buffer pH 5.6. Volumetric flows: Oil phase 8.8 l/hr; water phase 3.5 l/hr. Total residence time within the reactor sysem: 1.22 hr.

[a]Only mixer where the reaction mainly takes place due to the small interfacial area in the other parts of the reactor setup.
[b]Basis: 3.42 μmol fatty acid per g soybean oil.
[c]Derived from Figure 11.
[d]Corresponds to the amount of enzyme necessary to retain the level of hydrolysis in the steady state: 4.4 g/hr (2.2 g/hr with enzyme recovery from the water/glycerol phase after separator 1).
[e]Corresponds to the calculated enzyme losses within the mixer.

the reaction rate with increasing degree of hydrolysis (Fig. 2), it did not seem advantageous to use two mixers of the same volume. To achieve a high degree of hydrolysis, a mixer was chosen for the second stage, with three times the volume (9 liters) of the first-stage mixer (3 liters). The flow sheet and a photo of a part of the reactor display are shown in Figures 8 and 9. Some of its characteristic data are summarized in Table 1.

The reaction system was loaded with enzyme by adding, over a period of 4.25 hr, lipase-containing aqueous phase to reactor 2. The total residence time was 1.22 hr. After a steady state with 93% hydrolysis was reached, the dosage of enzyme was reduced to 25% (0.5 g/kg oil). The time

Continuous Use of Lipases in Fat Hydrolysis
Flow Sheet of the Two-Stage Mixer-Settler Setup

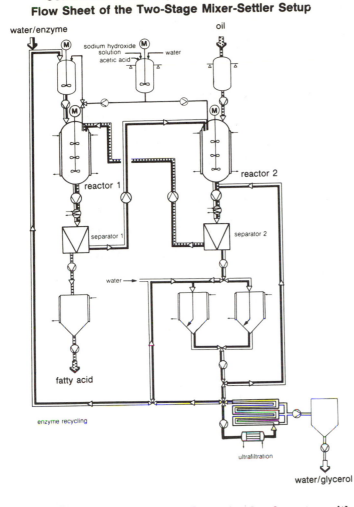

FIG. 8. Flow sheet of a cascade of two stirred tank reactors with enzyme recycling.

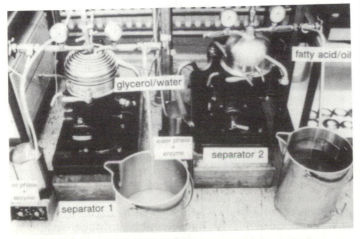

FIG. 9. Two-stage mixer-settler setup as used in the experiments described: view of the two separators (settlers).

course of the hydrolysis in the first and the second reaction stage is shown in Figure 10. Although a reduced amount of lipase was added, the degree of hydrolysis remained constant (about 65% in the first and 93% in the second stage). This indicates that the enzyme was recycled successfully.

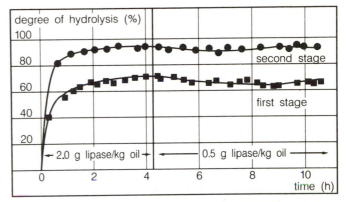

FIG. 10. Start-up of continuous hydrolysis of soybean oil in the two-stage mixer-settler setup (Table 1).

Mass Balances and Productivity

By calculating the mass flows of active enzyme from the volumetric enzyme activities determined at seven different positions around the reactor system and from the volumetric flows at these positions, a consistent mass balance for the enzyme within the whole reactor system was established (Fig. 11). The extraction of glycerol from the oil into the water phase was taken into account for the calculation of the volumetric flows. The flow of the interfacial layer containing the lipase was about 16% of the total flow of reactants entering the system. In the following discussion all enzyme flow data are given in g active enzyme/hr.

In the steady state, a flow of 26.8 g/hr leaves reactor 1, in which a loss of activity equivalent to 0.4 g active enzyme/hr occurs. In separator 1, a loss of 0.6 g/hr occurs while 2.2 g active enzyme/hr leave the system with the water phase. Together with fresh buffer, 4.4 g/hr are added to reactor 2, so that in all, 28.4 g/hr flow into reactor 2. Due to a loss of 0.4 g/hr within reactor 2, the enzyme flow at the inlet of the separator 2 thus decreases to 28.0 g/hr. Here a loss of 0.1 g/hr occurs during phase separation, while another 0.7 g active enzyme/hr leaves the system with the lipid phase, leading to an enzyme flow of 27.2 g/hr at the inlet of reactor 1.

The enzyme that is washed out together with the aqueous phase separated in separator 1 can be removed by ultrafiltration. Therefore, the total consumption of fresh enzyme that has to be added to reactor 2 to maintain this steady state can be reduced from 4.4 to 2.2 g/hr, providing there are no further losses during the recovery procedure. Taking into account the total quantity of active enzyme within the system (29.8 g) and the enzyme necessary to retain the degree of hydrolysis in the steady state (4.4 or 2.2 g/hr, respectively), the loss of enzyme was calculated as 14.8% (7.4%) per hr or 18.0% (9.0%) per total residence time. This means that in average, an active enzyme molecule can be recycled 5.5 (11.0) times.

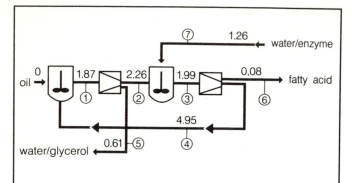

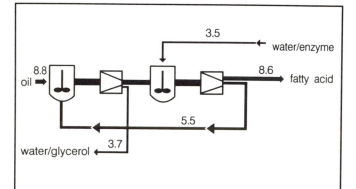

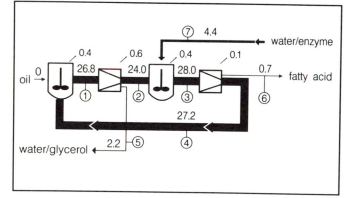

FIG. 11. Concentrations (A), volumetric flows (B) and mass flows (C) of active lipase within the two-stage mixer-settler setup. (A) Lipase concentrations (g active enzyme/liter), (B) Volumetric flows (l/hr), (C) Mass flows of active lipase (g active enzyme/hr). The numbers 1-7 indicate the positions of sampling where the volumetric activities (g active lipase/liter, Fig. 11A) were measured (see Experimental). From the volumetric activities (Fig. 11A) and the volumetric flows (Fig. 11B), the mass flows of active lipase (Fig. 11C) were calculated. Irreversible losses of enzyme activity occur in both of the mixers and settlers due to thermal and mechanical deactivation. These values were derived from the difference of the mass flows at the inlet and the outlet of the apparatus considered.

More data on the reaction engineering are summarized in Table 1. At a total degree of hydrolysis of 93%, the productivity was 171 kg fatty acid/d, and the space-time yield reaches 11.4 kg fatty acid/l.d. The enzymatic activity related to the effective residence time was calculated as 52.9 U/mg for 65% hydrolysis in reactor 1 or 7.2 U/mg at 93% hydrolysis in reactor 2. The average activity with

respect to the total enzyme in the system (29.8 g) was 14.3 U/mg. This takes into account that the reaction mainly takes place in the mixers, because the phase boundary in the other parts of the system is relatively low. The product-specific enzyme consumption is calculated as 0.62 g enzyme/kg fatty acid or 1.62 tons fatty acid/kg enzyme, respectively.

Alternative Reactor Systems

A relationship between reaction rate and degree of hydrolysis (Fig. 12A) can be derived from the initial reaction rate (maximal activity: 188 U/mg) and the activity at 65% and 93% hydrolysis (52.9 U/mg and 7.2 U/mg, respectively). This confirms that the reaction rate decreases with increasing degrees of hydrolysis (Fig. 2). In the plot of the inverse reaction rate vs degree of hydrolysis (Fig. 12B), two rectangles are drawn parallel to the axis and an intersecting point on the curve with an abscissa of 65% hydrolysis. The areas of both rectangles are equivalent to the product of individual residence time and enzyme concentration in each stage (24). At a constant enzyme concentration, the rectangles are then a measure for the volume of the reactors (mixers) in the two stages. Because the enzyme concentration in both mixers is nearly the same (Table 1), the only way to achieve a high degree of hydrolysis in the second stage is to increase the volume of reactor 2 leading to a prolonged residence time within this vessel. The data in Figure 12 confirm that it is more favorable to use mixers with a volume of 3 liters in the first and 9 liters in the second stage than to apply two reactors with an equal volume of 6 liters in each stage.

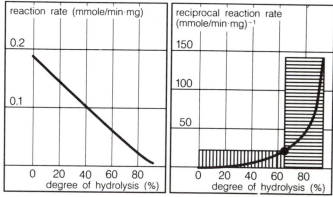

FIG. 12. Graphical procedure to estimate the relative volume of the mixers in the two reaction stages at a constant enzyme concentration. The curves have been graphically fitted to the activities at the following degrees of hydrolysis: 0% (initial rate), 65% (first stage) and 93% (second stage). Reaction rate (A) and reciprocal reaction rate (B) are depicted as a function of degree of hydrolysis. All values of activity are correlated to mg lipase preparation. See text for further details.

With a reactor device having the same mixer volume in the first and in the second stage (3 liters), a degree of hydrolysis of only 85% was achieved under otherwise identical reaction conditions. However, by substituting the stirred tank reactor in the second stage by a plug-flow reactor (3 liters) with static mixers, a steady degree of hydrolysis of 98% was measured. Although the specific interface area is clearly smaller than in a stirred vessel, the tubular flow behavior results in a higher degree of hydrolysis, especially when approaching equilibrium. These results are summarized in Figure 13.

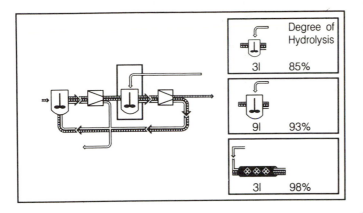

FIG. 13. Continuous hydrolysis of soybean oil by reactor setups with different mixers in the second stage of the process. The reactions were carried out under otherwise identical conditions and volumetric flows (Fig. 10, Table 1).

It has been shown in pilot scale that continuous use of a lipase is possible by exploiting the fact that the enzyme is enriched at the phase boundary. This principle should be applicable also to other two-phase systems in which hydrophilic and lipophilic substrates or products are involved. As the mass balance established for the lipase shows, it is possible to quantify all losses of enzyme from the recycling flow as deactivation within the various parts of the reactor system or as washout by the water or the oil phase, respectively. Thus, quantitative information can be obtained by a simple sampling technique during continuous operation. This seems to be an attractive approach to study the influence of process parameters on the product-specific enzyme consumption. Scaling-up probably is not going to be a problem because concentration gradients in the mixer and settler are very small. The operational stability of the lipase used in this study is not yet sufficient, but improvements might now be expected analogous to those achieved with amylases and proteases, after the technical feasibility has been demonstrated.

ACKNOWLEDGEMENTS

The authors appreciate the skillful technical assistance of G. Kasten, H. Rüth and R. Hermanns.

REFERENCES

1. Sonntag, N.O.V., in *Bailey's Industrial Oil and Fat Products*, 4th Edn., Vol. 2, edited by D. Swern, John Wiley & Sons, New York, NY, 1982, pp. 97–111.
2. Hartmann, H., in *Ullmann's Encyclopedia of Chemical Technology*, Vol. 11, Verlag Chemie, Weinheim, 1976, pp. 529–533.
3. Connstein, W., E. Hoyer and H. Wartenberg, *Berichte d. Deutschen Chemischen Gesellschaft 35*:3988 (1902).
4. Macrae, A.R., and R.C. Hammond, *Biotechnol. Genet. Eng. Rev. 3*:193 (1985).
5. Ishida, S., in *World Conference on Emerging Technologies in the Fats and Oils Industries*, Cannes 1985, edited by A.R. Baldwin, American Oil Chemists' Society, Champaign, IL, 1986, pp. 359–364.
6. Rattray, J.B.M., *J. Am. Oil Chem. Soc. 61*:1701 (1984).
7. Baumann, H., M. Bühler, H. Fochem, F. Hirsinger, H. Zoebelein and J. Falbe, *Angew. Chem.*, in press.
8. Brockmann, H.L., in *Lipases*, edited by B. Borgström and H.L. Brockmann, Elsevier, Amsterdam, New York, NY, Oxford, U.K., 1984, pp. 4–46.
9. Hirano, J., S. Ishida, T. Funada and S. Ikeda to Nippon Oils and Fats Co., Ltd., Jpn. 58, 126, 794 (1983).
10. Tanigaki, M., H. Wada and M. Sakata to Kao Corp., DE 36 19860 (1986).
11. Hoq, M.M., M. Koike, T. Yamane and S. Shimizu, *Agric. Biol. Chem. 49*:3171 (1985).
12. Liebermann, R.B., and D.F. Ollis, *Biotechnol. Bioeng. 17*:1401 (1975).
13. Bell, G., J.R. Todd, J.A. Blain, J.D.E. Patterson and C.E. Shaw, *Biotechnol. Bioeng. 23*:1703 (1981).
14. Kimura, Y., A. Tanaka, K. Sonomoto, T. Nihira and S. Fukui, *Eur. J. Appl. Microbiol. Biotechnol. 17*:107 (1983).
15. Tahoun, M.K., M.Y. El-Sayed and S. A. Abou Donia, *Z. Lebensm. Unters. Forsch. 183*:335 (1986).
16. Tamaura, Y., K. Takahashi, Y. Kodera, Y. Saito and Y. Inada, *Biotechnol. Lett. 8*:877 (1986).
17. Kwon, D.Y., and J.S. Rhee, *Kor. J. Chem. Eng. 1*:153 (1984).
18. Wisdom, R.A., P. Dunhill, M.D. Lilly and A. Macrae, *Enzyme Microb. Technol. 6*:443 (1984).
19. Taylor, F., C.C. Panzer, J.C. Craig, Jr. and D. J. O'Brien, *Biotechnol. Bioeng. 28*:1318 (1986).
20. Bühler, M., and C. Wandrey, *Fat Sci. Technol. 89*:156 (1987).
21. Shinota, A., H. Machita and T. Azuma to Meito Sangyo Co., Ltd., Jpn. 71, 16, 509 (1971).
22. Linfield, W.M., D.J. O'Brien, S. Serota and R.A. Barauskas, *J. Am. Oil Chem. Soc. 61*:1067 (1984).
23. Hoq, M.M., T. Yamane and S. Shimizu, *Enzyme Microb. Technol. 8*:236–240 (1986).
24. Wandrey, C., and E. Flaschel, in *Adv. Biochem. Eng.*, Vol. 12, edited by T.K. Ghose, A. Fiechter and N. Blakebrough, Springer, Berlin, 1979, pp. 147–218.

Economic Aspects of Lipid Biotechnology

N.K.H. Slater

Unilever Research Laboratorium, Vlaardingen, The Netherlands

The production of various lipids by microbial fermentation is now a technically feasible alternative to the continued dependence upon plant-derived oils and fats. Despite the particular attraction of this option to countries that face heavy import costs for lipids and recent developments in large-scale fermentation technology, commercial exploitation has been slow. Economic barriers exist for all but the specialty lipids that have high values. This paper examines the raw material costs and the various scale-related plant and processing costs for the batch production of microbial lipids. The sensitivity of production costs to the use of alternative carbohydrate sources, improvements in key process parameters and the utilization of fed-batch or continuous processing routes are considered to identify where developments in technology are required. Comparison of these costs with the current market values of lipids illustrates possible process opportunities.

Biotechnology offers novel opportunities to manufacture and modify lipids by fermentative and enzymic routes (1,2). The products of these processes will largely compete with materials derived from agricultural sources or chemical modifications of these products. In a related paper, the economic aspects of the enzymic modification of fats in membrane bioreactors have been considered (3). Here, I shall focus on the economic aspects that dictate whether lipid production by fermentation could be a feasible alternative to the current dependence upon agricultural feedstocks.

TABLE 1

Lipid Materials and their Modifications Derived by Fermentation (Examples and Market Values in Parentheses)

1. *Single-cell oils*
   - with range of saturated and unsaturated fatty acids (4).
2. *Fatty acids* (from hydrolyzed fats)
   - Short chain (i.e. caprylic acid, Dfl 4/kg)
   - Medium chain (lauric acid, Dfl 1.5/kg)
   - Long chain (erucic acid, Dfl 6/kg)
   - Polyunsaturated (linoleic acid, Dfl 6/kg)
   - Hydroxy (ricinoleic acid, Dfl 3.6/kg)
   - Dicarboxylic (azeloic acid, Dfl 10/kg)
3. *Wax Esters*
   - $C_{32}C_{40}$ (replacers for sperm whale and jojoba bean oil, Dfl 7-12/kg)
4. *Others*
   - Monoglycerides (glyceryl monooleate, Dfl 3.5/kg)
   - Surfactants (glycolipids)
   - Highly polyunsaturated fatty acids (eicosapentaenoic)

A wide range of lipid materials can be made by fermentation (Table 1); however, for the majority of materials little information is available on the process parameters necessary for fermentation plant design or the proper assessment of process costs. Fortunately, a wide body of data is available on the production of single-cell oil and the relevant literature has been reviewed extensively (4). Attention has been given to the contribution of raw material costs to single-cell oil production (4), but the magnitude of processing costs has not been assessed in detail. Sufficient data are available from laboratory studies for such an assessment of costs to be made. As a model study, and to illustrate where the costs lie in a biotechnological process for the production of lipids, a cost estimate is presented for the manufacture of single-cell oil by fermentation.

TABLE 2

Typical Oil Prices (5,6)

	Dfl/kg (1986)
Soybean	0.69
Cottonseed	1.02
Sunflowerseed	0.78
Peanut	1.16
Palm	0.50
Rapeseed	0.62
Coconut	0.46
Castor	3.90
Cocoa butter	10.00
Jojoba	33.33

Table 2 lists the market values of a range of commercially interesting oils. If fermentation is to compete with agricultural production then product costs must be comparable with these prices. Clearly, production of the higher value specialty oils will be attractive initially but it is interesting to question whether single-cell oil could be an alternative to lower cost products such as palm oil. If this is to be possible, then significant process development would be necessary to reduce manufacturing costs. The cost model indicates which parameters most influence production costs and could potentially assist such a development program.

NUTRIENT COSTS

Various assessments of the cost of carbon source upon lipid production by fermentation have been presented in the literature (4,5), and only a brief summary of the position is given here. The parameter that controls raw material costs is the yield of product (lipid) per unit weight of carbon source. Various carbon sources have been utilized for the production of single-cell oil although the majority of data refers to glucose for which, typically, 0.2 kg of lipid is produced per kg of glucose. At a price of Dfl 1.4/kg glucose, the carbon source contribution to lipid cost is at least Dfl 7/kg. Other nutrients such as yeast extract, a nitrogen source and trace metals also are required for a complete medium. Generally, the total nutrient costs would be 1.5–2 times the cost of carbon source. Total raw material costs with a glucose carbon source thus would amount to about Dfl 10.5-14/kg of oil. The manufacture of many of the commodity oils in Table 1 thus is not feasible purely on raw material costs. Other carbon sources could be considered. Typical yield coefficients for the production of lipids on various carbon sources are given in Table 3.

TABLE 3

Typical Yield Coefficients for the Production of Single-Cell Oil by Oleoginous Yeasts (7)

Substrate	Yield (kg SCO/kg Substrate)
Glucose	0.2–0.25
Whey	0.2–0.24
Starch	0.10
Molasses	0.24–0.44
Alkanes	—

Alkanes appear to be an economically attractive carbon source and would permit oil production at a total nutrient cost of about Dfl 5/kg depending upon yield coefficient. Bulk supplies of alkanes are available readily from petrochemicals, but concern arises over the presence of residual hydrocarbons in the cellular lipid whose removal may require unacceptable refining costs to render the oil suitable for consumption.

Whey also is an attractive nutrient source; as a by-product of cheese manufacture it is available at low cost, particularly in countries with a substantial dairy industry such as The Netherlands. Typically, for yield coefficients quoted in the literature (8), single-cell oil could be produced from whey with a total raw material cost of Dfl 7.5–10/kg; somewhat less than that of glucose. Whey would, however, only be suitable for the production of low tonnage or specialty oils due to its limited availability. The total annual European production of 185,000 tons (lactose) would be adequate for the manufacture of only 44,000 tons of oil. In practice, the use of lactose for other purposes would severely reduce its availability for specialty oils. Thus, the problems of supply may well detract from the benefits of reduced raw materials costs.

Starch has been used as a carbon source in laboratory trials, either as a hydrolyzed substrate (4) or as a nutrient for mixed-culture fermentation incorporating a hydrolytic organism (9). At present, starch appears to offer little benefit compared with glucose in terms of cost, but release of cereal-derived carbohydrates from EEC surplus stocks for use as industrial feedstocks could well permit a price reduction to Dfl 0.5/kg of carbohydrate (5).

Molasses may well prove to be the lowest cost practical substrate, although again, supply in adequate quantities may be a problem. Typically, beet molasses contains 54% saccharose and costs Dfl 0.19/kg. For a yield coefficient of 0.2 kg of oil produced per kg of saccharose, the total raw material costs would amount to Dfl 2.6–3.5/kg. It appears that little information is available on oil production with a molasses carbon source, although one study reports a yield coefficient as high as 0.44 (8). In view of the considerable nutrient cost reduction for molasses compared with glucose, a closer examination of this substrate would appear justified, particularly to see if the C:N ratio in molasses is satisfactory.

PROCESSING COSTS

Fermentation

Production of single-cell oils be fermentation involves two stages. First, a cell concentrate containing lipid as an intracellular product is prepared by fermentation and the resulting broth concentrated by either centrifugation or filtration to yield a cell slurry. This slurry is then treated in some downstream processing plant to extract the oil.

For the fermentation stage, the process economics are influenced by the mode of operation selected. For example, batch fermentations would typically be operated with a time-averaged oil productivity of 0.15 kg m^{-3} hr^{-1} (4,11). Alternatively, continuous fermentation may be operated with a productivity of 0.2 kg m^{-3} hr^{-1} (11,12). By contrast, investment costs invariably would be higher for a continuous fermentation plant, and its use may not be warranted by the small enhancement in productivity.

To illustrate the various contributors to processing

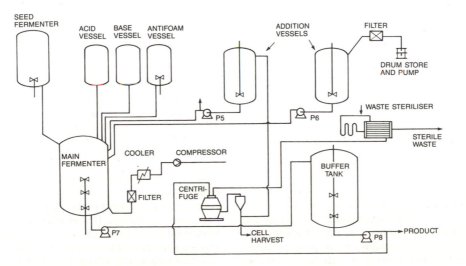

FIG. 1. Simplified flowsheet for batch fermentation facility to produce a cell slurry containing intracellular single-cell oil.

costs for the production of cell concentrate from fermentation, a base case is considered. Cells are assumed to be formed by batch fermentation in a simple dedicated facility operated at 100% capacity. Such a facility is illustrated in Figure 1 and contains as major equipment items a central aerobic fermenter and seed fermenter, medium make-up and feed vessels, a centrifuge(s) for cell concentration and a buffer vessel(s) for supply of cell concentrate to a downstream extraction facility. For large tonnage production, multiples of certain items may be required. Additional facilities would be required for plant control and sequencing of cleaning operations, provision of utilities (compressed air for the fermenter, process steam, cooling water and power), raw material storage and product packaging. Depending upon site location, treatment facilities also may be required for plant effluents, a particular problem with molasses as a nutrient because a high proportion of carbohydrates would not be fermented in the process.

Capital charges for the plant would accrue at typically 13% of total capital investment per annum (8% depreciation and insurance and 5% maintenance). Labor charges would accrue for the operation of a three-shift production schedule for seven days per week with at least two operators per shift for a fully automated process. Other costs also would accrue. Factory indirect costs would cover items such as factory management, staff and buying departments, accounts, etc. and would amount to typically 20% of factory operations expenses (capital-related costs + utilities + labor). Nonfactory indirect costs would cover central administration, sales and marketing and might typically account for about 5% of direct factory cots. Given these assumptions, the processing and labor-related costs per kilogram of lipid product depend upon the following parameters.

Production requirments. Increased production capacities enable large and more cost-effective plants to be employed as investment costs for plant automation, CIP and control are largely independent of the size of facility. Labor costs per unit volume of production capacity also are lower.

Production cycle time. For batch processes, reduction of the cycle time from medium make-up, cleaning and sterilization and fermentation to product harvesting enables greater plant utilization. Generally, at least 10 hours are required for cleaning and sterilization compared with 60–100 hours of fermentation time.

Final product concentration. Fermentation to a high-cell density with a high proportion of cell mass as lipid is a clear processing objective.

Product recovery. For a low-product recovery, the plant output per fermentation is reduced.

Typical unit plant and labor costs. Generally, such costs will vary with the degree of plant sophistication and construction materials as well as with plant location.

Table 4 gives a set of base-case parameters that have been selected from literature as typical values for the production of single-cell oil by yeasts. Using these parameters and estimates of plant investment costs, the scale-related processing and labor costs for the production of a con-

TABLE 4

Base-case Parameters for the Calculation of Scale-related Fermentation Processing Costs for Single-cell Oil Production

Production capacity: 10, 100, 1,000, 10,000 tons SCO/annum
Fermentation cycle time: 75 hr (batch)
Final product concentration: 15 kg/m³
Downstream processing yield: 70%
Plant utilization: 100%
Plant operation: 8,000 hours/annum

centrated slurry of lipid-containing cells are shown in Figure 2. Clearly, processing costs are a strong function of scale though even for an annual demand of 10,000 tons of lipid, the fermentation costs alone account for Dfl 9/kg.

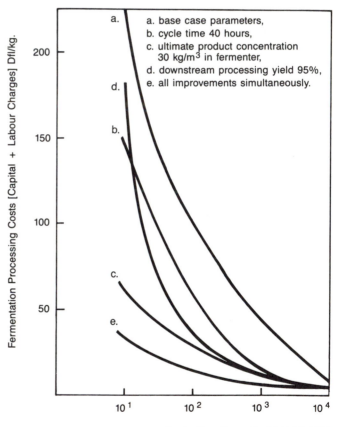

FIG. 2. Scale-related fermentation costs for the production of cell slurry containing intracellular single-cell oil. a., base case parameters; b., cycle time 40 hr; c., ultimate product concentration 30 kg/m³ in fermenter; d., downstream processing yield 95%; e., all improvements simultaneously. Fermentation Processing Costs [Capital + Labor Charges] Dfl/kg Production Capacity [tons/annum]

As a result of process optimization and scale-up, values of the base-case parameters may be varied to yield much better (or worse) process economics. The effects of changes in certain parameters are shown in Figures 2 and 3. In Figure 2, the effect of a fixed change in individual parameters upon scale-related costs is illustrated, and in Figure 3 the relationship between a range of parameter

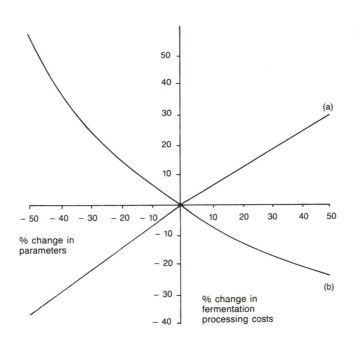

FIG. 3. Influence upon the fermentation processing costs of incremental variations in fermentation cycle time and ultimate product concentration and downstream processing yield. Production of 1,000 tons of oil/annum is assumed.

values and processing costs for a fixed production level of 1,000 tons oil/annum is shown.

Introduction of automated control of cleaning between batches and the development of rapidly fermenting hybrid yeast strains by protoplast fusion techniques may permit the total production cycle time to be reduced. Were this to be achieved, a smaller production plant could be employed with lower capital charges for a fixed output of oil. For the production of 1,000 ton/annum, a 10% reduction in cycle time permits a 7.5% reduction in fermentation costs (Fig. 3). With the most optimistic estimate, a cycle time of 40 hr may be attainable without influencing the other parameters. In this case, the scale-related costs could be reduced to the level shown in case (b) of Figure 2, yielding a minimum fermentation cost of Dfl 6/kg of oil at a production of 10,000 tons/annum. Figure 3 indicates the strong relationship between fermentation costs and the concentration of intracellular oil at the end of each production batch. For a 10% change in oil concentration from the base-case level, a 7% change in fermentation costs is anticipated. Efficient scale-up thus is essential.

The quantity of intracellular oil at harvest is fixed by both the cell density and the proportion of cell mass that is lipid. Improvements in cell density may well be achieved by optimization of processing conditions. For example, although single-cell oil production is promoted under conditions of carbon excess and nitrogen depletion, too-high levels of glucose commonly inhibit yeast growth. Fed-batch control of carbon and nitrogen source feed rates would enable inhibitory levels of glucose to be avoided more readily, and high cell densities to be reached before nitrogen exhaustion. In this case, it is likely that the oxygen transfer rate in the fermenter would become the growth-limiting factor. It is reasonable to assume that for

a fermenter with efficient aeration a cell density of at least 60 kg/ m^3 could be achieved, particularly as densities of 150 kg/m^3 are attained in baker's yeast production. Provided a lipid content of 50% of cell mass could be maintained as a conservative estimate, 30 kg of oil would be produced per m^3 of fermenter. Figure 2 (case c) shows that in this case, a minimum fermentation cost of Dfl 6/kg could be envisaged.

Finally, for the base-case calculations, an arbitrary downstream processing yield of 70% was chosen in view of the paucity of literature on oil recovery. As actual oil obtained from each fermentation is the product of oil produced in the fermenter and downstream processing yield, the sensitivity of costs to yield is the same as that to oil concentration in Figure 3. Efficient downstream processing may well enable a 95% recovery of oil to be achieved. For such a recovery, the base-case scale-related costs would be reduced to the level of case (d) in Figure 2, yielding a minimum cost of Dfl 7/kg.

Finally, if all the improvements in batch cycle time, product concentration and downstream processing yield could be achieved simultaneously, then the fermentation costs could be reduced to case (e) of Figure 2. Even with this optimistic target for process development the minimum fermentation cost is Dfl 3/kg of lipid at a production level of 10,000 tons/annum.

Fermentation costs for the production of another class of materials have been estimated. Wax esters can be produced by bacterial fermentation as replacements for jojoba waxes (3). Biological data from shake-flask cultures show waxes to be produced in a concentration of 15 g/m^3 by fermentation of an Acinetobacter strain. To achieve a final biomass concentration of 0.28 kg/m^3, a fermentation time of 24 hr was required. At this mean, productivity of 0.63 g/m^3/hr fermentation costs for the production of 1,000 tons wax/annum would be Dfl 800/kg. Although it is difficult to estimate what performance levels could be achieved by process development, it may be speculated that even if a biomass concentration of 40 kg/m^3 was achieved on a commercial scale, the fermentation cost would be Dfl 36/kg.

Costs for the provision of utilities also will accrue to the fermentation process. Assuming a plentiful supply of cooling water is available at negligible cost, the principal utility requirements are for steam and electricity. Steam is required to sterilize the fermentation broth. If direct steam injection into the fermenter is adopted with no heat recovery, steam costs would amount to about Dfl 0.9/kg of lipid. Electricity is required to drive a compressor to provide air to the fermenter. Typically, 1 kW of electricity would be utilized per m^3 of broth at a cost of Dfl 1.25/kg of lipid. These costs would accrue irrespective of fermentation capacity.

Downstream Processing

Estimation of downstream processing costs is more difficult because little data have been published on processing routes and design parameters for single-cell oil recovery. To provide a rough estimate of costs, the simple dry extraction procedure shown in Figure 4 has been considered.

The cell slurry from the centrifuge (say 10% solids) is

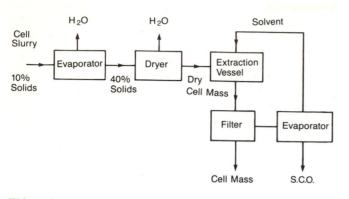

FIG. 4. Downstream processing scheme for the extraction of single-cell oil from yeast cells.

first concentrated to 40% solids in a multistage evaporator and then dried to leave the oil-containing cell mass. Extraction of oil with a solvent is accomplished in a single- stirred tank. The slurry is filtered to remove cell debris and the solvent removed from the extracted oil in a second evaporator. Solvent is recycled to the extraction tank with no losses. For the production of 10,000 tons of oil/annum, the total capital and labor charges (including a 20% contingency for equipment items) are Dfl 0.3/kg. Utilities account for a further Dfl 0.45/kg, yielding the total downstream processing cost of Dfl 0.75/kg. This price appears to be broadly in line with the cost of extracting oils from dry seeds where hexane extraction (preceded by pressing) is employed.

Discussion

Typical costs for the production of an intracellular lipid have been estimated taking as an example an unspecified single-cell oil. Representative values for process parameters were chosen from literature on the fermentation of yeasts. Cost data for fermentation equipment were obtained from estimates provided by suppliers and installation costs estimated according to standard engineering guidelines. Downstream processing costs were taken for a simple extraction scheme. In view of the various uncertainties involved in the estimation of installed equipment costs, and the lack of process data from pilot plant fermentations, the resulting cost estimate is accurate to only ±40%. However, it does indicate how various factors contribute to the overall cost and where process improvements may best be sought. A summary of base-case and optimistic production costs is given in Table 5.

For the most optimistic estimate, a total product cost of Dfl 18.4/kg is envisaged with a glucose nutrient or Dfl 10.6/kg with molasses as carbon source. Comparison with the current market values for various oils given in Table 1 shows that production of commodity oils that sell at less than Dfl 1/kg is unlikely to be economically feasible in the short term even if carbohydrate costs were to fall dramatically within the EEC due to the release of surplus grain for use as industrial feedstocks. Prices for oils do fluctuate substantially, depending upon agricultural and world economic conditions. For example, the price of palm oil fell from Dfl 1.4/kg in April 1985 to Dfl 0.9/kg in April 1986. Notwithstanding this, it is unlikely that the

TABLE 5

Cost Price Estimate for the Production of 10,000 Tons Single-cell Oil/Annum

		Optimistic Dfl/kg		Base Case Dfl/kg	
1.	Raw materials				
1.1	Glucose complex medium	10.5		14	
1.2	Molasses medium		(3)		(4)
2.	Fermentation	3		9	
3.	Fermentation utilities				
3.1	Steam	0.9		0.9	
3.2	Electricity	1.25		1.25	
4.	Downstream processing	0.75		0.75	
5.	Factory indirect costs	1.18		2.38	
6.	Non-factory indirect costs	0.82	(0.5)	1.41	(0.80)
	TOTAL	18.4	(10.58)	29.69	(19.69)

price of commodity oils would rise above the optimistic processing costs for the fermentation route (neglecting raw materials and indirect costs) of Dfl 5.9/kg. Higher value oils such as cocoa butter at Dfl 10/kg (5) may well be replaced by fermentation products, particularly at times when supplies are unreliable and market prices high. A cocoa butter replacement produced by fermentation additionally may be considered a natural product provided an approved solvent is used for the extraction stage. Thus, it could benefit from ready consumer acceptance although the fermentation route must compete with the enzymatic interesterification process on a cost basis. Other oils, such as a replacement for evening primrose oil at Dfl 115/kg appear even more attractive. This oil is rich in polyunsaturated γ-linolenic acid, which currently commands attention as a health food. The market for this product presently is unclear, although one fermentation company is embarking upon a large-scale process for its manufacture (5). Wax esters also command high prices (Dfl 7-12/kg), but the data available on productivity by fermentation are clearly suboptimal and further process development would be necessary before an evaluation of economic potential can be made.

It appears that the production of lipids by fermentation within European countries may well be appropriate for specialty oils, or during periods of uncertain supply and high prices for those oils that must be wholly imported. In this case, attention should be given to improving production economics. The cost estimate indicates that attention is required to reduce costs due to:

Raw materials. For a glucose nutrient, the carbon source accounts for more than 50% of production costs and as such is not a feasible process option. Molasses would be more economical, although little process data are available for this nutrient.

Fermentation. Clearly, productivity should be optimized by attaining high cell densities and shorter fermentation times. Strain selection and hybridization studies may well improve biological performance. Fed-batch fermentation may well overcome glucose inhibition for certain strains and enable cell growth under nitrogen limitation to proceed for longer periods. Additional investment in continuous-fermentation plants may be questioned in view of the extremely low dilution rates (<0.01/hr) that are necessary to achieve lipid productivities appreciably higher than in conventional batch culture.

Plant and labor requirements. Capital- and labor-related charges have been calculated for the conventional fermentation plant shown in Figure 1. Complete asepsis was considered essential and provision of sterility contributes a major share of fermenter costs. Development of protected fermentation processes would permit simpler plants to be used and process research might well be directed toward investigating combinations of pH and water activity at which bacterial contamination could be avoided. Plant sites could be selected so as to reduce labor costs or benefit from regional assistance programs.

Downstream processing. A high downstream- processing yield is essential to reduce fermentation costs. Extraction of oil from the wet-cell slurry would further reduce plant investment costs and would avoid the requirement of steam for drying.

REFERENCES

1. Rattray, J.B.M., *J. Am. Oil Chem. Soc. 61*:11, 1701 (1984).
2. Werdelmann, B.W., and R.D. Schmid, *Fette. Seifen. Anstrichmittel 84*:11, 436 (1982).
3. Kloosternam, J., P.D. van Wassenaar, and W.J. Bel, in *Proceedings of the World Conference on Biotechnology for the Fats and Oils Industry*, 1988, pp. xxx-xxx.
4. Ratledge, C., *Prog. Ind. Microbiol. 16*:119 (1982).
5. Sinden, K.W., *Enzyme Microb. Technol. 9*:124 (1987).
6. Anon., *J. Am. Oil Chem. Soc. 63*:944 (1986).
7. Ratledge, C., *Prog. Ind. Microbiol. 16*:119 (1982).
8. Moon, N.J., and E.G. Hammond, *J. Am. Oil Chem. Soc. 55*:683 (1978).
9. Dostalete, M., *Appl. Microbiol. Biotechnol. 24*:19 (1986).
10. Allen, L.A., N.H. Barnard, M. Fleming and B. Hollis, *J. Appl. Bact. 27*:27 (1964).
11. Yoon, S.H., and J.S. Rhee, *J. Am. Oil Chem. Soc. 60*:7, 1281 (1983).
12. Hansson, L., and M. Dostalek, *Appl. Microbiol. Biotechnol. 24*:187 (1986).
13. Neidleman, S.L., and J. Geigert, *J. Am. Oil Chem. Soc. 60*:7, 1281 (1984).

New Process for Purifying Soybean Oil by Membrane Separation and an Economical Evaluation of the Process

A. Iwama
Central Research Laboratory, Nitto Electric Industrial Co. Ltd., 1-chome, Shinohozumi, Ibaraki, Osaka 567, Japan

A feasibility study has been conducted on introducing membrane-separation technology into the refining process for edible oils. In general, vegetable oils for edible use such as soybean oil or rapeseed oil are purified and produced by chemical treatment with alkali agents. The conventional chemical process has the disadvantages of considerable loss of oil and highly contaminated effluents, and instead, a physical refining process is desired. Membrane separation can provide an answer to this need. Ultrafiltration makes it possible to manufacture purified oil by the following process: Degum crude miscella by ultrafiltration, obtain degummed oil by removing hexane from the permeated miscella by distillation and carry out bleaching and deodorization of the degummed oil. The solvent-resistant polyimide ultrafiltration membrane, designated as NTU-4200, which is available from Nitto Electric Industrial Co. Ltd., is effective for processing soybean miscella at 40-50 C, and the molecular weight cut-off value 20,000 of the polyimide membrane shows good rejection of gum materials. As a result, the complete process for the physical refining of soybean oil has proven practical. Based on an economic estimation with a bench-scale test plant, the physical refining of soybean oil by membrane separation is clearly profitable when compared with conventional chemical refining. This may have significance to the oil processing industry.

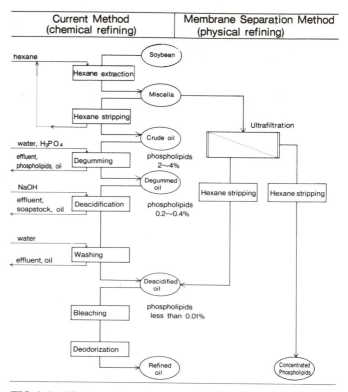

FIG. 1. Application of membrane separation method to the refining process of soybean oil.

Reverse osmosis or ultrafiltration are membrane separation processes that help save energy or resources due to no need for phase or chemical changes. Ultrafiltration (UF) is more effective than centrifugal separation and thus can give more precise separation than conventional methods.

Recently, in the food industries, UF has been investigated as a means of purifying oils and fats. The authors have been developing the membrane separation process for purifying soybean oil in cooperation with Rinoru Oil Mills Co., Ltd. of Japan.

This paper describes the results of the developmental work on the feasibility of membrane process for degumming soybean oil and the economic effectiveness of the process (1).

CONVENTIONAL PROCESS OF CHEMICAL REFINING OF SOYBEAN OIL AND ITS DISADVANTAGES

Most edible oils such as soybean oil or rapeseed oil are purified and produced by the chemical refining process shown in Figure 1. Chemical refining has the following disadvantages: 1. Oil losses due to the chemical treatments. Chemical damages are incurred by severe treatments with alkali solutions, resulting in approximately 2% loss of oil. 2. Chemical processing with water and chemicals results in an enormous volume of highly contaminated waste water, thus requiring large expenditures for disposing of the effluents.

In the oil industry, many attempts at solving these problems have been reported (2): dialysis, adsorption, UF (3), extraction with supercritical carbon dioxide (4) and the Alcon process (5) with the inactivation of soybean enzymes before hexane extraction. These approaches are used as a preamble to physical refining instead of conventional chemical refining.

Sen Gupta (6) reported the degumming of miscella by the use of UF, which may be valuable from an industrial viewpoint. However, in introducing UF separation into the degumming of miscella, UF membranes with heat and solvent resistance that are stable in the presence of hexane and oils had to be developed.

REFINING OF SOYBEAN OIL AND ULTRAFILTRATION

The Impurities of Crude Soybean Oil

Crude soybean oil (Fig. 1) contains impurities as shown in Table 1. These impurities are not desirable for the best quality of finished oils because of their decomposition or polymerization during storage, their discoloration on heating, their generation of unpleasant odors and their acceleration of oxidation or deterioration of oils. It is necessary, therefore, to remove these impurities from the crude oil as much as possible.

A major part of the impurities (7) consists of phospholipids, which are called gum material in the oil industry.

TABLE 1

Reported Components of Crude Soybean Oil

Components	Details	Composition
		%
Soybean oil	Glycerides	97~98
Phospholipids	Phosphatidyl —choline —ethanolamine —inositol	1.5~2
Free fatty acids		0.5~1
Other impurities	Tocopherols Carbohydrates Pigments Sterols	a small amount

Therefore, removal of the phospholipids results in substantial purification of soybean oil. That is, by reducing the phospholipid content to 100 ppm or less (as in Fig. 1 for deacidified oil), followed by bleaching and decoloring steps, purified oil is obtained. Thus, the whole process of physical refining will be completed (8).

Separating the Impurities from Soybean Oil by Ultrafiltration

Presumably, it is difficult to separate soybean oil (MW, ca. 900) and phospholipids (MW, ca. 800) by UF membranes because of the similarity of their molecular weights. However, in practice, they can be separated as shown in Figure 2.

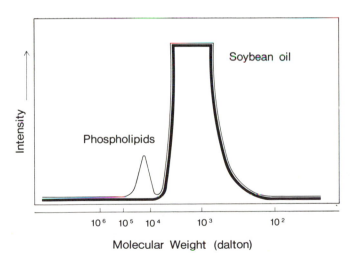

FIG. 2. Degumming effect by the ultrafiltration with 20000 MW cut-off membrane. GPC–analysis:
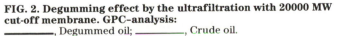
————, Degummed oil; ————, Crude oil.

The molecules of phospholipids tend to form micelles with imbedded hydrophilic ends inside in a nonpolar media of hexane or oil (9). The micelles thus formed can have a molecular weight of 20,000 or more. Thus, soybean oil and phospholipids can be separated by a UF membrane of MW cut-off value 20,000. As illustrated in Figure 3, degummed oil is recovered from the permeated mis-

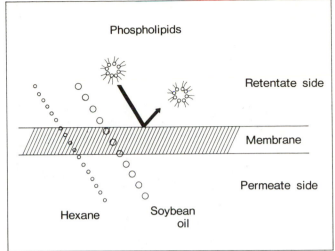

FIG. 3. Schematic diagram of ultrafiltration of crude soybean oil miscella with 20000 MW cut-off membrane.

cella. On the other hand, the phospholipids are concentrated within the retentate.

UF Membranes with Solvent Resistance

UF membranes for use in the processing of aqueous solutions have been developed and applied to industrial use. These include polysulfone, polyacrylonitrile or polyvinylalcohol. However, in treatment of soybean miscella with UF membranes, polymeric membranes must have chemical and physical stability in hexane or soybean oil, and, in addition, at elevated temperatures of 40–60 C. As indicated in Table 2, polyimide materials hold great promise in this regard (10).

TABLE 2

Solvent-resistant Materials that Could Be Applicable to Ultrafiltration Membranes for Organic Use

Polymer materials	Membrane available
Aromatic Polyamide (11)	
Aromatic polyamide-imide	Commercial membrane HFD available from Abcor Inc. (USA)
Aromatic imide	Commercial membrane NTU-4000 available from NITTO (Japan)
Polyvinylidene fluoride	Commercial membrane HFM available from Abcor Inc. (USA)
Polytetrafluoroethylene	Only microfiltration membrane available

Below, we describe an application of UF process to the purification of soybean oil with the polyimide UF membrane of MW cut-off value 20,000, which is available from Nitto and designated NTU-4220 (12).

BENCH-SCALE UF PLANT FOR THE PROCESS OF DEGUMMING MISCELLA

Crude miscella was piped from the extraction plant in a soybean oil factory to operate a bench-scale UF plant with the capacity of 500 kg oil/day. This UF plant is shown in Figure 4. The bench-scale plant first treats crude miscella by two-stage UF separation followed by evaporation of hexane from the permeated miscella to produce the degummed oil whose content of phospholipids ought to be 100 ppm or less as described in Figure 1.

FIG. 4. A photograph of the bench-scale UF plant for the process of degumming miscella.

Miscella Flux and Characteristics of the UF Treatment

Effect of pressure and temperature on miscella flux. The effects of pressure and temperature on miscella flux are depicted in Figures 5 and 6 respectively. Here, miscella flux is found to have a tendency to reach a balanced value in the region of high flux. This is a phenomenon in which the concentration polarization (13) of phospholipids that

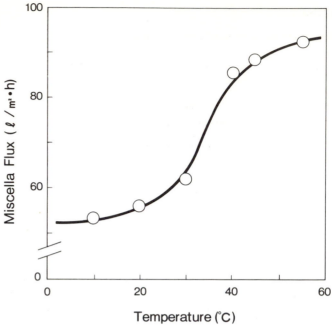

FIG. 6. Effect of temperature on miscella flux by the ultrafiltration with 20000 MW cut-off membrane. Miscella conc., 25 wt%; pressure, 3 kg/cm² ; feed velocity, 2 m/s.

are rejected onto the UF membrane surface grows with increasing flux to result in limiting miscella flux. Taking into account the feed velocity at the membrane surface, the system appears to be suitably operated at a pressure of 3 kg/cm² and at a temperature of 40–50 C.

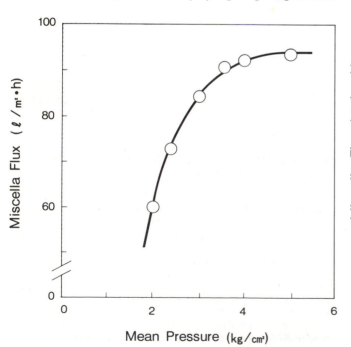

FIG. 5. Effect of pressure on miscella flux by the ultrafiltration with 20000 MW cut-off membrane. Miscella conc., 25 wt%; temperature, 40 C; feed velocity, 2 m/s.

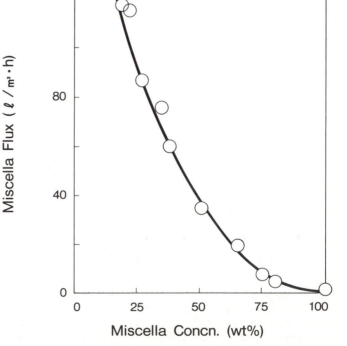

FIG. 7. Correlation of miscella flux and conc. by the ultrafiltration with 20000 MW cut-off membrane. Temperature, 40 C; pressure, 3 kg/cm² ; feed velocity, 2 m/s.

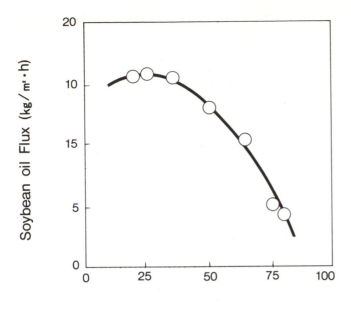

FIG. 8. Effect of miscella conc. on soybean oil flux by the ultrafiltration with 20000 MW cut-off membrane. Temperature, 40 C; pressure, 3 kg/cm²; feed velocity, 2 m/s.

Effect of miscella concentration on miscella flux. As shown in Figure 7, the lower the miscella concentration (i.e., principally oil concentration in miscella), the higher the miscella flux. However, considering the correlation shown in Figure 8, it is understandable from the viewpoint of oil productivity that the degumming of miscella can be most effectively processed at the miscella concentration of 20–30%. Currently, in soybean oil manufacturing, the extraction by hexane is as a rule carried out

within the range of miscella concentration of 20–30% because of the extraction efficiency. This fact is very advantageous for UF treatment.

Effect of miscella concentration on the phospholipids concentration in permeated oil. The concentration of phospholipids in permeated oil is related to the concentration of feed miscella as shown in Figure 9. To control the phospholipids concentration in the permeated oil to less than 100 ppm, UF treatment must be carried out at less than 50% of miscella concentration. In fact, as previously mentioned, the miscella concentration after extraction is processed in the range of 20–30%

Decreasing of miscella concentration by UF permeation. By permeation with an UF membrane of MW cut-off value 20,000, the concentration of permeated miscella is observed to decrease 4–5% as shown in Figure 10. Presumably, this is a phenomenon affected by the concentration polarization of phospholipids rejected at the membrane surface. However, it appears not to be of great disadvantage for practical engineering.

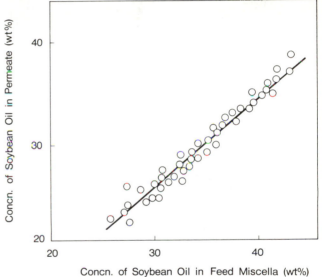

FIG. 10. Relationship between conc. of soybean oil in feed miscella and that in permeate by the ultrafiltration with 20000 MW cut-off membrane. Temperature, 40 C; pressure, 3 kg/cm²; feed velocity 2 m/s.

Long-Term Performance of the Bench-scale UF Plant and Recovery of Purified Oil

Because degummed and purified oil is the product of UF treatment, it is necessary to raise the concentration as high as possible to increase product output. In Figure 11, an example of the operation of the bench-scale plant is illustrated by a conceptual flow diagram showing continuous concentration of 40-fold.

Concentration ratio and the characteristics of concentrated miscella. As shown in Figure 12, the concentrations of some components in miscella change along with increasing concentration influencing UF permeation. Phospholipids, being rejected at the membrane surface, increase in concentration, but the oil concentration on the concentrated side makes no change due to the permeation of oil molecules through UF membrane. On the

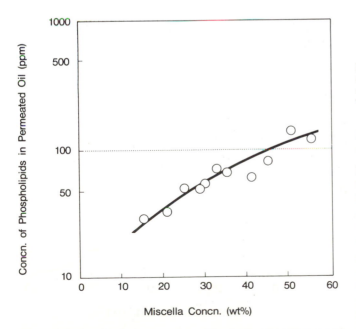

FIG. 9. Relationship between conc. of feed miscella and conc. of phospholipids in permeated oil. Temperature, 40 C; pressure, 3 kg/cm²; feed velocity, 2 m/s.

First Stage Concentration   Second Stage Concentration

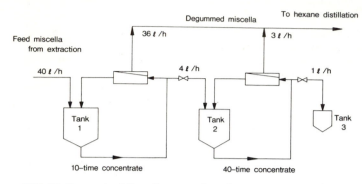

FIG. 11. Conceptual flow diagram of continuous concentration by 40 times.

other hand, water contaminated in feed miscella is rejected by the membrane remaining on the concentrated

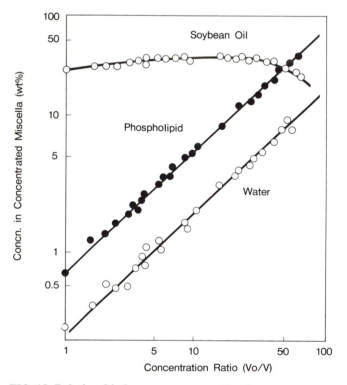

FIG. 12. Relationship between concentration ratio and conc. of components in concentrated miscella. Temperature, 40 C; pressure, 3 kg/cm2; feed velocity, 2 m/s.

side, and as it accumulated was found to severely influence miscella flux.

Concentration ratio and the recoveries of oil and phospholipids. As for the recovery of oil in the batchwise concentration, the results from the estimation of material balances are presented in Table 3. That is, 97 or 98% of oil was recovered as purified oil at a concentration of 40 or 50 times, respectively.

On the other hand, the balance of phospholipids is shown in Table 4. This balance indicates that phospholipids are almost completely found on the concentrated side of the membrane.

Long-term performance of the bench-scale plant. The results of the operation for 1,400 hours are depicted in Figure 13. Apart from the differential points due principally to the fluctuation of feed miscella concentration, miscella flux proved to be quite steady.

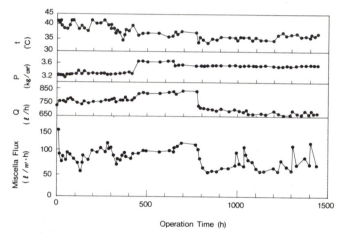

FIG. 13. Long-term membrane performance within treatment of crude miscella.

QUALITY OF THE SOYBEAN OIL PURIFIED BY MEMBRANE PROCESS

In the case of membrane process (physical refining) as shown in Figure 1, removal of hexane from the permeated miscella followed by bleaching and deodorizing provides oil acceptable for edible use. The comparison of the oil quality between the membrane process and the conventional process (chemical refining) is presented in Table 5. Thus, it was shown that the soybean oil purified by the membrane process is equal to or higher than the oil puri-

TABLE 3

Recovery of Soybean Oil by Batchwise Concentration

Concentration ratio[a]	Concentrated miscella (l)	Density of miscella (kg/l)	Soybean oil by concn. (wt %)	Soybean oil by weight (kg)	Recovery of soybean oil (%)
1	V_o[b]	0.718	25	$1.8 \times 10^{-1} V_o$	0
40	$V_o/40$	0.821	29	$6.0 \times 10^{-3} V_o$	96.7
56	$V_o/56$	0.862	24	$3.7 \times 10^{-3} V_o$	97.9

[a]Concentration ratio, initial volume V_o/concentrated volume V.
[b]Initial volume of crude miscella.

TABLE 4

Phospholipids Balance by Batchwise Concentration

		Miscella by volume (l)	Density of miscella (kg/l)	Phopholipids by concn. (wt %)	Phospholipids by weight (kg)	
Initial miscella		10,800	0.71	0.63		48.3
After concentration						
Concentrated miscella	1st stage	460	0.75	6.33	21.8	49.6
	2nd stage	155	0.82	21.8	27.8	
						50.7
Permeated miscella	1st stage	9,720	0.72	0.01	0.7	1.1
	2nd stage	460	0.78	0.10	0.4	

TABLE 5

The Comparison of Oil Quality between Membrane Process and Conventional Chemical Process

Analytical Items[a]	Acid value	Color[b]	Chloro-phyll	Phospho-lipids	Peroxide value	Flavor score	Odor by heating	Color by heating	Exposure test POV	Exposure test Odor	AOM[c]	Cold test
Crude soybean oil	1.82	Y35/R3.5	0.412	21000[d]	—	—	—	—	—	—	—	—
Purified oil by membrane process	0.03	Y4 /R0.5	0	21	0	5.0	A	Y10/R0.9	0.28	A	2.10	more than 60[e]
Purified oil by chemical process	0.03	Y4 /R0.4	0	24	0	5.0	A	Y 9/R0.9	0.64	A	1.80	25

[a]By standard of the method described in *J. Am. Oil Chem. Soc.* (1971).
[b]One-inch cell is used for crude soybean oil and 5.25 inch cell for purified oil.
[c]Six-hr value.
[d]Ppm.
[e]Hours.

fied by the conventional process at every item, which determines the quality of product oil. Also, it is emphasized that the membrane-purified oil made no deposit for the chilled storage of 60 hr or more, resulting in no need for dewaxing treatment (winterizing treatment) for salad oil use. Furthermore, in the long-term storage test the membrane-purified oil proved to be of equal or higher quality than the chemical refined oil.

Economic Efficiency of the UF Process for Purifying Soybean Oil

In Table 6, an estimate of the economic effectiveness of the membrane process is presented as compared with the conventional process. As shown here, in the case of the membrane process, a great deal of profit should be realized due to improved productivities and by eliminating the chemical process. Moreover, it is emphasized that the membrane process can be carried out without water and chemicals, resulting in no chemically contaminated

TABLE 6

The Comparison of Economic Effects between Conventional and Membrane Process in Refining Soybean Oil

	Conventional process	Membrane process
Refining method	Chemical	Physical
Productivity of purified oil	95.5%[a]	98%
Degumming	Chemical treatment	Membrane separation
Dewaxing	Wintering treatment	No need
Effluent and waste water	To be disposed of	No generating

[a]Treatment with chemical agents results in 2–2.5% loss of oil to be obtained.

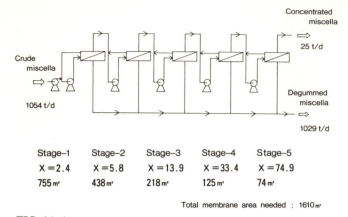

Stage-1 | Stage-2 | Stage-3 | Stage-4 | Stage-5
X = 2.4 | X = 5.8 | X = 13.9 | X = 33.4 | X = 74.9
755 m² | 438 m² | 218 m² | 125 m² | 74 m²

Total membrane area needed : 1610 m²

FIG. 14. A system of five-stage concentration of soybean miscella degumming (production of oil; 250 ton/day).

TABLE 7

Japanese Patent Applications on Purification of Oils and Fats by use of Membrane Processing[a]

Assignee	Numbers of patents applied	Details
Unilever N.V.	5	Purifications of miscellas
Asahi Chemical Co.	2	Purfications of oils and miscellas
Ajinomoto Co.	1	Purification of oil
Nisshin Oil Mills Co.	2	Purifications of miscellas
Nitto Electric Co.	5	Purifications of miscellas
Rinoru Oil Mills Co. Nitto Electric Co.	7	Purifications of miscellas
Asahi Chemical Co. Nisshin Oil Mills Co.	2	Purifications of oils

[a]As of February 1987.

effluent or waste water. This would reduce costs by eliminating installations for treating effluents. It also is meaningful that hexane removal by distillation after degumming results in the improvement of distillation effi-

ciency. Additionally, it would be profitable because the membrane process makes it possible to omit the dewaxing process and to recover phospholipids from the concentrated side.

In Figure 14, the results of a case study on the system of membrane process are presented for 250 ton/day of soybean oil production. Although multistage concentration is preferable due to the smaller membrane area needed from the viewpoint of flux, five-stage concentration is calculated to be most effective taking into account the investments for surrounding equipment needs. In this case, the membrane area needed is estimated to be 1400 m². Consequently, the membrane process could be paid back for the facility investments within about three years and would be economically effective.

As listed in Table 7, a number of patents have been applied for in Japan for the purification of edible oils by UF processes. This suggests that by introducing the UF process with solvent-resistant polyimide membrane for degumming soybean oil, the entire process of physical refining appears practical. This might become an epoch in the history of oil processing. In the future, the physical refining of soybean oil with UF separation can be shifted to cost-engineering.

REFERENCES

1. Iwama, A., *Yukagaku 34*:852 (1985).
2. Tadokoro, Y., *Ibid. 31*:820 (1982).
3. Lawhon, J.T., L.J. Manak, K.C. Rhee and E.W. Lusas, *J. Food Sci. 46*:391 (1981).
4. Totani, N., and H.K. Mangold, *Yukagaku 31*:411 (1982).
5. Lurgi GmbH (West Germany), Alcon process technical brochure: *Akzo-Lurgi-Conditioning.*
6. Sen Gupta, A.K., U.S. Patent 4062882.
7. Scholfield, C.R., *J. Am. Oil Chem. Soc. 58*:889 (1981).
8. Ishikawa, M., O. Okabe and M. Yamashita, *Yushi 36*:74 (1983).
9. Van Nieuwenhuyzen, W., *J. Am. Oil Chem. Soc. 58*:886 (1981).
10. Iwama, A., and Y. Kazuse, *J. Membrane Sci. 11*:297 (1982).
11. Zschocke, P., et al., *Desalination 34*:69 (1980).
12. Iwama, A., K. Tasaka and Y. Imamura, *Kagakukojo 27* (1983).
13. Kimura, S., and S. Sourirajan, *Am. Inst. Chem. Eng. J. 13*:497 (1967).

Justifying and Commercializing Biotechnology

Raymond H. Dull
Experience, Inc., Minneapolis, MN

Commercializing biotechnologically adapted products in the fats and oils industry requires careful and early attention to a range of production technology, product specification and marketing feasibility issues. Our experience indicates that many companies fail to adequately assess all facets of their new development plans on a timely basis. Resources committed to such evaluation are frequently minuscule relative to the costs of the development program and risks inherent in the marketplace. Evaluation often is initiated far too late in the development process to prevent substantial waste of resources in pursuing uneconomical business opportunities or inappropriate technological avenues. New competitive factors threaten to change the established industry patterns and may result in structural changes in the industry. New opportunity areas will be defined. The relative roles of patents and trade secrets may be altered. Marketing issues will become more important. Long-term commitment to technological strategies will become more important, even critical. Interindustry boundaries will be blurred, especially between fats and oils and seed. Higher product development risks will be required by biotechnology companies, fats and oils companies and fats and oils users, with highly exclusionary technology being the stake. The ante will go up. Those who wait and see may not survive. Premiums will accrue to market control and integrated downstream marketing savvy. There will be critical marketing success factors. Errors will be extremely costly, not only in economic terms but in customer confidence and supplier image. Some key questions are: Can established players adapt? Can small players survive? Will the lawyers get the whole pie? Is market research still valid or even possible? Can customers know what they need? Who will be left in 1999? Can we learn from analogs?

Commercializing biotechnology requires considerable effort long before the first order is taken; the order really is the last step. From the inception of the project idea, such considerations as sound market justification become integral to the commercialization process. I will discuss the process of commercializing biotechnology giving special attention to the importance of market research and applying basic marketing principles. My final points will cover the special considerations in commercializing biotechnology in the fats and oils industry.

During the last three years, I have talked with more than 150 firms involved in agricultural plant-related biotechnology. I also have monitored dozens of companies working in animal and human health biotechnology. Among the firms studied, there were significant differences in commercialization procedures. Companies with substantial experience in drug, human and animal concerns seemed to justify and document their opportunities more successfully and earlier than other companies. Firms that manufacture branded products, such as consumer food and ag chemicals, justify and document their opportunities well, except when dealing with biotechnol-

ogy not related to their core business.

Biotechnology companies, mostly start-up firms concerned with human health products, appear to be fairly thorough in justifying and documenting market opportunities. However, small companies working in plant sciences seem to have done little, if any, market research or strategizing. In several instances, I believe some companies deliberately distributed misleading or inappropriate justification information to investors in an effort to attract capital. Caveat emptor!

My studies also revealed a direct correlation between a company's success in commercialization and the qualifications of its executive staff. The firms most successful in assessing and fulfilling market needs are those with trained marketing people in key corporate positions. By contrast, firms poorest in market justification practices are those managed or directed by science-oriented executives. Whatever the rationale for a certain approach to commercializing biotechnology, there is no acceptable reason for omitting early market research, justification and documentation.

Incorporating justification measures into the earliest stage of project development is absolutely the most economical approach to running a new product program. In terms of total R&D costs relative to successful product introductions, the savings for programs identified as unfeasible before development can be 25 times the cost of the fundamental market research. It generally is easier and less costly to test market justification for a new technology before actually developing it than to develop and implement the product only to find it was a nonviable prospect. Investing in science without market rationale can be a costly, even bankrupting, mistake.

Four companies investigated that demonstrate relevant early market justification practices are DNA Plant Technology Corp., Calgene Inc., Plant Genetics, Inc., and Lubrizol Corp. All are headquartered in the U.S.

DNA Plant Technology (DNAP) provides an interesting example of how two very different products were justified in different, but effective, ways. The products are a high-solids tomato developed for Campbell, and Vegisnax™, a new snack food made from biotechnologically altered carrots and celery. In the case of high-solids tomatoes, a corporate sponsor (Campbell) internally justified production of the vegetables. That justification translated directly into a market justification for DNAP.

This manner of achieving market justification has been adopted by many biotechnology firms. Its appeal is that it provides an inexpensive means for the scientific company to identify a viable product, and it usually ensures successful results.

With Vegisnax™, a completely new venture, DNAP had to take a different course of justification. The product idea evolved from various assumptions about American consumers. With its key executives from big consumer goods companies, DNAP recognized that however exciting or promising the concept, the assumptions needed to be substantiated and other marketing issues addressed before scientific goals could be established. With the help of consultants, DNAP refined its product definition, combined quantitative research with qualitative research and defined the appropriate marketing strategy. The result was a well-conceived and apparently well-received product in test markets.

This phased approach allowed DNAP to risk only moderate sums in justifying the marketability of its product. Cost estimates for each phase might be approximately $35,000 to define market opportunities, $15,000 for consumer focus groups to target specific scientific development objectives and $100,000 for test market simulation. Whereas this entire justification process may have cost only $150,000, running the actual product development cycle could have easily cost $2.5 million—a hefty sum to lose should the product prove a failure. As you can see, the time and money spent in assessing and approving the market potential for Vegisnax™ was miniscule compared with what could have been lost through ill-defined product development.

Calgene Corporation's approach involves outlining potential opportunities in a general way, identifying possible R&D/marketing partners and then seeking agreement with the chosen partner. Once an agreement is arranged, Calgene further assesses the opportunity in terms of corporate objectives.

Calgene estimates that 7–10% of its research budget is spent examining opportunities in terms of marketing and commercializing issues. Furthermore, Calgene rejects 10 opportunities for each one accepted. Imagine the cost of implementing the projects rejected vs the cost of determining their unfeasibility in advance.

Another strategy at Calgene involves developing the necessary commercialization vehicle parallel with pursuing their scientific development program. In some instances, this has led to an acquisition, as in their recent purchase of Stoneville Pedigreed Seed Co., to provide entry into a business segment important to them.

In edible oils, Calgene's commercialization approach includes the use of production contracts and toll-crushing to begin carving a share in the market to support future developments.

Plant Genetics Corp. realized the importance of incorporating marketing into product development efforts. At the inception of its alfalfa program in 1983, Plant Genetics had one breeder and a part-time marketing consultant. By 1985, the scientific team numbered nine and the marketing consultant was full-time. Following the first sales in 1986, the 1987 staff at Plant Genetics consisted of six scientists and three marketing experts.

The fourth company I mentioned, Lubrizol, carefully assessed the potential of biotechnology before acquiring Agrigenetics.

In each of the four cases, relatively small costs for justification allowed the companies to pass milestones before investing significant money.

Because biotechnology is so new, clear operating standards have not yet been developed. Some of the big companies, such as Monsanto and Dupont, have transferred guidelines from related businesses and sciences. Most new companies do not have this advantage. However, I believe it is imperative to success that working guidelines be applied to R&D/product development and market justification. Otherwise, it is too easy to do research and too easy to omit justification.

My observations of the biotechnology field lead me to believe that about 10–20% of R&D budgets should be committed to market justification and documentation, commercialization, planning and market development.

I offer the following model as a way to justify product development and reduce wasteful R&D expenditures. The chain of events might be new project/product idea; initial market evaluation, general; initial science evaluation, general; decision point—go/no go; application analysis/definition (qualitiative concept evaluation); further market analysis; decision point—go/no go; concept review; project specification(s) from customer input; project/scientific goal setting and budgeting; scientific review success/failure probability; decision point—go/no go; critical path analysis; marketing development planning; prototype/pilot plant; price analysis; and market development implementation.

It is readily apparent that the emphasis is on generating confidence in the success of the project before large R&D expenditures are made.

Now you may be wondering what all of this has to do with commercializing biotechnology: everything, really. If you are conscientious about marketing justification and documentation and are responding to market needs, you have done the bulk of your commercialization. It becomes a relatively routine manner to add actual marketing development and promotional plans along with sales programs.

At this point, I will turn my focus toward the special considerations involved in commercializing biotechnology in the fats and oils industry.

What is special about fats and oils? A number of aspects set this business apart from others when it comes to commercialization. It is a complex, competitive business featuring a variety of vegetable oils and animal fats with high degrees of substitution involving a variety of applications. There are significant by-product or residual product markets that are interrelated or dominate the oil segment. In addition, the vast numbers of production localities are dispersed among many countries. Many factors come into play during the commercialization process: foreign exchange needs, livestock cycles, economic change, regulatory measures, government policy, and subsidy programs.

In light of such diverse and numerous issues, most new product concepts relating to biotechnological science cannot be assessed as simply as in the case of our earlier example involving high-solids tomato processing.

Making radical changes in the fatty acid composition of vegetable oils through seed selection is a long and tedious process. The development of a low erucic acid rapeseed by the Canadians is an excellent example of this process. The need for an edible oil that could be grown in Canada had been established, and commercialization of the new oil had sufficient time to develop in an orderly manner.

Biotechnology greatly reduces the time frame for changing the fatty acid composition and characteristics of vegetable oils, thereby shortening the time span for the development of markets and subsequent commercialization. Although it may be technically possible to change the composition of a vegetable oil, it may have no increased value in the marketplace. The potential volume and value of the oil must be established before laboratory work begins.

The development of a low erucic rapeseed and the subsequent regulations in Canada, and the EEC in production of a rapeseed containing under 2% erucic created a problem for the industrial users who require erucic acid as a raw material. This presents an opportunity for the development of a very high erucic rapeseed for the industrial market. The question that must be raised and answered by the biotechnical company is "can an 80% erucic rapeseed be developed and commercialized to compete with the 50% rapeseed that currently is produced in East Bloc countries and China?"

We also might consider the biotechnical modification of palm oil to produce higher oleic fractions, thus reducing or even eliminating the health concerns that now limit its acceptance in some countries. Doing so would increase its competitive edge with soy in many markets.

Barring a substantially higher price for the higher oleic palm, imagine its impact on the soy oil/meal market. As long as meal demand drives soy processing tonnage, the soy oil must be sold. At this point we must ask "how?" Lower prices to compete with "improved" palm oil? Higher soybean meal prices, which would make the meal less competitive to other oilseed meals? These are the sort of questions we must ask in order to design and tune commercialization strategies for success.

We believe there are opportunities for tailoring vegetable oils through biotechnology to meet specific edible and industrial needs. These specialty oils can demand a higher price in the market, but will their higher value offset the development and commercialization costs? An economic analysis of the markets must precede any in depth technical development.

It's becoming quite common to hear of outsiders proposing to use biotechnology to enter the business, and many of them think they will stand the vegetable oil industry on its ear. However accurate or inaccurate their assessment of their business opportunity, won't they surely impact the companies you represent?

What biotechnology strategy are you following? Proactive? Reactive? Head-in-the-sand? And, what marketing justification have you compiled to support your strategy? Are your customers being heard? How many food processing customers are reformulating to meet either health concerns or cost consideration? How many aren't?

Those of you who are working with animal fats surely must anticipate the fallout of plant-related biotechnology as a competitive factor, of bovine growth hormone as an impact on animal fat stocks, of leaner carcasses, and of work with ω-3 acids and oils.

I doubt there has been any decade that will be so significant to the fats and oils industry as the one ahead. Many strategic issues are facing you, and the issue that may set this decade apart from all others is that there may not be a free and economically viable choice about adopting new technology. You may be shut out by patent and plant protection barriers that severely limit alternatives.

So, today's R&D strategy and product planning may determine your ability to survive into the next century. I am a strong advocate of marketing-driven decision making. All marketing functions must be addressed: transportation, distribution, warehousing, grading, accumulating, packaging, financing, pricing, promotion and selling. The impact will vary between industries in relative importance and technique, but it will occur. And, industries will lean toward new biotechnically derived products.

However, technology now is more important than it has ever been. Guess wrong, and you are out of the game. The key requirement is identifying the best technological bets. Marketing research and strategic marketing are critical, and these must identify the right technologies or even the right paths within these technologies with only hypothesized product attributes to go on.

I believe this can be accomplished, but in order to do so a mindset must be altered. We need to ask different questions than before. We need to learn to make good decisions despite imprecise numerical documentation. This will be a traumatic change for some result-oriented companies because R&D costs continue to rise and paybacks are a long way off. Some managements may choose to exit from the industry as their only realistic course of action.

A larger part of our planning and analysis must include not just customer needs but end user evaluations as well. There is not an edible oil producer here today whose business will not be affected in some way by consumer attitudes toward health issues 5 and 10 years from now. Overall, the closer a company is to the end user, the more in tune it is with their needs, wishes and concerns. Those of you who are intermediate manufacturers must pay attention to your customer's future plans, strategies and research directions now, not later.

Break your work process down into smaller steps. Test project feasibility at every opportunity. Cutting off a loser early conserves cash and energy for more promising channels. Do your marketing homework first. Preproject feasibility studies are a good investment. As much as scientists may balk, you must be absolutely committed to assessing your probabilities for commercial success. You must do this to stay competitive.

Feasibility studies and pilot plant trials need to be conducted more often and earlier. If you can transfer a gene but cannot regenerate the plant, you are at a crucial decision point. Rethink your options. Maybe the time-frame requirement says "buy" rather than "develop." Look at the risks associated with failing and the rewards attendant to succeeding. We often see huge imbalances in this equation.

Earlier, I mentioned newcomers to the industry. You should be analyzing these potentially disruptive companies. Can they succeed? Why or why not? What does that mean for you? What is their strategy? What can you learn from them? Competitive intelligence now is a necessary tool. I do not mean illegal subversive measures; I mean paying careful attention to your competitors. Assess every piece of information you can legitimately acquire and continuously update that assessment.

What other industries are interfacing with yours? Certainly, seed companies are a major force impacting on your industry. How do their strategies affect you? Are their biotechnology programs a threat or an asset? Is any biotechnology company a potential future player in the oilseed business?

Explore ways of marketing your product more effectively. Perhaps you can sell a high oleic content palm oil, and then again you might discover that market development costs are too high. Ask yourself how you can differentiate your product in the customer's mind.

The personal computer industry is a good example of how drastically markets change direction, forcing players to adapt new strategies for survival. It was spawned from the mainframe industry, which had become relatively mature. Technological breatkthroughs, especially in chip design and manufacture allowed rapid expansion by small high-technology companies.

IBM responded by breaking with its traditional business methodology and was both spectacularly successful and badly bruised. A new round of industry shakeouts and consolidations occurred, and today fewer competitors are again vying for the control in the marketplace, this time with software as the key competitive issue.

By comparison, the oils and fats industry is at the mainframe stage, perhaps more classically mature as an industry but with the same vulnerability. Technology waits in the wings with small fast-moving competitors poised to play by new rules. Your mainframes, plants and basic products, may become obsolete. How will you respond?

And, over this entire panoply, hangs the spectra of a significant technology patent position or plant patents that could devastate the industry. In the end, the lawyers may have the only game in town.

Commercialization may require established companies to redefine their sales forces and re-examine the sales skills they need. Order takers may be out, missionary selling in. How do you make the change, by internal development, acquisition, joint venture, or by contract or partnerships?

To effectively move forward in the commercialization of biotechnology, the first partnership you need to make, if you haven't yet, is with your own marketing people. You need to move ahead together. You cannot open the lab door and hand out a product to marketing, saying, "it's yours now!"

Allow me to reiterate these points: consider commercialization as a parallel process to scientific development; commit funds at the outset for market research to justify your scientific opportunities; address all marketing requirements as early in the process as possible; address the special issues of your industry. Finally, "go get the order."

Discussion Session

In a lively start to these discussions, the panel was asked if anyone was using enzymes to modify/produce fats and oils. Unilever uses immobilized enzymes for interesterification, and up to 8,000 tons/yr of oil was treated by splitting fat with immobilized enzymes for soap making. Furthermore, it was said that phospholipase A-2 is used to produce a lysophospholipid. In response to a series of questions concerning the use of lipases, Mattias Buhler indicated that only degummed oil was employed in the studies, the lipase is rather specific, and a new set of conditions would be made for other fats such as tallow, coconut oil and corn oil because they all have different physical and structural characteristics. Several other questions were concerned with the economics of the process. Dr. Buhler stated that one should use the most simple reactor required for the process and optimize the process so it could compete with chemical processes. While the limiting factor is the price of the lipase, process improvement could result in 1.6 tons of fatty acid/kg of enzyme.

Akio Iwama's paper provoked several questions. In answer, he indicated that membrane-processed oil did not need winterization, and the chlorophylls remained in the oil, still requiring a bleaching step for their removal. He further indicated that the level of phosphorus remaining in the oil was about 30 ppm, and fatty acids were rejected by the membrane. Finally, he said very concentrated micella had to be further diluted by adding hexane.

Ray Dull, discussing commercializing biotechnology, received the question, "how should research in the public sector be managed?" His answer was it should be justified as the private-sector must. His second question requested discussion of strategies for successfully going public. These primary concerns appear to be the market viability of the product, and investment brokers and others are looking at the possible payoff. It cannot be done on the science alone; glamour also is needed. He emphasized the need for a seasoned business executive at the helm for success. The panel addressed the obstacles for implementation of biotechnology for the fat and oil industry and agreed it appeared to be funding.

Toxicological Evaluation of Biotechnology Products: Regulatory Viewpoint

Rolf Bass
Bundesgesundheitsamt, Berlin, Federal Republic of Germany

For drugs produced by biotechnological techniques, it is now understood that toxicity testing by routine protocols may be inadequate, unpredictive or even impossible to perform. Judgment of safe use of such products may depend, to a larger degree than for chemically synthesized compounds, on the availability of other data, e.g. obtained from clinical studies. On the other hand, quality, safety and efficacy will have to be established as for any other drug before registration. In the European Community, Council Directive 87/22/EEC has been enacted on July 1, 1987. To put this directive into action, two Notes for Guidance on quality aspects (monoclonal antibodies of murine origin; products derived by recombinant DNA technology) have been finalized, whereas the Note for Guidance on Pre-clinical Biological Safety Testing still awaits such finalization (Draft III/407/87-EN; revision 3-April 1987). In this Note for Guidance, proposals for testing requirements are outlined for the following product groups: hormones and cytokines and other regulatory factors, blood products, monoclonal antibodies, and vaccines. It is understood that quality (including problematic areas of identity, purity, determination of contaminants possibly encountered from the novel processes used in their manufacture and from the complex structural and biological characteristics of the products) be assessed, namely not by unsuitable toxicological techniques but by analytical tests. Depending on e.g., existing/non-existing pharmacological activity, the status of the patient and the disease to be treated, physiological or large nonphysiological doses expected to be used, foreseeable single or repeated administration of the product, the extent to which the biological effects of the substances are characterized, a flexible, case-by-case approach for safety testing will have to be sought. Major weight often will lie on (general) pharmacodynamic studies. No battery of safety tests can be prescribed. The usefulness of performing various test combinations, including toxicity and pharmacokinetic studies, has to take into account the development of immunological incompatibility of the product with the animal species used.

"Although the animal testing program outlined above can provide a core of knowledge about a particular product, this may only be of limited value in predicting problems in clinical practice. This has major implications regarding the exclusion of experimental studies in the investigation of a specific biotechnology product and on the clinicians' assessment of the relevance of experimental data to the therapeutic purposes of the compound in patients."

In the Institute for Drugs of the Federal Health Office we are responsible for the conditions of drug approval in general as well as for the risk-benefit decision to be made for each application—including those made for biotech-nology compounds.

It soon became obvious that toxicity testing by routine protocols may be inadequate, unpredictable or even impossible to perform. Judgments about safe use of such products may depend to a larger degree than for chemically synthesized compounds on the availability of other data, e.g. obtained from clinical studies. On the other hand, as for any drug, quality, safety and efficacy must be established before registration.

To address this problem, the European Economic Community (EEC) issued a directive (87/22/EEC) on July 1, 1987. This directive included two "Notes for Guidance: one on quality aspects (monoconal antibodies of murine origin and products derived by recombinant DNA technology). A third on pre-clinical biological safety still awaits finalization (Draft III/407/87-en; revision 3-April 1987). In these Notes for Guidance, proposals for testing requirements are outlined for the following product groups: hormones and cytokines and other regulatory factors; blood products; monoclonal antibodies; and vaccines.

It is understood that quality (including such problem areas as identity, purity and determination of contaminants possibly encountered from the novel processes used in their manufacture and from the complex structural and biological characteristics of the products) must be assessed by analytical tests and not by unsuitable toxicological techniques. However, the problems encountered in toxicity testing of biotechnology products often lie in the area of overlap between quality/purity and toxicological assessment.

Depending on existing or non-existing pharmacological activity, the status of the patient and the disease to be treated, physiological or large nonphysiological doses expected to be used, forseeable single or repeated administration of the product and the extent to which the biological effects of the substance are characterized, a flexible, case-by-case approach to safety testing must be sought. Major emphasis often lies on general pharmacodynamic studies. No battery of safety tests can be prescribed. The usefulness of performing various test combinations including toxicity and pharmacokinetic studies must take into account the development of immunological incompatibility between the product and the animal species tested.

This understanding by the European drug regulatory body CPMP has led to the activities of two working parties: one on Biotechnology quality and one on safety. Neither yet has had actual experience with a large number of applications for the marketing of biotechnology products. The requirements I shall describe, therefore, must be seen as rather theoretical. They will have to be adjusted in the near future as practical experience evolves.

How Can Such an Approach be Sought

It is easier to restrict the discussion to the first group of products: hormones, cytokines and other regulatory factors. The test philosophy for the other three product groups might be somewhat different. They are considered and described in the EEC paper that offers advice for the general and specific aspects to be considered for each product group.

Because toxicological testing for chemically synthesized drugs is regulated quite precisely in the EEC, we can use these test requirements as a matrix for the development of toxicological testing with group I products. We then have to look at the necessity of performing pharmacokinetic, pharmacodynamik and toxicological studies, including single and repeated dose toxicity, local tolerance, reproduction studies and mutagenic and cancinogenic potential. In addition to these well-regulated areas, we must concentrate on immunological aspects of toxicology including the possible usefulness of homologous animal systems.

All countries of the EEC and the EFTA as well as the U.S., Canada and Japan now agree that single-dose toxicity studies of the LDl5l0-type are inadequate. Assessment of the effects of a single high dose of a product on major physiological systems should be investigated using a wide range of techniques. The dose selected for this purpose and for other toxicity studies should be appropriate to the intended amount and duration of dosage in humans. To prevent repetition, such studies should combine the methodology available for both types into one study.

To select animal species, it is necessary to first perform experiments on one model species, then to perform pharmacokinetic investigations, comparing substance behavior in several animal species and humans and then decide if and which other species should be employed for additional toxicological studies. We must remember that even monkeys may not be appropriate and that it is necessary to test for many effects in the human species.

Studies Beyond Subacute Ones

This refers to chronic toxicology up to six-month, lifetime carcinogenicity studies. Reprotox studies seldom are needed. However, our inquisitiveness is rather great due to the fact that we know so little about physiology, pharmacology and immune response with such products, e.g., it is known that high interferon levels are associated with autoimmune disorders and allergies. It is unknown, however, how diseases and their outbreak are causally related to high interferon levels. Therefore, everybody would like to know more about the relationship between auto-immune-disease, naturally occurring interferon levels, and those used therapeutically.

This leads to a "case-by-case approach." Companies seeking advice discuss their problems together with us or with members of other drug regulatory agencies. The test approach and strategy to be followed after such discus-

sion is guided by pharmacological activity of the product, its mechanism of action, the intended dosage and route of administration, the clinical status of the recipient, the prophylactic, therapeutic, diagnostic indications and any results from preliminary clinical investigations. In addition to such criteria as pharmacological profile and the areas of clinical use, the test strategy will depend on the degree of biochemical similarity or identity of the product with its naturally occurring analog. The more distant the product from the natural one, the greater the likelihood of unexpected adverse reactions and the greater the necessity for stringent testing.

Particular problems may arise in the area of preclinical safety testing, especially in relation to toxicity testing in vivo. Safety testing therefore must take into account many factors. For example, certain proteins are highly species-specific and thus are much more pharmacologically active in humans than in animals. In addition, the amino acid sequences of human proteins are often significantly different from their natural counterparts in other species. Thus, these proteins frequently produce immunological responses in foreign hosts that ultimately will modify their biological effects and that may result in toxicity due to immune complex formation. Such toxicity, of course, would have little bearing on the safety of the product in humans, the intended natural hosts. Standard toxicological procedures applied to the preclinical safety testing of chemical drugs may be inappropriate because they will not result in a useful measure of the toxic potential of some protein products in humans. In such situations, in vitro biological tests, e.g., to ascertain specific activity, species specifity and immunological characteristics may aid the choice of test species.

Therefore, nobody here can expect that we or the Note for Guidance on preclinical testing of biotechnical products in preparation by the EEC can provide a firm network for testing. Whether this will develop in the future cannot be answered today. The draft of the note for guidance concludes, and I concur:

"At the present time, no battery of safety tests can be described, which would be applicable to all types of product groups and all biochemical groups available today. The usefulness of performing various combinations of tests should be ascertained by discussion among pharmacologists, toxicologists and clinicians, both from the pharmaceutical companies (including those making the product) and the regulatory agency concerned. The outcome of these discussions is expected to be reflected in the expert reports submitted for each product, which would explain the case-by-case approach employed.

"In all cases, toxicological investigations would be preceded by studies addressing the biotechnological problems of production quality control, contaminants and impurities."

Toxicological Evaluation of Biotechnology Products

W.E. Parish

Environmental Safety Laboratory, Unilever Research, Colworth House, Sharnbrook, Bedford, MK44 1LQ, U.K.

Biotechnology enables preparation of a diverse range of products, which may be end products for the user or processing agents. Toxicological evaluation requires greater attention to some potentially harmful properties but does not differ in essentials from standard considerations and test requirements. Products of biotechnological processing may be considered in terms of examination for potential hazards for the products user, protection of the environment from contamination by organisms derived by genetic engineering and public emotional responses to biotechnological products and processes. This presentation is mainly concerned with the first issue. Examination for potential hazards requires consideration of the product-forming microorganisms, whether or not the product is the recipient of genes derived from other organisms. If gene transfer has been made, detailed information is required on the identity of the gene, identity and properties of the source and recipient organisms, and transfer of the expression vector into the host. The examination of the fermenter product will be considered in detail, particularly with reference to the detection of any unknown toxins and potential mutagens. First considerations are the stability of the organisms used and the identity and purity of the gene product. Subsequent examination for mutagenic activity depends upon compatibility with in vitro assay. These include bacterial mutation, primary rat heptocytes (toxicity, DNA repair, inhibition of protein synthesis) and Chinese hamster V79 cells (toxicity and cytogenetics). Parallel tests may be made of digests of food products. If the test product is not compatible with in vitro assays, in vivo tests for micronucleus formation, metaphase chromosome analysis and DNA repair are available as substitutes. Apart from this screen for mutagenic activity, additional tests follow standard procedures for safe handling and feeding tests, pharmacokinetics and teratology, all supported by full pathology.

The three main considerations in toxicological evaluation of products or processing aids derived from microorganisms are the properties of the source organism, the safety of any processing aid, e.g. enzyme, and the safety and acceptability of the final product. A plan is presented of the bioassays used to determine that microbial enzymes are free from unknown toxins, or toxins possibly not detected by chemical analysis.

A representative production batch that becomes the model to set a product specification is subject to very critical examination. Much less is required to monitor routine production batches. The representative enzyme batch is first examined by chemical analysis for likely mycotoxins. Then, samples are prepared as crude solutions in appropriate media (DMSO or culture media). Further samples are extracted in water/chloroform and evaporated to dryness. Then the partially purified, concentrated extract is dissolved in media. The crude solution and concentrated extract are examined on four

strains of salmonella in the Ames test, on primary rat hepatocytes for cytotoxicity, DNA repair and inhibition of protein synthesis, and on Chinese Hamster V79 cells for cytotoxicity and chromosomal changes. If necessary, results may be supported in vivo by a micronucleus test and in vivo/in vitro DNA repair.

When monitoring routine production batches, conformity with the specification is a strong indication of process stability and freedom from toxins. The routine tests proposed are examination of the organism for purity, morphology, biochemistry and product identity. Tests for toxins may be nonessential but required. Toxicity tests on animals are the most effective but are undesirable. It is proposed that colony-growth inhibition tests on an established cell line with and without exogenous metabolism are appropriate and less susceptible artifacts than other cytotoxicity assays.

Biotechnology is defined as the application of living organisms or their products to the preparation of substances, among which are foods, or to controlled changes in the environment. Products of biotechnology may be derived from natural or "genetically modified" bacteria and yeasts that form the substance or the end product, or that produce an enzyme or similar processing agent. Toxicological evaluation requires greater attention to some potentially harmful properties but does not differ essentially from current, standard test requirements. An important requirement is to examine the product to detect the presence of unknown toxins or those that may have escaped detection by chemical analysis.

Investigation of the safety and acceptability of a biotechnological process may be considered in three phases as in the sequence of product preparation: (1.) The identity, properties and nonpathogenicity of the microorganisms used as the source of the product and, as in the case of genetically manipulated (engineered) organisms, the source organism of the transferred gene. (2.) The properties and safety of the crude product or of the enzyme, catalyst or additive used to make the final product. (3.) The final product to be supplied to the consumer.

It is the second phase of preparation, i.e., the safety of the processing enzyme or additive, that is of most current concern and stimulating research to define techniques sufficiently comprehensive to detect any acute or chronic toxic activity. Some enzymes, including a lipase of fungal origin, have been accorded GRAS status by the U.S. FDA (1). In the U.K., the Food Additives and Contaminants Committee (2) reviewed the toxicological data and classified enzymes in use according to the amount of evidence supporting safety. The Committee strongly recommended the use of a general nonspecific biological screening test for the detection of toxins that may be present in enzyme preparations. To achieve a comprehensive toxicological examination of microbial enzymes intended for food processing and to comply with the recommendations of the FACC (2), we have devised a program of biological tests effective in detecting small amounts of mycotoxins used to spike the enzyme samples.

In this report, the safety test considerations for the product-source microorganisms (phase 1) and of the final product (phase 3) are briefly summarized. The investigations of toxicological tests on enzymes used for processing (phase 2) with a proposal for a program of techniques to critically examine a representative first batch of product (enzyme) for freedom from toxins and to set a specification and for a test regimen to monitor each subsequent production batch are considered in more detail.

Safety Considerations for Microorganisms and Consumer Product

Safety considerations and requirements for the microorganism and for the final consumer products, represent phases 1 and 3 of product preparation.

Phase 1: The Properties of the Microorganism

The essential considerations for organisms not modified by genetic manipulation include the identity of the production microorganism, its properties, history of use, freedom from adventitious agents and a thorough examination of its potential pathogenicity, if any. It is desirable, whenever possible, that nonpathogenic organisms be used. It is necessary to determine the stability of the organism in large-volume culture conditions as well as the stability of the product formed.

If the product is derived from an organism that has been modified by recombinant DNA techniques, additional information is required, including the identity, properties and potential pathogenicity of the source organism, identity of the gene, nature of the vector and method of transfer of the gene. This evidence forms the basis of advice on safe handling and any containment necessary (3,4).

Phase 3: Examination of the Final Product

It has been proposed that the acceptability of biotechnology products be considered on a case-by-case basis (5). This is a rational approach as it is not possible to impose rigid guidelines for a diversity of products, even foods, with greater or lesser similarity with existing natural substances used widely and safely for many years.

The final product must not be harmful to the consumer over short or long periods of use. The manufacturer has the responsibility of ensuring that the product is safe. However, there are few directives for the requirements to demonstrate the safety of a food though there is much legislation controlling the acceptability of food additives.

Enzymes incorporated freely in products are food additives and subject to regulatory control (1,2,3). If bound to an insoluble matrix and not intended for release in the product, they become processing aids. However, as some enzyme or impurity may be desorbed from the matrix, the same considerations for safety apply.

Tests for Toxins

The objective is a thorough examination of a representative production batch for freedom from toxins. This is applied to a product specification so that subsequent batches will be safe for use. The program of tests is designed not only to detect the presence of a toxin but also to indicate the nature of the toxic activity whether acute as in membrane lysis and in cytoplasmic and metabolic poisons or chronic as in genotoxicity.

The microbial enzyme preparation must be examined by chemical analysis for those toxins known to be formed by the source organism and by other common organisms (2). Chemical analysis is much more sensitive than bioassays to detect mycotoxins (6), but being very specific detects only the toxin under investigation. Therefore a scheme of biological tests was devised to comply with the FACC (2) recommendation for detection of unknown toxins or those not included in the chemical analytical screen. The test program has proved effective in detecting small amounts of some mycotoxins used to spike enzyme samples. The tests were selected to satisfy defined objectives and criteria.

Objectives

The test program is sufficiently comprehensive to detect known and possibly unidentified toxins. The results of the tests will indicate if the toxins have acute or chronic activity and the nature of their effects, e.g. metabolic and cytoplasmic organelle toxicity or genotoxicity.

Criteria for the Selection of the Tests

(1.) The sequence of tests is designed to detect potential for both acute and chronic effects and to be sufficiently comprehensive to detect the common mycotoxins. (2.) The tests have known limits of detection for mycotoxins and give consistent, reproducible results during routine use. (3.) The techniques are readily useable in all laboratories without the introduction of a test species requiring special maintenance procedures and not known to have a unique property to detect microbial toxins.

Selection of tests for cytotoxicity. Cytotoxicity in this context refers to the killing of cells or more complex organisms, or severe derangement of cell function, e.g., protein synthesis. Procedures using many species ranging from bacteria, yeasts and cell culture to shrimps, fish and birds, have been used to detect mycotoxins. In a critical review, Watson and Lindsay (6) concluded that such assays are subject to interference by nonfungal agents and are less reproducible than chemical assays. They stated that the bioassays have revealed little information of value in the surveillance of food and foodstuffs.

Cytotoxicity assays whose end point is death of the test species have limitations in examination of crude fermenter products. (1.) Death of the test species is not selective evidence for the presence of a mycotoxin or other significant toxins. Substances harmless to man may be lethal to the test species, e.g. ingredients of culture media or somatic products of the bacteria or yeasts that do not appear in the final product or in insufficient concentration to be harmful. (2.) The concentration of the substance killing the test species is not a reliable guide to the presence of a potent toxin unless it is active in high dilution. The activity of small amounts of toxins will be masked by the activity of other components of the crude preparation. Determinations of the threshold limit of detection of mycotoxins often have been made on pure preparations, not on crude extracts of culture media or foodstuffs containing substances that may inhibit or mask the mycotoxin.

In a preliminary investigation, five mycotoxins were examined for their ability to induce release of lactate dehydrogenase (LDH) or to inhibit incorporation of [^{3}H]-leucine (protein synthesis) in primary cultures of rat hepatocytes. Release of LDH tended to be slow over 24 hr and the effective concentration for each mycotoxin varied much: aflatoxin B^1 (0.03 μg/ml) patulin (10.00 μg/ml), T2 toxin (0.03 μg/ml), vomitoxin (10.00 μg/ml) and citrinin (>10.00 μg/ml). As the activity of the mycotoxins was likely to be distorted in this test in the presence of crude culture preparations, this type of cytotoxic test was considered insufficiently applicable for routine use.

Therefore, in the study protocol described below evidence of cytotoxicity, i.e. acute cell degeneration and cell death, would be observed in the preliminary dose range-finding tests for DNA repair and cytogenetic studies together with the examination for inhibition of [^{3}H]-leucine incorporation into the cell protein.

Bioassay tests on representative production batch of enzyme. The program of tests devised to examine enzyme preparations is presented in the scheme (Table 1). Preparatory toxicity studies on salmonella and hepatocytes show if there is any physical property (e.g., osmolarity, mineral salt imbalance, pH, etc.) that may be incompatible with in vitro tests and correctable. Tests are made on solutions of the crude enzyme preparation (DMSO for the Ames test, and in culture media without serum for the DNA repair and with serum for the cytogenetic test) and filter sterilized. Tests also are made on an aqueous/chloroform extract, evaporated to dryness, reconstituted in the appropriate medium and filter sterilized. The extracts enable a concentrated preparation of a putative toxin and reduce the side effects of media products.

The solutions and concentrated extracts are examined by the Ames test with salmonella strains TA 1535, 100, 1537 and 98 by standard OECD procedure. Three investigations, each at five concentrations at half-log 10 concentration steps, are made on test samples with primary rat hepatocytes, their toxicity in the preliminary dose-range finding procedure and in DNA repair and their inhibition of protein synthesis determined by [^{3}H]-leucine incorporation. Finally, five concentrations of the text substance are examined for cytogenetic changes in Chinese Hamster (CHV79) cells; the preliminary test for dose range also being a cytotoxicity test (Table 1).

The results of the in vivo assays are assessed for their significance. If all are negative, it is very likely that no in vivo tests, as in Table 1, will be required. Should any of the in vitro tests show toxin activity, a judgment has to be made on further pursuing the project and any in vivo investigation.

Modified Program for Tests on a Representative Batch of a Foodstuff

If the product of the process is a foodstuff and not a processing aid or additive, a modification of the scheme (Table 1) is available for a more intensive investigation. A procedure has been reported (7-9) to reproduce gas-

TABLE 1

Program of Tests to Examine an Enzyme Preparation Derived from Microorganisms for Freedom from Unknown Toxins

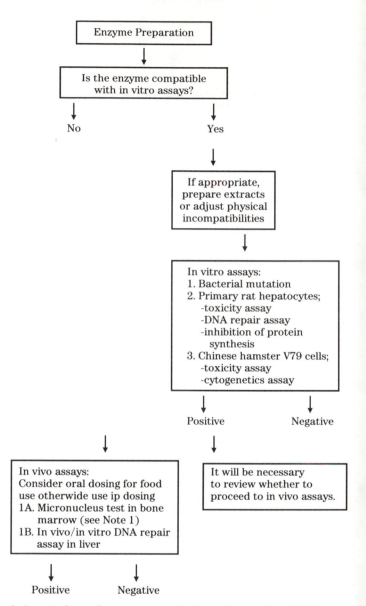

1. A metaphase chromosome analysis may be preferred if direct evidence of chromosome damage is required.

trointestinal digestion in vitro to examine the digest by the Ames or mammalian cell tests for evidence of formation of a mutagen. For this to be necessary, the foodstuff must be very different from its natural homologues.

Monitoring the Safety of Production Batches

The proposed full test procedure (Table 1) is intended as a detailed examination of the safety of a representative production batch. When the full program of tests on the first representative production batch is complete and there is no evidence of formation of any harmful toxins, it should be possible to define production conditions that preclude contamination by toxin in subsequent routine batches and to set a specification for the product. There should be no contamination by toxins if the approved procedure is followed meticulously. Furthermore, any formation of toxins should be readily detected as they could result from a mutation in the production organism, contamination by other organisms or contamination during subsequent processing (downstream processing) of the product in inadequately cleaned containers shared with other products.

Each production batch should be examined for (1.) compliance with the specification of the product and the amount of the particular ingredient, e.g. enzyme, (2.) immunoprecipitation or immunoelectrophoresis against antibodies specific for the components of the first representative production batch to confirm identity. (3.) examination of the production organism for freedom from contaminating organisms, and (4.) examination of the production organism for any deviation from normal colony morphology or deviation from its biochemical properties, e.g., sugar fermentation tests.

Any significant deviation from normal would be apparent by decreased yields, deviations from product specification or changes in the properties of the production organism. There should be no need for further tests for the presence of toxins. However, if further testing was required to obtain evidence of freedom from toxins, the most suitable method is chemical analysis for defined mycotoxins that are produced by common contaminants or possibly immunological methods for detection. Chemical analysis is much more sensitive than biological methods and is quicker and less costly once the equipment is available.

Should there be a regulatory requirement for a test for an unknown toxin, a biological method is required. One simple method is sufficient; there is little advantage in doing tests on several different species, e.g. fish, brine shrimp, germinating pea and mammalian cells. The most logical test is intraperitoneal injection into five mice and two guinea pigs as used to examine each batch of therapeutic antisera or into guinea pigs for tests on vaccines (10). A production enzyme with biological activity is neutralized with specific antibody before injection to enable sufficiently large amounts to be administered to detect any other harmful substance. This is the most comprehensive test for a substance toxic to vertebrates. There are, however, two limitations: the procedure requires access to animal house facilities and it is not desirable to add to requirements for tests on animals.

An in vitro test that appears appropriate is inhibition of colony formation of an established cell line, e.g. CHV79 or Hela cells. The method is simple. Exogenous metabolism may be provided using standard metabolizing systems, e.g. aroclor-induced rat liver S9, and the criterion, a colony count, is a clear endpoint determination. This reduces the difficulties of interpretation of results obtained using other criteria for toxicity, e.g., enzyme release, inhibition of synthesis of protein or DNA, morphological changes and dye exclusion. Personal experience has shown that the products resulting from incubation of some sterile bacteriological media induce changes in cultured cells indicating definite toxicity or an inconclusive result when the substances are harmless to vertebrates. Cell perturbation induced by the culture media may well give false positive results or mask small amounts of real toxins.

The colony count technique has much to recommend it if a bioassay for unknown toxins is required. However, investigations comparing the various methods are essential before selecting a screening technique.

ACKNOWLEDGMENTS

I thank M. Chamberlain and M. Richold for their advice and stimulating discussions.

REFERENCES

1. *Fed. Regist. 38*: 9256 (1973).
2. Food Additives Contaminants Committee U.K., *Report on the Review of Enzyme Preparations*, London, H.M.S.O., 1982.
3. *Fed. Regist. 51* (1986).
4. Organ. Econ. Co-op. Dev., *Recombinant DNA Safety Considerations*, Paris, 1986.
5. *Fed. Regist. 51*: 23310 (1986).
6. Watson, D.H. and D.G. Lindsay, *J. Sci. Food Agr. 33*:59 (1982).
7. Phillips, B.J. and P.S. Elias, *Food Cosmet. Toxicol. 16*:509 (1978).
8. Phillips, B.J., E. Kranz, and P.S. Elias, *Ibid 18*:371 (1980).
9. Phillips, B.J., E. Kranz, and P.S. Elias, *Ibid 18*:471 (1980).
10. *British Pharmacopoeia*, Vol. 2:861 (1980).

General Regulatory Aspects of Biotechnology: Europe

P.S. Elias
Bundesforschungsanstalt für Ernährung, Postfach 3640

Biotechnology is any technique that uses living organisms to make or to modify products, to improve plants or animals, or to develop microorganisms for specific uses. Like any other technology with implications for the health of the public and the integrity of the environment, it requires some form of regulatory control. The major areas of application of biotechnology, including genetic engineering, relate to novel foods, industrial enzymes, certain pharmaceuticals, some animal feeds and new processes. Within Europe, the Commission of the European Economic Communities is making efforts to develop a harmonized approach, acceptable to all member states, for the regulation of biotechnological processes and products. Certain countries, such as the United Kingdom and Germany, already have set up some national machinery to deal with the problems. For example, the United Kingdom has published guidelines on the procedures for approval of novel foods. In a similar manner, the Federal Republic of Germany has developed guidelines for the approval of single-cell proteins in animal feeds. Many national efforts are directed toward the creation of control systems for products of genetic engineering, particularly in the pharmaceutical areas. These controls also are concerned with the conditions under which research, experimental trial and the environmental releases of products or organisms resulting from recombinant DNA technology may be carried out. The need for greater uniformity in the regulatory approach to the control of biotechnology will be discussed.

Biotechnology has been defined as including "any technique that uses living organisms (or parts of organisms) to make or modify products, to improve plants or animals or to develop microorganisms for specific uses" (1). By providing an understanding of biological systems, recent advances in molecular biology have permitted more rapid, more diverse and more precise modifications of biological structures than have conventional techniques. The availability of methods to transfer genetic information including regulatory sequences by means of a vector DNA among diverse organisms in the laboratory, known as recombinant DNA (rDNA) techniques, plant cell and protoplast cultures, plant regeneration, somatic hybridization and embryo transfer, have provided society with new drugs, new biologicals, novel food products and novel food ingredients.

As with any other technology with implications for the health of the public and the integrity of the environment, biotechnology requires some form of regulatory control. The latter may rest entirely on existing legislative powers if these are sufficiently wide to ensure implementation of and compliance with any relevant regulations. Alternatively, it may be necessary to seek new powers and issue new regulations for adequate control of the new situation created by biotechnology in order to protect society, agriculture and the environment.

The major areas of application of biotechnology are

food products, agricultural commodities, pharmaceuticals and novel industrial processes using genetically modified organisms. Consequently, the regulatory aspects will differ according to the biotechnology products or processes concerned. There are many examples of novel food products and food ingredients developed through biotechnology using rDNA technology or other genetic modifications. Food ingredients such as citric acid, glutamic acid and other amino acids, vitamins, flavors and various gums will be manufactured by new processes in the future. Microorganisms used in fermentation cultures may be modified to improve production by altered affinity for the substrate, better resistance to noxious pH or temperatures, and modified control of metabolic functions. The cloning and expression of prokaryotic and eukaryotic genes in prokaryotes and lower eukaryotes now permit, for example, the production of high levels of active enzymes in a pure form from fermentation of inexpensive substrates. Single-cell proteins, peptides, amino acids and carotenoids now can be produced commercially by biotechnology.

The genetic manipulation of agricultural food organisms is probably the oldest form of biotechnology known to mankind. It is exemplified by saving seeds from the most productive plants for planting in subsequent years and by the selective mating within species of the most productive animals. By these manipulations of the agricultural gene pool the selection of genes for complex traits such as yield and environmental tolerance was achieved. Genetic engineering now allows the insertion of new genetic material from virtually any existing organism into agriculturally important organisms thus crossing the breeding barriers imposed by species. For example, cell fusion technology joins two cells from different species and combines their genetic material without the need for sexual compatibility. In the case of plants, the genetically engineered cells then are regenerated into whole plants for use in further breeding into novel food plants. With animals, new genes are inserted into very young embryos, and these then are transplanted into surrogate mothers for development and subsequent use in breeding programs for new animal mutants with particularly desirable characteristics resulting from the altered genetic make-up.

Biotechnological pharmaceutical products include cytokines, hormones, monoclonal antibodies, antigens, growth factors, diagnostic reagents and agents for gene therapy. Biotechnology products for use in industrial processes include microorganisms occurring either naturally or derived from familiar species endowed by gene transfer with additional fermentative capabilities. Other products are isolated enzymes specifically constructed to perform particular chemical reactions, to concentrate metals or for the treatment of agricultural and biological waste materials. In addition, others are completely new molecular entities such as DNA probes.

Whether considered by those responsible for the use, manufacture or legal regulation of biotechnology products, a risk assessment must be made in every case to serve as a basis for any action to be taken. This applies equally to the clinician and patient using a pharmaceutical product, to the consumer of a novel foodstuff, to the farmer and veterinarian producing a novel agricultural product and to the regulator considering the safety of a biotechnology product.

General Principles for a Regulatory Approach

Any regulatory policy to be developed by the European Economic Community (EEC) aimed at ensuring the safety of biotechnological research and biotechnology products must necessarily be uniform and comprehensive if barriers to trade are to be avoided within the EEC and elsewhere. Consequently, the Commission of the EEC will have to undertake the task not only of devising a regulatory policy acceptable to all member states but also of developing agreeable guidelines for assessing the safety aspects of biotechnology research, processes and products. Such efforts have been under way in the U.S. since 1984, where an Interagency Working Group proposed a coordinated framework for the regulation of biotechnology by the various federal agencies involved (2).

Traditional genetic modification techniques are applied broadly for enhancing characteristics of food (e.g. hybrid cereals, selective breeding), for food production (e.g. bread, cheese, meat products), for waste disposal (e.g. sewage treatment), for medicines (e.g. vaccines) and for pesticides and other products. Existing legislation has been adequate hitherto for assuring safety and compliance with existing national and community regulations. However, the newly emerging technologies of genetic engineering such as rDNA, rRNA and cell fusion raise concerns about the risks posed by these novel products. Although these newer methods are only extensions of traditional manipulations, they enable more precise genetic modifications in a much shorter span of time. In the food area, for example, genetic engineering may give rise to three distinct kinds of products requiring different regulatory approaches: (1.) Foods where the newly introduced DNA does not cause a substance novel to the food to be expressed in the part that is eaten, e.g. crops developed with different amino acid, vitamin or fructose content. (2.) Foods where the newly introduced DNA causes a substance novel to the food to be expressed in the part that is eaten, e.g. crops with herbicide resistance or with a new organoleptic property. (3.) Food ingredients produced by genetically engineered organisms or in vitro bioprocesses, e.g. flavors, colors.

Category 3 products are of less concern because they are no longer associated with the genetically engineered organisms and can be treated like any other chemical present as a food ingredient. This implies that existing regulations adequately cover this kind of biotechnology product. In the case of Category 1 and 2 products, people actually will ingest rDNA from different species. Because the building blocks of DNA are the same in all species, and the traditional rDNA produced through breeding has never been a food safety issue, this type of rDNA should not be a safety problem provided it is derived from a food or the gene product is common to food. Thus, category 1 products would not require toxicological studies but may need nutritional evaluation and evidence that no other genetic element foreign to food has been transferred.

Existing regulations may not be adequate to control the nature and safety of the novel food and additional guidelines may be required. In the case of category 2 foods where the new gene is strange to food, the possibility exists that new proteins or secondary metabolites not normally present in the diet may be expressed in some genetically engineered foods. This raises food safety problems and will require new regulations and guidelines.

The regulatory aspects of biotechnology include legislative control of the processes, the products and the environmental consequences. Some form of coordinating advisory mechanism also may have to be instituted. Although the actual regulatory measures may vary in detail depending on the area of application of biotechnology, there is also a considerable element of commonality insofar as the primary aim of the regulations must be the elimination of risks to humans, animals, plants and the various components of the environmental ecosystems. Any proposed regulatory framework must allow, however, for future scientific developments and must be sufficiently flexible to reflect better understanding in the future of the potential risks involved. These concepts should be applied when the necessary degree of control over novel commercial produce, new plants and animals and new microorganisms is being established.

Within agriculture, for example, new plants, animals or microorganisms have long been introduced routinely without regulatory approval except in the case of certain microorganisms. It should be noted that microorganisms play many essential roles in agriculture and the environment and therefore have been exploited and introduced into the environment for decades without apparent harm to the environment.

Jurisdiction over the varied biotechnology products and genetically engineered microorganisms probably will be determined by their use. However, there remains a need to adopt consistent definitions for genetically engineered organisms and to utilize comparable standards in scientific evaluations and assessments. Guidelines also will be needed for governing the release of new microorganisms into the environment ranging from highly contained facilities to the progressively lesser degrees of containment represented by greenhouse testing, small field trials and full field trials.

When considering the applicability of regulatory measures to biotechnology research and products, it should be noted that novel microorganisms may be formed in several ways. One approach uses the deliberate combinations of genetic material from sources in different genera. The other uses microorganisms containing genetic material from a pathogenic or nonpathogenic species. Whether novel organisms should be regulated will depend largely on the nature of the inserted genetic material with respect to the presence of coding regions for proteins, peptides and functional RNA molecules. Insertion of genetic material containing only noncoding regulatory regions such as operators, promoters, replication origins, terminators, ribosome binding regions, flanking sequences or recognition sites for nucleic acid synthesis may not require regulatory control. Similar considerations may apply to specially constructed nonpathogenic strains of species that also contain pathogenic strains.

OECD Guidelines

A group of experts called together by the Organization for Economic Cooperation and Development (OECD) has been studying the safety aspects of releasing rDNA-modified organisms into the environment. Their report, which appeared in 1986, laid down the following general concepts (3):

The vast majority of industrial rDNA large-scale applications probably will use organisms of intrinsically low risk. Therefore, these operations warrant only minimal containment satisfying Good Industrial Large-Scale Practice (GILSP). If rDNA organisms of higher risk are to be used, additional criteria for risk assessment must be identified. However, physical and biological containment of pathogenic organisms is not a new technique so that appropriate handling procedures can be regarded as adequate.

Experience in assessing the potential risks of organisms for environmental or agricultural applications is not as wide as that relating to industrial applications. However, the approach may be made analogous to the data base existing for traditionally modified organisms used generally in agriculture and the environment. If a step-by-step assessment is employed during the research and development stages, the potential risk could be minimized.

With regard to general recommendations, it has been suggested that a free exchange of information on guidelines for national regulations, developments in risk analysis and experience in risk management would be helpful for harmonizing approaches to rDNA technology. Because there is no basis for specific legislation, existing regulatory mechanisms should be reviewed for use as adequate controls in order to avoid undue burden that could hamper progress in this field. To facilitate data exchange and minimize trade barriers, there should be emphasis on international agreement on testing methods, equipment design and microbial taxonomy, taking into account work on standards by WHO, ISO, FAO, EC and the Microbial Strains Data Network. Special effort should be made to improve public understanding of the major aspects of rDNA technology. For certain industrial, environmental and agricultural applications, it may be advantageous to introduce a notification scheme provided appropriate means are developed to protect intellectual property and confidentiality while ensuring safety.

Specific recommendations for industry propose that large-scale applications of rDNA technology should utilize intrinsically low-risk microorganisms in compliance with GILSP. If this is not possible, additional physical containment corresponding to the risk assessment will be required. Monitoring and controlling the nonintentional release of rDNA organisms is an essential part of such containment.

Special recommendations for environmental and agricultural applications include the use of existing data on environmental and human health effects in making risk assessments. Potential risk evaluations must be made before any agricultural or environmental release. However, general international guidelines on such applications are not yet achievable so that independent case-by-case assessments remain, where necessary, the only option. The pathway to be followed in developing rDNA organisms should start at the laboratory where appropriate and move to growth chambers, greenhouses, limited field testing and finally large-scale field testing. Further research to improve prediction, evaluation and monitoring should be encouraged (3).

Ultimately, risk assessment will be the major factor in deciding the extent of regulatory control to be exercised over biotechnology products. Because this always involves an analysis of the hazards of the individual process or product, its biological impact on living organisms and an evaluation of the exposure, it is illustrative to consider the regulatory aspects of novel foods as an example of the general problems likely to be encountered with biotechnology products.

Biotechnology already shows the promise of successful delivery for marketing new foods and food components at a much faster rate than ever before. Current trends in the development of novel foods, both in agriculture and in food technology, are aimed at modest incremental changes in existing practices. Biotechnological efforts are directed toward speeding the progress and sharpening the focus of classical genetics to enhance the production of novel substances by fermentation technology and to stimulate the development of rapid and effective methods of on-line process control. Examples are the development of new and improved cultivars for consumption or processing; the development of improved animal strains and breeds with increased disease resistance, higher lean vs fat ratios of the carcasses and more efficient food utilization. The development of new sources of food ingredients such as flavors, enzymes, single-cell proteins or modified carbohydrates; the development of novel processes and packaging materials leading to novel foods and the development of major novel food ingredients as replacements for traditional ingredients. These new entities must be evaluated individually, probably initially on a case-by-case basis, for their safety, nutritional adequacy and environmental impacts. Benchmarks for these evaluations will have to be supplied by traditional foods, as these foods hitherto generally have been exempted from testing for safety or composition.

Novel Foods—Key Scientific and Regulatory Issues

In any attempts to tackle the safety problems of novel foods, certain key issues must be addressed. To the extent possible, regulatory resources should be employed commensurate with the effort necessary to contain the potential risks and should avoid becoming clogged by trivia or imposing economic burdens detrimental to commercialization or international trade in new products. The degree of novelty of a new product should be established in comparison with the spectrum of traditional counterparts and should determine the levels of nutritional and safety concerns. Specific toxicological methods and safety evaluation procedures that can be applied to traditional and novel foods need to be developed because the established methodologies and assessment procedures are designed principally to cater to single substances. If biotechnology products are to be compared with their traditional counterparts, one must avoid raising unnecessary safety, nutritional or other concerns over these traditional foods. It also must be realized that the key to understanding novel food systems is the existence of an adequate qualitative and quantitative data base on the components of natural and traditional foods. Because such a data base does not yet exist, its establishment is a matter deserving urgent priority. Finally, there is an overriding need to ensure that legislation and regulations do not result in the creation of nontariff barriers to international and community trade.

Societal Issues

The advent of biotechnology is bound to exacerbate the familiarity gap of the nonspecialist with agricultural and food technology. Similarly, the public needs to be more informed technologically and to be reassured about the safety and wholesomeness of novel foods. However, existing scientific and regulatory mechanisms are inadequate for such public reassurance. There also exists a potential for adverse economic impacts on various sectors of agriculture that would delay the acceptance of crops and foods produced by biotechnology.

Safety Issues

From the point of view of safety, it is immaterial whether novel foods derive from raw materials not previously considered as human food or from foods or food ingredients subjected to processes not previously used in food production. As novel foods may potentially substitute for traditional dietary components, their safety becomes paramount.

Safety considerations relate to two aspects: one concerns the exclusion of possible toxic and antinutritional factors; the other relates to the health effects of a lifetime ingestion of unusual constituents or unaccustomed ratios of basic food components. Novel technological procedures, e.g. irradiation and extrusion cooking, applied to traditionally consumed foods are just as liable to engender the same problems. Novel foods derived from vegetable and animal sources not previously considered as human food are equally subject to the same considerations.

Apart from the problems of integrating novel foods and food ingredients into existing compositional food standards and specifications for purity and wholesomeness, there is a general need to consider the effects of the inclusion of novel foods in the national diet of nutritionally vulnerable and captive subsections of the population, e.g. young children, pregnant women, the elderly and those with special dietary needs. Nutritional equivalence must be established with familiar and traditional foods likely to be replaced or supplemented by novel foods. The effects on the bioavailability of micronutrients also require consideration. If derived from microbial sources, their pathogenicity, possible metabolic byproducts, the immunogenicity of contaminating microbial protein and the effects on the environment of any surviving live organisms require attention. Similarly, in the case of novel products incorporated into animal feeds a check must be made for any abnormalities in the physiological state of the target species or in the composition of the carcasses and organs.

The U.K. Approach to Regulating Novel Foods

Interest in novel foods and particularly novel proteins has been prominent in U.K. regulatory bodies since the early 1970s when a worldwide protein shortage was perceived by international organizations such as FAO. As alternative sources of protein for use as food or animal feed were developed, the necessity arose for new powers to control the production, treatment and testing of these novel proteins and of the microorganisms employed in their manufacture. However, the need for flexibility in the approach to any controls over novel foods and the rapid progress in technology and safety evaluation procedures made it impractical to permit the marketing of novel protein foods only after evaluation of their safety. In practice, this would have entailed amending existing regulations each time a new product was developed and passed as safe. Therefore, the government decided on a voluntary notification scheme in cooperation with the food industry, whereby data on all novel foods would be submitted to the Ministry of Agriculture, Fisheries and Food (MAFF) for assessment and clearance before any such food was marketed. Adequate controls were already in place for proteins and other foods from traditional sources or for physically enriched foods. Thus, the general provisions of the Food and Drugs Act cover food hygiene, safety, quality and contaminants. Additional controls are provided by the existing compositional regulations and food standards which ensure a minimum supply of major nutrients, vitamins, minerals and trace elements in food for sale to the public. The labeling regulations in force are designed to furnish adequate information to the consumer and to prevent deception.

To enable the MAFF to grant formal ministerial clearance under the new arrangements, firms are invited to submit adequate data on their novel products. These data then are evaluated by the Advisory Committee on Irradiated and Novel Foods (ACINF), having regard, where appropriate, for the views of other relevant advisory bodies such as The Panel on Novel Foods, the Committee on Toxicity of Chemicals in Food, Consumer Products and the Environment (COT) and the Standing Panel on Hazards from Microbial Contamination of Food. The voluntary notification scheme has the additional advantage of leaving the door open for later introduction of legislation in accord with any future EEC directive on biotechnology products.

ACINF has defined novel foods as "foods or food ingredients produced from raw material that has not hitherto been used for human consumption or that has been consumed in only small amounts, or produced by new or extensively modified processes not previously used in the production of foods" (4). This definition includes foods consumed in only small amounts to cover possible adverse effects from the consumption of large quantities. Severe modifications are potentially just as capable of bringing about changes of toxicological or nutritional importance. If traditional processes have been used, extracted components, recipe changes or minor process modifications are not considered to be novel foods and therefore would not choke the evaluation process. This should avoid undue delays in the development of new biotechnology products and offers a measure of protection for the consumer and the manufacturer. In addition, official acceptability may help in developing markets abroad, although it will not absolve the manufacturer from his responsibilities to comply with current food regulations.

The degree of testing will depend largely on the degree of novelty of the new food and on its source, composition, processing and intended level of population exposure. Extreme novelty foods will need to be tested more extensively than foods subjected to minor changes. Tests would have to cover the effects of any future processing and storage. ACINF therefore suggested elaborate guidelines rather than rigid rules. The experience of the developer should suggest the initial amount of testing and information to be supplied.

Background information on the nature of the new food, its production process and potential market is essential for determining the extent of nutritional and toxicological testing. The source may indicate potential problems, e.g. novel foods from plant material should be examined for natural toxins or antinutritional factors. Novel fats and oils may contain unusual fatty acids and marine products may be contaminated with heavy metals and biotoxins. The use of microorganisms may indicate the need to look for bacterial or mycotoxins and pathogenic microbes. Chemical and harsh physical processing may cause the appearance of toxic products or result in significant damage to nutrients. It is also important to know the anticipated market size, the potential exposure, the likelihood of population groups with high intake and the food's probable effect on the nutrient composition of the existing national diet.

The product specifications should detail product variability and analytical methods to permit confirmation of the identity of the product marketed with that used in the safety testing program. Alternatively, complex products might be more easily specified by a process specification. Scaling up manufacture from pilot plant to full industrial production requires demonstrating the compliance of each batch with the pilot plant specification as well as consistency of the nutritional and toxicological properties.

Chemical analysis, both proximate and detailed analyses of the components, is an essential preliminary to any biological evaluation and might point to potential problems. Particular interest would center on the presence of unusual or toxic amino acids and aminoglycosides. A full fatty acid spectrum should reveal any cyclic fatty acids, toxic fatty acids, the presence of *cis* and *trans* double bonds, acids with chain lengths over 22 carbon atoms, peroxides and degradation products of polyunsaturated fatty acids. Carbohydrates should be examined for non-metabolizable fiber and chitin as well as for the presence of tannins. In addition, the level of toxic metals and nutritionally significant trace metals should be known. The vitamin content and the possible presence of naturally occurring or adventitious antinutritional factors as well as chemical and biological screening for mycotoxins and other toxins are essential analytical information.

Nutritional studies are needed to forecast the nutritional impact of the intended uses on the consuming public and the likely maximum consumption by particular population groups. Both animal and human studies are needed to establish nutritional equivalence in all important aspects with the natural food that is to be simulated. If necessary, fortification may be necessary. If the novel food is to replace a traditional food, its influence on the diet of children, the elderly and captive populations, such as hospital patients and school-age children, must be examined. The effects of further cooking, processing and storage on the components of a novel food and on its nutritional value should be established.

Novel foods likely to form a significant part of the human diet must be studied in vivo in animals for determination of metabolizable energy, protein quality and bioavailability of vitamins and minerals in the novel food and in the remainder of the diet as well as possible interactions with other dietary components. Nutritional studies are also a prerequisite to toxicity studies in laboratory animals to avoid nutrient imbalance and palatability problems when the novel food is incorporated in laboratory animal diets. Results of nutritional studies in animals may be extrapolated to human beings provided the availability of nutrients to man can be verified.

On the subject of toxicological studies to be carried out on novel foods, the U.K. guidelines underline the specific problems arising when testing novel foods by standard toxicological procedures. First, a hundred-fold safety margin cannot be left between the maximum amount of novel food likely to be consumed in the diet and the maximum amount in the laboratory animal diet that has no toxic effect, if the novel food constitutes more then 1% of the human diet. There are also practical limits to the amounts of food that may be incorporated into animal diets, if nonspecific adverse effects on the health and nutrition of test animals are to be avoided. Second, any novel food requires tolerance and allergenicity testing, including monitoring, in human volunteers under controlled conditions and medical supervision because predictive animal models for these effects are not available. Third, the stability of the novel food in the test diet requires investigation; palatability problems may require paired-feeding designs.

The complex composition of novel foods makes routine metabolic studies on their constituents impractical except if toxic contaminants or components are present or if the major component is a new chemical entity not normally occurring in food. In these cases, the toxicologically relevant components require a full metabolic investigation. Changes in excretory functions may be detected by analysis of urine and feces.

Acute studies are normally inappropriate with a novel food. Subchronic studies, usually extending over at least 90 days in one rodent and at least one nonrodent species are an essential part of the safety assessment of novel foods. The highest level studied should be the maximum practical level and should exceed the anticipated exposure level substantially. Only one intermediate dose level need be employed as the determination of a precise no-adverse-effect level (NEL) also is inappropriate. Such a design will help in distinguishing between dose-related toxic and nonspecific effects. Care must be taken that the test and control diets contain equivalent amounts of macronutrients and micronutrients. The usual toxicological parameters should be determined, including hormone, mineral and vitamin excretory levels. Full macroscopic and microscopic pathological evaluation at the termination of the experiment will be required.

The need for chronic toxicity and carcinogenicity studies will depend largely on the nature of the novel food. These may be omitted on the basis of the results of the subchronic and mutagenicity studies or they may replace subchronic studies if the food or process is particularly novel. These studies should comply with the OECD guidelines (5) and with good laboratory practice (GLP) in the protocols used and in their execution. Control groups should be larger than test groups and may be supplemented by historical data from animals of the same strain. The choice of dose levels should not be dictated by the need to establish an NEL but should follow the suggestions made for subchronic studies.

Studies on embryotoxicity, including teratogenicity and reproduction, may only be necessary if extremely novel processes or source materials are being used. If carried out, they must comply with the OECD guidelines (5). Mutagenicity tests will be required on novel foods irrespective of the availability of long-term studies although it is recognized that problems in interpretation may arise because of the presence of numerous mutagens in natural foodstuffs. Both germ cells and somatic cells will have to be investigated for possible mutagenic effects. In vitro mutagenicity testing of novel foods may present particular problems because of the interference of nutrients derived from the food with the growth media used in some of the tests. Specially modified test procedures may therefore have to be used or suitable extracts prepared for testing.

An important requirement of the U.K. guidelines is the need to carry out studies in humans. A minimum proviso before such studies are undertaken is the absence of adverse effects in subchronic studies in two animal species and in the mutagenicity tests. Provided women of child-bearing age are excluded, reproduction and teratogenicity studies need not be available for initial testing in volunteers. However, all studies on human subjects must comply with ethical and legal requirements and should be designed to include regular analyses of urine, feces, blood and plasma as well as renal, hepatic and other organ function tests if indicated. Initial human studies should use a single meal containing a known dose of the novel food given to one volunteer at a time and allow a sufficiently long intervening observation period for any gastrointestinal, allergic or intolerance reactions to appear. These tests may be followed by four-wk feeding studies with longer follow-up periods using different doses of the novel food in different groups of adequate numbers of individuals with a concurrent control group on the same diet without the novel food. Test and control groups should be matched for age, size, alcohol intake, smoking habits, sex distribution and food habits. If possible, a blind cross-over design should be used in which volunteers act as their own control. If the novel food is intended for special population groups, a representative test group should be included in the preliminary studies. If the novel food has been tolerated well, it may be administered ad libitum for a short period to assess acceptability.

The health of workers involved in the manufacture of novel foods should be monitored irrespective of whether exposure is to the raw or cooked food. Both volunteers and manufacturing staff should be examined for allergenicity to the novel food. Apart from overt evidence of allergy it may be useful to test individuals with a standard antigen from a component of the novel food. Detecting induced allergy in the general population requires the monitoring of very large numbers of exposed people.

Large-scale acceptability and marketing trials constitute the final step in evaluating the safety of a novel food. Supervision by local medical services in the test areas is needed to evaluate the occurrence of rare adverse reactions to the novel food among the local exposed population. To obtain all relevant information, one medical practitioner should be given overall responsibility for health monitoring. Special emphasis should be placed on monitoring vulnerable exposed groups such as pregnant women, children, the elderly, diabetics and other risk groups. It also is necessary to establish whether the novel food interferes with the action of commonly used drugs. If interference is detected, additional pharmacological studies will be needed.

The U.K. memorandum did not consider the complex problem of informative labeling of novel foods and food ingredients and the extent to which the consumer should be informed about the presence of novel constituents in the food supply.

The EEC Approach

Within the EEC, a harmonized approach to novel foods has not been developed yet. However, the future program of the EEC Commission announced in its white paper of June 14, 1985, and submitted in November 1985, to the Council of Minsters and the European parliament, includes among others a new strategy in the field of foodstuffs regulation. This strategy incorporates proposals for legally regulating the processes for the manufacture and treatment of foods in order to protect public health. The current state of industrial and technological development among the community members requires, in the view of the commission, the preparation of directives for the control of deep freezing, irradiation preservation of food and biotechnological processes. It leaves open the power to add other technological processes and treatments of food to this list.

The commission foresees the preparation of a general directive on new foodstuffs and on those obtained by a biotechnological process for submission to the Council of Ministers in 1987 and adoption by the Council in 1988. In anticipation of these future activities, the commission already has instituted two working groups composed of members of the Scientific Committee for Food. The commission is legally required to consult that committee on all matters relating to food safety. The working groups will be considering biotechnology on the one hand and food processing and nutrition on the other. It is to be expected that the guidelines already published by the U.K. as a member of the EEC will have considerable influence on the deliberations of the SCF and its working groups.

More precise regulations concerning certain biotechnology products to be used in animal nutrition already have been provided in a council directive adopted in 1982 (6). These relate specifically to protein products that are obtained from cultured microorganisms. These products were subject to extensive investigation with respect to their chemical composition, manufacturing process, purity and safety for the target animal as well as for their toxicological properties in laboratory animals. The data were required to be submitted in the form of a dossier in accordance with the provisions of a council directive adopted in 1983 (7) on the fixing of guidelines for the assessment of certain products used in animal nutrition. Dossiers for new substances to be added to the annex of the directive are reviewed by technical experts from member governments and subsequently by the Scientific Committee for Animal Nutrition of the EEC Commission. To date, no novel feeds produced by genetic engineering have been considered by the committee.

The Approach of JECFA

The Joint FAO/WHO Expert Committee on Food Additives (JECFA), established in 1956, has been concerned with the assessment and evaluation of chemicals used as food additives. It functions as an advisory body to the Codex Alimentarius Commission and its subordinate committees. Considerable efforts have been made recently by JECFA to update the toxicological methods and the system of evaluation of toxicological data in use since 1956. In its 30th session in 1986, JECFA adopted the final version of a revision document prepared by WHO, which for the first time includes a section on the safety evaluation of novel foods as distinct from the evaluation of food additives.

This section discusses the reasons for using a different approach to the evaluation of the safety of novel foods, already alluded to earlier, such as their potentially high level of consumption, similarity to natural foods in some cases and metabolism into normal constituents in most cases. Emphasis is laid again on the establishment of adequate specifications and the identification of potentially toxic impurities. A corollary to this requirement is the need to provide a full chemical analysis, particularly including components of nutritional importance. Among substances to be looked for are antinutritional factors, toxins and the products of interactions with other food components. The assessment of the nutritional value of the novel food using in vivo studies in experimental animals and complementary studies in man is considered essential before release of food for wider consumption. Although metabolic studies on each component of a novel food would be impractical, the influence of a novel food on the physiological processes of digestion, absorption, distribution and excretion of macro- and micronutrients of the normal diet requires clarification by appropriate studies. The transience or progression and reversibility of any observed changes must be established.

The document suggests a series of studies in experimental animals and man much in line with those listed in the U.K. memorandum. It proposes, however, to carry out the safety evaluation by the establishment of a traditional acceptable daily intake (ADI) for novel foods consumed at less than 1% of the total daily diet of 1500 g of solid food. For novel foods with higher consumption levels it still proposes to set ADIs using safety factors less than 100 or, as an alternative, suggests setting upper limits of inclu-

sion in the total daily diet. Appropriate labeling is recommended once the novel food is released generally. No attempt has been made to deal specifically with the data requirements for the evaluation of novel foods produced by genetic engineering.

The Approach of the Codex Alimentarius

No specific efforts have been made yet in the framework of the Codex Alimentarius to devise principles applicable to the general inclusion of novel foods in Codex food standards. Some work has been carried out on establishing general principles for the addition of nutrients to food. The definition of nutrients is such to include novel foods and food ingredients as a substance, consumed as a food constituent, which is needed for growth, development and the maintenance of healthy life, and cannot be synthesized in adequate amounts by the body. Some novel foods also could be covered under the provisions for substitute foods that are intended to be used as a complete or partial replacement for the foods they resemble. In these cases, the essential requirement is nutritional equivalence in terms of the essential nutrients of concern. Clearly, much remains to be done and little guidance on regulating biotechnology products is likely to emanate from the Codex Alimentarius in the near future.

The Approach of the FRG

No specific regulations exist in the Federal Republic of Germany for novel foods and food proteins. If these can be regarded legally as food, they are not subject to special legislative control except for compliance with general hygiene and labeling regulations. They may therefore be marketed freely. It is not clear from the existing legislation at what level of addition to food novel foods or food proteins become food additives. If they were to be defined as such they would be subject to the provisions applicable to food additives.

Discussion

There is as yet little uniformity in the regulatory approaches to biotechnology and its products within Europe. In the pharmaceutical area, the production by rDNA techniques of species-specific proteins with important pharmacological properties such as insulin, interleukines, interferons, tumor necrosis factor and others has led to the elaboration of draft recommendations for preclinical toxicity studies in animals (8–11). Present experience has shown, however, that animal models have failed to predict adverse effects observed when used in man nor are they as sensitive as chemical analysis for the detection of toxic impurities. Furthermore, the immunogenicity of these proteins restricts the usefulness of standard toxicological tests (12). With regard to novel foods and food ingredients, there are no EEC directives on their control in existence and only the U.K. memorandum published by ACINF attempts to lay down guidelines for assessing the safety of these products within the framework of a voluntary notification scheme.

No harmonized guidelines that deal with the information required to assess the safety of organisms and microorganisms modified by rDNA technologies and to evaluate their impact on the environment exist. Microbiological information on genetically engineered microorganisms has a pivotal role in assessments of the risks from exposure at work, from unintentional and from deliberate release into the environment. Similarly, chemical, toxicological and, where appropriate, nutritional information will have to be supplied for the gene products expressed by genetically modified organisms and microorganisms. Major efforts will be needed in the near future to arrive at a uniform approach to regulating biotechnology in all its fields of application to avoid hampering progress in this rapidly evolving field and to prevent the development of barriers to intercommunity and international trade.

REFERENCES

1. *Commercial biotechnology: an international analysis*, Office of Technology Assessment, Washington, D.C. (1984).
2. *Coordinated Framework for Regulation of Biotechnology*, Office of Science and Technology Policy, Washington, D.C. (1986).
3. *Recombinant DNA Safety Considerations*, OECD, Paris (1986).
4. Advisory Committee on Irradiated and Novel Foods, *Memorandum on the Testing of Novel Foods*, HMSO, London (1984).
5. *Guidelines for testing new chemicals*, OECD, Paris (1981).
6. EEC Council, *Directive concerning certain products used in animal nutrition*, 82/471/EEC, O.J. No. L213, 21/7/82, 1982, p. 8.
7. EEC Council, *Directive on the fixing of guidelines for the assessment of certain products used in animal nutrition.* 82/228/EEC, O.J. No. L126, 13/5/83, 1983, pp. 23-27.
8. *Control of quality and safety of active ingredients of drugs with polypeptide structure produced by recombinant DNA, cell fusion, fermentation and cell culture techniques*, Bundesverband der Pharmazeutischen Industrie, Sept. 1985, Frankfurt.
9. *Recommendation concernant le protocole toxicologique des interferons pour l'obtention d'une autorisation de mise sur le marche*, Syndicat National de l'Industrie Pharmaceutique, March 16, 1984, Paris.
10. *Proposed supplementary instructions for registration and clinical trial of biosynthetic drugs based on recombinant DNA technique*, National Board of Health and Welfare, April 1986, Sweden.
11. *Considerations for the standardization and control of the new generation of biological products*, National Institute for Biological Standards and Control, Draft 1984, London.
12. Teelmann, K., H. Hohbach, H. Lehmann and The International Working Group, *Arch. Toxicol. 59*:195 (1986).

General Regulatory Aspects for Biotechnology in Japan

M. Tamaki

Bioindustry Office, Basic Industries Bureau, Minister of International Trade and Industry, 3-1 Kasumijaseki, 1-chome, Chiyoda-Kw, Tokyo 100, Japan

r-DNA technology was developed as a means for improving organisms. It was pointed out that organisms produced by r-DNA technology have potential risks for environment and health soon after r-DNA technology was published. So, governments began setting guidelines for experimental work to secure the safety of r-DNA technology. The Japanese government also made guidelines, in 1979 estimating potential risk based on knowledge available at that time. Now that r-DNA technology has made rapid progress and has grown to be used in industrial processes, the Japanese Ministry of International Trade and Industries (MITI) proclaimed "Guidelines for Industrial Application of Recombinant DNA Technology" in June 1986. The guidelines provide that the organizer of a working organization can request the Minister of International Trade and Industry to authorize the industrial plan if it conforms with the present guidelines. The MITI already has authorized 68 plans through June of 1987. Application technology of r-DNA organisms in open areas, such as purification of wastewater, is developing now. We are researching to improve prediction, evaluation and monitoring of the outcome of applications of r-DNA organisms in these situations.

GENERAL ASPECTS OF BIOTECHNOLOGY POLICY IN JAPAN

Since the discoveries of recombinant DNA technology in 1973 and cell fusion technology in 1972, many countries have recognized the importance of these as key technologies in the development of future industries. Japan is, of course, not exceptional. Now, we recognize that biotechnology is becoming one of the most important high technologies for Japan. It is not an exaggeration to say that Japan's future success depends on whether Japan can develop this technology.

Japanese Government Offices in Biotechnology

Many government offices are actively involved in developing biotechnology in Japan. Table 1 shows the role of each government office in promoting research and development. There are five major government offices: Ministry of International Trade and Industry (MITI), Ministry of Agriculture, Forestry and Fishery (MAFF), Ministry of Health and Welfare (MHW), Ministry of Education (MOE) and Science and Technology Agency (STA). Of these government offices, MITI has taken the lead in promoting biotechnology as well as other high technologies. MITI is widely involved in numerous areas of biotechnology such as the consolidation of the infrastructure for bioindustry, promoting fundamental research and development, application for the production of chemicals and substitutional energy for petroleum, ensuring the safety in the application of recombinant DNA technology and so forth. STA is in charge of researching life sciences, especially developing the basic technology of biotechnology. MAFF promotes the application of biotechnology in the agricultural and food fields. MHW promotes the application of biotechnology in the medical and pharmaceutical fields. MOE is in charge of research in universities.

TABLE 1

Japanese Government Organization in the Biotechnology Field

Office	Ministry of International Trade and Industry (MITI)	Science and Technology Agency (STA)	Ministry of Agriculture, Forestry and Fishery (MAFF)	Ministry of Health and Welfare (MHW)	Ministry of Education (MOE)
Field	Mining, Industry and Energy	Development of basic technology	Agriculture and Foods	Public Health and Welfare	University's Research
Policy	Promotion of Bioindustry Promotion of Biotechnology Fuel Alcohol from Biomass Infrastructure of Bioindustry Ensure Safety of Biotechnology International Cooperation	Promotion of Life Science Development of Basic Technology in Biotechnology	Plant Breeding Bioreactor in Food Processing	10 Years Project of Cancer Research New Medicines by r-DNA Technology	Research in Universities Prevention of Biological Resources in Universities
Budget ('87)	ca ¥5,600 mil.	ca ¥12,800 mil.	ca ¥5,900 mil.	ca ¥4,000 mil.	ca ¥5,000 mil.

TABLE 2

MITI's Policy on Bioindustry

(1) Consolidation of the Infrastructure for Bioindustry
   1) Organization of private enterprises in bioindustry and reinforcement of industry-university cooperation
   2) Standardization
   3) Gene bank
   4) Tax incentives
   5) Supporting venture businesses
   6) Consolidation of data base

(2) Promotion Fundemental Research and Development in the Bioindustry Field
   1) R&D project on basic technology for future industry
     a. Utilization of recombinant DNA
     b. Large scale cell cultivation
     c. Bioreactor
     d. Bio-electronic
   2) R&D Project to obtain substitutional energy for pretroleum from biomass
   3) National research and development program (new water treatment system)
   4) Utilization of cohesive yeast for industrial ethanol production
   5) National research institutes

(3) Ensuring safety in the application of biotechnology
   1) Application of the "Guideline for Industrial Application of Recombinant DNA Technology"
   2) Study to find adaptable safety measures for the Use of Recombinant DNA Organisms in Open Environments

(4) Developing Regional Economies Utilizing Biotechnology
   1) Study the technological potential
   2) Study the cases of introducing biotechnology
   3) Holding the conference of regional bureau's officials in charge of bioindustry

(5) Promotion of International Cooperation in Bioindustry
   1) Promotion of research cooperation and industrial cooperation between advanced countries
   2) Promotion of technological assistance to LDC's
   3) Contribution to the Human Frontier Science Program

MITI's Policy on Biotechnology

MITI's policy on biotechnology consists of five major measures as shown in Table 2.

Consolidation the Infrastructure for Bioindustry. Organization of private enterprises in bioindustry and reinforcement of industry-university cooperation: Bioindustry Development Center (BIDEC) is an association that has a large membership of nearly 180 large enterprises plus nearly 160 small and medium enterprises with nearly 1,400 private members. As one easily can guess from these figures, BIDEC is the largest representative bioindustry organization in Japan. BIDEC promotes cooperation between the industrial and academic world, and carries out various activities to promote bioindustry as shown in Figure 1. BIDEC was reorganized last February with MITI's support. MITI carries out many activities through BIDEC. Standardization: MITI promotes standardization in the field of technical languages, capacities of reagent, machinery and equipment. Gene bank: MITI collects, preserves and distributes organisms through a patent depository system. Tax incentives: tax incentives are given to enterprises to encourage research and development in the bioindustry field. Supporting venture busi-

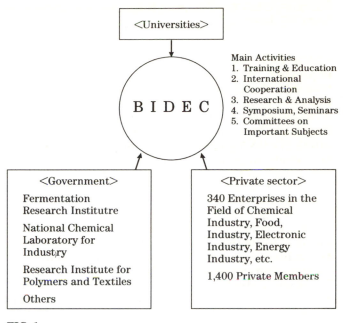

Main Activities
1. Training & Education
2. International Cooperation
3. Research & Analysis
4. Symposium, Seminars
5. Committees on Important Subjects

FIG. 1.

nesses: the Japan Key Technology Center gives financial support to venture businesses. Consolidation of data base: a data base is under development by BIDEC for efficient research and development in biotechnology.

Promotion Fundamental Research and Development in the Bioindustry Field. R&D project on basic technologies for future industries: long-term, about 10 years, research and development projects are in progress. They are utilization of recombinant DNA ('81FY–'90FY), large-scale cell cultivation ('81FY–'89FY), bioreactor ('81FY–'88FY), bioelectronic ('86FY–). R&D projects to obtain substitutional energy for petroleum from biomass: to obtain ethanol efficiently from biomass resources such as cellulose, agricultural and forestry waste, R&D projects utilizing biotechnology are in progress. National research and development program: to meet long-term water demand and supply, and to obtain methane, an R&D project to clean waste water utilizing biotechnology is in progress. Utilization of cohesive yeast for industrial ethanol production: to create a high productivity process utilizing cohesive yeast, an R&D project is in progress. National research institutes: MITI has 16 research institutes. The following research institutes are involved in studying biotechnology: the Fermentation Research Institute, the National Chemical Laboratory for Industry and the Research Institute for Polymers and Textiles.

Ensuring safety in the application of biotechnology. To ensure safety in the application of recombinant DNA technology, MITI announced the "Guideline for Industrial Application of Recombinant DNA Technology" last year and is involved in research aimed at finding adaptable safety measures for the use of recombinant DNA organisms in open environments.

Developing regional economies utilizing biotechnology. There are many types of fermented food in Japan, such as soybean paste, soy and rice wine. MITI is studying the technological potential of local enterprises and local universities related to biotechnology.

TABLE 3

Specification for Evaluating Equipment and Apparatuses

Items for evaluation:	Classification of recombinant / Evaluation item	GILSP	Category 1	Category 2	Category 3
(1) Extent to which equipment and apparatus can be sealed effectively.					
a. Visible organisms should be handled in a system that physically separates the process from the external environment (a closed system)	(1) a	Semi-closed[1]	Yes	Yes	Yes
b. Leakage form the closed system	b	Minimize release[2]	Minimize release	Prevent release	Prevent release
c. Performance of seals	c	Minimize release	Minimize release	Prevent release	Prevent
(2) The closed system should be located within a work site					
a. Designation of a work site	(2) a	Optional	Optional	Yes	Yes and purpose built
b. If a work site has been designated:					
(a) Biohazard signs shold be posted	b (a)	No	Optional	Yes	Yes
(b) An air lock controlling ingress and egress should be established	(b)	No	No	No	Yes
(c) Decontinuation and washing facilities should be provided for personnel	(c)	Optional	Yes	Yes	Yes
(d) Personnel should shower before leaving the work site	(d)	No	No	Optional	Yes
(e) Effluent from sinks and showers should be collected and inactivated before release	(e)	No	No	Optional	Yes
(f) Work site should be adequately ventilated to minimize contamination	(f)	Optional	Optional	Optional	Yes
(g) Work site should be maintained at a less-than-atmospheric air pressure	(g)	No	No	Optional	Yes
(h) Input air and extract air to work work site should be HEPA filtered	(h)	No	No	Optional	Yes
(i) Work site should be designed to contain spillage of the entire contents of the closed system	(i)	No	No	Optional	Yes
(j) Work site should be sealable to permit fumigation	(j)	No	No	Optional	Yes

[1] A system in which well-repaired equipment and apparatuses are used and a closed system is followed.
[2] Release shall be decreased to an appropriate level depending on the safety level of a recombinant.

Promotion International Cooperation in Bioindustry. The research and development of biotechnology has become important throughout the world. MITI is actively promoting international cooperation through various channels.

ENSURING SAFETY MEASURE

Recombinant DNA technology was championed as an innovative technology to improve cross-fertilization. Soon after recombinant DNA technology was discovered, it was pointed out that organisms produced by recombinant DNA technology posed a potential risk to health and the environment. Therefore, some countries established guidelines for experiments to ensure the safety of recombinant DNA technology. The Japanese government also established a guideline for experiments in 1979 estimating potential risks based on knowledge at that time.

Because recombinant DNA technology has made rapid progress and is used in the industrial process, MITI announced the "Guideline for Industrial Application of Recombinant DNA Technology" in June, 1986. The guideline specifies that the organizer of a working organization can request the Minster of International Trade and Industry to authorize that the industrial plan conforms with the present guideline. MITI authorized 68 proposals as of June 1987.

Application technology of recombinant DNA organisms in open environments, such as waste-water treatment, is being developed. MITI is studying the area to improve prediction, evaluation and monitoring of the outcome of application of recombinant DNA organisms.

Guideline for Industrial Application of recombinant DNA Technology

Background. On October 4, 1984, the Minister of International Trade and Industry submitted "How to keep safety

Table 4

A List of Authorized Proposals by MITI. The First Authorization (October 14, 1986)

Enterprise	Outline of the Proposal			
	Recipient Organism	Classification	Product	Number of proposals
Kyowa Hakko Kogyo Co., Ltd.	*E Coli* K-12	GILSP	Catalysts (Enzyme)	1
Showa Denko K.K.	*B. Amiloliquefaciens*	GILSP	Amino Acid	1
Takara Syuzo Co., Ltd.	*E. Coli* K-12	GILSP	Catalysts (Enzyme)	8
	E. Coli K-12	GILSP	Reagents	14
Toyobo Co., Ltd.	*Providencia Stuartii*	Category 1	Catalysts (Enzyme)	1
	E. Coli K-12	GILSP	Catalysts (Enzyme)	2
	E. Coli K-12	GILSP	Reagents	5
Nippon Gene Co., Ltd.	*E. Coli* K-12	GILSP	Catalysts (Enzyme)	3
	E. Coli K-12	GILSP	Reagents	9
Mitsui Touatsu Chemicals, Inc.	*E. Coli* K-12	GILSP	Amino Acid	1
Mitsubishi Petrochemical Co., Ltd.	*E. Coli* K-12	GILSP	Aminio Acid	1
Yakult Honsha Co., Ltd.	*E. Coli* K-12	GILSP	Reagent	1

Total   8 Enterprises   47 Proposals

The Second Authorization (December 19, 1986)

Enterprise	Recipient Organism	Classification	Product	Number of proposals
Sanraku, Inc.	*E. Coli* K-12	GILSP	Amino Acid	1
Takara Syuzo Co., Ltd.	*E. Coli* K-12	GILSP	Catalysts (Enzyme)	1
	E. Coli K-12	GILSP	Reagent	1
Yakult Honsha Co., Ltd.	*E. Coli* K-12	GILSP	Reagent	1
Wakunaga Pharmaceutical Co., Ltd.	*E. Coli* K-12	GILSP	Reagent	1

Total   4 Enterprises   5 Proposals

The Third Authorization (March 25, 1987)

Enterprise	Recipient Organism	Classification	Product	Number of proposals
Takara Syuzo Co., Ltd.	*E. Coli* K-12	GILSP	Catalyst (Enzyme)	1
	E. Coli K-12	GILSP	Reagents	13

Total   1 Enterprises   14 Proposals

The Fourth Authorization (May 27, 1987)

Enterprise	Recipient Organism	Classification	Product	Number of proposals
Novo Biochemical Industry Japan, Ltd.	*B. subtilis*	GILSP	Catalysts (Enzyme)	1
Amano Pharmaceutical Co., Ltd.	*B. stearotheromophilus*	GILSP	Catalysts (Enzyme)	1

Total   2 Enterprises   2 Proposals

Total amount   15 Enterprises   68 Proposals

for industrial application of recombinant DNA technology?" to the Chemical Product Council. The council vigorously examined this question, evaluating the arguments of the OECD/CSTP ad-hoc group. On May 30, 1986, the council submitted a report to the minister. The report stated that the Japanese Government should make a guideline for industrial application of recombinant DNA technology, evaluating the argument of the OECD/CSTP ad-hoc group. On June 19, 1986, the guideline was announced by the Minister of International Trade and Industry.

Contents. This guideline consists of the five chapters following: Chapter 1 is General Provisions, and it consists of two sections. Section 1 is the purpose of the guideline and it says "It is the purpose of the present guidelines to provide the basic conditions for ensuring adequate safety in the application of recombinant DNA technology to various industrial processes, including manufacturing and mining, thus providing complete safety and promoting

Working of the Guideline Investigation Revision of the Guideline

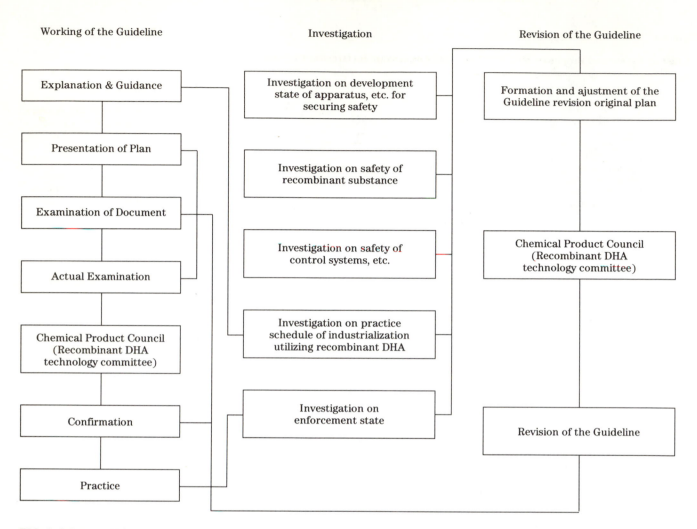

FIG. 2. Scheme of Bioindustry Safety Security Measure

appropriate use when applying recombinant DNA technology. Section 2 is a definition of terms. The guideline defines eight terms, recombinant DNA technology, recombinant DNA molecule and so forth. Chapter 2 is Evaluation of Recombinant's Safety, and it consists of three sections: section 1 is a general rule; section 2 contains items for evaluation of recipient organism, recombinant DNA molecules and recombinants; and section 3 is safety evaluation and classification for recipient organism and recombinants. There are four classifications, namely GILSP (Good Industrial Large-Scale Practice), Category 1, Category 2 and Category 3. Chapter 3 is Equipment, Apparatuses, Operations and Management for Recombinant Organisms and it consists of three sections: section 1 is a general rule; section 2 contains specifications for evaluating equipment and apparatuses as shown in Table 3; section 3 is about operation and control rules that the person in charge of a working organization shall obey. Chapter 4 is Management and Responsibility System, and it consists of eight sections including an organizer of working organizations, persons in charge of working oganizations, persons in charge of working organizations, a director of operations, operations personnel, a committee for safe operations, a manager for safe operations, a training system and a health-care sys-

tem. Chapter 5 is "Others," and it says in the first half as follows. "In order to ensure their safety, the organizer of a working organization can request the Minister of International Trade and Industry to authorize that the equipment, apparatuses, operations and management of the industrial application of recombinant DNA technology conforms with the present guidelines."

Application. The application scheme of the guideline is shown in Figure 2. There are three flowcharts. The left flowchart shows a procedure when an enterprise requests the Minister in conformity with Chapter 5 of the guideline. The inside flowchart shows a way to gain additional information related to the guideline. The right flowchart shows a procedure to revise the guideline. Because the minister announced the guideline, a total of 15 companies requested the minister to authorize 68 proposals, and the minister authorized all of them. A list of the authorized proposals is shown in Table 4.

Study to Find Adaptable Safety Measures for the Use of Recombinant DNA Organisms in Open Environments

There are some enterprises involving research recombinant DNA technology to utilize it in open environments

GENERAL REGULATORY ASPECTS FOR BIOTECHNOLOGY IN JAPAN

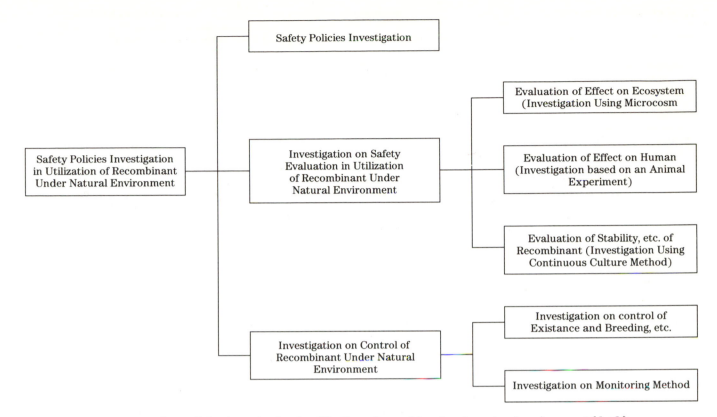

FIG. 3. System chart of safety policies investigation in utilization of recombinant under natural environment (draft)

such as waste water treatment. MITI is attempting to find adaptable safety measures for the use of recombinant DNA organisms in open environments through BIDEC from '86FY. Figure 3 shows the contents of the study. The study consists of three parts: safety measures investigation, investigation on safety evaluation in utilizing of recombinant in open environments, and investigation on control of recombinant in open environments. The annual program is shown in Table 5. MITI is planning to study the arguments of the OECD/CSTP ad-hoc group for four years.

Table 5

Yearly Program of Safety Policies Investigation in Utilization of Recombinant Under Natural Enviroment (draft)

Safety policies investigation in utilization of recombinant under natural environment	1986	1987	1988	1989
1. Safety policies investigation	The steering of following experiments 2 and 3 in view of the trend of a chemical product council and safety policies investigation in utilization of recombinant substance under a natural environment	The steering of following experiments 2 and 3 in view of the trend of a chemical product council and safety policies investigation in utilization of recombinant substance under a natural environment	The steering of following experiments 2 and 3 in view of the trend of a chemical product council and safety policies investigation in utilization of recombinant substance under a natural environment	The steering of following experiments 2 and 3 in view of the trend of a chemical product council and safety policies investigation in utilization of recombinant substance under a natural environment
2. Investigation on safety evaluation in utilizing of recombinant under a natural environment				
(1) Evaluation of effect on ecosystem	Effect exerted on ecosystem of microorganism, etc. of recombinant in single phase (soil) microcosm used is investigated	Effect exerted on microorganism ecosystem of recombinant in single phase (soil and aqueous system) microcosm used is investigated	Effect exerted on microorganism ecosystem of recombinant in single phase (soil and aqueous system) microcosm used is investigated	A method for evaluating effort on an ecosystem is established
(2) Evaluation of effect on human	———	Animal experiment is performed and effect on human is investigated	Animal experiment is performed and effect on human is investigated	An animal experiment is conducted and a technique for evaluating the effect on a human is established
(3) Evaluation of stability, etc. of recombinant substance	The stability and growth conditions, etc. are investigated by using the continuous culture apparatus	The stability and growth conditions, etc. are investigated by using a continuous culture apparatus	Continous experiment is conducted with respect to stability, etc. and evaluation technique is established	———
3. Investigation on control of recombinant under natural environment				
(1) Investigation on control of existance and breeding, etc.	A control method percipitated in existence and breeding, etc. of a recombinant liberated to the natural field is investigated in a solid phase (soil) by using chemical agent	Investigation is conducted on a method for controlling a recombinant liberated under natural environment	A method for controlling a recombinant liberated under a natural environment is established	———
(2) Investigation on monitoring method	———	Investigation is conducted on a method for monitoring a recombinant liberated under a natural environment	Investigation is conducted on a method for monitoring a recombinant liberated under a natural environment	A method for monitoring recombinant liberated under a natural environment is established

Toxicology of Technological Products—Dietary Fat, Cancer and Other Chronic Diseases

Kenneth K. Carroll

Department of Biochemistry, University of Western Ontario, London, Ontario, Canada, N6A 5C1

Consumption of high-fat diets is associated with increased risk from a number of chronic diseases including cancer and cardiovascular disease. The high caloric density of high-fat diets also is conducive to obesity. Studies on experimental cancer in animals have shown that polyunsaturated vegetable oils promote tumorigenesis more effectively than saturated fats or polyunsaturated fish oils, but cancer mortality data for human populations show the strongest positive correlations with total and/or saturated fat and little or no correlation with polyunsaturated fat. Substitution of polyunsaturated fat for saturated fat has been recommended to reduce the risk of cardiovascular disease, but there is some concern about high intakes of polyunsaturated fat, and monounsaturated fat may be a better substitute. The n-3 polyunsaturated fatty acids in fish oils differ in their effects on cancer and cardiovascular disease from the n-6 fatty acids in polyunsaturated vegetable oils; more studies are needed to assess the possible beneficial effects of fish oils on these and other chronic diseases. Effects of *trans* fatty acids formed during industrial hydrogenation of fats are of interest because relatively large amounts may be present in the diet, but there is little evidence of specific deleterious effects of these fatty acids. For some conditions such as obesity, total dietary fat intake is more relevant than the type of dietary fat. The diets of Western, industrialized nations have a much higher fat content than those of most other countries, and experimental and epidemiological evidence linking dietary fat to various chronic diseases has led to recommendations for reducing the fat content of the diet of these countries. Although dietary fats may differ in their effects on chronic diseases, reduction of total fat probably is easier to implement than altering the proportions of different types of fat, and may also be more appropriate in the present state of our knowledge.

Dietary fat is a major source of the energy required for normal functioning of the body; the remainder is provided mainly by carbohydrate and protein. Rough estimates of the relative amounts of these components in the food supplies of different countries can be obtained from data collected by the Food and Agriculture Organization of the United Nations (1).

Analysis of such data shows that the proportions of dietary fat and carbohydrate vary over a wide range in the diets of different nations, while protein is much more constant, averaging in most countries between 9 and 15% of total calories. In contrast, the percentage of calories from fat varies more than four-fold, from less than 10% to more than 40% of total calories (Table 1). Because the level of protein is fairly constant, carbohydrate tends to vary inversely with the fat content of the diet.

It can be seen from Table 1 that the diets of Western industrialized nations contain a very high proportion of fat compared with most of the rest of the world. Recently there has been increasing concern that these high-fat diets may be at least partly responsible for the high incidence of cardiovascular disease, cancer and other types of chronic disease. Such diets also may help to increase the prevalence of obesity since fat is a very concentrated source of calories.

In considering the possible role of dietary fat as a factor in the development of chronic diseases, attention has been focused on its composition and the total amount in the diet. Dietary fat consists mainly of triacylglycerols (triglycerides), but may also contain phospholipids, glycolipids and micronutrients such as sterols and fat-soluble vitamins. The overall composition of dietary fat is determined largely by the fatty acid components of the triacylglycerols. Some of these fatty acids, including the common saturated fatty acids such as palmitic acid and stearic acid and the monounsaturated oleic acid, can be synthesized in the body, whereas n-6 polyunsaturated fatty acids such as linoleic acid, and n-3 fatty acids such as linolenic acid and eicosapentaenoic acid, are derived solely from the diet. *Trans* fatty acids are produced by microorganisms in the rumen, and fatty acids with various branched-chain or ring structures can be synthesized by bacteria, but such fatty acids normally are not produced by animals. Levels of these fatty acids in the tissues of humans and non-ruminant animals thus are dependent on the amounts present in the diet (2).

The following discussion will be concerned with possible effects of variations in amount and type of dietary fat on cancer, obesity and some other diseases, with major emphasis on the amount of dietary fat and its fatty acid composition.

POSSIBLE ADVERSE EFFECTS OF HIGH-FAT DIETS

Dietary Fat and Cancer

Experiments on animals over the past 40-50 years have provided evidence that high-fat diets promote the development of tumors of the skin and mammary gland (3–6), colon (4–6), pancreas (6) and prostate (5–7). This evidence is complemented by epidemiological data showing a strong positive correlation between mortality from cancer at these and other sites and the amount of fat in the diets of different countries (3–6). This is illustrated in Figure 1, which shows the correlation between breast cancer mortality and the percentage of calories derived from dietary fat.

In studies on experimental cancer, the results are influenced by the type of fat in the diet, with polyunsaturated fats being more effective than saturated fats. This appears to be related to a requirement for n-6 polyunsaturated fatty acids (8,9). When adequate amounts of n-6 fatty acid are present, tumorigenesis can be enhanced further by increasing the total amount of fat in the diet (9). This may be why cancer mortality in human populations typically shows a better correlation with total dietary fat than with any particular type of fat (10).

TABLE 1
Percentage of Calories as Fat in the Diets of Different Countries by the Food and Agriculture Organization of the United Nations (1)

5-10%	10-15%	15-20%	20-25%	25-30%	30-35%	35-40%	40-45%
Bangladesh	Afghanistan	Albania	Belize	Antiqua	Argentina	England	Austria
Bhutan	Angola	Algeria	Benin	Barbados	Australia	Faeroe Islands	Belgium
Burundi	Bolivia	Brazil	Botswana	Bulgaria	Bahamas	Finland	Canada
Kampuchea	Burma	Brunei	Cameroon	French	Bermuda	France	Denmark
Democrat	China	Cape Verde	Central African	Polynesia	Cyprus	Iceland	Federal Republic
Korea,	Congo	Chad	Republic	Libya	Czechoslovakia	New Zealand	of Germany
Democratic	Ethiopia	Chile	Costa Rica	Macau	Greece	Spain	Netherlands
People's	Haiti	Columbia	Dominica	Malta	Hong Kong	Sweden	Norway
Republic of	India	Comoros	Dominican	Mongolia	Hungary	Vanuatu	Switzerland
Korea,	Indonesia	Cuba	Republic	Netherlands	Ireland	West Germany	U.S.A.
Republic of	Iraq	Egypt	Fiji	Antilles	Israel		
Rwanda	Lao	Equador	French Guiana	New Caledonia	Italy		
Thailand	Lesotho	Gabon	Gambia	Portugal	Poland		
	Madagascar	Ghana	Grenada	St. Kitts-Nevis-	Samoa		
	Malawi	Guatamala	Guadeloupe	Anguilla	Somalia		
	Malaysia	Guinea	Jamaica	St. Lucia	Uruguay		
	Mozambique	Guinea-Bissau	Japan	Sao Tome &			
	Nepal	Guyana	Liberia	Principe			
	Niger	Honduras	Martinique	Tonga			
	Phillipines	Iran	Mauritania	USSR			
	Tanzania	Ivory Coast	Mauritius				
	Uganda	Jordan	Mexico				
	Vietnam	Kenya	Nambia				
	Yeman Arab	Lebanon	Nicaragua				
	Republic	Maldives	Panama				
	Zaire	Mali	Paraguay				
		Morocco	Reunion				
		Nigeria	Romania				
		Pakistan	Saudia Arabia				
		Papua	Senegal				
		New Guinea	Sierre Leone				
		Peru	Singapore				
		San Salvador	Soloman Islands				
		St. Vincent	Sudan				
		South Africa	Surinam				
		Srilanka	Trinidad &				
		Swaziland	Tobago				
		Syria	Tunisia				
		Togo	Venezuela				
		Turkey	Yugoslavia				
		Upper Volta					
		Yemen, People's					
		Republic of					
		Zambia					
		Zimbabwe					

In contrast with the promoting effects of vegetable oils that contain polyunsaturated fatty acids belonging mainly to the n-6 family, fish oils containing n-3 fatty acids have a little effect on tumorigenesis and may be inhibitory at higher levels of intake (11). Ingestion of fish oils and other marine lipids may help to reduce the risk of cancer in some populations such as the Eskimos and Japanese, whose intake is relatively high, but in general it appears that the amount of fat in the diet is more closely related to cancer in humans than the type of fat.

Various mechanisms have been suggested to account for the effects of dietary fat on tumorigenesis (3–6). Polyunsaturated fatty acids may exert their effects by altering the composition and properties of cellular membranes, which in turn may affect the responses of cells to hormones, antigens or other stimuli of cellular proliferation. Polyunsaturated fatty acids are the precursors of many biologically active eicosanoids such as prostaglandins,

thromboxanes and leukotrienes. Promotion of mammary cancer by dietary fat has been shown to be counteracted by indomethacin, an inhibitor of prostaglandin biosynthesis, suggesting that the effect may be mediated by such compounds.

Different mechanisms may be involved in the promoting effects associated with total dietary fat, and these may not be the same for each type of cancer. For example, dietary fat may enhance the development of colon tumors by stimulating bile acid secretion because there is evidence that bile acids are tumor promoters (5).

As a general mechanism, high-fat diets may promote tumorigenesis by increasing caloric intake. It is known that tumorigenesis can be inhibited by restricting caloric intake (12,13), even when high-fat diets are used, but it is less certain that the increased tumor yields in animals fed high-fat diets can be explained entirely on the basis of overnutrition.

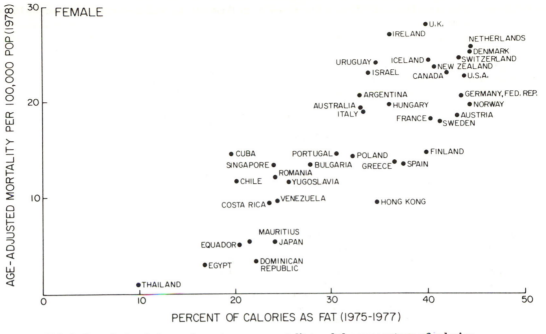

FIG. 1. Correlation between breast cancer mortality and the percentage of calories from dietary fatty. Reproduced from chapter by Carroll (6) with permission of the publisher.

Dietary Fat and Obesity

High-fat diets may contribute significantly to obesity, which is relatively common in the sedentary populations of industrialized countries. Because high-fat diets have a higher caloric density and may also be more appealing than low-fat diets, they encourage excessive intake of calories. When this is coupled with lower levels of physical activity, the excess energy is likely to be deposited as fat.

Data from life insurance companies have shown that overweight and obesity are associated with decreased life expectancy (14). People with excess weight are evidently at increased risk from a number of chronic diseases; maintenance of normal body weight is thus a desirable goal. It is difficult to limit one's food intake voluntarily in the midst of plenty, but this may be made easier by choosing foods that are lower in fat and higher in complex carbohydrates. Such foods are bulkier, and hunger may be satisfied with lower caloric intake. It is a challenge to industry to produce foods of this type that can compete in terms of sensory appeal and acceptability with others that have a higher caloric density.

Dietary Fat and Other Chronic Diseases

Heart attacks and strokes due to atherosclerotic vascular disease continue to be the major causes of death and illness in Western society. Dietary guidelines issued periodically by the American Heart Association to reduce the risk of cardiovascular disease have included recommendations to reduce dietary fat intake, with particular emphasis on saturated fat (15). Similar recommendations have been made by various other committees and organizations, both in the U.S. and in other countries (16–18). The idea of dietary intervention for whole populations, is however, not universally accepted (19).

Other chronic diseases are much more prevalent in Westernized countries than in developing or third-world countries (20); much of this difference has been attributed to diet. Dietary fat may have a significant effect on some of these so-called Western diseases such as hypertension, diabetes mellitus (type II) and cholesterol gallstones.

There is evidence that decreasing the level of dietary fat produces some lowering of blood pressure (21). Low-fat, high-carbohydrate, high-fiber diets also are accompanied by a marked reduction in the insulin requirements of diabetics as well as a lowering of plasma glucose and cholesterol levels (22,23).

Diet appears to be an important factor in the formation of cholesterol gallstones, but the details are far from clear (24–27). Although gallstones have been linked to obesity and excess energy intake, they also occur frequently in slim individuals, while many obese people are free of gallstones (24). In a case-control study, Scragg et al. (27) observed that fat intake was higher in young gallstone patients but was not a risk factor after middle age. There is some evidence that dietary polyunsaturated fats may promote the development of gallstones, but this has not been observed consistently (25,28).

EFFECTS OF SPECIFIC TYPES OF DIETARY FAT

The foregoing discussion has been concerned mainly with effects on health and diseases of the amount rather than the type of dietary fat. However, the relative merits and disadvantages of different types of dietary fat also have been the subject of much investigation and debate.

Concern over the high mortality from cardiovascular

disease has given rise to recommendations to replace saturated fat in the diet by polyunsaturated fat (15). These were based on association of high saturated fat intake with coronary heart disease in epidemiological studies and on evidence that saturated fats raise the level of serum cholesterol, whereas polyunsaturated fats reduce it as reflected in the equations of Keys and Hegsted (29,30). Monounsaturated fat was considered to be relatively neutral with respect to its effect on serum cholesterol.

More recent studies, however, have provided evidence that the monounsaturated fatty acid, oleic acid, lowers plasma levels of cholesterol in the atherogenic, low density lipoprotein (LDL) fraction as effectively as the polyunsaturated fatty acid, linoleic acid. Moreover, oleic acid is less likely to reduce the level of cholesterol in the protective high density lipoprotein (HDL) fraction (31). It also has been reported (32) that a diet high in monounsaturated fat is at least as effective in lowering LDL cholesterol as a low-fat diet and, unlike the low-fat, high-carbohydrate diet, it does reduce HDL cholesterol levels or increase plasma triglyceride levels.

These findings suggest that monounsaturated fat may be preferable to polyunsaturated fat as a means of reducing the risk of cardiovascular disease. It also may be more desirable for other reasons. As indicated earlier, polyunsaturated fats promote carcinogenesis in experimental animals and may increase the likelihood of gallstone formation. Polyunsaturated fatty acids also are more susceptible to oxidative changes than monounsaturated fatty acids. The consumption of relatively large quantities of olive oil with its high content of oleic acid in Mediterranean countries may help to account for the relatively low incidence of cardiovascular disease (33) and cancer (34) in that region.

The observation that Greenland Eskimos are relatively free of cardiovascular disease led to studies that showed that dietary fish oils effectively lower plasma lipids, particularly triglycerides (35,36), and inhibit blood clotting (37). The effects of dietary fish oils on experimental cancer in animals were referred to earlier.

The effects on blood clotting are thought to be due to competition between n-3 and n-6 fatty acids leading to differences in production of prostacyclins and thromboxanes (37,38). There are indications that the differing effects of dietary vegetable and fish oils on carcinogenesis also may be related to differences in eicosanoid production. Eicosanoids being implicated in a variety of inflammatory and autoimmune disorders has led to studies of the effects of dietary fish oils on chronic diseases such as multiple sclerosis and rheumatoid arthritis, which are rare in the Greenland Eskimo population, but more research is required to define the potential benefits. Meanwhile, the possibility of adverse effects should be kept in mind. The n-3 fatty acids in fish oils are even more susceptible to oxidative changes than the n-6 fatty acids in polyunsaturated vegetable oils (39), but it seems unlikely that the products of these oxidative changes promote carcinogenesis, because tumor yields in animals fed fish oils generally are lower than in those fed polyunsaturated vegetable oils (11). Whether fish oils have other deleterious effects, the odors and off-flavors due to autoxidation products present problems in their use as dietary components.

Trans-fatty acids are synthesized only by ruminants, so the diet is the only source for nonruminants. Since *trans* fatty acids are formed in large quantities during the commercial hydrogenation of fats, they can be a significant component of human diets, and concerns about their safety have been expressed from time to time. Most of the *trans* fatty acids are monoenoic although the diet may contain small amounts of dienoic *trans* acids (40).

In terms of their effects in serum cholesterol levels and atherosclerosis, *trans* fatty acids appear to be intermediate between saturated fatty acids and *cis* monoenoic fatty acids (40,41). *Trans* fatty acids do not seem to promote tumorigenesis in experimental animal models to any greater extent than their *cis* counterparts or saturated fatty acids (42,43). An ad hoc panel selected by the Federation of American Societies for Experimental Biology concluded that there is little reason for concern over the safety of *trans* fatty acids in the American diet but recommended further studies to understand better their properties and effects (44).

NUTRITIONAL RECOMMENDATIONS REGARDING DIETARY FAT

Nutritionists often recommend a varied diet as a means of avoiding excesses or deficiencies of specific nutrients that may have undesirable effects. The association of high-fat diets with a number of chronic diseases and that Western industrialized countries are at the high end of the spectrum in level of dietary fat (Table 1) suggest the desirability of altering the diets to reduce this excess of dietary fat. Essentially, this means replacing high fat foods in the diet by others that are lower in fat and higher in carbohydrate because these countries often already are at the high end of the scale in terms of protein intake (1).

Some of the dietary guidelines that have been issued have suggested that dietary fat be reduced to 30% or less of total calories (4,15–18). This is obviously an arbitrary figure that is still high in relation to the level in many national diets (Table 1). It perhaps can be viewed as an interim goal designed to achieve a gradual reduction in the fat content of the diet. Such changes should be introduced gradually to minimize the economic impact on agriculture and the food industry.

In attempting to reduce dietary fat, the question arises whether to aim for a general reduction or to concentrate on reduction of specific types of dietary fat. Those concerned with cardiovascular disease have emphasized reducing saturated fat and increasing polyunsaturated fat (15,16,18), but the recent report of Grundy (32) suggests that it might be better to replace saturated fat in the diet with monounsaturated fat. Organizations concerned with cancer risk have focused on reductions in total dietary fat without specifying particular types of fat (4,16,45). This also may help to prevent obesity and probably is easier to implement than reductions involving changes in the relative proportions of different types of dietary fat. Studies based on data from the Greenland Eskimo population have indicated that a high intake of fish oils containing n-3 polyunsaturated fatty acids can reduce the risk of various chronic diseases, but further investigation is needed to assess the effects of more moderate intakes of this type of fatty acids.

The major stimulus for recommending dietary change is the knowledge that morbidity and mortality from cardiovascular disease, cancer and other chronic diseases are much higher in Western, industrialized countries than in many other parts of the world. The fact that migrants to these countries show an increased incidence of such diseases indicates that the observed differences are related more to the environment than to heredity (46,47). Furthermore, experiments on animals in combination with epidemiological data have provided evidence that diet is one of the most important variables affecting the incidence of these chronic diseases.

Recommendations for dietary change thus have been aimed primarily at reducing the risk of chronic diseases. It is not possible in our present state of knowledge to be certain that the recommended changes will have the desired effects. However, diet is something that can be altered, and there is reasonable hope that the proposed changes will be beneficial. If other means of preventing premature death from chronic diseases can be identified, they certainly should be advocated but, in the meantime, it seems worthwhile to pursue the dietary option. To do nothing is to accept that death and disability from such diseases will continue to be much higher in Western industrialized nations than in other parts of the world.

REFERENCES

1. Food and Agriculture Organization of the United Nations, *Food Balance Sheets, 1975-77 Average and per Caput. Food Supplies 1961-65 Average, 1967 to 1977*, Rome, 1980.
2. Carroll, K.K., *J. Am. Oil Chem. Soc. 42*:516 (1965).
3. Carroll, K.K., and H.T. Khor, *Prog. Biochem. Pharmacol. 10*:308 (1975).
4. Committee on Diet, Nutrition, and Cancer, *Diet, Nutrition, and Cancer*, National Academy Press, Washington, D.C., 1982.
5. *Diet, Nutrition, and Cancer: A Critical Evaluation, Vol. I*, edited by B.S. Reddy and L.A. Cohen, CRC Press, Inc., Boca Raton, FL, 1986.
6. *Prog. Clin. Biol. Res. Vol. 222, Dietary Fat and Cancer*, edited by C.Ip, D.F. Birt, A.E. Rogers and C. Mettlin, Alan R. Liss, Inc., New York, 1986.
7. Pollard, M., and P.H. Luckert, *Cancer Lett. 32*:223 (1986).
8. Carroll, K.K., G.J. Hopkins, T.G. Kennedy and M.B. Davidson, *Prog. Lipid Res. 20*:685 (1981).
9. Ip, C., *Am. J. Clin. Nutr. 45*:218 (1987).
10. Carroll, K.K., L.M. Braden, J.A. Bell and R. Kalamegham, *Cancer 58*:1818 (1986).
11. Carroll, K.K., *Lipids 21*:731 (1986).
12. Kritchevsky, D., and D.M. Klurfeld, *Am. J. Clin. Nutr. 45*:236 (1987).
13. Pariza, M.W., *Am. J. Clin. Nutr. 45*:261 (1987).
14. Simopoulos, A.P., *Am. J. Clin. Nutr. 45*:271 (1987).
15. Nutrition Committee, American Heart Association, *Circulation 74*:1465A (1986).
16. O'Connor, T.P., and T.C. Campbell, in *Prog. Clin. Biol. Res. Vol. 222 Dietary Fat and Cancer*, edited by C. Ip, D.F. Birt, A.E. Rogers and C. Mettlin, Alan R. Liss, Inc., New York, 1986, pp. 731-771.
17. Consensus Conference, *J. Am. Med. Assoc. 253*:2080 (1985).
18. Study Group, European Atherosclerosis Society, *Eur. Heart J. 8*:77 (1987).
19. Olson, R.E., *J. Am. Med. Assoc. 255*:2204 (1986).
20. *Western Diseases: Their Emergence and Prevention*, edited by H.C. Trowell and D.P. Burkitt, Edward Arnold, London, 1981.
21. Puska, P., J.M. Iacono, A Nissinen, E. Vartianen, R. Dougherty, P. Pietinen, U. Leino, U. Uusitalo, T. Kuusi, E. Kostainen, T. Nikkari, E. Seppala, H. Vapaatalo and J.K. Huttunen, *Prev. Med. 14*:573 (1985).
22. Simpson, H.C.R., and J.I. Mann. *Adv. Nutr. Res. 7*:39 (1985).
23. Anderson, J.W., and N.J. Gustafson, *Geriatrics 41*:28 (1986).
24. Low-Beer, T.S., *Proc. Nutr. Soc. 44*:127 (1985).
25. Bennion, L.J., and S.M. Grundy, *New Engl. J. Med. 299*:1161, 1221 (1978).
26. Bouchier, I.A.D., *Gut. 25*:1021 (1984).
27. Scragg, R.K.R., A.J. McMichael and P.A. Baghurst, *Br. Med. J. 288*:1113 (1984).
28. Carroll, K.K., in *Adv. Exp. Med. Biol., Vol. 83, Function and Biosynthesis of Lipids*, edited by N.G. Bazan, R.R. Brenner and N.M. Giusto, Plenum Press, New York, 1977, pp. 535-546.
29. Keys, A., and R.W. Parlin, *Am. J. Clin. Nutr. 19*:175 (1966).
30. McGandy, R.B., and D.M. Hegsted in *The Role of Fats in Human Nutrition*, edited by A.J. Vergroesen, Academic Press, New York, 1975, pp. 211-230.
31. Mattson, F.H., and S.M. Grundy, *J. Lipid Res. 26*:194 (1985).
32. Grundy, S.M., *New Engl. J. Med. 314*:745 (1986).
33. *Circulation 41*, edited by A. Keys, Suppl. 1, 1970.
34. Cohen, L.A., D.O. Thompson, Y. Maeura, K. Choi, M.E. Blank and D.P. Rose, *J. Nat. Cancer Inst. 77*:33 (1986).
35. Goodnight, S.H. Jr., W.S. Harris, W.E. Connor and D.R. Illingworth. *Arteriosclerosis 2*:87 (1982).
36. Herold, P.M., and J.E. Kinsella, *Am. J. Clin. Nutr. 43*:566 (1986).
37. Dyerberg, *J. Nutr. Rev. 44*:125 (1986).
38. Bunting, S., S. Moncada and J.R. Vane, *Br. Med. Bull. 39*:271 (1983).
39. Cho, S-Y, K. Miyashita, T. Miyazawa, K. Fujimoto and T. Kanida, *J. Am. Oil Chem. Soc. 64*:876 (1987).
40. Gottenbos, J.J., in *Dietary Fats and Health*, edited by E.G. Perkins and W.J. Visek, American Oil Chemists' Society, Champaign, IL, 1983, pp. 375-390.
41. Beare-Rogers, J.L., *Adv. Nutr. Res. 5*:171 (1983).
42. Selenskas, S.L., M.M. Ip and C. Ip, *Cancer Res. 44*:1321 (1984).
43. Ip, C., M.M. Ip and P. Sylvester, in *Prog. Clin. Biol. Res., Vol. 222, Dietary Fat and Cancer*, edited by C. Ip, D.F. Birt, A.E. Rogers and C. Mettlin, Alan R. Liss, Inc., New York, 1986, pp. 283-294.
44. Anon. *J. Am. Oil Chem. Soc. 62*:1622 (1985).
45. ECP-IUNS Workshop on Diet and Human Carcinogenesis, *Nutr. Cancer 8*:1 (1986).
46. Haenszel, W., and M. Kurihara, *J. Nat. Cancer Inst. 40*:43 (1968).
47. Gori, G.B., *Nutr. Cancer 1*:5 (1978).

Dilemma of Patenting for Oilseed Breeders and Biotechnologists

R.K. Downey

Agriculture Canada, Research Station, 107 Science Crescent, Saskatoon, Saskatchewan S7N OX2 Canada

In most countries, the rapid biotechnological advances in crop plants, and oil crops in particular, has outpaced the present policy and legal framework designed to handle intellectual property rights. The present legal interpretation of what biotechnology innovations are patentable is still evolving and varies widely from country to country, with the U.S. and Japan having the most liberal interpretations. One of the major problem areas for crop plants is at the interphase between patent legislation and plant breeders' rights (PBR), the latter being the proprietary right to a crop variety as legislated under the International Union for the Protection of New Varieties of Plants (UPOV). Some have advocated replacing PBR with patent law. The vast majority of plant breeders believe that the patenting of sexually propagated plant varieties is neither appropriate nor practical. However, some system is required to provide proprietary protection for engineered genes inserted into crop plants. Similarly, biotechnology must recognize the value and rights associated with protected varieties that are the vehicle for the commercialization of the patented gene. Fortunately, the legal and moral dilemmas presented by biotechnology manipulation of plants have and are continuing to bring officials into direct contact with interested scientific circles, a process long overdue. These and other current plant gene patenting problems will be discussed, and possible solutions will be presented.

The application of biotechnology to oil-bearing plants has a tremendous potential to increase and diversify the type of oil available to the vegetable oil industry. However, to attract and sustain the very high level of investment required for biotechnology and to extend it over the 8-10 years needed to develop a new or improved oil-bearing plant, adequate worldwide intellectual property protection must be assured. It is clear that the field of biotechnology innovation has far outpaced regulations and policies of patent offices and governments worldwide. Thus, the question of adequate protection of biotechnology inventions is being examined on an urgent basis in almost every industrialized country.

In the medical field, most biotechnical innovations deal with specific proteins and other compounds that are either injected, added to the diet or used as diagnostics. As such, they can and are being accommodated under utility patent law. Microorganisms and plants, on the other hand, reproduce themselves and are less easily accommodated under patent protection. Prerequisites for patenting require that the invention be novel, have utility, not be obvious to those skilled in the art and be described in such a way that others would duplicate the invention (enabling disclosure).

Biological processes are not as accurate or predictable as chemical or physical processes so that the requirement for enabling disclosure normally cannot be met in living matter. This difficulty has been overcome for microorganisms by requiring the patenting applicant to place the organism in a national repository where the organism is available to others to test the claim(s) and build upon the invention. However, there are needs for safeguards against the misuses of deposited materials. Minimum protection should include provisions whereby the deposited materials are not available to third parties before effective protection is granted, if a patent is not granted, the culture must be returned to its depositor in its entirety, and those with access to the deposit are prevented from passing on such materials or their progeny to others or exporting them to foreign countries without the consent of the patent holder. Some even have recommended that for novel microorganisms, the burden of proof be reversed whereby an alleged infringer would have to establish that deposited materials or their offspring were not used in a process or manufacture. Thus, although the repository system appears to be workable, controls on distribution need to be refined.

With sexually propagated plants, a different system of protection has been devised under the convention entitled, "International Convention for the Protection of New Varieties of Plants." This convention, called "UPOV," establishes an international union of countries bound by the convention. National legislation relating to this convention often is called Plant Variety Protection (PVP) or Plant Breeders' Rights (PBR). The UPOV Convention for variety protection was needed first because the requirement for enabling disclosure could not be met. Even if the breeder revealed the exact parentage used to develop the new improved variety, it is extremely unlikely that any other breeder could exactly duplicate the new variety. In addition, the vast majority of varieties are developed using well-known techniques and procedures. Thus, the why and how of most variety improvements are obvious to someone skilled in the art. For these reasons, the granting of a proprietary right for crop plants is based on a different set of criteria; namely, each variety must be uniform, stable and distinct from all other varieties.

The convention also ensures that anyone has the legal right to unrestricted use of any variety as a source of breeding material. This rule (5[3]), which is basically a research exemption, is the basis for all plant improvement. There is a difference in understanding of "research" between plant breeders and patent experts. Patent holders only will condone research on their patents provided the patent item does not become incorporated into anything else. A plant breeder's research, however, is result-oriented (a new variety), and successful research incorporates without recompense the results of others' research. The starting point for any new variety is the presently existing varieties, since they embody the accumulation of all previous breeding improvements. The PVP protected varieties of today are the parents for the superior varieties of tomorrow. Although the free access principle is basic to future conventional breeding, the UPOV Rule (5[3]) in its present form tends to eliminate the required protection for biotechnology innovations incorporated into the protected plant varieties. The matter is further complicated because the UPOV Convention also contains a clause (Rule 2[1]) that prohibits a variety from being protected by both PVP and a patent(s).

PVP has worked well for conventional breeders, particularly small breeding firms. Although some larger firms have called for a stronger form of protection, PVP has numerous advantages. First, it is relatively inexpensive, as well as fast and easy, to obtain a plant variety right when compared with utility patent procedures. Second, protection is available almost immediately so that the holder of a right can utilize the granted monopoly privileges for the full term (17 or 18 years). Third, the access to a PVP right is not affected by prior disclosure.

The application of biotechnology to plant improvement has added a new dimension to variety protection. PVP in its present form offers little or no protection for genetically engineered plant modifications. However, the biotechnologist needs superior varieties in which to incorporate and market his innovations. Ideally, the biotech innovator would like to generate royalties through the licensed use of a patented innovation from as many varieties as would be needed to cover the globe. Indeed, some companies are attempting to do just that. However, under the present laws of most countries only the first transformed variety or varieties will yield a royalty. Upon release of the modified variety, the innovation will be freely accessible to other breeders for incorporation into their own new varieties. Since the average life span of the conventional variety is five to six years, such limited protection probably will be insufficient to attract and retain biotech research dollars for plant improvement.

There also is concern, particularly in Europe, that the first sale of a patented reproducing organism would exhaust the patent rights. It seems logical that if an oilseed, containing a patented gene, is sold for sowing and the harvested seed is used for oil extraction, the patent right is exhausted with the production of the oil and meal. However, the patent right is not or should not be exhausted if the harvested seed is used to obtain another generation.

In addition, it should not be possible to patent the phenotype of a trait, as has been done in the U.S. for the high oleic characteristic in sunflowers. Such phenotypic characteristics cannot be precisely defined and tend to close off whole areas of research. There are many reasons why plant breeders should be encouraged to find alternative genetic means of achieving the same plant expression, whether it be a new oil type or resistance to a disease or a pest.

Most biotechnology developments that one would want to incorporate into a plant variety are controlled by genes, nucleotide sequences or DNA segments that are novel, useful and can be described precisely. Thus, the majority of such innovations would meet the criteria for patenting. Therefore, it would appear that some means of accommodating gene patents and PVP is urgently required. Some groups, particularly in Europe, have called for the abandonment of PVP and UPOV so that plant varieties can be dealt with solely under patent legislation. From a plant breeder's view, such drastic action would be akin to cutting off your nose to spite your face. It would destroy one of the cornerstones of plant improvement, the free access to basic germplasm.

A much more workable solution would be to permit patenting of genes or gene combinations for a specific trait but not extend patenting to a host variety. The variety would be protected by the provisions of UPOV under a modified Rule (2[1]) that would allow patented genetic material to coexist in a PVP protected variety. However, to safeguard access to basic germplasm and to meet the needs of both the breeder and the biotechnologist, a system of automatic compulsory licensing in which the patent is made nonexclusive is required. Compulsory licensing could take effect either immediately upon the granting of a patent or after an initial period of exclusivity. Such a provision should ensure that all breeders have access to the new gene while giving the biotech innovator a broad base upon which to recover his research and development expenses. Compulsory licensing also might solve another potential problem, that of the possibility of the biotechnology developer superceding or usurping a PVP-protected variety by merely adding one patented gene. If this were to happen on a regular basis, the conventional plant breeder, who invests almost as much as the biotechnologist in developing a new superior variety, would receive little or no return for his efforts because the transformed variety, with the additional superior trait, would tend to dominate the market. Compulsory licensing would remove this real or imagined threat and provide the conventional breeder and developer of the genetically engineered trait the necessary protection and income.

These are but a few of the dilemmas facing the plant breeder and the biotechnologist in harnessing these two important disciplines to the betterment of the oilseed industry. I have tried to provide logical solutions to some of the major dilemmas. Whatever instruments are devised by the lawyers to overcome these barriers and concerns, they must be internationally based and accommodate the requirements of the plant breeding process. It should be noted that biotechnology is only one of many tools available to the plant breeder and is an adjunct to, but cannot replace, the conventional system. With this in mind, let's not kill the goose that's laying golden eggs. In addition, we must not lose sight of the fact that in the long run it is the agricultural producer who will have to pay the patent royalties. With the possible exception of producers in northern Europe, nearly all these clients or customers are operating on the narrowest of margins and, indeed, may not be able to afford the latest technology.

Those wishing additional information and insight into this urgent, complex question, may consult the following proceedings: Workshop on Plant Gene Patenting in Canada, Ottawa, Canada, March 22-25, 1987, a Canadian Agriculture Research Council Report. Symposium on the Protection of Biotechnological Inventions, Ithaca, NY, June 4 and 5, 1987, organized by the World Intellectual Property Organization (WIPO) and Cornell University, Department of Agriculture Economics. The WIPO Committee of Experts on Biotechnological Inventions and Industrial Property, Geneva, June 29-July 3, 1987.

Discussion of Session

W.E. Parrish mentioned that biotechnology can use bacteria, yeast or animals; he specifically did not cover plants. This was not to infer there were no potential hazards from genetically engineered plants. He further indicated that all microbial enzymes were treated the same way in toxicological evaluations regardless of their origin. A question to P.S. Elias was "is cocoa butter produced as single cell oil from whey allowed, and are products containing it called chocolate?" His response indicated that one probably could not get away with it. It probably would be treated as single cell protein is, as a food additive. A similar question to the panel in general asked "How would regulators view the production of an oil from a traditional plant source that had been modified to contain a nontraditional fatty acid?" The reply was that one probably would have to submit the oil to a complete toxicity testing protocol including mutagenicity and teratological studies because one also must worry about small amounts of other (minor) components in fats.

K.K. Carroll fielded the last question, which was that because saturated fatty acids are considered as C12:0, 14:0, 16:0 and 18:0, the Keys' equation, however expressly, excludes stearic acid (18:0). What were his thoughts on this? In answer, he stated that 18:0 certainly was a saturated fatty acid from a chemical standpoint. However, scientists always are trying to simplify matters; it probably was left out because it does not have the same cholesterol-elevating properties as the others.

Production of Eicosapentaenoic and Arachidonic acids by the Red Alga *Porphyridium Cruentum*

Zvi Cohen

Algal Biotechnology Laboratory, Jacob Blaustein Institute for Desert Research, Ben-Gurion University of the Negev, Sede-Boqer Campus 84993, Israel

The red microalga *P. cruentum* is a potential source for the pharmaceutically valuable fatty acids eicosapentaenoic acid and arachidonic acid. The conditions leading to a high content of either fatty acid were investigated. Strain 1380-1d produced high eicosapentaenoic acid content (2.4%, w/w) under conditions resulting in a high-growth rate. High arachidonic acid content (2.9%) was obtained under slow growth conditions in the stationary phase or under nitrogen starvation. Strain 1380-1a produced 1.9% w/w arachidonic acid under exponential conditions. By imposing nitrogen starvation, it was possible to obtain a lipid mixture enriched with arachidonic acid and eicosapentaenoic acid.

Arachidonic acid (AA, 20:4w6) and eicosapentaenoic acid (EPA, 20:5w3) are rare fatty acids of potential pharmaceutical value. AA was suggested as a precursor for biosynthetic production of prostaglandin PGE_2 (1). EPA was shown to be effective in preventing blood platelet aggregation (2) and is claimed to be useful for blood cholesterol reduction (3).

EPA is found in marine fish oils, yet the low EPA content and the presence of other fatty acids of less-desired properties have initiated several studies aimed at the production of EPA from microalgae. The marine alga *Chlorella minutissima* (4) and the freshwater alga *Monodus subterraneus* (5) were suggested as potential sources for EPA. AA was shown (6) as present in *Euglena gracilis*, *Ochromonas danica* and in the red microalga *Porphyridium cruentum*, which also contained EPA.

P. cruentum was studied as a source for sulfated polysaccharides (7,8), which could be utilized for enhanced oil recovery from oil wells. It also contained phycoerythrine, a red protein pigment used for immunofluorescent detection of tumors.

We have found that the fatty acid composition of *P. cruentum* was highly dependent on environmental conditions. Contrary to previous reports, EPA was the main polyunsaturated fatty acid (PUFA) found in cultures cultivated at optimal temperature and nonlimiting light conditions. When growth was slowed by increased light intensity, increased cell concentration, nonoptimal temperature, pH or salinity, the content of EPA decreased, and AA became the major PUFA. This work reports the potential of various strains of *P. cruentum* as a source of EPA and AA, and the growth conditions under which maximum productivity of each of these fatty acids can be obtained.

MATERIALS AND METHODS

P. cruentum strains 1380 1a to f and B113.80 were obtained from the Göttingen Algal Culture collection. Maintenance of stock cultures and inocula preparation were performed according to Vonshak (9). Cultures were grown on Jones' medium (10) in a temperature-regulated water bath illuminated with fluorescent lamps providing $170\ \mu E.m^{-2}.s^{-1}$ at the side of the bath. The cultures were mixed by bubbling an air-CO_2 mixture (99:1, v/v) through a sintered glass tube placed in the bottom of each culture tube. For the nitrogen starvation experiments, cultures in the exponential phase of growth (28 C) were centrifuged, washed and resuspended in a nitrogen-free medium. The cultures were kept for an additional three days under the same light and temperature conditions. Cultures were grown to the exponential stage and maintained at steady state by daily dilution for at least three days before the onset of the experiment. Growth rate was estimated by chlorophyll and turbidity measurements.

Fatty Acid Analysis

Gas chromatographic analysis was performed on a SP-2330 fused silica capillary column. Fatty acid methyl esters were identified by cochromatography with authentic standards (Sigma Chemical Co., St. Louis, MO) and by gas chromatography-mass spectrometry (GC-MS). Fatty acid contents were determined by comparing their peak areas with that of the internal standard. The data shown are mean values of at least two independent samples, each analyzed in duplicate.

RESULTS

Under optimal conditions, the seven strains of *P. cruentum* had similar growth rates, ca. 1.1 d^{-1}, (Table 1). The EPA content ranged from 1.85% to 2.44% of ash-free dry weight (AFDW). The range of AA/EPA values in the exponential phase (0.35-0.47) was lower than the range at the stationary phase (0.9-2.1) because AA increased, and the quantity of EPA decreased. The highest EPA content was found in strains 1380-1b and 1d. At 30 C, exponentially cultivated strains contained ca. 39.2% AA and 12.1% EPA on average (Table 2). However, total fatty acid content was in most cases the highest in the stationary phase at 30 C. In strain 113.80, total fatty acid content increased from 4.2% in the exponential phase to 7.7% (AFDW) in the stationary phase. The R-value (11.5) at the stationary phase in strain 113.80 was the highest value observed.

Nitrogen Starvation

When the exponentially cultivated *P. cruentum* strain 113.80 was transferred to a nitrogen-free medium, the total fatty acid content increased to 8.8% (AFDW) after four days. The fatty acid composition resembled that observed at the stationary phase. The content of AA increased to 2.9% and that of EPA decreased to 0.9% (AFDW).

The major effect of the nitrogen starvation on the lipid concentration of *P. cruentum* strain 113.80 was the increase in neutral lipids from 20.6% in the control cul-

TABLE 1

Fatty Acid Composition and Content in *P. cruentum* Strains Cultivated at 25 C

Strain	Growth phase[1]	μ^2 (d^{-1})	Total	AA	EPA	AA/EPA
		d^{-1}	\multicolumn{4}{l}{% w/w ash-free dry wt.}			
1380-1a	E	1.05	5.34	0.84	1.89	0.6
	S	—	5.83	1.91	1.12	1.7
1380-1b	E	1.13	5.61	0.86	2.44	0.4
	S	—	5.36	1.58	1.24	1.3
1380-1c	E	1.13	5.38	0.91	2.13	0.4
	S	—	5.35	1.27	1.48	0.9
1380-1d	E	1.15	5.47	0.81	2.41	0.3
	S	—	5.05	1.24	1.36	0.9
1380-1e	E	1.39	5.34	0.98	2.10	0.5
	S	—	5.65	1.89	0.92	2.1
1380-1f	E	1.10	5.15	0.91	2.06	0.4
	S	—	5.10	1.69	0.91	1.8
113.80	E	0.98	4.44	0.80	1.85	0.4
	S	—	5.10	1.39	1.30	1.1

[1]E, exponential phase, 4-5 mg-1^{-1} chlorophyll; s, stationary phase, 27-30 mg-l^{-1} chlorophyll.
[2]Growth rate, turbidity.

TABLE 2

Fatty Acid Composition and Content in *P. cruentum* Strains Cultivated at 30 C

Strain	Growth phase	μ	Total	AA	EPA	AA/EPA
		d^{-1}	\multicolumn{4}{l}{% w/w ash-free dry wt.}			
1380-1a	E	0.67	4.84	1.95	0.32	6.1
	S	—	6.23	2.18	0.60	3.7
1380-1b	E	0.79	3.78	1.49	0.54	2.8
	S	—	4.67	1.95	0.31	6.4
1380-1c	E	0.80	4.42	1.80	0.52	3.4
	S	—	5.42	2.01	0.82	2.3
1380-1d	E	0.70	5.39	1.78	1.02	1.7
	S	—	4.74	1.84	0.40	4.5
1380-1e	E	0.76	4.30	1.78	0.55	3.2
	S	—	6.48	2.45	0.26	9.4
1380-1f	E	0.49	4.46	1.85	0.55	3.3
	S	—	5.53	2.23	0.29	7.8
113.80	E	0.62	4.19	1.62	0.30	5.2
	S	—	7.67	2.48	0.22	11.5

TABLE 3

Effect of Nitrogen Starvation on the Fatty Acid Composition in *P. cruentum* Strains T13.80[1]

| | \multicolumn{6}{c}{Lipid} | | | | | |
| | \multicolumn{2}{c}{NL[2]} | | \multicolumn{2}{c}{GL[3]} | | \multicolumn{2}{c}{PC[4]} | |
Statistic	+N[5]	–N[6]	+N	–N	+N	–N
EE/APA	2.4	7.6	0.1	0.2	2.7	21.4
% of Total	20.6	61.0	66.6	21.8	7.7	8.7

[1]Cultivated at 28 C, 170 μE·m·s^{-1}.
[2]Neutral lpipids.
[3]Glycolipids monogalactosyldiacylglycerol, digalactosyldiacylglycerol plus sulfolipid.
[4]Phosphatidylcholine.
[5]Nitrogen sufficient.
[6]Three-day nitrogen starvation.

DISCUSSION

EPA Production

If one is to rate the strains of *P. cruentum* in terms of their potential to produce EPA, three factors should be primarily considered: the growth rate under optimal conditions, the EPA content and the R-values at the exponential and at the stationary phases. The combination of growth rate and EPA content yields the EPA production rate. In an unpublished study, it was shown that under optimal growth conditions both the EPA content and the growth rate are maximal leading to the highest EPA production rate. The R-value at the exponential phase reflects the extent of AA "contamination" and the degree of difficulty in EPA purification. The lower the R, the easier the separation of EPA from AA. The differences in the corresponding R-values for the exponential and stationary phases in the various strains could be utilized as an indicator of EPA contents under other growth conditions.

On the basis of these data, strain 1380-1d was judged to be the most promising as far as EPA production and relative purity are concerned. It had the highest content of EPA (2.4% of AFDW) and a low R-value (0.33) at the exponential phase, which increased to only 0.90 in the stationary phase.

AA Production

High AA contents were obtained by inducing light limitation (low light or high cell concentration) and under increased temperature, reduced pH or increased salinity (Z. Cohen, A. Vonshak and A. Richmond, unpublished data). The intensities of these effects was found to be maximal at the stationary phase. Cultivation under low pH or high salinity resulted however in sharp reductions in the total fatty acid content.

For large-scale production of AA in outdoor ponds of *Porphyridium*, cultivation at 30 C presents a favorable option, as one stage exponential cultivation is possible. Under these conditions, strain 1380.1a had the highest AA content (1.95%). However, cultivation at this temperature outdoors could be prohibitively expensive in the winter.

ture to 61.0% in the N-starved culture with a concurrent decrease in the glycolipids. A major increase in AA was noted in the neutral lipids and in phosphatidylcholine. Both AA and EPA decreased proportionally in the glycolipids. The combined fraction containing neutral and phospholipids was rich in AA (33.0% of total fatty acid). The glycolipids remained rich in EPA and low in AA.

AA enrichment also could be achieved by nitrogen starvation. In N-starved *P. cruentum* cultures, the content of EPA in the various lipids decreases, while that of AA increases in neutral lipids and in phospholipids but not in glycolipids. This is contrary to the situation found when growth is retarded by other means (i.e light, temperature, pH) and where the content of AA increases in all lipids. The practical outcome of this phenomenon is that while the increases in R-values in the neutral lipids and in PC result in a relatively pure AA in these lipids, the glycolipids are still low in AA and could be separated and used as a source of a high-quality EPA.

The cultivation of *P. cruentum* as a source for EPA or AA offers several advantages over other algal alternatives. The alga is cultivated on a marine medium, which is especially advantageous where freshwater is scarce. Moreover, the cultivation on this high salinity medium may provide an ecological niche, thus aiding the maintenance of monoalgal culture. Harvest is less problematic because the alga have a tendency for autofloculation, which could be enhanced by pH reduction (11). Another important factor in the assessment of the feasibility of large-scale *Porphyridium* production is the possibility to obtain several products from the biomass. The presence of sulfated polysaccharides and phycoerythrine also should encourage outdoor cultivation of *Porphyridium*.

ACKNOWLEDGMENTS

I thank Shoshana Didi for algal cultivation and chemical analyses. This study was supported by the Israeli National Council for Research and Development.

REFERENCES

1. Ahern, T.J., *J. Amer. Oil Chem. Soc. 61*:1754 (1984).
2. Dyerberg, J., *Nutr. Rev. 44*:125 (1986).
3. Kromhout, D., E.B. Bosschieter and C. de L. Coulander, *New Eng. J. Med. 312*:1205 (1985).
4. Seto, A., H.L. Wang and C.W. Hesseltine, *J. Amer. Oil Chem. Soc. 61*:892 (1984).
5. Iwamoto, H., and S. Sata, *Ibid. 63*:434 (1986).
6. Nichols, B.W., and R.S. Appelby, *Phytochemistry 8*:1907 (1969).
7. Anderson, D.B., and D.E. Eakin, *Biotech. and Bioeng. Symp. No. 15*:533 (1985).
8. Thepenier, C., and C. Gudin, *Biomass 7*:225 (1985).
9. Vonshak, A., in *Handbook of Microalgal Mass Culture*, edited by A. Richmond, CRC Press, Boca Raton, FL.
10. Jones, R.E., L. Speer and W. Kury, *Physiol. Plant 16*:636 (1983).
11. Vonshak, A., Z. Cohen and A. Richmond, *Biomass 8*:13 (1985).

Comparison of Biointeresterification and Conventional Processes for the Preparation of Vanaspati and Other Valuable Products

M.M. Chakrabarty, S. Ghose Chaudhuri, S. Khatoon and **A. Chatterjee**
Centre of Advanced Studies on Natural Products, Department of Chemistry, University Colleges of Science and Technology, 92, Acharyya Prafulla Chandra Road, Calcutta-700 009, India

Extracellular microbial lipases have been used in recent years to catalyze interesterification of oils and fats. Use of specific lipases gives products that are unobtainable by chemical interesterification methods. These processes also have been used for products normally produced by chemical methods in commerce. Vanaspati, a partially hydrogenated plastic fat, is used in India for cooking and as stearin. Margarine fat bases are produced by hydrogenation or chemical interesterification. It has been possible to prepare similar fats by biointeresterification using lipase from *Candida cylindracea*. Mowrah fat and a blend of palm stearin with soybean oil were interesterified. Substantial similarity in product characteristics were observed compared with those from chemical interesterification.

Biointeresterification has been gaining importance as an alternative to chemical interesterification processes for making plastic fats from vegetable oils. The present work describes the preparation of polyunsaturated fatty acid vanaspati and margarine base fat from selected vegetable oils by biointeresterification using *Candida cylindracea* lipase as a catalyst (1). The results of the biointeresterification process were compared with the commercial chemical interesterification processes.

MATERIALS AND METHODS

Biointeresterification was carried out by stirring 60 g of oil at 37.5 C with 5% w/w alumina, 0.85 gm lipase from *Candida cylindracea* obtained from Sigma Chemical Co. (St. Louis, MO) (1 mg enzyme equivalent to 700 units) and 0.6 ml of 1M Tris/HCL buffer at pH 6.0. Products were collected after 24 or 48 hr. Reaction products were dissolved in 40 ml of diethyl ether and filtered through Whatman No. 1 filter paper. The filtrate was dried with sodium sulphate. The filtrates containing the interesterified fats were analyzed for fatty acid and mono- and diacylglycerol content by standard methods (2). The material was then subjected to miscella refining with hexane and ethanolic caustic to remove fatty acids and mono-acylglycerol. Traces of diacylglycerol, however, remained. The refined fat then was analyzed for solid fat indices (3) and slip point by AOCS methods (4). All oil samples were gifts from M/s. Rasoi Vanaspati and Oil Industries, Calcutta, and were analyzed in the laboratory before use. Chemically catalyzed interesterification was carried out on refined oils with NaOMe (0.25% v/v) for 45 min (5). Original and interesterified fats and oils from mowrah (*Maduca latifolia*), palm (*Elaeis guineeneis*), and soybean (*Glycine max*) were analyzed in this study.

Results and Discussion

The initial fatty acid compositions of fats from mowrah, soybean and palm stearin are in Table 1. Mowrah fat

TABLE 1

Fatty Acid Composition of Fats and Oils

Fat	\%w/w 16:0	18:0	20:0	18:1	18:2	18:3
Mowrah	25.3	17.5	0.7	40.8	15.3	0.4
Palmstearin	60.2	3.3	0.0	32.1	4.4	0.0
Soybean	14.1	0.8	2.4	22.5	52.2	8.0

biointeresterified for 24 hr with *Candida cylindracea* lipase yielded products comparable with those produced by conventional interesterification with NaOMe for 45 min (Table 2). Bioesterification for 48 hr resulted in a higher melting point and significantly increased SFI, particularly at elevated temperatures. With a 4:6 (w/w) blend of palm stearin and soybean oil, bioesterification for 48 hr gave products with nearly identical properties as those from conventional methods. However, the time taken in the biointeresterification process (24-48 hr) compared with the conventional process (45 min) was a serious disadvantage. Even so, the physical properties of the products of biointeresterification simulated vanaspati fats and were suitable as a margarine fat base.

Therefore, the present work confirmed that biointeresterification of individual fats or blends of fats could be achieved with *Candida cylindracea* lipase. On the basis

TABLE 2

Solid Fat Indices and Slip Points of Mowrah Fat and Products Obtained by Convential Interesterification or Biointeresterification Candida using *Cylinderacea* Lipase

Treatment	Slip point C	Solid fat index (ml/g) 15 C	20 C	25 C	30 C	35 C	40 C
Control	30.0	27.4	8.0	1.2	0.7	ND	ND
Conventional interesterification with sodium methoxide (0.25% for 45 min)	34.0	19.1	6.1	2.4	1.3	0.1	ND
Biointeresterification (24 hr)	34.5	23.5	7.0	3.7	2.7	0.4	ND
Biointeresterification (48 hr)	35.2	19.1	15.2	10.4	7.7	4.0	1.3

TABLE 3

Solid Fat Indices and Slip Points of Palmstearin and Soybean Oil Blends (4:6, w/w) Obtained by Conventional Interesterification and Biointeresterification Using _Candida Cylindracea_ Lipase

Treatment	Slip point	Solid fat index					
		15 C	20 C	25 C	30 C	35 C	40 C
	C	ml/g					
Control	42.0	22.7	17.5	12.7	10.0	7.7	4.8
Conventional interesterified	38.0	19.0	15.0	10.7	8.5	6.4	5.0
Biointeresterified (48 hr)	36.2	19.3	15.3	10.4	7.7	4.0	1.3

of this work, it was concluded that biointeresterification can be useful in making plastic fats such as vanaspati and margarines. However, the replacement of conventional nonenzymatic processes with this technology will depend on the economics of the new process.

REFERENCES

1. Macrae, A.R., in _Biotechnology for the Oils and Fats Industry_, American Oil Chemists' Society, Champaign, IL, 1984, pp. 189.
2. Van Handel, E., and D.B. Zilversmit, _J. Lab. Clin. Med. 50_:152 (1957).
3. Booekenogen, H.A., in _Analysis and Characterization of Oils, Fats and Fat Products_, Interscience Publishers, New York, 1964, pp. 143.
4. _Indian Standard Methods of Sampling and Test for Oils and Fats_ (Revised), Fourth Reprint, May 1975, _Indian Standard Institution IS_:548 (Part 1), 1964, pp. 33.
5. Adhikari, S., J. Dasgupta, D.K. Bhattacharyya and M.M. Chakrabarty, _Fette Seifen Anstrichm. 83_:262 (1981).

Biochemical Characterization of a Genetic Trait for Low Palmitic Acid Content in Soybean

R.F. Wilson, P. Kwanyuen and **J.W. Burton**

USDA-ARS, Crop Science Department, North Carolina State University, Raleigh, NC 27695-7620

Recurrent mass and within half-sib selection methods have been used to develop soybean (*Glycine max* L. Merr.) germplasm (N79-2077) exhibiting low palmitic acid content. The palmitic acid content of N79-2077 seed at maturity was 2.3-fold lower than the cultivar "Dare", which is a typical commercial variety. This trait was expressed primarily in triacylglycerol. Palmitic acid accumulation in triacylglycerol was linear during seed development of both genotypes with a significantly greater rate of accumulation in Dare. However, genotypic differences in the content and concentration of palmitic acid among molecular species in triacylglycerol could not be explained completely by the different rates of palmitic accumulation. The theory that diacylglycerol acyltransferase (EC 2.3.1.20) might influence these compositional differences between genotypes as a result of inherent substrate specificities was tested with enzyme purified from both genotypes. Triacylglycerol synthetic activity was determined with various combinations of three species of *sn*-1,2 diacylglycerol and three species of acyl-CoA. Genotypic differences in the rate of substrate utilization were positively correlated with differences in the concentration of palmitic acid in given molecular species. Therefore, genetic regulation of palmitic acid synthesis plus inherent differences in substrate utilization for triacylglycerol synthesis were major factors governing the composition and deposition of palmitic acid in triacylglycerol produced by these soybean genotypes.

Palmitic acid (16:0) is a saturated 16-carbon fatty acid found in virtually all fats and oils. Genetic diversity for 16:0 concentration appears to be high among oilseed species, ranging from ca. 4% (w/w) in rapeseed to ca. 47% (w/w) in palm oil (1). However, within a species such as soybean, there is little natural genetic variation for this trait. Soybean oil typically contains 11-14% (w/w) 16:0.

Research on the genetic control of fatty acid composition in soybean has led to the selection of the first known germplasm (N79-2077) that exhibits a stable mutation affecting low (ca. 5.9% w/w) 16:0 concentration. The identification of this genetic resource has established a unique means to detect and describe the biochemical mechanisms responsible for control of 16:0 content in soybean oil. Investigations of lipid metabolism with N79-2077 should not only provide a better understanding of how lower amounts of 16:0 were affected but also should aid the future development of genetic technologies to manipulate 16:0 content in soybean oil as might be desired for specific industrial or food applications.

The ability to genetically "tailor" the content and composition of glycerolipids containing 16:0 could result in advantageous effects upon functional properties of soybean oil in foods by increasing 16:0 (2), potential health benefits by lowering saturated fatty acid levels in soybean oil (3), or more cost-effective utilization of soybean oil with low 16:0 for nonfood uses (4). Hence, development of

high or low 16:0 germplasm with other acceptable agronomic traits could significantly enhance the quality, nutritional value and economic worth of soybean oil. Progress made toward implementing biotechnological advances in this area are contingent upon knowledge of the interaction of enzymatic mechanisms governing 16:0 synthesis and 16:0 metabolism during N79-2077 seed development. Characterizations of lipid biochemistry and composition in N79-2077 have shown that genetic influences upon the rate of 16:0 synthesis plus inherent differences in the utilization of 16:0-CoA and DG species containing 16:0 for triacylglycerol synthesis were important factors governing the content and deposition of 16:0 in TG. Hence, genetic regulation of 16:0 in soybean oil may be determined by one or more structural genes encoding enzymes in the fatty acid synthetase I and/or II complexes (5) with additional modifier genes governing the biophysical characteristics of DGAT.

Experimental Procedures

Soybean (*Glycine max*, L. Merr. cv N79-2077) germplasm exhibiting low palmitic acid concentration was selected from a fifth cycle recurrent selection population for altered fatty acid composition and was inbred to the F_5 generation. Glycerolipids were isolated and analyzed from developing seed of N79-2077 and the cultivar "Dare" as described by Carver et al. (7). Triacylglycerol molecular species composition was determined as described by Carver and Wilson (8). Fatty acid and glycerolipid synthesis was evaluated via acetate saturation kinetics derived from incorporation of [2-^{14}C] acetate (5 μCi, 0.1 μmol) plus 0 to 1000 μmol potassium acetate with seed harvested at 30 days after flowering (DAF). Reactions were conducted for two hr at 25 C in 10 ml 0.2 N MES buffer, pH 5.5. DGAT was purified from germinating soybean cotyledons as described by Kwanyuen and Wilson (9). DGAT activity was determined by the method of Martin and Wilson (10) with the following modifications. Assays were conducted with 8.5 μM [1-^{14}C]acyl-CoA (55 Ci/mol), 1.6 mM *sn*-1,2 DG sonicated in 2 mg/ml Tween-20, 50 mM Bicine (pH 8.0), 10 mM mgCl$_2$ and 0 to 50 μg enzyme in a final volume of 200 μl at 30 C for 10 min. DGAT substrate specificity was demonstrated with various combinations of [1-^{14}C]16:0-CoA, [1-^{14}C]18:1-CoA or [1-^{14}C]18:2-CoA (50 Ci/mol) with *sn*-1,2 diolein, *sn*-1, 2 palmitoylolein or *sn*-1,2 palmitoyllinolein. Enzyme kinetics using diolein with varied concentrations of [1-^{14}C]16:0-CoA were determined by the same procedures. All data were reported as the mean of at least three replications.

Results and Discussion

N79-2077 is an experimental soybean germplasm selected for low 16:0 content. In comparison with a standard cultivar, the 16:0 content in total lipid extracts of mature seed was 2.3-fold greater in Dare than N79-2077 (Table 1). N79-2077 also contained a greater amount of 18:1, whereas Dare contained a greater amount of 18:2. How-

Table 1

Fatty Acid Content of Mature Soybean Seed

Genotype	Fatty acid					
	16:0	18:0	18:1	18:2	18:3	Total
	mmol Fatty acid/kg dry weight					
Dare	111.3	28.0	180.9	467.1	62.0	849.3
N79-2077	47.8	28.4	327.5	345.4	61.6	810.7
LSD 0.05	52.2	0.4	120.6	100.1	0.4	41.7

ever, 18:1 plus 18:2 content of these oils was not different between genotypes. Therefore, the higher 18:1 content in N79-2077 appeared to be a result of a slower rate of 18:1-desaturation. Although the 18:1 content of glycerolipids in N79-2077 seed hypothetically could be dependent upon 16:0 metabolism, the proposition that 16:0 and 18:1 levels were governed by separate and independent genetic mechanisms has been accepted during the initial stages of this investigation.

Given that assumption, the 16:0 content of N79-2077 most likely would be determined by the activities of fatty acid synthetases I and II (5). However, the work of Stobart and Stymne (11) has shown that regulation of the incorporation of acyl-CoA into phospholipids and the subsequent metabolism of phospholipids to provide DG and acyl-CoA for TG synthesis also could be a determining factor for glycerolipid composition. In addition, selective utilization of DG by the enzyme that catalyzes TG synthesis, diacylglycerol acyltransferase (DGAT), has been demonstrated in safflower (12) and maize (13). To initiate characterization of the biochemical basis for genetic control of 16:0, glycerolipids were isolated from developing seed of N79-2077 and the standard cultivar Dare. At comparable stages throughout seed development, the 16:0 concentration of total polar glycerolipids (TPL), diacylglycerol (DG) and triacylglycerol (TG), respectively, was significantly lower in N79-2077 than Dare (Table 2). Developmental changes in seed metabolism had little effect upon expression of the low 16:0 trait in TG, which accounted for ca. 97% of the total 16:0 in both genotypes at maturity (70 DAF).

The rate of fatty acid and glycerolipid synthesis in seed at 30 DAF was evaluated by in vivo incorporation of saturating levels of acetate. These experiments revealed a significantly lower rate of total 16:0 synthesis in N79-2077 seed (Table 3). In relative terms, the rate of 16:0 incorporation into TPL equalled 62.4% of the total in N79-2077 and 56.6% of the total in Dare. Because this percentage for N79-2077 was equal to or greater than that in Dare, restriction of 16:0 incorporation into TPL could be discounted as a potential regulatory factor. The percentage of total 16:0 radioactivity in TG was greater in Dare (34.5%) than N79-2077 (27.2%), which suggested a slower rate of 16:0 incorporation into TG by the low 16:0 germplasm. Indeed, the measured rate in N79-2077 was nearly half that in Dare. These data were directly comparable with the genotypic differences in actual 16:0 content of TG expressed on a seed basis at 30 DAF (Table 4). Therefore, in addition to reduced 16:0 synthesis, the availability

Table 2

Palmitic Acid Concentration in Glycerolipids from Developing Soybean Seed

Genotype	DAF[a]	Glycerolipid[b]			Total lipid
		TPL	DG	TG	
		mole %[c]			
Dare	30	25.8	19.5	13.3	15.1
	50	23.3	18.6	12.9	14.0
	70	18.7	17.0	12.1	12.4
N79-2077	30	14.8	10.2	7.9	8.8
	50	14.4	10.2	6.0	6.4
	70	13.5	9.8	5.7	5.9
LSD 0.05		3.0	2.7	2.0	2.3

[a]Days after flowering; 70 DAF, maturity.
[b]TPL, total polar lipid; DG, diacylglycerol; TG, triacylglycerol.
[c]Percentage of total mmol fatty acid/kg dry weight in each glycerolipid.

Table 3

Comparison of Palmitic Acid Metabolic Activities Among Glycerolipids form Soybeans at 30 DAF

Genotype	Glycerolipid			
	TPL	DG	TG	Total
	nmol [^{14}C]acetate/h/g/dry weight[a]			
Dare	394.3	62.0	240.3	696.6
N79-2077	252.5	42.1	110.0	404.6
LSD 0.05	116.6	16.4	107.2	240.2

[a]Maximal velocity derived from acetate saturation kinetics as described in Methods; [S]$_{90}$ for acetate incorporation into total lipid was 32.4 μmol for both genotypes.

and/or utilization of 16:0 substrates derived from TPL contributed to the low 16:0 trait.

Although 16:0 accumulation in TG was linear during development of both genotypes (Table 4), significant differences in 16:0 distribution were found among TG molecular species (Table 5). The extent to which this distribution could be attributed to differences in the endogenous levels of individual DG species containing 16:0 or 16:0-CoA available for TG synthesis has not yet been determined. However, inherent differences in the utilization of substrates by DGAT purified from these two genotypes was shown by the enzyme kinetics for dioleoylpalmitin (SM_2) synthesis (Fig. 1). These data established a V_{max} of 10.7 mU for this reaction with DGAT from N79-2077 and 4.4 mU from Dare. Apparent K_ms were 3.5 μM and 2.6 μM, respectively. Hence a greater capacity for SM_2 synthesis was demonstrated by DGAT purified from N79-2077. This observation was positively correlated with genotypic differences in 16:0 distribution in the SM_2 species extracted from mature seed. Without absolute knowledge of the endogenous level and metabolic flux of diolein or 16:0-CoA for TG synthesis in vivo, these data did show that

Table 4

Accumulation of Palmitic Acid in Triacylglycerol During Soybean Seed Development

Genotype	Days after flowering			Rate of accumulation in TG
	30	50	70	
	μmol 16:0/seed			μmol 16:0/day
Dare	3.2	8.9	14.5	0.283
N79-2077	1.6	5.5	9.5	0.197
LSD 0.05	1.3	2.8	4.1	0.071

Table 5

Relative Distribution of Palmitic Acid Among Triacylglycerol Molecular Species in Mature Soybean Seed

TG molecular species[a]	Genotype	
	Dare	N79-2077
	mol % 16:0[b]	
S_2M	16.7	10.2
SM_2	10.0	22.0
S_2D	8.3	15.2
SMD	20.8	22.0
SD_2	44.2	30.6
LSD 0.05	4.8	4.8

[a]S, 16:0; M, 18:1, D, 18:2; T, 18:3.
[b]The amount of 16:0 in each species relative to the total amount of 16:0 in TG at 70 DAF; no other TG species contained 16:0.

DGAT substrate specificity could effect 16:0 distribution, if not content, within TG. This premise was supported by additional evidence (Table 6) with similarity to corresponding data presented in Table 5. Hence, within the limits of this study, it was concluded that the low 16:0 trait

Table 6

Effect of Diaclyglycerol Acyltransferase Upon Synthesis of Triacylglycerol Species Containing Palmitic Acid

TG species	DGAT product[a]			Genotype	
	sn-1	sn-2	sn-3	Dare	N79-2077
				Relative maximal reaction velocity[b]	
S_2M	16:0	18:1	16:0	46.4	33.0
SM_2	18:1	18:1	16:0	100.0	100.0
S_2D	16:0	18:2	16:0	106.2	120.9
SMD	16:0	18:2	18:1	90.6	96.3
SD_2	16:0	18:2	18:2	138.3	148.3
LSD 0.05				11.0	11.0

[a]Product of specific sn-1,2 DG plus acyl-CoA (sn-3) substrate combinations with diacylglycerol acyltransferase (DGAT) purified from each genotype.
[b]Diolein plus 16:0-CoA = 100%.

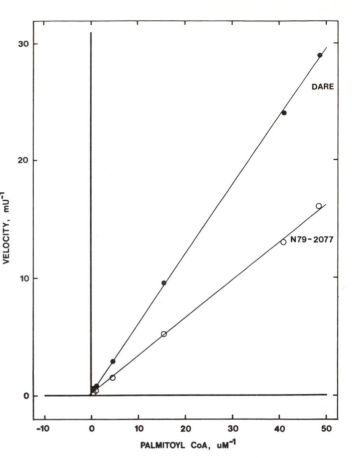

FIG. 1. **Lineweaver-Burk plot of DGAT kinetics with diolein and [1-^{14}C]16:0-CoA, One unit equals 1 mmol dioleoylpalmitin/min/mg enzyme. V_{max} (N79-2077), 10.7 mU; V_{max} (Dare)), 4.4 mU.**

in N79-2077 was affected by genetic regulation of 16:0 synthesis per se with additional effects contributed by inherent differences in substrate utilization for TG synthesis.

ACKNOWLEDGMENTS

These were cooperative investigations of the Agricultural Research Service/USDA and the North Carolina Agricultural Research Service (Raleigh, NC). Financial support was received in part by a grant from the Research Foundation of the American Soybean Association (St. Louis, MO).

REFERENCES

1. Weiss, T.J., in *Foods Oils and Their Uses*, 2nd ed., Avi Publishing Co., Inc., Wesport, CN, 1983, p. 35.
2. Ohlson, R., in *Dietary Fats and Health*, edited by E.G. Perkins and W.J. Visek, American Oil Chemists' Society, Champaign, IL, 1983, p. 124.
3. Herbert, P.N., *Ibid.*, 1983, p. 512.
4. Zilch, K.T., in *Proceedings World Soybean Research Conference II*, edited by F.T. Corbin, Westview Press, Boulder, CO, 1980, p. 693.
5. Stumpf, P.K., in *Proceedings World Conference on Emerging Technologies in the Fats and Oils Industry*, edited by A.R. Baldwin, American Oil Chemists' Society, Champaign, IL, 1986, p. 312.
6. Wilson, R.F., J.W. Burton and C.A. Brim, *Crop Sci. 21*:788 (1981).
7. Carver, B.F., R.F. Wilson and J.W. Burton, *Crop Sci. 24*:1016 (1984).

8. Carver, B.F., and R.F. Wilson, *Ibid.*, 1020 (1984).
9. Kwanyuen, P., and R.F. Wilson, *Biochim. Biophys. Acta. 877*:238 (1986).
10. Martin, B.A., and R.F. Wilson, *Lipids 18*:1 (1983).
11. Stobart, A.K., and S. Stymne, *Planta 163*:119 (1985).
12. Ichihara, K., and M. Noda, *Phytochemistry 21*:1895 (1982).
13. De la Roche, I.A., E.J. Weber and D.E. Alexander, *Lipids 6*:537 (1971).

Substrate Specificity of Diacylglycerol Acyltransferase Purified from Soybean

P. Kwanyuen, R.F. Wilson and **J.W. Burton**

USDA-ARS, Crop Science Department, North Carolina State University, Raleigh, NC, 27695-7620

Diacylglycerol acyltransferase (EC 2.3.1.20) catalyzes the synthesis of triacylglycerol, the predominate glycerolipid in soybean (*Glycine max*, L. Merr.) seed. The molecular species composition of triacylglycerol presumably is determined by the endogenous concentration and acyl composition of diacylglycerol and acyl-CoA substrates available for triacylglycerol synthesis. However, the degree and extent to which various substrate combinations may be utilized by diacylglycerol acyltransferase from soybean are unknown. The theory that diacylglycerol acyltransferase may exert an additional influence upon triacylglycerol molecular species composition, through inherent differences in substrate specificity, was tested with the experimental low linolenic acid genotypes N78-2245 and PI 123440. These genotypes exhibited genetic differences in the biochemical mechanisms affecting acyl desaturation and contained significant phenotypic differences in triacylglycerol composition. Triacylglycerol synthetic activity of the purified enzyme was determined with all possible combinations of nine species of *sn*-1,2 diacylglycerol and four species of acyl-CoA. Results revealed compelling evidence for genotypic differences in substrate utilization. These findings implied that triacylglycerol molecular species composition in soybean exhibiting low linolenic acid concentration could be influenced by selective utilization of specific substrates and suggested that genetic variability for the physical characteristics of diacylglycerol and acyltransferase may exist in these soybean germplasm.

Recent biotechnological advances affecting genetic alteration of fatty acid composition have been achieved with many crop species. Such technology has opened a new era in lipid research that should lead not only to a better understanding of why differences in glycerolipid composition exist among oilseed species but also to a more complete knowledge of the enzymatic mechanisms governing glycerolipid composition within species.

Current concepts on the synthesis of storage lipids by developing seeds have been reviewed by Slack and Browse (1). Triacylglycerol, the predominant glycerolipid in seed oils, is synthesized by diacylglycerol acyltransferase (EC 2.3.1.20), which utilizes diacylglycerol (DG), derived from phosphatidic acid (PA) or phosphatidylcholine (PC), and acyl-CoA as substrates (2). There is a growing body of information (3-5) indicating that the composition of endogenous DG and acyl-CoA pools in seed is determined by the rate and specificity of *sn*-glycerol 3-phosphate acyltransferase (EC 2.3.1.15) and acyl-CoA:lysophosphatidylcholine acyltransferase (EC 2.3.1.23). These two enzymes may influence positional acyl specificity within phospholipids and through acyl-exchange could provide a significant portion of the acyl-CoA species used in TG synthesis. Interconversions of DG and PC, catalyzed by

CDP-DG cholinephosphotransferase (EC 2.7.8.2), which reportedly shows no selectivity among DG species (6), probably determines the endogenous concentration and availability of DG for TG synthesis. Therefore, it generally is presumed that these enzymes are of major importance in determining the acyl composition of TG, particularly in germ plasm exhibiting genetically altered systems for fatty acid synthesis.

Another factor that also could influence or determine the amount and type of TG molecular species produced during seed development is the ability or degree to which DGAT per se may utilize certain substrate combinations. DGAT substrate specificity has been documented extensively in mammalian tissues (7). Although less well-described in plant tissues, evidence for selective utilization of DG by DGAT has been reported in safflower and maize seed (8,9). These investigations show that DGAT activity may influence TG composition; however, purified enzyme is required to evaluate critically the potential ramifications of such inherent differences in substrate specificity upon TG synthesis. Recently, methods have been developed for purification of DGAT from soybean (10). These procedures have been used to obtain DGAT from two experimental genotypes exhibiting significant differences in TG molecular species composition. Substrate specificities for these enzyme preparations were determined and related to phenotypic differences in TG composition.

EXPERIMENTAL PROCEDURES

Development, inheritance studies and biochemical characterization of the soybean (*Glycine max*, L. Merr) genotypes, N78-2245 and PI 123440, used in this work are referenced and described by Wilson and Burton (11). DGAT was purified from germinating N78-2245 and PI 123440 cotyledons as described by Kwanyuen and Wilson (10). DGAT activity was determined by the method of Martin and Wilson (12) with the following modifications. Assays were conducted with 8.5 μM [1-^{14}C]acyl-CoA (55 Ci/mol), 1.6 mM *sn*-1,2 DG sonicated in 0.2% (w/v) Tween-20, 50 mM Bicine (pH 8.0), 10 mM MgCl$_2$, and 0-50 μg enzyme in a final volume of 200 μl at 30 C for 10 min. Reactions were conducted with all combinations of [1-^{14}C]16:0-CoA, [1-^{14}C]18:0-CoA, [1-^{14}C]18:1-CoA, [1-P^{14}C]18:2- CoA (50 Ci/mol) with *sn*-1,2 dipalmitin, *sn*-1,2 palmitoyl-olein, *sn*-1,2 palmitoyl-linolein, *sn*-1,2 stearoyl-olein, *sn*-1,2 stearoyl-linolein, *sn*-1,2 oleoyl-palmitin, *sn*-1,2 diolein, *sn*- 1,2 dilinolein and *sn*-1,2 dilinolenin. TG synthesis with each substrate combination was determined as decribed (10). Enzyme kinetics using diolein with varied concentrations of [1-^{14}C]18:1-CoA were determined by the same procedures. TG molecular species composition and fatty acid analyses of glycerolipids extracted from mature seed were performed using the methods of Carver and Wilson (13) and Carver et al. (14). All data were reported as the mean of at least three replications.

RESULTS AND DISCUSSION

N78-2245 and PI 123440 represent two types of experimental soybean germplasm that contain a low concentration of 18:3 as compared with the standard cultivar "Dare" (Table 1). The low 18:3 trait in these genotypes is affected by two different genetic influences upon the enzyme systems responsible for desaturation of 18:1 to 18:3. Indirect analyses (Table 2) show that 18:1-desaturase activity (k_1) was significantly lower in N78-2245 than PI 123440 or Dare but not different between the latter two genotypes. The indirect estimate of 18:2-desaturase activity (k_2) is significantly lower in PI 123440 than N78-2245 or Dare and not different between the latter two genotypes. These observations are confirmed by a more detailed description of the biochemical and genetic systems involved (11,15). As diagrammed in Figure 1, the phenotype or unsaturated fatty acid composition of the respective oils is a function of the rates of these consequent desaturation reactions. Even so, recent work indicates that genetic alteration of the biochemical mechanisms of 18:1 and 18:2 desaturation may not singularly determine the acyl composition of glycerolipids, principally TG, which accounts for 90% (w/w) of most vegetable oils.

Table 1

Fatty Acid Composition of Triacylglycerol from Mature Soybean Seed

Genotype	Fatty acid						
	16:0	16:1	18:0	18:1	18:2	18:3	DBI
	mol %						
N78-2245	9.8	0.1	3.3	54.6	28.3	3.9	123.0
PI123440	10.7	0.1	3.8	29.6	51.7	4.1	145.4
Dare	12.0	0.1	2.9	21.8	55.8	7.4	155.7
LSD 0.05	1.3	0.0	0.2	19.9	17.2	0.6	19.4

DBI: Double bond index, Σ mol % [16:1 + 18:1 + 18:2(2) + 18:3(3)].

Table 2

Genotypic Differences in Acyl Desaturation Reactions Affecting the Linolenic Acid Concentration of Soybean Oil

Genotype	Desaturase reaction		Product	
	18:1 to 18:2[a]	18:2 to 18:3[b]	Calculated[c]	Expected[d]
	K_1	k_2	mol % 18:3	
N78-2245	37.1[e]	12.1	4.5	4.5
PI123440	65.3	7.3	4.8	4.8
Dare	74.3	11.7	8.7	8.7
LSD 0.05	11.2	1.5	1.3	1.3

[a]Mol % [(18:2 + 18:3)/(18:1 + 18:2 + 18:3)] = k_1.
[b]Mol % [(18:3)/(18:2 + 18:3)] = k_2.
[c]$k_1 k_2$.
[d]Mol % [(18:3)/(18:1 + 18:2 + 18:3)].
[e]Calculations based on date from Table 1.

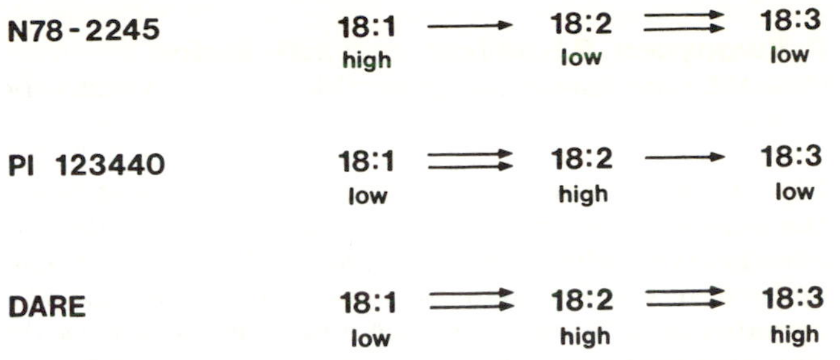

FIG. 1. Effect of genotypic differences in the rates of acyl-desaturation upon the phenotypic character of unsaturated fatty acid composition in soybean.

Research on high-18:1 and high-18:2 safflower genotypes (3) has shown that in addition to the rate of acyl desaturation in PC, the rate plus specificity of glycerolphosphate acyltransferases also may mediate the metabolic levels of DG and acyl-CoA molecular species utilized in TG synthesis. Hence, these enzymes apparently exert finite control over the concentration and availability of given DG and acyl-CoA species that may influence the products formed by diacylglycerol acyltransferase (DGAT) in safflower, and presumably soybean as well (Table 3). However, these current concepts on the biochemical regulation of TG composition in plants remain incomplete due to a lack of knowledge of DGAT specificities, which could alter the utilization of the endogenous substrates provided for TG synthesis.

Table 3

Molecular Species Composition of Triacylglycerol from Mature Soybean Seed

TG Species[a]	Genotype	
	N78-2245	PI123440
	mol %	
S_2M	2.6	1.6
SM_2	17.7	9.1
M_3	21.3	6.0
S_2D	1.5	3.5
SMD	6.7	9.5
M_2D	18.1	13.0
S_2T	1.0	0.0
SD_2	4.8	14.6
MD_2	9.6	15.8
M_2T	4.9	0.0
D_3	8.8	17.3
D_2T	0.8	6.7
DT_2	2.2	2.9

[a]S, 16:0 or 18:0; M, 18:1; D, 18:2; T, 18:3, LSD 0.05 among and between treatments, 1.6.

There are no definitive studies on DGAT substrate specificity in plants, but the best available work (8) has shown that DGAT from safflower exhibited broad specificity for both DG and acyl-CoA species. To determine the extent to which DGAT per se might contribute to genotypic differences in the concentration of individual TG molecular species in soybean, DGAT was purified (10) from N78-2245 and PI 123440 seed. Hyperbolic kinetics

P. KWANYUEN ET AL.

for 18:1-CoA with diolein were obtained with DGAT purified from both genotypes. Maximal reaction velocities (V_{max}) of 6.7 mU (N78-2245) and 3.4 mU (PI 123440) were estimated from Lineweaver-Burk plots of these data (Fig. 2); the $[S]_{90}$ was determined to be 44.5 μM (N78-2245) and 91 μM (PI 123440). Thus, it appeared that DGAT from N78-2245 had a greater capacity for production of triolein than PI 123440. Therefore, the actual difference in concentration of triolein (M_3) between genotypes (Table 3), could be attributed to more than the availability of endogenous diolein and 18:1- CoA.

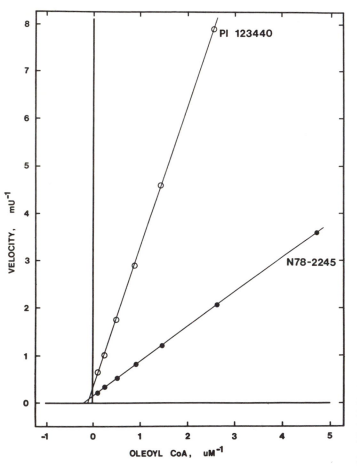

FIG. 2. Lineweaver-Burke plot of DGAT kinetics with diolein and [1-14C]18:1-CoA. One unit equals 1 mmol triolein/min/mg enzyme. V_{max} (N78-2245), 6.7 MU; V_{max} (PI 123440). 3.4mU.

One set of kinetic data, however, does not properly test the potential relationship between DGAT activity and TG molecular species composition. Additional tests for DGAT specificity were conducted with 35 other DG and acyl-CoA combinations under saturating conditions. The maximal velocities from each reaction relative to diolein plus 18:1-CoA within each genotype were compared in Table 4. Significant differences in DGAT substrate specificity were expressed among and within genotypes. A high order of discrimination against the utilization of dipalmitin with 16:0-CoA, 18:0-CoA, 18:1-CoA or 18:2-CoA, and also for 18:0-CoA with any of the nine DG species used was common to both genotypes. These data were independent of potential problems resulting from poor substrate solubil-

Table 4

Relative Comparison of Substrate Utilization by DGAT Purified from N78-2245 (N) and PI123440 (P).

sn-1,2 DG	16:0-CoA N	16:0-CoA P	18:0-CoA N	18:0-CoA P	18:1-CoA N	18:1-CoA P	18:2-CoA N	18:2-CoA P
				Relative Reaction Velocities[a]				
16:0/16:0	3.0	1.2	0.6	0.0	2.8	1.2	2.8	1.2
16:0/18:1	55.4	17.9	0.2	1.2	40.5	23.8	59.7	45.2
16:0/18:2	160.5	171.4	2.6	6.0	130.5	204.8	207.5	281.0
18:0/18:1	13.8	13.1	0.2	0.0	7.9	9.5	17.1	32.1
18:0/18:2	80.4	117.9	0.8	3.6	33.8	70.2	92.3	167.9
18:1/16:0	71.9	11.9	0.0	2.4	67.4	17.9	86.5	29.8
18:1/18:1	151.9	125.0	2.6	0.0	100.0	100.0	156.8	171.4
18:2/18:2	85.5	90.5	1.2	1.2	55.8	73.8	100.4	147.6
18:3/18:3	165.2	139.3	2.6	2.4	72.1	140.5	118.9	170.2

[a]Relative to the velocity for diolein plus 18:1-CoA within each genotype; LSD 0.05 among and between treatments, 10.2.

Table 5

Comparison of DGAT Reaction Velocities for Specific Triacylglycerol Products

TG species	DGAT product[a] sn-1	sn-2	sn-3	Genotype N78-2245	PI123440
				Relative reaction velocity[b]	
S_2M	16:0	18:1	16:0	55.4	17.9
SM_2	18:1	18:1	16:0	151.9	125.0
M_3	18:1	18:1	18:1	100.0	100.0
S_2D	16:0	18:2	16:0	160.5	171.4
SMD	16:0	18:2	18:1	130.5	204.8
SD_2	16:0	18:2	18:2	207.5	281.0
MD_2	18:2	18:2	18:1	55.8	73.8
D_3	18:2	18:2	18:2	100.4	147.6
DT_2	18:3	18:3	18:2	118.9	170.2

[a]Product of specific sn-1,2 DG plus acyl-CoA (sn-3) substrate combinations.
[b]Relative to the velocity for diolein plus 18:1-CoA within each genotype; LSD 0.05 among and between treatments, 10.2.

ity, and substrate levels in all reactions were below reported critical micelle concentrations (16). Excluding those respective treatments, statistically significant genotypic differences between the relative velocities of comparable reactions were found for 21 of the remaining 23 substrate combinations. These data were compelling evidence that product formation by DGAT was not merely a function of substrate availability but also depended upon the ability of DGAT to utilize specific substrate combinations. Comparison of the effect of acyl substitution of various sn-positions upon DGAT activity shows this more clearly (Table 5). In each case, the genotypic difference between comparable reactions was correlated positively with the genotypic difference between comparable reactions was correlated positively with the genotypic difference in concentration of the corresponding TG molecular species (Table 3). Therefore, it was concluded that DGAT substrate specificity could play an important role in determining TG molecular species composition of soybean. Furthermore, this work suggested genetic variability for DGAT specificities among soybean genotypes. Such variability may be affected by genetic alteration of structural or biophysical properties of the enzyme.

ACKNOWLEDGMENTS

These were cooperative investigations of the Agricultural Research Service/USDA and the North Carolina Agricultural Research Service (Raleigh, NC). Financial support was received in part by a grant from the Research Foundation of the American Soybean Association (St. Louis, MO).

REFERENCES

1. Slack, C.R., and J.A. Browse, in *Seed Physiology*, Vol. 1, edited by D.R. Murray, Academic Press Australia, Sidney, 1984, p. 209.
2. Slack, C.R., in *Proceedings of the 6th Annual Symposium in Botany, Riverside*, edited by W.W. Thompson, J.B. Mudd and M. Gibbs, University of California, Riverside, CA, 1983, p. 40.
3. Stymne, S., A.K. Stobart and G. Glad, *Biochim. Biophys. Acta* 752:198 (1983).
4. Stymne, S., and A.K. Stobart, *Biochem. J. 220*:481 (1984).
5. Stobart, A.K., and S. Stymne, *Planta 163*:119 (1985).
6. Justin, A.M., C. Demandre, A. Tremolieres and P. Mazliak, *Biochim. Biophys. Acta 836*:1 (1985).
7. Bell, R.M., and R.A. Coleman, in *The Enzymes*, Vol. XVI, Academic Press, Inc., New York, 1983, p. 87.
8. Ichihara, K., and M. Noda, *Phytochemistry 21*:1895 (1982).
9. De la Roche, I.A., E.J. Weber and D.E. Alexander, *Lipids 6*:537 (1971).
10. Kwanyuen,P., and R.F. Wilson, *Biochim. Biophys. Acta 877*:238 (1986).
11. Wilson, R.F., and J.W. Burton, in *Proceedings World Conference on Emerging Technologies in the Fats and Oils Industry*, edited by A.R. Baldwin, American Oil Chemists' Society, Champaign, IL, 1986, p. 386.
12. Martin, B.A., and R.F. Wilson, *Lipids 18*:1 (1983).
13. Carver, B.F. and R.F. Wilson, *Crop Sci. 24*:1020 (1984).
14. Carver, B.F., R.F. Wilson and J.W. Burton, *Ibid.*, 1016 (1984).
15. Carver, B.F., and R.F. Wilson, *Ibid.*, 1023 (1984).
16. Smith, R.H., and G.L. Powell, *Arch. Biochem. Biophys. 244*:357 (1986).

Oleic Acid Conversion to 10-Hydroxystearic Acid by *Nocardia* *Species*

S. Koritala, L. Hosie and **M.O. Bagby**
Northern Regional Research Center, ARS-USDA, Peoria, IL 61604

Five *Nocardia* species catalyzed the conversion of oleic acid to 10-hydroxystearic acid. Oleic acid disappeared from the reaction mixture in four hr, and the yield of the hydroxy acid was 50-75% based on GC analysis of methyl esters of reaction products with palmitate as the internal standard. 10-Ketostearic acid was a secondary product that appears to be formed by oxidation of the primary product. The ratio of 10-hydroxystearic to 10-ketostearic acids varied from run to run and from species to species. Oleic acid apparently was hydrated to form the main product as evidenced by the formation of 9-deuterio-10-hydroxystearic acid when the reaction was carried out in deuterated water. The enzyme system was specific for oleic acid because other substrates such as elaidic, palmitoleic, petroselenic and erucic acids and oleyl alcohol were unchanged. Triolein was unchanged, but a portion of the substrate was hydrolyzed followed by conversion to hydroxy acid. Little, if any, linoleic acid was transformed to hydroxyoctadecenoic acid.

Presently, castor oil is the only commercial source of ricinoleic acid. Some industrially important chemicals are derived from this hydroxy fatty acid (1). A saturated hydroxy acid is formed microbiologically by a *Pseudomonas* species (2). Subsequently, a soluble enzyme system capable of hydrating oleic acid to 10-hydroxystearic acid was prepared from a *Pseudomonas* sp. NRRL B-2994 (3). More recently, a patent (4) issued to Litchfield disclosed the formation of hydroxy-fatty acids and keto-fatty acids from oleic acid by *Rhodococcus rhodochorus*. Because hydroxystearic acid has industrial applications in lubricating greases (5), we have studied the conversion of oleic to hydroxy fatty acid with *Nocardia* species (6). The preliminary findings of this study form the basis for this report.

EXPERIMENTAL

Materials

Oleic acid (purity >99% by GC) was purchased from commercial sources. Other fatty acids and derivatives were either purchased or prepared in the laboratory.

Fermenations

All microorganisms used in this study were obtained from the ARS Culture Collection, Northern Regional Research Center. They were grown and maintained on yeast-malt agar plates. About five loopfuls of the organism were transferred to 1 liter of sterile medium that contained 4 g yeast extract, 10 g malt extract and 4 g glucose. The mixture was shaken in a flask (150 rpm) at 25 C for 17 hr, and then 0.5 ml of oleic acid was added as an inducer. The culture was then shaken for further 24 hr. The cells were isolated (yield 5-6 g wet cells) by centrifugation at 24,000

g for 10 min. About 1 g of wet cells was added to 19 ml sodium phosphate buffer (0.1M, pH 6.5) and 0.4 ml oleic acid (0.356 g) in a 125-ml flask and shaken at 25 C and 150 rpm for 24 hr. At the end of fermentation, 10 mg of palmitic acid was added as an internal standard, and the lipids were extracted with diethyl ether, freed of solvent and weighed.

Analysis

Fatty acids were converted to methyl esters with diazomethane and analyzed by capillary GC on a Hewlett-Packard 5890 instrument equipped with flame ionization detectors and a 15-m 0.32 mm i.d. fused silica column coated with SPB-1 (Supelco Inc., Bellefonte, PA). The peak areas were integrated with the aid of a ModComp Computer

RESULTS

Several strains of *Nocardia* and one *Actinomycete* sp. were tested for their ability to convert oleic acid to hydroxy derivative. As shown in Table 1, five different strains of *Nocardia* showed activity, while four others were unable to convert oleic to hydroxy acid. Two *Nocardia* strains showed slight activity, while the *Actinomycete* sp. was negative.

Table 1

Screening of Microorganisms for the Bioconversion of Oleic Acid to 10-Hydroxystearic Acid

Organism	NRRL #	Activity
Nocardia cholesterolicum		
(rough)	5767	+
(smooth)	5768	+
Nocardia sp.	5635	+
Nocardia sp.	5636	+
Nocardia aurantia	B3287	+
Nocardia minima	B5477	–
Nocardia sp.	B3068	–
Rhodococcus erythropolis	B1532	–
Rhodococcus corallinus	B5476	–
Nocardia amarae	B16281	slight
Nocardia amarae	B16282	slight
Actinomycete sp.	B16216	–

The GC compositions of typical reaction products are shown in Table 2. At the end of 24 hr, oleic acid nearly disappeared, and the product consisted essentially of keto- and hydroxystearic acids. The proportion of these two products differed from strain to strain and, within the same strain, from run to run. For example, NRRL 5767 produced 22% ketostearic acid as shown in Table 2, but in other experiments not shown, the content of keto acid was as low as 5%. The variable production of ketostearate

Table 2

GC Composition of Reaction Products[a]

NRRL No.	Oleate	Ketostearate	Hydroxystearate
B3287	2.6	21.4	75.9
5767	1.1	22.1	76.8
5768	3.1	16.5	80.4
5635	3.6	4.3	92.1
5636	2.2	12.4	85.5

[a]1 g wet cells + 19 ml sodium phosphate buffer (0.1M, pH 6.5) + 0.4 ml oleic acid. Reaction mixture shaken at 25 C and 150 rpm for 24 hr. Area percent.

is not understood. Based on palmitic acid content added as an internal standard, the yield of hydroxystearic acid was 50-75%.

The time study (Fig. 1) of NRRL 5767 revealed that the conversion was essentially complete in four hr. Ketostearic acid, which appeared to increase with time after four hr, is a secondary reaction product, probably formed by oxidation of the hydroxy acid.

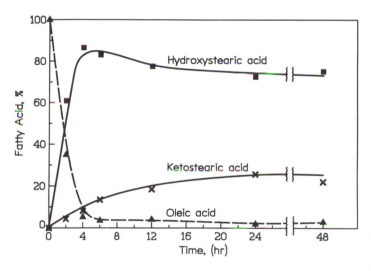

FIG. 1. GC analysis of products formed during fermentation of oleic acid by NRRL B-5767.

10-Ketostearic and 10-hydroxystearic acids were identified by GC-MS (2,7). Hydroxystearic acid was formed by hydration of the double bond as seen from the comparison of mass-spectra of the products formed in presence of either deuterated or normal water (Fig. 2). Because fragments 201 and 172 corresponded to 202 and 173, respectively, upon adding deuterated water the deuterium was presumed to be on carbon 9 of the hydroxystearic acid.

Various other fatty acids, esters and alcohols were used as substrates. As shown in Table 3, these were not converted to hydroxy acids. It appears that the enzyme system, unlike that in *Pseudomonas* (8,9), is specific for conversion of *cis*-9-octadeconoic acid.

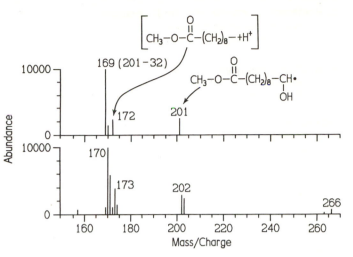

FIG. 2. A portion of the GC-MS of deuterated and protiated 10-hydroxystearic acid.

Table 3

Substrates Tested Negative for Hydroxy Acid Production

cis Fatty acids		
Petroselinic	18:1	Δ6
Palmitoleic	16:1	Δ9
Vaccenic	18:1	Δ11
Octadecenoic	18:1	Δ8
Linoleic	18:2	Δ9, 12
Erucic	22:1	Δ13
trans Fatty acids		
Elaidic	18:1	Δ9
Other Fatty acids		
cis-9, 10-epoxystearic		
10-hydroxystearic		
Ricinoleic		
Triglycerides		
Triolein		
Fatty alcohols		
Oleyl alcohol		

DISCUSSION

The primary emphasis of the new program at NRRC has been to identify reactions of industrial significance for the conversion of soybean oil or fatty acids and other vegetable oils into useful derivatives. To that end, the conversion of oleic acid to hydroxy acid was explored. In this study, we have successfully converted oleic to 10-hydroxystearic acid enzymatically, which may have industrial significance. 12-Hydroxystearic acid derived from castor oil is used in greases and lubricants (5). We hoped to convert linoleic acid to ricinoleic acid or its isomer, but the microorganisms were found to be very specific for oleic acid. The formation of ricinoleic from oleic acid by bacteria, fungi and yeast was reported recently (10), although none of the microorganisms were named in the abstract. The *Nocardia* sp. (NRRL 5767) most studied by us had a broad pH (5-7) for optimum growth. Good yields were obtained at temperatures of 15-38 C. The amount of ketostearic acid byproduct varied from experiment to experiment. It is hoped that this byproduct can be entirely eliminated by running the fermentation anaerobically. Such experiments are currently underway.

ACKNOWLEDGMENTS

Technical help was by A. McKnight and E.K. Hertenstein.

REFERENCES

1. Achaya, K.T., *J. Am. Oil Chem. Soc. 48*:748 (1971).
2. Davis, E.N., L.L. Wallen, J.C. Goodwin, W.K. Rohwedder and R.A. Rhodes, *Lipids 4*:356 (1969).
3. Niehaus,W.G. Jr., A. Kisic, A. Torkelson, D.J. Bednarczyk and G.J. Schroepfer Jr., *J. Biol. Chem. 245*:3790 (1970).
4. Litchfield,J.H., U.S. Patent 4,582,804 (1986).
5. McDermott, G.N., in *Bailey's Industrial Oil and Fat Products*, edited by Daniel Swern, John Wiley & Sons, New York, 1982, pp. 343-405.
6. Hosie, L., S. Koritala and C.W. Hesseltine, Abstract No. 18, presented at the 194th National Meeting of the American Chemical Society Division of Agricultural and Food Chemistry, New Orleans, LA, Aug. 30-Sept. 4 (1987).
7. Wallen, L.L., R.G. Benedict and R.W. Jackson, *Arch. Biochem. Biophys. 99*:249 (1962).
8. Schroepfer, G.J. Jr., and W.G. Niehaus Jr., *J. Biol. Chem. 245*:3798 (1970).
9. Wallen, L.L., E.N. Davis, Y.V. Wu and W.K. Rohwedder, *Lipids 6*:745 (1971).
10. Soda, K., *J. Am. Oil Chem. Soc. 64*:1254 (1987).

Molecular Biological Studies of Plant Lipoxygenases

G. Bookjans, M. Altschuler, J. Brockman, R. Yenofsky[a]**, J. Polacco**[b]**, R. Dickson**[c]**, G. Collins** and **D. Hildebrand**

Dept. of Agronomy, Univ. of Kentucky, Lexington, KY, [a]Phytogen Corp., Pasadena, CA, [b]Dept. of Biochemistry, University of Missouri, Columbia, MO, and [c]Dept. of Biochemistry, University of Kentucky, Lexington, KY.

Preliminary studies indicated that tobacco DNA contains sequences homologous to soybean lipoxygenase cDNA. Therefore, a genomic DNA library of *Nicotiana tabacum* (cv Samsun) cloned in the lambda phage *Charon 32* was screened with soybean cDNA coding for lipoxygenase. One *Charon 32* clone, which showed cross-hybridization with the cDNA clone for lipoxygenase-3 from soybean (pLX-10), was found to contain sequences very homologous to all soybean lipoxygenase cDNAs isolated to date. Most of this putative lipoxygenase gene from tobacco has been sequenced, and its amino acid sequence deduced is compared with the primary structures of three lipoxygenases expressed in soybean seed.

Lipoxygenases (Linoleate:oxygen Oxidoreductase, E.C. 1.13.11.12) are found in many plant and animal species. They catalyze the peroxidation of compounds containing a 1,4 pentadiene structure such as polyunsaturated fatty acids. The resulting lipid hydroperoxides are chemically unstable and convert either spontaneously or enzymatically to a variety of secondary oxidation products. In animals, lipoxygenases have been shown to catalyze the formation of precursors of potent physiological effectors. Any physiological role of lipoxygenases found in plants, however, remains to be elucidated (1,2).

At least three isoforms of lipoxygenase, designated LOX-1, LOX-2 and LOX-3, have been characterized in soybean seeds (3). All three isozymes are globular proteins consisting of a single polypeptide of about 96 Kd (4).

MATERIALS AND METHODS

Gene screen filters were purchased from New England Nuclear (Boston, MA) and used for colony hybridizations according to the manufacturer's recommendations. Screening of the genomic library of tobacco DNA (*Nicotiana tabacum* cv Samsun) in lambda *Charon 32* was done with the internal Bam HI/Eco RI cDNA fragment of pLX-10 (5). Nitrocellulose filters used for Northern blots were obtained from Millipore Corp. (Bedford, MA). RNA and DNA from soybean and tobacco plants, respectively, were isolated by the CTAB (cetyltrimethylammonium bromide) extraction procedure (6). Total cellular RNA (20 μg/lane) was separated electrophoretically in a 1% w/w agarose/formaldehyde gel, transferred to nitrocellulose and probed with the cDNA 3' to the Eco RI restriction site in pLX 10 (Eco-RI/Hin PI fragment) and with a HIN cII/xhoII fragment of the 3' end of pLX-65 (5). Both 3' cDNA probes were cloned in M13 vectors. All probes used in hybridization experiments were labeled with [³²P]-dCTP by the oligonucleotide labeling procedure (7). Eco-RI DNA fragments of *Charon 32* clone Cl were subcloned in the Bluescribe M13+ vector of Stratagene Cloning Systems (San Diego, CA). DNA nucleotide sequencing was done by the M13 dideoxynucleotide method (8).

RESULTS AND DISCUSSION

Recently, Start et al. (5) reported the isolation and characterization of three partial LOX cDNA clones (pAL-134, pLX-65 and pLX-10) from developing soybean seeds. Based on Northern blots of embryo transcripts derived from soybean genotypes lacking one of the three LOX isozymes and individually probed with all three LOX cDNA clones, pAL-134 as well as pLX-65 were tentatively designated as C lipoxygenase-1-like' while pLX-10 was tentatively identified as the clone encoding LOX-3. The data of Shibata et al. (9), however, strongly suggested that the clone pLX-65 does not encode LOX-1. Figure 1 shows the hybridization of the 3' untranslated sequences of pLX-65 (G-65) and pLX-10(I10) to mRNA isolated from soybean mutants missing one or two of the lipoxygenase isozymes. The hybridization pattern obtained with the 3' untranslated sequence of pLX-10 represents LOX-3. The Northern blot probed with the 3' untranslated DNA of pLX-65, however, is inconclusive (Fig. 1). The lipoxygenase isozyme encoded in the pLX-65 cDNA clone is unknown.

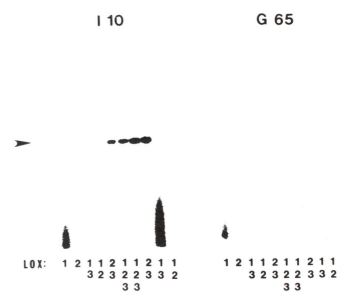

FIG. 1. Northern blots of total RNA from soybean genotypes lacking one or two lipoxygenase isozymes (10). The blots were probed with the 3' untranslated DNA sequences of clones pLX-10 (I10) and pLX-65 (G65). The arrow indicates the position of LOX transcripts. The LOX-isozymes present in the soybean genotypes used for the isolation of total RNA are indicated below each lane.

Preliminary studies indicated that tobacco DNA contained sequences homologous to soybean lipoxygenase cDNAs. Therefore, a genomic DNA library from *N. tabacum* cloned in lambda phage *Charon 32* (provided by R. Goldberg) was screened with the soybean cDNA coding for LOX-3 (pLX-10). Three clones containing putative lipoxygenase genes have been isolated from the genomic DNA library. A detailed restriction map for two of these

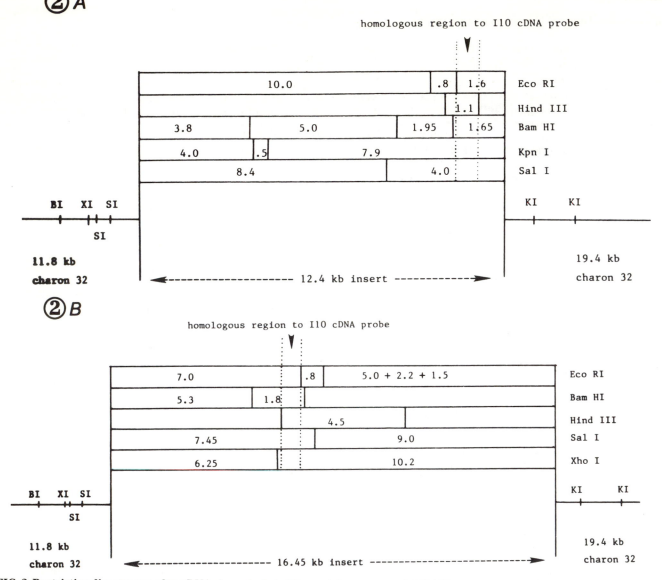

FIG. 2. Restriction digest maps of two DNA clones isolated from a tobacco genomic DNA library constructed in phage lambda *Charon 32* (supplied by R. Goldberg). The clones were isolated by screening the DNA library with an internal EI/BI 700 b.p. fragment of the cDNA clone pLX-10 (5). The numbers in the restriction fragments indicate their sizes in Kb. KI, Kpn-I; BI, Bam-HI; XI, XHO-I; S1, Sal-I; EI, Eco- RI.

clones has been established (clones C1 and L1). *Charon 32* clone L1 contains 12.4 kb, clone C1 about 16.45 kb tobacco DNA as an insert (Fig. 2a,b). The homologous DNA sequence to the cDNA probe of pLX-10 has been pinpointed in both of them. The orientation of clone C1 with respect to the 5′ and 3′ end of the putative lipoxygenase gene was determined by comparing the DNA sequence of pLX-10 (G. Bookjans, unpublished) with its homologous counterpart in clone C1. From this comparison, it was concluded that the entire putative LOX-gene of clone C1 is located on the Eco R1 restriction enzyme fragments of 7 and 0.8 Kb (Fig. 2B). Subsequently, it was decided to subclone both Eco R1 restriction enzyme fragments and to determine their DNA sequences. It was found that both Eco R1 fragments indeed contain the DNA sequence of a putative lipoxygenase gene. The amino acid sequences deduced for this putative LOX gene

(pGBC1) is presented in Table 1. Table 1 also contains the amino acid sequences of three soybean lipoxygenases that were deduced from the DNA sequences of the corresponding cDNA clones. The amino acid sequence of LOX-3 shown is partly encoded in pLX-10 (G. Bookjans, unpublished), a full-length cDNA clone for LOX-3 and its DNA sequence was obtained by R. Yenofsky (unpublished data). The amino acid sequence of LOX-1 was taken from Shibata et al. (9); the amino acid sequence deduced from pLX-65 is based on a corrected version (G. Bookjans, unpublished) of the DNA-sequence published by Start et al. (5). Evidently, the primary structures of the lipoxygenases exhibit high degrees of homology. Furthermore, it should be mentioned that the putative lipoxygenase gene from *N. tabacum* also contains the same apparent intron/exon splicing sites as the gene for LOX-3 in soybeans (data not shown), indicating a high degree of con-

TABLE 1

Lipoxygenase Amino Acid Sequence Comparisons

Lox 3 :	1	MLGGLLHRGH	KIKGTVVLMR	KNVLDVNSVT	SVGGIIGQGL	DLVGSTLDTL	TAFLGRPVSL
Lox 1 :		M FSAGH	KIKGTVVLMP	KNELEVNPDG	S	AV DNL	NAFLGRSVSL
pG65 :		————————	————————	————————	————————	————————	————————
pGBC1 :		————————	————————	————————	————————	————————	————————
	61	QLISATKADA	NGKGKLGKAT	FLEGIITSLP	TLGAGQSAFK	INFEWDDGSG	ILGAFYIKNF
		QLISATKADA	HGKGKVGKDT	FLEGINTSLP	TLGAGESAFN	IHFEWDGSMG	IPGAFYIKNY
		——————A	NDLQGKHSNPA	YLENWLTTIT	PLTAGESAYG	VTFDWDEEFG	LPGAFIIKNL
	121	MQTEFFLVSL	TLEDIPNHGS	IHFVCNSWIY	NAKLFKSDRI	FFANQTYLPS	ETPAPLVKYR
		MQVEFFLKSL	TLEAISNQGT	IRFVCNSWVY	NTKLYKSVRI	FFANHTYVPS	ETPAPLVSYR
		HFTEFFLKSV	TLEDVPNHGK	VHFVCNSWVY	PANKYKSDRI	LFANKTYLPS	ETPAPLRKYR
	181	EEELHNLRGD	GTGERKEWER	VYDYDVYNDL	GDPDKGENHA	RPVLGGNDTF	PYPRRGRTGR
		EEELKSLRGN	GTGERKEYDR	IYDYDVYNDL	GNPDKSEKLA	RPVLGGSSTF	PYPRRGRTGR
		————————	————————	————————	——————A	RPILGGSSTH	PYPRRGRTAR
		ENELLTLRGD	GTGKLEAWDR	VYDYAFYNDL	GDPDLGAQHAHV	RPILGGSSDY	PYP————
	241	KPTRKDPNSE	SR SNDVYLPR	DEAFGHLKSS	DFLTYGLKSV	SQNVLPLLQS	AFDLNFTPRE
		GPTVTDPNTE	KQGEV FYVPR	DENLGHLKSK	DALEIGTKSL	SQIVQPAFES	AFDLKSTPIE
		YPTRKDQNSE	NLGE VYVPR	DENFGHLKSS	DFLAYGIKSL	SQYVLPAFES	VFDLNFTPNE
		———DPESE	SRIPLLLSLDIYVPR	DERFGHLKLS	HFLTYALKSM	VQFILPELHA	LFD STPNE
	301	FDSFDEVHGL	YSGGIKLPTD	IISK SI PLPV	LKEIFRTDGE	QA LKFPPPKV	IQVSKSAWMT
		FHSFQDVHDL	YEGGIKLPRD	VIST II PLPV	IKELYRTDG	QHILKFPQPHV	VQVSQSAWMT
		FYSFQDVRDL	HEGGIKLPTE	VIST IM PLPV	VKELFRTDGE	QV LKFPPPHV	IQVSKSAWMT
		FDSFEVVLSI	YEGGIKLPQG	PLFKALISSIPLEM	VKELLRTDGE	G IMKFPTPLV	IKEDKTAWRT
	361	DEEFAREMLA	GVNPNLIRCL	KEFPPRSKLD	SQVYGDHTSP	ITKEHLEPNL	EGLTVDEAIQ
		DEEFAREMIA	GVNPCVIRGL	EEFPPKSNLD	PAIYGDQSSK	ITADSLD L	DGYTMDEALG
		DEEFAREMVA	GVNPCVIRGL	QEFPPKSNPD	PTIYGEQTSK	ITADALD L	DGYTVDEAHA
		DEEFGREMLA	GVNPVIIRNL	QEFPPKSKLD	PQVYGNQDST	ITIQHIEDRL	DGLTIDEAIK
	421	NKRLFLLGHH	DPIMPYLRRI	NA TSTKAYAT	RTILFLKNDG	TLRPLAIELS	LPHPQGDQSG
		SRRLFMLDYH	DIFMPYVRQI	NQLNSAKTYAT	RTILFLREDG	TLKPVAIELS	LPHSAGDLSA
		SRRLFMLDYH	DVFMPYIRRI	NQ TYAKAYAT	RTILFLRENG	TLKPVAIELS	LPHPAGDLSG
		SNRLFILNHH	DTIMPYLRRI	NA TTTKTYAS	RTLLFFQDNG	SLKPLAIELS	LPHPDGDQFG
	481	AFSQVFLPAD	EGVESSIWLL	AKAYVVVNDS	CYHQLVSHWL	NTHAVVEPFI	IATNRHLSVV
		AVSQVVLPAK	EGVESTIATS	KAYVIVNDS	CYHQLMSHWL	NTHAAMEPFV	IATHRHLSVL
		AVSQVILPAK	EGGESTIWLL	AKAYVVVNDS	CYHQLMSHWL	NTHALIEPFI	IATNRHLSAL
		AISKVYTPAG	EGVEGSIWEL	AKAYVAVNDS	GVHQLISHWL	NTHAVIEPFV	IATNSELSVI
	541	HPIYKLLHPH	YRDTMNINGL	ARLSLVNDGG	VIEQTFLWGR	YSVEMSAVVY	KDWVFTDQAL
		HPIYKLLTPH	YRNNMNINAL	ARQSLINANG	IIETTFLPSK	YSVEMSSAVY	KNWVFTDQAL
		HPIYKLLTPH	YRDTMNINAL	ARQSLINADG	IIEKSFLPSK	HSVEMSSAVY	KNWVFTDQAL
		HPIHKLLHPH	FRYTMNINAM	ARQILINAGG	VLESTVFPTK	CAMEMSAVVY	KNWIFPDESL
	601	PADLIKRGMA	IEDPSCPHGI	RLVIEDYPYA	VDGLEIWDAI	KTWVHEYVFL	YYKSDDTLRE
		PADLIKRGVA	IKDPSTPHGV	RLLIEDYPYA	ADGLEIWAAI	KTWVQEYVPL	YYARDDDVKN
		PADLIKRGVA	IKDPSAPHGL	RLLIEDYPYA	VDGLEIWAAI	KTWVQEYVSL	YYARDDDVKP
		PTDLLKRGMA	VEDSSSPHGI	RLLIQDYPYA	VDGLEIWSAI	KSWVTEYCSF	YYKSDDSILK
	661	DPELQACWKE	LVEVGHGDKK	NEPWWPKMQT	REELVEACAI	IIWTASALHA	AVNFGQYPYG
		DSELQHWWKE	AVEKGHGDLK	DKPWWPKLQT	LEDLVEVCLI	IIWIASALHA	AVNFGQYPYG
		DSELQQWWKE	AVEKGHGDLK	DKPWWPKLQT	IEELVEICTI	IIWTASALHA	AVNKGQYPYG
		DNELQAWWKE	LREAGHGDLK	DEPWWPKMEN	CQELIDSCTI	IIWTTSALHA	AVNFGQYPYA
	721	GLILNRPTLS	RAFMPEKGSA	EYEELRKNPQ	KAYLKTITPK	FQTLIDLSVI	EILSRHASDE
		GLIMNRPTAS	RRLLPEKGTP	EYEEMINNK	KAYLRTITSK	LPTLISLSVI	EILSTHASDE
		GFIQNRPTSS	RRLLPEKGSP	EYEEVVKSHQ	KAYLRTITSK	FQTLVDLSVI	EILSRHASDE
		GYLPNRPTVS	RRLMPEPGTS	EYELLKTNPD	KAFLRTITAQ	LQTLLGVSLI	EILSRHTSDE
	781	VYLGDRDNPN	WTSDTRALEA	FKRFGNKLAQ	IENKLSERNN	DEKLR NRCGP	VQMPYTLLLP
		VYLGQRDNPH	WTSDSKALQA	FQKFGNKLKE	IEEKLVRRNN	DPSLQGNRLGP	VQLPYTLLYP
		VYLGQRDNPH	WTSDSKALQA	FQKFGNKLKE	IEEKLARKNN	DQSLS NRLGP	VQLPYTLLHP
		IYLGQRDSPK	WTNDEVPLAA	FERFGNKLSD	IENRIIEMNG	DQIWR NRSGP	IKAPYTLLFP
	841	SSKEGLTFRG	IPNSISI				
		SSEEGLTFRG	IPNSISI				
		NSEEGLTCRG	IPNSISI				
		TSEGGLTGKG	ILVFDFEGFT	HGFHQVTVED	GTIISDN		

F, phe; V, val; A, ala; ;N, asn; C, cys; L, leu; S, ser; Y, tyr; ;K, lys; W, trp; I, ilu; P, pro; H, his; D, asp, R, arg, M, met; T, thr; Q, gln; E, glu; G, gly.

servation of the soybean and tobacco LOX genes.

To elucidate the physiological roles of LOX, we have undertaken the task of cloning the full-length cDNA of the LOX-3 gene from soybean in the "sense" and "antisense" orientation into a vector suitable for the transformation of plants (pKYLX7). Tobacco plants transformed with both constructs will be evaluated for levels of LOX mRNAs, proteins and levels of enzymatic activity. Those plants with altered LOX activity will be investigated for changes in volatile lipid peroxidation product formation in different tissues and for changes in growth and development and pest resistance.

REFERENCES

1. Mack, A.J., T.K. Peterman and J.N. Siedow, *Current Topics in Biological and Medical Research 13*:127 (1987).

2. Needleman, P., J. Turk, B.A. Jakschik, A.R. Morrison and J.B. Lefkowith, *Ann. Rev. Biochem. 55*:69 (1986).

3. Christopher, J.P., E.K. Pistorius and B. Axelrod, *Biochim. Biophys. Acta 284*:54 (1972).

4. Axelrod, B., T.M. Cheesbrough and S. Laasko, *Meth. Enzymol. 71*:441 (1981).

5. Start, W.G., Y. Ma, J.C. Polacco, D.F. Hildebrand, G.A. Freyes and M. Altschuler, *Plant Mol. Biol. 7*:11 (1987).

6. Taylor, B., and A. Powell, *Bethesda Res. Lab. Focus 4*:4 (1982).

7. Feinberg, A.P., and B. Vogelstein, *Anal. Biochem. 132*:6 (1983).

8. Sanger, F., S. Nicklin and A.R. Coulson, *Proc. Natl. Acad. Sci. 74*:5463 (1977).

9. Shibata, D., J. Steczko, J.E. Dixon, M. Hermodson, R. Yazdanparast and B. Axelrod, *J. Biol. Chem. 262*:10080 (1987).

10. Davies, C.S., and N.S. Nielsen, *Crop Sci. 27*:370 (1987).

Synthesis of Fatty Acids in Cocoa Beans at Different Stages of Maturity

C. Foster[a], M. End[b], R. Leathers[a], G. Pettipher[c], P. Hadley[b] and A.H. Scragg[a]

[a]Wolfson Institute of Biotechnology, University of Sheffield, Sheffield S10 2TN, U.K., [b]School of Horticulture, University of Reading, Whiteknights, Reading RG6 2LA, U.K., [c]Cadbury-Schweppes PLC, The Lord Zuckerman Research Centre, University of Reading, Whiteknights, Reading RG6 2LA, U.K.

Cocoa beans from the tree *Theobroma cacao* L. produce the main ingredient of chocolate. The fat, cocoa butter, also is used in the pharmaceutical and cosmetics industry. Fatty acid accumulation and synthesis were studied during cocoa bean development. These data indicated that cocoa butter synthesis occurred late in the development of the bean. In addition, the fatty acid profile of *T. cacao* suspension cultures was determined to be similar to that of the immature embryo. These data indicated that production of cocoa butter by suspension culture methods was not a viable alternative.

The cocoa bean, the seed of the tree *Theobroma cacao* L., is the main ingredient of chocolate and a source of cocoa butter, which is used in the cosmetics and pharmaceutical industry (1). The cocoa bean contains 55-60% oil w/w, of which 96% w/w is triacylglycerol. The triacylglycerol primarily is comprised of three molecular species: 2-oleopalmitostearin, 2-oleodipalmitin and 2-oleodistearin. These triacylglycerols account for the unique physical properties of cocoa fat (cocoa butter), and hence chocolate.

The demand for cocoa butter is growing, but not all cocoa butters have the correct triacylglycerol composition. In some cases, soft cocoa butter results, which requires the addition of a hardening agent. Detailed knowledge of the process of fat synthesis in cocoa beans is lacking, hence an understanding of the reasons for soft cocoa butter formation is not possible. Using cocoa beans of various degrees of maturity, we have followed lipid metabolic changes during development to investigate triacylglycerol synthesis in cocoa beans.

EXPERIMENTAL PROCEDURES

Suspension cultures of *T. cacao* L. were developed from callus cultures initiated from mature beans of the Forastero-type. The callus was initiated on Murashige and Skoog medium (2) containing 1 mg/l zeatin, 1 mg/l indole butyric acid and 1% agar. Suspension cultures were grown at 25 C with continuous shaking (150 rpm) on the same medium minus agar and subcultured every two wk using a 1:5 inoculation density.

Cocoa pods of known age, measured as days after pollination, were obtained from *Theobroma cacao* trees grown under controlled heat and humidity conditions similar to Ghanayan, Malaysian and Brazilian climates.

Lipids were extracted using the method of Bligh and Dyer (3), and fatty acid methyl esters (FAME) were prepared using boron trifluoride (4). The FAME were analyzed using a Shimadzu GC-8A gas chromatograph equipped with a flame ionization detector. Separations were carried out in a glass column 2 m x 4 mm, packed with 10% diethylene glycol succinate on Chromosomal W

HP 80-100 mesh support. Operating conditions were Nitrogen carrier gas, 30 ml/min; oven temperature, 155 C; injection port and flame ionization detector, 250 C; sample size, 1 μl.

For the uptake of ^{14}C-sucrose, 2-10 beans, depending on size, were incubated in 50 ml sterile water containing 0.5 mM sucrose and 2 μCi ^{14}C-sucrose for 24 hr. The lipids were extracted according to the method of Bligh and Dyer (3), and phenacyl bromide esters were prepared (5) for analysis by HPLC. Two Gilson model 303 pumps were used in conjunction with a Gilson variable UV detector measuring absorbance at 254 nm. Separations were carried out on two, 10 cm x 8 mm, Z Modules (Millipore-Waters) in series containing C18 microbondpak. The acetonitrile:water (80:20, v/v) mobile phase was run isocratically for 60 min and then linearly increased to 100% acetonitrile during another 10 min. The flow rate was 2 ml/min. Samples (50 μl) were applied to the column through a Rheodyne 7125 injection port and 2 ml fractions were collected. Before derivitization, 10 fatty acid standards were added to the sample to visualize the fatty acid esters on the HPLC. Aliquots of the collected fractions were mixed 1:9 with Labscint (Raytest Instruments). Radioactivity was determined with a Philips PW 4700 liquid scintillation counter.

RESULTS

Fatty Acid Composition of Beans of Various Ages

At 100 days post-pollination, the developing cocoa embryos were approximately 4-8 mm in length with recognizable features such as cotelydons. The embryos matured

Table 1

Fatty Acid Composition of Developing Cocoa Beans

Fatty acid	Stage of development				
	White embryo	White/pink embryo	Pink bean	Purple bean	Cocoa butter
	% w/w				
12:0	1.2	0.5	0.00	0.0	0.0
14:0	2.0	0.6	0.5	0.2	0.0
16:0	36.2	33.7	35.9	27.4	29.2
18:0	6.9	7.7	7.2	22.4	35.9
18:1	14.6	29.0	21.3	41.1	31.9
18:2	39.1	26.3	31.7	7.8	3.0
18:3	0.0	0.7	0.8	0.6	0.0
20:0	0.0	1.5	2.6	0.5	0.0

rapidly, increasing in size and becoming pink in color due to the synthesis of anthocyanins. The color continued to darken until the bean was fully mature at 180 days after pollination, giving a dark purple bean 2 cm in length and weighing about 1-2 grams. Although the age in days after pollination was known for the beans used in the experiments, some variation was noted in bean development such that the visual appearance of the bean was a more accurate estimate of maturity as found by Lehrian and Keeney (6). Four bean types: white embryo, white/pink embryo, fully pink bean and purple bean were analyzed for their fatty acid content and composition. Figure 1 shows that the fatty acid content of the bean increased rapidly from the pink bean to purple bean (6,7). The changes in the relative proportions of the fatty acids are shown in Table 1. The emergence of a fatty acid profile representative of cocoa butter can only be seen in the last stage, purple bean, with a significant increase in stearic acid and a decrease in linoleic acid.

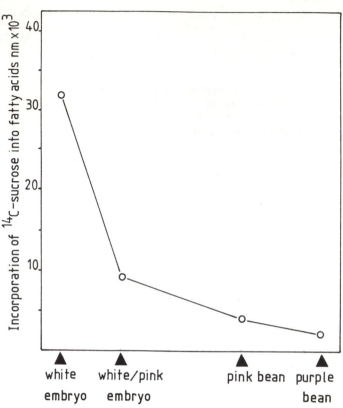

FIG. 2. Fatty acid synthetic rates during cocoa bean development.

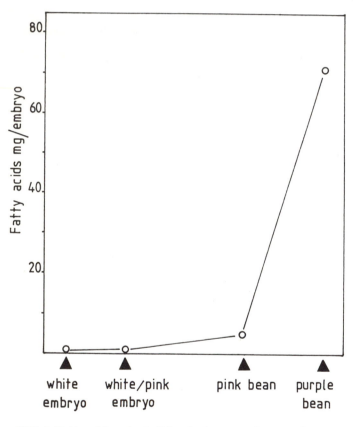

FIG. 1. Fatty acid content of developing cocoa bean embryos.

The Rate of Fatty Acid Synthesis in Beans of Various Ages

Beans of the same age as those assayed for fatty acid content were incubated with ^{14}C-sucrose for 24 hr. The fatty acids were extracted from the beans and the incorporation of radioactivity into the fatty acids determined by HPLC. Figure 2 shows that the rate of incorporation of sucrose into fatty acids was greatest at the white embryo stage. The distribution of incorporated radioactivity among 16:0, 18:0, 18:1 and 18:2 is shown in Table 2. Although 16:0 and 18:1 were not separated by HPLC, the trend over the whole age range indicated a reduction in linoleic acid and increase in stearic acid synthesis.

Table 2

Distribution of Radioactivity Among Major Cocoa Bean Fatty Acids During Seed Development

	Stage of development			
Fatty acid	White embryo	White/pink embryo	Pink bean	Purple bean
	% of radioactivity			
16:0+18:1	53.0	71.5	57.1	64.0
18:0	15.7	1.1	18.6	29.6
18:2	31.3	13.4	24.3	6.4

Fatty Acid Composition and Synthesis in Suspension Cells

The growth of *T. cacao* L. cells in culture has been proposed as an alternative source of cocoa butter (8). We have developed a suspension culture system for *T. cacao*. However, the fatty acid composition of the suspension culture was not similar to that of cocoa butter. The fatty acid composition (%, w/w) of the suspension culture contained 12:0, 0.9%; 14:0, 1.7%; 16:0, 28.4%; 18:0, 5.7%; 18:1, 6.4%; 18:2, 52.2%; 18:3, 3.3%; and 20:0, 1.4%.

DISCUSSION

Using beans at various stages of maturity, from the small white embryo stage (100-110 days after pollination) to deep-purple bean (160-170 days after pollination), we have studied the formation of lipid in cocoa beans. The accumulation of fatty acid was rapid between the pink bean and purple bean stages, 120-160 days after pollination. A similar case has been demonstrated by Lehrian and Keeney (6) and Wright et al. (7). The fatty acid profile of developing cocoa beans became similar to that of cocoa butter between the pink bean and purple bean stages of development.

The metabolism of ^{14}C-sucrose into fatty acids, however, was greatest at the white embryo stage. Lipid synthesis declined rapidly at the white/pink embryo stage. These data indicated that the rate of fatty acid synthesis was most rapid at the early stages. Although lower radioactivities at later stages may be due to permeability problems, the incorporation of ^{14}C-sucrose into various fatty acids showed a trend towards cocoa butter synthesis between the pink bean and purple bean stage. These data corresponded with the accumulation of cocoa butter during the maturation of the cocoa bean.

Suspension cells of *T. cacao* L. Forastero type produced a fatty acid profile similar to that found in the youngest embryo. A similar result was found for the Criollo *T. cacao* suspension culture by Tsai and Kinsella (8). A study of the enzymes involved in lipid synthesis in both beans and suspension culture is needed to explain why the suspension cells were unable to synthesize cocoa butter.

ACKNOWLEDGMENT

The authors wish to thank Cadbury Schweppes PLC for their support.

REFERENCES

1. Wood, G.A.R., *Biologist 33*:99 (1986).
2. Murashige, T., and F. Skoog, *Physiol. Plant 15*:431 (1962).
3. Bligh,E.G., and W.J. Dyer, *Can. J. Biochem. Physiol. 37*:911 (1959).
4. Kinsella, J.E., J.L. Shimp, J. Mai and J.L. Weihrauch, *J. Am. Oil Chem. Soc. 54*:424 (1977).
5. Borch, R.F., *Anal. Chem. 47*:2437 (1975).
6. Lehrian, D.W., and P.G. Keeney, *J. Am. Oil Chem. Soc. 57*:61 (1980).
7. Wright, D.C., W.D. Park, N.R. Leopold, P.M. Hasegawa and J. Janick, *J. Am. Oil Chem. Soc. 59*:475 (1982).
8. Tsai, C.H., and J.E. Kinsella, *Lipids 16*:577 (1981).

Enzymatic Acidolysis Reaction of Some Fats

D.K. Bhattacharyya and **S. Bhattacharyya**
Oil Technology Division, Chemical Technology Department, University Colleges of Science and Technology, 92, Acharyya Prafulla Chandra Roy Road, Calcutta 700 009, India

Acidolysis reactions with sal (*Shorea robusta*), mango (*Mangifera indica*), mowrah (*Madhuca latifolia*) and palm (*Elaeis guineensis*) fats were conducted with *Mucor miehei* lipase and selected exogenous fatty acid substrates. Sal and mango kernel fats were enriched in palmitic acid content by this acidolysis reaction. Mowrah and palm fats were enriched with stearic acid by the same procedure. Physical properties of these fats such as the slip point and the Solid Fat Index (SFI) were significantly changed in the resulting products.

The incorporation of a desired fatty acid specifically at the 1,3-positions of natural triglycerides by acidolysis reactions with the aid of a 1,3-specific microbial lipase has become a very important area of recent research investigations. Such a selective acidolysis reaction has been used primarily in making natural triacylglycerols (TG) of composition and properties more suited to the preparation of specialty products like confectionery, vanaspati and margarine fats. Some pioneering work has been published in this area by Macrae (1), Hansen and Eigtved (2) and Neidleman and Geigert (3).

This study describes the acidolysis reaction of some fats with a microbial lipase. Fats like sal (*Shorea robusta*), mango (*Mangifera indica*), mowrah (*Madhuca latifolia*) and palm (*Elaeis guineensis*) are distinguished by their typical fatty acid profiles. Sal and mango fats naturally are rich in stearic acid (18:0) and oleic acid (18:1) but are deficient in palmitic acid. Mowrah fat contains about 50% saturated acids, primarily 16:0 and 18:0. Palm oil has low 18:0 but has high levels of 16:0 and 18:1. This work relates to the preparation of palmitic acid-rich sal and mango fat, and stearic acid-rich mowrah fat and palm oil by enzymatic acidolysis with *Mucor miehei*. The products of these reactions were characterized in terms of fatty acid composition, slip point and Solid Fat Index (SFI) by standard methods of analysis.

EXPERIMENTAL

Acidolysis reactions were conducted with immobilized *Mucor miehei lipase* (Lipozyme®) by the procedure of Hansen and Eigtved (2). Equimolar proportions of triacylglycerol from each oil species and exogenous fatty acid (16:0 or 18:0) were dissolved in a minimum quantity of petroleum ether (Glaxo, A.R. grade). The reaction was carried out at 40 C with enzyme (10%, w/w) and an equal amount of water. Reaction products were recovered from the reaction mixture with diethyl ether. The etheral solution was dried over anhydrous sodium sulphate, and the product was isolated by desolventization at atmospheric pressure and then under vacuum. Excess exogenous fatty acids were removed from the product by mixed solvent refining (4).

Thermal characteristics determined were slip point (5), SFI (6) and fatty acid (7) composition of each reaction product.

RESULTS AND DISCUSSION

The fatty acid compositions of the original fats and acidolysis reaction products are shown in Table 1. After acidolysis, the concentration of 16:0 in sal fat and mango kernel fat was enriched 6.1- and 3.7-fold, respectively; the 18:0 concentration in mowrah and palm oil was increased 1.8- and 6.5-fold, respectively.

Apparently, 18:0 was replaced by 16:0 in sal and mango fat specifically at the 1,3-positions. In mowrah oil after the acidolysis reaction, the concentration of 18:0 was increased at the expense of 18:1 and 18:2. However, the amount of 16:0 remained virtually unaltered. In palm oil, 16:0 again remained virtually unaltered. In palm oil, 16:0 was decreased as 18:0 increased after reaction. Although the amount of 18:1 increased slightly, 18:2 was completely eliminated. Further analysis suggested that the lipase displaced 16:0 with 18:0 primarily from the saturated TG molecular species PPP, PPSt and POP, in which P, 16:0; St, 18:0; O, 18:1.

The effect of exogenous fatty acid substitution in TG

Table 1

Fatty Acid Composition of Original and Acidolyzed Fats with *Mucor miehei* Lipase

Fat	Exogenous fatty acid	Reaction time (hr)	Major fatty acids (% w/w)					
			14:0	16:0	18:0	18:1	18:2	20:0
Sal original	none	0.0	0.1	4.2	43.8	48.0	0.0	3.8
acidolyzed	16:0	3.5	1.1	25.6	29.3	42.3	0.0	1.6
Mango original	none	0.0	0.0	7.0	40.7	47.2	2.6	2.4
acidolyzed	16:0	4.0	0.0	25.7	28.1	42.8	1.6	1.8
Mowra original	none	0.0	0.2	23.3	19.0	46.2	11.2	0.0
acidolyzed	18:0	4.0	1.9	22.5	34.6	33.9	7.2	0.0
Palm original	none	0.0	1.5	45.2	5.6	40.0	7.7	0.0
acidolyzed	18:0	4.0	2.2	17.7	36.4	43.6	0.0	0.0

Table 2

Slip Point and SFI Characteristics of Enzyme-acidolyzed Products of Sal, Mango, Mowrah and Palm Oil

Fat	Slip point (C)	SFI at (C)						
		15	20	25	30	35	40
Sal	30.5	69.1	64.0	45.5	29.5	12.0	0.0
Acidolyzed sal	24.0	45.4	38.9	11.7	4.5	0.7	0.0
Mango	33.0	54.8	51.4	42.2	37.7	13.8	0.0
Acidolyzed mango	34.5	31.8	26.6	8.5	5.6	2.4	0.0
Mowrah	27.0	20.4	2.8	0.0	0.0	0.0	0.0
Acidolyzed mowrah	36.5	29.1	16.6	13.0	10.1	5.9	0.0
Palm	36.5	41.2	30.7	22.8	8.0	3.5	0.0
Acidolyzed palm	35.0	36.5	31.2	19.6	11.1	6.3	0.0

upon the slip point and SFI characteristics of the four fats are shown in Table 2. Acidolyzed sal fat exhibited a lower melting point, and SFI values were greatly decreased. This could be explained by significant reduction in the amount of 18:0 in sal fat after acidolysis. Mango fat after the reaction had a slightly greater slip point, while the SFI was significantly decreased. Such a drastic decrease in the SFI content could not be explained easily, even though the amount of stearic acid was lowered by about 12% w/w. Mowrah fat showed significant increase in both slip point and SFI characteristics due to higher 18:0 after acidolysis. The incorporation of 18:0 in palm oil had little effect on slip point and the SFI characteristics, but at higher temperatures the SFI was slightly greater for the product.

A comparison with the previous report (8) reveals that the direct acidolysis of a fat with a 1,3 specific lipase is better than the process of first splitting it by pancreatic lipase and then reesterifying it with the desired fatty acid. Such an approach invariably leads to a random pattern of incorporation of the desired and other fatty acids and much higher slip points and SFIs at higher temperatures. The selective incorporation of fatty acids such as 16:0 or 18:0 in specific positions of triglycerides can be achieved by the use of *Mucor miehei* lipase. These products also show significant changes in thermal properties.

ACKNOWLEDGMENTS

This work was supported by a grant from the Council of Scientific and Industrial Research, New Delhi, India. The enzyme used was provided by Novo Industri A/S, Denmark, The Netherlands.

REFERENCES

1. Macrae, A.R., *J. Am. Oil Chem. Soc. 60*:291 (1983).
2. Hansen, T.T., and P. Eigtved, *Proceedings World Conference on Emerging Technologies in the Fats and Oils Industry*, edited by A.R. Baldwin, American Oil Chemists' Society, Champaign, IL, 1986, pp. 365–369.
3. Neidleman, S.L., and J. Geigert, *J. Am. Oil Chem. Soc. 61*:290 (1984).
4. Bhattacharyya, D.K., and A.C. Bhattacharyya, *J. Oil Tech. Assoc. India 17*:31 (1985).
5. *Official and Tentative Methods of the American Oil Chemists' Society*, Vol. 1, 3rd edn., American Oil Chemists' Society, Champaign, IL, 1974, pp. CA 9a–52.
6. Boekenoogen, H.A., *Analysis and Characterization of Oils, Fats and Fat Products*, Vol. 1, Interscience Publishers, London, England, 1964, p. 143.
7. Luddy, F.E., R.A. Barford, P. Magidman and R.W. Riemenschneider, *J. Am. Oil Chem. Soc. 41*:693 (1964).
8. Bhattacharyya, D.K., S. Majumdar and S. Khatoon, *Proceedings World Conference on Emerging Technologies in the Fats and Oils Industry*, edited by A.R. Baldwin, American Oil Chemists' Society, Champaign, IL, 1986, pp. 414–417.

Oil Body Formation in Developing Rapeseed, *Brassica napus*

Denis J. Murphy and **Ian Cummins**

Department of Botany, University of Durham, Department of Botany, Science Laboratories, South Road, Durham DH1 3LE, U.K.

The biosynthesis and ultrastructure of oil bodies in the embryos of developing rapeseed have been followed. The mature oil bodies were made up of about 95% (w/v) triacylglycerol, 2% (w/w) phospholipid and 2% (w/w) protein. Oil synthesis commenced three wk after anthesis, and the maximal rate of oil accumulation was between three and six wk. The oil body proteins were not synthesized until six wk after anthesis, and the maximal rate of synthesis occurred between 10 and 12 wk. Immature oil bodies differed from mature oil bodies by having no electron-dense limiting membrane. A membrane only was seen around the oil bodies in the later stages of seed maturation, and this coincided with a large increase in the amount of oil body protein. The oil body polypeptide composition was unchanged during seed development and resembled that of the mature seed. These data were inconsistent with proposals that the endoplasmic reticulum was the origin for oil bodies.

Oilseeds are an increasingly important agricultural resource for both edible and nonedible oil products, and as possible sources of catalysts such as lipases for biotechnological applications. Despite this, many aspects of oilseed metabolism remain obscure. This is particularly true of the mechanism and regulation of storage oil and protein synthesis during seed development. The most widespread oilseed crop in the northern temperate regions of America, Europe and Asia is rapeseed. The dried seeds of rape typically contain 40-50% (w/w) storage oil and 20-25% (w/w) storage protein. This oil and protein is synthesized during a six to nine wk period, beginning at about three wk after flowering in the case of oil synthesis and several weeks later in the case of protein synthesis. There is some dispute concerning the subcellular localization of the enzymes responsible for storage oil synthesis and the mechanism by which oil bodies are assembled in developing oilseeds. It often has been assumed that triacylglycerol biosynthesis is exclusively confined to the endoplasmic reticulum (1,2), although this recently has been challenged (3). Various models of oil body formation have been proposed, including budding off from the endoplasmic reticulum (4-6) and direct synthesis in the cytoplasm of developing embryo cells (7-10). In this study, the mechanism and localization of oil body synthesis in developing rapeseed embryos is examined.

EXPERIMENTAL PROCEDURES

Isolation of Subcellular Fractions

Rapeseed plants (var Mikado) were collected at the stage of maximum oil accumulation. Intact embryos were dissected carefully out of the seeds and homogenized in 50 mM HEPES-KOH, pH 7.4, 1 mM EDTA, 1 mM DTT and 0.5M sucrose. Microsomal, oil body and soluble fractions were prepared as described (3).

Oleoyl-CoA Incubations

Incubations of subcellular fractions were performed in 55 μl 0.1M potassium phosphate, pH 7.2, 10 mM mgCl$_2$, 20 μM [1-^{14}C] oleoyl-CoA and 20 μM glycerol 3-phosphate. Lipid products were extracted and separated by thin layer chromatography as described (3).

Oil and Protein Estimation

Oil was estimated as either chloroform:methanol-soluble material or by quantitative gas chromatography of methyl esters. Protein was determined in fatted oil bodies by the method of Markwell (11).

Electron microscopy

Tissue slices (1 mm^3) were excised from the cotyledon under a pool of fixative. The tissue then was fixed in 1.5% paraformaldehyde plus 2.5% glutaraldehyde in 50 mM sodium cacodylate buffer, pH 7.0 for 16 hr at 4 C. After two washes in buffer and one in distilled water, the tissue was post-fixed in 1% w/w osmium tetroxide for 16 hr at 4 C. The specimen then was washed in distilled water for 16 hr at 4 C, dehydrated through an alcohol series, and embedded in Spurr's resin (12). Ultra-thin sections (gold/silver interference colors, 150-60 nm) were cut on an LKB ultratome (Type 4801A). Sections were collected on 200 mesh formvar-coated copper/rhodium grids and stained for 15 min in saturated aqueous uranyl acetate, followed by 15 min in alkaline lead citrate. Grids were viewed in a Phillips EM 400 at 80 kV and images recorded on Kodak electron image film 4489.

Table 1

Incorporation of [1-^{14}C] Oleoyl-CoA by Subcellular Fractions of Developing Rapeseed

Subcellular fraction	Total oleoyl-CoA incorporation	Lipid products[a]				
		Monoacylglycerol	Diacylglycerol	Triacylglycerol	Phosphatidylcholine	Phosphatidic acid
Oil bodies	30.3	1.0	26.1	23.1	18.8	0.7
Microsomes	24.5	0.8	26.1	20.0	20.8	0.8
Soluble	73.1	4.9	54.7	1.0	19.8	1.0

[a]Percentage total incorporation.

RESULTS AND DISCUSSION

The presence of the biosynthetic activities leading to triacylglycerol formation was tested by incubating oil body microsomal and soluble fractions of six wk-old developing rapeseed embryos with [1-^{14}C] oleoyl-CoA. In each case, as shown in Table 1, there was an active incorporation of the oleoyl-CoA into complex lipids. When considering total oleoyl-CoA incorporation, the soluble fraction was more active, on a per tissue basis, than the oil body and microsomal fractions together. The major product of the soluble fraction was diacylglycerol with little or no triacylglycerol being synthesized. In contrast, the oil body and microsomal fractions contained high percentages of radioactive triacylglycerol. These results and other studies (3) demonstrated that triacylglycerol biosynthesis in developing rapeseed was not confined to the microsomal fraction.

The accumulation of oil during the development of rapeseed embryos is shown in Figure 1. For the first three wk, there is no detectable synthesis of storage products. After about three wk, cell division ceases, and the cells become oleogenic. The most rapid rate of oil synthesis occurs from weeks 3-6, after which there is a gradual decrease in rate until the oil content of the mature seed is reached.

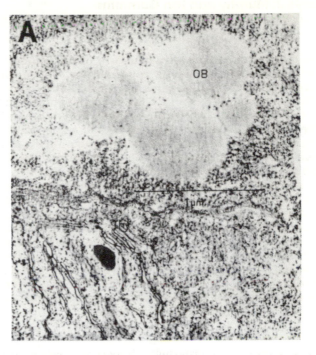

FIG. 2. Electron micrographs of different stages of oil body formation in embryos of developing rapeseed. (A) Eight wk after anthesis. Small (0.7 μm) oil bodies are scattered throughout the cytoplasm. No limiting membrane comparable to other cellular membranes can be seen around the newly formed oil bodies, but their contents are slightly stained.

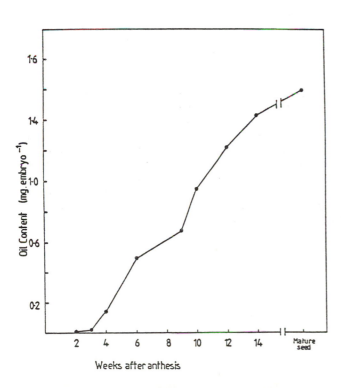

FIG. 1. Accumulation of oil in embryos of developing rapeseed. Oil content was measured either as chloroform:methanol-soluble material or by quantitative gas chromatography of fatty acid methyl ester derivitives.

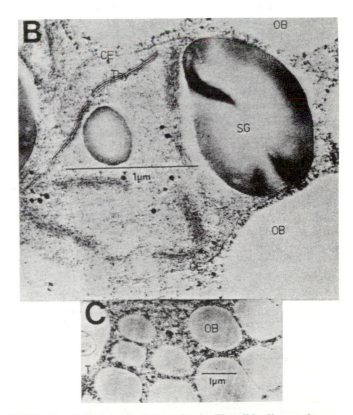

FIG 2. (B and C) Ten wk after anthesis. The oil bodies are larger and more numerous but still lack a definite electron-dense membrane. The contents have become less stained.

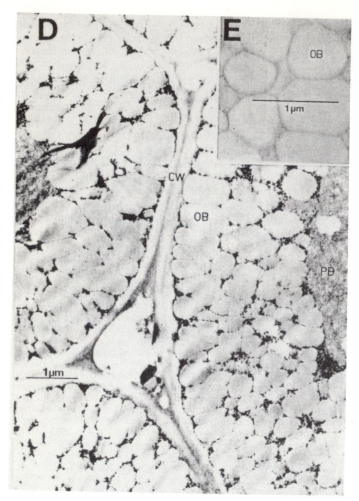

The above ultrastructural evidence is at variance with the hypothesis that oil bodies arise directly from the endoplasmic reticulum (4-6). Another way to test this hypothesis is to follow the protein composition of oil bodies during seed development. If the oil bodies bud off the endoplasmic reticulum, oil bodies should contain endoplasmic reticulum polypeptides.

The protein content of purified oil bodies from developing rapeseed embryos is shown in Figure 3. Little protein is associated with the oil bodies during the time when oil is accumulating at its maximum rate, i.e. between weeks 3-6. The first significant increase in oil body protein content is found between weeks 8 and 10, and the maximum increase is between weeks 10 and 12. The final composition of the mature oil bodies is about 95% w/w triacylglycerol, 2% w/w phospholipid and 2% w/w protein. These results are consistent with the initial formation of naked, protein-free oil bodies, which only acquire a protein coat at a later stage of development.

The polypeptide composition of the oil body proteins is shown in Fig. 4A. This gel electrophoretogram shows the lack of protein until 8 wk. It also is noteworthy that the polypeptide composition of the oil body proteins does not change during development. This was confirmed by applying an equal protein loading to each lane (Fig. 4B). The results show that when protein is present, the polypeptide composition is essentially identical at all developmental stages. The oil body protein fraction contains an 18.5 kDa polypeptide that constitutes 15% of the total protein in the mature seed. The 18.5 kDa polypeptide may be analagous to the 16.5 kDa polypeptide recently characterized in oil bodies from maize scutellum (13). The 18.5

FIG. 2. (D) Twelve wk after anthesis. The oil bodies are densely packed and occupy most of the cell apart from the central protein body. (E) Twelve wk after anthesis. Each oil body now has a strongly staining outer membrane and the contents are unstained. OB, oil body; Thy, thylakoids; SG, starch grain; CE, chloroplast envelope; T, tonoplast; CW, cell wall; PB, protein body.

The formation of oil bodies during seed development is shown in Figure 2. After three to five wk (Fig. 2A), the newly formed oil bodies are relatively small (0.5-0.8 μm) and are scattered throughout the cell. No limiting membrane is apparent around the oil bodies, although other membranes, e.g. thylakoids, readily are seen in the section. No continuity was ever observed between the surface of the oil bodies and components of the endomembrane system, such as the endoplasmic reticulum. After 8-10 wk (Fig. 2B), the oil bodies enlarge to about 1.0 μm but still are lacking a distinct electron-dense limiting membrane. Such a membrane is found only at the later stages of oil body development, at about 10-12 wk, which also coincides with the maximum rate of membrane protein accumulation in the oil body fraction (Fig. 2C,2D,2E).

This sequence of events was very similar to that seen in developing embryos of mustard seeds (7). The mustard oil bodies were not seen to arise from the endoplasmic reticulum and lacked a distinct membrane, until the final stages of oil body maturation.

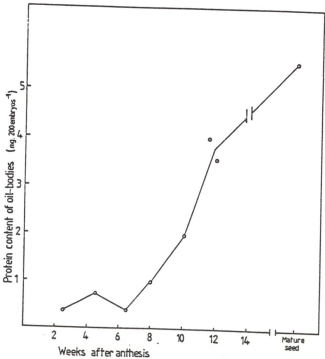

FIG. 3. Protein content of oil bodies from embryos of developing rapeseed. Proteins were estimated by a modified Lowry assay of the delipidated oil bodies or by scanning densitometry of the separated polypeptides following polyacrylamide gel electrophoresis.

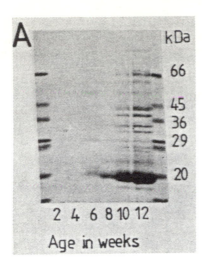

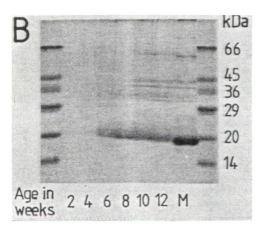

FIG. 4 Polypeptide composition of oil body proteins in embryos of developing rapeseed. Gels were run as described. (A) protein normalized to a per embryo basis. (B) equal loading of protein in each well.

kDa polypeptide is absent from all other subcellular fractions, including the endoplasmic reticulum. None of the characteristic endoplasmic reticulum polypeptides were found to be associated with the oil body fraction at any

developmental stage. These data do not support the endoplasmic reticulum theory for origin of oil bodies, at least in the case of rapeseed but indicate direct formation of oil bodies in the cytoplasm. In this regard, safflower microsomes incubated with acyl-CoAs and glycerol 3-phosphate also produce triacylglycerol in the form of naked oil droplets (14).

Therefore, it is proposed that storage oil in developing rapeseed embryos initially is synthesized as naked oil droplets. Several subcellular compartments may cooperate in this synthesis. The oil droplets continue to grow until, at the later stages of oil deposition, a protein-containing membrane is formed around the mature oil bodies. This proteinaceous membrane is largely made up of a single 18.5 kDa polypeptide, which persists through seed desiccation and remains around the oil bodies during germination and oil mobilization. We now are characterizing this 18.5 kDa polypeptide in more detail.

ACKNOWLEDGMENT

This work was supported by grants from the Agriculture & Science Research Council, U.K., the NATO Science Council and the British Council. We are grateful to N. Harris and the staff of the Durham Botany Department Electron Microscopy Unit for their assistance and advice.

REFERENCES

1. Gurr, M.I., in *The Biochemistry of Plants*, Vol. 4, edited by P.K. Stumpf and E.E. Conn, Academic Press, New York, 1980, p. 205.
2. Roughan, P.G., *Ann. Rev. Plant Physiol. 33*:97 (1982).
3. Murphy, D.J., and K.D. Mukherjee, *Lipids 22*:293 (1987).
4. Frey-Wyssling, A., E. Grieshaber and K. Muhlethaler, *J. Ultrastruct. Res. 8*:506 ((1963).
5. Schwarzenbach, A.M., *Cytobiologie 4*:145 (1971).
6. Wanner, G., and R.R. Theimer, *Planta 140*:163 (1978).
7. Rest, J.A., and J.G. Vaughan, *Ibid. 105*:245 (1972).
8. Smith, C.G., *Ibid. 119*:125 (1974).
9. Bergfeld, R., Y.N. Hong, T. Kuhnl and P. Schopfer, *Ibid. 143*:297 (1978).
10. Ichihara, K., *Agric. Biol. Chem. 46*:1767 (1982).
11. Markwell, M.A.K., S.M. Haas, N.E. Tolbert and L.L. Bieber, *Meth. Enz. 72*:296 (1981).
12. Spurr, A.R., *J. Ultrastruct. Res. 26*:31 (1969).
13. Wang, H., and A.H.C. Huang, *J. Biol. Chem. 262*:11275 (1987).
14. Stobart, A.K., S. Stymne and S. Hoglund, *Planta 169*:33 (1986).

Improvement of the Economic Feasibility of Microbial Lipid Production

H. Smit, A. Ykema, E.C. Verbree, M.M. Kater and **H.J.J. Nijkamp**

Department of Genetics, Biological laboratory, Vrije Universiteit, P.O. Box 7161, 1007 MC Amsterdam, The Netherlands

The oleaginous yeast *Apiotrichum curvatum* was used to investigate whether the economic feasibility of microbial lipid production could be improved. It was observed that the C/N ratio of the growth medium had a strong influence on lipid yield and the lipid production rate. In general, the optimum C/N ratio for lipid production was 30/40. The lipid production rate also was dependent on the cultivation technique used. Lipid production rates were highest in partial recycling cultures, followed by continuous cultures, fed-batch cultures and batch cultures respectively. Whey permeate was found to be a suitable growth medium for lipid production by *A. curvatum*. Lipid yields and lipid production rates generally were higher than in a semidefined medium with glucose as a carbon source. The fatty acid composition of the lipids of *A. curvatum* could be improved by the development of strains that were defective in the conversion of stearic acid to oleic acid.

Some species of bacteria, yeasts and molds have the potential to accumulate lipid up to more than 60% on a dry weight basis. Lipids accumulation is greater when conditions for growth are unfavorable, for instance when nitrogen becomes limiting and the carbon source is in excess.

The microbial storage lipids usually consist of 80-90% w/w triacylglycerols with a fatty acid composition similar to many plant seed oils. However, microbial lipid production compared with oil production by green plants is economically feasible. Costs in microbial lipid production may be minimized by optimizing the efficiency of converting carbohydrate substrates to lipid with inexpensive growth media, preferentially waste materials with zero or negative value. Also, the fermentor operational costs may be decreased by producing more lipid in a shorter time. The quality of the microbial lipid is also of great importance. Microbial lipids should be intended for human consumption and have unique and desirable properties to command a high price. Thus far, the application of microbial lipid accumulation on an industrial scale has been restricted to the fungal production of γ-linolenic acid (1). This compound is applied in the health food industry, and very high prices can be obtained for it. The production of cocoa butter equivalents by oleaginous yeasts has been patented, but the economic feasibility is uncertain (23).

We have studied the oleaginous yeast *Apiotrichum curvatum* to improve the economic feasibility of microbial lipid production. The objectives were to optimize lipid yield (unit weight lipid produced/unit weight carbon in growth medium consumed), to optimize the lipid production rate, to test whey permeate as growth medium for lipid production, and to improve the quality of the yeast lipids.

RESULTS AND DISCUSSION

Optimization of Lipid Yield

It is well-established that the lipid yield of oleaginous yeasts is high when these organisms are grown in a medium with a high carbon to nitrogen ratio (C/N ratio). After nitrogen is exhausted, oil droplets are accumulated in the yeast cells.

The influence of the C/N ratio of the growth medium on the lipid yield was studied in continuous cultures of *A. curvatum* using a semidefined medium with glucose or whey permeate with lactose as the main carbon source. The results (Fig. 1) showed that lipid yield increased with increasing C/N ratio. At the most favorable conditions for lipid accumulation, the production of 1 kg of lipid required about 5 kg of carbon substrate. The cost of the carbon substrate was an important element in the process, especially when a semidefined medium was used. The influence of the C/N ratio on the substrate costs ($_s$) could be calculated from the equation: $\$_s$ (price of semidefined medium per kg carbon source)/(lipid yield). The influence of the C/N ratio on substrate costs in a semidefined medium containing glucose is shown in Figure 2. The lowest substrate costs occurred when the C/N ratio was 40-80.

When whey permeate was used as a growth medium, the influence of the C/N ratio on substrate costs depended on the geographical location of the fermentation plant. In Europe, the lactose from whey might cost $0.25-$0.30 per kg. In the U.S. and New Zealand, whey is a real waste with zero or negative value. In either case, the influence of the C/N ratio on substrate costs was less important than with the semidefined medium.

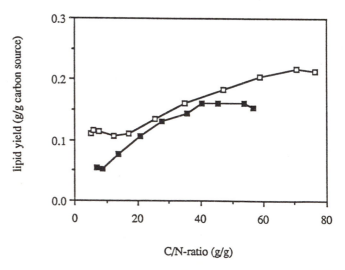

FIG. 1. Influence of the C/N ratio of the growth medium on lipid yield of *A. curvatum* in whey permeate (□) and semidefined growth medium (■). *A. curvatum* was grown in continuous cultures. NH$_4$Cl was added in the appropriate amount to obtain the desired C/N ratio. The composition of semidefined growth medium is described (5).

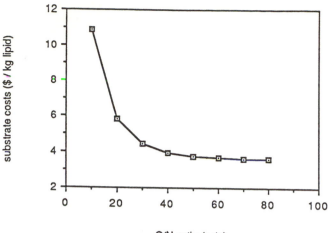

FIG. 2. Influence of the C/N ratio of the growth medium on substrate costs in a semidefined growth medium with glucose as a carbon source. A price of $0.60 per kg carbon source was assumed for the semidefined medium.

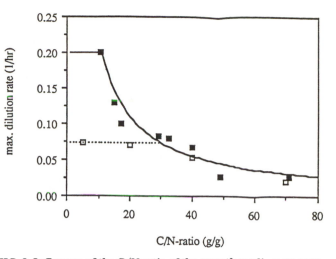

FIG. 3. Influence of the C/N ratio of the growth medium on maximal dilution rates of continuous cultures of *A. curvatum* determined in whey permeate (□) and semidefined growth medium (■). The lines drawn are based on a mathematical model constructed by Ykema et al. (5).

Optimization of the Lipid Production Rate

The rate of lipid production is low in oleaginous yeasts. This means that prolonged times are required in batch fermentations to obtain a substantial amount of lipid. In continuous cultures, lipid production is not efficient at dilution rates higher than 0.04-0.06 hr^{-1} (4). Consequently, fermentation operational costs (electricity, steam, labor, etc.) per kg lipid produced will be high and an extensive fermentor capacity is needed to obtain lipid production of some significance. To optimize this process, we investigated the influence of the C/N ratio of the growth medium and the cultivation technique on the lipid production rate.

Using continuous cultures of *A. curvatum*, the effect of the C/N ratio on the maximal dilution rate at which the carbon substrate is still fully consumed is depicted in Figure 3 for semidefined medium with glucose or whey permeate. In general, the dilution rate decreased with increasing C/N ratios. At low C/N ratios (<30), a marked difference was observed between the two growth media. In whey permeate, the maximal dilution rate appeared to be about 0.07 hr^{-1} and 0.20 hr^{-1} in semidefined medium. An explanation for this most probably is related to the difference in lipid production when nitrogen was in excess (Fig. 1).

Lipid production rates can be calculated as a function of the C/N ratio when lipid yield and maximal obtainable dilution rates are known using the equation $Q_1 = L.D. [S]$, in which Q_1 = the lipid production rate (g lipid produced/ l. hr); L = the lipid yield (g lipid produced/g carbon source consumed); D = dilution rate(hr^{-1}) and [S] = the carbon substrate concentration (g/l). The influence of the C/N ratio on lipid production rates of *A. curvatum* in semidefined medium and whey permeate is shown in Figure 4.

Another important factor that affects the lipid production rate is the cultivation technique. Lipid production rates are faster in continuous cultures than in batch cultures (4,6). However, high lipid production rates will be achieved in culture modes with very high cell densities like fed-batch cultures as reported for *Lipomyces star-*

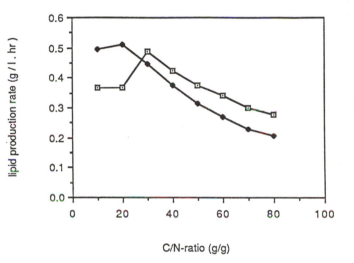

FIG. 4. Influence of the C/N ratio of the growth medium on maximal lipid production rates of *A. curvatum* calculated from experimental data of continuous cultures in whey permeate (□) and semidefined growth medium (■). Carbon substrate concentration was assumed to be 45 g/l. Production rates of total lipid.

keyi (7) and for *Rhodotorula glutinis* (8). We have determined the lipid production rate of *A. curvatum* in batch cultures, fed-batch cultures, continuous cultures and partial recycling cultures (Fig. 5). Whey permeate with NH$_4$Cl was added to obtain a C/N ratio of 40 in all cases. Fed-batch cultures and continuous cultures yielded higher lipid production rates than batch cultures, but the highest lipid production rates were obtained in partial recycling cultures. We obtained a lipid production rate of nearly 1 g/l/hr in a partial recycling culture, which to our knowledge is the highest production rate reported. Based on theoretical grounds, even much higher lipid production rates can be reached, provided that cell densities are maintained at 150 g dry weight/l (Fig. 5). We observed that oxygen limitation occurred at cell densities exceeding about 90 g dry weight/l.

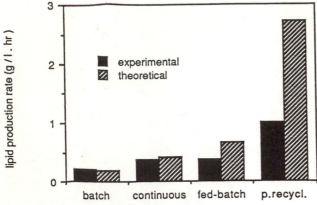

FIG. 5. Influence of the cultivation technique on total lipid production rate of *A. curvatum*. Lipid production rates were determined from experimental data of batch, fed-batch, continuous and partial recycling cultures or from theoretical calculations that were based on a mathematical model that was constructed by Ykema et al. (5) with the assumption that cell densities of 150 g dry wt./l can be obtained. Whey permeate with a C/N ratio adjusted to 40 was used as a growth medium in all cases. The fed-batch mode was simulated by a batch culture in which the cell dry weight was increased by operating in the recycling mode because of difficulties with concentrating the whey permeate. Therefore, the chemostat was modified by attaching one or two 0.1 μm polycarbonate membrane filtration unit(s) that continuously returned cells to the culture vessel while removing filtered broth at a rate equal to the rate of fresh medium input (14).

From the data in Figure 5, it is obvious that in partial recycling cultures superior lipid production rates can be obtained in comparison with the other cultivation techniques used. Although the advantages of the partial recycling technique have been reported (9–12), it should be stated that this cultivation mode requires very sophisticated equipment, including expensive filter systems to separate the yeast cells from the consumed medium. Because partial recycling is a continuous process, special care has to be taken to avoid contamination with other micro-organisms. These aspects have to be considered seriously before this technique can be applied on a large scale.

Potentials of Whey Permeate as a Growth Medium for Microbial Lipid Production

When comparing whey permeate and semidefined growth medium based on glucose, it was demonstrated that lipid yield (Fig. 1) and lipid production rate (Fig. 4) were generally higher in whey permeate. Because whey is an inexpensive growth medium, this medium seems very suitable for the production of microbial lipids especially when nitrogen is added to increase the lipid production rate.

Improvement of the Fatty Acid Composition of the Yeast Lipids

The fatty acid composition of the oil produced by wild type *A. curvatum* approximates to that of palm oil. The economic feasibility of microbial lipid production should be far more favorable when a higher-priced lipid could be produced. In several publications, the use of the desaturase inhibitor sterculic acid has been described as a tool to obtain microbial lipid resembling cocoa butter (3,13). The limited availability, however, and the additional costs involved in obtaining this compound may prevent its application on an industrial scale. The development in our laboratory of special yeast strains that are defective in the conversion of stearic acid to oleic acid may be a more promising approach. A detailed description of the construction of these strains and the analysis of the lipids they produce will be published separately.

REFERENCES

1. Suzuki, O., *J. Am. Oil Chem. Soc. 64*:1251 (1987).
2. Fuji Oil Co. Ltd., U.K. Patents 1,501,355 (1976); 1,555,000 (1978); U.S.Patent 4,032,405 (1977).
3. CPC International Inc., U.K. Patent Applications 2,091,285 and 2,091,282 (1982); European Patent Application 0,005,277 (1979); U.S. Patent 4,485,173 (1984).
4. Evans, C.T., and C. Ratledge, *lipids 18*:623 (1983).
5. Ykema, A., E.C. Verbree, H.W. van Verseveld and H. Smit, *Antonie van Leeuwenhoek 52*:491 (1986).
6. Floetenmeyer, M.D., B.A. Glatz and E.G. Hammond, *J. Dairy Sci. 68*:633 (1985).
7. Yamauchi, H., H. Mori, T. Kobayashi and S.Shimizu, *J. Ferment. Technol. 61*:275 (1983).
8. Pan, J.G., M.Y. Kwak and J.S. Rhee, *Biotechnol. Lett. 8*:715 (1986).
9. Damiano, D., C.S. Shin, N. Ju and S.S. Wang, *Appl. Microbiol. Biotechnol. 21*:69 (1985).
10. Maiorella, B.L., H.W. Blanch and C.R. Wilke, *Biotechnol. Bioeng. 26*:1003 (1984).
11. Mehaia, M.A., and M. Cheryan, *Enzyme Microb. Technol. 8*:289 (1986).
12. Enzminger, J.D., and J.A. Asenjo, *Biotechnol. Lett. 8*:7 (1986).
13. Moreton, R.S., *Appl. Microbiol. Biotechnol. 22*:41 (1985).
14. Chesbro,W.R., T. Evans and R. Eifert, *J. Bact. 139*: 625 (1979).

Effect of Temperature and pH on the Saturation of Lipids Produced by *Rhodosporidium Toruloides* ATCC 10788

G. Turcotte[a,b] and **N. Kosaric**[a]

[a]Department of Chemical and Biochemical Engineering, the University of Western Ontario, London, Ontario, Canada, and
[b]Department of Chemical Engineering, University of Ottawa, Ottawa, Ontario, Canada

Lipid accumulation and composition was studied in an oleaginous strict aerobe yeast *Rhodosporidium toruloides* strain ATCC 10788. Temperature and pH effects on lipid composition were investigated using a factorial design approach in a batch culture. Lipid composition profiles in all treatments were very similar in nitrogen-limited media; however, changes in the fatty acid distribution occurred before nitrogen depletion. The main treatment effects on lipid unsaturation were found at 35 C and pH 7.5 when more saturated lipids were produced.

The effect of temperature on lipid synthesis in fat-producing (oleaginous) yeasts is similar to that of most living organisms (1). The lipid content rises as the environmental temperature is lowered (2,3), and low temperatures bring about an increase in the unsaturation of the lipids. Although these effects occur in yeasts containing less than 15% lipids, lipids produced in mature oleaginous yeasts seem little affected by changes in temperature (3-6).

Varying pH also might play a role in altering lipid composition from certain oleaginous yeasts. Goulet (3) found a constant lipid content of 20% w/w in *Rhodotorula glutinis* cultivated at pH 3-8, whereas the lipid content of *Rhodotorula gracilis* decreased from 63 to 40% w/w when pH was increased from 3 to 6 (7). Furthermore, cells of *Candida stellatoidea* were shown to respond differently in unbuffered solutions, and higher pH reportedly increased the unsaturation of the lipids (7,8).

The goal of this study was to investigate the possible synergistic effects of temperature and pH on the amount and saturation of the lipids produced by *Rhodosporidium toruloids*, a strict aerobe oleaginous yeast.

EXPERIMENTAL PROCEDURES

Studies were performed using *Rs. toruloides* ATCC 10788. This obligate aerobe oleaginous yeast produced cells and intracellular lipids with a deep orange color. Preconditioned cells were cultivated in 1 l Bellco fermenters until mid-exponential phases when cells contained ca. 5-7% w/w lipid (10). Sufficient amounts of this inoculum solution were added to semi-pilot plant batch bioreactors to obtain an initial cell concentration of about 0.1 g/l (dry wt). The aeration rate was 0.6 volume of air/volume of medium/min (vvm), and the agitation speed was 300 rpm. The dissolved oxygen (DO) concentration was monitored after calibration in the freshly inoculated medium. The pH was held constant at 4.5, 6 or 7.5 for each temperature treatment (25 C, 30 C, 35 C). The total of nine treatments represented a three-level factorial design with two variables. The foam was controlled by a mechanical antifoam system or by the chemical Antifoam B, diluted to 1/10.

The dry wt of the cells was measured by filtration through cellulose nitrate membranes of 0.8 μm pore size. Total lipids were obtained by extraction of wet centrifuged cells with a choloroform:methanol (2:1, v/v) coupled with glass beads and glass powder. The lipid extracts were washed according to the method of Bligh and Dyer (11). The lipids were transformed into methyl esters of fatty acids following the method of Morrison and Smith and analyzed by gas chromatography on a 10% DEGS-stabilized column (on Chromosorb W-AW 100/120) at 190 C.

RESULTS AND DISCUSSION

The temporal patterns for the accumulation of intracellular lipids by *Rs. toruloides* ATCC 10788 were observed in three different temperature/pH treatments. Cells placed at pH 4.5 and 25 C proliferated until the NH_4^+ concentration became limiting (process time of ca/20 hr). Thereafter, excess glucose was converted into storage lipids in the absence of protein synthesis. The lipid content of mature cells reached 35% (dry wt basis). At 30 C and pH 6, nitrogen limitation occurred at about 16 hr, and the final lipid content was 35% w/w (Fig. 1). When cells were cultivated in the medium at pH 7.5 and 35 C, NH_4^+ could not be detected after 15 hr; the lipid content of the cells after 50 hr was around 30% w/w. Growth-associated production of lipids was present in all treatments (Table 1). The fatty acid composition of the total intracellular lipids at pH 6 and 30 C changed drastically before nitrogen-limitation but returned to a constant distribution thereafter (Fig. 2). This pattern was recurrent in all experiments performed. The unsaturation of the total lipids, as

Table 1

Ratio of Cell Growth Rate (h^{-1}) to Lipids Production Rate (h^{-1}) Before Nitrogren Limitation in the Medium

	pH 4.5	pH 6	pH 7.5
25 C	1.42	0.63	1.05
30 C	1.04	0.70	1.00
35 C	0.79	0.97	0.99

Table 2

Double Bond Index of the Total Lipids Before Nitrogen Limitation in the Medium

	pH 4.5	pH 6	pH 7.5
25 C	1.23	1.59	1.40
30 C	1.50	1.44	1.18
35 C	1.21	1.30	1.20

Batch Growth of <u>Rs.toruloides</u> at pH 6 and 30 C

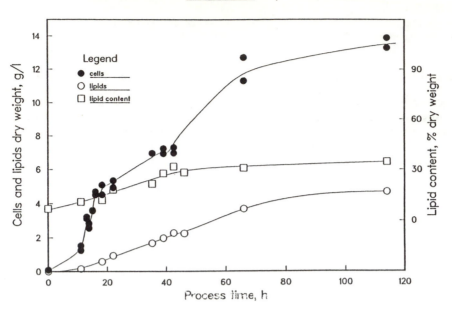

FIG. 1. Batch Growth of *Rs. toruloides* ATCC 10788 at pH 6 and 30 C.

Fatty Acid Composition of the Total Lipids from <u>Rs.toruloides</u> ATCC 10788 at pH6 and 30 C

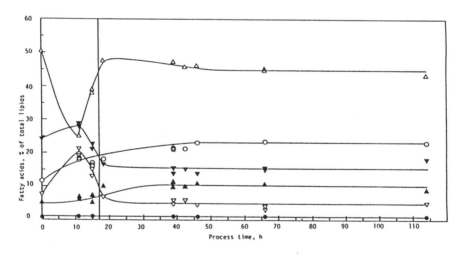

FIG. 2. Fatty Acid Composition of the total intracellular lipids from *Rs. toruloides* ATCC 10788 at pH 6 and 30 C. ○, C 14:0; ●, C 16:0; △, C 18:0; ▲, C 18:1; ▽, C 18:2; ▼, C 18:3.

calculated by the Double Bond Index (DBI) value (13), was found to be correlated to the sum of the polyunsaturated in the total lipids (Fig. 3). As shown in Figure 4, pH did not have a consistent effect on the unsaturation of the lipids. This is exemplified in Table 2, in which the DBI fluctuated widely among the treatments. However, the DBI value after 30 hr of cultivation (Table 3) was similar in most cases studied (14). The only exception occurred when cells were subjected to 35 C and pH 7.5, where the lipids were significantly more saturated than at other conditions.

Therefore, the DBI for lipids from batch cultures of

Rs.toruloides ATCC 10788, in the range 25-35 C, had no consistent trend when grown in nitrogen-sufficient conditions. However, the degree of unsaturation of the lipids of mature cells increased long after nitrogen limitation. The same was true when the pH ranged from 4.5 to 7.5 or when the combined effect of temperature and pH was considered. Although mature cells at 35 C and Ph 7.5 contained a significantly more saturated oil, it seemed improbable to alter the degree of saturation of the intracellular lipids produced by mature cells when temperature and pH alone are changed.

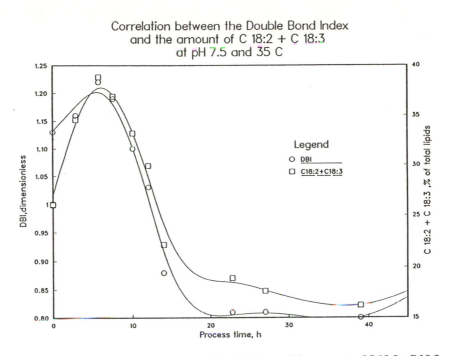

FIG. 3. Correlation between the Double Bond Index and the amount of C 18:2 + C 18:3 at pH 7.5 and 35 C.

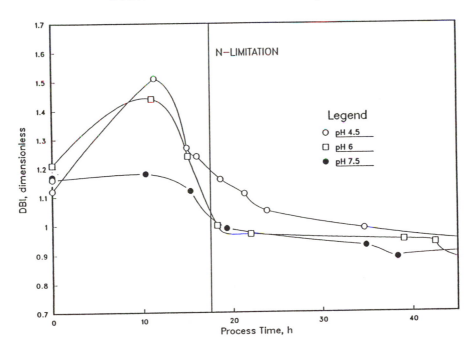

FIG. 4. Double Bond Index of the total lipids at 30 C.

Table 3

Degree of Unsaturation of the Total Lipids in Mature Cells

	pH 4.5	pH 6	pH 7.5
25 C	1.80	1.93	2.00
30 C	1.83	1.85	1.82
35 C	1.87	1.88	1.32

REFERENCES

1. Stokes, J.L., in *The Yeasts* Vol. 2, edited by A.H. Rose and J.S. Harrison, Academic Press, New York, 1971, p. 118.
2. Kates, M., and M. Paradis, *Can. J. Biochem. 51*:184 (1973).
3. Goulet, J., Ph.D. Thesis, McGill University, Canada (1975).
4. Hall, M.J., and C. Ratledge, *Appl. Environ. Microbiol. 33*:577 (1977).
5. Rattray, J.B.M., A. Schibeci and D.K. Kidby, *Bacteriol. Rev. 39*:197 (1975).
6. Hansson, L., and M. Dostalek, *J. Am. Oil Chem. Soc. 63*:1179 (1986).
7. Kessel, R.H.J., *J. Appl. Bacteriol. 31*:220 (1968).
8. Zalashko, M.V., E.S. Gurinovich, N.V. Obraztsova, I.F. Koroleva, Zh. N. Bogdanovskaya, V.D. Andreevskaya and S.S. Andriyashina, *Khimiya Tverdogo Topliva 3*:146 (1974).
9. Yoon, S.H., and J.S. Rhee, *J. Am. Oil Chem. Soc. 60*:1281 (1983).
10. Turcotte, G., Ph.D. Thesis, The University of Western Ontario, Canada (1987).
11. Bligh, E.G., and W.J. Dyer, *Can. J. Biochem. Physiol. 37*:911 (1959).
12. Morrison, W.R., and L.M. Smith, *J. Lipid Res. 5*:600 (1964).
13. Kates, M., and R.M. Baxter, *Can. J. Biochem. Physiol. 40*:1213 (1962).
14. Moreton, R.S., *Appl. Microbiol. Biotechnol. 22*:41 (1985).

A Non-Destructive Method for Seed Phenotype Identification in Plant Breeding

R. Garcés, J.M. García and **M. Mancha**
Instituto de la Grasa (C.S.I.C.), Aptdo. 1078, E- 41080 Sevilla, Spain

A new nondestructive technique that has advantages over the half-seed method was developed to identify phenotypes of germinating seeds. This method allowed selection of seeds with special characters by analyzing fatty acid composition in a part of one sprouting cotyledon. Oil seeds with different fatty acid composition were tested and had successful results with no viability problems. This method also could be used for selection of other traits such as seed protein content and quality.

In inheritance studies, it is necessary to distinguish the phenotype of FL_1 generation seed before the material is advanced to the next generation. The half-seed method, widely used for this purpose, involves cutting dry seed such that a small part is used for chemical analysis, and the remainder is stored for germination and seedling growth. This method has been very useful with oil seeds like safflower (1), rapeseed (2) and sunflower (3,4). However, the viability of cut seeds may be decreased due to microbial contaminations and physiological injury. Additionally, small, hard-coated seeds are difficult to cut. In this paper, a new method in which part of one germinating cotyledon is analyzed to determine the seed phenotype has been developed.

EXPERIMENTAL

Two species of Brassica with high (*B. napus*) and zero (*B. carinata*) erucic acid content and two sunflower (*Helianthus annuus*) lines with high and low oleic acid content, respectively, were used. The seeds were germinated in vermiculite moistened with tap water, in a growth chamber (14-hr photoperiod; 23 C/18 C). At the indicated times, samples were taken from three different seedlings of each genotype. The samples were ground in test tubes by using sand and a glass rod with up to 2 mg of heptadecanoic acid as the internal standard. The total lipids were extracted with petroleum ether-isopropanol (5). Lipids were converted into methyl esters (6), and the fatty acid composition and lipid content were determined by GLC.

RESULTS AND DISCUSSION

The fatty acid composition of seeds from *B. carinata* contained high levels of eicosenoic (20:1) and erucic (22:1) acids. These fatty acids were absent in *B. napus*, and consequently the oleic acid content was increased. At the beginning of germination, the fatty acid composition of the cotyledon was unchanged in spite of the rapid degradation of reserve lipids (Table 1). This allowed the identification of the seed phenotype by analyzing the fatty acid composition in a portion of one germinating cotyledon. The sample should be taken before the fourth day after germination.

Similar results were obtained with soil-germinated seeds as soon as the cotyledons emerged. In *B. carinata*, the fatty acid composition was 18.9% 18:1, 7.0% 20:1, and 35.5% 22:1. *B.napus* seed contained 60.2% 18:1.

Similar experiments were made with low and high oleic acid sunflower germplasm (Table 2). Once again, the fatty acid composition of cotyledon remained stable during the first days of germination. As before, the sample should be taken before the fourth day after germination.

There was no seedling survival problems with this method because the cotyledons were not exposed to microbial contamination as in the half-seed method. Therefore, this method could be used for selection of special seed characters such as fatty acid content, proteins or other interesting compounds.

Table 1

Fatty Acid Composition and Lipid Content of Germinating Brassica Seed

	B. carinata				B. napus		
	18:1	20:1	22:1	Total lipids %, w/w	18:1	Total lipids %, w/w	
DAG[a]			mol %		mol %		Stage of development
0	13.2	9.2	40.6	31.9	68.6	37.2	Dormant seed
1	11.9	7.0	43.0	n.d.[b]	72.1	n.d.	Germination
2	15.2	10.1	36.6	30.9	69.1	30.9	
3	18.2	7.9	38.1	15.1	66.6	24.7	
4	19.0	7.9	29.6	7.2	68.0	16.8	Aerial cotyledons
5	12.6	8.2	28.8	7.5	64.6	8.4	
6	24.2	4.8	22.7	3.0	51.0	4.5	Green cotyledons
7	23.4	5.0	22.8	1.8	49.9	2.9	
9	16.4	5.3	23.4	0.9	49.8	2.8	First leaves

[a]Days after germination.
[b]Not determined.

Table 2

Fatty Acid Composition and Lipid Content of Germinating Sunflower Seed

	High oleic			Low oleic			
	18:1	18:2	Total lipids	18:1	18:2	Total lipids	
DAG[a]	mol %		%, w/w	mol %		%, w/w	Stages of development
0	90.1	1.9	47.5	28.9	59.9	46.8	Dormant seed
2	89.5	2.7	29.1	29.8	59.3	41.4	Germination
4	87.4	4.1	11.3	33.3	55.3	27.1	Aerial cotyledons
6	69.9	15.6	0.9	37.5	50.5	12.7	Green cotyledons
8	33.3	28.4	0.3	37.0	48.5	2.7	
10	10.1	22.2	0.2	26.4	48.3	0.5	First leaves

[a]Days after sowing.

ACKNOWLEDGMENTS

Thanks are due to M.C. Ruiz for technical assistance and to M.D. Garcia for typing the manuscripts. This work was supported by a grant from the Junta de Andalucia and C.S.I.C.

REFERENCES

1. Ladd, S.L., and P.F. Knowles, *Crop Sci. 10*:525 (1970).
2. Röbbelen, G., and A. Nitsch, *Z. Pflanzenzuchty 75*:93 (1975).
3. Urie, A.L., *Crop Sci. 25*:986 (1985).
4. Fernández-Martinez, J., A. Jiménez-Ramirez, J. Dominguez-Gimenez and M. Alcántara, *Grasas y Aceites 37*:326 (1986).
5. Hara, A., and N.S. Radin, *Anal. Biochem. 90*:420 (1978).
6. Mancha, M., and J. Sánchez, *Phytochem. 20*:2139 (1981).

Candida cylindracae Lipase-catalyzed Interesterification of Butter Fat

Paavo Kalo[a], **Asmo Kemppinen**[b] and **Matti Antila**[b]

[a]Department of General Chemistry, University of Helsinki, SF-00710 Helsinki, Finland, and [b]Department of Dairy Science, University of Helsinki, SF-00710, Finland

Butter fat was subjected to *Candida cylindracae* lipase catalyzed interesterification reactions. The lipase was immobilized by adsorption on Celite, and the enzyme was activated with glycerol or with variable amounts of water to produce fats with different contents of lipolysis products. To follow the nonspecific lipase catalyzed interesterification reaction, a method for the quantitative determination of triacylglycerols separated on capillary columns according to their acyl carbon number and level of saturation was developed. To overcome the problems of irreversible adsorption, polymerization and degradation, prior silylation was combined with an on-column injection technique. Immobilized SE-54 and OV-17 capillary columns were studied under different analytical conditions. The effect of silylation on the degradation of the higher triacylglycerols was evaluated on the basis of the empirical correction factors. Coefficients of variation were high, however the analysis on capillary columns was regarded as a useful method.

The possibility of altering the physical properties of fats through interesterification reactions catalyzed by lipase enzymes (triacylglycerol acylhydrolases E.C. 3.1.1.3.) has aroused considerable interest recently. Lipase enzymes can be divided into three groups according to their specificity. One group of lipases specifically catalyzes acyl exchange at *sn*-1 and 3. A second group of lipases catalyzes interesterification of acyl groups with certain structures. The third class, nonspecific lipases (e.g. extracellular lipase of *Candida cylindracae*), does not show positional or substrate specificity. The interesterification with a nonspecific lipase as a catalyst, as does the interesterification with a chemical catalyst, leads to a random distribution of fatty acyl groups in the triacylglycerols (TAG) (1).

Changes in physical properties induced by interesterification can be followed by dilatometry, calorimetric methods, and by nuclear magnetic resonance (NMR). The changes in TAG composition can be followed by mass spectrometry, enzymatic deacylation, stereospecific analysis, argentation thin layer chromatography (TLC) (2), high performance liquid chromatography (HPLC) (3) and gas liquid chromatography (GLC).

In GLC on packed and short capillary columns, TAG are separated into groups of TAG with the same number of carbon atoms. Apolar (OV-1, SE-52, SE-54) 8- to 20-m immobilized capillary columns separate triacylglycerols according to acyl carbon number and level of unsaturation. Recent developments in the coating technique enable the preparation of chemically bonded, thermally stable capillary columns of medium polarity (OV-17). Using such columns, Geeraert and Sandra (4) separated TAG with the same acyl carbon number and level of saturation into several isomer peaks.

Problems reported in the quantitation of TAG include irreversible adsorption, degradation and polymerization. Grob (5) has suggested that the loss of TAG in capillary columns are mainly the result of degradation to free fatty acids. To offset the problem, Grob and coworkers (6) introduced the use of a cold on-column injection technique. D'Alonzo et al. (7) have preferred prior silylation of the free fatty acids and mono- and diacylglycerols (MAG, DAG) Kalo et al. (8-10) have studied the quantitation of TAG separated on SE-54 columns according to acyl carbon number and level of saturation.

In this work, butter fat was interesterified with *Candida cylindracae* lipase as the catalyst. The enzyme was activated with glycerol or with different amounts of water to produce fats with variable content of hydrolysis products. Changes in the TAG composition and the MAG and DAG content of reaction products induced by interesterification were studied from silylated samples on an immobilized OV-17 silica capillary column using an on-column injection technique. Resolution, degradation of higher TAG, and repeatability of quantitations were compared with previous determinations (8-10) of butter fat and butter fat/rapeseed oil mixtures. Melting properties were studied by measurement of solid fat content (SFC) at different temperatures using pulsed NMR.

EXPERIMENTAL

Reagents

Butter fat was supplied by Valio Finnish Co-operative Dairies' Association. MAG, DAG and TAG standards were purchased from NU Check Prep (Elysian, MN), *Candida cylindracae* lipase and TES-buffer from Sigma Chemical Co. (St. Louis, MO), HPLC grade hexane from Rathburn Chemicals Ltd., Bond Elut columns and Vac Elut Accessory from Analytichem International Inc. (Harbor City, CA), and BSTFA and p.a. grade solvents from Merck, Sharp and Dome (Weste Pointe, PA).

Lipase-catalyzed Interesterification

Lipase from *Candida cylindracae* was immobilized by adsorption on Celite Hyflo Super Gel as described (7). Butter fat was interesterified according to the procedure described earlier (7), with the exception that batches containing 20 g fat and 200 mg enzyme were activated with 100, 200 or 400 mg TES buffer solution (pH 6.5) or 400 mg glycerol.

Determination of Free Fatty Acids

The FFA contents of the original butter fat and the reaction products were determined by titration of the fat sample in ethanol/diethyl ether solution (1:1, v/v) with 0.1 M methanolic sodium hydroxide solution, using thymol blue (0.04% in ethanol) as indicator.

Gas Chromatography

A Carlo Erba 5300 Gas Chromatograph with on-column injector, flame-ionization detector and constant pressure-constant flow cp-cf 516 control module was used. The calculation of chromatograms was carried out with an IBM-compatible microcomputer, connected to the EL-480 electrometer output through a Chrom-1 data acquisition board (Metrabyte) using Chrom+ software (Laboratory Technologies Corporation). To derivatize the FFA and MAG and DAG, the fat samples were treated with BSTFA as described (10).

The chromatograms were run on a 25-m 0.32 mm i.d. immobilized OV-17 column (Carlo Erba) with 0.15-μm film thickness. Injection was made with constant carrier gas (hydrogen) pressure mode. After the injection the chromatogram was run with constant carrier gas flow (linear velocity 62 cm/s). The temperature programs were Program A, in which after an isothermal period of one min at 160 C, the column temperature was raised 15 C/min to 345 C; and Program B, in which after an isothermal period of one min at 80 C, the column temperature was raised 10 C/min to 245 C, a further 3 C/min to 300 C and ballistically to 345 C.

A calibration mixture containing tricaprin, trilaurin, trimyristin, tripalmitin, tristearin, triolein, diolein, dinonadecanoin, monoolein and monononadecanoin was used for determination of the empirical correction factors. The correction factors of other TAG peaks were calculated from those of the calibration TAG. The average value of the empirical correction factors of MAG and DAG of the calibration mixture was used for MAG and DAG peaks. A sample of 5 μg interesterified butter fat was analyzed six to seven times with temperature programs.

Identification of peaks was based on the retention times of calibration compounds. Natural fats (coconut oil, cocoa butter, raw palm kernel oil, olive oil, linseed oil, groundnut oil) of known composition (11) were analyzed by GLC. The MAG, DAG, TAG, FFA, cholesterol (C) and cholesteryl ester (CE) fractions obtained from fractionation of interesterified butter fat also were analyzed on a Bond Elut aminopropyl column (12).

Pulsed NMR Measurements

A Bruker Minispec pc 120 pulsed NMR instrument was used in SFC determinations (13). All samples were subjected to the following thermal pretreatment: the sample was melted at 80 C, held at 60 C for 30 min and at 0 C for 90 min. The spin-spin relaxation signal was measured 12 and 70 μs after irradiation with radio frequency radiation. The instrument was calibrated using a stable reference standard whose solids content was accurately known. The repeatability of the SFC measurement was studied by repeating the analysis of untreated butter fat 11 times.

Calculation of Random Triacylglycerol Composition

Random TAG composition was calculated from the fatty acid composition of butter fat expressed in mol % (4:0, 9.1; 6:0, 4.1; 8:0, 1.4; 10:0, 2.3; 12:0, 2.7; 14:0, 10.6; 14:1, 1.5; 16:0, 29.1; 16:1, 1.8; 18:0, 11.1; 18:1, 19.5; 18:2, 1.0; 18:3, 3.0 mol %) disregarding fatty acids below 1 mol %. The calculation

was carried out by microcomputer according to the equations % AAA $a^3/10000$, % AAB $3a^2b/10000$, % ABC $6abc/10000$, in which a, b and c are the concentrations of fatty acids expressed in mol percent and AAA, AAB and ABC are TAG comprised of one, two and three fatty acids, respectively.

RESULTS

Separation of Fat Components

The OV-17 column (program B) separated the TAG of butter with different acyl carbon number and level of unsaturation (Fig. 1). In addition, the TAG in the acyl carbon number range 26-46 were separated into several isomer peaks. All silylated DAG and MAG were separated completely from TAG. FFA with 8-18 acyl carbons, cholesterol and most cholesteryl esters were separated from other peaks.

Quantitative Considerations

The empirical correction factors (Table 1) revealed degradation of TAG containing more than 36 acyl carbons with both temperature programs. The degradation of TAG also was confirmed by the presence of FFA (14-18) and by high proportions of DAG in the interesterification products. The degree of degradation of these TAG also was shown by the large coefficient of variation for TAG 54:3 (23.9% on program B and 6.6% on program A).

The standard deviations of saturated and monoene TAG determined for interesterified butter fat analyzed with the temperature program A varied between 0.078 and 0.776, with a coefficient of variation in the range

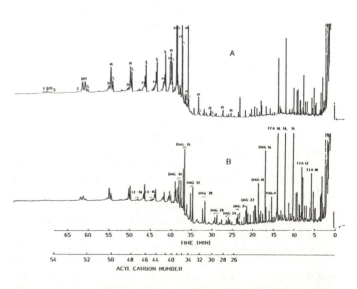

FIG. 1. Chromatograms of 0.5 μg butter fat run on a 25-m OV-17 column with temperature program B. (A) Untreated butter fat; (B) Interesterified with lipase as catalyst (2.0% water as activator). Peak designations: S, saturated TAG; M, monoene TAG; D, diene TAG; T, triene TAG; FFA, free fatty acid; MAG, monoacylglycerol; DAG, diacylglycerol; TAG, triacylglycerol; C, cholesterol; CE, cholesteryl ester. Numbers indicate the acyl carbon number of the compounds.

Table 1

Empirical Correction Factors for GC Quantification of Mono-, Di- and Triacylglycerols Separated on a 25-m OV-17 Column

	Component correction factor	
	Temperature program A	Temperature program B
MAG 18:1	0.948 ± 0.037	0.998 ± 0.127
MAG 19:0	0.904 ± 0.032	0.991 ± 0.128
DAG 36:2	0.834 ± 0.013	0.876 ± 0.096
DAG 38:0	0.856 ± 0.019	0.844 ± 0.111
TAG 30:0	1.092 ± 0.023	1.034 ± 0.023
TAG 36:0	1.000 ± 0.000	1.000 ± 0.000
TAG 42:0	1.225 ± 0.213	1.083 ± 0.023
TAG 48:0	1.311 ± 0.051	1.315 ± 0.049
TAG 54:0	2.067 ± 0.170	2.174 ± 0.244
TAG 54:3	2.237 ± 0.148	2.672 ± 0.638

3.5-23.5% for TAG more abundant than 2% w/w (Table 2). For chromatograms run on program B, the standard deviations determined for single peaks varied between 0.065 and 1.1, with a coefficient of variation in the range of 3.1-21.8% for peaks more abundant than 2% w/w.

Table 2

Variation Coefficients (%) of Saturated and Monoene Triacylglycerols

Acyl carbon number	Program A		Program B	
	Saturated	Monoene	Saturated	Monoene
26	n.d.	—	n.d.	n.d.
28	80.2	20.5	n.d.	n.d.
30	40.3	5.8	n.d.	n.d.
32	36.8	6.5	n.d.	31.9
34	25.5	11.0	n.d.	13.8
36	15.6	23.5	6.8-11.4	6.3-8.8
38	18.6	5.4	n.d.	4.8-15.7
40	3.5	7.6	10.7	4.6-13.4
42	8.9	10.1	4.9	21.6
44	7.1	12.1	3.1-8.2	n.d.
46	5.8	12.5	13.5-8.2	n.d.
48	5.0	3.8	6.6	3.9
50	10.5	4.6	10.0	4.3
52	6.5	6.7	n.d.	18.3
54	19.0	22.5	n.d.	—

Determined for interesterified butter fat on a 25-m OV-17 column as described in Methods.
C.V. was calucated for the sum of peaks with program A and for single peaks with program B.
n.d., not determined.

Changes in the Composition of Fat

The compositions of saturated TAG of untreated butter fat and interesterified products are compared with the calculated random TAG composition, in Figure 2. Interesterification increased the proportion of saturated TAG with 34 and 36 acyl carbons and in the acyl carbon number range 46-52. In general and especially in the high acyl carbon number range, the compositions of saturated TAG of products of reactions activated with glycerol and the lowest amount of water showed the best resemblance to the calculated composition.

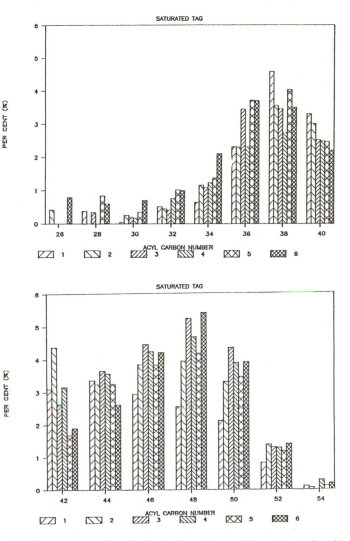

FIG. 2. Proportion of saturated triacylglycerols in untreated and interesterified butter fat. 1, untreated butter fat; 2, interesterified with 2.0% glycerol; 3, interesterified with 0.5% water; 4, interesterified with 1.0% water; 5, interesterified with 2.0% of water; 6, calculated random distribution.

Interesterification decreased the proportion of monoene TAG in the acyl carbon number range 36-42 and increased that in the acyl carbon number range 46-52 (Fig. 3). The compositions of monoene TAG of the products of reactions activated with glycerol and the lowest amount of water were most similar to the calculated compositions.

DAG compositions of reaction products and untreated butter fat were calculated from the gas chromatograms run on program B (Fig. 4). The interesterification product of the reactions activated with glycerol showed the lowest proportion of DAG, whereas the reaction activated with the highest amount of water yielded the highest level of DAG. Table 3 shows the TAG, DAG and MAG contents calculated from gas chromatographic analysis; and FFA content was determined by titration with methanolic sodium hydroxide solution. The DAG and FFA content of treated butter fat was high relative to the control, content, most probably because of the degradation of higher TAG. High DAG and FFA contents in the reaction products was an undesirable result.

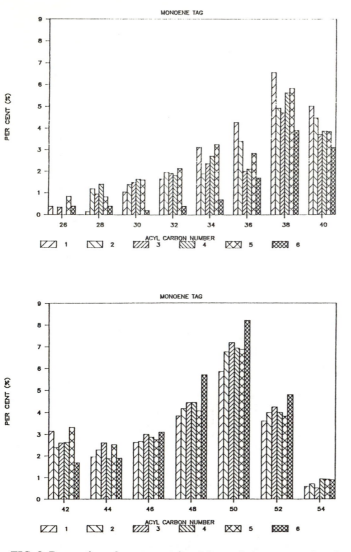

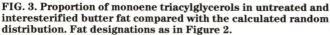

FIG. 3. Proportion of monoene triacylglycerols in untreated and interesterified butter fat compared with the calculated random distribution. Fat designations as in Figure 2.

Solid Fat Content

SFC of reaction products and untreated butter fat were determined by pulsed NMR at nine temperatures between 0 and 40 C. Determination of the repeatability of the measurement showed standard deviations of SFC to vary between 0.09 and 0.21. The determinations at low temperatures showed better repeatability than those at high temperatures. In general, SFC was lower below 20 C and higher above 20 C in the interesterified than in the untreated butter fat. The SFC was lower in the fats with a high content of hydrolysis products than in those with low content.

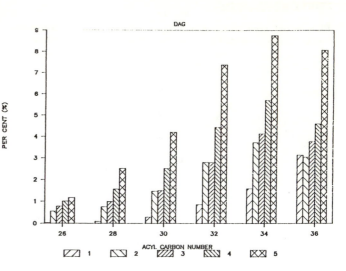

FIG. 4. Proportion of diacylglycerols in untreated and interesterified butter fat. Fat designations as in Figure 2.

DISCUSSION

Separation Efficiency

The OV-17 column showed superior separation power to the SE-54 columns used earlier (10). The ability of the OV-17 column to separate TAG, DAG and MAG made it the superior choice for following the changes induced by interesterification. In addition, the ability to separate FFA, cholesterol and most cholesteryl esters allowed a broader characterization of interesterification products than the apolar column.

Quantitation of TAG

The empirical correction factors for TAG determined with a mixture of standards reveal a higher degree of degradation of high molecular weight TAG. This is at least partly due to the long isothermal period at the maximum temperature. The higher degree of degradation also led to a greater deviation for TAG with acyl carbon number 54. However, the repeatability of the determination of the major TAG was the same as in the separations on SE-54 column with 0.2 mm i.d. Although the overall accuracy of the determination of TAG composition was only fair, the change in the TAG composition due by interesterification was so marked that this method still could provide valuable information for product development.

Interesterification-induced Changes

Comparison of the TAG compositions determined in this study with the determinations on SE-54 columns (8-10) showed general agreement. Owing to the ability of the OV-17 column to separate DAG, the determinations made in this study in the acyl carbon number range 36-40 showed better agreement with the calculated composition than those of earlier studies.

ACKNOWLEDGMENTS

This work was supported by research grants from the National Research Council for Agriculture and Forestry, the Academy of Finland and from the Finnish Culture Foundation.

REFERENCES

1. Macrae, A.R., *J. Am. Oil Chem. Soc. 60*:291 (1983).
2. Sreenivasan, B., *Ibid. 55*:796 (1978).
3. Frede, E., and H. Thiele, *Ibid. 64*:521 (1987).
4. Geeraert, E., and P. Sandra, *J. High Resolut. Chromatogr. Chromatogr. Commun. 7*:431 (1984).
5. Grob, K. Jr., *J. Chromatogr. 205*:289 (1981).
6. Grob, K. Jr., H.P. Neukom and R. Battaglia, *J. Am. Oil Chem. Soc. 57*:282 (1980).
7. D'Alonzo, R.P., W.J. Kozarek and H.W. Wharton, *Ibid. 58*:215 (1981).
8. Kalo, P., P. Parviainen, K. Vaara, S. Ali-Yrkkö and M. Antila, *Milchwissenschaft 41*:82 (1986).
9. Kalo, P., K. Vaara and M. Antila, *Fette-Seifen-Anstrichm. 88*:362 (1986).
10. Kalo, P., K. Vaara and M. Antila, *J. Chromatogr. 368*:145 (1986).
11. Padley, F.B., in *The Lipid Handbook*, edited by F.D. Gunstone, J.L. Harwood and F.B. Padley, Chapman and Hall, London, 1986, pp. 55-112.
12. Kaluzny, M.A., L.A. Duncan, M.V. Merritt and D.E. Epps, *J. Lipid Res. 26*:136 (1985).
13. Van den Enden, J.C., J.B. Rossell, L.F. Vermaas and D. Waddington, *J. Am. Oil Chem. Soc. 59*:433(1982).

Interesterification of Lipids by an *sn*-1, 3-Specific Triacylglycerol Lipase

Ricardo Schuch and **Kumar D. Mukherjee**

Bundesanstalt für Fettforschung, Institut für Biochemie und Technologie, H.P. Kaufmann Institut, Piusallee 68, D-4400 Münster, Federal Republic of Germany

The rates of interesterification of medium-chain (C_{12} and C_{14}) triacylglycerols with various compounds using an immobilized *sn*-1,3-specific triacylglycerol lipase (Lipozyme®) of the following order: long-chain alcohol> triacylglycerol> fatty acid> methyl ester> polyhydric alcohol. In all these reactions, exchange (transfer) of acyl moieties occurred almost exclusively at the *sn*-1,3- positions of the glycerol backbone.

Lipase-catalyzed reactions, such as hydrolysis, esterification and interesterification of lipids, have received considerable attention because of possible applications in biotechnological processes (1). We have studied interesterification reactions catalyzed by an immobilized triacylglycerol lipase (EC 3.1.1.3) from *Mucor miehei*, which hydrolyzes triacylglycerols specifically at the *sn*-1,3-positions.

EXPERIMENTAL

Medium-chain triacylglycerols from ucuhuba (*Virola surinamensis*) having the composition given in Table 1 were used in all interesterification reactions. The medium-chain triacylglycerols (0.5 mmol) were interesterified at 45 C in the presence of hexane with 0.5 mmol of heptadecanoic acid, methyl heptadecanoate, trioleoylglycerol, octadecyl alcohol or glycerol. The lipase preparation Lipozyme® (Novo Industrie GmbH, Mainz, Federal Republic of Germany) was used as catalyst at a concentration of 10% of the reacting substances. Details of the experimental conditions and analytical procedures are given elsewhere (2).

RESULTS AND DISCUSSION

The positional specificity of interesterification reactions catalyzed by the *sn*-1,3-specific lipase has been shown in Table 1. It was quite evident from the data that heptadecanoyl moieties, not present in the medium-chain triacylglycerols of ucuhuba, are incorporated almost exclusively into the *sn*-1,3-positions of the traicylglycerols upon interesterification with heptadecanoic acid. Concomitantly, lauroyl, myristoyl and palmitoyl moieties from the *sn*-1,3-positions are liberated as fatty acids. Similarly, interesterification of the medium-chain triacylglycerols with methyl heptadecanoate using Lipozyme® as a catalyst resulted in the incorporation of heptadecanoyl moieties into the *sn*-1,3- positions with concomitant release of the acyl moieties from these positions as methyl esters.

Interesterification of the saturated medium-chain triacylglycerols of ucuhuba with trioleoylglycerol using Lipozyme® resulted in the transfer of oleoyl moieties to the *sn*-1,3-positions of ucuhuba oil. Thus, during the course of interesterification, molecular species of triacylglycerols containing one and two double bonds per molecule were

Table 1

Positional Distribution of Major Acyl Moieties in Medium-Chain Triacylglycerols of Ucuhuba (*Virola surinamensis*) Seed and in the Product Formed by Their Interesterification with Heptadecanoic Acid Using Lipozyme® for Four Hr

	Positional distribution			
	Ucuhuba		Interesterified product	
Acyl moieties	*sn*-1,3	*sn*-2	*sn*-1,3	*sn*-2
12:0	18.4	2.8	12.7	3.5
14:0	68.9	90.2	42.5	83.0
16:0	6.9	0.7	2.8	2.9
17:0	0.0	0.0	40.4	0.5
18:1	2.5	4.3	1.6	3.7

formed. The initial reaction mixture was almost exclusively composed of triacylglycerols having none or three double bonds per molecule. Analysis of the molecular species of triacylglycerols containing one and two double bonds by hydrolysis with pancreatic lipase revealed that the exchange of oleoyl moieties against the saturated medium-chain acyl moieties occurred specifically at the *sn*-1,3-positions.

Medium-chain triacylglycerols also were converted upon interesterification with octadecyl alcohol to the octadecyl esters (wax esters) of the medium chain fatty acids; concomitantly, *sn*-2-monoacylglycerols and *sn*-1,2(2,3)-diacylglycerols were formed. Composition of the acyl moieties of the wax esters (12:0, 20.4%; 14:0, 71.9%; 16:0, 4.5%; 18:1, 3.0%) closely resembled that of the acyl moieties at the *sn*-1,3-positions of the medium-chain triacylglycerols (Table 1). These date further demonstrated the positional specificity of this interesterification reaction.

Interesterification of the medium-chain triacylglycerols with glycerol using Lipozyme® yielded a mixture of *sn*-1(3)-monoacylglycerols and *sn*-1,3-diacylglycerols by transfer of medium-chain acyl moieties to glycerol. Concomitantly, *sn*-2-monoacylglycerols and *sn*-1,2(2,3)-diacylglycerols were formed from the starting mixture of triacylglycerols. Positional specificity of this interesterification reaction were established conveniently by radiochemical techniques (R. Schuch and K.D. Mukherjee, unpublished data).

Table 2 shows the extent of interesterification of the medium-chain triacylglycerols with various compounds after four hr. Longer reaction periods gave no further increase in the extent of interesterification (2). The rates of interesterification with the various classes of compounds were of the general order: long-chain alcohol > triacylglycerol > fatty acid > methyl ester > polyhydric alcohol.

Table 2

Extent of Interesterification of Medium Chain Triacylglycerols of Ucuhuba with Various Reaction Partners Using Lipozyme® for Four Hr

Reaction partner	Extent of interesterification (μmol reaction partner interesterified)	
	Total	per mg Lipozyme®
Heptadecanoic acid	249.1	5.0
Methyl heptadecanoate	158.2	3.2
Trioleoylglycerol	310.8	4.1
Octadecyl alcohol	449.6	9.0
Glycerol	30.5	0.8

ACKNOWLEDGMENT

A research fellowship provided to Ricardo Schuch by the German Academic Exchange Service (Deutscher Akademischer Austauschdienst, DAAD), Bonn, Federal Republic of Germany, is gratefully acknowledged.

REFERENCES

1. Macrae, A.R., in *Biotechnology for the Oils and Fats Industry*, edited by C. Rattledge, P. Dawson, and J. Rattray, American Oil Chemists' Society, Champaign, IL, 1984, p. 189.
2. Schuch, R., and K.D. Mukherjee, *J. Agric. Food Chem.*, in press.

Radiochemical Techniques for the Assay of Lipase-catalyzed Reactions

Ricardo Schuch and **Kumar D. Mukherjee**
Bundesanstalt für Fettforschung, Institut für Biochemie und Technologie, H.P. Kaufmann Institut, Piusallee 68, D-4400 Münster, Federal Republic of Germany

Lipase-catalyzed reactions such as hydrolysis, interesterification and esterification of lipids can be assayed conveniently by using radioactively labeled substrates. A method for analyzing the products of these reactions by thin layer chromatography and liquid scintillation counting is described.

Chromatographic techniques such as thin layer chromatography (TLC) in conjunction with gas chromatography (GC) or high performance liquid chromatography (HPLC) commonly are used to assay the products of lipase-catalyzed hydrolysis, interesterification and esterification of lipids (1-4). We recently have shown that radiochemical techniques are highly suitable for a simple and highly sensitive assay of lipase-catalyzed interesterification reactions (R. Schuch and K.D. Mukherjee, unpublished data). This communication describes application of radiochemical techniques in the assay of rates and positional specificity of lipase-catalyzed interesterification reactions. These techniques involve the use of radioactively labeled substrates in various reactions and the analysis of the reaction products by TLC and/or liquid scintillation counting (LSC).

EXPERIMENTAL

Radiolabeled fatty acids and triacylglycerols (uniformly labeled in the acyl moieties) were commercially available or were prepared from commercially available intermediates. Triacylglycerols containing radioactive acyl moieties exclusively at the sn-1,3 positions were prepared by interesterification of an unlabeled triacylglycerol with a radioactively labeled fatty acid using a sn-1,3-specific lipase. Hydrolysis of triacylglycerols using lipases (5) and fractionation of the hydrolysis products by TLC on Silica Gel containing boric acid (6) were carried out according to established procedures. Radioactivity in various lipid fractions was measured by LSC using suitable scintillation liquids. Interesterification and esterification of lipids using immobilized lipases were carried out with stirring in the presence of hexane; the products were analyzed by TLC and LSC.

RESULTS AND DISCUSSION

Radiochemical techniques have been found to be highly suitable for the assay of the rates and positional specificity of various lipase-catalyzed reactions. These techniques were applied to follow the lipase-catalyzed interesterification of triacylglycerol.

Interesterification of an unlabeled triacylglycerol with [1-14C]oleic acid yielded radioactive triacylglycerols with concomitant release of the unlabeled fatty acids. The amount of fatty acid interesterified was determined from the radioactivity in the triacylglycerol fraction by taking into account the specific radioactivity of the reaction sub-

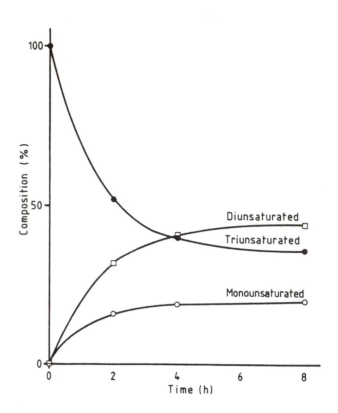

FIG. 1 Composition (%) of molecular species of 14C-labeled triacylglycerols formed during interesterification of saturated triacylglycerols (tristearoylglycerol) with [carboxyl-14C] trioleoylglycerol using an sn-1,3-specific triacylglycerol lipase (Lipozyme®).

strate. Lipase-catalyzed interesterification of an unlabeled saturated triacylglycerol, tristearoylglycerol, with [carboxyl-14C]trioleoylglycerol, yielded a mixture of triacylglycerols containing one, two and three double bonds per molecule. The amount of trioleoylglycerol interesterified was determined from the total radioactivity in the triacylglycerol fractions containing one and two double bonds per molecule by taking into account the specific radioactivity of [carboxyl-14C]trioleoylglycerol (Fig. 1).

In addition, unlabeled triacylglycerols may be interesterified with a radioactively labeled long-chain alcohol, e.g. [1-14C]octadecenol, to yield a mixture of radioactive wax esters, or they may be interesterified with [U-14] glycerol to yield a mixture of radioactive mono-, di- and triacylglycerols.

The radiochemical techniques described above also can be used to determine the positional specificity of various lipase-catalyzed reactions. In such reactions, positional specificity of lipases may be assayed using triacylglycerols with radioactive acyl moieties exclusively at the sn-1 and sn-3 positions. Such triacylglycerols may be prepared conveniently by interesterification catalyzed by sn-1,3-specific lipases. Positional specificity of an enzyme suspected to be a *sn*-1,3-specific lipase would be revealed if radioactive *sn*-1,3-diacylglycerol was not found in the lipolysis products. In this case, radioactivity should be associated with free fatty acids and *sn*-1,2(2,3)-diacylglycerols. Virtually no radioactive *sn*-2-monoacylglycerols should be found either.

REFERENCES

1. Tsujisaka, Y., S. Okumura and M. Iwai, *Biochim. Biophys. Acta* *489*:415 (1977).
2. Yokozeki, K., S. Yamanaka, K. Takinami, Y. Hirose, A. Tanaka, K. Sonomoto and S. Fukui, *Eur. J. Appl. Microbiol. Biotechnol.* *14*:1 (1982).
3. Nielsen, T., *Fette Seifen Anstrichmittel 87*:15 (1985).
4. Kalo, P., K. Vaara and M. Antila, *Ibid. 88*:362 (1986).
5. Christie, W.W., *Lipid Analysis*, 2nd edn., Pergamon Press, Oxford, England, 1982, p. 156.
6. Thomas, A.E. III, J.E. Scharoun and H. Ralston, *J. Am. Oil Chem. Soc. 42*:789 (1965).

Differential Determination of Free Fatty Acids Using Acyl-CoA Synthetase and Acyl-CoA Oxidase Enzymes Purified from Microorganisms

Sakayu Shimizu and **Hideaki Yamada**
Department of Agricultural Chemistry, Kyoto University, Sakyo-ku 606, Japan

Two fatty acyl-CoA synthetases (EC 6.2.1.2 and EC 6.2.1.3) and an acyl-CoA oxidase (EC 1.3.3.x) were purified from *Pseudomonas aeruginosa* and *Candida tropicalis*, respectively. Crystalline preparations of a long-chain fatty acyl-CoA synthetase activated saturated and unsaturated free fatty acids (FFA) with chain lengths of 6-18 carbons. Crystalline short-chain fatty acyl-CoA synthetase activated free fatty acids with chain lengths of three and four carbons. Crystalline acyl-CoA oxidase catalyzed conversion of fatty acyl-CoAs with chain lengths of 4-18 carbons to the corresponding enoyl-CoAs. Based on the substrate specificities of these three enzymes, a new method was developed for FFA determination in which activation of FFA was measured by production of H_2O_2. Results obtained by this method agreed with those obtained by gas chromatographic analysis.

Free fatty acids (FFA) found in biological fluids and tissues usually are derived from the hydrolysis of triglycerides and phospholipids. Conventional methods for quantitative determination of FFA are based on titrimetric determination of total acidity (1) or colorimetric analyses based on the transfer of metal soaps from a copper or cobalt triethanolamine reagent (2,3). However, these methods have limited sensitivity and accuracy. Therefore, an improved and simplified method was developed for determination of FFA in biological samples using microbial enzymes. The principal reactions for this method are (1) FFA + CoA + ATP → acyl-CoA + AMP + PPi and (2) acyl-CoA + O_2 → enoyl-CoA + H_2O_2. These reactions are catalyzed by long-chain fatty acyl-CoA synthetase (EC 6.2.1.3) and a acyl-CoA oxidase (EC 1.3.3.x), which are abundantly produced in *Pseudomonas aeruginosa* (4) and *Candida tropicalis* (5), respectively. It also was found that *P. aeruginosa* produced a short-chain fatty acyl-CoA synthetase (EC 6.2.1.2). The long-chain fatty acyl-CoA synthetase, acyl-CoA oxidase and the short-chain fatty acyl-CoA synthetases have been crystallized. Characterization of these enzymes and a new method for FFA analysis is described.

METHODS

Acyl-CoA Synthetases

Fatty acyl-CoA synthetases were assayed in a variety of microorganisms. Stock cultures were grown at 28 C in a medium containing 1% sodium palmitate, 0.2% peptone, 0.1% KH_2PO_4, 0.1% K_2HPO_4, 0.05% $MgSO_4$ • 7 H_2O and 0.3% yeast extract, pH 6.0 or 7.0 (4) Palmitoyl-CoA synthetase activity was assayed in *Pseudomonas, Citrobacter, Escherichia, Klebsiella, Enterobacter, Serratia, Candida, Saccharomycopis, Gibberela, Fusarium, Cylindrocarpon* and other strains of basidiomycetes. Pseudomonads showed detectable activity. Further investigation re-

vealed that *P. aeruginosa* IFO 3919 was the most promising producer of the enzyme.

Acyl-CoA Oxidase

This enzyme activity was assayed in microorganisms depending on the presence of *n*-alkane or fatty acid in the growth medium (5). Most fatty acid-utilizing fungal strains, but none of the bacterial strains, showed this enzyme activity. *Candida tropicalis* AKU 4617 was selected as the preferred source of this enzyme.

Method for FFA Determination

After optimization of the reaction conditions, the following simple procedure was developed. The reaction mixture (0.42 ml) contained μmol of Tris-HCl buffer, pH 8.0, 0.34 μmol of CoA, 0.9 μmol of ATP, 0.23 μmol of $MgCl_2$, 0.20 μmol of KCl, 0.045 of Triton X-100, 138 mU of acyl-CoA synthetase, and 4 U of myokinase. The sample solution (30 μl) containing 10-50 nmol of FFA was added to the reaction mixture. After incubation of the mixture at 37 C for five min, 0.05 ml of *N*- ethylmaleimide (12.5 mmol phenol in 100 ml 12 mM *N*-ethylmaemide) was added to stop the reaction. Then, 0.1 ml of color reagent containing 500 μmol of 4-aminoantipyrine, 200 U of acyl-CoA oxidase and 3000 U of peroxidase in 100 ml of 300 mM potassium phosphate buffer, pH 7.4, was added. After further incubation of this solution at 37 C for 10 min, absorbance of the solution was measured at 505 nm against a reagent blank.

RESULTS AND DISCUSSION

Characterization of Long-chain Fatty Acyl-CoA Synthetase

A constitutive long-chain fatty acyl-CoA synthetase in *P. aeruginosa* was purified to a homogenous state and crystallized from cells grown with succinate as a major carbon source (Fig. 1A). Molecular mass of the enzyme was 166 kd. The enzyme was comprised of four identical 41-kd subunits. This enzyme catalyzed the stoichiometric conversion of saturated and unsaturated fatty acids with 6-18 carbon atoms plus CoA to acyl-CoAs in the presence of

FIG. 1. Crystals of (A) long-chain fatty acyl-CoA synthetase, (B) medium-chain fatty acyl-CoA synthetase and (C) Acyl-CoA oxidase.

Table 1

Properties of Acyl-CoA Synthetase and Acyl-CoA Oxidase

Property	Long-chain acyl-CoA synthetase	Short-chain acyl-CoA synthetase	Acyl-CoA oxidase
Source	*P. aeruginosa*	*P. aeruginosa*	*C. tropicalis*
Mr (Kd)	166	146	600
Subunit Mr (Kd)	41	37	80
Subunit No.	4	4	8
S(20,w)	11.3 *S*	10.5 *S*	16.2 *S*
C-terminal	Ala	Leu	Ser
pI	4.8	4.3	5.5
Substrate	C-6 - C-28	C-3 - C-5	C-4 - C-28
Kmn (μM)			
fatty acid	20 (C-16)	10 (C-4)	34 (C-16)
CoA	45	75	N.D.
ATP	560	970	N.D.

N.D., not determined.

ATP, Mg^{2+} and K$^+$ (6). Apparent *Km*-values for myristic, palmitic, oleic, α-linolenic and arachidonic acid were 10-30 μM. The enzyme gave maximum activity at pH 8.0 and at 42 C. Other properties of the enzyme have been summarized in Table 1.

Short-chain Fatty Acyl-CoA Synthetase

A short-chain fatty acyl-CoA synthetase was purified and crystallized (Fig. 1B) from the cells of *P. aeruginosa*. The enzyme was formed inducibly when the bacterium was grown with butyrate as a major carbon source. Molecular mass of the enzyme was estimated to be 146 kd with four identical 37-kd subunits. The enzyme catalyzed the stoichiometric conversion of butyric acid and CoA to butyryl-CoA in the presence of ATP, Mg^{2+} and K$^+$. It also activated propionic acid and *n*-valeric acid but was inactive toward fatty acids with chain lengths of more than four carbons. Other fatty acids with four carbon atoms, such as isobutyric acid, DL-3-hydroxybutyric acid, 2-ketobutyric acid and *trans*-crotonic acid also were found to serve as substrates. Therefore, this enzyme appeared to be a butyryl-CoA synthetase (7). Properties of the enzyme are summarized in Table 1.

Acyl-CoA Oxidase

This enzyme was crystallized from *C. tropicalis* (Fig. 1C). The enzyme was an octamer with a molecular mas of 600 kd and contained eight mol of FAD per mol of enzyme (5). The enzyme catalyzed the stoichiometric conversion of palmitoyl-CoA and O$_2$ into 2-hexadecenoyl-CoA and H$_2$O$_2$. Acyl-CoAs with carbon chain lengths of 4-18 carbons also were oxidized, however, acetyl and succinyl-CoAs were not oxidized (5,6). Maximum activity was achieved at pH 8.0 and 50 C. The enzyme was stable in the pH range of 5.5-9.0 at 35 C for 60 min but was completely inactivated at 65 C for 10 min. Properties of this *C. tropicalis* enzyme are summarized in Table 1.

Substrate specificities of the fatty acyl-CoA synthetases and acyl-CoA oxidase are compared in Table 2. The long-

Table 2

Relative Substrate Specificities of Acyl-CoA Synthetases and Acyl-CoA Oxidase

Carbon chain length of substrate	Long-chain acyl-CoA synthetase	Short-chain acyl-CoA synthetase	Acyl-CoA oxidase
	Relative reaction velocity (%)		
3:0	0	53	N.D.
4:0	0	100	10
6:0	189	0	14
8:0	257	0	33
10:0	121	0	210
12:0	149	0	256
12:1 (Δ11)	153	N.D.	N.D.
14:0	100	N.D.	105
16:0	100[a]	N.D.	100[b]
18:0	73	N.D.	36
18:1 (*cis*-9)	76	N.D.	121
18:1 (*trans*-9)	80	N.D.	N.D.
18:2 (*cis*-9,12)	30	N.D.	104
18:3 (*cis*-9,12,15)	57	N.D.	N.D.

[a]41.5 μmol/min/mg.
[b]22.2 μmol/min/mg.
N.D., not determined.

chain fatty acyl-CoA synthetase shows the widest substrate specificity. The short-chain fatty acyl-CoA synthetase (butyryl-CoA synthetase), specifically activated butyric acid and propionic acid. Because the acyl-CoA oxidase also was found to have broad substrate specificity for acyl-CoAs, coupling of this enzyme with the fatty acyl-CoA synthetases made it possible to quantify FFA. The hydrogen peroxide generated by the acyl-CoA oxidase could be measured by the oxidative coupling of 4-aminoantipyrine and phenol by peroxidase (5): H$_2$O$_2$ + 4-aminoantipyrine and phenol red quinoimine dye. When the necessary substrates and enzymes were supplied in excess, the rate of the overall reaction was limited by FFA concentration (8-10).

This method gave a quantitative response for fatty acids with carbon chain lengths of 6-18, when the long-chain fatty acyl-CoA synthetase was used and also for equimolar amounts of butyric acid with butyryl-CoA synthetase. In all cases, the degree of saturation did not affect the response. The values obtained by this method well-agreed with those obtained by gas-liquid chromatographic analysis.

The FFA reacted with this fatty acyl-CoA synthetase, and the acyl-CoA oxidase system was stoichiometrically and virtually completely converted to hydrogen peroxide. Therefore, these enzymes proved to be very useful for total FFA analysis. Based on FFA substrate specificities, FFA may be quantified differentially. Furthermore, this method has the advantage of being rapid and simple and having good sensitivity.

ACKNOWLEDGMENTS

This work was supported in part by a Grant-in-Aid for Scientific Research from the Ministry of Education, Science and Culture of Japan.

REFERENCES

1. Dole, V.P., *J. Clin. Invest. 35*:150 (1956).
2. Duncombe, W.G., *Clin. Chim. Acta 9*:122 (1964).
3. Falholt, K., B. Lund and W. Halholt, *Ibid. 46*:105 (1973).
4. Shimizu, S., H. Morioka, K. Inoue, K. Yasui, Y. Tani and H. Yamada, *Agric. Biol. Chem. 44*:2659 (1980).
5. Shimizu, S., K. Yasui, Y. Tani and H. Yamada, *Biochem. Biophys. Res. Commun. 91*:108 (1979).
6. Yamada, H., S. Shimizu, Y. Shinmen, H. Kawashima and K. Akimoto, *Proceedings of World Conference on Biotechnology for the Fats and Oils Industry*, edited by T.H. Applewhite, American Oil Chemists' Society, Champaign, IL, 1988.
7. Shimizu, S., K. Inoue, Y. Tani and H. Yamada, *Biochem. Biophys. Res. Commun. 103*:1231 (1981).
8. Shimizu, S., Y. Tani, H. Yamada, M. Tabata and T. Murachi, *Anal. Biochem. 107*:193 (1980).
9. Shimizu, S., Y. Tani and H. Yamada, in *Enzyme Engineering*, Vol. 6, edited by I. Chibata, S. Fukui and L.B. Wingard Jr., Plenum Publishing, New York, 1982, p. 467.
10. Shimizu, S., and H. Yamada, in *Methods of Enzymatic Analysis*, Vol. 8, edited by H.U. Bergmeyer, VCH Verlagsgesellschaft, Weinheim, 1985, p. 19.

Immunological Characterization of Lipases from a Wide Range of Oilseed Species

Denis J. Murphy[a] and **Matthew J. Hills**[b]

[a]Department of Botany, University of Durham, Science Laboratories, South Road, Durham DH1 3LE, England, U.K., and
[b]Bundesanstalt für Fettforschung, Piusallee 68/76, D4400 Münster, Federal Republic of Germany

A monospecific polyclonal antibody raised against a glyoxysomal lipase purified from castor bean was used to study the lipases of a wide variety of oilseed species. In each of the 12 oilseeds tested, the antilipase cross-reacted with a single polypeptide of apparent molecular mass, 62 kDa. The subcellular localization of the immunoreactive 62 kDa polypeptide varied considerably in different oilseeds but was similar to the localization of lipase activity in each species. The antilipase specifically inhibited lipase activity in all species tested. No antilipase-binding proteins were present in the nonoilseeds lima bean and kidney bean, which also lacked lipase activity. In germinating rapeseed cotyledons, the extent of lipase binding closely followed both the rise and fall of lipase activity and the amount of a 62 kDa polypeptide. These data suggested that many oilseeds possess lipases made up of structurally similar 62 kDa polypeptides.

The storage oils of oilseeds are mobilized during germination by a specific triacylglycerol lipase (EC 3.1.1.3). This enzyme catalyzes the sequential formation of diacylglycerols, monoacylglycerols, glycerol and free fatty acids. Lipases now are used widely as catalysts for such reactions as transesterification, interesterification, hydrolysis and stereoselectivity (1–4). Most of these lipases hitherto have been obtained from microbial sources, but oilseed lipases are potential sources of such industrial catalysts (5). Despite the increasing interest in oilseed lipases, relatively little is known about their synthesis, localization and regulation during germination. In this study, these questions have been addressed by immunological techniques using a specific antilipase antibody.

EXPERIMENTAL PROCEDURES

Plant Material

Seed of castor bean (*Ricinus communis* L. cv. Hale), pea (*Pisum sativum* L. cv alaskan), kidney bean (*Phaseolus vulgaris* L.), lima bean (*Phaseolus lunatus* L.), cotton (*Gossipium hirsutum*), peanut (*Arachis hypogaea* L.) and garbonzo bean (*Cicer arietinum*) were soaked for one day and germinated in moist vermiculite at 30 C for four days. Seed of three corn varieties (*Zea mays* L. cv. Golden Bantam T-51, cv. Trojan and cv. Hopi white) were germinated in moist vermiculite for six days. Erucic acid-free rape (*Brassica napus* L. cv. Andor) and zucchini (*Cucurbita pepo* cv. Burpee hybrid zucchini) were germinated for three days. Soybean (*Glycine max* L.) and sunflower (*Helianthus annus* L.) seed were germinated on moist filter paper in deep petri dishes for five and seven days, respectively. All seedlings except castor bean were grown in the dark at 25 C. In addition, seedlings of corn, rape, sunflower, cotton and castor bean were grown for six wk in a greenhouse.

Subcellular Fractionation and Density Gradient Centrifugation

These were performed as described (6).

Enzyme Assays

Lipase activity was determined by measuring fatty acids produced by triacylglycerol hydrolysis. All assays were performed in triplicate in a total volume of 100 μl. Reactions were carried out for up to one hr in a shaking water bath at 30 C. Fatty acid release was linear for more than one hr. The fatty acids were assayed colorimetrically as copper soaps using 1,5 diphenylcarbazide (7,8). The purity of the triacylglycerol substrates was verified by TLC, and no breakdown products were detected. The marker-enzymes CDP-choline diacylglycerol transferase and catalase were assayed as described (9).

Protein Analysis

Proteins were assayed in the presence of SDS by a modified Lowry method using BSA as standard (10). Protein extracts were solubilized before SDS gel electrophoresis by incubation at 90 C for five min in 1.7% w/w SDS, 1% w/w mercaptoethanol, 48% w/w urea, 16% w/w sucrose and 0.1M Tris-HCl at pH 6.8. Electrophoresis was performed using slab gels containing 15% w/w polyacrylamide. Gels were normally run at 8 mA for 16 hr to achieve good resolution without excessive heating effects. Polypeptides were stained with Coomassie-brilliant blue R. The relative proportions of polypeptide bands on stained polyacrylamide gels were estimated using an LKB 2222-010 Ultroscan XL Laser Densitometer.

Immunological Procedures

The antilipase antibody used in this study was prepared in a rabbit by M. Maeshima and was raised against a purified castor bean lipase. Polypeptides were transferred onto nitrocellulose membranes by electroblotting for two hr at 0.8 mA cm^{-2} using a Sartorius Model II Sartoblot dry blotter. The nitrocellulose membrane was blocked by incubation for 30 min in 3% w/w dried milk powder in a saline buffer comprising 20 mM Tris-HCl at pH 7.4 with 150 mM NaCl. The membrane was incubated overnight in 20-50 ml blocking buffer containing a 1:1500 dilution of antilipase and was washed twice in saline buffer containing 0.5% w/w Triton X- 100 and once with saline buffer. The membrane then was incubated for two hr with a 1:2000 dilution of alkaline phosphatase conjugated goat anti-rabbit antibody (Sigma Chemical Co., St. Louis, MO), washed as described above and stained for alkaline phosphatase activity using the fast red/napthol assay (11). Antilipase-stained bands were localized with reference to prestained molecular weight markers on the same electrophoretic gels and blotted onto the nitrocellulose membranes along with the rapeseed proteins. In some cases, gels were blotted onto two nitrocellulose membranes; one for antibody probing, the other for amido black staining of polypeptides. This allowed for the exact localization of the antibody-binding polypeptide.

For the detection of lipase by enzyme-linked immunosorbent assay (ELISA), samples were diluted in 0.2 M Tris, 0.3 M NaCl, pH 9.5 to 1 μg/ml protein, and 200 μl was loaded to the wells of the microtitration plate (Nunc, Immunoplate 1, Gibco, Europe) and incubated 18 hr at 4 C. Plates were washed six times with 10 mM Na_2HPO_4/ NaH_2PO_4 (pH 7.2), 1.38 mM NaCl, 2.7 mM KCl (PBS) containing 0.05% w/w Tween 20 (PBST) using a Nunc Immunowash 8 hand-held plate washer and rinsed twice with water. The sample wells were postcoated by incubating 300 μl PBS containing 0.1% w/w BSA (98-99% Albumin, Sigma Chemical Co.) for one hr at ambient temperature and then washed as before. Rabbit antilipase was diluted (1/10,000) in PBST, 200 μl was added to the wells and incubated at ambient temperature for two hr, after which the washing procedure was repeated, followed by the addition of goat anti-rabbit IgG horse radish peroxidase conjugate (Sigma Chemical Co.) diluted (1/1000) in PBST at 200 μl/well. After a further incubation of two hr at ambient temperature, the plate was washed and the colorimetric substrate 1.1 mM azinobis (3-ethylbenzthiazol:sulphonic acid) 11 mM citrate buffer, pH 4.0, 0.017% v/v H_2O_2 was added at 200 μl/well. After an incubation of 15-n30 min at ambient temperature, the optical density at 414 nm of each well was determined using a Titertek Multiscan MCC (Flow Laboratory U.K.) microtitration plate reader. Optical densities of the wells were directly proportional to the lipase content of the samples.

RESULTS AND DISCUSSION

A wide range of oilseed and nonoilseed plants was screened for immunological cross-reactivity with the castor bean antilipase. Twelve oilseeds were probed with the antibody. In each case, a single immunoreactive polypeptide of 61-63 kDa was observed on Western blots of proteins separated from germinating tissue. The two nonoilseed species, kidney bean and lima bean exhibited no immunological cross-reactivity with the antilipase. No cross-reactivity was detected in any of the oilseed or nonoilseed species when pre-immune serum was used. In addition to specifically binding to a single ca. 62-kDa polypeptide, the antilipase inhibited lipase activity in all five oilseeds tested for this, i.e. castor bean, rapeseed, maize, peanut and soybean. Nonspecific inhibition of lipases by some proteins that bind to the emulsified substrate has been reported. This was ruled out following the demonstration that neither pure defatted BSA nor pure rabbit IgG significantly affected lipase activity over the concentration range in which activity was completely inhibited by antilipase.

The subcellular distribution of antilipase binding in a range of oilseeds is shown in Figure 1. These data are derived from immunoblots of extracts from germinating oilseeds at the stage of maximum lipase activity. These results clearly demonstrate the variability in apparent lipase distribution in different oilseed species. The results generally correlate with those of other studies of lipase

subcellular distribution. For example, the oil bodies of peanut or soybean seedlings have been shown to have no lipolytic activity over the physiological pH range (13,14) but do have an active glyoxysomal lipase. In Figure 1, both peanut and soybean lack any immunoreactive 62 kDa polypeptide in their oil body fractions. In each case, however, most of the antibody binding was in the 10,000 G pellet that would contain the glyoxysomal fraction (15). Castor bean, rapeseed, cotton and corn are known to have oil body lipases, and in all four cases the antilipase was bound to a 62 kDa polypeptide of their oil body fractions. The only appreciable antilipase binding to the soluble (100,000 G supernatant) fraction occurred with the three maize varieties, most notably with sweet corn (var T51). It has been reported that the oil body lipase of maize is easily detached from the membrane and therefore may be recovered from the soluble fraction (16,17). The oil body lipase purified from maize oil bodies had an apparent molecular mass of 65 kDa, which is very similar to the value reported in this study.

In addition to the subcellular fractions of germinating oilseeds, three other tissues also were tested for antilipase binding, i.e. mature seeds, leaves and roots. There was no antilipase binding with any of the leaf and root extracts in the range of oilseeds tested. Neither was binding detected with mature seeds of soybean, peanut, cotton and corn. Very low levels of binding were observed

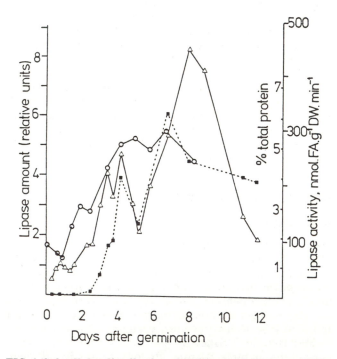

FIG. 1. Subcellular distribution of antilipase binding to a 62 kDa polypeptide in a range of oilseeds. Oilseed homogenates were fractionated and 7 μg protein from each fraction separated by PAGE, transferred to nitrocellulose, probed with antilipase and stained as described in Experimental Procedures. The extent of antilipase binding to the 62 kDa polypeptides was assessed by scanning densitometry. The relative staining in each fraction is given as a percentage of the total for that species. Subcellular and whole organ fractions are (A) oil body, (B) 10,000 G pellet, (C) 100,000 G pellet, (D) 100,000 G supernatant, (E) leaf homogenate, (F) root homogenate.

with mature seeds of castor bean and rapeseed. No binding was found in any tissues or subcellular fractions of kidney bean or lima bean. Neither of these plants is an oilseed, and no lipase activity was detected in any of their extracts.

The reason for the variations in apparent subcellular localization of oilseed lipases is unclear. To better elucidate this question, organelles were separated by SDS-PAGE and immunoblotted. In Figure 2, the results of such an immunoblot of fractions from a sucrose gradient of germinating sunflower can be seen. There are three main sites of antilipase binding in these gradient fractions, i.e. the oil body fraction, the fractions at about 1.15 g/l and the fractions at about 1.24 g/l. The latter two fractions roughly corresponded with marker enzymes for endoplasmic reticulum and glyoxysomes, respectively. Essentially similar results were obtained with the immunoblots of castor bean and rapeseed sucrose gradient fractions.

It is difficult to conclude with certainty that the lipases from these oilseeds were partially localized in the endoplasmic reticulum. One reason for this caution is that the peak of antilipase binding, which coincided exactly with lipase activity (as assayed by the copper soaps method), was invariably shifted slightly with respect to endoplasmic reticulum marker enzymes, such as CDP-choline diacylglycerol acyltransferase. It has been suggested that the microsomal lipase of germinating rapeseed may derive from depleted oil bodies or fragments of oil body membranes (18). This suggestion is, however, not supported by recent studies in which oil body membranes were largely made up of a single 18.5 kDa polypeptide that was com-

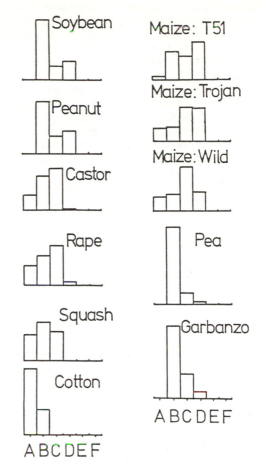

FIG. 3. Changes in lipase activity, antilipase binding and amount of 62 kDa polypeptide during germination. Lipase activity was determined by the copper soaps method. Antilipase binding was estimated using an ELISA. The relative amount of the 62 kDa immunoreactive polypeptide was assessed by scanning densitometry of gel electrophoretograms. (sketch), lipase activity; (sket), antilipase binding; (skt), amount of kDa polypeptide.

pletely absent from microsomal fractions having high lipase activities (19). These results appear to rule out the accumulation of depleted oil body membranes in the microsomal fraction. Another possibility that currently is being tested is that the lipase in the microsomal fraction is derived from glyoxysomal membrane fragments. Preliminary results from immunocytochemical studies using antilipase and antibodies specific to known glyoxysomal proteins suggest that much of the rapeseed lipase may originate from glyoxysomes, as is the case in oilseeds such as soybean and peanut (13,14).

Lipase activity in oilseeds is known to rise during the early stages of germination and subsequently to decline (13,18–21). In an attempt to correlate lipase activity with antilipase binding to the 62 kDa polypeptide, these parameters were followed over the first 12 days of germination. In Figure 3, the change in lipase activity is compared with the amount of antilipase binding, as determined by an ELISA, and with the relative amount of the immunoreactive 62 kDa polypeptide, as determined by scanning densitometry. The results, which were repeatedly reproducible, show a good correlation of the extent

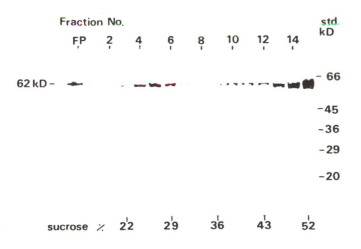

FIG. 2. Immunoblot of fractions from sucrose density gradient separation of organelles from germinating sunflower cotyledons. Cotyledons (3 g) from 10-day-old sunflower seedlings were chopped with razor blades, filtered, centrifuged at 1000 G for 10 min and the supernatant added to a 15-60% sucrose density gradient. The gradient was centrifuged at 100,000 G for three hr and fractionated. The proteins from 2 µl aliquots of each fraction then were subjected to SDS-PAGE and the separated polypeptides transferred to nitrocellulose. The nitrocellulose blots were probed with antilipase and finally stained using the alkaline phosphatase/fast red system. The immunoblots show strong staining in the oil body (FP), light membrane (equilibrium density 1.11 g/ml) and microbody (equilibrium density 1.23 g/ml fractions.

of immunobinding with lipase activity. The changes in the amount of 62 kDa immunoreactive polypeptide also are consistent with its involvement in lipase activity. The data in Figure 3 are consistent with the de novo synthesis of rapeseed lipase during the first few days of germination rather than the activation of an enzyme or precursor form already present in dry seeds. It also has been found that the maize scutellum lipase is synthesized de novo as a 65 kDa protein during the early stages of germination (16,17). These data seem to preclude, at least in the case of maize and rapeseed, an important regulatory role for the soluble proteinaceous inhibitors of lipase found in some oilseeds (23-25).

The evident ease of cross-reaction of an antibody raised against a castor bean glyoxysomal lipase with a single 62 kDa polypeptide from a wide variety of oilseed species and subcellular fractions poses interesting questions regarding structural similarities between oilseed lipases. Similarities already have been pointed out between the active sites of many microbial lipases and serine hydrolases, which can catalyze the addition of nucleophiles to ester bonds (26). The three oilseed lipases so far purified to homogeneity, i.e. castor bean, maize (16), and rapeseed (Theimer, R.R., personal communication), all have a subunit molecular weight of 62-65 kDa. As yet, no sequence information is available for oilseed lipases, although this is being actively sought both in this laboratory and elsewhere. However, we would predict on the basis of the above evidence that there should be extensive structural homologies between the various oilseed lipases.

ACKNOWLEDGMENTS

This work was supported by grants from the Agricultural Research Council, U.K., the NATO Science Council and the British Council. We are grateful to L.M. Thistal for reading the manuscript and U.-P. Shyminder for helpful discussions.

REFERENCES

1. Macrae, A.R., *J. Am. Oil Chem. Soc. 60*:291 (1983).
2. Macrae, A.R., and R.C. Hammond, *Biotech. Gen. Eng. Rev. 3*, 193 (1985).
3. Schuch, R., and K.D. Mukherjee, *J. Agric. Food Chem.*, in press.
4. Tanaka, A., *J. Am. Oil Chem. Soc. 64*:1251 (1987).
5. Hassanien, F.R., and K.D. Mukherjee, *Ibid. 63*:893 (1986).
6. Hills, M.J., and H. Beevers, *Plant Physiol.*, in press.
7. Huang, A.H.C., in *Modern Methods of Plant Analysis*, Vol. 1, edited by H.F. Linskens, and J.F. Jackson, Springer Verlag, Berlin, 1986, pp. 141-151.
8. Nixon, M., and S.H.P. Chen, *Anal. Biochem. 97*:403 (1979).
9. Lord, J.M., T. Kagawa, T.S. Moore and H. Beevers, *J. Cell Biol. 57*:659 (1973).
10. Markwell, M.A.K., S.M. Haas, N.E. Tolbert and L.L. Bieber, *Meth. Enz. 72*:296 (1981).
11. Mason, D.Y., and R.E. Simmons, *J. Histochem. Cytochem. 27*:832 (1979).
12. Kang, A.S., N. Harris, R.R.D. Croy, J.R. Ellis, L.A. O'Shea and D. Boulter, *FEBS Lett.*, in press.
13. Huang, A.H.C., and R.A. Moreau, *Planta 141*:111 (1978).
14. Lin, Y.-H., R.A. Moreau and A.H.C. Huang, *Plant Physiol. 70*:108 (1982).
15. Cooper, T.G., and H. Beevers, *J. Biol. Chem. 244*:3507 (1969).
16. Lin, Y.-H., and A.H.C. Huang, *Plant Physiol. 76*:719 (1984).
17. Wang, S.-M, and A.H.C. Huang, *J. Biol. Chem. 262*:2270 (1987).
18. Theimer, R.R., and I. Rosnitschek, *Planta 139*:249 (1978).
19. Hills, M.J., and D.J. Murphy, *Biochem. J.*, in press.
20. Lin, Y.-H., L.T. Wimer and A.H.C. Huang, *Plant Physiol. 73*:460 (1983).
21. Lin, Y.-H, and A.H.C. Huang, *Archs. Biochem. Biophys. 225*:360 (1983).
22. Satouchi, K., T. Mori and S. Matschushita, *Agric. Biol. Chem. 38*:97 (1974).
23. Satouchi, K., and S. Matschushita, *Ibid. 40*:888 (1976).
24. Wang, S., and A.H.C. Huang, *Plant Physiol. 76*:929 (1984).
25. Estell, D., T.P. Graycar, J. Von Beilen, A.J. Poulose, M.J. Repsin and M.V. Arbidge, *J. Am. Oil Chem. Soc. 64*:1253 (1987).

Characteristics of Spiculisporic Acid as a Polycarboxylic Biosurfactant

Yutaka Ishigami[a], **Yasuo Gama**[a], **Shinsuke Yamazaki**[a] and **Shergeru Suzuki**[b]

[a]National Chemical Laboratory for Industry, Tsukuba, Ibaraki, Japan, and [b]Tokyo Pearl Co., Ltd., Chuo-ku, Tokyo

Several kinds of amphiphiles were derived from spiculisporic acid, a biosurfactant of microbial origin. The surface-active properties and functional properties of these compounds were compared. Alkylamine salts exhibited multifunctional surface-active, liquid crystal-, gel- and liposome-forming properties. Furthermore, a new rhodamine-type dye and a liposaccharide containing glucosamine residue were produced from spiculisporic acid.

Trace amounts of spiculisporic acid are a membranous constituent of microorganisms and lichens (1,2). Recently, Tabuchi et al. (3,4) reported a new bioindustrial process for the mass production of this compound (up to 110 g/l culture broth) using *Penicillium spiculisporum* Lehman No. 10-1. This has allowed the efficient production of spiculisporic acid. The application of high-priced surfactants with safe, biodegradable products has been discussed in previous papers (5–8). This paper reports the functional properties of compounds derived from spiculisporic acid.

EXPERIMENTAL PROCEDURE

Materials

Spiculisporic acid (S-acid) was supplied by Iwata Chemical Industrial Co., Ltd. and purified by the procedure of Tabuchi et al. (3,4). Open-ring acid, 3-hydroxy-1,34-tetradecanetricarboxylic acid (O-acid) was prepared by ethyl ether extraction from the neutralized product with hydrochloric acid after the saponification of both the carboxylic and the lactone groups with 3N NaOH. N-(5-fluoresceinthiocarbamoyl) dipalmitoyl-L-α-phosphatidylethanolamine was purchased from Molecular Probes; Aerosol OT (A. OT), Wako Pure Chemical Ind., was purified by eliminating insoluble fractions with acetone and petroleum ether. Polyoxyethylated (11) nonylphenylether (NP-11) was supplied from Lion Corp. as Liponox NCK; α-copper phthalocyanine blue (α-PC), Sumimoto Chemical Ind., Cyanine blue HB, av. 2.0 μm; γ-Fe$_2$O$_3$ and Fe$_3$O$_4$ (magnetic material), Toda Ind., MX-450 and KN-320, respectively (7). Water was redistilled in a glass-tip packed Pyrex column (3 cm x 60 cm) and deionized in the presence of alkaline potassium permanganate. Other chemicals were reagent grade.

Methods

Surface tension measurements were carried out with a Wilhelmy-type surface tensiometer (Shimadzu ST-1) at 30 ± 0.5 C. NMR spectra were obtained using a Hitachi R-90 (90 mHZ). Liposome structure was observed using an electron microscope (TEM), Hitachi HU-12A, after negative staining with 1% phosphotungstic acid solution (9). A Shimadzu UF-265FS spectrophotometer and a Shimadzu RF-540 spectrofuorophotometer were used. Emulsifying action was measured by observing oil (kerosene), emulsion and water layers with time (5-120 min) at 95 C using stoppered 30-ml volumetric test tubes. The emulsions

contained 2:3, v/v sample solution and kerosene. Emulsifying rate was calculated by measuring changes in the ratio of emulsified area in the oil layer to total oil layer area with time (10). Suspensions containing dispersoid were allowed to stand for four hr after mixing of the aqueous sample solution with dispersoid. The degree of dispersing action was evaluated from the turbidity of the solution at 445 nm. The stability of dispersion was determined by the appearance of the sedimented boundary in the 30-ml suspension as described elsewhere (10). Liposome-forming material (7-30 mmol) was dissolved in 30 ml of CHCl$_3$-CH$_3$OH (2:1) in a 100 ml round-bottomed flask. The thin film was attached to the inner wall of the flask by evaporating the solvent to dryness under vacuum and drying for one hr in a vacuum dessicator over P$_2$O$_5$ at 25 C. Thereafter, 10 ml of water or M/15 phosphate buffer (pH 5.8 to 6.0) was added to the flask and vortexed for 20 min at 50 C. Fluorescent microscopic observations were conducted with N-(5-fluoresceinthiocarbamoyl) dipatlmitoyl-L-α-phosphatidylethanolamine added to the CHCl$_3$-CH$_3$OH solution.

RESULTS AND DISCUSSION

Functional chemicals derived from spiculisporic acid (I) are shown in Figure 1. Spectral data for these compounds are listed in Table 1. Spiculisporic acid (50 g) was dehydrated by heating it gradually under vacuum (1-2 mm Hg) to temperatures above the melting point of the acid (146 C). After one hr at 170-180 C, the light-brown liquid was cooled and purified by crystallization from hexane-ether. The yield was 45.3 g (96%), mp 40-41 C, of the anhydride (III).

Synthesis of D-glucosamine N-acyl derivative (IV)

Acid anhydride (III) (5 g) was refluxed in dry benzene with freshly distilled thionyl chloride for five hr. After removal of excess thionyl chloride, the residual oil was dissolved in dry terahydrofuran. The solution was added drop-wise over a 15 min period with vigorous stirring to a mixture of D-glucosamine hydrochloride and sodium bicarbonate in water at 0 C. After one hr, the product was precipitated by addition of 6 N HCl and purified by washing by water and crystallization from dioxane. The yield was 3.6 g (49%), mp 194-196 C.

Synthesis of rhodamine-type derivative (V)

A 1:2 molar ratio mixture of III and 3-diethylaminophenol was heated gradually in the presence of a trace amount of sulfuric acid. The reaction mixture, which melted at 65 C, was reacted for one hr at 145-150 C. The resulting product was a purple-colored solid. The solid was dissolved in ethyl acetate, washed with excess of

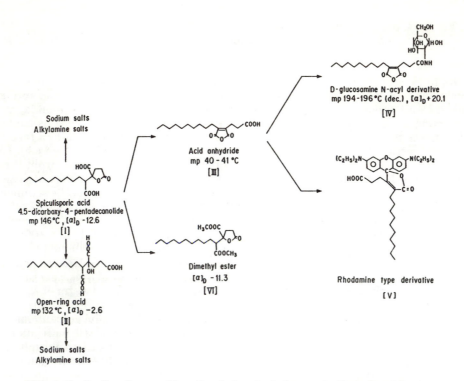

FIG. 1. Synthetic scheme of functional chemicals from spiculisporic acid.

diluted alkaline solution and evaporated. The resulting solid was purified by silica gel chromatography. The final product was eluted from a silica gel (Wakogel C-100) column, preconditioned with hexane and methanol.

Synthesis of dimethyl ester (VI)

Spiculisporic acid (10 g) dissolved in ether was reacted with etheral diazomethane at room temperature for one hr. The product, a colorless oil, was purified by silica gel chromatography. The yield was 10.1 g (93%).

Surface tension and micelle formation

In general, sodium salts of both S-acid and O-acid had small surface activities. However, the lower hydrophilicity of monosubstituted sodium salts of S-acid (S-1Na) and O-acid (O-1Na) resulted in high surface activities (Fig. 2). Alkylamine salts of both S-acid and O-acid exhibited high surface activities due to equal hydrophobic and lipophilic properties (Fig. 3). Disubstituted n-hexylamine salt of S-acid (S-2n-HA) had a critical micelle concentration (CMC) of 1.0×10^{-2}M with an equilibrium surface tension at the CMC (γcmc) of 27 mNm. The disubstituted cyclohexylamine salt of S-acid (S-2c-HA) had a CMC of 1.3×10^{-2}M and γcmc of 36 mN/m. Therefore, the absorption behavior of S-acid appeared to be effected to a greater extent by the cyclohexylamine derivative. In addition, the lactone group of S-2n-HA gave greater hydrophobic properties, due to the low Pγcmc and the large CMC, in comparison with O-2n-HA.

Plots of surface tension against concentration of three amphipathic derivatives (IV, V and VI) from I, and the corresponding water-soluble derivatives obtained by

their neutralization are illustrated in Figure 4. V-2HCl was the most surface-active of all compounds due to the contribution of the rhodamino group to the molecular hydrophobicity. IV-2 Na had a large Pγcmc because of the increase in the hydrophilic properties due to the glucosamino moiety. The CMC of VI-1Na was smaller than the I-2Na because of the difference in the molecular orientation of the hydrophilic group.

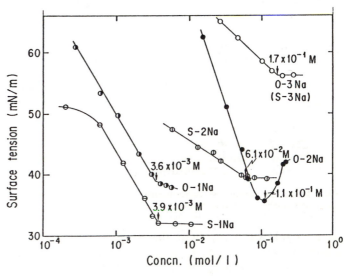

FIG. 2. Surface tension vs concentration plots of sodium salts of S-acid and O-acid at 30 C.

Dispersability

Alkylamine salts of I and II exhibited large dispersing

Table 1

Spectral Data for the Derivatives of Spiculisporic Acid [(I)]

Sample	IR (cm^{-1})	NMR (ppm)
III	1836, 1760 (anhydride); 1707 (COOH); 1665 (C=C)	0.88 (t, CH$_3$); 1.28 [m, (CH$_2$)]; 9.15 (vs. COOH)
IV	3200-3600 (OH); 3280 (NH); 1821, 1769 (anhydride); 1617 (C=C); 1636, 1556 (CONH)	
IV-2Na	3200-3600 (OH); 1600 (CONH); 1640 C=C); 1540-1580 (CONH and COO$^-$)	1.35 (t, CH$_3$); 1.78 [m, (CH$_2$)$_n$]; 2.5-3.2, 3.7-4.0, 4.2-4.4 (m, CH$_2$ and CH)
V	1810, 1770 (lactone); 1720 (COOH)	1.32 (t,CH^3); 1.75 [m, (CH^2)$_n$], 1.55 (t, NCH$_2$CH$_3$); 4.01 (q.NCH$_2$CH$_3$); 7.2-8.0 (aromatic H) in DC1-D$_0$)
VI	792 (lactone); 1734 (COOCH$_3$)	0.85 (s, CH$_3$); 2.25 [m, (CH$_2$)$_n$]; 2.3-2.7 (m, CH$_2$); 2,9-3.1 (m, CH); 3.72, 3.60 (s, CH$_3$)
VI-1Na	1720 (COOCH$_3$); 1580 (COO$^-$)	4.10, 4.22 (s, CH$_3$)

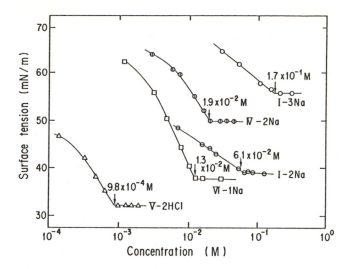

FIG. 4. Surface tension vs concentration plots of functional derivatives from spiculisporic acid at 30 C.

ethyl acrylate and methyl methacrylate (7:3) with potassium persulfate as the polymerization initiator demonstrated the effect of 2% w/v solutions of neutralized salts of I and II (Table 4). Polymerization proceeded smoothly for at least 33 min, yielding stable latices. However, the large particle size of these polymer latices reflected a low solubilizing action of the emulsifiers concerned (11). The transparency and the water-resisting quality of the films were improved by heating or UV-irradiation (366 nm). This phenomenon was attributed to cross-linking of the polyfunctionality of S-acid.

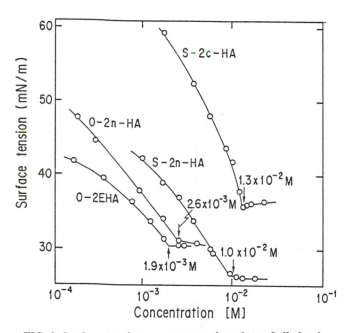

FIG. 3. Surface tension vs concentration plots of alkylamine salts of S-acid and O-acid at 30 C.

actions for water-repellent dispersoids such as α,-PC (Table 2).

Emulsifiability

Table 3 shows the emulsifying actions of sodium and alkylamine salts of I, IV and VI. Monosubstituted salts (especially S-1n-HA) generally were good emulsifiers as determined by surface activity. Emulsion copolymerization of

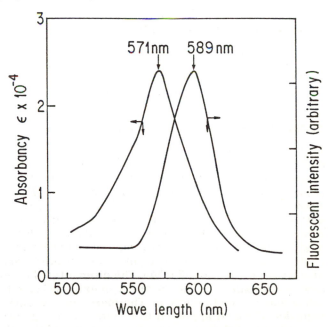

FIG. 5. Absorption and fluorescent spectra excited at 571 nm of V-HCl.

Characteristics of fluorescent dye (V)

Absorption spectrum of V-HCl determined in water peaked at 571 nm, while the maximum fluorescent inten-

Table 2

Dispersing Actions for α-Copper Phthalocyanine Blue (α-Pc)

Dispersant	Concn. (%)	Dispersing actions (%)	Stability of dispersion 4 hr	Stability of dispersion 24 hr	Dispersant	Concn. (%)	Dispersing actions (%)	Stability of dispersion 4 hr	Stability of dispersion 24 hr
S-1n-HA	0.1	29	17.5	17.0	O-1n-HA	0.1	0	—	—
S-1n-HA	0.5	53	30.0	24.8	O-1n-HA	0.5	0	—	—
S-2n-HA	0.1	66	27.0	—	O-2n-HA	0.1	24	6.2	—
S-2n-HA	0.5	50	28.3	28.3	O-2n-HA	0.5	48	28.2	28.0
S-2c-HA	0.5	19	6.5	—	O-2c-HA	0.5	0	—	—
S-2EHA	0.05	0	—	—	O-2EHA	0.05	62	30.0	30.0
A. OT	0.1	46	24.3	22.3	O-2EHA	0.1	67	30.0	30.0
NP-9	0.1	56	23.8	22.0	O-2EHA	0.5	69	30.0	30.0
S-1Na	0.5	4	0	—	O-1Na	0.5	3	0	—
Water		7	—	—					

TABLE 4

Emulsion Copolymerization of Ethyl Acrylate and Methyl Methacrylate

Emulsifier	Concn. (%)	Polymerization time (min)	Particle size (A)	Properties of polymer films Haze (%)	Properties of polymer films Tensile strength (kg/cm^2)	Properties of polymer films Elongation (5)	Properties of polymer films Gel content (%)
S-1Na	2.0	45	553	57.4			0
S-2Na	2.0	40	1107	20.6	71	840	19
O-3Na	2.0	70	2507	42.6			66
S-1NH$_4$	20	40	836	51.5	60	780	0
S-2NH$_4$	20	40	1601	47.3			
S-3NH$_4$	20		2582				
S-1morpholine	2.0	45	550	13.7			0

TABLE 3

Emulsifying Actions for Kerosene

Emulsifier	0.1%	0.5%	1.0%
I-1Na	0	0	5
I-2Na	0	0	0
IV-2Na	10	15	15
VI-1Na	30	45	70
S-1n-HA	17	77	89
S-2n-HA	0	0	0
SDS	10	90	95

sity was found at 589 nm (Fig. 5). Compound (V), a new type of amphipathic fluorescent dye, exhibited a wine color rather than the typical bright-red color of rhodamine dyes.

Liposome formation

It was observed with polarized microscopy that the aqueous suspensions of octylamine salts of I and II formed lipidic spherulites 6-50 μm in diameter. Aqueous suspensions of octylamine salts of I and II with N-(5- fluo-resceinthiocarbamoyl) dipalmitoyl-L-α-phosphatidyl-ethanolamine gave spheres having 1-10 nm diameter under fluorescent microscopy. Negatively stained suspensions containing octyl or hexylamine salts were observed using TEM. Liposomal sizes averaged 170 nm in diameter.

REFERENCES

1. Bodo, B., M. Massias, L. Mohlo and S.Combrisson, *Bull. Mus. Natl. Hist. Nat., Sci. Phys.-Chim.* 11:53 (1976).
2. Asano, M., and Y. Kameda, *J. Pharm. Soc., Japan* 61:80 (1941).
3. Tabuchi, T., I. Nakamura and T. Kobayashi, *J. Ferment. Technol.* 55:37.
4. Tabuchi, T., I. Nakamura and E. Higashi, *Ibid.* 55:43 (1977).
5. Ishigami, Y., S. Yamazaki and Y. Gama, *J. Colloid Interface Sci.* 94:141 (1983).
6. Yamazaki, S., Y. Ishigami and Y. Gama, *Kobunshi Ronbunshu* 40:569 (1983).
7. Ishigami, Y., S. Yamazaki and Y. Gama, *J. Natl. Chem. Lab. Ind. 80 (223)*:231 (1985).
8. Ishigami, Y., Y. Gama and S. Yamazaki, *J. Japan Oil Chem. Soc.* 36:490 (1987).
9. Lukac, S., A. Perovic, *J. Colloid Interface Sci.* 103:586 (1985).
10. Ishigami, Y., S. Suzuki, S. Yamazaki and H. Suzuki, *J. Japan Soc. Colour Material* 54:671 (1981).
11. Ishigami, Y., and S. Yamazaki, *J. Japan Oil Chem. Soc.* 32:735 (1983).

Engineering Parameters for the Application of Immobilized Lipases in a Solvent-free System

T. Luck, **T. Kiesser** and **W. Bauer**

Fraunhofer Institut für Lebensmitteltechnologie und Verpackung, Schragenhofstr. 35, D-8000 München 50, Federal Republic of Germany

The interesterification of lipids catalyzed by immobilized lipases in a continuous solvent-free process was studied. The important engineering parameters that are necessary for the design of a bioreactor were examined. The pressure drop across the catalyst bed was determined as a function of the fluid temperature, linear superficial velocity, height of the biocatalyst bed and average particle diameter of the biocatalyst. The overall biocatalyst effectiveness factor was calculated numerically. The dependence of its value on different ratios of film to pore diffusion are discussed. The influence of external transport on the conversion of the substrate was tested in a fixed bed reactor. External mass transfer limitations can be neglected for flow rates more than 8.0×10^{-5} m/s.

The application of immobilized microbial lipases for the interesterification of fats and oils has been reported (1-3). The immobilization of the lipase on a support offers the possibility of using it in a continuous process involving fixed-bed reactors. The development of an immobilized lipase system for commercial application requires a systematic analysis of engineering parameters and reaction kinetics. Investigations require the inclusion of the reaction rate, external and internal mass transfer limitations, backmixing of the fluid in the reactor, and the pressure drop across the biocatalyst bed. These aspects need to be taken into consideration when a mathematical model of the bioreactor is developed. Mathematical modeling is an important factor in scaling-up an immobilized enzyme system.

EXPERIMENTAL

Pressure Drop Experiments

The experiments were performed in thermostated glass reactors ($L=1.0$ m, $D_R=1.0\times10^{-2}$ m). Pure olive oil (Fa. H. Lamotte Bremen, FRG) was used as the fluid. Carrier materials were Duolite ES 568 (Fa. Diamont Shamrock, Chem. Co., Cleveland; 100 μ m$<$ dp $<$ 500 μm, $\overline{dp}$ =360 μm), two sieve fractions of Duolite ES 568 (200 μm $< \overline{dp}$ $<$ 315 μm, $\overline{dp}$ = 260 μm and 400 μm $< \overline{dp}$ $<$ 500 μm, $\overline{dp}$ = 450 μm), and glass beads (Fa. Braun, Melsungen, FRG; 250 μm $< \overline{dp}$ $<$ 300 μm, $\overline{dp}$ = 270 μm). The following parameters were varied: bed height: H_B: 0.3 m, 0.5m, 0.7m; temperature: T: 60 C, 70 C, 80 C; and flow rate: u: 1.0×10^{-4} m, 1.1×10^{-3} m. The pressure drop across the biocatalyst bed was detected by a pressure sensor (142 PC 30 G; Honeywell, Co., Freeport, IL).

Conversion Experiments

The interesterification was studied in a thermostated glass reactor ($L=1.0$ m; $D_R = 5\times10^{-3}$m). The immobilized enzyme used was obtained from Novo, Copenhagen (Lipozyme; ρbulk 36 kg/m^3). Triolein (purity$>$70%, w/w) and

trimyristin (purity$>$97%, w/w/) were obtained from Fluka (Neu Ulm, FRG). The substrate feed consisted of 92.5% triolein and 7.5% trimyristin (w/w). The reaction was performed at 60 C. The conversion was determined as a function of the residence time. The influence of external transport on the conversion was examined by the operation of reactors with different bed heights at flow rates that gave the same residence time. The bed height was varied from 0.05 m to 0.8 m and the flow rate from 8.0×10^{-5} m/s to 1.3×10^{-3} m/s. The triacylglycerol composition was analyzed by RP-HPLC using the following equipment and conditions: Gynkotek isocratic pump 300 C (Fa. Gynkotek, Germering, FRG); two 250 mm Nucleosil 5 μm columns in series (Fa. Macherey-Nagel, Düren, FRG); Shodex SE 51 refractive index detector (Showa Denko Co., Tokyo, Japan); Shimadzu C-R 3 A integrator (Shimadzu Co., Kyoto, Japan); solvent system: acetonitrile/acetone/chloroform; (33.2/61.8/5.0 v/v/v); and samples were solved in acetone (5% v/v), sample size 20 μm. The conversion factor X was calculated by: X = 100 x (area of dimyristoyl-monooleyl glycerol)/(total area of triacylglycerols). The equilibrium concentration of MMO (X_{eq} = 3.90%) was determined in a stirred tank reactor. The modified fixed bed conversion X' is given by: X' = ($X_{fixed\ bed}$)/($X_{equilibrium}$).

RESULTS AND DISCUSSION

Pressure Drop Experiments

The experimental results for three different carrier materials and the calculated data are presented in Figure 1. The results are expressed by the friction factor ϕ as a function of the Reynolds number, Re. Theoretical data were calculated by the correlation: $\phi = [22.4 \times (1-\epsilon) \times \epsilon^{-4.55}]/$Re, which describes the flow of a fluid through a particle bed (4). As can be seen from Figure 1, the results for each carrier material can be represented by a straight line possessing a slope of -1, respectively. Each straight line is characterized by one value of the fractional voidage ϵ. The pressure drop across a fixed bed of Duolite ES 568 ($\overline{dp}$ μm, D_R = 1.0×10^{-2} m, H_B = 0.7 m, T = 60 C, u = 7.1×10^{-4} m/s) was 0.5 bar. This low value allows for the performance of a continuous process without using a solvent.

Influence of Mass Transfer Effects on the Effectiveness Factor

The effectiveness factor η_o is the ratio of the apparent reaction rate to the reaction rate in the absence of mass transfer limitations (5). For η_o= 1 the reaction rate is kinetically controlled. The effectiveness factor will decrease ($\eta_o<$1) if diffusion effects reduce the reaction rate. For an immobilized enzyme system, the value of η_o should be near unity for efficient operation. The numerically calculated overall effectiveness factor is presented in Figure 2 as a function of the Thiele modulus ϕ, which

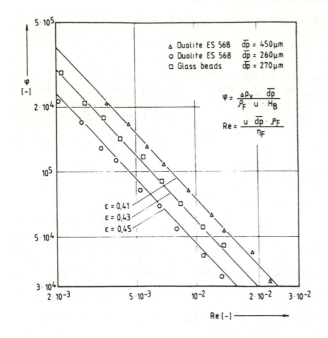

FIG. 1. Calculated and experimentally determined friction factor ϕ as a function of the Reynolds number Re (experimental results for different carrier materials).

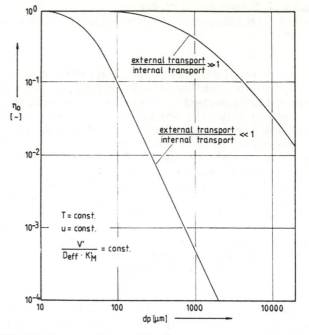

FIG. 3. Influence of the particle diameter dp on the overall effectiveness factor η_o for different ratios of film to pore diffusion.

characterizes the intraparticle diffusion and the reaction kinetics. With increasing particle diameter dp, ϕ increases, and the effectiveness factor will decrease. Different ratios of film to pore diffusion are expressed by the Biot number Bi. In the range of $0.01 < Bi < 100$, film and pore diffusion effects reduce the reaction rate. For high Biot numbers, $Bi > 100$ the influence of external mass transport limitations can be neglected as the reaction rate is pore-diffusion controlled. For values of $Bi < 0.01$, the reaction rate is film-diffusion controlled. In Figure 3, η_o is plotted as a function of the particle diameter dp. The value of η_o for constant dp is strongly influenced by the ratio of external to internal transport.

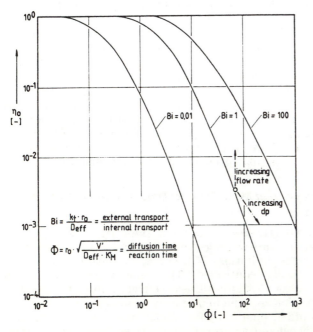

FIG. 2. Calculated overall effectiveness factor η_o as a function of the Thiele modulus ϕ for different values of the Biot number Bi.

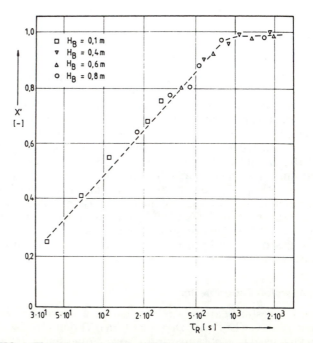

FIG. 4. Fixed bed conversion X' for different residence times τ_R (particle diameter and pore size are constant).

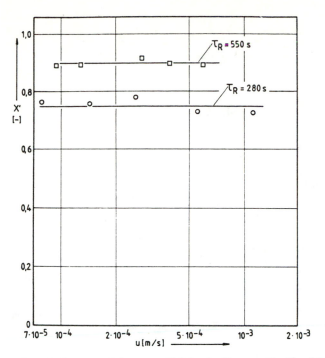

FIG. 5. Influence of the superficial velocity u on the fixed bed conversion X' for constant residence times τ_R (particle diameter and pore size are constant).

Conversion experiments

Figure 4 shows the modified fixed bed conversion X' for different residence times τ_R. The fixed bed conversion for residence times more than 1000 sec approaches equilibrium. Significant external mass transfer limitations can be neglected if the conversion does not change while the flow rate is increased (6). As can be seen from Figure 5, the conversion at two different residence times did not change for flow rates over 8.0×10^{-5} m/s, respectively. The influence of internal mass transfer limitations on the reaction rate need to be studied in further investigations.

REFERENCES

1. Macrae, A.R., *J. Am. Oil Chem. Soc. 60*:219 (1983).
2. Nielsen, T., *Fette, Seifen, Anstrichm. 87*:15 (1985).
3. Hansen, T.T., and P. Eigtved, *Proceedings of the World Conference of Emerging Technologies in the Fats and Oils Industry*, edited by A.R. Baldwin, American Oil Chemists' Society, Champaign, IL, 1986, p. 336.
4. Molerus, O., M. Pahl and H. Rumpf, *Chem. Eng. Techn. 43*:376 (1971).
5. Beck, M., T. Kiesser, M. Perrier and W. Bauer, *Can. J. of Chem. Eng. 64*:553 (1986).
6. Weetall, H.H., and W.H. Pitcher, *Science 232*:1396 (1986).

*Acceptance address as recipient of the 1987 Normann Award of Deutsche Gesellschaft für Fettwissenschaft.

Nutritional Attributes of Fatty Acids*

Joyce Beare-Rogers

Bureau of Nutritional Sciences, Food Directorate, Department of National Health and Welfare, Ottawa K1A OL2

When Wilhelm Normann hydrogenated a double bond of a fatty acid, who would have dreamed that the position of double bonds would be of utmost concern in human nutrition? The essential fatty acids were unkown, and at the end of the last century, fat was not regarded as essential because it could be synthesized in the body from carbohydrate (1).

The various families of unsaturated fatty acids now are designated by the biochemist and nutritionist according to the number of carbon atoms from the last double bond to the terminal methyl group, the (n-3) and (n-6) series of polyunsaturated fatty acids have become recognized as essential for normal development and health. The question of quantities of these fatty acids that should be present in the human diet has been hotly debated.

Successes in manipulating the fatty acid composition of oilseeds have highlighted the need for better information about nutritionally desirable levels of different fatty acids. No one source of fat is ideal for humans but the mix of fats and oils in the total diet determines the nature of the dietary fat. Where one type of oil has a prominent position in the food supply, there is a need to examine its contribution to dietary fatty acids and under some circumstances to modify the situation.

For countries with a high intake of fat and a prevalence of cardiovascular disease, it has been proposed that dietary fat should be comprised of equal quantities of saturated, monounsaturated and polyunsaturated fatty acids. The last group has to provide at least two essential nutrients to represent the (n-3) and (n-6) families.

ESSENTIALITY OF (n-3) FATTY ACIDS

Decades after the recognition of the role of linoleic acid in growth and reproduction, (n-3) linolenic acid appeared not to meet the traditional criteria for essentiality. The failure to find any effect of a linolenic deficiency on growth of experimental animals posed an uncertainty about the requirement in mammals (2). Life was maintained with no more than trace amounts of linolenic acid obtained as a contaminant from casein while the level of linoleic acid was about 300 times higher than that of linolenic acid. It appeared that if dietary linolenic acid or other (n-3) fatty acids were essential to life, the minimum level would be exceedingly low.

The distribution of (n-3) fatty acids and their concentration in particular tissues and phospholipids indicates that a strict metabolic control must be in operation. Docosahexaenoic acid, the last in the (n-3) series

ESSENTIAL FATTY ACIDS

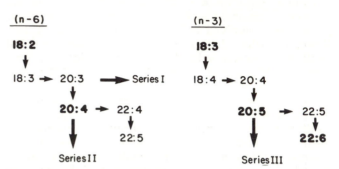

FIG. 1. The (n-6) and the (n-3) families of essential fatty acids and the steps at which eicosanoid series begin.

(Fig. 1), is also the most abundant member. It occurs most prominently in the retina, cerebral cortex, spermatozoa and testes and is suspected of having specific roles. The fine regulation of this fatty acid was suggested by its fairly constant level in the brain of different animal species (3).

The photoreceptor membranes of the retina are particularly rich in docosahexaenoic acid (4). With its six *cis* methylene-interrupted double bonds that cannot rotate, this fatty acid has a rather rigid structure. Its specific molecular roles in close proximity to rhodopsin of the retina, in cerebral grey matter or in reproductive cells have yet to be elucidated and remain a challenge for investigators.

The demonstration of the essentiality of (n-3) fatty acids in the retina was achieved in rhesus monkeys (5). When the only oil in the diet was safflower oil, containing 150 times more (n-6) than (n-3) fatty acids, and this was supplied before the female monekys conceived, during pregnancy and again in the infant formula the young monkeys, which were deficient in (n-3) fatty acids, exhibited impaired visual acuity as compared with those fed soybean oil. Because graduated amounts of the (n-3) fatty acids were not fed, the dietary requirement for normal visual acuity is yet to be determined. Nutrition experiments have tended to test extreme situations. In the deficient monkeys tested by Connor et al., the feeding of marine oil enhanced the level of (n-3) docosahexaenoic acid in the brain, particularly in the phosphatidyl ethanolamine (6). A reciprocal change in the (n-6) docosapentaenoic occurred. It is not known what functional changes might be associated with such substitution of membrane fatty acids.

In young rats, low levels of docosahexaenoic acid were produced by depriving females of (n-3) fatty acids during pregnancy and lactation (7). The dietary fats employed have been reminiscent of some primitive attempts to feed human infants with diets containing high levels of linoleic acid and no (n-3) fatty acids.

Results of discrimination learning tests in rats suggested that (n-3) fatty acids might be involved in cerebral functions (8). After two generations of rats were fed either soybean oil or safflower oil (no blends), the learning of correct responses was significantly better with the soybean oil that furnished (n-3) fatty acids. It is not known if visual acuity influenced this result from discrimination learning tests.

Human data on a (n-3) fatty acid deficiency are meager. A growing girl with a shortened bowel who parenterally received a lipid emulsion containing sunflower oil, which has a high level of linoleic acid but negligible linolenic acid, developed neurological symptoms and blurring of vision (9). The serum fatty acids were low in (n-3) fatty acids as compared with controls. This picture and the neurological symptoms disappeared when a soybean oil emulsion replaced the sunflower one. The report that a human linolenic acid deficiency had been produced caused skepticism because lecithin, a vehicle for choline known to affect neural transmission, was a constituent of the emulsion and the patient was being treated with long-term parenteral nutrition that can cause other metabolic disturbances (10). All of these caveats do not negate the apparent (n-3) deprivation that developed and that was corrected.

DIETARY SOURCES AND REQUIREMENTS OF (n-3) FATTY ACIDS

The shortest member of the (n-3) series, α-linolenic acid, is a constituent of soybean oil and also is found in rapeseed or canola oil. The chloroplasts in green vegetables supply (n-3) linolenic acid as a major fatty acid in low-fat foods. Examples of this in vegetables from Ottawa, Canada are shown in Table 1. Even cereal-based foods in which linoleic acid predominates provide appreciable amounts of (n-3) linolenic acid. (Table 2). The long-chain (n-3) fatty acids are provided by cold-water fish, which are rich sources of (n-3) eicosapentaenoic and docosahexaenoic acids, fatty acids that

TABLE 1

(n-3) Fatty Acids in Vegetables

	m/100g
Celery	5
Iceberg lettuce	6
Spinach	10
Cabbage	17
Chicory	29
Romaine lettuce	55
Shallots	70
Parsley	84
Brussels sprouts	150
Broccoli	162

TABLE 2

Bread Fatty Acids (FA)

Bread	Total FA	(n-6) 18:2	(n-3) 18:3
	mg/g	%	%
Rye	14.5	21.4	1.4
Whole wheat	35.8	19.3	1.3
Bran	17.2	29.5	2.7
Pumpernickel	11.8	50.3	5.2
White	21.9	22.8	1.3

also are found in neural membranes. That humans can synthesize eicosapentaenoic and docosahexaenoic acids from dietary (n-3) linolenic acid is established since strict vegetarians do not consume the long-chain derivatives.

The requirement for (n-3) fatty acids may change during a lifespan. For a six-year-old, it was found that 0.54% energy was sufficient (9) but half of that intake met the need of immobilized elderly (11). The amount of (n-3) fatty acids ingested must be considered along with the amount of (n-6) fatty acids because the two families affect the metabolism of each other.

The ratio of the (n-6) to (n-3) fatty acid in the human fetus appeared to be 8:1 (12). In human milk, this ratio was found to be approximately 5:1 (13). It is recognized that (n-3) linolenic acid has a competitive advantage over linoleic acid for desaturation (14,15). The selectivity of acyltransferase ensures that arachidonic acid, a preferred substrate, is incorporated into membrane lipids (16).

SUBSTITUTION OF (n-3) FOR (n-6) FATTY ACIDS

The extent to which eicosapentaenoic acid from fish oil can substitute for the (n-6) arachidonic acid in biomembranes is of considerable interest. The many studies on platelets confirm that the long chain (n-3) fatty acids from the diet increase in the membranes (17–19). They can be incorporated into all tissues and cells so far investigated. Red blood cells that are synthesized in bone marrow can be used as an index of their prolonged intake. When monkeys were fed 30% of their energy as fat with an equal distribution among saturated, monounsaturated and polyunsaturated fatty acids, the erythrocyte fatty acids were followed for 15 weeks (personal communication). The ratio of (n-6) to (n-3) fatty acids, including all derivatives, increased with a high linoleic, low (n-3) diet but with the diet containing linoleic and linolenic acids the ratio remained at a relatively constant level at about 2.5 and appeared metabolically regulated. In contrast, the longer chain (n-3) fatty acids from the fish oil, which did not have to be desaturated or elongated, increased progressively in the erythrocytes. The enhanced incorporation of the marine-type compared to the plant-type (n-3) fatty acids has been observed in human platelets (19,20) and rat tissues, notably the heart (7–21). A supply of pre-formed fatty acid substrates for acyltransfer reactions permits an enhanced concentration of them in membrane lipids. This may be advantageous in some situations but does circumvent some metabolic control steps.

EICOSANOIDS

Large quantities of (n-6) and (n-3) fatty acids are incorporated into the 2-position of glycerol phospholipid but only a small amount of these polyunsaturated fatty acids may be used in the production of the potent eicosanoids. A thrombotic tendency was thought to be regulated by the balance between the proaggregatory thromboxane A_2 in the platelets and the antiaggregatory prostacyclin in the vessel endothelium (22,23). Thromboxane A_3 from eicosapentaenoic acid was found to be lacking in the proaggregatory properties that characterized thromboxane A_2 (24).

The low incidence of ischemic heart disease in the Greenland Eskimos (25) was attributed to the consumption of (n-3) fatty acids and the inhibition of eicosanoid production from arachidonic acid (26,27). Eicosapentaenoic acid was proposed as an antithrombotic agent. Among the hypotheses proposed for its action were competitive inhibition for cyclo-oxygenase (24,28) and enhanced synthesis of both thromboxane A_3 and prostacyclin I_3. In the in vivo situation, eicosapentaenoic acid was not incorporated into phosphatidylinositol but into phosphatidylcholine and phosphatidylethanolamine (29). As phosphatidylinositol, from which precursors for eicosanoid synthesis arise (30) appeared not to play a role, the metabolic pathway became questionable. Also there was a failure to find parallel effects between the uptake of eicosapentaenoic acid into platelet membranes and a diminished production of thromboxane B_2 (18). The mechanism by which dietary eicosapentaenoic acid alters haemostasis still is not clear but many findings have been well confirmed.

Arachidonic acid and, more particularly, eicosapentaenoic acid also are substrates for lipoxygenase. From studies on neutrophils, it was postulated that the (n-3) fatty acids of marine origin may have anti-inflammatory effects by inhibiting the 5-lipoxygenase pathway by which arachidonic acid is converted to 5-hydroperoxyeicosatetraenoic acid (5 HPETE) and subsequently to leukotriene A_4 and its more stable derivatives (31).

INFLUENCES OF MARINE FATTY ACIDS

A fall in blood pressure has accompanied the ingestion of marine oil (20,32). In Japan, a fishing population consuming more eicosapentaenoic acid than a farming population exhibited a lower blood viscosity (33). The addition of 10–20 ml per day of cod liver oil to a normal western diet increased bleeding time, decreased production of TXB_2, platelet aggregation and blood pressure (34). PGI_3 was detected within one day after the ingestion of marine oil (35). This also cast doubt on the original concept that the eicosanoid precursors must arise from the membrane phospholipids.

When the marine oils or concentrates of them have been administered to humans, the most consistent result has been lowering of the serum triacylglycerol levels (36–38). Some investigators also have found lowering of serum cholesterol. The effects obtained cannot be related easily to dose and duration because protocols have differed greatly. The greatest changes appear to have occurred in hyperlipidemic patients (39).

In a 20-year mortality study in Holland, the consumption of even one or two fish dishes per week has been claimed to prevent coronary heart disease (40). It appears that the level of intake of (n-3) fatty acids need not be above that easily obtained from food.

The mechanism for the action of (n-3) fatty acids in reducing platelet aggregation requires further study. It even has been suggested that levels of arachidonic acid that give rise to thromboxane A_2 may be determined genetically (41). For example, Indians living on an island off the West Coast of British Columbia who consume a traditional diet rich in eicosapentaenoic acid similar to the Eskimos maintain low levels of arachidonic acid in plasma even when they switched to nonmarine foods. Eskimos and Indians may be more adapted to high fish diets than Hugh Sinclair (42), whose bleeding time increased abnormally on a diet of only marine foods.

Another manifestation of possible effects from a high intake of marine fat has been investigated in the Faroe Islands where birthweights are among the highest in the world and gestation periods are prolonged (43). It was postulated that there is an interference with the normal production of prostaglandins required to induce parturition. However, the diet on the Faroe Islands did not provide relief from cardiovascular disease, which was more prevalent than in Denmark (44).

MEMBRANE STRUCTURE

The major noneicosanoid function of essential fatty acids relates to their being integral compounds of membrane bilayers, in which they control the effectiveness of membranes through the conformational changes in the lipid surrounding intrinsic enzymes or transport proteins (45). In this regard, there appear to be distinctive functions in different membranes for the (n-3) and the (n-6) fatty acids. It has been suggested that mobility of insulin receptors on membranes is enhanced by many double bonds (46) but the mechanism by which blood glucose levels are elevated by the administration of fish oil is not known. This phenomenon appeared as a postprandial increase of plasma glucose in pigs fed mackerel oil (47) and in humans treated with a fish oil concentrate (48). If a change in the reactive sites for insulin occurs through ingestion of long chain (n-3) fatty acids, this apparent noneicosanoid influence may precipitate or worsen diabetes.

OTHER PRECAUTIONS

Another aspect of the ingestion of the (n-3) fatty acids is that they readily oxidize and when heated also cyclize and form geometric isomers (49). More information therefore is required on the composition of fatty acids in foods as eaten, particularly those rich in (n-3) fatty acids.

Elevated levels of docosahexaenoic acid in cardiac tissue may not be advantageous. Studies by Gudbjarnason and Hallgrisson (50) indicated that an elevated level of this acid in the heart might be an indication of cardiac lesions. With diets containing different amounts of (n-6) and (n-3) fatty acids, the level of docosahexaenoic acid in the tissue of experimental animals did not correlate with that in the diet, but did correlate the frequency of cardiac lesions (51).

Yellow fat disease developed in growing pigs receiving about 100 g of mackerel oil per day for four weeks, despite a supplement of 0.1% α-tocopheryl acetate in the oil (52). The production of yellow fat disease with the consumption of the highly unsaturated fatty acids in mackerel oil has been attributed to the number of double bonds and not to their position in the fatty acid molecule (53). A disctinction could not be made on the basis of (n-3) or (n-6) fatty acids but only on their degree of unsaturation. The increased requirements for tocopherol or other anti-oxidants in the presence of various (n-3) fatty acids have yet to be determined quantitatively. Caution has been urged with respect to the widespread use of diets high in (n-3) fatty acids (54).

It sometimes is not realized that fish oils contain cholesterol, a factor that usually has not been taken into account in either the design or the interpretation of experiments. The oils high in (n-6) fatty acids from vegetable oils have been hypocholesterolemic whereas the (n-3) fatty acids from marine oils have been consistently hypotriglyceridemic.

RANGE OF INTAKE

There is a need to determine the safe range of intake for the (n-3) fatty acids that is between inadequate intakes producing deficiency signs in membrane composition and an excessive intake that produces deleterious effects. If a concentrate with eicosapentaenoic acid is used as a drug to lower blood triglyceride levels, the patient must be monitored to ensure that blood glucose levels are not dangerously high. For normal nutrition, the challenge is to provide an appropriate proportion of (n-3) to (n-6) fatty acids and those in a suitable mix with other fatty acids.

A prime nutritional attribute of fatty acids is the energy value of the hydrocarbon chain. No other food component makes such a contribution; a situation that both comforts us in face of energy deficiency and alarms us in face of energy excess. What are the upper limits of safe intake of total fatty acids? A high level of saturates in maternal milk provides energy for the infant but in the diet of a mature adult increases blood cholesterol levels and the risk of cardiovascular disease. The level of various fatty acids should be such as to provide long-term health benefits.

The nutritional attributes of fatty acids can be expected to increase in importance as our understanding of them grows.

REFERENCES

1. McCollum, E.V., *A History of Nutrition*, Riverside Press, Cambridge, Massachusetts, (1957).
2. Tinoco, J., R. Babcock, I. Hincenberg, B. Medwadowski, P. Miljanick and M.A. Williams, *Lipids 14*:166 (1979).
3. Crawford, M.A., N.M. Casperd and A.J. Sinclair, *Comp. Biochem. Physiol. 54B*:395 (1976).
4. Anderson, R.E., R.M. Benolken, P.A. Dudley, D.J. Landis and T.G. Wheeler, *Exp. Eye Res. 18*:205 (1974).
5. Neuringer, M., W.E. Connor, C. Van Petten and L. Barstad, *J. Clin. Invest. 73*:272 (1984).
6. Connor, W.E., M. Neuringer and D. Lin, *Am. J. Clin. Nutr. 41*:874 (1985).
7. Roshanai, F., and T.A.B. Sanders, *Ann. Nutr. Metab. 29*:189 (1985).
8. Lamptey, M.S., and B.L. Walker, *J. Nutr. 106*:86 (1976).
9. Holman, R.T., S.B. Johnson and T.F. Hatch, *Am. J. Clin. Nutr. 35*:617 (1982).
10. Bozian, R.C., and S.N. Moussavian, *Am. J. Clin. Nutr. 36*:1253 (1982).
11. Bjerve, K., I.L. Mostad and L. Thoresen, *Am. J. Clin. Nutr. 45*:66 (1987).
12. Clandinin, J.T., J.E. Chappell, T. Heim, P.R. Swyer and G.W. Chance, *Early Human Devel. 5*:355 (1981).
13. Budowski, P., and M.A. Crawford, *Prog. Lipid Res. 25*:615 (1986).
14. Brenner, R.R., and R. Peluffo, *J. Biol. Chem. 241*:5213 (1966).
15. Arens, V.M., S. Konker, G. Werner and U. Petersen, *Fette. Seifen. Antrichmittel 84*:89 (1984).
16. Iritani, N., Y. Ideda and H. Kajitani, *Biochim. Biophys. Acta. 793*:416 (1984).
17. Dyerberg, J., and H.O. Bang, *Lancet ii*:433 (1979).
18. Thorngren, M., and A. Gustafson, *Lancet ii*:1190 (1981).
19. Ahmed, A.A., and B.J. Holub, *Lipids 19*:617 (1984).
20. Sanders, T.A.B., M. Vickers and A.P. Haines, *Clin. Sci. 61*:317 (1981).
21. Holmer, G., and J.L. Beare-Rogers, *Nutr. Res. 5*:1011 (1985).
22. Hamberg, M., J. Svenson and B. Samuelsson, *Proc. Nat. Acad. Sci. U.S.A. 72*:2994 (1975).
23. Moncada, S., R. Gryglewski, S. Bunting and J.R. Vane, *Nature 263*:663 (1976).
24. Needleman, P., A. Raz, M.S. Minkes, J.A. Ferrendelli and H. Sprecher, *Proc. Nat. Acad. Sci. U.S.A. 76*:944 (1979).
25. Bang, H.O., and J. Dyerberg, *Acta Med. Scand. 192*:85 (1972).
26. Dyerberg, J., H.O. Bang, E. Stofferson, S. Moncada and J.R. Vane, *Lancet ii*:117 (1978).
27. Dyerberg, J., and H.O. Bang, *Scand. J. Clin. Lab. Invest. 40*:589 (1980).
28. Culp, B.R., W.E.M. Lands, B.R. Lucchesi, B. Pitt and J. Romson, *Prostaglandins Med. 20*:1021 (1980).
29. Galloway, J.H., I.J. Cartwright, B.E. Woodcock, M. Greaves, R.G.G. Russell and F.E. Preston, *Clin. Sci. 68*:449 (1985).
30. Lefkowith, J.B., H. Sprecher and P. Needleman, *Prog. Lipid Res. 25*:111 (1986).
31. Lee, T.H., R.L. Hoover, J.D. Williams, R.I. Sperling, J. Ravalese, B.W. Spur, D.R. Robinson, E.J. Corey, R.A. Lewis and K.F. Austen, *New Engl. J. Med. 312*:1217 (1985).
32. Mortensen, J.Z., E.B. Schmidt, A.H. Nielsen and J. Dyerberg, *Thromb. Haemostas. 50*:543 (1983).
33. Tamura, Y., *Proceedings of the Second International Congress on Essential Fatty Acids and Prostaglandins*, London (1985).
34. Siess, W., B. Scherer, B. Bohlig, P. Roth, I. Kirzmann and P.C. Weber, *Lancet ii*:441 (1980).
35. Fischer, S., and P.C. Weber, *Nature 307*:165 (1984).
36. von Lossonczy, T.O., A. Ruiter, H.C. Bronsgeest-Schoute, C.M. van Gent and R.J.J. Hermus, *Am. J. Clin. Nutr. 31*:1340 (1978).
37. van Gent, C.M., J.B. Luten, H.C. Bronsgeest-Schoute and A. Ruiter, *Lancet ii*:1249 (1979).
38. Saynor, R., D. Verel and T. Gillott, *Atherosclerosis 50*:3 (1984).
39. Phillipson, B.E., D.W. Rothbrock, W.E. Connor, W.S. Harris and D.R. Illingworth, *New England J. Med. 312*:1210 (1985).
40. Kromhout, D., E.B. Bosschieter and C.L. Coulander, *New Engl. J. Med. 312*:1205 (1985).
41. Bates, C., C. van Dam, D.F. Horrobin, N. Norse, Y-S Huang

and M.S. Manku, *Prostaglandins Leukotrienes & Medicine* *17*:77 (1985).

42. Sinclair, H.M., *Postgrad. Med. J. 56*:579 (1980).

43. Olsen, S.F., H.S. Hansen, T.I.A. Sorensen, B. Jensen, N.J. Secher, S. Sommer and L.B. Knudsen, *Lancet ii*:367 (1986).

44. Joensen, H.D., *Ungeskrift for Laeger*, 2781 (1985).

45. Mead, J.F., *J. Lipid Res. 25*:1517 (1984).

46. Ginsberg, B.H., T.J. Brown, I. Simon and A.A. Spector, *Diabetes 30*:773 (1981).

47. Hartog, J.M., J.M.J. Lamers, A. Montfoort, A.E. Becker, M. Klompe, H. Morse, F.J. ten Cate, L. van der Werf, W.C. Hulsmann, P.G. Hugenholtz and P.D. Verdouw, *Am. J. Clin. Nutr. 46*:258 (1987).

48. Glauber, H.S., P. Wallace and G. Brechtel, *Clin. Res. 35*:504A (1987).

49. Grandgirard, A., J.L. Sebedio and J. Fleury, *J. Am. Oil Chem. Soc. 61*:1563 (1984).

50. Gudbjarnason, S., and J. Hallgrimsson, *Acta Med. Scand. Suppl. 587*:17 (1976).

51. Beare-Rogers, J.L., and E.A. Nera, *Lipids 12*:769 (1977).

52. Ruiter, A., A.W. Jongbloed, C.M. van Gent, L.H.J.C. Danse and S.H.M. Metz, *Am. J. Clin. Chem. 31*:2159 (1978).

53. Danse, L.H.J.C., and H. Nederbragt, *Int. J. Vit. & Nutr. Res. 51*:319 (1981).

54. Hornstra, G., E. Haddeman and F. ten Hoor, *Lancet ii* 1080, (1979).

Conversions of Lipophilic Substances by Encapsulated Biocatalysts

Saburo Fukui

Kyoto University, Kyoto, Japan

It is a great honor and pleasure to be selected as the laureate of the prestigious Normann Medal of the Deutsche Gessellschaft für Fettwissenschaft (DGF) in 1987. I am deeply grateful to Karl Gander, DGF president, Bruno Werdelmann, former DGF president, the members of the selecting committee, and all the members of DGF.

This paper is comprised of two parts. The first deals with the metabolism of alkanes and fatty acids in yeast cells by the action of the enzyme system localized in the organelles, particularly the function of peroxisomes and, namely, bioconversions of these lipophilic compounds in vivo by the enzymes biologically encapsulated. The second part is concerned with bioconversions of lipophilic and slightly water-soluble substances in organic solvent media by using artificially encapsulated biocatalysts.

These works were carried out in my laboratory in collaboration with A. Tanaka, my successor, K. Sonomoto, and my former coworkers Numa and Mishina, and others in the Medical Chemistry Department of Kyoto University and Professor Osumi, Japan Women's University, Tokyo, and some Japanese industries.

METABOLISM OF ALKANES AND FATTY ACIDS BY YEAST (1,2,3)

One specific feature of alkane-utilizing yeasts and higher fatty acid-utilizing yeasts is conspicuous appearance of peroxisomes in the cells. The subtle diversity in the metabolism of alkanes and fatty acids is mediated by subcellular localization of enzymes. My paper describes the metabolism of alkanes and fatty

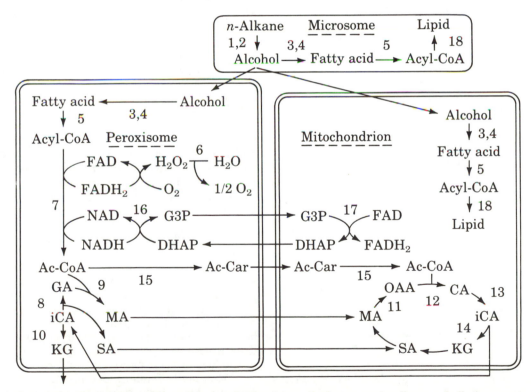

FIG. 1. Presumptive roles of peroxisomes, mitochondria and microsomes in alkane-assimilating yeasts. Enzymes: 1, cytochrome P-450; 2, NADPH-cytochrome P-450 (cytochrome c) reductase; 3, long-chain alcohol dehydrogenase; 4, long-chain aldehyde dehydrogenase; 5, acyl-CoA synthetase; 6, catalase; 7, β-oxidation system; 8, isocitrate lyase; 9, malate synthase; 10, NADP-linked isocitrate dehydrogenase; 11, malate dehydrogenase; 12, citrate synthase; 13, aconitase; 14, NAD-linked isocitrate dehydrogenase; 15, carnitine acetyl transferase; 16, NAD-linked glycerol-3-phosphate dehydrogenase; 17, FAD-linked glycerol-3-phosphate dehydrogenase; 18, glycerol-3-phosphate acyl transferase.

Abbreviations: Ac-Car, acetyl carnitine; Ac-CoA, acetyl-CoA; CA, citrate; DHAP, dihydroxyacetone-phosphate; GA, glyoxylate; G3P, glycerol-3-phosphate; iCA, isocitrate; KG, α-keto-glutarate; MA, malate; OAA, oxalacetate; SA, succinate.

acids in yeasts with special emphasis on the physiological function of peroxisomes.

In microsomes, alkanes are hydroxylated to the corresponding fatty alcohols which are further oxidized to fatty acids via aldehydes in microsomes, mitochondria and peroxisomes, respectively. Degradation of fatty acids to acetyl-CoA via β-oxidation pathways is carried out exclusively in peroxisomes while fatty acids formed in microsomes and mitochondria are incorporated into cellular lipids, each after being activated to acyl-CoAs. Acetyl-CoA produced in peroxisomes is converted to C_4-compounds by the cooperative action of peroxisomes and mitochondria. The existence of acyl-CoA synthetases of different subcellular localization has been demonstrated. The distinctly localized acyl-CoA synthetases play an important role to supply acyl-CoAs, which will be utilized for chain elongation and intact incorporation, or for degradation yielding acetyl-CoA, the substrates for the de novo synthesis system of cellular components. Figure 1 shows the presumable roles of peroxisomes, mitochondria and microsomes in alkane (or fatty acid)-assimilating yeasts. Figure 2 depicts fatty acid metabolism in alkane (or fatty acid)-utilizing yeasts with special reference of the distinct functions of acyl-CoA synthetase I and acyl-CoA II, which are localized exclusively in peroxisomes. The mutants lacking acyl-CoA synthetase I and those lacking acyl-CoA synthetase II were obtained by Numa and his group. When mutants lacking acyl-CoA synthetase I were cultivated on alkanes of odd-chain carbon skeletons, the proportion of odd-chain fatty acids to total cellular fatty acids was completely different from that of the wild strains (Table 1).

As mentioned above, peroxisomes of alkane- and fatty acid-grown yeast cells contain various enzymes,

TABLE 1

Ratios of Odd-Chain Fatty Acids to Total Cellular Fatty Acids in *C. lipolytica* Wild and Mutant Strains

Carbon source	Proportion of odd-chain fatty acids (%)	
	Wild	Mutants
Glucose	3.3	0.7– 0.8
n-Undecane	9.2	1.5– 1.7
n-Tridecane	62.0	2.1– 2.4
n-Pentadecane	97.8	8.9–11.4
n-Heptadecane	98.6	11.7–12.4
Oleic acid	0.8	0.2– 0.3

especially peroxisome-associated enzymes. As shown in Table 2, the levels of these peroxisome-associated enzymes are enhanced markedly when cultivated on these special substrates. Practical applications of these enzymes have been commercialized already.

BIOCONVERSIONS OF LIPOPHILIC AND SLIGHTLY WATER-SOLUBLE COMPOUNDS IN ORGANIC SOLVENT MEDIA BY USING ARTIFICIALLY ENCAPSULATED BIOCATALYSTS (4,5,6)

In the bioconversion of highly lipophilic or slightly water-soluble compounds, it is desirable to carry out the enzymatic reaction in a mixture of water and a suitable organic cosolvent or in an adequate organic solvent that contains a small amount of water. The use of an organic solvent will improve the poor solubility of the substrate and other reaction components that are hydrophobic. Moreover, for the utilization of hydrolytic enzymes for synthetic or group transfer reactions, the water fraction in the reaction mixture should be reduced by replacing water with appropriate organic solvents. Furthermore, along with the recent development of biotechnology much wider applications of biocatalysts are demanded, e.g. bioconversions of xenobiotic compounds such as aliphatic and aromatic compounds, having hydrophobic characters. In such bioprocesses, substrates and products themselves are organic solvents that are unconventional for biocatalysts.

Biocatalysts (enzymes, microbial cells) traditionally have been used in aqueous systems. It generally has been considered that biocatalysts are liable to be denatured in the presence of organic solvents, resulting in the loss of their catalytic abilities.

Attempts to render biocatalysts resistant to organic solvents have been made through different lines of approaches: chemical, biochemical and genetic. Lately, site-specific mutations of enzyme molecules by gene manipulation have attracted worldwide interests. Of these approaches, immobilization of biocatalysts on or in suitable supports seem to be the most general and promising.

We have developed convenient methods for entrapping biocatalysts inside gel matrices formed from

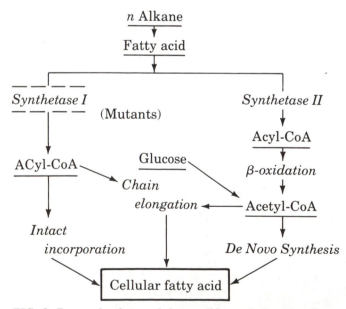

FIG. 2. Proposed scheme of fatty acid metabolism in alkane (fatty acid)-utilizing yeasts. If the mutant of *Candida lipolytica* lacking acyl-CoA synthetase I is grown on alkanes or fatty acids, only the pathway in the peroxisomes is operative, showing the different pattern of the cellular fatty acid from that of the wild strain (Table 1).

TABLE 2

Levels of Peroxisomal Enzymes in *Candida Tropicalis* pK 233

	Enzyme	
Enzyme activity of cells grown on	Glucose (nmol • min^{-1} • mg protein^{-1})	Alkanes (C$_{10}$–C$_{13}$)
Long-chain alcohol dehydrogenase	6	80
Long-chain aldehyde dehydrogenase	9	98
Acyl-CoA synthetase	37	70
β-Oxidation system	ca. 0	75
Catalase	206 × 10^3	5,540 × 10^3
Carnitine acetyl transferase	376	7,430
NAD-linked glycerol-3-phosphate dehydrogenase	10	169
Isocitrate lyase	40	81
Malate synthase	33	170
NADP-linked isocitrate dehydrogenase	242	491
Uricase	42	182
d-Amino acid oxidase	112	73

synthetic prepolymers. Figures 3 and 4 show the structures of the prepolymers of photo cross-linkable resins (ENT, hydrophilic and ENTP, hydrophobic) and the prepolymers of urethane resin (PU), respectively. Photo cross-linkable resin prepolymers have photosensitive functional groups such as acryloyl groups at both terminals of the linear main chain. The chain length of prepolymers can be adjusted by using poly(ethylene glycol) or poly(propylene glycol) of optional chain length as the starting material for synthesis. Thus, ENT-4000, for instance, means that the prepolymer is formed with poly(ethylene glycol)-4000 (average Mw, ca. 4000; the chain length, ca. 40 nm). When the main skeleton consists of poly(ethylene oxide), the prepolymer and accordingly the gels formed from the prepolymers should have a hydrophilic character. On the other hand, the prepolymers containing poly(propylane oxide) in the main skeleton (ENTP) and the gels formed from ENTP should be hydrophobic. In the case of urethane prepolymers, the hydrophilic or hydrophobic character of prepolymers can be adjusted by changing the ratio of the poly(ethylene oxide) part and the poly(propylene oxide) part in the polyether moiety of the main skeleton. Thus, PU-3 with a high content of poly(propylene oxide) gives more hydrophobic gels while PU-6 and PU-9 with a high content of poly(ethylene oxide) hydrophilic gels.

Gelation of photo cross-linkable resin prepolymers can be completed easily by illuminating the mixture of prepolymer solution, a small amount of photosensitizer, e.g. benzoin ethyl ether or benzoin isobutyl ether, and enzyme solution or microbial cell suspensions by near-ultraviolet irradiation for three to five min. Entrapment of biocatalysts with urethane prepolymers is much

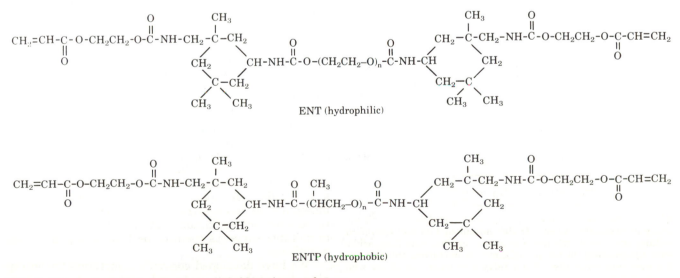

FIG. 3. Structures of typical photo-crosslinkable resin prepolymers.

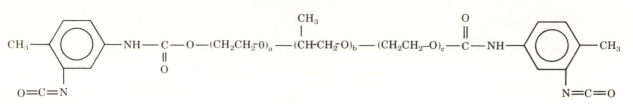

FIG. 4. Structure of polyurethane resin prepolymers (PU).

simpler. When liquid prepolymers are mixed with aqueous solutions of enzymes or aqueous suspensions of microbial cells, prepolymers react each other, being cross-linked by forming urea linkages with liberation of carbon dioxide.

The conformational structure of an immobilized enzyme will be more resistant to the distortion caused by organic solvents as compared with free native enzyme. In the cases of biotransformations of biological substances in vivo, many enzymes, particularly those catalyzing transformations of lipophilic biological compounds, function in membrane-bound states, and the stability of such membrane-associated enzymes is in general greater than that of enzymes released from the membrane.

Inclusion of enzymes or microbial cells within suitable gels would give an environment analogous to that in vivo. Multi-point interactions between entrapped enzymes (and microbial cells including enzymes) and gel matrices will give stabilizing effects. If the gels have desired hydrophobic characters and a suitable network structure, the environment around the biocatalysts entrapped in the gels will be more similar to that in vivo. However, the situation of immobilized biocatalysts in vitro is more complicated. For application of such gel-entrapped biocatalysts for conversions of lipophilic compounds in organic solvent systems, affinity of lipophilic substrates for the gels entrapping biocatalysts and diffusion of reactants through gel matrices are important factors. Low affinity of hydrophilic gels for lipophilic substrates will lower the apparent activity of the gel-entrapped biocatalysts. Thus, use of suitably hydrophobic gels with

an adequate network structure will be preferable depending on hydrophobicity of substrates and polarity of solvents to be used.

In this article, we would like to report comprehensively on our experimental results using biocatalysts immobilized with prepolymers of photo cross-linkable resins and urethane resins. The biocatalyst-entrapping gels had the desired hydrophobicity-hydrophilicity balances and network structures. They were used for bioconversions of a variety of highly lipophilic or slightly water-soluble substrates carried out in homogeneous reaction systems composed of water-water miscible cosolvents or water-containing organic solvents. The effects of solvents and water content in reaction systems also were investigated.

Bioconversion in water-organic cosolvent systems. Water-water miscible organic cosolvent systems have been employed widely to dissolve water-insoluble, lipophilic compounds to prepare homogeneous reaction systems and to shift reaction equilibrium in a desired direction, especially in the synthetic direction with hydrolyzing enzymes.

We have studied bioconversions extensively by biocatalysts entrapped by our prepolymer methods in appropriate water-water miscible organic solvent systems for carrying out the reactions continuously. Table 3 shows our results on bioconversions of various lipophilic or slightly water-soluble compounds carried out in water-water miscible organic solvent systems by the use of microbial cells entrapped with prepolymers.

Bioconversion in organic solvent systems. Although water-immiscible organic solvents have been used for bioreactions to increase the solubility of substrates and

TABLE 3

Bioconversions in Water-Organic Cosolvent Systems by Microbial Cells Entrapped with Prepolymers

Microorganisms (Condition)	Organic Cosolvent		Application
Arthrobacter simplex (acetone-dried)	10%	Methanol	Δ^1-Dehydrogenation of hydrocortisone
Curvularia lunata (living)	2.5%	Dimethyl sulphoxide	11β-Hydroxylation of cortexolone
Rhizopus stolonifer (living)	2.5%	Dimethyl sulphoxide	11α-Hydroxylation of progesterone
Sepedonium ampullosporum (living)	0.65%	Dimethyl formamide	16α-Hydroxylation of estrone
Corynebacterium sp. (living)	15%	Dimethyl sulphoxide	9α-Hydroxylation of 4-androstene-3,17-dione
Enterobacter aerogenes (thawed)	40%	Dimethyl sulphoxide	Synthesis of adenine arabinoside

TABLE 4

Bioconversions in Organic Solvent Systems by Biocatalysts Entrapped with Prepolymers

Biocatalysts	Organic solvent (water-saturated)	Application
Nocardia rhodocrous (cells)	Benzene-n-Heptane (1:1)[a]	$\triangle^1$-Dehydrogenation of ADD[b]
	Benzene-n-Heptane (4:1)	$\triangle^1$-Dehydrogenation of TS[c]
	Benzene-n-Heptane (1:1)	3β-Hydroxysteroid dehydrogenation
	Benzene-n-Heptane (4:1)	17β-Hydroxysteroid dehydrogenation
	Chloroform-n-Heptane (1:1)	3β-Hydroxysteroid dehydrogenation
Rhodotorula minuta (cells)	n-Heptane	Stereoselective hydrolysis of *dl*-menthyl ester
Candida cyclindracae (lipase)	Cyclohexane or Isooctane	Stereoselective esterification of *dl*-menthol
Rhizopus delemar (lipase)	n-Hexane	Interesterification of triglycerides

[a]Mixed ratio by volume.

[b]4-Androstene-3,17-dione.

[c]Testosterone.

water-organic solvent two-phase systems. Only a limited number of papers have described the use of organic solvents alone (mostly, water-saturated organic solvents) in bioreactions.

We have investigated extensively bioconversions of lipophilic compounds by immobilized biocatalysts in organic solvent systems (Table 4).

The activities of immobilized biocatalysts were found to be affected significantly by the hydrophilicity-hydrophobicity balance of gels, the hydrophobicity of substrates, and the polarity of reaction solvents.

REFERENCES

1. Fukui, S., and A. Tanaka, *Trends in Biochemical Sciences* 4:246 (1979).

2. Fukui, S., and A. Tanaka, *Adv. Biochem. Engin.-Biotechnology* ,edited by A. Fiechter, Vol. 19, Springer-Verlag, 1981, pp. 217–237.

3. Tanaka, A., M. Osumi and S. Fukui, *Ann. New York Acad. Sci.*, 1982, pp. 183–199.

4. Fukui, S., and A. Tanaka, *Annual Rev. Microbiol.*, *36*:145 (1982).

5. Fukui, S., and A. Tanaka, *Adv. Biochem. Engin.-Biotechnology*, edited by A. Fiechter, Vol. 29, 1984, pp. 1–33.

6. Fukui, S., A. Tanaka and T. Iida, *Biocatalysis in Organic Media*, edited by C. Laane, J. Tramper and M.D. Lilly, Elsevier Science Publishers, Amsterdam, 1986, pp. 21–41.

Registrants

AUSTRALIA	Frank Annison	Sydney University	Sydney NSW
	Allan Green	C.S.I.R.O.	Canberra City
	Keith R. Harris	Mabule Pty. Ltd.	Miranda NSW
AUSTRIA	Rudolf Franzmair	Pharmaz. Industrie	Linz
	Manfred Frenzl	Oesterr. Agrar-Industrie GmbH	Wien
	Rudolfine Kolmer	Vereinigte Fettwarenindustrie GmbH	Vienna
	Hans Georg Lotter	Krems-Chemie GmbH	Krems
	Johann Marimart	Oesterr. Agrar-Industrie Ges.M.B.H.	Wien
	Fritz Paltauf	Dept. of Biochem.	Graz
	Wilhelm Scheiblauer	Oesterr. Agrar-Industrie Ges.m.b.H.	Wien
	Peter Weiland	Schaerdinger O. Holkereiverband	Enns
BELGIUM	Dirk R.B. De Buyser	N.V. Vandemoortele Coordin Center	Izegem
	Pieter De Geus	Plant Genetic Systems N.V.	Gent
	Jacqueline Destain	Faculte Sciences Agronomiques	Gembloux
	Albert Dijkstra	N.V. Vandemoortele Coordin. Center	Izegem
	Etienne Deffense	SA Fractionnement Tirtiaux	Fleurus
	Jean-Paul Perraudin	Synfina-Oleofina SA	Brussels
	Lea Tirtiaux	Synfina-Oleofina SA	Brussels
	Dirk Verhaeghe	State University Ghent	Ghent
	Bernhard Von Wullerstorff	Kommission Europe Gemeinschaft	Brussels
BRAZIL	Anna Chimeri Battaglia	Sanbra Soc. Alg. Nordeste Bras. SA	Sao Paulo
	Walter Esteves	Unicamp-Fea	Campinas
	Gerhard Fred Plonis	Unicamp - FEA	Campinas
	Otto Rohr	Miracema Nuodex	Campinas
CANADA	Kenneth Carroll	University of Western Ontario	London, Ontario
	R. Keith Downey	Agric. Canada Research Station	Saskatoon
	Daniel Guerra	Biotechnica Canada	Calgary
	Sean Hemmingsen	National Research Council PBI	Saskatoon
	Naim Kosaric	University of Western Ontario	London, Ontario
	Samuel L. Mackenzie	National Research Council of Canada	Saskatoon
	Ted Mag	Canada Packers Inc.	Toronto
	J. Mikle	POS Pilot Plant Corporation	Saskatoon
	M. Keith Pomeroy	Agriculture Canada	Ottawa
	Edward Underhill	Plant Biotechnology Institute	Saskatoon
	Terry S. Walker	National Research Council Canada	London, Ontario
CHILE	Oscar Contreras	Watt's Alimentos S.A.	Santiago
CHINA	Zu-Hua Yi	Institute of Microbiology	Peking
DENMARK	Jens Birk Lauridsen	Grindsted Products A/S	Brabrand
	Peter Eigtved	Novo	Bagsvaerd
	Ulrik Engelrud	Grindsted Products A/S	Brabrand
	Tomas Hansen	Novo A/S	Bagsvaerd
	Finn Hansted	Atlas Industries A/S	Ballerup
	Henrik Hoeg-Petersen	Atlas Industries A/S	Ballerup
	Jesper Korning	Aarhus Olie Ltd.	Aarhus C
	Poul Moller	Consultant Agency	Aarhus
	Peder Holk Nielsen	Novo Industri A/S	Bagsvaerd
	Tommy Pedersen	Aarhus Oliefabrik A/S	Aarhus C
	Inger Plum	Danisco Biotechnology	Glostrup
	Morten Poulsen	Breeding Station	Holeby
	Soeren Rasmussen	Aarhus Oliefabrik A/S	Aarhus C
	Vijai Shukla	Aarhus Oliefabrik A/S	Aarhus
ENGLAND	David Allen	Pura Foods Group	Merseyside
	Susan Armfield	Laboratory of The Government Chemist	London

ENGLAND (cont'd)

Barry Arnold	Mars Confectionery	Slough
Melanie J. Brown	QMC Industrial Research Ltd.	London
Stephen J. Bungard	Imperial Chemical Industries Plc.	Billingham, Clev
Dr. G.G. Connor	Albright & Wilson Limited	Cumbria
Stephen Crook	Castro Ltd.	Reading Berks
Anthony Fentem	Plant Biotechnology Iciple	Runcorn
Caroline Foster	Wolfson Sheffield University	Sheffield
Roger Hammond	Unilever Research	Bedford
Neil Graham Hargreaves	Loders And Nucoline	London
Ronald Harris	Tropical Dev. and Res. Institute	Abingdon, Oxon
John Harrison	ICI International Seed Business	Surrey
John Harwood	Department of Biochemistry	Cardiff
Michael Haughton	Albright & Wilson Limited	Cumbria
Andrew Hepher	Shell Research Ltd.	Kent
Leo Hepner	L. Hepner	London
John Hodgson	Elsevier Publications Cambridge	Cambridge
Jane Holosworth	Rowntree Plc.	York
Laurence Jones	Unilever Research	Bedford
Haydn Jones	Tate and Lyle	Reading
A. James Lambie	Albright & Wilson Limited	Cumbria
Nigel Langley	Croda International, Universal Div.	Humberside
Barry Law	AFRC Institute of Food Research	Reading Berks
Francois Loury	Cargill Europe	London
Alastair Macrae	Unilever Research	Bedford
George W.J. Matcham	Shell Research Ltd.	Kent
Denis Murphy	University of Durham	Durham
William Parish	Unilever Research	Bedford
Robert Pearson	Procter & Gamble Ltd.	Hayes, Middlesex
Colin Ratledge	University of Hull	Hull
Renton Righelato	Tate & Lyle Group R&D	Reading Berks
Richard Safford	Unilever Research	Sharnbrook Bed.
David Steer	Unilever Research Laboratory	Wirral
Jim C. Taylor	Imperial Chemical Industries Plc.	Billingham, Clev

FEDERAL REPUBLIC OF GERMANY

Erik Aalrust	Lurgi Gmbh, R&D Division	Frankfurt/Main
Kurt Aitzetmuller	Fed. Center for LIpid Research	Muenster
Hans-Joachim Asmer	T.U., Inst. F. Biochem	Bravnschweig
Volkbert Bade	Th. Goldschmidt AG	Essen 1
Michael Bahn	Henkel KGaA	Dusseldorf
Horst Baumann	Henkel KGaA	Dusseldorf
Roland Bernerth	Universitat Hamburg	Hamburg
Siegfried Billenstein	Hoechst AG Werk Gendorf	Burgkirchen
Jacob Geert Bindels	Milupa AG	Friedrichsdorf
Dr. Boelsche	Walter Rau Lebesmittelwerke GmbH	Hilteri
Horst Brinkmann	Huels Aktiengesellschaft	Marl
Nils Brolund	Akzo GmbH	Dueren
Karlheinz Brunner	Westfalia Separator AG	Oelde
W. Matthias Buhler	Henkel KGaA	Dusseldorf
Hans Peter Bulian	Noblee & Thoerl GmbH	Hamburg
Ulrich Buller	Redaktion Tenside	Peterhausen
Wolfgang Burk	Bruker Analyt Messtechnik GMbH	Rheinstetten 4
Guenther Demmering	Henkel	Solingen
Christian Duve	Deutsche Snia Vertriebs GmbH	Wuppertal 1
Helmut Eicke	Sud Chemie AG	Munchen
Horst Eierdanz	Henkel	Dusseldorf
Peter S. Elias	Federal Research Centre	Karlsruhe
Friedrich Elstner	Extraktionstechnik	Hamburg 76
Jorgen Erbe	Sud-Chemie AG	Munchen
Sonsoles Espinosa	University of Goettingen	Goettingen
Jurgen Falbe	Henkel KGaA	Dusseldorf 13
Karl-W. Fangauf	American Soybean Association	Hamburg
Karl Gander		Hamburg
Klaus Gottmann	Roehm GmbH	Darmstadt
Friedrick Gotz	Tu-Munchen Lehrstuhl Mik Robiol.	Munchen

FEDERAL REPBULIC OF GERMANY (cont'd)	Walter Gresch	Bucher-Guyer Ltd.	Niederweningen
	Franco Griselli	W.R. Grace	Worms
	Burghard Gruening	Th. Goldschmidt AG	Essen 1
	Guenter Eberhard	Enka AG Research Institute	Obernburg
	Achinto Sen Gupta	Walter Rau Neusserol & Fette AG	Neuss
	Enst Haase	Eastman Kodak Company	Stuttgart
	Dr. Haberiettl	German Patent Office	Munich
	Wolfgang Hares	Institute for General Botany	Hamburg
	Hedtrich Hartmann	Munzing Chemie GmbH	Heilbronn
	Joachim Haselbach	Chemische Fabrik Stockhausen GmbH	Krefeld
	Masahide Hayakawa	Mitsubishi International GMBH	Hamburg
	Arnold Heins	Henkel KGaA	Duesseldorf
	E. Heinz	Universitat Hamburg	Hamburg
	Klaus-Peter Heise	Institut fur Biochemie D. Pflanze	Gottingen
	Horst Hilpert	Deutsche Unilever GmbH	Hamburg 36
	Frank Hirsinger	Henkel KGaA	Duessldorf
	Ulrich Horcher	Basf Aktiengesellschaft	Ludwigshafen
	Karl-Werner Jach	Att-Verfahrenstechnik GmbH	Muenster
	Finn Jacobsen	Novo Industrie GmbH	Nierstein
	Torsten Kiesser	Fraunhofer-Institut fur Leben.	Munchen 50
	Petra Klagge	Walter Rau neusserol & Fett AG	Neuss
	Helmut Klimmek	Chemische Fabrik Stockhausen GmbH	Krefeld
	Gunter Knauel	Chemische Fabrik Stockhausen GmbH	Krefeld
	Dr. Manfred Knuth	Broekelmann & Company	Hamm
	Konrad Kraeling	University of Goettingen	Goettingen
	Wolfram Krieger	Lehmann & Voss & Company	Hamburg 36
	Walter Kuehns	Schmidding-Werke	Koeln
	Berthold Kulp	Koerting Hannover AG Buro Nord	Wulfsen
	Dieter Kundrun	American Soybean Association	Hamburg
	Wolfgang Ladner	BASF	Ludwigshafen
	Siegmund Lang	T.U., Inst. F. Biochem	Braunschweig
	Albrecht Laufer	Ruetgerswerke AG	Castrop-Rauxel
	Walter Lindorfer	Wintershall AG	Kassel
	Thomas Luck	Fraunhofer-Institut fur Leben.	Munchen 50
	Christoph Martin	BASF AG	Ludwigshafen
	Robert Merkle	Hobum Oele Und Fette AG	Hamburg 90
	Franz Meussdoerffer	Henkel KGaA	Duesseldorf
	Wolfgang Meyer-Ingold	Beiersdorf AG	Hamburg
	Otto Moeller	Consultant	Kleve 1
	Kumar D. Mukherjee	Bundesanstalt fur Fettforschung	Munster
	Alice Nasner	Lucas Meyer GmbH & Co.	Hamburg 28
	Wolfgang Neelsen	Franz Kirchfeld GmbH & Co. KG	Duesseldorf
	Theodor Neher	R.P. Scherer GmBH	Eberbach
	Karl-Heinz Neureiter	Coesfeld GmbH Mess-Technik	Dortmund 1
	Hans Pfeiffer	Henkel KGaA	Dusseldorf
	Edwin Pilepp	Krupp Forschungsinstitut	Essen
	Fritz Pipa	Amandus Kahl Nachf.	Hamburg
	Hans-Jurgen Rehm	University Inst. fur Mikrobiologie	Munster
	Gerhard Roebbelen	Georg-August Universitat	Goettingen
	Wulf Ruback	Huels AG	Marl
	Angelos N. Sagredos	Natec Institut	Hamburg 50
	Manfred Salzer	Union Deutsche Lebensmittelwerke	Hamburg 36
	Juan Sanchez-Garcia	Institut Biochemie der Pflanze	Gottingen
	Andreas Sander	Chemische Fabrik Grunau	Illertissen
	Wolfgang Schatzle	Redaktion Bioengineering	Muenchen 5
	Klaus Schloter	Akzo GmbH.	Dueren
	Dr. Bernd Schlotthauer	August Wolff Chem. Pharm. Fabrik	Bielefeld 1
	Armin Schmeichel	T.U., Inst. F. Biochem.	Braunschweig
	Rolf G. Schmid	Gesellschaft fur Biotechnologische	Braunschweig
	Hermann Schmidt	Institut Allg. Botanik	Hamburg
	Heribert Schmitz	Munzing Chemie GmbH	Heilbronn
	Wolfgang Schnabel	Mallinckrodt GMBH	Neunkirchen
	Michael Schneider	Lucas Meyer GmbH & Co.	Hamburg 28
	Rulf Schneider	Oelmuhle Hamburg AG	Hamburg 93

FEDERAL REPUBLIC OF GERMANY (cont'd)	Olaf Schreurs	Laves Chemie	Bad Soden
	Jaap Schuddeboom	Deutsche Snia Vertiebs GmbH	Wuppertal 1
	Heinz Schumacher	Heinz Schumacher V.D.I.	Hamburg 80
	Eckhart Schweizer	University of Erlangen	Erlangen
	Artur Seher	Deutsche Ges. Fettwissenschaft	Muenster
	Klaus Sommermeyer	Fresenius AG	Oberursel
	Kurt Spanier	Universitaet	Goettingen
	Friedrich Spener	Inst. Fur Biochemie, Univ. Munster	Munster
	Hermann Stage	Att-Verfahrenstechnik GmbH	Muenster
	Herbert Stilken Baeumer	Diessel GmbH & Co.	Hildesheim
	Helmut L. Stoehr	W.R. Grace	Worms
	Thomas Teucher	University Hamburg	Hamburg 52
	Roland Theimer	Bergische Universitaet	Wuppertal
	Werner Thies	Universitat Goettingen	Goettingen
	Hans-Jurgen Treede	Institut Biochemie D. Pflanze	Goettingen
	Eduard Von Boguslawski		Ebsdorfergrund 4
	Sabine Von Witzke	University of Goettingen	Goettingen
	Fritz Wagner	Technical University	Braunschweig
	Christian Wandrey	Institute of Biotechnology	Juelich
	Wilhelm Wanke	Deutsche Gelatine-Fabriken	Eberbach
	Sabine Weber	Institut F. Allgemeine Botriu	Hamburg 52
	Klaus Weber	Extraktionstechnik	Hamburg 76
	Gerhard Weilandt	Gebruder Bauermeister & Co.	Norderstedt.
	Freidrich Wengenmayer	Hoechst AG	Frankfurt/Main 8
	Bruno W. Werdelmann	DGF	Ratingen 1
	Cornelius Weser	Kruss GmbH	Hamburg 61
	Theophil Wieske	Consultant	Hamburg 61
	Betlef Wilke	Kalichemie AG	Hannover
	Martin Woelk	Tintometer	Dortmund 1
	Gerhard Wohner	Hoechst AG	Frankfurt
	Rudiger Ziegelitz	Lucas Meyer GmbH & Co. KG	Hamburg
	Werner Zschau	Sud Chemie AG	Munchen
FINLAND	Pirkko Foassell	Technical Research Ctr. of Finland	Espoo 15
	Ben Gronlund	Valio Finnish Coop Diaries	Helsinki
	Jukka Hollo	Raision Tehtaat	Raisio
	Paavo Kalo	Dept. of General Chemistry	Hensinki
	Veikko Kankare	State Institute fur Dairy Research	Jokioinen
	Asmo Kemppinen	Dept. of Dairy Science	Helsinki
	Rune Nystrom	Oljynpuristamo Oy	Helsinki
	Unto Tulisalo	Oljynpuristamo Oy	Helsinki
FRANCE	Daniel Auriol	ITERG	Paris
	Daniel Ballerini	Institut Francais Du Petrole	Rueil Malm Aison
	Jean-Bernard Chazan	Institut Des Corps Gras - ITERG	Paris
	Rene Duterte	S.I.O.	St Laurent Blang
	Chone Emile	Cetiom	Paris
	Jacques Evrard	Cetiom	Pessac
	Jean-Louis Fribourg	Lesieur-Alimentaire	Boulogne/Seine
	Christian Gancet	GRL/ELF Aquitaine	Lacq Artix BP34
	Gerard Gellf	Roussell UCLAF	Romainville
	Lamberet Gilles	INRA	Jouy en Josas
	Michel Goldberg	Universite De Compiegne	Compiegne
	Rene Guillaumin	Institut Des Corps Gras - ITERG	Paris
	Olivier Midler	Genencor Inc.	Le Vesinet
	Frederic Monot	Institut Francais Du Petrole	Rueil Malmaison
	Bernard Prilleux	S.I.O.	St Laurent Blang
	Aldo Uzzan	Revue Francaise Corps Gras	Paris
GERMAN DEMOCRATIC REPUBLIC	Gebauer Herbert	Kernforschungsanlage Juelich	Juelich
	Gerhard Konetzke	VEB Deutsches Hydrierwerk Rodleben	Rosslau
	Hans-Georg Muller	Central Institute of Moleclarbiol.	Berlin-Buch
	Wolfgang Paul	Veb Kombinatoel U Margarine	Magdeburg

GREECE	Ioannis Lazaridis	173 Sygrou Avenue	Athens
HUNGARY	Tibor Farkas	Ist. of Biochem Biol Res. Center	Szeged
ICELAND	Gudmundur G. Haraldsson	Science Inst., Univ. of Iceland	Reykjavik
	Baldur Hjaltason	Lysi H.F.	Reykjavik
INDIA	Dipak Bhattacharyya	Calcutta University Chemical Tech.	Calcutta
	Monindra Chakrabarty		Calcutta
	Nadir Godrej	Godrej Soaps Pvt. Ltd.	Bombay
IRAN	Alireza Koroor	ORAC	Tehran
IRELAND	Diarmuid Kilduff	Kerry Group Plc.	Listowell
	Gerald McNeill	Agricultural Institute	Fermoy Co. Cork
ISRAEL	Zvi Cohen	Ben Gurio U. Desert Res. Inst.	Sde Boqer
	Shimona Geresh	Institutes for Applied Research	Beer Sheva
ITALY	Alessandro Boggiani	Salga SpA	Trecate
	Bruno Boggiani	Salga SpA	Trecate
	Enzo Fedeli	Staz. Sper. Olie. Grassi	Milano
	Vannozzi Gianpaolo	Universita Di Pisaistitutoagronomia	Pisa
	Ambrogio Ricco	Caffaro S.p.A.	Milano
	Giacomo Uccelli	Star S.p.A.	Milan
JAPAN	Saburo Fukui	Kyoto University	Kyoto
	Tadashi Funada	Nippon Oil & Fats Co., Ltd.	Ibaraki
	Shinobu Gocho	T. Hasegawa Co., Ltd.	Kawasaki-shi
	Yukio Hashimoto	Fuji Oil Co., Ltd.	Izumisano
	Hisao Hidaka	Meisei University	Hino, Tokyo
	Hibino Hidehiko	Nippon Oil & Fats Co., Ltd.	Tokyo
	Jiro Hirano	Nippon Oil & Fats Co., Ltd.	Tsukuba, Ibaraki
	Shigeyuki Imamura	Toyo Jozo Co., Ltd.	Shizuoka
	Shigeo Inoue	Kao Corporation	Kashima-Gun
	Yutaka Ishigami	National Chem. Lab. Industry	Ibaraki
	Akio Iwama	Nitto Electric Industrial Company	Osaka
	Isao Karube	Tokyo Institute of Technology	Yokohama
	Masayasu Kawashima	Henkel Hakusui Corporation	Ibaraki Pref.
	Takao Kawata	Nikkei Biotechnology	Tokyo
	Hideo Kikutsugi	Idemitsu Petrochemical Co., Ltd.	Chiodaku
	Hideo Kikutugi	National Chemical Lab. for Industry	Tsukuba, Ibaraki
	Yoshihito Koizumi	Riken Vitamin Co., Ltd.	Hirakata, Osaka
	Yoshitaka Kokusho	Meito Sangyo Co., Ltd.	Tokyo
	Satoshi Kudo	Yakult Honsha Co., Ltd.	Tokyo
	Jun Kurashige	Ajinomoto Co., Inc.	Yokohama City
	Wataru Matsumoto	Asahi Denka Kogyo K.K.	Tokyo
	Hiroyuki Mori	Fuji Oil Co., Ltd.	Osaka Pref.
	Masafumi Moriya	Miyoshi Oil & Fat Co., Ltd.	Tokyo
	Wataru Murayama	Sanki Engineering Limited	Kyoto
	Tomizo Niwa	R&D Assoc. For Bioreactor System	Chuo-ku, Tokyo
	Tadashi Numata	Shibaura Engineering Works Co., Ltd.	Yokohama
	Kan-Ichi Nunogaki	Sanki Engineering Limited	Kyoto
	Kenkichi Oba	Lion Corporation	Odawara City
	Takamasa Ohki	Nigata Engineering Co., Ltd.	Yokohama
	Jungo Okada	Lion Corporation	Odawara
	Takashi Satoh	Nippon Oil & Fats Co., Ltd.	Hyogo
	Norio Sawamura	Fuji Oi Co., Ltd.	Osaka
	Hajime Seino	Kitasato Univ., Eiseigakubu	Sagamihara-Shi
	Sakayu Shimizu	Kyoto University	Kyoto
	Yoshifumi Shinmen	Suntory Research Center	Osaka
	Kenji Soda	Institute of Chemical Research	Uji, Kyoto-fu

JAPAN (cont'd)	Yukio Sugimura	Kao Corporation	Tochigi
	Joji Takahashi	University of Tsukuba	Ibaraki
	Mitsuyoshi Tamaki	MITI	Tokyo
	Atsuo Tanaka	Kyoto University	Kyoto
	Kazuyuki Torii	Suntory Ltd.	Osaka
	Namio Uemura	Nippon Mining Co., Ltd.	Toda Saitama
	Shun Wada	Tokyo University of Fisheries	Tokyo
	Tomiaki Yamada	JGC Corporation	Yokohama
	Hideaki Yamada	Kyoto University	Kyoto
	Yutaka Yamada	The Nisshin Oil Mills, Ltd.	Yokohama
	Tsuneo Yamane	Nagoya University	Nagoya
	Shin Yoshida	Mitsubishi Corporation	Tokyo
	Tamao Yoshida	Tokyo University Fisheries	Tokyo
KENYA	John Gitau Mark	Elianto Kenya Limited	Nakuru
KOREA	Sue Kwon Kim	Lucky, Ltd.	Seoul
MALAYSIA	Suan-Choo Cheah	Palm Oil Research Inst. Malaysia	Kuala Lumpur
	Ismail Hamzah	HMPB	Banting
	Vinod K.Lal	Pan Century Edible Oils Sdn Bhd	Pasir Gudang
	Raton Lal Parakh	Pan Century Edible Oil Sdn Bhd	Pisar Gudang
	Brian John Wood	Sime Darby Plantations	Kuala Lumpur
MEXICO	Eduardo Collingnon	Sappsa	Guadalajara
	Fernando Puig	Enmex SA de CV	Mexico
	Jorge Rios	La Higienica SA de CV	Guadalajara
NEW ZEALAND	Gregory Brown	Mirinz	Hamilton
	Julian Davies	Industrial Process Division DSIR	Petone
NORWAY	Hans A. Blom	A/S Denofa Og Lilleborg Fabriker	Fredrikstad
	Ivan Burkow	FTFI	Tromso
	Geir Heimdal	Peter Moller A/S	Oslo 5
	Aksel Kleppe	A/S Johan C. Martens & Company	Bergen
	Odd Magnar Krog	A/S Denofa Og Lilleborg Fabriker	Fredrikstad
	Einar Mork	Norsk Hydro A.S.	Porsgrunn
	Kristin Saarem	Peter Holler A/S	Oslo 5
	Olav Thorstad	Norsk Hydro A.S.	Porsgrunn
PHILIPPINES	Ernesto Lozada	Phil Coconut R&D Foundation	Diliman QC
	Hubertus E.R. Wriedt	United Coconut Chemicals Inc.	Manila-Makati
POLAND	Bronislan Drozdowski	Tech. Univ. of Gdansk/Chemistry	Gdansk
	Augustyn Jakubowski	Fats Oils and Meat Institute	Warsaw
	Stefan Mitosz	Fats And Oils Industry Association	Warsaw
	Hieskawa Walisiewicz	Ind. Chemistry Research Institute	Warszawa
PORTUGAL	Jose Empis	Sociedade Nacional De Saboes	Lisboa
	Jose Lopes	Sociedade Nacional De Saboes	Lisboa
	Jose Lopes	Fabrica Nacional De Margarina	Sacavem
SCOTLAND	Frank Gunstone	University of St. Andrews	St. Andrews
	R. James Henderson	University of Stirling	Stirling
SOUTH AFRICA	L. Du Plessis	Nat. Food Research Inst. CSIR	Pretoria
SPAIN	Jose Maria Boya	Union Derivan S.A.	Barcelona
	F. Xavier Freixa	Starlux, S.A.	Barcelona
	Puyuelo Galiano	Glyco Iberica S.A.	Gava (Barcelona)

SPAIN (cont'd)	Rafael Garces	Instituto De La Grasa	Sevilla
	Manuel Pacheco	CEPSA	Madird
SWEDEN	Bo Ekstrand	Sik-The Swedish Food Institute	Goteborg
	Rolf Fornhammar	Ellco Food AB	Klippan
	Gareth Griffiths	Swedish Univ. of Agric. Sciences	Uppsala
	Ulf Hakansson	Kabivitrum AB	Stockholm
	Jan-Olaf Lidefelt	Karlshamns AB	Karlshamn
	Anncharlotte Lundgren	Karlshamns AB	Karlshamn
	Lars Moberger	Margarinbolaget AB	Stockholm
	Ragnar Ohlson	Karlshamns AB	Karlshamn
	Hans Olofsson	Margarinbolaget AB	Helsingborg
	Lembitu Reio	Food Administration	Uppsala
	Haj-Britt Stark	Institute of Neurobiology	Goteborg
	Keith Stobart	University of Bristol	Uppsala
	Sten Stymne	Swedish Univ. of Agric. Sciences	Uppsala
	Lennart Svensson	Kabivitrum AB	Stockholm
	Peter Svensson	Kabivitrum AB	Stockholm
	Bengt Uppstrom	Svalof AB	Svalov
	Bob Verduyckt	Norba-Nerlander AB	Zerum
SWITZERLAND	Ernst Braendli	Sulzer AG	Winterthur
	Luc De Bry	Jacobs Suchard	Neuchatel
	Urs Hengartner	F. Hoffmann-La Roche & Co. AG	Basel
	Ruediger A. Hoeren	Prognos AG	Basle
	Othmar Kaeppeli	Motur-Columbus Consulting Eng.	Baden
	Victor Krasnobajew	Givaudan Research Company	Duebendorf
	Juerg P. Marmet	F. Hoffmann-La Roche & Co AG	Basel
	Gerard Moine	F. Hoffmann-La Roche & Co. AG	Basel
	Rodney Moreton	Nestec Ltd.	Vevey
	Marc Olivier Perret	Ciba-Geigy Limited	Basle
	Ernest Romann	Cantonal Laboratory	Regensberg
	Hans-Jurgen Wille	Nestec S.A.	Vevey
THE NETHERLANDS	Marianne Backx	De Melkindustrie Veghel BV	Veghel
	Teun Biere	Friwessa B.V.	Zaandijk
	Ellen Bloksma	Friwessa B.V.	Zaandijk
	Koos Brandenburg	Croklaan B.V.	Wormerveer
	Farrokh Farin	Gist-Brocades N.V.	Dlft
	Marco Giuseppin	Unilever Research Laboratory	Vlaardingen
	Hans Robert Kattenberg	Cacao De Zaan BV	Koob, Zaan
	Jos Keurentjes	Wageningen Agricultural University	Wageningen
	Jan Kloosterman	Unilever Research Laboratorium	Vlaardingen
	Pieter D. Meyer	Suiker Unie Research	Roosendaal
	H. John Nijkamp	Free University Genetics Dept.	Amsterdam
	Wouter Pronk	Agricultural University	Wageningen
	Berend A. Rijkens	Inst. Storage Process. Agric. Prod.	Wageningen
	Robert Schijf	Unilever Research	Vlaardingen
	Nigel Slater	Unilever Research Laboratorium	Vlaardingen
	Henk Smit	Vrye Universiteit Amsterdam	Amsterdam
	Martin Staber	DSM	Heerlen
	Toon Stuitje	Vrye Universiteit Amsterdam	Amsterdam
	Graham Taylor	Unichema International Division	Gouda
	August M. Trommelen	Unilever Research Laboratory	Vlaardingen
	Klaas Van 't Riet	Wageningen Agricultural University	Wageningen
	Jankees Van Der Have	Wessanen Nederland B.V.	Wormerveer
	Albert Van Der Padt	Wageningen Agricultural University	Wageningen
	Leo Van Dorp	Agricultural University	Wageningen
	Jan Van Ee	Gist Brocades	Delft
	Hendrik Vreeman	Nizo	Ede
	D.H. Vuyk	Dir. Agricult. Research M.L.V.	Wageningen
	P.A. Th. J. Werry	Dir. Agricult. Research M.L.V.	Wageningen
	Helmuth Willinger	Hewin International Inc.	Amsterdam
	Gerara Zijlema	Gebr. Smilde BV	Heerenveen

THE NETHERLANDS (cont'd)	Peter Zuurendonk	De Melkindustrie Veghel BV	Veghel
TURKEY	Ali Kulcay	Komili Chemical Industries, Inc.	Istanbul
UNITED STATES	George Abraham	U.S. Dept. of Agriculture	New Orleans, LA
	Terry Andreasen	Sungene Technologies Corporation	Palo Alto, CA
	Thomas H. Applewhite		Prospect Heights, IL
	Marvin Bagby	Northern Regional Research Center	Peoria, IL
	G.R. Beecher	Beltsville Human Nutrition Res. Ctr.	Beltsville, MD
	Gerhard Bookjans	Dept. of Agronomy	Lexington, KY
	Catherine Brady	Akzo Chemie America	McCook, IL
	Glenn Brueske	Crown Iron Works Company	Minneapolis, MN
	Edward Campbell	ADM Company	Decatur, IL
	Joan Dixon	American Oil Chemists' Society	Champaign, IL
	Paul Drzewiecki	Procter and Gamble	Cincinnati, OH
	Michael Dueber	Kraft, Inc.	Glenview, IL
	Raymond Dull	Experience, Inc.	Minneapolis, MN
	D. Estell	Genencor, Inc.	S. San Francisco, CA
	William Finnerty	University of Georgia	Athens, GA
	Eddy R. Hair	Procter & Gamble	Cincinnati, OH
	John Hunter	The Procter & Gamble Company	Cincinnati, OH
	Peter Kalustian	Peter Kalustian Associates, Inc.	Boonton, NJ
	Edward Kiggins	Lubrizol Enterprises Inc.	Wickliffe, OH
	David Kingsbury	National Science Foundation	Washington, DC
	Waldemar Klassen	Agricultural Research Service	Beltsville, MD
	Robert Kleiman	USDA/ARS NRRC	Peoria, IL
	Karl Klein	Centrico	Northvale, NJ
	William Klopfenstein	Kansas State University	Manhattan, KS
	Paul Knowles	University of California/Davis	Blaine, WA
	Sambasivarao Koritala	Northern Regional Research Center	Peoria, IL
	Michael Kotick	Miles Laboratories	Elkhart, IN
	R.G. Krishnamurthy	Kraft, Inc.	Glenview, IL
	Clem Kuehler	Frito Lay, Inc.	Dallas, TX
	David Kyle	Martex Corporation	Columbia, MD
	Steven Laning	Capital City Products Company	Columbus, OH
	E. Charles Leonard	Humko Chemicals	Memphis, TN
	Warner Linfield		Oreland, PA
	Joel Livingston	Exxon Chemical Company	Annandale, NJ
	Norman Lloyd	Nabisco Brands, Inc.	East Hanover, PA
	Willie H.-T. Loh	DNA Plant Technology Corporation	Cinnaminson, NJ
	Charles Martin	Rutgers University -Bio. Sci.	Piscataway, NJ
	David McGee	Sungene Technologies Corporation	Palo Alto, CA
	Jonathan Mielenz	Henkel Research Corporation	Santa Rosa, CA
	Kazunori Miyagawa	Palmco, Inc.	Potland, OR
	Timothy Mounts	Northern Regional Research Ctr.	Peoria, IL
	John-Brian Mudd	The Plant Cell Research Institute	Dublin, CA
	Saul Neidleman	Cetus Corporation	Emeryville, CA
	Niels Nielsen	USDA/ARS/Purdue University	West Lafayette, IN
	Max Norris	Durkee	Strongsville, OH
	John Ohlrogge	Michigan State University	East Lansing, MI
	Fred Paulicka	Durkee	Strongsville, OH
	Edward Perkins	University of Illinois	Urbana, IL
	L.H. Princen	Northern Regional Research Center	Peoria, IL
	Raghav Ram	Sungene Technologies Corp.	Palo Alto, CA
	Lee J. Romanczyk	M&M Mars	Hackettstown, NJ
	W. Jan Rowe	Agrigenetics Advanced Science	Madison, WI
	Ronald Sleeter	ADM Company	Decatur, IL
	Paul K. Stumpf	University of California	Davis, CA
	Ilona Taussky	Consultant	New York, NY
	Gregory Thompson	Calgene Inc.	Davis, CA
	Ruxton Villet	USDA-ARS-NPS	Beltsville, MD
	Richard Wilson	USDA/ARS	Raleigh, NC
	Randall Wood	Texas A&M University	College Station, TX

UNITED STATES (cont'd)	Narendra Yadav David Yang	Dupont Procter and Gamble	Wilmington, DE Cincinnati, OH
VENEZUELA	Antonio Ganzalez Manuel Sucre	Mavesa Mavesa S.A.	Caracas Caracas
YUGOSLAVIA	Simic Dragoslav Stanka Lupsa	Agricultural Kom. Bglgrad Tovarna Oilja Slov Bistrica	Belgrade Slov Bistrica